Major Histocompatibility Complex (MHC) in Health and Disease

Major Histocompatibility Complex (MHC) in Health and Disease

Special Issue Editors

Jerzy K. Kulski
Takashi Shiina
Johannes M. Dijkstra

MDPI • Basel • Beijing • Wuhan • Barcelona • Belgrade

Special Issue Editors

Jerzy K. Kulski
Tokai University
Japan

Takashi Shiina
Tokai University
Japan

Johannes M. Dijkstra
Fujita Health University
Japan

Editorial Office
MDPI
St. Alban-Anlage 66
4052 Basel, Switzerland

This is a reprint of articles from the Special Issue published online in the open access journal *Actuators* (ISSN 2076-0825) in 2019 (available at: https://www.mdpi.com/journal/cells/special_issues/major_histocompatibility_complex).

For citation purposes, cite each article independently as indicated on the article page online and as indicated below:

LastName, A.A.; LastName, B.B.; LastName, C.C. Article Title. *Journal Name* **Year**, *Article Number*, Page Range.

ISBN 978-3-03928-072-8 (Pbk)
ISBN 978-3-03928-073-5 (PDF)

Contents

About the Special Issue Editors

Jerzy K. Kulski graduated from the University of Western Australia with honors in science and a Ph.D. (1977) for his research on the biochemical and endocrine changes in human milk composition during normal and abnormal lactation. He has published more than 180 peer-reviewed papers and chapters in the medical and biological fields of lactation, virology, cancer, microbiology, genetics, genomics, and immunology and has been a staff and freelance researcher at a number of universities, institutes, and hospitals in Australia, the USA, and Japan. He is the author of the crime novel China Heist and the historical novel *Leonardo da Vinci: The Melzi Chronicles* (2017). *Leonardo and the Pacioli Code* (2019) is his first non-fiction book on science and the art history of Leonardo da Vinci. He also guest edited and contributed chapters to two scientific books: *Next Generation Sequencing: Advances, Applications and Challenges* (InTech Publishing 2015) and *The Major Histocompatibility Complex in Health and Disease* (MDPI Books, 2019).

Takashi Shiina graduated from Tokyo University of Agriculture, Graduate School of Agriculture and received a Ph.D. (1993) for his research on the molecular genetic analysis of major histocompatibility complex class I genes in quail. He has published more than 150 peer-reviewed papers in the medical and biological fields, and most of the papers include various contents related to the major histocompatibility complex.

Johannes M. Dijkstra graduated from the University of Groningen in the Netherlands with a degree in molecular genetics and obtained his Ph.D. (1998) at the University of Utrecht in the Netherlands for virology studies performed at the Friedrich Loeffler Institute, Insel Riems, Germany. He has published more than 60 peer-reviewed papers and is one of the leading experts on the evolution of the MHC and cytokines. He, with co-authors, has published the most extensive genetic analyses to date of fish MHC regions, fish MHC class I genes, and fish MHC class II genes and the evolution of classical and nonclassical MHC class I genes in tetrapod species (*Immunogenetics* 2018, 70:459-476). His analysis of cytokine genes has been critical for understanding that type 1 ("Th1") versus type 2 ("Th2") immunity organization is, essentially, conserved throughout jawed vertebrate species [*Nature* 2014, 511:E7; *Biology* (Basel) 2015, 4:814]; this ended a scientific controversy and cemented the use of fish (zebrafish) as an animal model for immune functions.

Editorial

Genomic Diversity of the Major Histocompatibility Complex in Health and Disease

Jerzy K. Kulski [1,2,*], Takashi Shiina [2] and Johannes M. Dijkstra [3]

[1] Faculty of Health and Medical Sciences, UWA Medical School, The University of Western Australia, Crawley, WA 6009, Australia

[2] Department of Molecular Life Sciences, Division of Basic Medical Science and Molecular Medicine, Tokai University School of Medicine, 143 Shimokasuya, Isehara, Kanagawa 259-1193, Japan; tshiina@is.icc.u-tokai.ac.jp

[3] Institute for Comprehensive Medical Science, Fujita Health University, Toyoake, Aichi 470-1192, Japan; dijkstra@fujita-hu.ac.jp

* Correspondence: yurek.kulski@uwa.edu.au or kulski@me.com; Tel.: +61-047-799-9443

Received: 11 October 2019; Accepted: 14 October 2019; Published: 17 October 2019

1. Introduction

The human Major Histocompatibility Complex (MHC) genes are part of the supra-locus on chromosome 6p21 known as the human leukocyte antigen (HLA) system. This genomic complex consists of more than 250 annotated genes and expressed pseudogenes usually partitioned into three distinct regions known as Classes I, II and III. Some of these MHC genes are located closely together in diverse haplotype blocks or clusters that are involved in encoding proteins for cellular and extracellular antigen presentation to circulating T cells, inflammatory and immune-responses, heat shock, complement cascade systems, cytokine signalling, and the regulation of various aspects of cellular development, differentiation, and apoptosis. In addition, there are hundreds of putative microRNA, long noncoding RNA (lncRNA) and antisense RNA non-protein coding loci within the HLA genomic region that may be expressed by different cell types and play important roles in the regulation of immune-response genes and in the aetiology of numerous diseases [1–6]. Since about 2010, the next generation sequencing revolution has been contributing slowly to a better understanding of human MHC gene diversity in worldwide populations, non-coding region variation of HLA loci, the effect of regulatory variation on HLA expression, diversity and polymorphisms in shaping lineage-specific expression, and the impact of HLA expression on disease susceptibility and transplantation outcomes [7]. There is considerable diversity of the MHC genomic region within and between different jawed vertebrate species and much of this diversity is found in the large structural and architectural differences in the genomic organisation of the MHC Class I, II and III genes [8–11]. The MHC of all jawed vertebrate species is characterised specifically by two primary classes of glycoproteins that bind peptides derived from intracellular or extracellular antigens to present to circulating T-cells and play an integral role in adaptive and innate immune systems [12]. Because of the MHC Class I and II gene sequences, duplications and functional diversity, the use of animal experimental models such as macaque, mice, quail, fish, etc., to evaluate the importance of the structure, diversity, expression and function of these genes in immunity, reproduction, mate choice, health, disease, transplantation and vaccination is invaluable [13–15].

This Special Issue on the "Genomic Diversity of the MHC in Health and Disease" consists of eighteen papers with one commentary [16], five reviews [17–21], eleven research articles [22–32] and one communication [33]. These papers cover a broad range of topics on the genomic diversity of the MHC regulatory system in various vertebrate species in health and disease including structure and function; MHC Class I, II and III genes; antigen presentation; innate and adaptive immunity; neurology; transplantation; haplotypes; alleles; infectious and autoimmune diseases; fecundity; conservation;

lineage; and evolution. Although this Special Issue is largely limited to the MHC of mammals, birds and fish, with no expert paper provided on the MHC of monotremes/marsupials, reptiles or amphibians, taken together, these articles demonstrate the immense complexity and diversity of the MHC structure and function within and between different vertebrate species.

2. MHC Genomics, Functions and Diseases from Humans to Fishes

Ten of the 18 papers in the Special Issue are human related, starting with a commentary by Dawkins and Lloyd who provided an overview of the history of the discovery of the association between HLA Class I, II and III gene alleles and certain human autoimmune diseases such as ankylosing spondylitis, systematic lupus erythematosus, myasthenia gravis, and type-1 diabetes from the perspective of conserved population (ancestral) haplotypes [16]. The authors were critical of the modern genome-wide association studies that are based solely on SNP typing and recommended that all MHC genomics and SNP typing results associated with phenotypes or disease be defined as haplotypes, preferably through segregation in extensive family studies for a better understanding of the mechanisms and concepts between HLA genetics, function and phenotypes. A similar sentiment about segregation analysis was extended recently to the study and sequencing of two MHC Class I loci in European barn owls in an investigation of allele segregation patterns in families, showing that family studies not only help to improve the accuracy of MHC genotyping and haplotyping, but also contribute to enhanced analyses in the context of MHC evolutionary ecology [34,35].

Shiina and Blancher provided an extensive review on the use of Old World monkeys in experimental medicine to study the role of MHC polymorphisms in allograft transplantation of organs and stem cells, immune response against infectious pathogens and to vaccines, and various biological systems including reproduction [17]. They compared and expanded on the essential differences and similarities between the human and monkey genomic organisation of the MHC following from their previous comprehensive review comparing the MHC genomics of humans, macaques and mice [36]. They also pointed out the difficulties of reconstructing the complex MHC haplotypes in Old World monkeys by whole genome sequencing using short reads because of the complexity and large number of MHC gene duplications in these animals.

O'Connor and co-authors reviewed the current concepts of avian MHC evolution in the era of next generation sequencing and genomics, focussing on the use of MHC Class I and II sequences to evaluate their associations with fitness, ecological effects, mating preferences, and parasite resistance [18]. Their review refers to the MHC genes of many bird species rather than focusing solely on the chicken MHC, which is an avian MHC model reference that is not wholly representative of most birds. The authors discussed the phylogeny of MHC structural evolution across the avian tree of life, highlighting the enormous diversity between MHC Class I and II gene copy numbers in over 200 species. They concluded that, despite the many inroads made in the last 20 years with the advent of high-throughput sequencing in understanding MHC structure, diversity and evolution, significant improvements still are needed in assembling complete MHC regions with long-read sequencing to establish robust genetic and physical maps in exemplar lineages of birds and to provide anchor points for MHC studies in diverse species.

The MHC Class I and II antigen presentation systems probably emerged in the gnathostome (jawed vertebrates) because these two particular adaptive immune systems are absent in agnathans (jawless vertebrates such as the lamprey and hagfish) and invertebrates [37]. The cartilaginous sharks are elasmobranch fish and the earliest extant representatives of jawed vertebrates with a functional MHC antigen presentation system already established before the emergence of the teleost (modern bony fish) [9,10]. In this Special Issue, Yamaguchi and Dijkstra provided a critical review of classical MHC Class I and II functional analyses and disease resistance in teleost (modern bony fish) and a detailed account of MHC polymorphism and haplotype variation [19]. The authors were critical of many MHC-specific genotype-phenotype association reports in teleost fish, especially of those that claimed an association between MHC Class II haplotypes and mating preferences. Concerning disease-resistance association studies, they only considered whole genome quantitative trait loci (QTL) analyses that were based on statistical reliability. The authors concluded that the teleost classical MHC Class I allelic variations cannot be explained only by selection for different peptide binding properties, and they hypothesised that the extremely divergent alleles may have been selected to induce a more rigorous allograft rejection. In addition, in this Special Issue, Grimholt and co-authors communicated their discovery of a new nonclassical MHC Class I lineage that was found in Holostei (primitive bony fish) and as a new, sixth lineage in Teleostei (modern bony fish) [33].

While three reviews of the MHC structure and function focus mostly on the MHC classical and nonclassical Class I and II genes [17–19], one review [20] and a research article [22] in this Special Issue specifically describe some of the genes in the MHC Class III region that are associated with the innate immune system, complement activation, inflammation and regulation of immunity [1–4]. Zhou and co-authors reviewed a cluster of four genes NELF-E, SKIV2L, DXO and STK19 (the NSDK cluster) in the human MHC Class III region that are involved in RNA metabolism and surveillance during the transcriptional and translational processes of gene expression [20]. These four genes seem to engage in the surveillance of host RNA integrity, in the destruction and turnover of faulty or expired RNA molecules or RNA viruses, and in the fine-tuning of innate immunity. The NSDK cluster is located between the complement gene cluster that codes for constituents of complement C3 convertases (C2, factor B and C4) and the humoral effector functions for immune response. The authors regarded these four genes as highly under-rated because the genetic, biochemical and functional properties for the NSDK cluster in the MHC have remained relatively unknown to many immunologists. Some related gene sequences were found in *Drosophila*, *C. elegans* and zebrafish, but their important roles in human carcinogenesis, infectious and autoimmune diseases are only starting to emerge.

Plasil and co-workers provided a synopsis of the emerging genomic sequencing data for the tumour necrosis factor (TNF) gene and the lymphocyte antigen 6 (LY6G6) multicopy gene family in the MHC Class III region of camels [22]. The LY6 proteins that also are encoded by the MHC Class III region of humans and mice contain a cysteine-rich domain, and they are attached to the cell surface by a glycophosphatidylinositol (GPI) anchor, which is involved in signal transduction. In a comparative and phylogenetic analysis of these gene sequences, the authors found that the camel TNFA and LY6G6 genes mostly resemble those of pigs and/or cattle, as part of their continuing contribution to constructing and improving the genomic map of the entire MHC region of Old World camels.

The human MHC genomic Class I, II and III regions spanning ~4 Mbp from the telomeric myelin oligodendrocyte glycoprotein (MOG) gene to the centromeric collagen type XI alpha 2 chain (COLL11A2) gene also harbour numerous putative microRNA, lncRNA and antisense RNA non-protein coding loci that receive little or no investigative attention [5,6]. Kulski reviewed the origin and structure of the HCP5 gene located between the MICA and MICB genes of the MHC Class I region [21]. This lncRNA gene is a hybrid structure carrying the MHC Class I promoter sequences for the expression of a fossilised endogenous viral sequence ERV16, a repeat sequence that is widely distributed across the genomes of primates and some other mammals. Kulski also found that the HCP5 gene probably expresses the small protein PMSP that binds to the capsid protein of human papillomaviruses. Although the PMSP amino acid sequence appeared to be limited mainly to humans, its homologue was found recently in the baboon (Madrillus genomic sequencing project, UniprotKB: A0A2K5XZB9). Many recent studies have shown that HCP5 SNP sequences are strongly associated with various chronic and infectious diseases including HIV and that the HCP5 RNA interacts with genes inside and outside the MHC genomic region especially with microRNA in the regulation of different cancers. This review highlights the importance of gaining more information and a better understanding of the many noncoding RNA genes expressed by the MHC region that can affect health and disease in association with or independently of the MHC classical Class I and II genes.

3. MHC Classical and Nonclassical Class I and Class II Genomic Diversity (Haplotypes) and Peptide Presentation in Health and Disease

Five research papers are specifically on the topic of MHC antigen presentation and/or interactions with receptors of T cells or killer cells in health or disease [23–25,27,28]. One research paper focusses on haplotyping Class II genes using SNPs associated with disease [26], whereas another examines the importance of MHC Class I gene expression on spinal motoneuron survival and glial reaction following a spinal ventral root crush in wild type and beta2-microglobulin knockout mice [29].

The interaction between T-cell receptors (TCRs) and antigenic peptides presenting major histocompatibility complexes (pMHCs) is a crucial step in adaptive immune response. It triggers the generation of cell-mediated immunity to pathogens and other antigens. The response is driven by TCRs specifically recognising antigenic peptides bound to and presented by the MHC molecules of infected or transformed cells [12,13]. In this Special Issue, Karch and co-workers presented a molecular dynamics simulation study of bound and unbound TCR and pMHC proteins of the LC13-HLA-B*44:05-pEEYLQAFTY complex to monitor differences in relative orientations and movements of domains between bound and unbound states of TCR-pMHC [23]. They found decreased inter-domain movements in the simulations of bound states when compared to unbound states; and increased conformational flexibility was observed for the MHC alpha-2-helix, the peptide, and for the complementary determining regions of the TCR in TCR-unbound states as compared to TCR-bound states. In this regard, Tedeschi and co-workers showed for the first time using a combination of a computer molecular dynamics simulation and in vitro experimentation that HLA-B*27:05, the strongest risk factor for the immune-mediated disorder ankylosing spondylitis (AS), was able to elicit anti-viral CD8+ T cell immune-responses even when the binding groove seemed to be only partially occupied by the Epstein Barr Virus epitope (pEBNA3A-RPPIFIRRL) [24]. In contrast, the non-AS-associated B*27:09 allele, distinguished from the B*27:05 by the single His116Asp polymorphism, was unable to display this peptide and therefore did not unleash specific CD8+ T cell responses in healthy subjects. The authors suggested that even partially filled grooves involved in peptide binding and presentation to CD8+ T cell receptors should be considered as part of the B27 immunopeptidome in evaluating viral immune-surveillance and autoimmunity.

HLA-DQA1*05 and -DQB1*02 alleles encoding the DQ2.5 molecule and HLA-DQA1*03 and -DQB1*03 alleles encoding DQ8 molecules are strongly associated with celiac disease (CD) and type 1 diabetes (T1D). Farina and co-workers demonstrated previously that DQ2.5 genes showed a higher expression with respect to non-CD associated alleles in heterozygous DQ2.5 positive (HLA DR1/DR3) antigen presenting cells of CD patients. They showed that the HLA-DQA1*05 and -DQB1*02 alleles were co-ordinately regulated and expressed as a haplotype at significantly higher levels than non-predisposing alleles [25]. A different study of HLA DQ in T1D by Vadva and co-workers reports on a pedigree-based method for the haplotype analysis of the SNPs in and around the HLA-DR, DQ region using an optimised selection of SNP data to test whether SNPs inside and outside the gene regions are as useful for haplotyping as using HLA-typed alleles [26]. This new pedigree-based methodology for generating edited, non-ambiguous SNP haplotype phasing of minor allele frequency variation as found in the T1DGC pedigree resource might be useful in HLA SNP typing for association with various genetic phenotypes including autoimmune diseases such as T1D.

Experimental allergic encephalomyelitis (EAE) models are being developed in the rhesus monkey and cynomolgus macaque to elucidate the role of Epstein Barr Virus and MHC-E molecules in the presentation of encephalitogenic MOG peptides in multiple sclerosis [17]. The nonclassical HLA-E Class Ib molecules exhibit regulatory functions in both innate and adaptive immune responses and act as indicators for "missing-self" by continuously presenting peptides derived from signal sequences from HLA classical Class Ia molecules. HLA-E presents a 9-mer peptide derived from the signal sequences of HLA-A, -B, -C, and -G proteins to the CD94/NKG2 receptor that transduce an inhibitory signal to NK cells. In addition, it can bind and present antigenic peptides derived from bacterial and viral pathogens to HLA-E restricted CD8+ T cells that secrete antiviral cytokines and kill infected cells [17].

Rohm and co-workers reported in this Special Issue that, although limited, HLA-E polymorphism is associated with susceptibility to BK polyomavirus nephropathy (PyVAN) after a living-donor kidney transplant [27]. Their statistically significant findings suggest that a predisposition based on a defined HLA-E marker is associated with an increased susceptibility to developing PyVAN, and that assessing HLA-E polymorphisms may enable physicians to identify patients who are at an increased risk of this viral complication.

Yao and co-authors reported on the distribution of killer-cell immunoglobulin-like receptor genes and combinations of their HLA ligands in 11 ethnic populations in China [28]. The KIR and its HLA ligands exhibited diverse distribution and characteristics, where each group had its specific KIR and KIR–HLA pair profile. These findings could be expanded on in future population studies on the differential role of these receptors in health and disease.

Neuronal MHC-I has a role in synaptic plasticity, brain development, axonal regeneration, neuroinflammatory processes, and immune-mediated neurodegeneration. In the spinal cord, the MHC-I and beta-2 microglobulin (B2M) transcripts and proteins are upregulated after generating a peripheral motoneuronal lesion. In this Special Issue, Cartarozzi and co-workers presented their experimental findings that, after a ventral root crush, synaptic stripping and neuronal loss occurred more severely in B2M knockout (B2M-KO) mice than wild type mice [29]. Enhanced synapse detachment in B2M-KO mice was attributed to a preferential removal of inhibitory terminals, and the authors concluded that MHC-I molecules are important for a selective maintenance of inhibitory synaptic terminals after lesion formation, and that, with the absence of functional MHC-I expression in the B2M-KO mice, glial inflammatory reactions resulted in a more pronounced synaptic detachment in and around the lesion.

4. Breeding and Conservation: MHC Association with Reproductive Traits, Mate Choice and Fitness

Thirty-six years ago, Jones and Partridge suggested that the MHC is a system primarily for sexual selection and avoidance of inbreeding with histocompatibility fulfilling a secondary role [38]. However, to this day, the evidence for a role of the MHC as a life history gene complex with pleiotropic actions affecting reproduction and other fitness components such as mate selection, fecundity and survival remains relatively inconsistent and debatable. Some controversial aspects of the role of the MHC sexual selection and reproduction in primates [17], birds [18] and fish [19] are reviewed in this Special Issue. Three research papers specifically report on the MHC association with reproductive traits and kin selection (MHC-based mate choice) and fitness [30–32]. Ando and co-authors examined the association between Class II haplotypes and reproductive performances such as fertility index, gestation period, litter size, and number of stillbirths in the highly inbred population of Microminipigs [30]. They found statistically significant differences between haplotypes and the fertility index of dams, litter size at birth, litter size at weaning of dams, and body sizes of adult animals. Their findings suggest that MHC Class II genes of Microminipigs can affect some aspects of reproduction and therefore could be used as differential genetic markers for further haplotype and epistatic studies of reproductive traits and for improving selective breeding and fitness programmes.

Lan and co-workers described the use of MHC haplotypes as adaptive markers in their study of the relative roles of selection and genetic drift in seven populations of the endangered crested ibis [31]. They concluded that genetic drift had a predominant role in shaping the genetic variation and population structure of MHC haplotypes in bottlenecked populations, although some populations showed elevated differentiation of the MHC due to limited gene flow. The seven populations were significantly differentiated into three groups with some groups showing genetic monomorphism attributed to founder effects. The MHC haplotype results allowed the authors to propose various strategies for future conservation and management of the endangered crested ibis.

Zhu and co-workers used ten MHC loci as haplotypes and seven microsatellites outside the MHC region to test three hypotheses of female mate choice in a 17-year study of the giant panda [32].

They found female-choice for heterozygosity and disassortative mate choice at the inter-individual recognition level and that the MHC haplotypes were the mate choice target and not any of the seven microsatellite markers outside the MHC genomic region. They concluded from their long-term field, behavioural and genetic study that the MHC genes of giant pandas should be included when studying MHC-dependent reproductive studies. In this regard, the giant pandas [32] and the minimicropigs [30] appear to be two unique inbred mammalian models for investigating the correlation between the MHC and reproduction.

5. MHC Genomic Alleles (SNPs) and Haplotypes

An important subtheme to emerge from this Special Issue is that the association between MHC genomic SNP sequences and diseases, infections and phenotypes should be examined more often in the context of haplotypes (phased) rather than just genotypes (unphased). Two of the pioneers of human MHC haplotype research, Roger L. Dawkins who coined the term "Ancestral Haplotypes" and Chester Alper (and colleagues) who originated the term "Conserved Extended Haplotypes", both published articles in this Special Issue showing that human population variation studied at the MHC haplotype level is a key requirement to better understanding the role that the MHC and its various genes and subregions may have in human traits including those of health and disease [16,26]. It is noteworthy that, apart from SNPs at gene loci, HLA interspersed indels such as the Alu, SVA, HERV and LTR retroelements also are useful MHC haplotype markers for differentiating between worldwide populations and for case-control stratification in disease association studies [39–41]. The benefits and disadvantages of assessing haplotypes as phased combinations of multilocus alleles instead of genotypes, single locus alleles or diplotypes were considered also in the reviews of MHC genetic diversity of primates, birds and fish [17–19]. In regard to the research articles, Farina and co-workers highlighted the importance of analysing the coordinated haplotypic expression of HLA-DQA and -DQB to better understand susceptibility to the autoimmune diseases T1D and CD [25]. Ando and colleagues used the MHC Class II haplotypes determined from breeding records of highly inbred Microminipigs to investigate their association with reproductive traits [30]. Lan and co-workers described the use of MHC haplotypes as adaptive markers in their study of the relative roles of selection and genetic drift in seven populations of the endangered crested ibis [31]. Zhu and co-workers used ten MHC loci as haplotypes and seven microsatellites outside the MHC region to test three hypotheses of female mate choice in a 17-year study of the giant panda [32]. Many of the reviews and research articles in this Special Issue demonstrate that there is a growing trend towards MHC haplotype analysis rather than simply limiting most genetic/phenotypic associations to only alleles or SNPs.

6. Conclusions

The 18 papers gathered together in this Special Issue highlight the enormous genetic diversity and broad complexity of the MHC regulatory system and why its genomic structure and function is continuously under scientific investigation. These articles provide new insights as well as confirm some of the more tenuous and/or established beliefs about the genetic and biological roles of the MHC [16–33]. More importantly, many of these articles point MHC researchers and scholars in new directions where technical developments and research can greatly improve our knowledge and concepts of the structure and function of the MHC genomic region, especially as functional haplotypes in humans and all the other vertebrate species on the planet that thrive or are in danger of extinction. Some endangered species already need the assistance of researchers, breeders, and conservationists to use informative MHC genetic markers to help establish outbred colonies and families for their conservation and survival.

Acknowledgments: We thank all the authors, reviewers, editors and assistant editors for their efforts and timely submissions and patience during the review process for this Special Issue, the staff of Cells' editorial office, and especially Daniela Zhang for her friendly and accommodating editorial assistance.

Conflicts of Interest: The authors declare no conflict of interest.

References

1. Shiina, T.; Inoko, H.; Kulski, J.K. An update of the HLA genomic region, locus information and disease associations: 2004. *Tissue Antigens* **2004**, *64*, 631–649. [CrossRef] [PubMed]
2. Shiina, T.; Hosomichi, K.; Inoko, H.; Kulski, J.K. The HLA genomic loci map: Expression, interaction, diversity and disease. *J. Hum. Genet.* **2009**, *54*, 15–39. [CrossRef] [PubMed]
3. Horton, R.; Gibson, R.; Coggill, P.; Miretti, M.; Allcock, R.J.; Almeida, J.; Forbes, S.; Gilbert, J.G.R.; Halls, K.; Harrow, J.L.; et al. Variation analysis and gene annotation of eight MHC haplotypes: The MHC Haplotype Project. *Immunogenetics* **2008**, *60*, 1–18. [CrossRef] [PubMed]
4. Matzaraki, V.; Kumar, V.; Wijmenga, C.; Zhernakova, A. The MHC locus and genetic susceptibility to autoimmune and infectious diseases. *Genome Biol.* **2017**, *18*, 76. [CrossRef] [PubMed]
5. Clark, P.M.; Chitnis, N.; Shieh, M.; Kamoun, M.; Johnson, F.B.; Monos, D. Novel and Haplotype Specific MicroRNAs Encoded by the Major Histocompatibility Complex. *Sci. Rep.* **2018**, *8*, 3832. [CrossRef]
6. Gensterblum-Miller, E.; Wu, W.; Sawalha, A.H. Novel Transcriptional Activity and Extensive Allelic Imbalance in the Human MHC Region. *J. Immunol.* **2018**, *200*, 1496–1503. [CrossRef]
7. Petersdorf, E.W.; O'hUigin, C. The MHC in the era of next-generation sequencing: Implications for bridging structure with function. *Hum. Immunol.* **2019**, *80*, 67–78. [CrossRef]
8. Flajnik, M.F.; Kasahara, M. Comparative genomics of the MHC: Glimpses into the evolution of the adaptive immune system. *Immunity* **2001**, *15*, 351–362. [CrossRef]
9. Kulski, J.K.; Shiina, T.; Anzai, T.; Kohara, S.; Inoko, H. Comparative genomic analysis of the MHC: The evolution of Class I duplication blocks, diversity and complexity from shark to man. *Immunol. Rev.* **2002**, *190*, 95–122. [CrossRef]
10. Kaufman, J. Unfinished Business: Evolution of the MHC and the Adaptive Immune System of Jawed Vertebrates. *Annu. Rev. Immunol.* **2018**, *36*, 383–409. [CrossRef]
11. Abduriyim, S.; Zou, D.; Zhao, H. Origin and evolution of the major histocompatibility complex Class I region in eutherian mammals. *Ecol. Evol.* **2019**, *9*, 7861–7874. [CrossRef] [PubMed]
12. Kelly, A.; Trowsdale, J. Genetics of antigen processing and presentation. *Immunogenetics* **2019**, *71*, 161–170. [CrossRef] [PubMed]
13. Doherty, P.C.; Zinkernagel, R.M. Enhanced immunological surveillance in mice heterozygous at the H-2 gene complex. *Nature* **1975**, *256*, 50–52. [CrossRef] [PubMed]
14. Shimizu, S.; Shiina, T.; Hosomichi, K.; Takahashi, S.; Koyama, T.; Onodera, T.; Kulski, J.K.; Inoko, H. MHC Class IIB gene sequences and expression in quails (Coturnix japonica) selected for high and low antibody responses. *Immunogenetics* **2004**, *56*, 280–291. [CrossRef] [PubMed]
15. Ganeshpurkar, A.; Saluja, A.K. Experimental animal models used for evaluation of potential immunomodulators: A mini review. *Bull. Fac. Pharm. Cairo Univ.* **2017**, *55*, 211–216. [CrossRef]
16. Dawkins, R.L.; Lloyd, S.S. MHC Genomics and Disease: Looking Back to Go Forward. *Cells* **2019**, *8*, 944. [CrossRef] [PubMed]
17. Shiina, T.; Blancher, A. The Cynomolgus Macaque MHC Polymorphism in Experimental Medicine. *Cells* **2019**, *8*, 978. [CrossRef]
18. O'Connor, E.; Westerdahl, H.; Burri, R.; Edwards, S.V. Avian MHC evolution in the era of genomics: Phase 1.0. *Cells* **2019**, *8*, 1152. [CrossRef]
19. Yamaguchi, T.; Dijkstra, J.M. Major Histocompatibility Complex (MHC) Genes and Disease Resistance in Fish. *Cells* **2019**, *8*, 378. [CrossRef]
20. Zhou, D.; Lai, M.; Luo, A.; Yu, C.-Y. An RNA Metabolism and Surveillance Quartet in the Major Histocompatibility Complex. *Cells* **2019**, *8*, 1008. [CrossRef]
21. Kulski, J.K. Long Noncoding RNA HCP5, a Hybrid HLA Class I Endogenous Retroviral Gene: Structure, Expression, and Disease Associations. *Cells* **2019**, *8*, 480. [CrossRef] [PubMed]
22. Plasil, M.; Wijkmark, S.; Elbers, J.P.; Oppelt, J.; Burger, P.A.; Horin, P. The major histocompatibility complex of Old World camels—A synopsis. *Cells* **2019**, *8*, 1200. [CrossRef] [PubMed]
23. Karch, R.; Stocsits, C.; Ilieva, N.; Schreiner, W. Intramolecular Domain Movements of Free and Bound pMHC and TCR Proteins: A Molecular Dynamics Simulation Study. *Cells* **2019**, *8*, 720. [CrossRef] [PubMed]

24. Tedeschi, V.; Alba, J.; Paladini, F.; Paroli, M.; Cauli, A.; Mathieu, A.; Sorrentino, R.; D'Abramo, M.; Fiorillo, M.T. Unusual Placement of an EBV Epitope into the Groove of the Ankylosing Spondylitis-Associated HLA-B27 Allele Allows CD8+ T Cell Activation. *Cells* **2019**, *8*, 572. [CrossRef] [PubMed]
25. Farina, F.; Picascia, S.; Pisapia, L.; Barba, P.; Vitale, S.; Franzese, A.; Mozzillo, E.; Gianfrani, C.; Del Pozzo, G. HLA-DQA1 and HLA-DQB1 Alleles, Conferring Susceptibility to Celiac Disease and Type 1 Diabetes, are More Expressed Than Non-Predisposing Alleles and are Coordinately Regulated. *Cells* **2019**, *8*, 751. [CrossRef]
26. Vadva, Z.; Larsen, C.E.; Propp, B.E.; Trautwein, M.R.; Alford, D.R.; Alper, C.A. New Pedigree-Based SNP Haplotype Method for Genomic Polymorphism and Genetic Studies. *Cells* **2019**, *8*, 835. [CrossRef]
27. Rohn, H.; Michita, R.T.; Schramm, S.; Dolff, S.; Gäckler, A.; Korth, J.; Heinemann, F.M.; Wilde, B.; Trilling, M.; Horn, P.A.; et al. HLA-E Polymorphism Determines Susceptibility to BK Virus Nephropathy after Living-Donor Kidney Transplant. *Cells* **2019**, *8*, 847. [CrossRef]
28. Yao, Y.; Shi, L.; Yu, J.; Liu, S.; Tao, Y.; Shi, L. Distribution of Killer-Cell Immunoglobulin-Like Receptor Genes and Combinations of Their Human Leucocyte Antigen Ligands in 11 Ethnic Populations in China. *Cells* **2019**, *8*, 711. [CrossRef]
29. Cartarozzi, L.P.; Perez, M.; Kirchhoff, F.; de Oliveira, A.L.R. Role of MHC-I Expression on Spinal Motoneuron Survival and Glial Reactions Following Ventral Root Crush in Mice. *Cells* **2019**, *8*, 483. [CrossRef]
30. Ando, A.; Imaeda, N.; Matsubara, T.; Takasu, M.; Miyamoto, A.; Oshima, S.; Nishii, N.; Kametani, Y.; Shiina, T.; Kulski, J.K.; et al. Genetic Association between Swine Leukocyte Antigen Class II Haplotypes and Reproduction Traits in Microminipigs. *Cells* **2019**, *8*, 783. [CrossRef]
31. Lan, H.; Zhou, T.; Wan, Q.-H.; Fang, S.-G. Genetic Diversity and Differentiation at Structurally Varying MHC Haplotypes and Microsatellites in Bottlenecked Populations of Endangered Crested Ibis. *Cells* **2019**, *8*, 377. [CrossRef] [PubMed]
32. Zhu, Y.; Wan, Q.-H.; Zhang, H.-M.; Fang, S.-G. Reproductive Strategy Inferred from Major Histocompatibility Complex-Based Inter-Individual, Sperm-Egg, and Mother-Fetus Recognitions in Giant Pandas (*Ailuropoda melanoleuca*). *Cells* **2019**, *8*, 257. [CrossRef] [PubMed]
33. Grimholt, U.; Tsukamoto, K.; Hashimoto, K.; Dijkstra, J.M. Discovery of a Novel MHC Class I Lineage in Teleost Fish Which Shows Unprecedented Levels of Ectodomain Deterioration while Possessing an Impressive Cytoplasmic Tail Motif. *Cells* **2019**, *8*, 1056. [CrossRef] [PubMed]
34. Gaigher, A.; Burri, R.; Gharib, W.H.; Taberlet, P.; Roulin, A.; Fumagalli, L. Family-assisted inference of the genetic architecture of major histocompatibility complex variation. *Mol. Ecol. Resour.* **2016**, *16*, 1353–1364. [CrossRef] [PubMed]
35. Gaigher, A.; Roulin, A.; Gharib, W.H.; Taberlet, P.; Burri, R.; Fumagalli, L. Lack of evidence for selection favouring MHC haplotypes that combine high functional diversity. *Heredity* **2018**, *120*, 396–406. [CrossRef]
36. Shiina, T.; Blancher, A.; Inoko, H.; Kulski, J.K. Comparative genomics of the human, macaque and mouse major histocompatibility complex. *Immunology* **2017**, *150*, 127–138. [CrossRef]
37. Flajnik, M.F. Re-evaluation of the Immunological Big Bang. *Curr. Biol.* **2014**, *24*, R1060–R1065. [CrossRef]
38. Jones, J.S.; Partridge, L. Population genetics: Tissue rejection: The price of sexual acceptance? *Nature* **1983**, *304*, 484–485. [CrossRef]
39. Kulski, J.K.; Dunn, D.S. Polymorphic Alu insertions within the Major Histocompatibility Complex Class I genomic region: A brief review. *Cytogenet. Genome Res.* **2005**, *110*, 193–202. [CrossRef]
40. Kulski, J.K.; Shigenari, A.; Inoko, H. Genetic variation and hitchhiking between structurally polymorphic Alu insertions and HLA-A, -B, and -C alleles and other retroelements within the MHC Class I region. *Tissue Antigens* **2011**, *78*, 359–377. [CrossRef]
41. Kulski, J.K.; Mawart, A.; Marie, K.; Tay, G.K.; AlSafar, H.S. MHC Class I polymorphic *Alu* insertion (POALIN) allele and haplotype frequencies in the Arabs of the United Arab Emirates and other world populations. *Int. J. Immunogenet.* **2019**, *46*, 247–262. [CrossRef] [PubMed]

 *cells*

Commentary

MHC Genomics and Disease: Looking Back to Go Forward

Roger L. Dawkins * and Sally S. Lloyd

Centre for Innovation in Agriculture, Murdoch University and C Y O'Connor ERADE Village Foundation, North Dandalup 6207, Western Australia, Australia
* Correspondence: rldawkins@cyo.edu.au

Received: 14 June 2019; Accepted: 16 August 2019; Published: 21 August 2019

Abstract: Ancestral haplotypes are conserved but extremely polymorphic kilobase sequences, which have been faithfully inherited over at least hundreds of generations in spite of migration and admixture. They carry susceptibility and resistance to diverse diseases, including deficiencies of CYP21 hydroxylase (47.1) and complement components (18.1), as well as numerous autoimmune diseases (8.1). The haplotypes are detected by segregation within ethnic groups rather than by SNPs and GWAS. Susceptibility to some other diseases is carried by specific alleles shared by multiple ancestral haplotypes, e.g., ankylosing spondylitis and narcolepsy. The difference between these two types of association may explain the disappointment with many GWAS. Here we propose a pathway for combining the two different approaches. SNP typing is most useful after the conserved ancestral haplotypes have been defined by other methods.

Keywords: MHC; ancestral haplotype; autoimmune disease

---

1. Introduction

It has been very nearly 50 years since Terasaki, Brewerton, and their colleagues discovered the extraordinary association between ankylosing spondylitis (AS) and HLA B27 [1,2]. With a few caveats of great interest to clinicians, all patients with AS have this allele, justifying the idea that B27 is an essential requirement for the disease–effectively a sine qua non [3]. However, the allele is much more frequent than the disease, and is therefore not itself sufficient. Penetrance is low [4,5]. Other known requirements are male sex and adult age, indicating that the mechanisms of susceptibility and pathogenesis may be quite complex and difficult to unravel. This is still the case today, although the fundamental finding has been confirmed to exhaustion [6].

It has been nearly 40 years since it was established that other HLA associations are completely different [3,7]. For example, in Caucasoids, the more severe forms of systemic lupus erythematosus (SLE) and myasthenia gravis with thymic hyperplasia (MG) are associated with the 8.1 ancestral haplotype [8,9]. This sequence includes B8, but extends over more than a megabase from HLA A to HLA DP. The genetic factors responsible for susceptibility and severity must be numerous and widely spaced throughout the MHC, although most evidence implicates the central MHC, including the *C2*, *Bf*, *C4*, *HSP*, and *TNF* genes, together with their associated non-coding or regulatory sequences. In contrast to AS, female sex is important for these conditions [10].

By 1983, these two different types of association were well recognized [8]. Over subsequent decades, many more examples have been added. Narcolepsy is another example of the allelic form in that the DQB1 0602 association is not restricted to one ethnic group [11,12]. Therefore, explanations for a direct role of a single allele are sought, in this case, and may ultimately appear to be disarmingly simple as for haemochromatosis [13,14] and some drug hypersensitivities [15].

Deficiencies of C4, C2, and 21-hydroxylase (21OH) are examples of associations with extensive ancestral haplotypes (8.1, 18.1, and 47.1, respectively) [16,17]. In each case, there is a plausible explanation, in that the particular sequence includes a missing or defective gene. The observations by Alper et al. [7,18,19] were important in leading to concepts of conserved population haplotypes, which have been faithfully inherited over thousands of generations and are best illustrated by 57.1, which is represented to various degrees in multiple ethnicities.

As different ancestral groups formed and migrated out of Africa and beyond, they carried conserved MHC sequences, which were fixed at each of the alpha, beta, gamma, and delta polymorphic frozen blocks (PFB) (see Figure 1 and Table 1). These sequences were shuffled laterally somewhat as populations mixed, and new combinations appeared, but the more polymorphic regions survive to this day [20–22]. In the case of these deficiencies, the defective or missing *C4*, *C2*, or *21OH* genes remain within the frozen sequence.

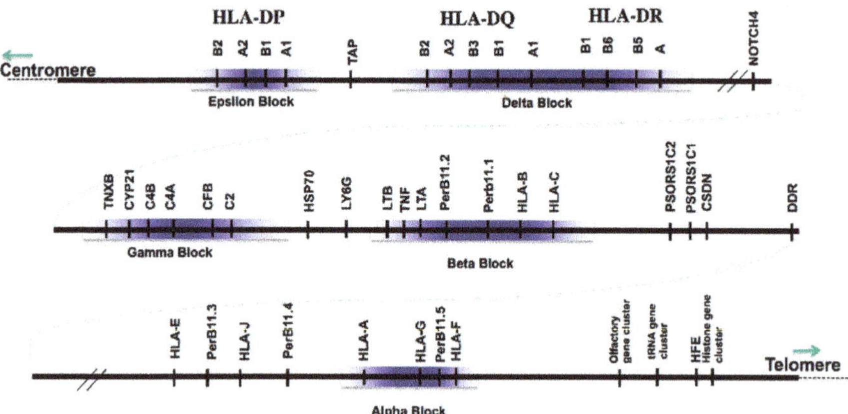

Figure 1. Polymorphic frozen blocks (PFB) in the MHC. Reproduced with permission from [5], and adapted from [16].

The organization of the MHC provides a model for the genome. Each ancestral haplotype has its own map. Polymorphic frozen blocks are shaded. Not all genes are shown. PerB11 is now designated *MIC*. Frozen refers to the freezing of the sequence by inhibition of recombination and mutations, whether

1. unequal crossing-over,
2. double recombination,
3. nucleotide replacement,
4. insertions and deletions,
5. duplication,
6. other.

It was only through extensive family studies that it became clear that recombination occurred outside polymorphic frozen blocks (Figure 2). Indeed, the frozen blocks must have been inherited faithfully over many generations, since identical haplotypes occur in subjects with extensive family trees showing no known relationship, and even in populations that were widely scattered geographically, implying only very remote common ancestry.

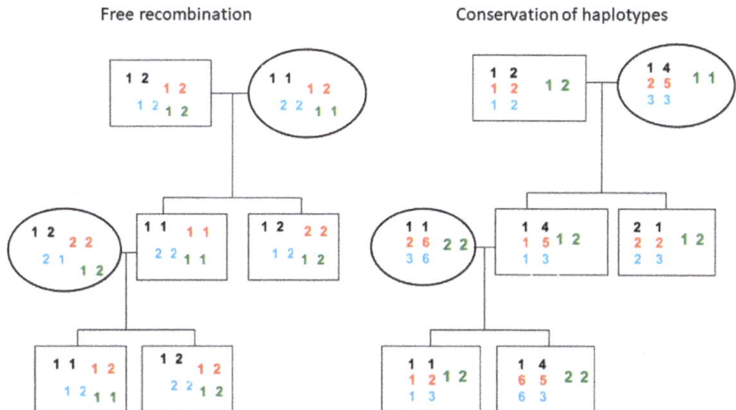

Figure 2. Stable segregation of ancestral haplotypes through three generation family trees. The left describes the inheritance pattern, where the loci are bi-allelic and segregate independently. The right describes the inheritance pattern within PFB, where the alleles are highly polymorphic and the four ancestral haplotypes have segregated predictably.

The reality is that the frozen blocks occupy only a limited proportion of the whole MHC region of a megabase or more, and we have not been able, as suggested in Figure 1, to define hard boundaries between frozen blocks and areas subject to recombination. This difficulty leads us to conclude that there are degrees of freezing as well as specific hotspots. Carrington [23] has contributed by identifying some of the regions which do recombine, but we suspect that such regions are dependent on the genomic environment rather than distance. For example, B8, which is Caucasoid specific by any measure, occurs in non- caucasoids with MG but hardly otherwise [24]. This observation implies that lateral transfer between haplotypes of very different ancestry may lead to thawing as a consequence of differences in the cis, trans, and epistatic interactions. Using multiethnic mapping is powerful [25–27] as a way to find which components carry susceptibility. In other words, ancestral haplotypes are preserved in ethnicities, but eventually fall apart in multiracial combinations.

Susceptibility to autoimmune diseases, such SLE, MG, and insulin-dependent diabetes mellitus (IDDM) must be similarly frozen [8,28–32]—although, to date, there is no mechanistic explanation for the susceptibility [20,32]. Largely, for this reason, we have suggested a dominant role for non-coding regulatory sequences associated with duplicons, indels, and retroviral-like elements (Table 1). We also implicate *epistatic, trans*, and *cis* interactions, with their potential to increase the degree of functional polymorphism exponentially [5,16]. *Epistatic* refers to sequences on different chromosomes, which segregate independently. *Trans* refers to the sequence of the second alternative chromosome. *Cis* refers to the same chromosome.

Table 1. Lessons from MHC genomics.

- Human diversity is inherited from ancestors, rather than created by recent mutation.
- Diversity is regenerated at speciation and maintained by meiotic recombination between the ancestral haplotypes within polymorphic frozen blocks.
- The unit of inheritance is the ancestral haplotype.
- Such sequences carry specific alleles, duplicons, indels and retroviral-like elements (RLEs), which together regulate gene expression.

Reproduced with permission from [5].

These lessons could not have been predicted from the prevailing concepts, which still underlie the thinking behind SNPs and GWAS (Table 2). This SNP-based thinking promotes an overemphasis on the role of ongoing mutation, compared to conservation of ancient polymorphism [33].

Table 2. Alternative dogma.

- Genetic diversity or inherited variation requires ongoing mutation.
- Diversity accumulates through errors in the copying of DNA.
- The unit of inheritance is the allele.
- At each locus alleles may be deleterious, beneficial, or neutral.
- Meiotic recombination scrambles maternal and paternal alleles.
- Linkage disequilibrium between SNPs defines haplotypes.

Reproduced with permission from [5].

Clearly, any understanding of MHC genomics leading to disease must take account of the pragmatic observations implicating age- and sex-dependent penetrance. In Figure 3, we illustrate the potential importance of cis and trans interactions by proposing that ancestral haplotypes can be represented by meshing polymorphic cogs. The degree of meshing or interaction, whether cis or trans, is affected by size and density (expression), as well as by competition with similar cogs, including those representing paralogous sequences encoded on other chromosomes. Thus, MHC, genotypically identical subjects may be affected to different degrees (Figure 3).

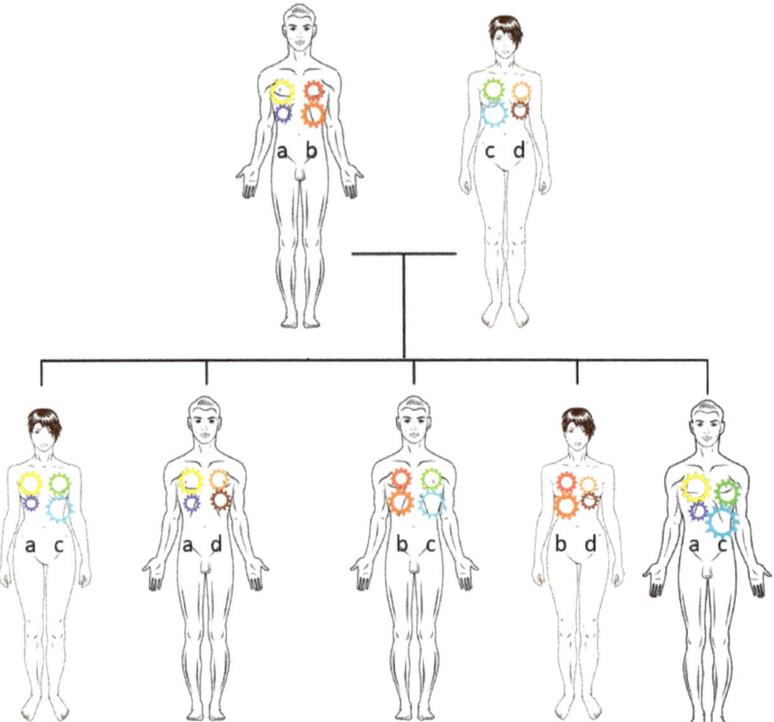

Figure 3. Segregating ancestral haplotypes preserve cis interactions. Ancestral haplotypes are represented by two cogs meshing vertically (in *cis*). The size and shape of cogs represent polymorphism

or variant forms of the blocks. For example, the father has yellow–blue (a) and red–orange (b) haplotypes while the mother has green-cyan (c) and orange-brown (d) haplotypes. Haplotype (a) has been transmitted to the eldest daughter, eldest son, and youngest son. The middle two children have inherited haplotype (b). Reactive meshing is dependent on hormonal and other environmental influences. The oldest and youngest offspring are genotypically identical, but the interactions are different as a consequence of the sexual environment. Adapted with permission from [5].

These models have led to the concept of multifunctional polymorphic control of metabolic and other pathways. Cascades involving stepwise activation of related products, such as the complement system, are promising targets for further study.

2. Recognition of Conserved Ancestral Haplotypes

Table 3 illustrates how ancestral haplotypes were recognized initially. Today there are many more haplotypic markers, including noncoding sequences between the loci shown. As the number of haplotypic markers increased, including SNPs, it became more and more obvious that very few are haplo-specific. It follows that linkage disequilibrium (LD) cannot define such haplotypes. For example, 2×2 delta values cannot identify 18.1 and 18.2, because the alleles are shared by other haplotypes.

Table 3. Haplotype definitions and frequencies in an Australian population.

Ancestral Haplotype	A	Cw	B	C2	Bf	C4A	C4B	DR	DQ	Frequency
7.1	3	7	7	C	S	3	1	2	6	12.9%
8.1	1	7	8	C	S	0	1	3	2	13.2%
13.1			13		S	3	1	7		2.6%
18.1	25	-	18	0	S	4	2	3	6	1.1%
18.2	30	5	18	C	F1	3	0	3	2	1.7%
18.3			18		S	3	1	5		5.2%
35.1		4	35		S	3	1	5		6.9%
35.2	3	4	35	C	F	3 + 2	0	1	5	0.9%
35.3	11	4	35		S	3	0	1	5	2.3%
44.1	2	5	44	C	S	3	0	4	7	5.5%
44.2	29	4	44	C	F	3	1	7	2	2.6%
47.1	3	6	47	C	F	1	0	7	2	<0.6%
57.1	1	6	57	C	S	6	1	7	9	2.6%
65.1		8	65	C	S	2	1 + 2	1	5	0.6%

Adapted from [34,35].

3. Genome-Wide Association Studies and Single Nucleotide Polymorphisms

We believe that the above concepts and models of ancestral haplotypes suggest an alternative approach to the conduct and analysis for genome-wide association studies. When applied to the MHC, some results of commercial SNP typing have been disappointing, even to the point that a recent review by Kennedy et al. [36] essentially dismisses the many classic studies, and very unwisely blames "HLA typing errors, disregard of population structure and lack of replication" [36]. The same authors cite a paper which promotes a "focus on haplotypes ... first suggested in 1987", thereby ignoring important previous contributions on haplotypes—including the original use of the term in 1967 by Ruggero Ceppellini [37,38]—and a huge body of careful observations that have been confirmed repeatedly and rediscovered, without attribution, in the past few decades.

The International Hapmap Project [39–41] is a potentially valuable resource of high-resolution SNP-typed individuals, and includes samples of cell lines used to define the genomic sequence of conserved, extended ancestral haplotypes, which segregate faithfully through families. Surprisingly, the proponents and users [42] of Hapmap have ignored the opportunities for reconciliation with earlier studies, which addressed the shortcomings of LD analysis and focused on haplotypic sequences, including RLEs; indels; duplications; and single nucleotide polymorphism in the literal sense, used by Gaudieri et al. [43] and Longman et al. [44] to map regions of extensive, interrupted sequence differences. The importance of PFB [16] or fixity [45] was also ignored until rediscovered [46].

A more balanced review by Petersdorf and O'hUigin [47] begins the daunting process of integrating population genetics, classic HLA associations, ancestral haplotypes, polymorphic frozen blocks, SNP typing, and gene expression [47]. The authors hope for "the study of haplotype-associated phenotypic differences" and for "haplotype-matching" in transplantation. Indeed, there is already great encouragement for each of these ambitions. The functional differences conferred by ancestral haplotypes, such as 8.1, have been well known for more than 30 years, and include *TNF* and *IgA* concentrations, even though the latter is not encoded within the MHC [16,48,49]. The benefit of haplotype matching in renal and bone marrow transplantation was established decades ago [50,51].

4. Reconciling MHC Genomics and SNP Typing

While we recognize that the disconnect between classical MHC and later SNP genomics will decrease [47,52], we hope this happens quickly. To this end, we summarize the issues as follows:

1. There is no possibility of a single reference sequence. Rather, there are numerous ancestral haplotypes, each with its own very extensive and specific sequence [35,53–55].
2. These sequences are characteristic of ancestral populations or ethnicities. SNP typing on heterozygous mixed populations cannot reveal ancestral haplotype—or at least, must be extremely inefficient [20].
3. MHC complexity is best managed by defining haplotypes through segregation in extensive family studies [56]—not trios—since the power of segregation increases with the number of copies in different heterozygous combinations. Recombination can be demonstrated given sufficient generations to study.
4. Ancestral haplotypes include specific duplications, indels, etc. [54]. Fortunately, there are now many panels of homozygous cell lines and libraries of their sequences available [57,58]. These should be the references and should replace allele and SNP databases.
5. Penetrance is crucial but complex, and depends on age, sex, and a multitude of environmental factors that will vary from time to time and in different settings [53]. Cis and trans interactions are well known contributors to susceptibility and severity [5] These need to be included in the experimental design, but will be difficult to understand until the relevant pathways are defined. We recommend careful consideration of the concept of whole genome duplications resulting in paralogous sequences [16,59], which may compete with and modify the effect of any sequence implicated through SNP analysis.
6. Linkage disequilibrium (LD) is an incomplete reflection of conserved haplotypes and their relative frequencies. Many highly conserved haplotypes cannot be detected by delta values, because even though haplotypic (i.e., present on all examples), individual SNPs are *not* haplo-specific. Different haplotypes share nucleotides and alleles at different locations and relative frequencies. LD between SNPs is misleading if used to define functional haplotypes. Haplospecific recombination creates further complexity [23,60,61].
7. There are at least two types of association with disease, as described above. No doubt, there will be further categories, especially as cis and trans interactions are defined. IDDM is an example of the need to address the mode of inheritance and multiple interactions, as described by Alper et al. [62,63]. Epistatic interactions may also be important.
8. The low positive predictive values of a genetic marker for a disease will remain so until the pathogenic pathways are understood. Fortunately, for clinical purposes, there are examples where the absence of an allele or sequence can be useful for the exclusion of a diagnosis. However, the presence of the same allele does not permit confirmation of the diagnosis—take, for example,

HLA B27 [4]. Thus, those designing future studies should consider how the results will be of practical value, and at the same time, be useful in defining the biology.

9. Many regions of biological and statistical importance are not included in commercial SNP typing. In fact, duplications, indels, RLEs, and ambiguities may be very informative [16].

10. The MHC is a useful model for other genomic regions with polymorphic frozen blocks [64].

11. An understanding of synteny and paralogy is very valuable [65].

12. Many MHC associations, such as narcolepsy, 21-OH deficiency, and haemochromatosis are *not* immunologically mediated. There is no justification for prejudice in interpreting results. Current limited understanding of pathological processes may lead to confusion. In fact, as illustrated by the value of informative clinic-to-laboratory studies, there is potential to elucidate these processes.

13. A promising approach includes understanding the processes and genetics responsible for autoimmune diseases induced by immune checkpoint inhibitors, vaccines, and drugs, such as D-Penicillamine. The value of genomics increases when the inducing agent is known [15].

5. Conclusions

One lesson might be that there are incompatible concepts and terminology, as well as a very patchy understanding of the incontrovertible facts of MHC associations and structure. Another might be that research is constrained by the limitations of commercial platforms.

We hope our commentary is helpful in providing background for new discoveries. Hence the subtitle: "Looking Back to Go Forward". Surely, it is important not to dismiss the history as "HLA typing errors". In fact, the International HLA workshops provide a useful resource for haplotype association with disease, as well as for matching donors and hosts for transplantation.

SNP typing is most useful after the conserved ancestral haplotypes have been defined by other methods.

Funding: This research received no external funding.

Acknowledgments: We would like to thank Chester Alper, Jerzy Kulski, Michael Chorney, Ted Steele, Sue Lester, and the late Jon van Rood, for their encouragement and contributions to our understanding of this field. This is publication 1901 of the C Y O'Connor ERADE Village Foundation. Reference [5] is available on request.

Conflicts of Interest: The authors declare no conflict of interest.

References

1. Brewerton, D.A.; Hart, F.D.; Nicholls, A.; Caffrey, M.; James, D.C.O.; Sturrock, R.D. Ankylosing Spondylitis and HL-A 27. *Lancet* **1973**, *301*, 904–907. [CrossRef]

2. Schlosstein, L.; Terasaki, P.I.; Bluestone, R.; Pearson, C.M. High association of an HL-A antigen, W27, with ankylosing spondylitis. *N. Engl. J. Med.* **1973**, *288*, 704–706. [CrossRef] [PubMed]

3. Dawkins, R.L.; Christiansen, F.T.; Zilko, P.J. Musculoskeletal Disease and D-penicillamine, including reports from the D-Pen-HLA 82 Workshop. *Immunogenet. Rheumatol.* 1982.

4. Christiansen, F.T.; Hawkins, B.R.; Dawkins, R.L.; Owen, E.T.; Potter, R.M. The prevalence of ankylosing spondylitis among B27 positive normal individuals—A reassessment. *J. Rheumatol.* **1979**, *6*, 713–718. [PubMed]

5. Dawkins, R.L. *Adapting Genetics*; Near Urban Publishing: Dallas, TX, USA, 2015; ISBN 978-0-9864115-1-9.

6. Brown, M.A.; Kenna, T.; Wordsworth, B.P. Genetics of ankylosing spondylitis—Insights into pathogenesis. *Nat. Rev. Rheumatol.* **2016**, *12*, 81–91. [CrossRef]

7. Alper, C.A.; Awdeh, Z.L.; Raum, D.D.; Yunis, E.J. Extended major histocompatibility complex haplotypes in man: Role of alleles analogous to murine t mutants. *Clin. Immunol. Immunopathol.* **1982**, *24*, 276–285. [CrossRef]

8. Dawkins, R.L.; Christiansen, F.T.; Kay, P.H.; Garlepp, M.; McCluskey, J.; Hollingsworth, P.N.; Zilko, P.J. Disease associations with complotypes, supratypes and haplotypes. *Immunol. Rev.* **1983**, *70*, 5–22. [CrossRef]

9. Rigby, R.J.; Dawkins, R.L.; Wetherall, J.D.; Hawkins, B.R. HLA in systemic lupus erythematosus: Influence on severity. *Tissue Antigens* **1978**, *12*, 25–31.

10. Christiansen, F.T.; Dawkins, R.L.; Uko, G.; McCluskey, J.; Kay, P.H.; Zilko, P.J. Complement allotyping in SLE: association with C4A null. *Aust. N. Z. J. Med.* **1983**, *13*, 483–488. [CrossRef]

11. Chabas, D.; Taheri, S.; Renier, C.; Mignot, E. The Genetics of Narcolepsy. *Annu. Rev. Genom. Hum. Genet.* **2003**, *4*, 459–483. [CrossRef]

12. Hollingsworth, P.N.; Dawkins, R.L.; Peter, J.B. HLA and narcolepsy. *Neurology* **1993**, *43*, 1444–1445. [CrossRef]

13. Gerhard, G.S.; Ten Elshof, A.E.; Chorney, M.J. Hereditary haemochromatosis as an immunological disease. *Br. J. Haematol.* **1998**, *100*, 247–255. [CrossRef]

14. Feder, J.N.; Gnirke, A.; Thomas, W.; Tsuchihashi, Z.; Ruddy, D.A.; Basava, A.; Dormishian, F.; Domingo, R., Jr.; Ellis, M.C.; Fullan, A.; et al. A novel MHC class I-like gene is mutated in patients with hereditary haemochromatosis. *Nat. Genet.* **1996**, *13*, 399–408. [CrossRef]

15. Mallal, S.; Nolan, D.; Witt, C.; Masel, G.; Martin, A.M.; Moore, C.; Sayer, D.; Castley, A.; Mamotte, C.; Maxwell, D.; et al. Association between presence of HLA-B*5701, HLA-DR7, and HLA-DQ3 and hypersensitivity to HIV-1 reverse-transcriptase inhibitor abacavir. *Lancet* **2002**, *359*, 727–732. [CrossRef]

16. Dawkins, R.L.; Leelayuwat, C.; Gaudieri, S.; Tay, G.; Hui, J.; Cattley, S.; Martinez, P.; Kulski, J. Genomics of the major histocompatibility complex: haplotypes, duplication, retroviruses and disease. *Immunol. Rev.* **1999**, *167*, 275–304. [CrossRef]

17. McCluskey, J.; Kay, P.H.; Stickey, M.; Christiansen, F.T.; Dawkins, R.L.; Wilson, G. MHC "supratype" predicting heterozygous 21-hydroxylase deficiency. *Lancet* **1983**, *1*, 764–765. [CrossRef]

18. White, P.C.; New, M.I.; Dupont, B. HLA-linked congenital adrenal hyperplasia results from a defective gene encoding a cytochrome P-450 specific for steroid 21-hydroxylation. *Proc. Natl. Acad. Sci. USA* **1984**, *81*, 7505–7509. [CrossRef]

19. Raum, D.; Awdeh, Z.; Yunis, E.J.; Alper, C.A.; Gabbay, K.H. Extended major histocompatibility complex haplotypes in type I diabetes mellitus. *J. Clin. Investig.* **1984**, *74*, 449–454. [CrossRef]

20. Alper, C.A.; Larsen, C.E. Major Histocompatibility Complex: Disease Associations. In *eLS*; John Wiley & Sons, Ltd.: Chichester, UK, 2015; pp. 1–8. ISBN 9780470015902.

21. Gaudieri, S.; Leelayuwat, C.; Tay, G.K.; Townend, D.C.; Dawkins, R.L. The major histocompatibility complex (MHC) contains conserved polymorphic genomic sequences that are shuffled by recombination to form ethnic-specific haplotypes. *J. Mol. Evol.* **1997**, *45*, 17–23. [CrossRef]

22. Lam, T.H.; Tay, M.Z.; Wang, B.; Xiao, Z.; Ren, E.C. Intrahaplotypic variants differentiate complex linkage disequilibrium within human MHC haplotypes. *Sci. Rep.* **2015**, *5*, 1–16. [CrossRef]

23. Carrington, M. Recombination within the human MHC. *Immunol. Rev.* **1999**, *167*, 245–256. [CrossRef]

24. Christiansen, F.T.; Pollack, M.S.; Garlepp, M.J.; Dawkins, R.L. Myasthenia gravis and HLA antigens in American blacks and other races. *J. Neuroimmunol.* **1984**, *7*, 121–129. [CrossRef]

25. Todd, J.A.; Mijovic, C.; Fletcher, J.; Jenkins, D.; Bradwell, A.R.; Barnett, A.H. Identification of susceptibility loci for insulin-dependent diabetes mellitus by trans-racial gene mapping. *Nature* **1989**, *338*, 587–589. [CrossRef]

26. Alper, C.A.; Larsen, C.E.; Dubey, D.P.; Awdeh, Z.L.; Fici, D.A.; Yunis, E.J. The Haplotype Structure of the Human Major Histocompatibility Complex. *Hum. Immunol.* **2006**, *67*, 73–84. [CrossRef]

27. Degli-Esposti, M.A.; Andreas, A.; Christiansen, F.T.; Schalke, B.; Albert, E.; Dawkins, R.L. An approach to the localization of the susceptibility genes for generalized myasthenia gravis by mapping recombinant ancestral haplotypes. *Immunogenetics* **1992**, *35*, 355–364. [CrossRef]

28. Kelly, H.; McCann, V.J.; Kay, P.H.; Dawkins, R.L. Susceptibility to IDDM is marked by MHC supratypes rather than individual alleles. *Immunogenetics* **1985**, *22*, 643–651. [CrossRef]

29. Fernando, M.M.; Stevens, C.R.; Sabeti, P.C.; Walsh, E.C.; McWhinnie, A.J.; Shah, A.; Green, T.; Rioux, J.D.; Vyse, T.J. Identification of two independent risk factors for lupus within the MHC in United Kingdom families. *PLoS Genet.* **2007**, *3*, e192. [CrossRef]

30. Pugliese, A.; Gianani, R.; Moromisato, R.; Awdeh, Z.L.; Alper, C.A.; Erlich, H.A.; Jackson, R.A.; Eisenbarth, G.S. HLA-DQB1*0602 is associated with dominant protection from diabetes even among islet cell antibody-positive first-degree relatives of patients with IDDM. *Diabetes* **1995**, *44*, 608–613. [CrossRef]

31. Schloot, N.C.; Roep, B.O.; Wegmann, D.; Yu, L.; Chase, H.P.; Wang, T.; Eisenbarth, G.S. Altered immune response to insulin in newly diagnosed compared to insulin-treated diabetic patients and healthy control subjects. *Diabetologia* **1997**, *40*, 564–572. [CrossRef]

32. Rewers, M.; Norris, J.M.; Eisenbarth, G.S.; Erlich, H.A.; Beaty, B.; Klingensmith, G.; Hoffman, M.; Yu, L.; Bugawan, T.L.; Blair, A.; et al. Beta-cell autoantibodies in infants and toddlers without IDDM relatives: Diabetes autoimmunity study in the young (DAISY). *J. Autoimmun.* **1996**, *9*, 405–410. [CrossRef]

33. Dawkins, R.L.; Williamson, J.F.; Lester, S.; Dawkins, S.T. Mutation versus polymorphism in evolution. *Genomics* **2013**, *101*, 211–212. [CrossRef]

34. Dawkins, R.L.; Leaver, A.; Cameron, P.U.; Martin, E.; Kay, P.H.; Christiansen, F.T. Some disease-associated ancestral haplotypes carry a polymorphism of TNF. *Hum. Immunol.* **1989**, *26*, 91–97. [CrossRef]

35. Degli-Esposti, M.A.; Leaver, A.L.; Frank, T.; Witt, C.S.; Abraham, L.J.; Dawkins, R.L. Ancestral Haplotypes: Conserved Population MHC Haplotypes. *Hum. Immunol.* **1992**, *34*, 242–252. [CrossRef]

36. Kennedy, A.E.; Ozbek, U.; Dorak, M.T. What has GWAS done for HLA and disease associations? *Int. J. Immunogenet.* **2017**, *44*, 195–211. [CrossRef]

37. Ceppellini, R.; Curtoni, E.S.; Mattuiz, P.L.; Miggiano, V.; Scudeller, G.; Serra, A. Genetics of Leukocyte Antigens: A Family Study of Segregation and Linkage. In *Histocompatibility Testing 1967*; Curtoni, E.S., Mattiuz, P.L., Tosi, R.M., Eds.; Williams & Wilkins: Munksgaard, Copenhagen, Denmark, 1967; pp. 149–187.

38. Petersdorf, E.W. In celebration of Ruggero Ceppellini: HLA in transplantation. *HLA* **2017**, *89*, 71–76. [CrossRef]

39. Manolio, T.A.; Brooks, L.D.; Collins, F.S. A HapMap harvest of insights into the genetics of common disease. *Fournal Clin. Investig.* **2008**, *118*, 1590–1605. [CrossRef]

40. International HapMap Consortium. A haplotype map of the human genome. *Nature* **2005**, *437*, 1299–1320. [CrossRef]

41. International HapMap Consortium. The International HapMap Project. *Nature* **2003**, *426*, 789–796. [CrossRef]

42. Ka, S.; Lee, S.; Hong, J.; Cho, Y.; Sung, J.; Kim, H.N.; Kim, H.L.; Jung, J. HLAscan: Genotyping of the HLA region using next-generation sequencing data. *BMC Bioinform.* **2017**, *18*, 1–12. [CrossRef]

43. Gaudieri, S.; Dawkins, R.L.; Habara, K.; Kulski, J.K.; Gojobori, T. SNP profile within the Human Major Histocompatibility Complex reveals an extreme and interrupted level of nucleotide diversity. *Genome Res.* **2000**, *10*, 1579–1586. [CrossRef]

44. Longman-Jacobsen, N.; Williamson, J.F.; Dawkins, R.L.; Gaudieri, S. In Polymorphic Genomic Regions Indels Cluster with Nucleotide Polymorphism: Quantum Genomics. *Gene* **2003**, *312*, 257–261. [CrossRef]

45. Romero, V.; Larsen, C.E.; Duke-Cohan, J.S.; Fox, E.A.; Romero, T.; Clavijo, O.P.; Fici, D.A.; Husain, Z.; Almeciga, I.; Alford, D.R.; et al. Genetic fixity in the human major histocompatibility complex and block size diversity in the class I region including HLA-E. *BMC Genet.* **2007**, *8*, 14. [CrossRef]

46. Gabriel, S.B.; Schaffner, S.F.; Nguyen, H.; Moore, J.M.; Blumenstiel, B.; Higgins, J.; Defelice, M.; Lochner, A.; Faggart, M.; Liu-cordero, S.N.; et al. The Structure of Haplotype Blocks in the Human Genome. *Science* **2002**, *296*, 2225–2229. [CrossRef]

47. Petersdorf, E.W.; O'hUigin, C. The MHC in the era of next-generation sequencing: Implications for bridging structure with function. *Hum. Immunol.* **2019**, *80*, 67–78. [CrossRef]

48. Abraham, L.J.; French, M.A.; Dawkins, R.L. Polymorphic MHC ancestral haplotypes affect the activity of tumour necrosis factor-alpha. *Clin. Exp. Immunol.* **1993**, *92*, 14–18. [CrossRef]

49. Wilton, A.N.; Cobain, T.J.; Dawkins, R.L. Family studies of IgA deficiency. *Immunogenetics* **1985**, *21*, 333–342. [CrossRef]

50. Tay, G.K.; Witt, C.S.; Christiansen, F.T.; Charron, D.; Baker, D.; Herrmann, R.; Smith, L.K.; Diepeveen, D.; Mallal, S.; McCluskey, J.; et al. Matching for MHC haplotypes results in improved survival following unrelated bone marrow transplantation. *Bone Marrow Transplant.* **1995**, *15*, 381–385.

51. Wilton, A.N.; Christiansen, F.T.; Dawkins, R.L. Supratype matching improves renal transplantation survival. *Transplant. Proc.* **1985**, *17*, 2211–2216.

52. Vadva, Z.; Larsen, C.E.; Propp, B.E.; Trautwein, M.R.; Alford, D.R.; Alper, C.A. A New Pedigree-Based SNP Haplotype Method for Genomic Polymorphism and Genetic Studies. *Cells* **2019**, *8*, 835. [CrossRef]

53. Lloyd, S.S.; Steele, E.J.; Dawkins, R.L. Analysis of Haplotype Sequences. In *Next Generation Sequencing—Advances, Applications and Challenges*; Kulski, J.K., Ed.; BoD–Books on Demand: Norderstedt, Germany, 2016; pp. 345–368. ISBN 978-953-51-2240-1.

54. Horton, R.; Gibson, R.; Coggill, P.; Miretti, M.; Allcock, R.J.; Almeida, J.; Forbes, S.; Gilbert, J.G.R.; Halls, K.; Harrow, J.L.; et al. Variation analysis and gene annotation of eight MHC haplotypes: the MHC Haplotype Project. *Immunogenetics* **2008**, *60*, 1–18. [CrossRef]

55. Jensen, J.M.; Villesen, P.; Friborg, R.M.; Mailund, T.; Besenbacher, S.; Schierup, M.H. Assembly and analysis of 100 full MHC haplotypes from the Danish population. *Genome Res.* **2017**, *27*, 1597–1607. [CrossRef]

56. Alper, C.A.; Larsen, C.E. Pedigree-Defined Haplotypes and Their Applications to Genetic Studies. In *Haplotyping: Methods and Protocols, Methods in Molecular Biology*; Humana Press: New York, NY, USA, 2008; Volume 1551, pp. 113–127. ISBN 9781493967506.

57. International Histocompatibility Working Group. Available online: http://www.ihwg.org/ (accessed on 19 August 2019).

58. Norman, P.J.; Norberg, S.J.; Guethlein, L.A.; Nemat-Gorgani, N.; Royce, T.; Wroblewski, E.E.; Dunn, T.; Mann, T.; Alicata, C.; Hollenbach, J.A.; et al. Sequences of 95 human MHC haplotypes reveal extreme coding variation in genes other than highly polymorphic HLA class i and II. *Genome Res.* **2017**, *27*, 813–823. [CrossRef]

59. Kasahara, M. The chromosomal duplication model of the major histocompatibility complex. *Immunol. Rev.* **1999**, *167*, 17–32. [CrossRef]

60. van Oosterhout, C. Trans-species polymorphism, HLA-disease associations and the evolution of the MHC. *Commun. Integr. Biol.* **2009**, *2*, 408–410. [CrossRef]

61. Ahmad, T.; Neville, M.; Marshall, S.E.; Armuzzi, A.; Mulcahy-Hawes, K.; Crawshaw, J.; Sato, H.; Ling, K.L.; Barnardo, M.; Goldthorpe, S.; et al. Haplotype-specific linkage disequilibrium patterns define the genetic topography of the human MHC. *Hum. Mol. Genet.* **2003**, *12*, 647–656. [CrossRef]

62. Aly, T.A.; Eller, E.; Ide, A.; Gowan, K.; Babu, S.R.; Erlich, H.A.; Rewers, M.J.; Eisenbarth, G.S.; Fain, P.R. Multi-SNP analysis of MHC region: remarkable conservation of HLA-A1-B8-DR3 haplotype. *Diabetes* **2006**, *55*, 1265–1269. [CrossRef]

63. Hutton, J.C.; Eisenbarth, G.S. A pancreatic β-cell-specific homolog of glucose-6-phosphatase emerges as a major target of cell-mediated autoimmunity in diabetes. *Proc. Natl. Acad. Sci. USA* **2003**, *100*, 8626–8628. [CrossRef]

64. McLure, C.A.; Hinchliffe, P.; Lester, S.; Williamson, J.F.; Millman, J.A.; Keating, P.J.; Stewart, B.J.; Dawkins, R.L. Genomic evolution and polymorphism: Segmental duplications and haplotypes at 108 regions on 21 chromosomes. *Genomics* **2013**, *102*, 15–26. [CrossRef]

65. Dawkins, R.L.; Berry, J.; Martinez, P.; Gaudieri, S.; Hui, J.; Cattley, S.; Longman, N.; Kulski, J.; Carnegie, P. Potential for Paralogous Mapping to Simplify the Genetics of Diseases and Functions Associated with MHC Haplotypes. In *The Major Histocompatibility Complex: Evolution, Structure, and Function*; Kasahara, M., Ed.; Springer-Verlag: Tokyo, Japan, 2000; pp. 146–157.

 *cells*

Review

The Cynomolgus Macaque MHC Polymorphism in Experimental Medicine

Takashi Shiina [1] and Antoine Blancher [2,3,*]

[1]   Department of Molecular Life Sciences, Division of Basic Medical Science and Molecular Medicine, Tokai University School of Medicine, 143 Shimokasuya, Isehara, Kanagawa 259-1193, Japan
[2]   Centre de Physiopathologie Toulouse-Purpan (CPTP), Université de Toulouse, Centre National de la Recherche Scientifique (CNRS), Institut National de la Santé et de la Recherche Médicale (Inserm), Université Paul Sabatier (UPS), 31000 Toulouse, France
[3]   Laboratoire d'Immunologie, CHU de Toulouse, Institut Fédératif de Biologie, Hôpital Purpan, 330 Avenue de Grande Bretagne, TSA40031, CEDEX 9, 31059 Toulouse, France
*   Correspondence: blancher.a@chu-toulouse.fr or blancher.antoine@neuf.fr; Tel.: +33-5-61-77-61-43; Fax: +33-5-67-69-03-27

Received: 27 June 2019; Accepted: 22 August 2019; Published: 26 August 2019

Abstract: Among the non-human primates used in experimental medicine, cynomolgus macaques (*Macaca fascicularis* hereafter referred to as *Mafa*) are increasingly selected for the ease with which they are maintained and bred in captivity. Macaques belong to Old World monkeys and are phylogenetically much closer to humans than rodents, which are still the most frequently used animal model. Our understanding of the *Mafa* genome has progressed rapidly in recent years and has greatly benefited from the latest technical advances in molecular genetics. Cynomolgus macaques are widespread in Southeast Asia and numerous studies have shown a distinct genetic differentiation of continental and island populations. The major histocompatibility complex of cynomolgus macaque (*Mafa* MHC) is organized in the same way as that of human, but it differs from the latter by its high degree of classical class I gene duplication. Human polymorphic MHC regions play a pivotal role in allograft transplantation and have been associated with more than 100 diseases and/or phenotypes. The *Mafa* MHC polymorphism similarly plays a crucial role in experimental allografts of organs and stem cells. Experimental results show that the *Mafa* MHC class I and II regions influence the ability to mount an immune response against infectious pathogens and vaccines. MHC also affects cynomolgus macaque reproduction and impacts on numerous biological parameters. This review describes the *Mafa* MHC polymorphism and the methods currently used to characterize it. We discuss some of the major areas of experimental medicine where an effect induced by MHC polymorphism has been demonstrated.

Keywords: cynomolgus macaque; *Macaca fascicularis*; MHC polymorphism; experimental medicine; nonhuman primate models

---

1. Introduction

Cynomolgus macaques (*Mafa*) have been increasingly used over the past decades due to the limited availability of Indian rhesus macaques following the export ban in 1978 [1]. After the ban, researchers tried to find alternative models by using rhesus monkeys from China as well as other species of macaques such as pig-tailed macaque and the cynomolgus macaque. Because of the extensive distribution of *Mafa* throughout South East Asia, animals of different origins were used in experimental medicine. Numerous companies, primarily in the Philippines, Vietnam, China, and Malaysia, specialized to breed cynomolgus macaques in captivity. An original population of *Mafa* can also be found on Mauritius. Macaques were introduced on this island around 400 years ago by

Dutch and/or Portuguese sailors [2]. Recent molecular work confirmed that these animals, which were released in the wild, originated from either Java [3] or Sumatra [4]. The current population on Mauritius is derived from a small number—the estimate is around 20—founding animals [5,6]. The founding effect and total isolation of the Mauritius macaque produced a severe population bottleneck, resulting in a relatively poor genetic polymorphism when compared to the wild *Mafa* population of South East Asia [6]. The geographical distribution of cynomolgus macaques in South East Asia overlaps with that of rhesus monkeys north of the Kra isthmus. In this specific region, genetic studies demonstrated an introgression leading to gene flow from rhesus monkey males to *Mafa* [7–10].

Nonhuman primate models, including macaques, are frequently used in clinical and non-clinical testing because of their greater phylogenetic proximity to humans, when compared to mice or rats. The cynomolgus macaque is one of the nonhuman primate species used in clinical testing of organ allo-transplantations [11], stem cell allografts [12,13], the transfer of genes into stem cells [14], innovative vaccines [15] and immunotherapies [16], experimental infectious diseases [17], degenerative diseases and aging [18]. In all these fields, the use of a nonhuman primate model is justified by the absence of alternative animal models such as mouse or rat. Cynomolgus macaques are also used in non-clinical testing such as safety testing (toxicity, dependency, and reproductive and developmental toxicity testing), pharmacokinetic testing, and pharmacological efficacy testing. The proximity of the macaque and human immune systems has been a significant advantage in terms of the pertinence to extrapolate to humans the results obtained in macaques. For example, monoclonal antibodies for human therapy frequently cross-react with the macaque equivalent of the targeted human antigen [16]. However, decades of experimental data also demonstrate that specific aspects of the macaque immune system differ significantly from that of human. The example of the CD28 superagonist monoclonal antibody (TGN1412) is a paradigm of the dangers of extrapolating observations made in the macaque model to humans [19]. Despite the fact that this review focuses on the description of the *Mafa* MHC and highlights differences with its human counterpart, it is important to note that macaque immune related genes also differ from their human equivalents in many other loci, such as killer cell immunoglobulin-like receptors (KIRs) and leukocyte immunoglobulin-like receptors (LILRs), which coevolved with the MHC genes [20].

A detailed understanding of *Mafa* MHC polymorphism is essential for numerous experimental protocols where inter-individual histocompatibility must be met, or which involve antigenic peptide presentation by MHC class I or class II proteins. Indeed, numerous human studies have shown that MHC polymorphism is pivotal in allo-rejection of cells or tissues, vaccine responses, the control of infectious diseases, the development of autoimmune diseases and immune regulation during pregnancy. In all these areas, experimental medicine using the *Mafa* model has to take into account the MHC genetic variability of the animals used in the protocols. For example, in most therapeutic trials, it is crucial that animals in the treated and control groups are matched as closely as possible for their MHC type. In other cases, when an innovative immunosuppressive regimen is being tested to block allogeneic rejection, it is crucial to use donor/recipient pairs which are as incompatible as possible. It has been shown that the DRB incompatibility of macaques is correlated with the results of mixed lymphocyte cultures and that DRB genotyping facilitates the choice of donor recipient pairs [21].

The main problem which compromises the interpretation of medical research results based on animal models such as the *Mafa*, is that the impact of genetic polymorphism is not always taken into account. In captive breeding programs, genetic control usually refers to planned crosses that prevent the progression of consanguinity and promote the maintenance of genetic diversity. However, in order to improve the power of animal experiments while keeping the number of individuals used as low as possible, it is necessary to select animals sharing a common geographical origin and to systematically select animals with the experimentally appropriate polymorphic alleles in loci known to influence immune-related responses. The MHC genotype plays a key role in the selection of animals in all fields of medical research involving immune responses.

This review focusses on the macaque cynomolgus MHC and examines its genomic similarities and differences with human MHC regions, the history of techniques developed to study the MHC polymorphism of *Mafa* and their application to various fields such as the study of regenerative medicine and experimental infectious diseases.

2. Brief Review of Experimental MHC Polymorphism Milestones

In the early 70s, Hans Balner and his collaborators started the study of histocompatibility antigens in rhesus monkeys and chimpanzees [22–29]. The costs of breeding animals in captivity and the reduction in the number of feral animals that could be sourced, limited the overall use of chimpanzee and rhesus monkey models. These difficulties motivated the search for alternatives. In the early 70s, cynomolgus macaques could be sourced from Malaysia, Indonesia, and the Philippines. This triggered a renewed research interest in this species and Carolyn Keever and Eugene Heise characterized the Class I and Class II allo-antigens from alloimmune sera by working on a group of 277 feral Malaysian and Indonesian monkeys and 250 colony bred offspring for family studies [30–32]. In total, these serological studies led to the definition of three loci called A, B and C, consisting of 14, 10 and 6 alleles, respectively. The fact that in family studies the mixed lymphocyte culture (MLC) reactivity and survival times of skin grafts depended on donor-recipient CyLA compatibility, lead to the hypothesis that the serologically defined alleles were MHC gene protein products.

Development of MHC genotyping started with the study of the DRB exon 2 polymorphism by DGGE (denaturing gradient gel electrophoresis) separating allelic amplified fragments, followed by direct Sanger sequencing [33], a technique first applied to the study of rhesus monkeys [34]. An alternative approach used microsatellites to characterize the polymorphic MHC region in cynomolgus from various geographical origins [35]. A BAC-based contig map of the *Mafa* MHC genomic region was published by Watanabe et al. (2007) [36]. Wiseman et al. published the use of massive parallel pyrosequencing of cDNA amplified fragments to genotype MHC of cynomolgus macaque as well as other species of macaques [37]. Alice Aarnink et al. characterized MHC class I transcripts of a Malaysian cynomolgus macaque from EST libraries [38]. Westbrook et al. reported on the use of full-length MHC class I transcripts using single-molecule real-time (SMRT) sequencing and described *Mafa* alleles in various populations [39]. The main steps in the MHC characterization of the cynomolgus macaque are presented in Figure 1.

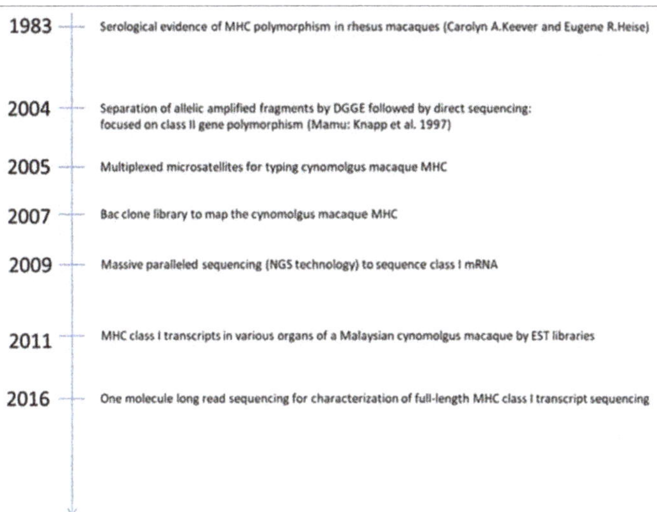

Figure 1. Milestones of cynomolgus macaque MHC studies.

3. Genomic Structure of the *Mafa* MHC Region and MHC Polymorphism

3.1. Length of the Mafa Region

Figure 2 shows a comparative MHC genomic map, based on the nucleotide length of each of the MHC class I, class II and class III regions in human (derived from genomic information of the current genome NCBI database (GRCh38.p2)) and *Mafa* taken from the previously mentioned Watanabe et al. (2007) study [36]. The nucleotide lengths of the whole *Mafa* and HLA regions from the class I region to the class II region are 3.92 Mb and 3.41 Mb, respectively. The *Mafa* and HLA class II and class III regions are well conserved and extend over a 0.93 Mb and 0.69 to 0.71 Mb region, respectively. In contrast, the *Mafa* and HLA class I regions extend over 2.28 Mb and 1.79 Mb nucleotide lengths, respectively, and the specific lengths of the MHC-A and MHC-B/I subregions which include redundant gene duplication events differ significantly between the two species.

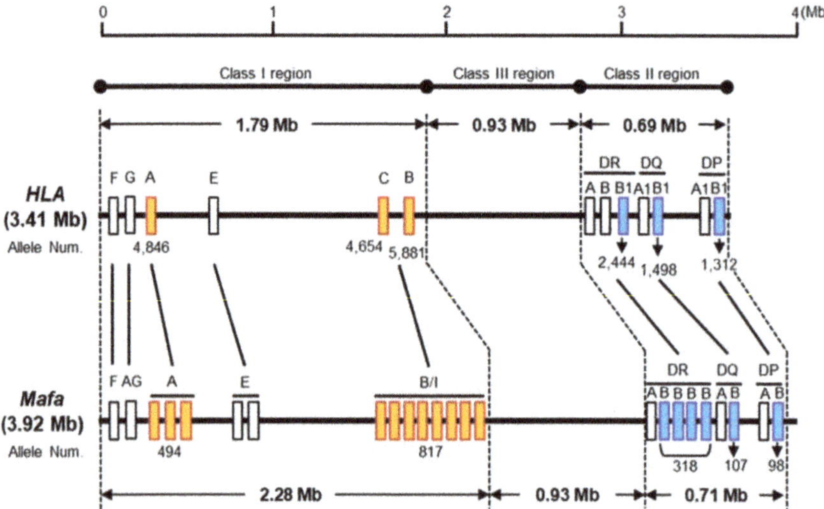

Figure 2. Comparative genome map of the HLA and *Mafa* regions. The number of *Mafa*-A, *Mafa*-B, and *Mafa*-DRB loci varies from one haplotype to another (*Mafa*-A from 1 to 3 loci, *Mafa*-B/I from 2 to 9, *Mafa*-DRB from 2 to 4). The numbers below the boxes indicate the number of alleles based on the IPD-IMGT/HLA version 3.35 and IPD-MHC version 3.2.0.0 database. The numbers of *Mafa*-A, *Mafa*-B and *Mafa*-DRB alleles correspond to the allele numbers of all *Mafa*-A loci (*Mafa*-A1 to -A6 and *Mafa*-A8), all *Mafa*-B loci (*Mafa*-B, -B11L -B12, -B16, -B17, -B20 and -B21), as well as all *Mafa*-DRB loci (*Mafa*-DRB*W, -DRB1, -DRB3, -DRB4, -DRB5 and -DRB6). For more details on allele numbers, see Table 1.

3.2. MHC Genes

The names of all thirty-three MHC loci and allele groups (18 class I: *Mafa*-A1, A2, A3, A4, A5, A6, A8, AG, B, B11L, B16, B17, B20, B21, E, F, G and I, and 15 class II: *Mafa*-DMA, DMB, DOA, DOB, DPA1, DPB1, DQA1, DQB1, DRA, DRB, DRB1, DRB3, DRB4, DRB5 and DRB6) in the *Mafa* region have been adopted by the Comparative MHC Nomenclature Committee as established by the International Society for Animal Genetics (ISAG) who are affiliated to the International Union of Immunological Societies (IUIS)-Veterinary Immunology Committee (VIC). Seven of these loci or allele groups (B11L, B16, B17, B20, B21, G and DRB6) do not express any RNAs.

The *Mafa* equivalents of the classical HLA-A and HLA-B loci and the non-classical HLA-E, HLA-F, and HLA-G loci are designated *Mafa*-A, *Mafa*-B, *Mafa*-E, *Mafa*-F and *Mafa*-G, respectively. *Mafa*-G is considered a pseudogene and the functions associated with HLA-G may have been taken over by

Mafa-AG, which shares similar features to HLA-G which is expressed in human placenta. In addition, eight splice variants have been observed in *Mafa*-AG [40]. To date, no ortholog of the HLA-C locus has yet been identified in the cynomolgus monkey or any of the other kinds of Old-World monkeys.

The MHC class II loci are well conserved between the *Mafa* and HLA class II regions, with the exception of the MHC-DRB, MHC-DQA2 and MHC-DQB2 loci that were generated by different evolutionary processes [41]. The gene order within the *Mafa* and HLA class II orthologous gene regions is essentially identical.

The MHC haplotypes of the Mauritian and Filipino populations, were characterized by nucleotide sequencing of the MHC genes and consisted of seven major *Mafa*-class I haplotypes and class II haplotypes in the Mauritian population [42–44] and 84 major *Mafa*-class I haplotypes and 16 *Mafa*-class I haplotypes in the Filipino population [45–47]. One to three *Mafa*-A loci, one to eight *Mafa*-B and one to three *Mafa*-DRB loci were located in the duplicated subregions. However, information on MHC haplotypes in populations other than Mauritius and the Philippines is still lacking.

3.3. Genes of MHC Class I Polypeptide-Related Sequences (MIC)

Three protein coding *Mafa*-MICA, *Mafa*-MICB and *Mafa*-MICB/A genes have been identified on the class III side of the *Mafa* class I region [36]. The *Mafa*-MICA and *Mafa*-MICB are orthologs of the human MICA and MICB genes, respectively, but *Mafa*-MICB/A is a hybrid of MICA and MICB generated by a crossing-over event with one breakpoint in the intron 3 region [48]. The MIC genes are polymorphic like the human MICA and MICB genes.

3.4. Non-MHC Genes

One hundred and twenty-two protein-coding non-MHC loci were identified in the HLA gene region, between GABBR1 and KIFC1 (Supplementary Table S1). The corresponding *Mafa* region contains 117 loci with 95.9% identity to the HLA sequences based on a comparison of the human GRCh38.p12 (GCF_000001405.38) and Macaca_fascicularis_5.0 (GCF_000364345.1) assemblies. In contrast, five HLA loci (PSORS1C1, LY6G6F, HSPA1A or HSPA1B, C4A and BTNL2) were not detected in the *Mafa* MHC genomic region.

3.5. Nomenclature of Mafa Class I and Class II Alleles

The *Mafa* class I and class II allele names were assigned by the IPD-MHC NHP database following classical rules [49]. An example of the allele nomenclature is *Mafa*-A1*001:01:01:01N where *Mafa*-A1 is the MHC allele of the *Mafa* and encoded by the A1 locus. The first field after the asterisk defines the lineage number, the second field after the first colon describes a non-synonymous substitution between two sequences, the third field after the second colon describes a synonymous substitution between two sequences, and the fourth field after the third colon describes a substitution in the non-coding region between two sequences. The final letter 'N' indicates an altered level of expression such as for a null allele in the example given above.

3.6. Characteristics of Mafa Class I and Class II Allele Numbers

Table 1 shows the latest HLA and *Mafa* published allele numbers dating from 30 May 2019. A total of 22,140 different allele sequence variations are annotated for HLA, with over 1000 different alleles published for the classical HLA loci, HLA-A, B, C, DRB1, DQB1 and DPB1. The class II α chain loci, HLA-DRA, DQA1 and DPA1, in contrast, are less polymorphic than the β chain loci, while the non-classical HLA loci, HLA-E, F, G, DOA, DOB, DMA and DMB, are not very polymorphic at all. In the case of *Mafa* only a total of 2196 different MHC allele sequence variants have been published to date. The MHC allele numbers in the macaque is expected to increase further because many novel alleles are still regularly identified in Vietnamese and Cambodian populations. In the *Mafa*-A1, B and DRB subregions, which result from gene duplication events, more allelic variants are observed for *Mafa*-A1, B and DRB*W than the other class I loci. The non-classical MHC loci, *Mafa*-E, F, G, DOA,

DOB, DMA and DMB, like the human equivalent region, are not very polymorphic at all. Unlike the HLA orthologs, *Mafa*-DRA, DQA1 and DPA1 have similar allele numbers as the β chain loci (Table 1).

Table 1. Allele numbers of MHC genes in human and cynomolgus macaque.

Region	Allele Group	HLA Locus	Allele Num.	*Mafa* Locus	Allele Num.
Class I	MHC-F	HLA-F	44	*Mafa*-F	8
	MHC-G	HLA-G	68	*Mafa*-G	10
	MHC-AG	X	X	*Mafa*-AG	36
	MHC-A	HLA-A	5018	*Mafa*-A1	320
				Mafa-A2	83
				Mafa-A3	30
				Mafa-A4	38
				Mafa-A5	8
				Mafa-A6	14
				Mafa-A8	1
	MHC-E	HLA-E	30	*Mafa*-E	16
	MHC-C	HLA-C	4852	X	X
	MHC-B	HLA-B	6096	*Mafa*-B	717
				Mafa-B11L	5
				Mafa-B16	11
				Mafa-B17	4
				Mafa-B20	2
				Mafa-B21	2
	MHC-I	X	X	*Mafa*-I	76
Class II	MHC-DR	HLA-DRA	7	*Mafa*-DRA	52
		HLA-DRB1	2403	*Mafa*-DRB*W	160
		HLA-DRB3	217	*Mafa*-DRB1	82
		HLA-DRB4	108	*Mafa*-DRB3	13
		HLA-DRB5	77	*Mafa*-DRB4	10
				Mafa-DRB5	18
				Mafa-DRB6	35
	MHC-DQ	HLA-DQA1	149	*Mafa*-DQA1	100
		HLA-DQB1	1560	*Mafa*-DQB1	107
	MHC-DO	HLA-DOA	12	*Mafa*-DOA	15
		HLA-DOB	13	*Mafa*-DOB	16
	MHC-DM	HLA-DMA	7	*Mafa*-DMA	11
		HLA-DMB	13	*Mafa*-DMB	7
	MHC-DP	HLA-DPA1	106	*Mafa*-DPA1	91
		HLA-DPB1	1360	*Mafa*-DPB1	98
Total			22,140		2196

The HLA and *Mafa* MHC allele numbers refer to IPD-IMGT/HLA Release 3.36.0 and IPD-MHC Release 3.2.0.0 (2018-12-18) build 780, respectively.

4. Diversity of Cynomolgus Macaque Populations

The polymorphic MHC region has been studied in several cynomolgus macaque populations (Table 2). Differences among these populations were clearly highlighted by microsatellite studies [35]. Genetic population studies of four distinct populations confirmed that the Mauritian cynomolgus macaque (MCM) population has the most restricted MHC polymorphism [50]. The Philippine macaque population also showed a restricted polymorphism compared to the Indonesian or Vietnamese *Mafa* populations [50]. The study of 36 autosomal microsatellites (18 spread across the MHC region and 18 outside MHC) in a sample of 254 individuals from four populations (Vietnam, Java, the Philippines, and Mauritius) suggested that the class III region was subjected to selection in the Philippine population [51]. This potential selection event was deduced from the difference between the observed

F(st) and that deduced from a neutral model using two methods based on contrasting demographic models. The two approaches showed a signal of positive selection in the MHC class III region. This departure from a neutral model in the class III region was much more significant than the one previously reported for the DRACA marker, which may have hitchhiked due to its proximity to the class III region [50].

4.1. Study of MHC Class I Alleles in Various Populations

Many investigators have examined the MHC class I region of *Mafa* from various origins [38,47,52–63], but only a limited number of studies have looked at *Mafa* from Chinese breeding facilities [39,64,65]. Because no feral *Mafa* are found in China, Chinese breeding facilities are based on animals sourced from a variety of geographical locations, most likely from Thailand, Laos, Vietnam, Cambodia, and Malaysia, but also possibly from Indonesia and the Philippines. Cynomolgus macaques from Chinese breeding facilities therefore are genetically heterogeneous, and this heterogeneity may be particularly biologically relevant in highly diverse genomic regions like the MHC.

The study of class I cDNA sequences in various populations confirmed the complexity of *Mafa* Class IA and class IB haplotypes. Indeed, as in other *Cercopithecidae*, *Mafa* MHC haplotypes are characterized by a number of classical class I genes which are much larger than the corresponding human HLA haplotypes, the latter only encompassing three classical class I genes (HLA-A, -B, -C). The study of class IA and class IB alleles present in various *Mafa* populations indicated that a limited number of alleles and haplotypes are shared between populations. These studies also revealed sharing of several class I alleles with *Macaca mulatta*, or *M. nemestrina* [49].

4.2. MHC Class II Allele Sharing

Comparative study of DRB exon 2 in various *Mafa* populations confirmed that Mauritius and the Philippines have a restricted DRB polymorphism when compared to Vietnamese or Indonesian macaques [45]. The Vietnamese and Indonesian populations have a high diversity of DRB alleles [66,67]. Extensive interspecies DRB allele sharing was demonstrated by comparison *with Macaca mulatta* and *Macaca nemestrina* [49,68].

About one third of the *Mafa*-DRB exon 2 sequences are identical to rhesus macaque orthologs and one half of the *Mafa*-DPB1, *Mafa*-DQA1, and *Mafa*-DQB1 alleles are identical to *Mamu* orthologs [68]. Allele sharing between *M. fascicularis*, *M. mulatta* and *M. nemestrina* was also observed for the MHC DRA gene [69]. Despite extensive allele sharing, rhesus and cynomolgus monkeys differ in terms of MHC class II haplotypes, suggesting that after speciation, recombination processes generated species-specific haplotypes [68].

Table 2. Cynomolgus macaque MHC polymorphism in various populations.

Reference	Year	Population of Macaques	Class I E	Class I F	Class I A	Class I B	Class II	DRA	DRB Exon 2	DRB cDNA	Class II Others	Microsat.
Boyson et al. [70]	1995	Not specified	X									
Gaur and Nepom [71]	1996	Not specified							X			
Alvarez et al. [72]	1997	Not specified	X									
Otting et al. [73]	2002	Not specified									DQB	
Leuchte et al. [33]	2004	Mauritius							X			
Uda et al. [62]	2004	Not specified			X							
Krebs et al. [65]	2005	Vietnam, Mauritius, Chinese breeding facilities			X (RSCA) (1)	X (RSCA)						
Uda et al. [61]	2005	Not specified			X	X						
Blancher et al. [21]	2006	Mauritius, The Philippines							X			
Sano K et al. [74]	2006	Indonesia, Vietnam, The Philippines									DPB1	
Doxiadis et al. [68]	2006	Indonesia Malaysia Thailand Java Sumatra Vietnam, Chinese breeding facilities							X		DPB1, DQA1, DQB1	
O'Connor et al. [43]	2007	Mauritius						X	X	X	DPA, DPB, DQA, DQB,	X
Wiseman et al. [75]	2007	Mauritius			X (RSCA)	X (RSCA)			X			X
Pendley et al. [59]	2008	Indonesia			X	X						X
Wiseman et al. [37]	2009	Mauritius			X	X						
Campbell et al. [53]	2009	The Philippines			X	X	X					
Kita et al. [54]	2009	Indonesia, Vietnam, The Philippines			X							
Aarnink et al. [69]	2010	Indonesia, Mauritius, The Philippines, Vietnam						X				X
Craeger et al. [66]	2011	Indonesia and Vietnam						X		X	DPA, DPB, DQA, DQB, cDNA	
Ling et al. [56]	2011	Vietnam									DOB, DPB1, DQB1	

Table 2. *Cont.*

Reference	Year	Population of Macaques	Class I E	Class I F	Class I A	Class I B	Class I I	DRA	DRB Exon 2	DRB cDNA	Class II Others	Microsat.
Blancher et al. [45]	2012	The Philippines Java, Vietnam, Mauritius						X	X	X		
Ling et al. [76]	2012	Vietnam				X			X		DQB1	
Mitchell et al. [77]	2012	Indonesia			X (RSCA)	X (RSCA)						X
de Groot et al. [78]	2014	Indonesia, Indochina			X	X			X		DQA, DQB	X
Shiina et al. [47]	2015	The Philippines	X	X	X	X	X					
Karl et al. [64]	2017	Chinese breeding facilities			X	X	X					

RSCA: [1] Reference strand conformational analysis.

5. Comparison of MHC Polymorphism among Populations

As mentioned above, the Mauritius *Mafa* MHC is not very polymorphic when compared to the other *Mafa* populations [43,52]. Table 3 compares MHC polymorphism among the other three populations (Vietnam, Indonesia, Philippines).

Vietnamese and Indonesian populations are better suited for biomedical research and non-clinical testing, in particular for exploring the impact of genetic polymorphisms. In contrast, the Mauritian and Filipino populations are better suited to biomedical research addressing genetic homogeneity because these populations have fewer alleles than other production areas [45,54,69,74]. Detection and breeding of MHC homozygotes is necessary for the development of vaccines and immunosuppression protocols, and for evaluating the usefulness of organs and regenerated cells derived from induced pluripotent stem (iPS) cells and embryonic stem (ES) cells [79,80]. It is for this reason that, Mauritius and Filipino cynomolgus macaques are considered to be the most suitable populations for use in biomedical research [3,4,81].

Table 3. Example of inter-population comparison based on MHC polymorphism.

Locus	Total Allele Num.	Vietnamese	Indonesian	Filipino
Mafa-A1	76	34	32	13
Mafa-DRA	22	17	–	12
Mafa-DRB	134	84	60	35
Mafa-DPB1	40	26	23	12

Polymorphism analysis results are based on 30 animals per population. The numbers indicate the number of MHC alleles detected in each population. The table shows that the Filipino population is less polymorphic than any of the other populations for any of the MHC genes tested.

6. Identification of MHC Region Homozygotes

MHC genotyping of animals originating from Mauritius using microsatellites was able to identify MHC homozygous animals [82,83]. In the case of Filipino animals, MHC genotyping based on high throughput sequencing of MHC class I and II cDNA (Ion PGM system Thermo Fisher Scientific) was used to identify MHC homozygous animals [47,84]. So far, a total of 207 MHC alleles have been identified from approximately 5,500 Filipino animals genotyped using this method, and we were able to detect 38 homozygous animals using any one of the 15 most frequently encountered MHC haplotypes (HT1 to HT15). Examples of Filipino animal MHC genotypes used for transplantation of induced pluripotent stem cells (iPS) are given in Figure 3 (see thereafter and Figure 4).

(A) MHC homozygote (e.g. HT1/HT1)

Mafa-F	F-like1	F-like1
Mafa-A	A1*089:03 A2*05:50 A3*13:03:01	A1*089:03 A2*05:50 A3*13:03:01
Mafa-E	E-like3 E-like10	E-like3 E-like10
Mafa-B/I	B*104:03 B*144:03N B*057:04 B*060:02 B*046:01:02 B*050:08 B*114:02 B*072:01 I*01:12:01	B*104:03 B*144:03N B*057:04 B*060:02 B*046:01:02 B*050:08 B*114:02 B*072:01 I*01:12:01
Mafa-DRB	DRB*W1:08 DRB*W3:01 DRB*W36:01	DRB*W1:08 DRB*W3:01 DRB*W36:01
Mafa-DQA1	DQA1*26:03	DQA1*26:03
Mafa-DQB1	DQB1*18:07:02	DQB1*18:07:02
Mafa-DPA1	DPA1*02:15:02	DPA1*02:15:02
Mafa-DPB1	DPB1*10:01	DPB1*10:01

(B) MHC heterozygote (e.g. HT1/Other)

Mafa-F	F-like1	F-like7
Mafa-A	A1*089:03 A2*05:50 A3*13:03:01	A1*094:01 A2*05:04
Mafa-E	E-like3 E-like10	E-like7
Mafa-B/I	B*104:03 B*144:03N B*057:04 B*060:02 B*046:01:02 B*050:08 B*114:02 B*072:01 I*01:12:01	B*159:01 B*007:01:01 B*158:01 B*085:01 B*079:02:02 B*098:08 I*01:13:01
Mafa-DRB	DRB*W1:08 DRB*W3:01 DRB*W36:01	DRB1*03:21 DRB1*10:07
Mafa-DQA1	DQA1*26:03	DQA1*01:07:01
Mafa-DQB1	DQB1*18:07:02	DQB1*06:08
Mafa-DPA1	DPA1*02:15:02	DPA1*02:05
Mafa-DPB1	DPB1*10:01	DPB1*15:04

(C) Animal with MHC haplotypes excepting HT1-type (Other/Other)

Mafa-F	F-like4	F-like4
Mafa-A	A1*052:02 A4*01:04	A1*071:02 A4*14:14
Mafa-E	E-like5 E-like11	E_like4 E_like9
Mafa-B/I	B*033:02 B*096:01 B*098:10	B*056:02:01 B*017:01 B*157:01 B*050:08 B*060:03:01 B*089:01:02 B*116:01 I*01:11 I*91:01:01
Mafa-DRB	DRB1*03:21 DRB1*10:07	DRB1*03:18 DRB*W1:04 DRB*W2:02
Mafa-DQA1	DQA1*01:07:01	DQA1*01:13
Mafa-DQB1	DQB1*06:08	DQB1*06:17:02
Mafa-DPA1	DPA1*02:05	DPA1*07:04
Mafa-DPB1	DPB1*15:04	DPB1*21:01

Figure 3. Examples of useful animals for non-clinical transplantation studies. Colored background shows MHC haplotype composition of HT1-type. (**A**) genotype of an HT1 homozygous animal, (**B**) genotype of an HT1 heterozygous animal (**C**) genotype of a heterozygous animal which has two haplotypes differing form HT1.

7. Characterization of MHC Class I Expression Levels in Various Tissues and Conditions

As detailed elsewhere in this review, the macaque MHC is characterized by its multiplicity of functional MHC class I genes. This is due to multiple rounds of duplications that resulted in the accumulation of classical class IA and class IB genes. Using high-throughput sequencing of exon 2 from amplified genomic DNA (190 bp), we characterized 23 classical class I genes from an MHC homozygous Mauritian animal (H2/H2) and 38 classical class I genes from a Malaysian cynomolgus monkey which was a putative MHC heterozygote [38]. Among these 38 classical MHC class I exon 2 variants only 25 had an intact open reading frame (ORF). Only 12 of these ORFs were associated with functional RNA sequences [38]. The availability of Expressed Sequence Tag (EST) libraries derived from mRNA extracted from six tissues of a Malaysian macaque (thymus, spleen, bone marrow, liver, heart and pancreas) [85], allowed the number of MHC class I sequences expressed in the various tissues to be estimated. MHC class I sequences deduced from EST libraries were compared to those obtained by massive pyrosequencing of MHC class I cDNA amplified from a lymph node RNA sample [38]. The study of the six EST libraries from various organs of the Malaysian animal studied revealed 16 MHC classical class I transcripts, 12 of which were associated with ORFs. The relative frequencies of classical class I transcripts in the lymphoid organs largely exceeded those observed in other non-lymphoid tissues. The thymus expressed the greatest number of transcript variants with relative frequencies of the various class I transcripts displaying a significant difference from those observed in the spleen (chi-square test, $p = 0.004$). The relative frequencies of the MHC class I transcripts in the lymph node differed significantly from those observed in the spleen and in the thymus (chi-square test, $p < 0.0001$ and $p = 0.002$, respectively). No MHC classical class I gene transcript was detected from the EST library derived from pancreatic mRNA. The number of MHC class I functional transcripts defined in the Malaysian animal was within the estimated range determined using high-throughput sequencing, with between 17–23 of classical MHC class I genes transcribed in the MHC heterozygote cynomolgus macaques [37,52].

The evolutionary advantage of maintaining the efficient transcription of such a large variety of classical MHC class I genes is difficult to understand. From a classical point of view, the expression of MHC genes may have subtly adapted, to on the one hand avoid expressing an excessive variety of alleles leading to disproportionate levels of negative thymic selection and, on the other hand, an allelic diversity too weak to ensure the presentation of a large variety of exogenous peptides which is essential for mounting immune responses to the huge variety of microbial pathogens potentially encountered [86]. According to theoretical models, the ideal number of alleles seems to be around three MHC class I genes and three MHC class II genes [86]. One possible functional advantage to expressing such a variety of class I genes per haplotype is to allow the differential tissue expression of these genes. Differential MHC class I expression in distinct leukocyte subsets was studied by Green et al. (2011) using high-throughput pyrosequencing [87]. Transcription of certain MHC class I genes species varied significantly between different cynomolgus macaque leukocyte subsets. For example, the *Mafa*-B*134:02 RNA was virtually undetectable in CD4+ T cells, while it represented over 45% of class I transcripts in monocytes [87]. The authors also analyzed the expression of MHC proteins at the cell surface with fluorescent peptides capable of accessing the peptide grove of certain cynomolgus MHC class I proteins [87]. This demonstrated that distinct leucocyte subsets expressed MHC proteins differentially. A parallel study of human leukocytes revealed that expression of human HLA class I proteins does not significantly vary in the human cell subsets [87].

Another advantage of having a large number of class I genes per haplotype could be substantiated by the ability to induce expression of specific genes during particular infections. Changes in global gene expression were assessed for the brain, lungs, and spleen after aerosol exposure to the Venezuelan equine encephalitis viruses (VEEV) [88]. At day 3 post inoculation, the study revealed the induction of major histocompatibility complex (MHC) class I transcripts in the brain with no effect in the lungs or spleen [88]. Systematic study of the virus-dependent induction of MHC class I expression in cynomolgus macaque tissues remains to be explored. As it stands, we do not currently understand the overall functional advantages of expressing such a large variety of MHC class I genes per haplotype.

8. MHC and Experimental Infectious Diseases in Cynomolgus Macaque

Because the sets of peptides presented by MHC class I and class II proteins are allele dependent, the MHC polymorphism is associated with the control numerous infectious diseases. The influence that MHC polymorphism exhibits in controlling the SIV infection has been extensively studied and will be expanded on below. The peptide repertoire presented to macaque CD4+ T lymphocyte was also investigated in the case of tuberculosis [89]. Although this study, concluded that the immune repertoire of *Mafa* and of rhesus monkey largely overlap with that of humans, the association between immune responses to tuberculosis and the MHC class II polymorphism was not explored. Any future studies of innovative therapeutics or vaccines in the cynomolgus macaque have to take into account the genetic variability of *Mafa* MHC. Indeed, despite the phylogenetic proximity of macaques and humans, the fact that some macaques present a given set of peptides derived from a given infectious agent is not a guarantee that humans will present the same set of peptides.

9. MHC and Control of SIV Infection

The years following the initial discovery of the AIDS epidemic in humans, were marked by reports of some individuals spontaneously controlling the infection over extensive periods of time. Multiple genetic association studies of the phenomenon demonstrated that the controller status is associated with specific HLA alleles, in particular the HLA-B57, HLA-B*58:01 and HLA-B27 alleles (for review see Martin and Carrington [90]). In contrast, the HLA-B35 allele is associated with poor control of viral infection and rapid progression to full-blown AIDS. The presence of protective HLA alleles was shown to be associated with a strong response of cytotoxic CD8+ T lymphocytes specific for particular HIV virus peptides, called dominant peptides. Multiple studies established that the HLA-B*57 B*58:01 and B*27 alleles present a set of peptides preferentially derived from gag protein, which become the dominant targets of cytotoxic CD8+ T cell responses. The proof of the effectiveness of the cytotoxic responses in controlling the infection comes from the demonstration of the progressive accumulation of viruses exhibiting non-synonymous mutations mapping to the regions encoding the dominant viral peptides presented by the protective HLA alleles, in controller patients. These mutated viruses circumvent the immune response because CD8+ T lymphocytes are now incapable of recognizing the mutated peptides since these can no longer be presented by the class I MHC molecules of the patient. These mutated viruses have a reduced replicative capacity (VRC viral replicative capacity) and the selection pressure exerted on the viruses by the controller's immune response results in selecting viruses that accumulate compensatory mutations to help restore viral replication capacities. The gradual accumulation of these mutated viruses containing reversal mutations correlates with progression of the disease. Viruses encoding mutations which impart resistance to protective HLA alleles are transmissible and accumulate in populations affected by the epidemic so that the protective effect of HLA alleles decreases over time. This was first reported in Japan, where the protective role of allele HLA-B*51 disappeared with the rapid spread of mutated viruses [91]. In summary, in this arms race, the adaptive power of the virus far exceeds that of humans so that the population of viruses that prevails in a given infected human population gradually adapts to the HLA alleles associated with controlling the infection [91–94].

After the discovery of the SIV virus, the rhesus macaque of Indian origin became the animal model for virally induced AIDS. This animal model is by far the most frequently investigated and the best characterized non-human primate model. It reflects the observations in humans, with some rhesus monkeys also able to control the SIV infection for extended periods of time. Because the MHC polymorphism of rhesus monkeys was so well characterized, it was possible to demonstrate that in Indian rhesus monkeys the SIV controller phenotype was associated with the *Mamu*-A*01, *Mamu*-B*08, and *Mamu*-B*17 alleles. After SIVmac251/SIVmac239 challenge, rhesus macaques expressing these alleles presented with slow disease progression [95–98]. Interestingly, the *Mamu*-B*08 allele has very similar peptide presentation restrictions to those of the human HLA-B*27 allele [99]. As for the peptides derived from SIV, the peptides presented by the *Mamu*-B*08 allele include the homologs of the HIV peptides presented by the human allele HLA-B*27 [99,100]. In a manner, similar to that observed in humans, viruses carrying mutations mapping to the regions of the dominant peptides presented by *Mamu*-B*08, evade immune control when injected into *Mamu*-B*08 expressing animals [100]. Although viruses with multiple mutations have reduced replicative capacity than non-mutated viruses, they are nevertheless capable of inducing experimental AIDS in *Mamu*-B*08 animals [101].

Due to the increasing difficulty of accessing sufficient numbers of rhesus monkeys of Indian origin, researchers have attempted to use rhesus monkeys from China, but these animals differed too much from animals of Indian origin in terms of SIV-induced AIDS susceptibility. Since then, the cynomolgus monkey has started to be used more extensively as an alternative animal model for SIV infection, with the Mauritius cynomolgus macaque (MCM) being the most frequently used animal source. As previously mentioned, the MCM population experienced a sharp genetic bottleneck, as only a relatively small number of animals were initially introduced onto the island. With time, the total isolation of this macaque population favored a decrease in the total number of MHC alleles by genetic drift and/or local adaptations. Extensive investigation of MHC polymorphism in the current Mauritian macaque population has found that a mere seven of the ancestral MHC haplotypes have persisted. Due to recombination events accumulated over a period of 400 years on the island, a high percentage (about 31%) of recombinant haplotypes can be detected in this specific population [102]. The percentage of recombinant haplotypes allowed us to estimate the rate of recombination in the MCM population to be between 0.4 and 0.8% [102].

Many studies have reported an association between MHC polymorphism and control of the SIV infection in the MCM model. However, the conclusions drawn differ from one study to the other. Wiseman et al. 2007 demonstrated that MCM with identical MHC haplotypes mounted comparable cellular immune responses and maintained similar viral loads following infection with the widely used virus isolate SIVmac239 [75]. However, further studies found that MCM sharing MHC genotypes did not mount similar CD8 immune responses [103]. Moreover, M3 and M6 class IB and class II haplotypes were reported to be associated with resistance to chimeric SHIV89.6P challenge, while H2 and H5 class IB and class II haplotypes were associated with susceptibility to infection [104].

In an earlier study, Burwitz et al. demonstrated that the three most frequently found Mauritian MHC haplotypes share a pair of MHC class IA alleles, *Mafa*-A*25 and *Mafa*-A*29, which condition the presentation of SIV peptides to CD8+ lymphocytes [105]. They also reported on the accumulation of substitutions in the targeted peptides consistent with an immune evasion as a result of the selective pressure exerted by the cellular immune response [105]. Mee et al., showed that the MHC M6 haplotype in MCM is associated with a sustained control of the SIVmac251 infection [106]. Mee et al. subsequently demonstrated that the M3 haplotype was associated with a rapid control of SHIVsbg infection (significant differences in viral load apparent at 28 days p.i. despite comparable viral loads at day-14 p.i.) [107]. Aarnink et al. reported on the influence of the MHC genotype on the progression of experimental SIV infection (SIVmac239 strain), in the MCM [108]. The study of 44 animals allowed to detect an association between the plasma viral load at the set point and markers located in the MHC class IB region. Three MHC class IB haplotypes were significantly associated with lower (M2 and M6) or higher (M4) set point PVL values [108]. Two recent studies based on the same animal cohort highlighted evidence that several genetic factors outside the MHC are potentially involved in controlling SIV infections in the Mauritian macaque model [109,110]. In 2014, Alessandra Borsetti showed that MHC could influence the secretion of cytokines such as IL10 [111]. In the Mauritian macaque model, the MHC Class II M3 haplotype is associated with a lower acute and post-acute IL-10 level. Moreover, the class IA M3 haplotype is associated with a lower level of alpha defensins in the post-acute phase of infection [111]. The Mauritian MHC haplotypes associated with controlling the SIV infection vary from one study to the next. There are multiple causes to explain these discrepancies.

Firstly, these studies are based on results from a limited number of animals that, in most cases, did not allow an analysis of all Mauritian MHC haplotypes. Secondly, the SIV strain used, the inoculation dose and the route of inoculation differ from one study to the other, so there are many confounding factors which likely bias results obtained in MHC association studies. Despite the discrepancies between the definition of the haplotypes associated with controlling SIV infection in the Mauritian macaque model, all published studies are in agreement with regards to the important role that the MHC class IB region plays in this mechanism. The host MHC class IB genetic background must be taken into consideration when designing and interpreting the results of any future SIV vaccine and therapeutic efficacy studies [112].

The beneficial effects of MHC heterozygosity in the control of experimental SIV infections were explored in the MCM model [113]. A statistically significant heterozygous advantage was demonstrated in HIV infected patients but it was difficult to explore the mechanism of this advantage because HIV strains are different for each individual HIV+ patient. In addition, individuals homozygous for a given HLA locus may also include different alleles at that locus and may also differ extensively from each other at other HLA loci. The advantage of the MCM model is that animals homozygous for the most frequent MHC haplotypes are not excessively rare. It is therefore possible to compare the evolution of the SIV infection in MCM homozygous for M1, M2 and M3 MHC haplotypes, with that in the heterozygous animals. Results from the numerous studies using this approach are nevertheless not so easy to synthesize because some conflicting observations were reported. In 2010, Shelby O'Connor et al. reported that MHC heterozygote MCM (five animals) have a lower plasma viral load than homozygous animals (eight animals) after experimental inoculation with the SIV239 strain [113]. In 2012, Budde et al. in a study on a larger number of MCMs (27 animals out of which 12 were homozygous) did not confirm the heterozygote advantage when all animals were considered together [42]. Restricting the comparison to M3/M3 ($N = 6$) and M1/M3 heterozygous animals ($N = 6$), showed that homozygous animals had a higher viral load at the set point than heterozygous animals [42]. Surprisingly, in 2012, Greene et al. reported that the T CD8+ lymphocytes from M1/M3 animals predominantly targeted SIV peptides restricted to the M1 haplotype [114]. This is not in favor of the theory arguing that the advantage of heterozygous individuals is imparted by the larger array of peptides presented and by heterozygotes who possess two different sets of MHC molecules instead of the one (as is case for homozygous animals). Although control of SIV infection differs between M1/M3 individuals and M3/M3 and M1/M1 individuals, it is not certain that the advantage of MHC heterozygosity results from an extension of the panel of viral peptides targeted by CD8 T lymphocytes. In 2016, the same group reported on immune "escaped" virus conditioned by MHC haplotypes [115]. They assessed the appearance of variants at the chronic phase of SIVmac239 infection in homozygous (M1/M1 or M3/M3) and heterozygous (M1/M3) MCM. The study demonstrated that a category of variants accumulated preferentially in M3/M3 individuals. By contrast, CD8+ lymphocytes specific for wild type (wt) peptides became more abundant in M1/M3 individuals. These results suggested that the accumulation of mutant escaped virus for a given peptide is associated with a progressive decrease of CD8+ lymphocytes specific for the wt peptide [115]. This association does not imply a cause and effect relationship between the two observations. Moreover, these observations are not consistent with the hypothesis of a more diversified CD8 response in heterozygous animals. Therefore, despite the possibilities offered by the MHC singularities of the MCM macaques, the detailed mechanisms of the heterozygous advantage in the fight against SIV infection remains elusive.

In a more recent study, Li et al. reported that the protective effect of the M3 haplotype could be restricted to M3/M4 heterozygous animals [116]. However, this study was based on a very small number of animals ($n = 12$) among these were two animals showing the lowest plasma viral load at the set point which had an MHC M3/M4 genotype. Another recent study by Bruel et al. reported that long term control of SIV infection in MCM is not associated with an efficient SIV-specific CD8+ T lymphocyte response [117]. This contradicts a previous study reporting that a CD8+ specific response correlated with restriction of SIV replication in the MCM model [42].

10. Allogenic Lymphocyte Transfer

The limited MHC diversity in the Mauritian macaque population, allows one to investigate adoptive transfers of allogenic lymphocytes between fully MHC matched animals (i.e., animals sharing two common MHC haplotypes). Indeed, the adoptive transfer of lymphocytes is indispensable for the study of the cellular immune response and is a promising strategy in particular clinical circumstance [118,119]. Initial experiments in MCM demonstrated that allogenic lymphocytes isolated from blood or various lymphoid organs did not subsist beyond 14 days in the full MHC-matched unrelated recipients [82,120]. Persistence of transferred lymphocytes was slightly prolonged in MHC-matched macaque siblings although it did not equal the persistence of autologous lymphocytes [83]. In the macaque model, only autologous adoptive transfer ensured the prolonged persistence of transferred lymphocytes in various organs [121]. However, neither allogenic nor autologous lymphocyte transfers of CD8+ T lymphocytes specific for SIV succeeded in controlling the SIV infection of the recipient. There are certainly multiple and complex reasons to explain this inefficacy. One can evoke the difficulty of transferred lymphocytes to localize to organs specifically affected by intense replication of SIV (i.e., the mucosae). Another difficulty in evaluating lymphocyte persistence in the recipient is that all experiments are based on labelling transferred cells with fluorescent dyes such as CFDA-SE (carboxyfluorescein diacetate succidimyl ester) or PKH67 (a compound characterized by a highly aliphatic alkyl tail attached to a fluorochrome). In both cases, labelling cells prior to transfer is susceptible to modify their cell surface properties as well as their metabolism. Alternatives to these fluorescent labels could be explored in order to avoid introducing CFDA-SE or PKH67 toxicity artefacts [122,123].

11. Role of MHC in Induced Pluripotent (iPS) Stem Cell Allografts

In Japan, an iPS cell stock project is underway to collect HLA haplotype homozygous iPS cells for treating HLA-matched patients [124]. The transplantation of differentiated cells from patient's autologous iPS cells encompasses three major problems: it is expensive, it requires time-consuming processing for the preparation of differentiated cells and runs the risk of unmasking dormant inherited diseases. Ready-to-use HLA homozygous iPSCs are expected to solve these problems. In order to investigate the efficacy of ready-to-use MHC homozygous iPS cells, we established an iPS cell transplantation model system in cynomolgus macaque in which regenerative cells differentiated from iPS cells derived from MHC homozygotes are transplanted into MHC heterozygotes (Figure 4A).

This macaque transplantation system is used for transplantation of differentiated iPS cells such as retinal pigment epithelial cells [125,126], dopaminergic neuron cells [127,128], cardiomyocyte cell sheets [129] and cardiomyocytes [84]. Differentiated cells from MHC homozygous iPS cells were functional in vivo and minimal rejection was observed in MHC heterozygote recipients after transplantation in any of the cases. In contrast, transplantation of MHC mismatched animals often resulted in severe rejection (Figure 4B). In addition, the dose of immunosuppressant can be reduced in *Mafa* matched transplantations relative to *Mafa* mismatched transplantations [84,125–130]. This transplantation model can therefore be applied to the development of a therapy for suppressing graft rejection and is an efficient protocol for studying the effects of immunosuppressants in a nonclinical context.

We established a reproductive technique using intracytoplasmic sperm injection (ICSI) to maintain the necessary number of MHC-controlled cynomolgus macaques [131]. Namely, this is a technique for injecting MHC homozygous sperm cells into MHC heterozygous oocytes by microinjection. We have so far produced several MHC homozygotes and more than 10 MHC heterozygous animals using this technology.

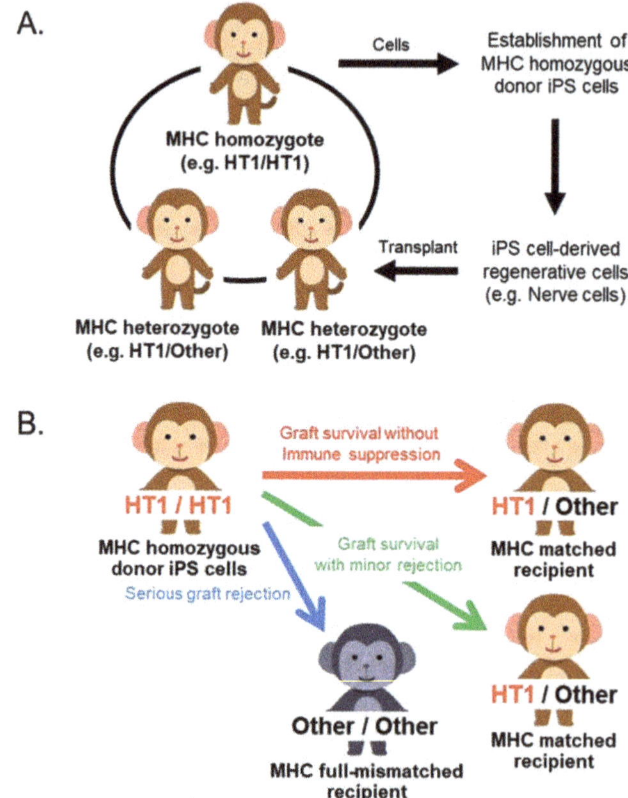

Figure 4. The iPS cell transplantation model and application for iPS cell transplantation study using MHC defined cynomolgus macaques. (**A**) This monkey model is used for analyses of engraftment rate of transplanted cells and inflammatory responses and used for evaluating safety and usefulness of iPS cell transplantation. (**B**) Summary of transplantation outcome between MHC matched recipient and mismatched recipient. MHC type mismatch in transplantation causes severe rejection. This transplantation model can be applied to the development of a therapy for suppressing graft rejection and an efficient protocol of an immunosuppressant.

12. Impact of MHC Polymorphism on Blood Counts

We investigated the impact of MHC polymorphism on various parameters obtained by complete blood count and flow cytometric analysis of lymphocyte populations in 200 unrelated cynomolgus macaques born in captivity from animals originating from the Philippines [132]. The MHC polymorphism was characterized using 14 microsatellites markers distributed across the MHC. The DRB locus was genotyped by denaturing gradient gel electrophoresis and sequencing. Among all cell count parameters analyzed, only two were associated with MHC polymorphism: CD4+ T lymphocytes and platelets [132,133]. The CD4+ T lymphocyte blood count was significantly associated with a DRACA-DRB haplotype ($p < 8 \times 10^{-7}$) [132], while the platelet count was significantly associated with two markers in the proximal class I region and class III region [133]. Examination of a Japanese cohort (14,967 subjects from the BioBank Japan Project) revealed several single nucleotide polymorphisms (SNPs) potentially associated with platelet count in a region homologous to the one we identified in the macaque model. It remains to be determined whether these associations are valid in other *Mafa* populations such as that of Mauritius.

13. Macaque MHC and Experimental Autoimmune Diseases

The cynomolgus macaque is currently used as an animal model for several human autoimmune diseases [134–136]. However, until recently, only one study evoked a possible association between the development of the experimental disease and presentation of peptides by MHC proteins [137]. This preliminary study was based on only twelve animals originating from Mauritius (six class II M6 haplotypes and six without) [137]. After immunization with citrullinated peptides, the authors observed that the T-cell response was specifically directed against citrullinated peptides and not against arginine peptides. They reported that the presence of a valine in position 11 of the DRB alleles, and to a lesser extent of phenylalanine in position 13, led to an increased T-cell response to citrullinated peptides. It is important to note that the intensities of both the cellular (T cells) and humoral (production of antibodies) responses against citrullinated peptides were unrelated to the development of experimental arthritis in this model [138]. Indeed, a combined systemic and intra-articular immunization with citrullinated peptides is needed to induce experimental arthritis in the *Mafa* model. Cynomolgus macaques are also reported to develop collagen-induced arthritis, but any potential association between the development of this disease and MHC polymorphism remains to be investigated [136,139].

14. Impact of MHC on Cynomolgus Macaque Reproduction

In humans, several studies have suggested that fetuses that are MHC-compatible with their mothers may have a selective survival disadvantage compared to fetuses that inherit paternal MHC antigens which differ from maternal antigens (MHC semi-compatible pregnancies) [140–142]. The reproductive effects of feto-maternal histocompatibility (i.e., HLA sharing) is difficult to explore in humans due to the large number of HLA class I (HLA-A, HLA-B and HLA-C) and class II (HLA-DR, HLA-DQ and HLA-DP) alleles, requiring the study of many small subgroups and a loss of statistical power. Unlike the human population, the MCM population has a very limited MHC polymorphism (only seven founder haplotypes) as a consequence of the severe founder bottleneck experienced by the population. This limited polymorphism is of great value for exploring the impact of MHC compatibility on reproduction. We took advantage of this characteristic to explore the impact of feto-maternal MHC compatibility on reproduction in a captive *Mafa* colony of Mauritian descent [143].

We studied the MHC polymorphism in 42 macaque trios (male, female, and offspring) for whom the identity of the father was ascertained and with an equal theoretical probability of producing a totally compatible or semi-compatible offspring [144]. Animal MHC genotypes were deduced from the study of 17 microsatellite markers distributed across the MHC. Of the 42 offspring obtained, 11 were fully compatible and 31 were semi-compatible with their respective mothers. These proportions were clearly outside the 99% confidence interval when equiprobability is assumed for compatible and semi-compatible offspring. This departure from equiprobability suggests that *Mafa* offspring that are fully MHC-compatible have a selective survival disadvantage compared to offspring inheriting a paternal MHC haplotype distinct from the maternal haplotypes. By restricting the study of MHC feto-maternal compatibility to the MHC class II region, or the regions outside the latter, we found that departure from equiprobability of compatible/semi-compatible offspring only implicated the class I-III region [144]. The mechanisms at work in the negative selection of mother-compatible offspring in the MHC class I-III region remains to be identified. Due to the low density of microsatellite markers in the class I-III regions and the low frequency of recombinant haplotypes, the genomic region involved in the selection of offsprings has not been able to be mapped more precisely. The class I to III regions include numerous candidate genes. For example, in the Hutterite population, Ober et al. demonstrated that fetuses with identical HLA-B as their mothers were at increased risk of death in utero [145]. Similarly, to observations in humans, one could also evoke the impact of particular class I molecules, which can be expressed on the surface of trophoblastic macaque cells (a hypothetical homolog of human HLA-C). Genes in the MHC class III region (e.g., TNF alpha and TNF beta) play a role in reproduction [146,147] and could also explain our observations. The exact moment when such negative selection occurs (before or after fertilization or after implantation) remains an open question. The MCM represents an animal model that is accessible as well as phylogenetically close to humans, and amenable to experimental investigation of this issue.

15. Association of MHC and Drug Related Adverse Effects

Only one study has so far explored the association between delayed drug hypersensitivity and MHC in the *Mafa* model. The authors observed that nine out of 62 MCM exhibited clinical and histopathologic skin symptoms consistent with the drug-induced skin reactions observed in humans [148]. They studied the MHC genotype of the 62 animals and found an association between the drug induced skin reaction and the class IB region of Mauritian haplotype M3. The authors reported a high percentage of similarity (from 86% to 93%) between the class IB cDNA sequences associated with the Mauritian macaque M3 haplotype and the cDNA sequences of HLA-B alleles (HLA-B*57:01, B*15:02, B*58:01 and B*35:05) that had been previously reported to be associated with drug-induced hypersensitivity reactions in humans [149]. Obviously, a similarity at the cDNA level alone is not enough to draw conclusions about the mechanism of drug induced adverse reactions observed in the macaque model. A comparison of the amino acid sequences is also necessary. Moreover, the 62 macaques included in this study also received different drug treatments: either Metabotropic Glutamate Receptor 5 (mGluR5), Negative Allosteric Modulators or an unknown drug being tested by the authors [148]. Considering that only very few animals actually had adverse reactions to any of the drugs investigated, the study was unable to draw any significant findings when each of the drugs was considered separately. Notwithstanding these limitations, the Mauritian haplotype M3 could still be associated with a high risk of developing idiosyncratic drug-induced skin reactions.

16. Interactions between MHC Proteins and Various Receptors: Killer Cell Immunoglobulin-Like Receptors (KIRs) and Leukocyte Immunoglobulin-Like Receptors (LILRs)

As well as presenting peptides to the T cell receptor (TCR) of CD8+ T cells, classical MHC class I proteins also interact with KIRs and LILRs. KIRs are involved in the regulation of natural killer (NK) cell functions [150], while LILRs are involved in the regulation of antigen presentation by dendritic cells (DC) [151]. In the case of LILRs, it has also been shown that the strength of the human LILRB2 receptor and HLA-B protein interaction, contributes to control HIV-1 infections [151,152]. The strength of interaction is influenced by the HLA-B polymorphism, and strong LILRB2-HLA binding impairs the capacity of myeloid DCs to present antigens to T cells [152]. To our knowledge, only one study has to date explored the possible impact of LILR2 on the control of SIV infection in the MCM [153]. A longitudinal study of LILR2 expression after SIV inoculation revealed an up-regulation of LILRB2 and its MHC-I ligands on DCs in the early phase of SIV infection. The impact of this up-regulation on the capacity of DCs to present antigens to T cells remains to be explored [153]. The *Mafa* model offers the possibility of assessing the efficacy of anti-LILRB2 monoclonal antibodies aimed at suppressing this putative checkpoint in order to restore the full extent of DC's capacity to present viral antigens to T cells.

As for KIRs, the human KIR locus encodes between 9 to 16 cell surface receptors expressed on NK and T cells, which interact with HLA molecules to transduce either inhibitory or activating signals. Inhibitory KIRs repress the NK cell-mediated killing of normal endogenous cells. In contrast, the downregulation of classical MHC class I protein expression at the surface of virally infected or malignant cells was shown to induce NK cell cytotoxicity, a process referred to as "missing self-recognition". In this killing mechanism, NK cells detect, and subsequently destroy, aberrant cells that fail to express classical MHC class I molecules. Activating KIRs also interact with MHC class I molecules and may recognize cell surface alterations on unhealthy cells leading to their destruction by NK cells. Another well-characterized function of a specific type of activating human KIR receptor (KIR2DS1) is the activation of decidual NK cells through their interaction with HLA-C proteins [154–156]. This activation conditions the secretion of various cytokines (interferon gamma, vascular endothelial growth factor (VEGF) and others) by uterine NK cells. These cytokines are indispensable for the development of spiral arteries in the uterus as well as invasion of decidua by placental trophoblasts. One mystery which still requires resolution is the identification of the equivalent mechanism in macaques, as they do not have HLA-C gene orthologs. We have shown that *Mafa* reproduction is influenced by the MHC class I region [144]. It remains to be explored whether KIR-MHC interactions are also at work in the development of an efficient feto-maternal interface.

The KIR and LILR receptors interact not only with conventional MHC-class I proteins but also with non-classical MHC class I proteins (HLA-E, F, G). The latter play a crucial role in the regulation of immune functions, and particularly in the activation of NK cells by their interaction with KIR and LILR receptors. The functional roles of these non-classical class I HLA proteins in the development of pregnancy, cancers, infectious diseases have recently been reviewed [157–160]. We have already indicated that the *Mafa*-E, *Mafa*-F genes, are found in all MHC macaque haplotypes and have low polymorphism, similarly to all other non-human primates. The functional role of *Mafa*-E (see below) has been particularly well established because of its involvement in infectious diseases and the success of innovative CMV (cytomegalovirus) vector-based vaccines, this is in contrast to the functional role of *Mafa*-F which remains to be investigated. As mentioned earlier in this review, there is no functional ortholog of HLA-G in macaques. However, the *Mafa*-AG genes, present in variable numbers in all *Mafa* MHC haplotypes, encode proteins that have tissue expression and functions similar to those of HLA-G.

Until recently, the complexity of the KIR locus in macaques has been detrimental to genetic association studies. Recent advances in molecular genetics now make it possible to genotype the macaque KIR locus. In the case of Mauritian macaque cynomolgus, a complete characterization of KIR haplotypes has been performed [20,161]. Although MHC/KIR epistatic interaction has been extensively studied in experimental SIV infection in the rhesus monkey model [162], it is still not understood in the cynomolgus macaque model. The rhesus monkey model has revealed a complex combined effect of MHC and KIR polymorphism in controlling SIV infections [162], and it is highly likely that the same epistatic interaction is at work for controlling SIV infections in the MCM model.

17. Roles of *Mafa*-E in Experimental Medicine

Similar to other non-classical MHC-Ib molecules, human HLA-E exhibits limited polymorphism (the two predominant HLA-E alleles differ by a single amino acid at position 107) [163]. HLA-E protein is known to present a 9-mer peptide derived from the signal sequences of HLA-A, B, C, and G proteins [164,165]. The HLA-E/signal peptide complexes are ligands of the CD94/NKG2 receptor which transduce an inhibitory signal to NK cells [166–168]. In addition to its essential function to protect healthy cells from NK cell mediated attack, HLA-E has also been demonstrated to present bacterial (*Mycobacterium tuberculosis* [169–172], *Salmonella enterica* [173]) and viral (EBV [174], CMV [175,176], and hepatitis C virus [177,178]) derived peptides to HLA-E restricted CD8+ T cells. The HLA-E–restricted CD8+ T cells recognize the pathogen-derived peptides by their TCR, secrete antiviral cytokines and kill infected cells [170,171,173,175,178], but the impact of these unconventional MHC-E– restricted CD8+ T cell responses on the control of infections remains elusive.

Recently, the induction of SIV-specific MHC-E–restricted CD8+ T cell responses was demonstrated in the rhesus monkey after vaccination with an innovative anti-SIV vaccine based on a rhesus CMV (RhCMV) vector [179]. The *Mamu*-E–restricted CD8+ T cells response results in a robust control of SIV infection in approximately 50% of vaccinated Indian rhesus monkeys [180], confirming the role of *Mamu*-E in the presentation to CD8+ T cells of pathogen-derived peptides. More recently, MHC-E of Indian rhesus monkey, Mauritian cynomolgus monkey and human were compare for their expression at the surface of various cell types and their capacity to present virus-derived peptides to CD8+ T cells [181]. As is observed in the case of human CD4+ T cells infected with HIV, SIV infection of cynomolgus macaque CD4+ T cells induces an increase in MHC-E expression and the disappearance of classical MHC-I protein expression [181]. The macaque MHC-E molecules present identical SIV-derived peptides to MHC-E–restricted CD8+ T cells, leading to allogeneic and cross-species peptide recognition of MHC-E presented epitopes [181]. The functional similarity between human MHC-E and their macaque orthologs reinforces the argument supporting the relevance of macaque models (the rhesus monkey and the cynomolgus macaque) in the study of HLA-E in the fight against infectious pathogens [160].

In addition to its role as a non-classical MHC presenting protein, the MHC-E protein is also able to block NK cell activation by interacting with the inhibitory NKG2A receptor. Recently, a monoclonal antibody called monalizumab has been developed as a ckeck-point inhibitor that neutralizes the HLA-E/NKG2A interaction, thereby promoting NK cell activation against tumor cells and virally infected cells [182]. Despite the fact that the macaque model may be a suitable model for non-clinical trials, the cross reactivity of anti-human NKG2A with macaque NKG2A and NKG2C proteins will probably impede the use of this animal model in the future [183].

Another potential role of MHC-E peptide presentation to CD8+ effector T cells was reported in experimental allergic encephalomyelitis (EAE) in the marmoset model. In this model, EAE progression involves MOG-specific MHC-E restricted CD8+ CD56+ cytotoxic T cells activated by B cells infected with the EBV-related lymphocryptovirus virus CalHV3. The MOG (myelin oligodendrocyte glycoprotein) peptide (residues 40–48) is presented by the Caja-E molecules [184]. An in vitro study revealed that rhesus monkey MHC-E protein expression is enhanced by EBV infection of B cells [185]. Rhesus monkey EBV-infected B cells adopt a phenotype compatible with the presentation of the MOG 40–48 peptide [185]. In EBV free marmoset and macaque B cells, the MOG peptide is degraded by the endolysosomal protease cathepsin G and this destruction was interpreted as a tolerogenic mechanism. In contrast, in B cells infected with EBV-related lymphocryptoviruses, the induction of peptidyl arginine deaminase (PAD2 and PAD4) expression led to the citrullination of the MOG 35–51 peptide which protects the embedded MOG 40–48 peptide from the destruction by cathepsin G, leading in turn to the presentation of the citrullinated MOG 40–48 peptide by MHC-E [185,186]. Infection of B cells by EBV or EBV-related lymphocryptoviruses transforms these cells into APCs. This plays an essential role in the immunopathogenic process and is crucial to clarify our understanding of the therapeutic effects of anti-CD20 mAbs in multiple sclerosis [187]. The development of EAE models in the rhesus monkey and cynomolgus macaque would certainly help elucidate the role of EBV and MHC-E in the presentation of encephalitogenic MOG peptides in multiple sclerosis [186].

18. Conclusions

In conclusion, the study of *Mafa* MHC polymorphism has benefited from recent developments in molecular genetics. However, MHC gene duplication makes this region inaccessible to whole genome sequencing based on short reads. Techniques based on sequencing of very long fragments are required to reconstruct the complex cynomolgus MHC haplotypes. Moreover, many questions remain to be investigated, specifically in relation to the epistasis between MHC genes and other polymorphic loci such as KIR or LILR. In addition to the MHC, KIR and LILR loci, many other loci interfere with the immune response and their polymorphic variants need to be considered in the design of experimental medicine protocols [109,110,188]. The MHC polymorphism of macaque presents significant advantages for protocols aimed at establishing genetic associations. However, in most cases, experimental medicine in macaques aims at testing innovative treatments in this animal model. In that context, genetic variability of animals is a strong disadvantage and numerous efforts have been made to address this problem [131]. The recent developments in macaque cloning by somatic cell nuclear transfers will certainly play a determinant role in many protocols, for which the influence of the genetic background on the experimental results obtained needs to be limited [189].

Supplementary Materials: The following is available online, Table S1: Comparison of non-MHC loci in the MHC genomic region between human and cynomolgus macaque.

Author Contributions: T.S. and A.B. equally contributed to the writing and validation of the manuscript.

Funding: We received no funds of this review work.

Acknowledgments: A.B. thanks the technicians of Laboratory of Immunogenetics of Toulouse who help him in the study of cynomolgus macaque MHC.

Conflicts of Interest: The authors declare no conflict of interest.

References

1. Southwick, C.H.; Siddiqi, M.F. Population Status of Nonhuman-Primates in Asia, with Emphasis on Rhesus Macaques in India. *Am. J. Primatol.* **1994**, *34*, 51–59. [CrossRef]

2. Sussman, R.W.; Tattersall, I. Distribution, Abundance, and Putative Ecological Strategy of Macaca-Fascicularis on the Island of Mauritius, Southwestern Indian-Ocean. *Folia Primatol.* **1986**, *46*, 28–43. [CrossRef]

3. Kawamoto, Y.; Kawamoto, S.; Matsubayashi, K.; Nozawa, K.; Watanabe, T.; Stanley, M.A.; Perwitasari-Farajallah, D. Genetic diversity of longtail macaques (Macaca fascicularis) on the island of Mauritius: An assessment of nuclear and mitochondrial DNA polymorphisms. *J. Med. Primatol.* **2008**, *37*, 45–54. [CrossRef] [PubMed]

4. Tosi, A.J.; Coke, C.S. Comparative phylogenetics offer new insights into the biogeographic history of Macaca fascicularis and the origin of the Mauritian macaques. *Mol. Phylogenet Evol.* **2007**, *42*, 498–504. [CrossRef] [PubMed]

5. Bonhomme, M.; Blancher, A.; Cuartero, S.; Chikhi, L.; Crouau-Roy, B. Origin and number of founders in an introduced insular primate: Estimation from nuclear genetic data. *Mol. Ecol.* **2008**, *17*, 1009–1019. [CrossRef] [PubMed]

6. Osada, N.; Hettiarachchi, N.; Babarinde, I.A.; Saitou, N.; Blancher, A. Whole-Genome Sequencing of Six Mauritian Cynomolgus Macaques (Macaca fascicularis) Reveals a Genome-Wide Pattern of Polymorphisms under Extreme Population Bottleneck. *Genome Biol. Evol.* **2015**, *7*, 821–830. [CrossRef] [PubMed]

7. Bonhomme, M.; Cuartero, S.; Blancher, A.; Crouau-roy, B. Assessing Natural Introgression in 2 Biomedical Model Species, the Rhesus Macaque (Macaca mulatta) and the Long-Tailed Macaque (Macaca fascicularis). *J. Hered.* **2009**, *100*, 158–169. [CrossRef] [PubMed]

8. Bunlungsup, S.; Kanthaswamy, S.; Oldt, R.F.; Smith, D.G.; Houghton, P.; Hamada, Y.; Malaivijitnond, S. Genetic analysis of samples from wild populations opens new perspectives on hybridization between long-tailed (Macaca fascicularis) and rhesus macaques (Macaca mulatta). *Am. J. Primatol.* **2017**, *79*. [CrossRef] [PubMed]

9. Osada, N.; Uno, Y.; Mineta, K.; Kameoka, Y.; Takahashi, I.; Terao, K. Ancient genome-wide admixture extends beyond the current hybrid zone between Macaca fascicularis and M-mulatta. *Mol. Ecol.* **2010**, *19*, 2884–2895. [CrossRef]

10. Trask, J.A.S.; Garnica, W.T.; Smith, D.G.; Houghton, P.; Lerche, N.; Kanthaswamy, S. Single-Nucleotide Polymorphisms Reveal Patterns of Allele Sharing Across the Species Boundary Between Rhesus (Macaca mulatta) and Cynomolgus (M. fascicularis) Macaques. *Am. J. Primatol.* **2013**, *75*, 135–144. [CrossRef]

11. Anderson, D.J.; Kirk, A.D. Primate Models in Organ Transplantation. *Csh Perspect Med.* **2013**, *3*. [CrossRef] [PubMed]

12. Chen, Y.C.; Niu, Y.Y.; Li, Y.J.; Ai, Z.Y.; Kang, Y.; Shi, H.; Xiang, Z.; Yang, Z.H.; Tan, T.; Si, W.; et al. Generation of Cynomolgus Monkey Chimeric Fetuses using Embryonic Stem Cells. *Cell Stem Cell* **2015**, *17*, 116–124. [CrossRef] [PubMed]

13. Honda, A.; Kawano, Y.; Izu, H.; Choijookhuu, N.; Honsho, K.; Nakamura, T.; Yabuta, Y.; Yamamoto, T.; Takashima, Y.; Hirose, M.; et al. Discrimination of Stem Cell Status after Subjecting Cynomolgus Monkey Pluripotent Stem Cells to Naive Conversion. *Sci. Rep.* **2017**, *7*, 45285. [CrossRef] [PubMed]

14. Hanazono, Y.; Terao, K.; Ozawa, K. Gene transfer into nonhuman primate hematopoietic stem cells: Implications for gene therapy. *Stem Cells* **2001**, *19*, 12–23. [CrossRef] [PubMed]

15. Rivera-Hernandez, T.; Carnathan, D.G.; Moyle, P.M.; Toth, I.; West, N.P.; Young, P.R.; Silvestri, G.; Walker, M.J. The Contribution of Non-human Primate Models to the Development of Human Vaccines. *Discov. Med.* **2014**, *18*, 313–322. [PubMed]

16. Iwasaki, K.; Uno, Y.; Utoh, M.; Yamazaki, H. Importance of cynomolgus monkeys in development of monoclonal antibody drugs. *Drug Metab. Pharmacok.* **2019**, *34*, 55–63. [CrossRef]

17. Gardner, M.B.; Luciw, P.A. Macaque models of human infectious disease. *ILAR J.* **2008**, *49*, 220–255. [CrossRef]

18. Verdier, J.M.; Acquatella, I.; Lautier, C.; Devau, G.; Trouche, S.; Lasbleiz, C.; Mestre-Frances, N. Lessons from the analysis of nonhuman primates for understanding human aging and neurodegenerative diseases. *Front. Neurosci.* **2015**, *9*, 64. [CrossRef]

19. Eastwood, D.; Findlay, L.; Poole, S.; Bird, C.; Wadhwa, M.; Moore, M.; Burns, C.; Thorpe, R.; Stebbings, R. Monoclonal antibody TGN1412 trial failure explained by species differences in CD28 expression on CD4(+) effector memory T-cells. *Brit. J. Pharmacol.* **2010**, *161*, 512–526. [CrossRef]

20. Prall, T.M.; Graham, M.E.; Karl, J.A.; Wiseman, R.W.; Ericsen, A.J.; Raveendran, M.; Alan Harris, R.; Muzny, D.M.; Gibbs, R.A.; Rogers, J.; et al. Improved full-length killer cell immunoglobulin-like receptor transcript discovery in Mauritian cynomolgus macaques. *Immunogenetics* **2017**, *69*, 325–339. [CrossRef]

21. Blancher, A.; Tisseyre, P.; Dutaur, M.; Apoil, P.A.; Maurer, C.; Quesniaux, V.; Raulf, F.; Bigaud, M.; Abbal, M. Study of Cynomolgus monkey (Macaca fascicularis) MhcDRB (Mafa-DRB) polymorphism in two populations. *Immunogenetics* **2006**, *58*, 269–282. [CrossRef]

22. Balner, H. Current knowledge of the histocompatibility complex of rhesus monkeys. *Transplant. Rev.* **1973**, *15*, 50–61.

23. Balner, H.; D'Amaro, J.; Toth, E.K.; Dersjant, H.; van Vreeswijk, W. The histocompatibility complex of rhesus monkeys. Relation between RhL and the main locus controlling reactivity in mixed lymphocyte cultures. *Transplant. Proc.* **1973**, *5*, 323–329.

24. Roger, J.H.; van Reeswijk, W.; D'Amaro, J.; Balner, H. The major histocompatibility complex of Rhesus monkeys IX. Current concepts of serology and genetics of Ia antigens. *Tissue Antigens* **1978**, *11*, 163–180. [CrossRef]

25. van Es, A.A.; Balner, H. The major histocompatibility complex of Rhesus monkeys. XII: Cellular typing for D locus antigens in families. *Tissue Antigens* **1979**, *13*, 255–272.

26. Van Rood, J.J.; Balner, R.; Gabb, B.W.; Dersjant, H.; Van Vreeswijk, W. Major histocompatibility locus of rhesus monkeys (RhL-A). *Nat. New Biol.* **1971**, *230*, 177–180.

27. Balner, H.; Dorf, M.E.; de Groot, M.L.; Benacerraf, B. The histocompatibility complex of rhesus monkeys. 3. Evidence for a major MLR locus and histocompatibility-linked Ir genes. *Transplant. Proc.* **1973**, *5*, 1555–1560.

28. Balner, H.; Gabb, B.W.; Toth, E.K.; Dersjant, H.; van Vreeswijk, W. The histocompatibility complex of rhesus monkeys. I. Serology and genetics of the RhL-A system. *Tissue Antigens* **1973**, *3*, 257–272. [CrossRef]

29. Balner, H.; Toth, E.K. The histocompatibility complex of rhesus monkeys. II. A major locus controlling reactivity in mixed lymphocyte cultures. *Tissue Antigens* **1973**, *3*, 273–290. [CrossRef]

30. Keever, C.A.; Heise, E.R. The major histocompatibility complex (CyLA) of the cynomolgus monkey. I. Serologic definition of 21 specificities. *Hum. Immunol.* **1983**, *7*, 131–149. [CrossRef]

31. Keever, C.A.; Heise, E.R. The major histocompatibility complex of the cynomolgus monkey: Absorption analysis of 24 CyLA antisera. *Hum. Immunol.* **1985**, *12*, 75–90. [CrossRef]

32. Keever, C.A.; Heise, E.R. The major histocompatibility complex of the cynomolgus monkey. II. Polymorphism at three serologically defined loci and correlation of haplotypes with stimulation in MLC and skin graft survival. *Hum. Immunol.* **1985**, *12*, 143–164. [CrossRef]

33. Leuchte, N.; Berry, N.; Kohler, B.; Almond, N.; LeGrand, R.; Thorstensson, R.; Titti, F.; Sauermann, U. MhcDRB-sequences from cynomolgus macaques (Macaca fascicularis) of different origin. *Tissue Antigens* **2004**, *63*, 529–537. [CrossRef]

34. Knapp, L.A.; Cadavid, L.F.; Eberle, M.E.; Knechtle, S.J.; Bontrop, R.E.; Watkins, D.I. Identification of new Mamu-DRB alleles using DGGE and direct sequencing. *Immunogenetics* **1997**, *45*, 171–179. [CrossRef]

35. Bonhomme, M.; Blancher, A.; Crouau-Roy, B. Multiplexed microsatellites for rapid identification and characterization of individuals and populations of Cercopithecidae. *Am. J. Primatol.* **2005**, *67*, 385–391. [CrossRef]

36. Watanabe, A.; Shiina, T.; Shimizu, S.; Hosomichi, K.; Yanagiya, K.; Kita, Y.F.; Kimura, T.; Soeda, E.; Torii, R.; Ogasawara, K.; et al. A BAC-based contig map of the cynomolgus macaque (Macaca fascicularis) major histocompatibility complex genomic region. *Genomics* **2007**, *89*, 402–412. [CrossRef]

37. Wiseman, R.W.; Karl, J.A.; Bimber, B.N.; O'Leary, C.E.; Lank, S.M.; Tuscher, J.J.; Detmer, A.M.; Bouffard, P.; Levenkova, N.; Turcotte, C.L.; et al. Major histocompatibility complex genotyping with massively parallel pyrosequencing. *Nat. Med.* **2009**, *15*, 1322–1326. [CrossRef]

38. Aarnink, A.; Apoil, P.A.; Takahashi, I.; Osada, N.; Blancher, A. Characterization of MHC class I transcripts of a Malaysian cynomolgus macaque by high-throughput pyrosequencing and EST libraries. *Immunogenetics* **2011**, *63*, 703–713. [CrossRef]

39. Westbrook, C.J.; Karl, J.A.; Wiseman, R.W.; Mate, S.; Koroleva, G.; Garcia, K.; Sanchez-Lockhart, M.; O'Connor, D.H.; Palacios, G. No assembly required: Full-length MHC class I allele discovery by PacBio circular consensus sequencing. *Hum. Immunol.* **2015**, *76*, 891–896. [CrossRef]

40. Bondarenko, G.I.; Dambaeva, S.V.; Grendell, R.L.; Hughes, A.L.; Durning, M.; Garthwaite, M.A.; Golos, T.G. Characterization of cynomolgus and vervet monkey placental MHC class I expression: Diversity of the nonhuman primate AG locus. *Immunogenetics* **2009**, *61*, 431–442. [CrossRef]

41. Bontrop, R.E.; Otting, N.; de Groot, N.G.; Doxiadis, G.G. Major histocompatibility complex class II polymorphisms in primates. *Immunol. Rev.* **1999**, *167*, 339–350. [CrossRef]

42. Budde, M.L.; Greene, J.M.; Chin, E.N.; Ericsen, A.J.; Scarlotta, M.; Cain, B.T.; Pham, N.H.; Becker, E.A.; Harris, M.; Weinfurter, J.T.; et al. Specific CD8+ T cell responses correlate with control of simian immunodeficiency virus replication in Mauritian cynomolgus macaques. *J. Virol.* **2012**, *86*, 7596–7604. [CrossRef]

43. O'Connor, S.L.; Blasky, A.J.; Pendley, C.J.; Becker, E.A.; Wiseman, R.W.; Karl, J.A.; Hughes, A.L.; O'Connor, D.H. Comprehensive characterization of MHC class II haplotypes in Mauritian cynomolgus macaques. *Immunogenetics* **2007**, *59*, 449–462. [CrossRef]

44. Wiseman, R.W.; Karl, J.A.; Bohn, P.S.; Nimityongskul, F.A.; Starrett, G.J.; O'Connor, D.H. Haplessly hoping: Macaque major histocompatibility complex made easy. *ILAR J.* **2013**, *54*, 196–210. [CrossRef]

45. Blancher, A.; Aarnink, A.; Tanaka, K.; Ota, M.; Inoko, H.; Yamanaka, H.; Nakagawa, H.; Apoil, P.A.; Shiina, T. Study of cynomolgus monkey (Macaca fascicularis) Mhc DRB gene polymorphism in four populations. *Immunogenetics* **2012**, *64*, 605–614. [CrossRef]

46. Blancher, A.; Aarnink, A.; Yamada, Y.; Tanaka, K.; Yamanaka, H.; Shiina, T. Study of MHC class II region polymorphism in the Filipino cynomolgus macaque population. *Immunogenetics* **2014**, *66*, 219–230. [CrossRef]

47. Shiina, T.; Yamada, Y.; Aarnink, A.; Suzuki, S.; Masuya, A.; Ito, S.; Ido, D.; Yamanaka, H.; Iwatani, C.; Tsuchiya, H.; et al. Discovery of novel MHC-class I alleles and haplotypes in Filipino cynomolgus macaques (Macaca fascicularis) by pyrosequencing and Sanger sequencing: Mafa-class I polymorphism. *Immunogenetics* **2015**, *67*, 563–578. [CrossRef]

48. Meyer, A.; Carapito, R.; Ott, L.; Radosavljevic, M.; Georgel, P.; Adams, E.J.; Parham, P.; Bontrop, R.E.; Blancher, A.; Bahram, S. High diversity of MIC genes in non-human primates. *Immunogenetics* **2014**, *66*, 581–587. [CrossRef]

49. de Groot, N.G.; Otting, N.; Robinson, J.; Blancher, A.; Lafont, B.A.; Marsh, S.G.; O'Connor, D.H.; Shiina, T.; Walter, L.; Watkins, D.I.; et al. Nomenclature report on the major histocompatibility complex genes and alleles of Great Ape, Old and New World monkey species. *Immunogenetics* **2012**, *64*, 615–631. [CrossRef]

50. Bonhomme, M.; Blancher, A.; Jalil, M.F.; Crouau-Roy, B. Factors shaping genetic variation in the MHC of natural non-human primate populations. *Tissue Antigens* **2007**, *70*, 398–411. [CrossRef]

51. Aarnink, A.; Bonhomme, M.; Blancher, A. Positive selection in the major histocompatibility complex class III region of cynomolgus macaques (Macaca fascicularis) of the Philippines origin. *Tissue Antigens* **2013**, *81*, 12–18. [CrossRef]

52. Budde, M.L.; Wiseman, R.W.; Karl, J.A.; Hanczaruk, B.; Simen, B.B.; O'Connor, D.H. Characterization of Mauritian cynomolgus macaque major histocompatibility complex class I haplotypes by high-resolution pyrosequencing. *Immunogenetics* **2010**, *62*, 773–780. [CrossRef]

53. Campbell, K.J.; Detmer, A.M.; Karl, J.A.; Wiseman, R.W.; Blasky, A.J.; Hughes, A.L.; Bimber, B.N.; O'Connor, S.L.; O'Connor, D.H. Characterization of 47 MHC class I sequences in Filipino cynomolgus macaques. *Immunogenetics* **2009**, *61*, 177–187. [CrossRef]

54. Kita, Y.F.; Hosomichi, K.; Kohara, S.; Itoh, Y.; Ogasawara, K.; Tsuchiya, H.; Torii, R.; Inoko, H.; Blancher, A.; Kulski, J.K.; et al. MHC class I A loci polymorphism and diversity in three Southeast Asian populations of cynomolgus macaque. *Immunogenetics* **2009**, *61*, 635–648. [CrossRef]

55. Lawrence, J.; Orysiuk, D.; Prashar, T.; Pilon, R.; Fournier, J.; Rud, E.; Sandstrom, P.; Plummer, F.A.; Luo, M. Identification of 23 novel MHC class I alleles in cynomolgus macaques of Philippine and Philippine/Mauritius origins. *Tissue Antigens* **2012**, *79*, 306–307. [CrossRef]

56. Ling, F.; Wei, L.Q.; Wang, T.; Wang, H.B.; Zhuo, M.; Du, H.L.; Wang, J.F.; Wang, X.N. Characterization of the major histocompatibility complex class II DOB, DPB1, and DQB1 alleles in cynomolgus macaques of Vietnamese origin. *Immunogenetics* **2011**, *63*, 155–166. [CrossRef]

57. Otting, N.; de Vos-Rouweler, A.J.; Heijmans, C.M.; de Groot, N.G.; Doxiadis, G.G.; Bontrop, R.E. MHC class I A region diversity and polymorphism in macaque species. *Immunogenetics* **2007**, *59*, 367–375. [CrossRef]

58. Otting, N.; Doxiadis, G.G.; Bontrop, R.E. Definition of Mafa-A and -B haplotypes in pedigreed cynomolgus macaques (Macaca fascicularis). *Immunogenetics* **2009**, *61*, 745–753. [CrossRef]

59. Pendley, C.J.; Becker, E.A.; Karl, J.A.; Blasky, A.J.; Wiseman, R.W.; Hughes, A.L.; O'Connor, S.L.; O'Connor, D.H. MHC class I characterization of Indonesian cynomolgus macaques. *Immunogenetics* **2008**, *60*, 339–351. [CrossRef]

60. Saito, Y.; Naruse, T.K.; Akari, H.; Matano, T.; Kimura, A. Diversity of MHC class I haplotypes in cynomolgus macaques. *Immunogenetics* **2012**, *64*, 131–141. [CrossRef]

61. Uda, A.; Tanabayashi, K.; Fujita, O.; Hotta, A.; Terao, K.; Yamada, A. Identification of the MHC class I B locus in cynomolgus monkeys. *Immunogenetics* **2005**, *57*, 189–197. [CrossRef]

62. Uda, A.; Tanabayashi, K.; Yamada, Y.K.; Akari, H.; Lee, Y.J.; Mukai, R.; Terao, K.; Yamada, A. Detection of 14 alleles derived from the MHC class I A locus in cynomolgus monkeys. *Immunogenetics* **2004**, *56*, 155–163. [CrossRef]

63. Zhang, G.Q.; Ni, C.; Ling, F.; Qiu, W.; Wang, H.B.; Xiao, Y.; Guo, X.J.; Huang, J.Y.; Du, H.L.; Wang, J.F.; et al. Characterization of the major histocompatibility complex class I A alleles in cynomolgus macaques of Vietnamese origin. *Tissue Antigens* **2012**, *80*, 494–501. [CrossRef]

64. Karl, J.A.; Graham, M.E.; Wiseman, R.W.; Heimbruch, K.E.; Gieger, S.M.; Doxiadis, G.G.; Bontrop, R.E.; O'Connor, D.H. Major histocompatibility complex haplotyping and long-amplicon allele discovery in cynomolgus macaques from Chinese breeding facilities. *Immunogenetics* **2017**, *69*, 211–229. [CrossRef]

65. Krebs, K.C.; Jin, Z.; Rudersdorf, R.; Hughes, A.L.; O'Connor, D.H. Unusually high frequency MHC class I alleles in Mauritian origin cynomolgus macaques. *J. Immunol.* **2005**, *175*, 5230–5239. [CrossRef]

66. Creager, H.M.; Becker, E.A.; Sandman, K.K.; Karl, J.A.; Lank, S.M.; Bimber, B.N.; Wiseman, R.W.; Hughes, A.L.; O'Connor, S.L.; O'Connor, D.H. Characterization of full-length MHC class II sequences in Indonesian and Vietnamese cynomolgus macaques. *Immunogenetics* **2011**, *63*, 611–618. [CrossRef]

67. Wang, W.; Lu, Y.E.; Zhuo, M.; Ling, F. Identification of five novel MHC class II alleles in cynomolgus macaques of Vietnamese origin. *HLA* **2016**, *88*, 61–62. [CrossRef]

68. Doxiadis, G.G.M.; Rouweler, A.J.M.; de Groot, N.G.; Louwerse, A.; Otting, N.; Verschoor, E.J.; Bontrop, R.E. Extensive sharing of MHC class II alleles between rhesus and cynomolgus macaques. *Immunogenetics* **2006**, *58*, 259–268. [CrossRef]

69. Aarnink, A.; Estrade, L.; Apoil, P.A.; Kita, Y.F.; Saitou, N.; Shiina, T.; Blancher, A. Study of cynomolgus monkey (Macaca fascicularis) DRA polymorphism in four populations. *Immunogenetics* **2010**, *62*, 123–136. [CrossRef]

70. Boyson, J.E.; Mcadam, S.N.; Gallimore, A.; Golos, T.G.; Liu, X.M.; Gotch, F.M.; Hughes, A.L.; Watkins, D.I. The-Mhc-E Locus in Macaques Is Polymorphic and Is Conserved between Macaques and Humans. *Immunogenetics* **1995**, *41*, 59–68. [CrossRef]

71. Gaur, L.K.; Nepom, G.T. Ancestral major histocompatibility complex DRB genes beget conserved patterns of localized polymorphisms. *Proc. Natl. Acad. Sci. USA* **1996**, *93*, 5380–5383. [CrossRef]

72. Alvarez, M.; MartinezLaso, J.; Varela, P.; DiazCampos, N.; GomezCasado, E.; VargasAlarcon, G.; GarciaTorre, C.; ArnaizVillena, A. High polymorphism of Mhc-E locus in non-human primates: Alleles with identical exon 2 and 3 are found in two different species. *Tissue Antigens* **1997**, *49*, 160–167. [CrossRef]

73. Otting, N.; de Groot, N.G.; Doxiadis, G.G.; Bontrop, R.E. Extensive Mhc-DQB variation in humans and non-human primate species. *Immunogenetics* **2002**, *54*, 230–239. [CrossRef]

74. Sano, K.; Shiina, T.; Kohara, S.; Yanagiya, K.; Hosomichi, K.; Shimizu, S.; Anzai, T.; Watanabe, A.; Ogasawara, K.; Torii, R.; et al. Novel cynomolgus macaque MHC-DPB1 polymorphisms in three South-East Asian populations. *Tissue Antigens* **2006**, *67*, 297–306. [CrossRef]

75. Wiseman, R.W.; Wojcechowskyj, J.A.; Greene, J.M.; Blasky, A.J.; Gopon, T.; Soma, T.; Friedrich, T.C.; O'Connor, S.L.; O'Connor, D.H. Simian immunodeficiency virus SIVmac239 infection of major histocompatibility complex-identical cynomolgus macaques from mauritius. *J. Virol.* **2007**, *81*, 349–361. [CrossRef]

76. Ling, F.; Zhuo, M.; Ni, C.; Zhang, G.Q.; Wang, T.; Li, W.; Wei, L.Q.; Du, H.L.; Wang, J.F.; Wang, X.N. Comprehensive identification of high-frequency and co-occurring Mafa-B, Mafa-DQB1, and Mafa-DRB alleles in cynomolgus macaques of Vietnamese origin. *Hum. Immunol.* **2012**, *73*, 547–553. [CrossRef]

77. Mitchell, J.L.; Mee, E.T.; Almond, N.M.; Cutler, K.; Rose, N.J. Characterisation of MHC haplotypes in a breeding colony of Indonesian cynomolgus macaques reveals a high level of diversity. *Immunogenetics* **2012**, *64*, 123–129. [CrossRef]

78. de Groot, N.; Doxiadis, G.G.M.; Otting, N.; de Vos-Rouweler, A.J.M.; Bontrop, R.E. Differential recombination dynamics within the MHC of macaque species. *Immunogenetics* **2014**, *66*, 535–544. [CrossRef]

79. Klimanskaya, I.; Chung, Y.; Becker, S.; Lu, S.J.; Lanza, R. Human embryonic stem cell lines derived from single blastomeres. *Nature* **2006**, *444*, 481–485. [CrossRef]

80. Takahashi, K.; Tanabe, K.; Ohnuki, M.; Narita, M.; Ichisaka, T.; Tomoda, K.; Yamanaka, S. Induction of pluripotent stem cells from adult human fibroblasts by defined factors. *Cell* **2007**, *131*, 861–872. [CrossRef]

81. Wiseman, R.W.; Wojcechowskyj, J.A.; Greene, J.M.; Blasky, A.J.; O'Connor, D.H. MHC identical cynomolgus macaques from Mauritius. *J. Med. Primatol.* **2007**, *36*, 296.

82. Greene, J.M.; Burwitz, B.J.; Blasky, A.J.; Mattila, T.L.; Hong, J.J.; Rakasz, E.G.; Wiseman, R.W.; Hasenkrug, K.J.; Skinner, P.J.; O'Connor, S.L.; et al. Allogeneic lymphocytes persist and traffic in feral MHC-matched mauritian cynomolgus macaques. *PLoS ONE* **2008**, *3*, e2384. [CrossRef]

83. Mee, E.T.; Stebbings, R.; Hall, J.; Giles, E.; Almond, N.; Rose, N.J. Allogeneic lymphocyte transfer in MHC-identical siblings and MHC-identical unrelated Mauritian cynomolgus macaques. *Plos ONE* **2014**, *9*, e88670. [CrossRef]

84. Shiba, Y.; Gomibuchi, T.; Seto, T.; Wada, Y.; Ichimura, H.; Tanaka, Y.; Ogasawara, T.; Okada, K.; Shiba, N.; Sakamoto, K.; et al. Allogeneic transplantation of iPS cell-derived cardiomyocytes regenerates primate hearts. *Nature* **2016**, *538*, 388–391. [CrossRef]

85. Osada, N.; Hirata, M.; Tanuma, R.; Suzuki, Y.; Sugano, S.; Terao, K.; Kusuda, J.; Kameoka, Y.; Hashimoto, K.; Takahashi, I. Collection of Macaca fascicularis cDNAs derived from bone marrow, kidney, liver, pancreas, spleen, and thymus. *BMC Res. Notes* **2009**, *2*, 199. [CrossRef]

86. Takahata, N. MHC diversity and selection. *Immunol. Rev.* **1995**, *143*, 225–247. [CrossRef]

87. Greene, J.M.; Wiseman, R.W.; Lank, S.M.; Bimber, B.N.; Karl, J.A.; Burwitz, B.J.; Lhost, J.J.; Hawkins, O.E.; Kunstman, K.J.; Broman, K.W.; et al. Differential MHC class I expression in distinct leukocyte subsets. *BMC Immunol.* **2011**, *12*. [CrossRef]

88. Koterski, J.; Twenhafel, N.; Porter, A.; Reed, D.S.; Martino-Catt, S.; Sobral, B.; Crasta, O.; Downey, T.; DaSilva, L. Gene expression profiling of nonhuman primates exposed to aerosolized Venezuelan equine encephalitis virus. *FEMS Immunol. Med. Microbiol.* **2007**, *51*, 462–472. [CrossRef]

89. Mothe, B.R.; Lindestam Arlehamn, C.S.; Dow, C.; Dillon, M.B.C.; Wiseman, R.W.; Bohn, P.; Karl, J.; Golden, N.A.; Gilpin, T.; Foreman, T.W.; et al. The TB-specific CD4(+) T cell immune repertoire in both cynomolgus and rhesus macaques largely overlap with humans. *Tuberculosis (Edinb)* **2015**, *95*, 722–735. [CrossRef]

90. Martin, M.P.; Carrington, M. Immunogenetics of HIV disease. *Immunol. Rev.* **2013**, *254*, 245–264. [CrossRef]

91. Kawashima, Y.; Pfafferott, K.; Frater, J.; Matthews, P.; Payne, R.; Addo, M.; Gatanaga, H.; Fujiwara, M.; Hachiya, A.; Koizumi, H.; et al. Adaptation of HIV-1 to human leukocyte antigen class I. *Nature* **2009**, *458*, 641–645. [CrossRef]

92. Avila-Rios, S.; Carlson, J.M.; John, M.; Mallal, S.; Brumme, Z.L. Clinical and evolutionary consequences of HIV adaptation to HLA: Implications for vaccine and cure. *Curr. Opin. HIV AIDS* **2019**, *14*, 194–204. [CrossRef]

93. Chikata, T.; Carlson, J.M.; Tamura, Y.; Borghan, M.A.; Naruto, T.; Hashimoto, M.; Murakoshi, H.; Le, A.Q.; Mallal, S.; John, M.; et al. Host-specific adaptation of HIV-1 subtype B in the Japanese population. *J. Virol.* **2014**, *88*, 4764–4775. [CrossRef]

94. Kloverpris, H.N.; Leslie, A.; Goulder, P. Role of HLA Adaptation in HIV Evolution. *Front. Immunol.* **2016**, *6*. [CrossRef]

95. Loffredo, J.T.; Maxwell, J.; Qi, Y.; Glidden, C.E.; Borchardt, G.J.; Soma, T.; Bean, A.T.; Beal, D.R.; Wilson, N.A.; Rehrauer, W.M.; et al. Mamu-B*08-positive macaques control simian immunodeficiency virus replication. *J. Virol.* **2007**, *81*, 8827–8832. [CrossRef]

96. Mothe, B.R.; Weinfurter, J.; Wang, C.; Rehrauer, W.; Wilson, N.; Allen, T.M.; Allison, D.B.; Watkins, D.I. Expression of the major histocompatibility complex class I molecule Mamu-A*01 is associated with control of simian immunodeficiency virus SIVmac239 replication. *J. Virol.* **2003**, *77*, 2736–2740. [CrossRef]

97. Muhl, T.; Krawczak, M.; Ten Haaft, P.; Hunsmann, G.; Sauermann, U. MHC class I alleles influence set-point viral load and survival time in simian immunodeficiency virus-infected rhesus monkeys. *J. Immunol.* **2002**, *169*, 3438–3446. [CrossRef]

98. Yant, L.J.; Friedrich, T.C.; Johnson, R.C.; May, G.E.; Maness, N.J.; Enz, A.M.; Lifson, J.D.; O'Connor, D.H.; Carrington, M.; Watkins, D.I. The high-frequency major histocompatibility complex class I allele Mamu-B*17 is associated with control of simian immunodeficiency virus SIVmac239 replication. *J. Virol.* **2006**, *80*, 5074–5077. [CrossRef]

99. Marcilla, M.; Alvarez, I.; Ramos-Fernandez, A.; Lombardia, M.; Paradela, A.; Albar, J.P. Comparative Analysis of the Endogenous Peptidomes Displayed by HLA-B*27 and Mamu-B*08: Two MHC Class I Alleles Associated with Elite Control of HIV/SIV Infection. *J. Proteome Res.* **2016**, *15*, 1059–1069. [CrossRef]

100. Loffredo, J.T.; Sidney, J.; Bean, A.T.; Beal, D.R.; Bardet, W.; Wahl, A.; Hawkins, O.E.; Piaskowski, S.; Wilson, N.A.; Hildebrand, W.H.; et al. Two MHC class I molecules associated with elite control of immunodeficiency virus replication, Mamu-B*08 and HLA-B*2705, bind peptides with sequence similarity. *J. Immunol.* **2009**, *182*, 7763–7775. [CrossRef]

101. Valentine, L.E.; Loffredo, J.T.; Bean, A.T.; Leon, E.J.; MacNair, C.E.; Beal, D.R.; Piaskowski, S.M.; Klimentidis, Y.C.; Lank, S.M.; Wiseman, R.W.; et al. Infection with "escaped" virus variants impairs control of simian immunodeficiency virus SIVmac239 replication in Mamu-B*08-positive macaques. *J. Virol.* **2009**, *83*, 11514–11527. [CrossRef]

102. Blancher, A.; Aarnink, A.; Savy, N.; Takahata, N. Use of Cumulative Poisson Probability Distribution as an Estimator of the Recombination Rate in an Expanding Population: Example of the Macaca fascicularis Major Histocompatibility Complex. *G3 (Bethesda)* **2012**, *2*, 123–130. [CrossRef]

103. Cain, B.T.; Pham, N.H.; Budde, M.L.; Greene, J.M.; Weinfurter, J.T.; Scarlotta, M.; Harris, M.; Chin, E.; O'Connor, S.L.; Friedrich, T.C.; et al. T cell response specificity and magnitude against SIVmac239 are not concordant in major histocompatibility complex-matched animals. *Retrovirology* **2013**, *10*, 116. [CrossRef]

104. Florese, R.H.; Wiseman, R.W.; Venzon, D.; Karl, J.A.; Demberg, T.; Larsen, K.; Flanary, L.; Kalyanaraman, V.S.; Pal, R.; Titti, F.; et al. Comparative study of Tat vaccine regimens in Mauritian cynomolgus and Indian rhesus macaques: Influence of Mauritian MHC haplotypes on susceptibility/resistance to SHIV89.6P infection. *Vaccine* **2008**, *26*, 3312–3321. [CrossRef]

105. Burwitz, B.J.; Pendley, C.J.; Greene, J.M.; Detmer, A.M.; Lhost, J.J.; Karl, J.A.; Piaskowski, S.M.; Rudersdorf, R.A.; Wallace, L.T.; Bimber, B.N.; et al. Mauritian Cynomolgus Macaques Share Two Exceptionally Common Major Histocompatibility Complex Class I Alleles That Restrict Simian Immunodeficiency Virus-Specific CD8(+) T Cells. *J. Virol.* **2009**, *83*, 6011–6019. [CrossRef]

106. Mee, E.T.; Berry, N.; Ham, C.; Sauermann, U.; Maggiorella, M.T.; Martinon, F.; Verschoor, E.; Heeney, J.L.; Le Grand, R.; Titti, F.; et al. Mhc haplotype H6 is associated with sustained control of SIVmac251 infection in Mauritian cynomolgus macaques. *Immunogenetics* **2009**, *61*, 327–339. [CrossRef]

107. Mee, E.T.; Berry, N.; Ham, C.; Aubertin, A.; Lines, J.; Hall, J.; Stebbings, R.; Page, M.; Almond, N.; Rose, N.J. Mhc haplotype M3 is associated with early control of SHIVsbg infection in Mauritian cynomolgus macaques. *Tissue Antigens* **2010**, *76*, 223–229. [CrossRef]

108. Aarnink, A.; Dereuddre-Bosquet, N.; Vaslin, B.; Le Grand, R.; Winterton, P.; Apoil, P.A.; Blancher, A. Influence of the MHC genotype on the progression of experimental SIV infection in the Mauritian cynomolgus macaque. *Immunogenetics* **2011**, *63*, 267–274. [CrossRef]

109. de Manuel, M.; Shiina, T.; Suzuki, S.; Dereuddre-Bosquet, N.; Garchon, H.J.; Tanaka, M.; Congy-Jolivet, N.; Aarnink, A.; Le Grand, R.; Marques-Bonet, T.; et al. Whole genome sequencing in the search for genes associated with the control of SIV infection in the Mauritian macaque model. *Sci. Rep.* **2018**, *8*, 7131. [CrossRef]

110. Shiina, T.; Suzuki, S.; Congy-Jolivet, N.; Aarnink, A.; Garchon, H.J.; Dereuddre-Bosquet, N.; Vaslin, B.; Tchitchek, N.; Desjardins, D.; Autran, B.; et al. Cynomolgus macaque IL37 polymorphism and control of SIV infection. *Sci. Rep.* **2019**, *9*, 7981. [CrossRef]

111. Borsetti, A.; Ferrantelli, F.; Maggiorella, M.T.; Sernicola, L.; Bellino, S.; Gallinaro, A.; Farcomeni, S.; Mee, E.T.; Rose, N.J.; Cafaro, A.; et al. Effect of MHC Haplotype on Immune Response upon Experimental SHIVSF162P4cy Infection of Mauritian Cynomolgus Macaques. *PLoS ONE* **2014**, *9*. [CrossRef]

112. Antony, J.M.; MacDonald, K.S. A critical analysis of the cynomolgus macaque, Macaca fascicularis, as a model to test HIV-1/SIV vaccine efficacy. *Vaccine* **2015**, *33*, 3073–3083. [CrossRef]

113. O'Connor, S.L.; Lhost, J.J.; Becker, E.A.; Detmer, A.M.; Johnson, R.C.; MacNair, C.E.; Wiseman, R.W.; Karl, J.A.; Greene, J.M.; Burwitz, B.J.; et al. MHC Heterozygote Advantage in Simian Immunodeficiency Virus-Infected Mauritian Cynomolgus Macaques. *Sci. Transl. Med.* **2010**, *2*. [CrossRef]

114. Greene, J.M.; Chin, E.N.; Budde, M.L.; Lhost, J.J.; Hines, P.J.; Burwitz, B.J.; Broman, K.W.; Nelson, J.E.; Friedrich, T.C.; O'Connor, D.H. Ex Vivo SIV-Specific CD8 T Cell Responses in Heterozygous Animals Are Primarily Directed against Peptides Presented by a Single MHC Haplotype. *PLoS ONE* **2012**, *7*. [CrossRef]

115. Gellerup, D.D.; Balgeman, A.J.; Nelson, C.W.; Ericsen, A.J.; Scarlotta, M.; Hughes, A.L.; O'Connor, S.L. Conditional Immune Escape during Chronic Simian Immunodeficiency Virus Infection. *J. Virol.* **2016**, *90*, 545–552. [CrossRef]

116. Li, H.Z.; Omange, R.W.; Czarnecki, C.; Correia-Pinto, J.F.; Crecente-Campo, J.; Richmond, M.; Li, L.; Schultz-Darken, N.; Alonso, M.J.; Whitney, J.B.; et al. Mauritian cynomolgus macaques with M3M4 MHC genotype control SIVmac251 infection. *J. Med. Primatol.* **2017**, *46*, 137–143. [CrossRef]

117. Bruel, T.; Hamimi, C.; Dereuddre-Bosquet, N.; Cosma, A.; Shin, S.Y.; Corneau, A.; Versmisse, P.; Karlsson, I.; Malleret, B.; Targat, B.; et al. Long-term control of simian immunodeficiency virus (SIV) in cynomolgus macaques not associated with efficient SIV-specific CD8+ T-cell responses. *J. Virol.* **2015**, *89*, 3542–3556. [CrossRef]

118. Kaeuferle, T.; Krauss, R.; Blaeschke, F.; Willier, S.; Feuchtinger, T. Strategies of adoptive T -cell transfer to treat refractory viral infections post allogeneic stem cell transplantation. *J. Hematol. Oncol.* **2019**, *12*, 13. [CrossRef]

119. Ottaviano, G.; Chiesa, R.; Feuchtinger, T.; Vickers, M.A.; Dickinson, A.; Gennery, A.R.; Veys, P.; Todryk, S. Adoptive T Cell Therapy Strategies for Viral Infections in Patients Receiving Haematopoietic Stem Cell Transplantation. *Cells* **2019**, *8*. [CrossRef]

120. Greene, J.M.; Lhost, J.J.; Hines, P.J.; Scarlotta, M.; Harris, M.; Burwitz, B.J.; Budde, M.L.; Dudley, D.M.; Pham, N.; Cain, B.; et al. Adoptive transfer of lymphocytes isolated from simian immunodeficiency virus SIVmac239Deltanef-vaccinated macaques does not affect acute-phase viral loads but may reduce chronic-phase viral loads in major histocompatibility complex-matched recipients. *J. Virol.* **2013**, *87*, 7382–7392. [CrossRef]

121. Mohns, M.S.; Greene, J.M.; Cain, B.T.; Pham, N.H.; Gostick, E.; Price, D.A.; O'Connor, D.H. Expansion of Simian Immunodeficiency Virus (SIV)-Specific CD8 T Cell Lines from SIV-Naive Mauritian Cynomolgus Macaques for Adoptive Transfer. *J. Virol.* **2015**, *89*, 9748–9757. [CrossRef]

122. Chapuis, A.G.; Desmarais, C.; Emerson, R.; Schmitt, T.M.; Shibuya, K.C.; Lai, I.P.; Wagener, F.; Chou, J.; Roberts, I.M.; Coffey, D.G.; et al. Tracking the fate and origin of clinically relevant adoptively transferred CD8(+) T cells in vivo. *Science Immunol.* **2017**, *2*. [CrossRef]

123. O'Neil, R.T.; Saha, S.; Veach, R.A.; Welch, R.C.; Woodard, L.E.; Rooney, C.M.; Wilson, M.H. Transposon-modified antigen-specific T lymphocytes for sustained therapeutic protein delivery in vivo. *Nat. Commun.* **2018**, *9*. [CrossRef]

124. Center for iPS Cell Research and Application, Kyoto University. Available online: https://www.cira.kyoto-u.ac.jp/e/research/stock.html (accessed on 23 August 2019).

125. Sugita, S.; Iwasaki, Y.; Makabe, K.; Kamao, H.; Mandai, M.; Shiina, T.; Ogasawara, K.; Hirami, Y.; Kurimoto, Y.; Takahashi, M. Successful Transplantation of Retinal Pigment Epithelial Cells from MHC Homozygote iPSCs in MHC-Matched Models. *Stem Cell Rep.* **2016**, *7*, 635–648. [CrossRef]

126. Sugita, S.; Makabe, K.; Fujii, S.; Iwasaki, Y.; Kamao, H.; Shiina, T.; Ogasawara, K.; Takahashi, M. Detection of Retinal Pigment Epithelium-Specific Antibody in iPSC-Derived Retinal Pigment Epithelium Transplantation Models. *Stem Cell Rep.* **2017**, *9*, 1501–1515. [CrossRef]

127. Morizane, A.; Doi, D.; Kikuchi, T.; Okita, K.; Hotta, A.; Kawasaki, T.; Hayashi, T.; Onoe, H.; Shiina, T.; Yamanaka, S.; et al. Direct comparison of autologous and allogeneic transplantation of iPSC-derived neural cells in the brain of a non-human primate. *Stem Cell Rep.* **2013**, *1*, 283–292. [CrossRef]

128. Morizane, A.; Kikuchi, T.; Hayashi, T.; Mizuma, H.; Takara, S.; Doi, H.; Mawatari, A.; Glasser, M.F.; Shiina, T.; Ishigaki, H.; et al. MHC matching improves engraftment of iPSC-derived neurons in non-human primates. *Nat. Commun.* **2017**, *8*. [CrossRef]

129. Kawamura, T.; Miyagawa, S.; Fukushima, S.; Maeda, A.; Kashiyama, N.; Kawamura, A.; Miki, K.; Okita, K.; Yoshida, Y.; Shiina, T.; et al. Cardiomyocytes Derived from MHC-Homozygous Induced Pluripotent Stem Cells Exhibit Reduced Allogeneic Immunogenicity in MHC-Matched Non-human Primates. *Stem Cell Rep.* **2016**, *6*, 312–320. [CrossRef]

130. Sugita, S.; Iwasaki, Y.; Makabe, K.; Kimura, T.; Futagami, T.; Suegami, S.; Takahashi, M. Lack of T Cell Response to iPSC-Derived Retinal Pigment Epithelial Cells from HLA Homozygous Donors. *Stem Cell Rep.* **2016**, *7*, 619–634. [CrossRef]

131. Yamasaki, J.; Iwatani, C.; Tsuchiya, H.; Okahara, J.; Sankai, T.; Torii, R. Vitrification and transfer of cynomolgus monkey (Macaca fascicularis) embryos fertilized by intracytoplasmic sperm injection. *Theriogenology* **2011**, *76*, 33–38. [CrossRef]

132. Aarnink, A.; Garchon, H.J.; Puissant-Lubrano, B.; Blancher-Sardou, M.; Apoil, P.A.; Blancher, A. Impact of MHC class II polymorphism on blood counts of CD4+T lymphocytes in macaque. *Immunogenetics* **2011**, *63*, 95–102. [CrossRef]

133. Aarnink, A.; Garchon, H.J.; Okada, Y.; Takahashi, A.; Matsuda, K.; Kubo, M.; Nakamura, Y.; Blancher, A. Comparative analysis in cynomolgus macaque identifies a novel human MHC locus controlling platelet blood counts independently of BAK1. *J. Thromb. Haemost.* **2013**, *11*, 384–386. [CrossRef]

134. Bagavant, H.; Sharp, C.; Kurth, B.; Tung, K.S. Induction and immunohistology of autoimmune ovarian disease in cynomolgus macaques (Macaca fascicularis). *Am. J. Pathol.* **2002**, *160*, 141–149. [CrossRef]

135. Massacesi, L.; Joshi, N.; Lee-Parritz, D.; Rombos, A.; Letvin, N.L.; Hauser, S.L. Experimental allergic encephalomyelitis in cynomolgus monkeys. Quantitation of T cell responses in peripheral blood. *J. Clin. Invest.* **1992**, *90*, 399–404. [CrossRef]

136. Shimozuru, Y.; Yamane, S.; Fujimoto, K.; Terao, K.; Honjo, S.; Nagai, Y.; Sawitzke, A.D.; Terato, K. Collagen-induced arthritis in nonhuman primates: Multiple epitopes of type II collagen can induce autoimmune-mediated arthritis in outbred cynomolgus monkeys. *Arthritis Rheum.* **1998**, *41*, 507–514. [CrossRef]

137. Bitoun, S.; Roques, P.; Maillere, B.; Le Grand, R.; Mariette, X. Valine 11 and phenylalanine 13 have a greater impact on the T-cell response to citrullinated peptides than the 70-74 shared epitope of the DRB1 molecule in macaques. *Ann. Rheum. Dis.* **2019**. [CrossRef]

138. Bitoun, S.; Roques, P.; Larcher, T.; Nocturne, G.; Serguera, C.; Chretien, P.; Serre, G.; Grand, R.L.; Mariette, X. Both Systemic and Intra-articular Immunization with Citrullinated Peptides Are Needed to Induce Arthritis in the Macaque. *Front. Immunol.* **2017**, *8*, 1816. [CrossRef]

139. Choi, E.W.; Lee, K.W.; Park, H.; Kim, H.; Lee, J.H.; Song, J.W.; Yang, J.; Kwon, Y.; Kim, T.M.; Park, J.B.; et al. Therapeutic effects of anti-CD154 antibody in cynomolgus monkeys with advanced rheumatoid arthritis. *Sci. Rep.* **2018**, *8*, 2135. [CrossRef]

140. Ho, H.N.; Yang, Y.S.; Hsieh, R.P.; Lin, H.R.; Chen, S.U.; Chen, H.F.; Huang, S.C.; Lee, T.Y.; Gill, T.J., 3rd. Sharing of human leukocyte antigens in couples with unexplained infertility affects the success of in vitro fertilization and tubal embryo transfer. *Am. J. Obstet. Gynecol.* **1994**, *170*, 63–71. [CrossRef]

141. Jin, K.; Ho, H.N.; Speed, T.P.; Gill, T.J., 3rd. Reproductive failure and the major histocompatibility complex. *Am. J. Hum. Genet.* **1995**, *56*, 1456–1467.

142. Kirby, D.R. The egg and immunology. *Proc. R. Soc. Med.* **1970**, *63*, 59–61.

143. Mee, E.T.; Badhan, A.; Karl, J.A.; Wiseman, R.W.; Cutler, K.; Knapp, L.A.; Almond, N.; O'Connor, D.H.; Rose, N.J. MHC haplotype frequencies in a UK breeding colony of Mauritian cynomolgus macaques mirror those found in a distinct population from the same geographic origin. *J. Med. Primatol.* **2009**, *38*, 1–14. [CrossRef]

144. Aarnink, A.; Mee, E.T.; Savy, N.; Congy-Jolivet, N.; Rose, N.J.; Blancher, A. Deleterious impact of feto-maternal MHC compatibility on the success of pregnancy in a macaque model. *Immunogenetics* **2014**, *66*, 105–113. [CrossRef]

145. Ober, C. Current Topic—Hla and Reproduction—Lessons from Studies in the Hutterites. *Placenta* **1995**, *16*, 569–577. [CrossRef]

146. Parimi, N.; Tromp, G.; Kuivaniemi, H.; Nien, J.K.; Gomez, R.; Romero, R.; Goddard, K.A. Analytical approaches to detect maternal/fetal genotype incompatibilities that increase risk of pre-eclampsia. *BMC Med. Genet.* **2008**, *9*, 60. [CrossRef]

147. Ziegler, A.; Santos, P.S.; Kellermann, T.; Uchanska-Ziegler, B. Self/nonself perception, reproduction and the extended MHC. *Self Nonself* **2010**, *1*, 176–191. [CrossRef]

148. Wu, H.; Whritenour, J.; Sanford, J.C.; Houle, C.; Adkins, K.K. Identification of MHC Haplotypes Associated with Drug-induced Hypersensitivity Reactions in Cynomolgus Monkeys. *Toxicol. Pathol.* **2017**, *45*, 127–133. [CrossRef]

149. Pompeu, Y.A.; Stewart, J.D.; Mallal, S.; Phillips, E.; Peters, B.; Ostrov, D.A. The structural basis of HLA-associated drug hypersensitivity syndromes. *Immunol. Rev.* **2012**, *250*, 158–166. [CrossRef]

150. Campbell, K.S.; Purdy, A.K. Structure/function of human killer cell immunoglobulin-like receptors: Lessons from polymorphisms, evolution, crystal structures and mutations. *Immunology* **2011**, *132*, 315–325. [CrossRef]

151. Hudson, L.E.; Allen, R.L. Leukocyte Ig-Like Receptors—A Model for MHC Class I Disease Associations. *Front. Immunol.* **2016**, *7*, 281. [CrossRef]

152. Bashirova, A.A.; Martin-Gayo, E.; Jones, D.C.; Qi, Y.; Apps, R.; Gao, X.; Burke, P.S.; Taylor, C.J.; Rogich, J.; Wolinsky, S.; et al. LILRB2 interaction with HLA class I correlates with control of HIV-1 infection. *PLoS Genet.* **2014**, *10*, e1004196. [CrossRef]

153. Alaoui, L.; Palomino, G.; Zurawski, S.; Zurawski, G.; Coindre, S.; Dereuddre-Bosquet, N.; Lecuroux, C.; Goujard, C.; Vaslin, B.; Bourgeois, C.; et al. Early SIV and HIV infection promotes the LILRB2/MHC-I inhibitory axis in cDCs. *Cell Mol. Life Sci.* **2018**, *75*, 1871–1887. [CrossRef]

154. Chazara, O.; Xiong, S.; Moffett, A. Maternal KIR and fetal HLA-C: A fine balance. *J. Leukoc. Biol.* **2011**, *90*, 703–716. [CrossRef]

155. Colucci, F. The role of KIR and HLA interactions in pregnancy complications. *Immunogenetics* **2017**, *69*, 557–565. [CrossRef]

156. Parham, P.; Moffett, A. Variable NK cell receptors and their MHC class I ligands in immunity, reproduction and human evolution. *Nat. Rev. Immunol.* **2013**, *13*, 133–144. [CrossRef]

157. Garcia-Beltran, W.F.; Holzemer, A.; Martrus, G.; Chung, A.W.; Pacheco, Y.; Simoneau, C.R.; Rucevic, M.; Lamothe-Molina, P.A.; Pertel, T.; Kim, T.E.; et al. Open conformers of HLA-F are high-affinity ligands of the activating NK-cell receptor KIR3DS1. *Nat. Immunol.* **2016**, *17*, 1067–1074. [CrossRef]

158. Lin, A.; Yan, W.H. The Emerging Roles of Human Leukocyte Antigen-F in Immune Modulation and Viral Infection. *Front. Immunol.* **2019**, *10*, 964. [CrossRef]

159. Morandi, F.; Rizzo, R.; Fainardi, E.; Rouas-Freiss, N.; Pistoia, V. Recent Advances in Our Understanding of HLA-G Biology: Lessons from a Wide Spectrum of Human Diseases. *J. Immunol. Res.* **2016**, *2016*, 4326495. [CrossRef]

160. Sharpe, H.R.; Bowyer, G.; Brackenridge, S.; Lambe, T. HLA-E: Exploiting pathogen-host interactions for vaccine development. *Clin. Exp. Immunol.* **2019**, *196*, 167–177. [CrossRef]

161. Bimber, B.N.; Moreland, A.J.; Wiseman, R.W.; Hughes, A.L.; O'Connor, D.H. Complete characterization of killer Ig-like receptor (KIR) haplotypes in Mauritian cynomolgus macaques: Novel insights into nonhuman primate KIR gene content and organization. *J. Immunol.* **2008**, *181*, 6301–6308. [CrossRef]

162. Walter, L.; Ansari, A.A. MHC and KIR Polymorphisms in Rhesus Macaque SIV Infection. *Front. Immunol.* **2015**, *6*, 540. [CrossRef]

163. Grimsley, C.; Kawasaki, A.; Gassner, C.; Sageshima, N.; Nose, Y.; Hatake, K.; Geraghty, D.E.; Ishitani, A. Definitive high resolution typing of HLA-E allelic polymorphisms: Identifying potential errors in existing allele data. *Tissue Antigens* **2002**, *60*, 206–212. [CrossRef]

164. Braud, V.; Jones, E.Y.; McMichael, A. The human major histocompatibility complex class Ib molecule HLA-E binds signal sequence-derived peptides with primary anchor residues at positions 2 and 9. *Eur. J. Immunol.* **1997**, *27*, 1164–1169. [CrossRef]

165. Lee, N.; Goodlett, D.R.; Ishitani, A.; Marquardt, H.; Geraghty, D.E. HLA-E surface expression depends on binding of TAP-dependent peptides derived from certain HLA class I signal sequences. *J. Immunol.* **1998**, *160*, 4951–4960.

166. Borrego, F.; Ulbrecht, M.; Weiss, E.H.; Coligan, J.E.; Brooks, A.G. Recognition of human histocompatibility leukocyte antigen (HLA)-E complexed with HLA class I signal sequence-derived peptides by CD94/NKG2 confers protection from natural killer cell-mediated lysis. *J. Exp. Med.* **1998**, *187*, 813–818. [CrossRef]

167. Braud, V.M.; Allan, D.S.J.; O'Callaghan, C.A.; Soderstrom, K.; D'Andrea, A.; Ogg, G.S.; Lazetic, S.; Young, N.T.; Bell, J.I.; Phillips, J.H.; et al. HLA-E binds to natural killer cell receptors CD94/NKG2A, B and C. *Nature* **1998**, *391*, 795–799. [CrossRef]

168. Lee, N.; Llano, M.; Carretero, M.; Ishitani, A.; Navarro, F.; Lopez-Botet, M.; Geraghty, D.E. HLA-E is a major ligand for the natural killer inhibitory receptor CD94/NKG2A. *Proc. Natl. Acad. Sci. USA* **1998**, *95*, 5199–5204. [CrossRef]

169. Caccamo, N.; Pietra, G.; Sullivan, L.C.; Brooks, A.G.; Prezzemolo, T.; La Manna, M.P.; Di Liberto, D.; Joosten, S.A.; van Meijgaarden, K.E.; Di Carlo, P.; et al. Human CD8 T lymphocytes recognize Mycobacterium tuberculosis antigens presented by HLA-E during active tuberculosis and express type 2 cytokines. *Eur. J. Immunol.* **2015**, *45*, 1069–1081. [CrossRef]

170. Heinzel, A.S.; Grotzke, J.E.; Lines, R.A.; Lewinsohn, D.A.; McNabb, A.L.; Streblow, D.N.; Brand, V.M.; Grieser, H.J.; Belisle, J.T.; Lewinsohn, D.M. HLA-E-dependent presentation of Mtb-derived antigen to human CD8(+) T cells. *J. Exp. Med.* **2002**, *196*, 1473–1481. [CrossRef]

171. Joosten, S.A.; van Meijgaarden, K.E.; van Weeren, P.C.; Kazi, F.; Geluk, A.; Savage, N.D.L.; Drijfhout, J.W.; Flower, D.R.; Hanekom, W.A.; Klein, M.R.; et al. Mycobacterium tuberculosis Peptides Presented by HLA-E Molecules Are Targets for Human CD8(+) T-Cells with Cytotoxic as well as Regulatory Activity. *PLoS Pathog.* **2010**, *6*. [CrossRef]

172. van Meijgaarden, K.E.; Haks, M.C.; Caccamo, N.; Dieli, F.; Ottenhoff, T.H.M.; Joosten, S.A. Human CD8(+) T-cells Recognizing Peptides from Mycobacterium tuberculosis (Mtb) Presented by HLA-E Have an Unorthodox Th2-like, Multifunctional, Mtb Inhibitory Phenotype and Represent a Novel Human T-cell Subset. *PLoS Pathog.* **2015**, *11*. [CrossRef]

173. Salerno-Goncalves, R.; Fernandez-Vina, M.; Lewinsohn, D.M.; Sztein, M.B. Identification of a human HLA-E-restricted CD8+ T cell subset in volunteers immunized with Salmonella enterica serovar Typhi strain Ty21a typhoid vaccine. *J. Immunol.* **2004**, *173*, 5852–5862. [CrossRef]

174. Jorgensen, P.B.; Livbjerg, A.H.; Hansen, H.J.; Petersen, T.; Hollsberg, P. Epstein-Barr virus Peptide Presented by HLA-E is Predominantly Recognized by CD8(bright) Cells in multiple Sclerosis Patients. *PLoS ONE* **2012**, *7*. [CrossRef]

175. Mazzarino, P.; Pietra, G.; Vacca, P.; Falco, M.; Colau, D.; Coulie, P.; Moretta, L.; Mingari, M.C. Identification of effector-memory CMV-specific T lymphocytes that kill CMV-infected target cells in an HLA-E-restricted fashion. *Eur. J. Immunol.* **2005**, *35*, 3240–3247. [CrossRef]

176. Pietra, G.; Romagnani, C.; Mazzarino, P.; Falco, M.; Millo, E.; Moretta, A.; Moretta, L.; Mingari, M.C. HLA-E-restricted recognition of cytomegalovirus-derived peptides by human CD8(+) cytolytic T lymphocytes. *Proc. Natl. Acad. Sci. USA* **2003**, *100*, 10896–10901. [CrossRef]

177. Nattermann, J.; Nischalke, H.D.; Hofmeister, V.; Ahlenstiel, G.; Zimmermann, H.; Leifeld, L.; Weiss, E.H.; Sauerbruch, T.; Spengler, U. The HLA-A2 restricted T cell epitope HCV core 35-44 stabilizes HLA-E expression and inhibits cytolysis mediated by natural killer cells. *Am. J. Pathol.* **2005**, *166*, 443–453. [CrossRef]

178. Schulte, D.; Vogel, M.; Langhans, B.; Kramer, B.; Korner, C.; Nischalke, H.D.; Steinberg, V.; Michalk, M.; Berg, T.; Rockstroh, J.K.; et al. The HLA-E-R/HLA-E-R Genotype Affects the Natural Course of Hepatitis C Virus (HCV) Infection and Is Associated with HLA-E-Restricted Recognition of an HCV-Derived Peptide by Interferon-gamma-Secreting Human CD8(+) T Cells. *J. Infect. Dis.* **2009**, *200*, 1397–1401. [CrossRef]

179. Hansen, S.G.; Wu, H.L.; Burwitz, B.J.; Hughes, C.M.; Hammond, K.B.; Ventura, A.B.; Reed, J.S.; Gilbride, R.M.; Ainslie, E.; Morrow, D.W.; et al. Broadly targeted CD8(+) T cell responses restricted by major histocompatibility complex E. *Science* **2016**, *351*, 714–720. [CrossRef]

180. Hansen, S.G.; Piatak, M., Jr.; Ventura, A.B.; Hughes, C.M.; Gilbride, R.M.; Ford, J.C.; Oswald, K.; Shoemaker, R.; Li, Y.; Lewis, M.S.; et al. Immune clearance of highly pathogenic SIV infection. *Nature* **2013**, *502*, 100–104. [CrossRef]

181. Wu, H.L.; Wiseman, R.W.; Hughes, C.M.; Webb, G.M.; Abdulhaqq, S.A.; Bimber, B.N.; Hammond, K.B.; Reed, J.S.; Gao, L.; Burwitz, B.J.; et al. The Role of MHC-E in T Cell Immunity Is Conserved among Humans, Rhesus Macaques, and Cynomolgus Macaques. *J. Immunol.* **2018**, *200*, 49–60. [CrossRef]

182. Andre, P.; Denis, C.; Soulas, C.; Bourbon-Caillet, C.; Lopez, J.; Arnoux, T.; Blery, M.; Bonnafous, C.; Gauthier, L.; Morel, A.; et al. Anti-NKG2A mAb Is a Checkpoint Inhibitor that Promotes Anti-tumor Immunity by Unleashing Both T and NK Cells. *Cell* **2018**, *175*, 1731–1743.e13. [CrossRef]

183. Walter, L.; Petersen, B. Diversification of both KIR and NKG2 natural killer cell receptor genes in macaques—Implications for highly complex MHC-dependent regulation of natural killer cells. *Immunology* **2017**, *150*, 139–145. [CrossRef]

184. 't Hart, B.A.; Gran, B.; Weissert, R. EAE: Imperfect but useful models of multiple sclerosis. *Trends Mol. Med.* **2011**, *17*, 119–125. [CrossRef]

185. Jagessar, S.A.; Holtman, I.R.; Hofman, S.; Morandi, E.; Heijmans, N.; Laman, J.D.; Gran, B.; Faber, B.W.; van Kasteren, S.I.; Eggen, B.J.L.; et al. Lymphocryptovirus Infection of Nonhuman Primate B Cells Converts Destructive into Productive Processing of the Pathogenic CD8 T Cell Epitope in Myelin Oligodendrocyte Glycoprotein. *J. Immunol.* **2016**, *197*, 1074–1088. [CrossRef]

186. Haanstra, K.G.; Wubben, J.A.M.; Jonker, M.; 't Hart, B.A. Induction of Encephalitis in Rhesus Monkeys Infused with Lymphocryptovirus-Infected B-Cells Presenting MOG(34-56) Peptide. *PLoS ONE* **2013**, *8*. [CrossRef]

187. Jagessar, S.A.; Heijmans, N.; Bauer, J.; Blezer, E.L.A.; Laman, J.D.; Hellings, N.; 't Hart, B.A. B-Cell Depletion Abrogates T Cell-Mediated Demyelination in an Antibody-Nondependent Common Marmoset Experimental Autoimmune Encephalomyelitis Model. *J. Neuropath. Exp. Neur.* **2012**, *71*, 716–728. [CrossRef]

188. Shen, S.; Pyo, C.W.; Vu, Q.; Wang, R.; Geraghty, D.E. The essential detail: The genetics and genomics of the primate immune response. *ILAR J.* **2013**, *54*, 181–195. [CrossRef]

189. Liu, Z.; Cai, Y.; Wang, Y.; Nie, Y.; Zhang, C.; Xu, Y.; Zhang, X.; Lu, Y.; Wang, Z.; Poo, M.; et al. Cloning of Macaque Monkeys by Somatic Cell Nuclear Transfer. *Cell* **2018**, *172*, 881–887.e7. [CrossRef]

 *cells*

Review

Avian MHC Evolution in the Era of Genomics: Phase 1.0

Emily A. O'Connor [1], Helena Westerdahl [1], Reto Burri [2] and Scott V. Edwards [3],*

[1] Department of Biology, Lund University, SE-223 62 Lund, Sweden; emily.o_connor@biol.lu.se (E.A.O.); helena.westerdahl@biol.lu.se (H.W.)

[2] Department of Population Ecology, Institute of Ecology & Evolution, Friedrich Schiller University Jena, 07737 Jena, Germany; burri@wildlight.ch

[3] Department of Organismic and Evolutionary Biology and Museum of Comparative Zoology, Harvard University, Cambridge, MA 02138, USA

* Correspondence: sedwards@fas.harvard.edu; Tel.: +1-617-384-8082; Fax: +1-617-495-5667

Received: 12 August 2019; Accepted: 20 September 2019; Published: 26 September 2019

Abstract: Birds are a wonderfully diverse and accessible clade with an exceptional range of ecologies and behaviors, making the study of the avian major histocompatibility complex (MHC) of great interest. In the last 20 years, particularly with the advent of high-throughput sequencing, the avian MHC has been explored in great depth in several dimensions: its ability to explain ecological patterns in nature, such as mating preferences; its correlation with parasite resistance; and its structural evolution across the avian tree of life. Here, we review the latest pulse of avian MHC studies spurred by high-throughput sequencing. Despite high-throughput approaches to MHC studies, substantial areas remain in need of improvement with regard to our understanding of MHC structure, diversity, and evolution. Recent studies of the avian MHC have nonetheless revealed intriguing connections between MHC structure and life history traits, and highlight the advantages of long-term ecological studies for understanding the patterns of MHC variation in the wild. Given the exceptional diversity of birds, their accessibility, and the ease of sequencing their genomes, studies of avian MHC promise to improve our understanding of the many dimensions and consequences of MHC variation in nature. However, significant improvements in assembling complete MHC regions with long-read sequencing will be required for truly transformative studies.

Keywords: MHC genes; birds; disease resistance; orthology; life history; gene duplication; long-read sequencing; high-throughput sequencing; concerted evolution; ecology

1. Introduction

Birds and the major histocompatibility complex (MHC) have a long and special relationship. The domestic chicken (*Gallus gallus domesticus*) was among the first species to have its MHC characterized at the functional level, and the chicken MHC has been the major non-mammalian vertebrate model for MHC structure and genomic organization [1–5]. The chicken MHC has also yielded some of the clearest associations between disease resistance and genotype, a link speculated to be associated with the relatively simple and compact structure of the chicken MHC when compared with that of mammals [1,5,6]. However, despite decades of immunological work on the chicken, interest in the MHC among ornithologists with a focus on ecology and evolution came not from studies on chickens but from mammals. The possibility that MHC variation might influence mating preferences as well as disease resistance were revealed for the first time outside of laboratory strains by the landmark studies of Wayne Potts [7,8]. These studies in semi-natural populations of mice, alongside a growing interest in the role of MHC in mate choice and kin recognition [9], catalyzed the first explorations of MHC variation in natural populations of birds [10–15]. The long-term goal of these avian MHC studies was to find the genes influencing fitness and behavior in the wild [11–15], and to aid the conservation of biodiversity through the analysis of genes with important functions in conservation biology [16]. Currently, there is a strong interest among evolutionary biologists in genes and regulatory regions underlying phenotypic variation, with many recent exciting examples from birds [17–19]. Genes of the MHC were the first candidates for variation in fitness and fitness-related phenotypes in birds, including disease resistance, plumage brightness, as well as mating preferences and success [11,13,14]. The extraordinarily rich body of theory relating to signals of superior disease resistance, sexual selection, mating patterns, and behavior [20–22] was inspired and compellingly exemplified to a large extent by birds; this body of theory also provided a fertile foundation for the flood of avian MHC studies beginning in the latter half of the 1990s [13,14,23,24].

This review provides an update of studies of MHC evolution in birds, including recent advances in our understanding of structural evolution of the avian MHC, as well as new insights into aspects of avian ecology and evolution provided by the MHC. Our update is timely because the field of avian MHC is poised for advances in our understanding of the structural genetic variation of MHC due to the adoption of affordable and accessible long-read sequencing technology: phase 1.0 of the genomic era. Our focus is on species that are models in ecology and evolutionary studies, such as songbirds (oscine passerine birds), and less so on avian species that are immunological models, such as the chicken, other gamebirds (Galliformes), and waterfowl, which have been reviewed elsewhere [2–4]. An assumption of our overall perspective is that little can be learned about the evolution of MHC genes in birds without comparisons among species. The publication of genome sequences for a large number of bird species has opened the door to comparative analyses of genomic regions of interest [25]. However, the incomplete assembly of the MHC region in genomes generated using short-read sequencing technology has limited the scope for whole genome data to be used in comparative analyses of avian MHC.

Recent phylogenetic studies of MHC evolution using data from targeted MHC genotyping studies, which we review here, have revealed exciting long-term patterns of gene duplication and divergence. The quality of non-chicken avian genomes lags behind that of the chicken. Nevertheless, we can start to see some clear trends in the evolution of MHC structure and gene number across the avian tree of life [3]. Even from early rudimentary studies using genomic cloning [23,24,26–31], it was evident that the paradigm put forward by the chicken MHC, with less than 20 genes and spanning less than 100 kb, did not represent birds generally; indeed, the zebra finch MHC genomic region(s) exceeds 700 kb [32,33]. The emerging picture of MHC evolution in birds is that species within the Galliformes appear to have smaller MHC genomic regions than most other species of birds [5,28,32,34–38] (but Shiina et al. 2004 provides evidence of more extensively duplicated MHC genes in the Japanese quail (*Coturnix japonica*) [39]). In this respect, the chicken, with its 'minimal essential MHC', may be the 'odd duck' when it comes to their genomic organization of the MHC. Moreover, expression of MHC genes in non-galliform birds can be highly complex, with multiple class I and II genes expressed in some songbirds [23,24,40–42], suggesting that the chicken may also be an outlier with regard to MHC expression (but see Drews et al. 2017 [43]). There is conclusive evidence for classical and non-classical MHC class I (MHC-I) or MHC class IIB (MHC-IIB) in Galliform species such as the chicken, the turkey (*Meleagris gallopavo*), the black grouse (*Tetrao tetrix*), and the golden pheasant (*Chrysolophus pictus*) [5,36,44–49]. Possessing both classical and non-classical MHC genes appears to be a taxonomically widespread phenomenon because it has been observed in primates, fish, amphibians, and reptiles [50–54]. Classical MHC genes are usually highly expressed, polymorphic, and have a well-established function in presenting antigens to T-cells [55]. On the other hand, non-classical MHC genes usually exhibit lower polymorphism, are only weakly expressed, and may have functions beyond classic antigen presentation [56–58]. The detailed studies required to confirm the presence of non-classical MHC genes outside of Galliformes are currently lacking, but evidence from allelic polymorphism and expression patterns suggests that a number of species within the Charadriiformes, Pelecaniformes, and Passeriformes may also possess both classical and non-classical MHC-I and MHC-II genes [42,43,59,60]. In songbirds, these putatively non-classical

MHC-I genes are not orthologous to MHC-Y, a second polymorphic MHC region also on chicken chr 16 [4], making it likely that non-classical MHC genes have arisen on multiple independent occasions across the evolution of birds [43]. Furthermore, the non-classical MHC genes described in mammals, fish, and birds all appear to have independent origins [51,61]. We use the terms MHC-I and MHC-II throughout this review to refer only to classical MHC genes unless otherwise stated.

We also discuss recent efforts to integrate studies of MHC variation in long-term ecological studies, for which birds provide many excellent examples. Studies on wild birds reveal how MHC evolves in nature and impacts fitness. The detailed mechanistic understanding of MHC, generated by studies on chickens, help us to interpret the biological relevance of the patterns of MHC variation we see in nature. In the last ten years, ornithologists and evolutionary biologists have gone to great lengths to understand the evolutionary forces driving MHC polymorphism. Here, we hope to provide an overview of this quiet revolution, which we believe will foster a second renaissance of MHC studies in birds.

2. The Avian Major Histocompatibility Complex (MHC) Enters the Genomic Era

2.1. Technical Advances: Large-Scale MHC Structure

Twenty years have passed since the first genomic structure of an avian MHC, that of the chicken, was characterized [5]. However, despite tremendous technological advances, strategies for the genomic characterization of bird MHCs have barely changed. Most of the recent genomic characterizations of bird MHCs, such as those of the black grouse (*Lyrurus tetrix*), crested ibis (*Nipponia nippon*), and oriental stork (*Ciconia boyciana*), have employed sequencing of MHC-containing fosmids or BAC clones as the preferred strategy [35,36,62,63]. Only in a very few instances have MHCs been characterized from *de novo* genome assemblies; even in these instances MHC structures were supported by sequencing of BAC clones [32]. Today, the cost of BAC-library construction, screening for MHC-containing clones, and subsequent sequencing likely exceed the costs of a typical *de novo* assembly of a bird genome. This begs the question why research on bird MHCs has made only limited use of existing genome assemblies [25]. Thus far, high-throughput sequencing and traditional genome assembly approaches have not enabled proper assembly of highly repetitive genomic regions such as the MHC [64], and existing tools for this purpose have not been applied extensively [65]. This deficit is particularly evident for bird species that have highly duplicated MHC genes. So far, *de novo* genome assembly in birds has predominantly relied on short-read sequencing technology and traditional assembly approaches, which results in the collapse of repeated sequences into a single location in the assembly. Bird MHCs are GC-rich, further hampering proper assembly [64] due to downward-biased sequence coverage.

Long-read sequencing technologies are quickly being adopted by the field of *de novo* genome assembly. Sequencing reads obtained through Pacific Biosciences Single Molecule Real-Time sequencing or Oxford Nanopore sequencing, for instance, now reach average lengths of several dozens of kilo base pairs or longer, dramatically improving the assembly of repetitive parts of the genome by anchoring repeats and duplicates to unique sequence content within the long reads [64]. In a number of primate species, long-read sequencing technology has already been employed to improve the characterization of complex gene families [66] including the MHC [67,68]. Similarly, in the newest chicken genome assembly, long-read sequencing has considerably improved the assembly of the MHC-Y region, extending it 1.5-fold in overall length [69]. Recent improvements in the accessibility of long-read sequencing technology are likely to result in the availability of many more high quality bird genomes, which offers promise for future comparative studies of MHC gene regions between species.

In addition to long-read sequencing, linked-read sequencing, such as offered by the 10x Genomics Chromium platform and other technologies in development [70], offer promising avenues for improving MHC assemblies. These technologies tag DNA molecules individually during library preparation and sequences them using short-read approaches in a cost-efficient manner, facilitating more efficient assembly and phasing, thereby reducing the assembly problem relative to phase-unaware approaches [71]. However, the best approach so far to document the catalog of MHC genes in any given species is likely to come from a combination of long-read sequencing and exhaustive short-read amplicon sequencing [72] (Westerdahl, unpublished). We expect such combined approaches and linked-read sequencing to complement and significantly assist the assembly of MHC regions based on long-read sequencing in the near future.

Nevertheless, avian MHC research has made use of short-read genome assemblies, which, at the same time, has revealed their limitations. A major PCR-screening of the avian tree of life for the presence of two ancestral MHC-IIB lineages was in part complemented by sequences retrieved from short-read bird genomes [34]. For several clades, the results from PCR screening of MHC-IIB genes were indeed confirmed in the genome assemblies, but only for species in which sequences of the two ancestral MHC-IIB lineages are highly divergent. In species with less divergent ancestral MHC-IIB lineages, the two regions were collapsed into a single assembled region (Burri, unpublished). Thus, genome assemblies based on short-reads can help confirm results obtained by other means, but the structure inferred from such genome assemblies may be biased toward models of multigene-family evolution that predict highly similar gene copies, such as concerted evolution, over others [73]. In birds, MHC-IIA genes are less well characterized than MHC-IIB, although initial evidence suggests that Pekin ducks (*Anas platyrhynchos*) and chickens only have a single MHC-IIA gene whereas the crested ibis has several sequentially repeated pairs of MHC-IIA and MHC-IIB genes [35,74,75]. Overall, it seems likely that MHC-IIA is less polymorphic than MHC-IIB in birds; however, further work is required to establish the general pattern for MHC-IIA genes across birds

2.2. Technical Advances: MHC Genotyping within Populations

As seen in Table 1, technological advances over the last two decades have brought about massive improvements in MHC genotyping methodology, resulting in an unforeseen resolution at the level of the individual [40,76,77]. In the 1990s, MHC genotyping was generally conducted with some form of fragment analysis, such as restriction fragment analyses (RFLP) and Southern blots using class I or IIB probes to visualize fragments [11,15,23]. However, with such approaches, a single allele could be represented by one or two bands, and the genetic polymorphisms indicated by the bands were not necessarily associated with coding DNA. PCR-based methods, whereby specific fragments of MHC genes are targeted and amplified, saw a focus on MHC-I exon 3 and MHC-IIB exon 2, which represent the most polymorphic exons of each class [27]. PCR-based amplicon sequencing, using degenerate or specific primers, has been the dominant method used to study MHC diversity in birds since. Meanwhile, methods used to characterize the PCR products have evolved, starting with cloning and Sanger sequencing, followed by various fragment conformation techniques (SSCP, DGGE, RSCA) [78–80] and, most recently, by high-throughput sequencing (Roche 454, Ion Torrent and Illumina) [81,82]. High-throughput amplicon sequencing catalyzed an enormous leap forward in terms of possible sequencing depth, enabling high-resolution MHC genotyping of individuals. This advance was especially advantageous for the songbird research community because many species within this clade have highly duplicated MHC genes.

Table 1. The development of major histocompatibility complex (MHC) genotyping methods in birds over four decades, with the advantages (Pros) and disadvantages (Cons) associated with different methods and future perspectives.

	Past		Present	Near Future	
	(1990s)	**(2000s)**	**(2010+)**	**(2019+)**	**(2019+)**
Method	Fragment analysis of genomic DNA (RFLP & Southern blot) [15,23]	Fragment analyses of PCR products (e.g., DGGE) [83–85]	High-throughput sequencing of PCR products (e.g., Illumina amplicon sequencing) [40,86,87]	Long-read sequencing (PacBio & Oxford nanopore)	Amplification of specific gene copies (High-throughput sequencing & qPCR)
Pros	No PCR artifacts Robust gene copy numbers	Moderate resolution	High resolution High coverage	No PCR artifacts Very long reads (100,000 bp possible) Gene synteny	Expression data Gene-specific amplification
Cons	Low resolution No sequence data	Artifactual alleles possible No sequence data	Artifactual alleles possible Short reads (up to 300 bp)	Low coverage Sequencing errors	Limited data on MHC diversity

The first generation of high-throughput MHC studies has revealed remarkably high levels of MHC diversity in many bird species [40,87–92], especially among passerine birds (Passeriformes). Despite early evidence from traditional MHC genotyping techniques that MHC diversity was higher in passerine than in non-passerine clades [24,93–97], the extreme levels of MHC diversity detected with high-throughput sequencing in some passerines was unprecedented and unexpected. For example, recent studies have found evidence for at least 33 MHC-I genes in sedge warblers (*Acrocephalus schoenobaenus*) [40] and 23 MHC-IIB genes in common yellowthroats (*Geothlypis trichas*) [76], numbers that have not been detected with traditional approaches (although these numbers may include both classical and non-classical MHC genes). Such numbers provide clear evidence that passerine MHCs are more gene-rich than those of galliforms, including the chicken [5,15,45]. Interestingly, thorough characterizations of the MHC genomic regions in some mammals and fish also report highly duplicated MHC-I and MHC-IIB regions [50,51].

An additional consideration for MHC genotyping using high-throughput amplicon sequencing is the bioinformatic processing of the sequence data. High-throughput sequencing produces vast amounts of data, but this comes at the cost of a relatively high error-rate [98,99]. Thus, it is necessary to remove artifactual reads via filtering to accurately estimate the number of MHC alleles detected. Although there is substantial variation in filtering approaches [100], a recent comparison of some of the most common filtering methods on the same sedge warbler dataset suggests that they yield similar estimates of the number of alleles per individual retained in the final dataset [101]. It is likely that the errors that are hardest to resolve for a typical MHC genotyping dataset are those arising during the PCR process [100,101]. Artifactual variants arising during the early stages of PCR may be highly represented in the final dataset and can appear very similar to true alleles, making them difficult to remove through filtering. Genotyping more than one preparation from the same individual can help identify PCR-generated errors [102]. Primer design can also have a dramatic effect on the accuracy of MHC genotyping. Unbalanced amplification efficiency across MHC alleles can lead to over-amplification of certain groups of genes and alleles while others may be missed—so-called 'allelic dropout' [30,102]. Thus, the use of multiple primer sets and the running of all samples in duplicates is advisable to gain the most complete and accurate survey of MHC alleles within and between individuals [30,88].

Since the arrival of high-throughput amplicon sequencing, there has been a huge increase in the number of bird species genotyped for MHC-I and -IIB genes, as seen in Figure 1: currently at more than 78 species for MHC-I and over 220 species for MHC-IIB [103]. This wealth of data has enabled comparative studies seeking to understand the extraordinary variation in MHC diversity we see across bird species [88,93,103]. When employing a purely bioinformatic approach to compile datasets for these analyses, such as harvesting the data from GenBank, it is important to exercise caution over the comparability of such data across species. As discussed above, there are many factors influencing the number of alleles detected in a typical MHC genotyping study, making it challenging to make between-species comparisons of MHC diversity across different studies using different amplification primers and methods [88]. However, in the largest study of this kind to-date, Minias and colleagues [103] implemented statistical approaches to account for variation in MHC diversity estimates generated by methodological differences between studies, demonstrating that it may be possible to remove some of the noise in comparative approaches. A wet-lab approach, whereby the MHC diversity estimate are generated within the same lab using comparably-designed primers and employing the same filtering steps, may result in more robust comparisons, but has the drawback of a more limited sample size [88,104].

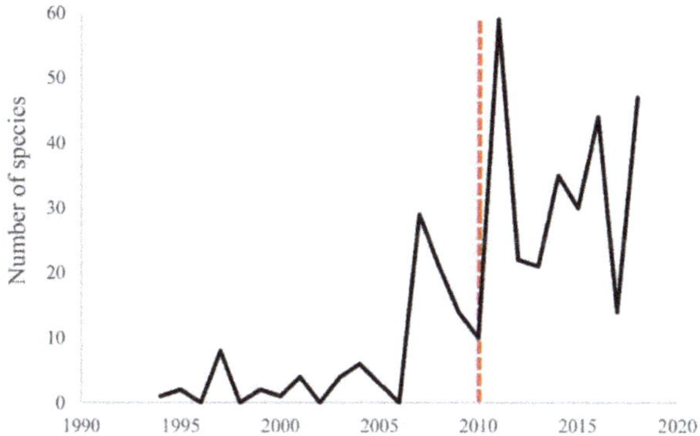

Figure 1. Evidence for the impact of high-throughput sequencing on the number of bird species genotyped for MHC, as demonstrated by the increase in the number of bird species genotyped for MHC each year since 2010 (red dashed line) when the first studies using high-throughput sequencing for avian MHC genotyping were published. Figure prepared using data from Minias et al. 2018 [103] with updated information added for studies until the end of 2018 using identical methods to those described within Minias et al 2018 [103].

3. Avian MHC Spreads Its Wings

3.1. High-throughput Studies of Avian Ecology and MHC

The ease with which individuals can be MHC-genotyped using high-throughput amplicon sequencing facilitates the search for associations between MHC variation and ecological traits in wild birds. The central role played by MHC molecules in adaptive immunity means that much of this work has focused on linking MHC variation or specific MHC haplotypes with resistance to pathogens [87,105,106]. One of the most well-studied pathogens in wild birds over the last two decades is avian malaria; the human studies reporting associations between specific MHC haplotypes and resistance to malaria have inspired such work in birds [21,107]. Qualitative resistance, in which the MHC haplotype is associated with the absence or clearance of malaria, and quantitative resistance, in which the MHC haplotype is associated with lower infection intensity of avian malaria parasites in the blood, have been studied in a wide range of passerine species, as seen in Table 2. There is not only evidence for particular MHC alleles or overall MHC diversity conferring qualitative and quantitative resistance to malaria but also evidence for potentially antagonistic effects, whereby certain alleles may confer resistance to one malaria strain but susceptibility to another strain [84,86,87,105,106,108–113]. A challenge with studies of avian malaria in the wild is that it is often difficult to distinguish between birds that have cleared the parasite after it has risen to high levels within individuals and birds that have resisted any accumulation of the parasite. Such uncertainties could be overcome by examining additional features of the immune response, such as testing birds for antibodies to malaria [114], which may be less likely to be detected if the pathogen never rose to high levels.

Table 2. Results of studies that have investigated the association between MHC and avian malaria. Resistance may be qualitative, whereby the malaria infection is cleared (prevalence), or quantitative, whereby the infection intensity is suppressed (intensity). Associations have been found between malaria and either specific MHC alleles, groups of MHC alleles that share similar antigen binding properties ('supertypes') or overall MHC diversity.

Species	Country	Class	Resistance	MHC Association	Reference
Cyanistes caeruleus	Sweden	MHC-I	Intensity	Alleles	Westerdahl et al. 2013 [110]
Cyanistes caeruleus	Spain	MHC-I	Prevalence Intensity	Alleles	Rivero-de Aguilar et al. 2016 [115]
Parus major	UK	MHC-I	Prevalence Intensity	Supertypes	Sepil et al. 2013 [87]
Passer domesticus	France	MHC-I	Prevalence	Alleles	Bonneud et al. 2006 [116] Loiseau et al. 2008; 2010 [84,108]
Acrocephalus arundinaceus	Sweden	MHC-I	Prevalence Intensity	Diversity Alleles	Westerdahl et al. 2005 [106] Westerdahl et al. 2011 [112]
Acrocephalus schoenobaenus	Poland	MHC-I	Prevalence	Supertypes	Biedrzycka et al. 2018 [105]
Ficedula albicollis	Sweden	MHC-IIB	Prevalence	Diversity	Radwan et al. 2012 [109]
Geothlypis trichas	US	MHC-IIB	Prevalence	Alleles Diversity	Dunn et al. 2013 [111] Whittingham et al. 2018 [86]
Melospiza melodia	Canada	MHC-I	Prevalence	Diversity	Slade et al. 2016 [113]

The above-mentioned associations between MHC and malaria resistance are correlational because the studies have been conducted on natural populations. Although there are significant advantages to studying wild animals subject to natural selection, this approach has its limitations, because it is difficult to be certain that MHC genes have a direct relationship to the pathogen/disease in question. However, in the case of Marek's disease in the chicken, which is caused by an oncogenic herpesvirus, more detailed knowledge is available. Individuals with low cell-surface expression of generalist MHC molecules do not get sick (MHC-I haplotypes B21, B2, B6 and B14), whereas individuals with high cell-surface expression of specialist MHC molecules (B4, B12 and B15) often die from the disease [117,118]. Moreover, for mycoplasmal conjunctivitis in house finches, individuals with high MHC-IIB diversity have been experimentally shown to be more resistant to *Mycoplasma gallisepticum*, the bacteria responsible for this infection [119]. This study also found evidence for greater MHC-IIB diversity in wild house finch populations exposed to this disease when compared with naïve populations, further supporting the important role played by MHC [119]. However, resistance to *Mycoplasma* in house finches, paradoxically, induces reduced expression of MHCIIB genes and is governed by additional immune genes, including genes of the innate immune system [120,121].

3.2. Fitness Assocations with MHC from Ecological Studies

Beyond MHC associations with specific traits, the aim of most ecological studies of avian MHC is ultimately to understand the relationship between MHC variation and fitness. This question is best addressed using data from longitudinal studies that follow a population of wild birds over many generations. This type of data is extremely labor-intensive and time-consuming to collect, but there are a number of such projects which have produced valuable insights. Von Schantz was a pioneer in this field and reported that ring-necked pheasants with specific MHC haplotypes had longer spurs, were more attractive as mates, and had increased survival rates [13,14]. Higher survival and lifetime reproductive success have been linked to particular MHC variants in great tits [91], whereas no such patterns were seen in a natural population of collared flycatchers [109]. A study of yellowthroats (*Geothlypis trichas*) found that males with higher MHC-IIB diversity had greater survival in the wild [111]. A long-term study of great reed warblers has demonstrated possible sex differences in the fitness effects of MHC-I diversity: offspring recruitment success was higher in males with higher MHC-I diversity, whereas females with lower MHC diversity had higher offspring recruitment success [90]. This last study raises the intriguing possibility that there are sexually antagonistic effects of selection on MHC diversity.

It is possible that MHC-based mate choice plays an important role in maintaining the high degree of MHC diversity seen in wild populations [122,123], although results vary among studies. MHC-based mate choice can maintain MHC genetic diversity, not only by the selective advantage of specific MHC haplotypes but also through mate choice for partners with more diverse MHC genes or partners with MHC genes that are compatible with the chooser's set of MHC genes, a type of disassortative mating leading to offspring with more divergent MHC genes [123,124]. Several studies have found evidence for each form of MHC-based mate choice in wild birds [85,125–129]. However, a recent meta-analysis investigating mate-choice driven by either MHC diversity or dissimilarity concluded that there is little evidence overall to support such mate-choice in birds [130]. MHC-based mate choice is still an active and ongoing field of research: 129 articles were published that matched the search terms "MHC", "birds", and "mate choice" in 2018 alone. Thus, it is likely that a clearer picture of the relationship between MHC and mate choice in birds will emerge in the future.

The high MHC polymorphism and the associations between MHC genes and fitness measures have led to interest in these genes from a conservation perspective, and it has been suggested that measures to maintain or promote MHC diversity should play a role in conservation efforts in endangered species [131–135]. Birds are no exception, with a number of studies on endangered or near threatened species, such as Hawaiian honeycreepers (*Drepanidinae*), New Zealand black robins (*Petroica traversi*), Seychelles warblers (*Acrocephalus sechellensis*), and crested ibis, focusing on MHC gene diversity [89,95,136,137].

3.3. Next Steps in Ecological Studies of MHC Variation

The ease of MHC genotyping using high-throughput amplicon sequencing is unfortunately not matched by our knowledge of the genomic structure of the MHC region in most birds or functionally important details such as expression differences between genes or even how sequence variation translates to differences in the antigen binding properties of MHC molecules. This knowledge deficit hampers our ability to interpret the biological relevance of the patterns revealed by MHC genotyping studies on most bird species.

A major restriction to estimating MHC diversity using high-throughput amplicon sequencing is an inability to assign alleles to specific genes in species with highly duplicated MHC genes. In these cases, differences in the number of MHC alleles between genotyped individuals will partly reflect heterozygosity or copy number variation. This ambiguity undermines the accuracy of counting the number of MHC alleles as an estimate of MHC diversity. Long-read sequencing technology will facilitate characterization of the MHC region across a wider range of bird species. A better understanding of the structure of MHC genes, haplotypes, linkage relationships, and the extent of recombination and interlocus gene conversion will lead to greater detail in studies of MHC correlates with fitness and disease in birds. In humans, different MHC genes (HLA-A, -B, and -C) encode different sets of MHC molecules, and genomic regions containing classical MHC genes with known immune function are more likely to be associated with disease resistance than regions containing non-classical MHC genes, which play a less well-characterized role in the immune system [56,138,139].

Ecological studies of avian MHC diversity are also hampered by a lack of knowledge of whether all MHC alleles are equally expressed. This is pertinent, because expression profiles relate directly to the function of MHC molecules. For example, classical MHC genes can have high or low expression, whereas non-classical MHC genes are defined by having generally low levels of expression [57]. Some basic information on expression, measured as transcription at the level of RNA, of MHC genes in birds beyond chickens is available [11,23,24,29,46,120,140–142], but less than a handful of studies have used high-throughput amplicon sequencing to address this question, and few studies have used specific organs, such as spleen [142–145], to measure expression, likely a more reliable guide to expression than whole blood, although further study is needed. Initial evidence suggests that, even in the case of passerines with numerous MHC genes, a high proportion of MHC-I gene copies are expressed [40,42,43,146], implying that the number of expressed MHC genes correlate with the number of MHC gene copies in the genome. The number of expressed MHC-I gene copies are thus likely to be higher in passerines than in galliforms. However, the degree of expression among passerine MHC genes may still vary. Although there is some evidence from house sparrows and tree sparrows (*Passer montanus*) of a similar single-locus dominance as chickens [43], Eurasian siskins (*Spinus spinus*) do not have a single dominantly expressed class I locus [42]. Thus, the question of whether single-locus dominance in MHC gene expression is generally the case across birds has yet to be conclusively answered. Differences in MHC gene expression may also occur depending on the sample used (blood versus various tissues, especially spleen) as well as the infection status of the individual sampled, although some studies have reported highly similar MHC-I expression profiles across tissues [35,41,42,120,147]. Further work is required to better understand how MHC gene expression differs across bird species and across organs. Adding MHC gene expression would add a novel dimension to the observed relationships between ecological traits and MHC diversity at the genomic level.

Finally, knowledge of how sequence differences between MHC genes actually affect the peptide-binding repertoire of MHC molecules is an overlooked area of research in most birds beyond the chicken. The existence of a crystal structure for chicken MHC-I molecules enables accurate predictions of the functional impact of MHC sequence variation in chickens because it demonstrates which amino acids are in close contact with the antigen and thus where MHC-I variation is likely to affect the peptide-binding repertoire [148]. This information led to the discovery of two distinct categories of MHC-I molecules in the chicken: those with promiscuous binding properties that are capable of binding a wide-range of antigenic peptides and those with fastidious binding properties that bind a narrow range of antigenic peptides [117]. In other bird species, in which there is limited knowledge of the structure of the MHC molecule and for which the precise peptide biding sites are unknown, it is more challenging to interpret the biological relevance of variation in MHC sequences. *In silico* prediction, whereby computer simulations attempt to model how sequence variation is likely to alter the peptide binding repertoire, opens up opportunities in this area [149] as it has in humans [150,151]. However, knowledge of the crystal structure of MHC in a wider variety of bird species would greatly improve the accuracy of such approaches.

4. Macroevolution of the MHC Across the Avian Tree of Life

4.1. Variation in the Number of MHC Genes Across the Avian Tree of Life

The study of MHC evolution across the avian tree of life provides promising insights into macroevolutionary links between genome structure, long-term forces molding gene number, and variation in life history traits. The evolutionary dynamics of the MHC are extraordinary, involving a combination of gene duplication, gene loss, gene conversion and different forms of recombination [152]. These mechanisms contribute to rapid expansions and contractions of MHC gene number and to high rates of sequence exchange within and between MHC duplicates, creating substantial variation in MHC sequence and haplotype structure [153–156]. As we have seen, variation in numbers of MHC genes across the avian tree of life is substantial [40,77,88,157,158], although still poorly known and likely to suffer from some degree of inaccuracy in birds outside of galliforms [30] with less well-characterized MHC genes.

Work spurred by the discovery of orthologous MHC-IIB genes across owls [159] led to the unveiling of two ancient MHC-IIB lineages in birds that originated before the radiation of extant birds [34,160]. This discovery was a major improvement in our understanding of MHC-IIB evolution across birds and has allowed researchers to ascribe new MHC-IIB genes to one of these two deep lineages in cases in which the sequence signatures that distinguish them have escaped concerted evolution [34]. Balasubramaniam et al. [104] found complex repertoires of MHC-IIB genes and high polymorphism in basal oscine songbirds of the largely Australian Corvides clade, suggesting that complex MHC structures are ancestral in oscine songbirds. A major gap in our knowledge of MHC evolution in passerine birds is in the suboscines [93], a diverse and largely tropical clade that deserves further study with regard to the immune system in general.

High-throughput sequencing approaches have accelerated surveys of MHC copy number across bird species, providing the first large-scale studies of MHC macroevolution. Studying 12 Passerida species, O'Connor et al. [88] suggested that MHC-I gene number was dominated by fluctuating selection and/or drift. However, in the most extensive survey of MHC gene copy number variation in birds thus far, Minias and colleagues [103] suggested that the evolutionary processes governing duplicate numbers differ between the two MHC classes. MHC-I gene number was found to be shaped by accelerated evolution and stabilizing selection (Figure 2, [103]), whereas MHC-IIB gene number appears to be shaped by fluctuating selection and genetic drift (Figure 3 [103]). The study by O'Connor et al. [88] generated MHC sequences in-house, using identical methods across species, thereby allowing more reliable comparisons across species; however, the larger sample size of Minias et al. (78 species) might have yielded a more robust result. Both studies confirm the extraordinary variation in MHC duplicate numbers across birds.

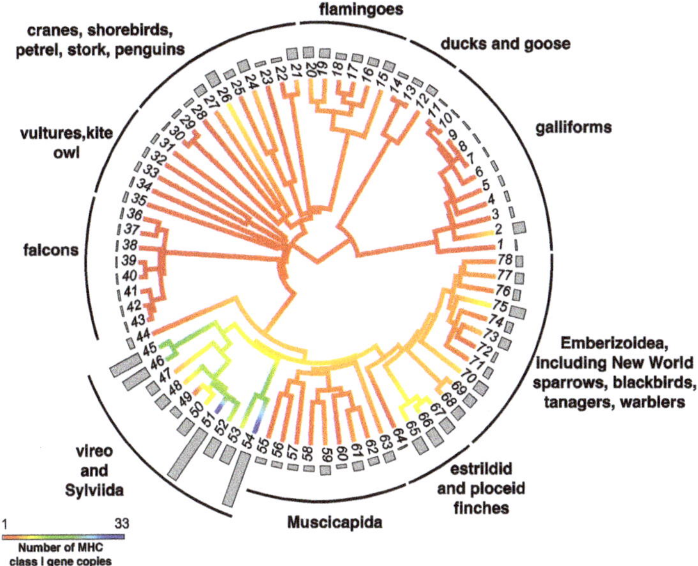

Figure 2. Ancestral character estimation of gene copy number at MHC class I genes along the branches and nodes of a tree for birds. Bars associated with each terminal node indicate the estimated number of MHC gene copies. Figure modified from Minias et al. 2018 [103]. Key to species names is supplied in Supplementary Table S1.

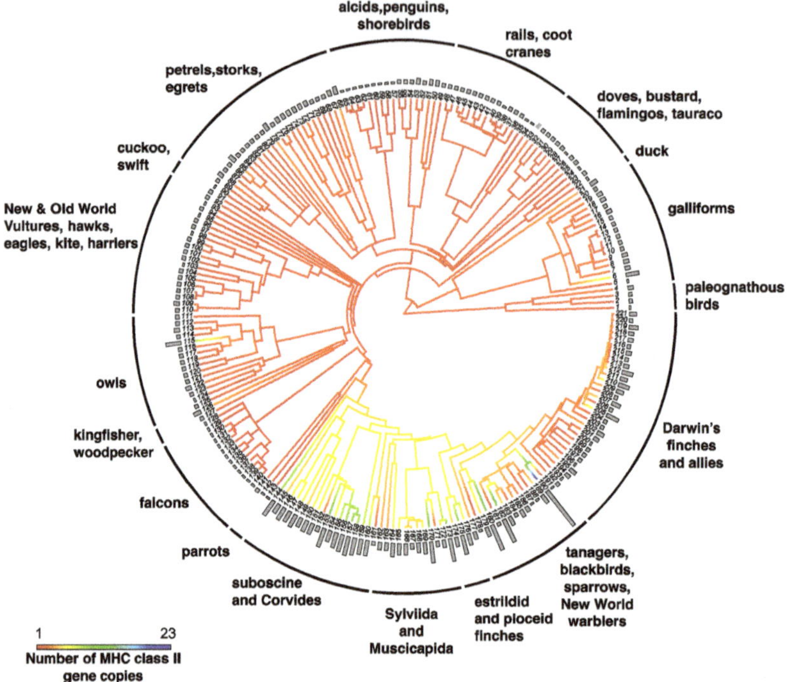

Figure 3. Ancestral character estimation of gene copy number at MHC class IIB genes along the branches and nodes of a tree for birds. Bars associated with each terminal node indicate the estimated number of MHC gene copies. Figure modified from Minias et al. 2018 [103]. Key to species names is supplied in Supplementary Table S1.

Many questions regarding the duplication history of the avian MHC remain open. First, the timing of duplications and the role of new duplication versus gene loss in governing MHC gene numbers are yet to be determined. Due to variable and often high rates of concerted evolution, sequence similarity alone provides no information on the age of MHC duplicates [161]. Highly similar duplicates may be recent and not diverged yet, or may be old, with gene conversion erasing sequence divergence in coding regions [30,31,34]. Only comparative genomic analyses based on comprehensive assemblies of the MHC region that include positional information on synteny, flanking sequences and orthology for each duplicate will provide the required resolution. Furthermore, confirmation of different modes of selection on MHC-I and MHC-IIB [103] will require additional data from less biased *de novo* genome-based approaches. With the data and sequencing methods used thus far, it is unclear whether evidence for different modes of selection on gene copy number between MHC-I and MHC-IIB could be driven by different rates of concerted evolution that determine the success with which MHC duplicates can be distinguished from one another.

4.2. MHC Gene Duplication and Life History

Some of the most pertinent questions regarding the evolution of the MHC multigene family concern the link between modes and histories of MHC evolution and the diversification of species and life histories. Have particular genomic structures of the MHC facilitated the exploitation of more diverse ecological niches and thus contributed to species radiations? For instance, might the high MHC gene numbers in passerine birds have played a role in the radiation of this most species-rich group of birds? Or enabled the evolution of life histories that risk exposure to larger or more diverse pathogen communities, such as encountered by species that perform seasonal migrations or breed in colonies?

A recent series of comparative analyses has investigated links between the MHC and life history evolution [103,162,163]. Minias et al. [162] suggested that evolution at pathogen-binding sites of MHC-IIB genes was faster and stronger in migratory and colonially breeding non-passerine birds in comparison with sedentary and solitary species. Additionally, using all available published data on avian MHC, Minias et al. [103] demonstrated that the number of MHC duplicates in 250 bird species correlated positively with lifespan and migratory behavior. O'Connor et al. [163] genotyped MHC-I loci in 39 species of Afro-Palearctic passerine birds and demonstrated that shifts in the number of MHC-I alleles are associated with the colonization of new environments as well the evolution of long-distance migration. Specifically, passerines that have escaped pathogen pressure, either by permanently colonizing the less pathogen-rich Palearctic from Africa, or by seasonally leaving Africa to breed in the Palearctic, show a reduction in number of MHC alleles over evolutionary time. The study by O'Connor et al. [163] also suggests that changes in life-history strategies in host species that affect the diversity of pathogens they encounter are strongly linked to MHC evolution. It is likely that, even with high rates of gene duplication, changes in MHC number probably do not evolve as fast as migratory behavior, which appears to change rapidly across the avian tree of life, sometimes over the span of a few decades [164,165]. However, several studies have reported substantial intra-species variation in the number of MHC alleles, and, by extension, copies per individual [40,86,90], raising the possibility that selection on standing genetic variation could potentially enable rapid adaptive shifts in MHC diversity. As with other areas of MHC evolution, more precise determination of MHC structure and gene number will help clarify relationships between MHC diversity and life history traits.

In summary, there is evidence for a relationship between MHC evolution and life history evolution in birds. Yet, the data that are available so far are restricted to a small number of studies which apply rudimentary approaches to studying MHC diversity. *De novo* genome sequencing using the latest long-read technology promises to provide crucial information on the actual number of MHC genes within genomes and, through information on gene orthology, insights into the timing of duplication. Data collected this way across the avian tree of life will provide unprecedented power to study macroevolutionary links between the MHC and life history evolution and speciation. Such studies might ultimately uncover links between MHC complexity and adaptive radiation across birds.

5. Conclusions

The avian MHC will continue to provide a major link between avian ecology and genetics, and harbors primary candidate genes for many behaviors of interest to ornithologists, including mate choice and disease resistance. The intersection between genetics and the ecology of birds is perhaps best captured by studies on the MHC. At the same time, the number of researchers pushing the boundaries to improve our understanding of avian MHC structure, expression, and evolution remains small. Although high-throughput PCR and sequencing surveys of MHC diversity have increased dramatically in recent years, such studies are unlikely to improve our understanding of the genomic structure of the MHC, and correlations with life history, mate choice, and disease resistance will remain tentative and uncertain due to our poor understanding of MHC gene number, linkage relationships, and other details.

We hope this update on MHC evolution in birds will spur further investigations of MHC structure, expression, function, and evolution in birds. What is sorely needed are robust genetic and physical maps of the MHC region in a handful of exemplar lineages of birds to provide anchor points for MHC studies in diverse species. Careful re-analysis and annotation of the wealth of short-read data, combined with comparisons between species, might alone yield substantial improvements in our understanding of MHC structure and evolution. Personal observations suggest that a substantial amount of gene structure data, including synteny information, can be gleaned from traditional short-read assemblies of bird genomes, particularly those with high coverage [166]. But, as this review has addressed, long-read data will accelerate and greatly improve our understanding of avian MHC evolution. Working both phylogenetically across species, as well as in-depth within key species with long-term ecological data sets, will undoubtedly provide new insights into the dimensions and consequences of MHC evolution in birds.

Supplementary Materials: The following is available online at http://www.mdpi.com/2073-4409/8/10/1152/s1, Table S1: Data for Figures 2 and 3, modified with permission from Minias et al. 2018. Genome Biology and Evolution 2018, 11 (1), 17–28 (ref. 103),

Funding: Funding was provided by the Swedish Research Council (2015-05149 (H.W.)); the European Research Council (ERC) under the European Union's Horizon 2020 research and innovation programme (grant 679799 (H.W.); and National Science Foundation grant DEB-1355343 (S.V.E.). Publishing fees were covered by a grant from the Wetmore Colles Fund of the Museum of Comparative Zoology, Harvard University.

Acknowledgments: We would like to thank the editors for inviting us to write this review and, along with three anonymous reviewers, for helpful comments on the manuscript. We would also like to thank Piotr Minias for encouragement and for kindly providing us with the original images for Figures 2 and 3.

Conflicts of Interest: The authors declare no conflict of interest.

References

1. Kaufman, J. Antigen processing and presentation: Evolution from a bird's eye view. *Mol. Immunol.* **2013**, *55*, 159–161. [CrossRef] [PubMed]
2. Kaufman, J. Generalists and Specialists: A New View of How MHC Class I Molecules Fight Infectious Pathogens. *Trends Immunol.* **2018**, *39*, 367–379. [CrossRef] [PubMed]
3. Kaufman, J. Unfinished Business: Evolution of the MHC and the Adaptive Immune System of Jawed Vertebrates. *Annu. Rev. Immunol.* **2018**, *36*, 383–409. [CrossRef] [PubMed]
4. Miller, M.M.; Taylor, R.L., Jr. Brief review of the chicken Major Histocompatibility Complex: The genes, their distribution on chromosome 16, and their contributions to disease resistance. *Poult. Sci.* **2016**, *95*, 375–392. [CrossRef] [PubMed]
5. Kaufman, J.; Milne, S.; Gobel, T.W.F.; Walker, B.A.; Jacob, J.P.; Auffray, C.; Zoorob, R.; Beck, S. The chicken B locus is a minimal essential major histocompatibility complex. *Nature* **1999**, *401*, 923–925. [CrossRef] [PubMed]
6. Wallny, H.J.; Avila, D.; Hunt, L.G.; Powell, T.J.; Riegert, P.; Salomonsen, J.; Skjodt, K.; Vainio, O.; Vilbois, F.; Wiles, M.V.; et al. Peptide motifs of the single dominantly expressed class I molecule explain the striking MHC-determined response to Rous sarcoma virus in chickens. *PNAS* **2006**, *103*, 1434–1439. [CrossRef]
7. Potts, W.K.; Manning, C.J.; Wakeland, E.K. Mating patterns in seminatural populations of mice influenced by MHC genotype. *Nature* **1991**, *352*, 619. [CrossRef]
8. Manning, C.J.; Wakeland, E.K.; Potts, W.K. Communal nesting patterns in mice implicate MHC genes in kin recognition. *Nature* **1992**, *360*, 581. [CrossRef]
9. Brown, J.L.; Eklund, A. Kin Recognition and the Major Histocompatibility Complex: An Integrative Review. *Am. Nat.* **1994**, *143*, 435–461. [CrossRef]
10. Brown, J.L. A theory of mate choice based on heterozygosity. *Behav. Ecol.* **1997**, *8*, 60–65. [CrossRef]
11. Edwards, S.V.; Grahn, M.; Potts, W.K. Dynamics of Mhc evolution in birds and crocodilians: Amplification of class II genes with degenerate primers. *Mol. Ecol.* **1995**, *4*, 719–729. [CrossRef] [PubMed]
12. Edwards, S.V.; Wakeland, E.; Potts, W. Contrasting histories of avian and mammalian Mhc genes revealed by class II B sequences from songbirds. *PNAS* **1995**, *92*, 12200–12204. [CrossRef] [PubMed]
13. Von Schantz, T.; Wittzell, H.; Bose, N.; Grahn, M.; Persson, K. MHC genotype and male ornamentation: Genetic evidence for the Hamilton-Zuk model. *Proc. R. Soc. Lond. Ser. B Biol. Sci.* **1996**, *263*, 265–271.
14. Von Schantz, T.; Wittzell, H.; Göransson, G.; Grahn, M. Mate choice, male condition-dependent ornamentation and MHC in the pheasant. *Hereditas* **1997**, *127*, 133–140. [CrossRef]
15. Wittzell, H.; Vonschantz, T.; Zoorob, R.; Auffray, C. Molecular characterization of 3 MHC class-II B-haplotypes in the Ring-necked Pheasant. *Immunogenetics* **1994**, *39*, 395–403. [CrossRef]
16. Edwards, S.V.; Potts, W.K. Polymorphism of Mhc genes: Implications for conservation genetics of vertebrates. In *Molecular Genetic Approaches to Conservation*; Smith, T.B., Wayne, R.K., Eds.; Oxford University Press: New York, NY, USA, 1996; pp. 214–237.

17. Lamichhaney, S.; Fan, G.; Widemo, F.; Gunnarsson, U.; Thalmann, D.S.; Hoeppner, M.P.; Kerje, S.; Gustafson, U.; Shi, C.; Zhang, H. Structural genomic changes underlie alternative reproductive strategies in the ruff (Philomachus pugnax). *Nat. Genet.* **2016**, *48*, 84. [CrossRef]

18. Bosse, M.; Spurgin, L.G.; Laine, V.N.; Cole, E.F.; Firth, J.A.; Gienapp, P.; Gosler, A.G.; McMahon, K.; Poissant, J.; Verhagen, I.; et al. Recent natural selection causes adaptive evolution of an avian polygenic trait. *Science* **2017**, *358*, 365–368. [CrossRef]

19. Lamichhaney, S.; Han, F.; Berglund, J.; Wang, C.; Almén, M.S.; Webster, M.T.; Grant, B.R.; Grant, P.R.; Andersson, L. A beak size locus in Darwin's finches facilitated character displacement during a drought. *Science* **2016**, *352*, 470–474. [CrossRef]

20. Emlen, S.T.; Oring, L.W. Ecology, sexual selection, and the evolution of mating systems. *Science* **1977**, *197*, 215–223. [CrossRef]

21. Hill, A.V. HLA associations with malaria in Africa: Some implications for MHC evolution. In *Molecular Evolution of the Major Histocompatibility Complex*; Klein, J., Klein, D., Eds.; Springer: Berlin/Heidelberg, Germany, 1991; pp. 403–420.

22. Loye, J.E.; Zuk, M. *Bird-parasite Interactions: Ecology, Evolution and Behavior (Ornithology Series)*, 1st ed.; Oxford University Press: Oxford, UK, 1991.

23. Westerdahl, H.; Wittzell, H.; von Schantz, T. Polymorphism and transcription of *Mhc* class I genes in a passerine bird, the great reed warbler. *Immunogenetics* **1999**, *49*, 158–170. [CrossRef]

24. Westerdahl, H.; Wittzell, H. k.; von Schantz, T. r. Mhc diversity in two passerine birds: No evidence for a minimal essential Mhc. *Immunogenetics* **2000**, *52*, 92–100. [CrossRef] [PubMed]

25. Jarvis, E.D.; Mirarab, S.; Aberer, A.J.; Li, B.; Houde, P.; Li, C.; Ho, S.Y.W.; Faircloth, B.C.; Nabholz, B.; Howard, J.T.; et al. Whole-genome analyses resolve early branches in the tree of life of modern birds. *Science* **2014**, *346*, 1320–1331. [CrossRef] [PubMed]

26. Edwards, S.V.; Gasper, J.; Garrigan, D.; Martindale, D.A.; Koop, B.F. A 39-kb sequence around a blackbird Mhc class II B gene: Ghost of selection past and songbird genome architecture. *Mol. Biol. Evol.* **2000**, *17*, 1384–1395. [CrossRef] [PubMed]

27. Gasper, J.; Shiina, T.; Inoko, H.; Edwards, S.V. Songbird genomics: Analysis of 45-kb upstream of a polymorphic Mhc class II gene in Red-winged Blackbird (*Agelaius phoeniceus*). *Genomics* **2001**, *75*, 26–34. [CrossRef] [PubMed]

28. Hess, C.M.; Gasper, J.; Hoekstra, H.; Hill, C.; Edwards, S.V. MHC class II pseudogene and genomic signature of a 32-kb cosmid in the House Finch (*Carpodacus mexicanus*). *Genome Res.* **2000**, *10*, 613–623. [CrossRef] [PubMed]

29. Westerdahl, H.; Wittzell, H.; von Schantz, T.; Bensch, S. MHC class I typing in a songbird with numerous loci and high polymorphism using motif-specific PCR and DGGE. *Heredity* **2004**, *92*, 534. [CrossRef] [PubMed]

30. Burri, R.; Promerová, M.; Goebel, J.; Fumagalli, L. PCR-based isolation of multigene families: Lessons from the avian MHC class IIB. *Mol. Ecol. Resour.* **2014**, *14*, 778–788. [CrossRef] [PubMed]

31. Hess, C.M.; Edwards, S.V. The evolution of the major histocompatibility complex in birds. *Bioscience* **2002**, *52*, 423–431. [CrossRef]

32. Balakrishnan, C.; Ekblom, R.; Volker, M.; Westerdahl, H.; Godinez, R.; Kotkiewicz, H.; Burt, D.; Graves, T.; Griffin, D.; Warren, W.; et al. Gene duplication and fragmentation in the zebra finch major histocompatibility complex. *BMC Biol.* **2010**, *8*, 29. [CrossRef] [PubMed]

33. Ekblom, R.; Stapley, J.; Ball, A.D.; Birkhead, T.; Burke, T.; Slate, J. Genetic mapping of the major histocompatibility complex in the zebra finch (*Taeniopygia guttata*). *Immunogenetics* **2011**, *63*, 523–530. [CrossRef] [PubMed]

34. Goebel, J.; Promerová, M.; Bonadonna, F.; McKoy, K.; Strandh, M.; Yannic, G.; Burri, R.; Fumagalli, L. 100 million years of multigene family evolution: Origins and evolution of the avian MHC class IIB. *BMC Genom.* **2017**, *18*, 460. [CrossRef] [PubMed]

35. Chen, L.-C.; Lan, H.; Sun, L.; Deng, Y.-L.; Tang, K.-Y.; Wan, Q.-H. Genomic organization of the crested ibis MHC provides new insight into ancestral avian MHC structure. *Sci. Rep.* **2015**, *5*, 7963. [CrossRef] [PubMed]

36. Wang, B.; Ekblom, R.; Strand, T.M.; Portela-Bens, S.; Höglund, J. Sequencing of the core MHC region of black grouse (Tetrao tetrix) and comparative genomics of the galliform MHC. *BMC Genom.* **2012**, *13*, 553. [CrossRef] [PubMed]

37. Eimes, J.; Reed, K.; Mendoza, K.; Bollmer, J.; Whittingham, L.; Bateson, Z.; Dunn, P. Greater prairie chickens have a compact MHC-B with a single class IA locus. *Immunogenetics* **2013**, *65*, 133–144. [CrossRef] [PubMed]

38. Ye, Q.; He, K.; Wu, S.-Y.; Wan, Q.-H. Isolation of a 97-kb Minimal Essential MHC B Locus from a New Reverse-4D BAC Library of the Golden Pheasant. *PLoS ONE* **2012**, *7*, e32154. [CrossRef] [PubMed]

39. Shiina, T.; Shimizu, S.; Hosomichi, K.; Kohara, S.; Watanabe, S.; Hanzawa, K.; Beck, S.; Kulski, J.K.; Inoko, H. Comparative Genomic Analysis of Two Avian (Quail and Chicken) MHC Regions. *J. Immunol.* **2004**, *172*, 6751–6763. [CrossRef] [PubMed]

40. Biedrzycka, A.; O'Connor, E.; Sebastian, A.; Migalska, M.; Radwan, J.; Zajac, T.; Bielanski, W.; Solarz, W.; Cmiel, A.; Westerdahl, H. Extreme MHC class I diversity in the sedge warbler (Acrocephalus schoenobaenus); selection patterns and allelic divergence suggest that different genes have different functions. *BMC Evol. Biol.* **2017**, *17*, 159. [CrossRef] [PubMed]

41. Ekblom, R.; Wennekes, P.; Horsburgh, G.J.; Burke, T. Characterization of the house sparrow (Passer domesticus) transcriptome: A resource for molecular ecology and immunogenetics. *Mol. Ecol. Resour.* **2014**, *14*, 636–646. [CrossRef]

42. Drews, A. Avian MHC: Characterization and expression patterns of classical and non-classical MHC-I genes. Ph.D. Thesis, Lund University, Scania, Sweden, 2018.

43. Drews, A.; Strandh, M.; Raberg, L.; Westerdahl, H. Expression and phylogenetic analyses reveal paralogous lineages of putatively classical and non-classical MHC-I genes in three sparrow species (Passer). *BMC Evol. Biol.* **2017**, *17*, 152. [CrossRef]

44. Kaufman, J.; Jansen, J.; Shaw, I.; Walker, B.; Milne, S.; Beck, S.; Salomonsen, J. Gene organisation determines evolution of function in the chicken MHC. *Immunol. Rev.* **1999**, *167*, 101–117. [CrossRef]

45. Strand, T.; Westerdahl, H.; Höglund, J.V.; Alatalo, R.; Siitari, H. The Mhc class II of the Black grouse (*Tetrao tetrix*) consists of low numbers of B and Y genes with variable diversity and expression. *Immunogenetics* **2007**, *59*, 725–734. [CrossRef] [PubMed]

46. Reed, K.M.; Bauer, M.M.; Monson, M.S.; Benoit, B.; Chaves, L.D.; O'Hare, T.H.; Delany, M.E. Defining the Turkey MHC: Identification of expressed class I- and class IIB-like genes independent of the MHC-B. *Immunogenetics* **2011**, *63*, 753. [CrossRef] [PubMed]

47. Zeng, Q.-Q.; Zhong, G.-H.; He, K.; Sun, D.-D.; Wan, Q.-H. Molecular characterization of classical and nonclassical MHC class I genes from the golden pheasant (Chrysolophus pictus). *Int. J. Immunogenet.* **2016**, *43*, 8–17. [CrossRef] [PubMed]

48. Chaves, L.; Krueth, S.; Reed, K. Characterization of the turkey MHC chromosome through genetic and physical mapping. *Cytogenet. Genome Res.* **2007**, *117*, 213–220. [CrossRef] [PubMed]

49. Briles, W.E.; Goto, R.M.; Auffray, C.; Miller, M.M. A polymorphic system related to but genetically independent of the chicken major histocompatibility complex. *Immunogenetics* **1993**, *37*, 408–414. [CrossRef]

50. Shiina, T.; Blancher, A. The Cynomolgus Macaque MHC Polymorphism in Experimental Medicine. *Cells* **2019**, *8*, 978. [CrossRef] [PubMed]

51. Yamaguchi, T.; Dijkstra, J.M. Major Histocompatibility Complex (MHC) Genes and Disease Resistance in Fish. *Cells* **2019**, *8*, 378. [CrossRef]

52. Glaberman, S.; Du Pasquier, L.; Caccone, A. Characterization of a Nonclassical Class I MHC Gene in a Reptile, the Galápagos Marine Iguana (Amblyrhynchus cristatus). *PLoS ONE* **2008**, *3*, e2859. [CrossRef] [PubMed]

53. Miller, H.C.; Belov, K.; Daugherty, C.H. Characterization of MHC class II genes from an ancient reptile lineage, Sphenodon (tuatara). *Immunogenetics* **2005**, *57*, 883–891. [CrossRef] [PubMed]

54. Flajnik, M.F.; Kasahara, M.; Shum, B.P.; Salter-Cid, L.; Taylor, E.; Du Pasquier, L. A novel type of class I gene organization in vertebrates: A large family of non-MHC-linked class I genes is expressed at the RNA level in the amphibian Xenopus. *EMBO J.* **1993**, *12*, 4385–4396. [CrossRef] [PubMed]

55. Murphy, K.; Janeway, C.A.J.; Travers, P.; Walport, M.; Ehrenstein, M. *Janeway's Immunobiology 7th Ed*; Garland Science: New York, NY, USA, 2008.

56. Shawar, S.M.; Vyas, J.M.; Rodgers, J.R.; Rich, R.R. Antigen presentation by major histocompatibility complex class IB molecules. *Annu. Rev. Immunol.* **1994**, *12*, 839–880. [CrossRef] [PubMed]

57. Afanassieff, M.; Goto, R.M.; Ha, J.; Sherman, M.A.; Zhong, L.; Auffray, C.; Coudert, F.; Zoorob, R.; Miller, M.M. At Least One Class I Gene in Restriction Fragment Pattern-Y (Rfp-Y), the Second MHC Gene Cluster in the Chicken, Is Transcribed, Polymorphic, and Shows Divergent Specialization in Antigen Binding Region. *J. Immunol.* **2001**, *166*, 3324–3333. [CrossRef] [PubMed]

58. Hunt, H.D.; Goto, R.M.; Foster, D.N.; Bacon, L.D.; Miller, M.M. At least one YMHCI molecule in the chicken is alloimmunogenic and dynamically expressed on spleen cells during development. *Immunogenetics* **2006**, *58*, 297–307. [CrossRef] [PubMed]

59. Cloutier, A.; Mills, J.A.; Baker, A.J. Characterization and locus-specific typing of MHC class I genes in the red-billed gull (Larus scopulinus) provides evidence for major, minor, and nonclassical loci. *Immunogenetics* **2011**, *63*, 377–394. [CrossRef] [PubMed]

60. Buehler, D.M.; Verkuil, Y.I.; Tavares, E.S.; Baker, A.J. Characterization of MHC class I in a long-distance migrant shorebird suggests multiple transcribed genes and intergenic recombination. *Immunogenetics* **2013**, *65*, 211–225. [CrossRef] [PubMed]

61. Hughes, A.L.; Nei, M. Evolution of the major histocompatibility complex: Independent origin of nonclassical class I genes in different groups of mammals. *Mol. Biol. Evol.* **1989**, *6*, 559–579. [PubMed]

62. Taniguchi, Y.; Matsumoto, K.; Matsuda, H.; Yamada, T.; Sugiyama, T.; Homma, K.; Kaneko, Y.; Yamagishi, S.; Iwaisaki, H. Structure and Polymorphism of the Major Histocompatibility Complex Class II Region in the Japanese Crested Ibis, Nipponia nippon. *PLoS ONE* **2014**, *9*, e108506. [CrossRef]

63. Tsuji, H.; Taniguchi, Y.; Ishizuka, S.; Matsuda, H.; Yamada, T.; Naito, K.; Iwaisaki, H. Structure and polymorphisms of the major histocompatibility complex in the Oriental stork, Ciconia boyciana. *Sci. Rep.* **2017**, *7*, 42864. [CrossRef]

64. Peona, V.; Weissensteiner, M.H.; Suh, A. How complete are "complete" genome assemblies?—An avian perspective. *Mol. Ecol. Resour.* **2018**, *18*, 1188–1195. [CrossRef]

65. Dilthey, A.; Cox, C.; Iqbal, Z.; Nelson, M.R.; McVean, G. Improved genome inference in the MHC using a population reference graph. *Nat. Genet.* **2015**, *47*, 682. [CrossRef]

66. Larsen, P.A.; Heilman, A.M.; Yoder, A.D. The utility of PacBio circular consensus sequencing for characterizing complex gene families in non-model organisms. *BMC Genom.* **2014**, *15*, 720. [CrossRef] [PubMed]

67. Westbrook, C.J.; Karl, J.A.; Wiseman, R.W.; Mate, S.; Koroleva, G.; Garcia, K.; Sanchez-Lockhart, M.; O'Connor, D.H.; Palacios, G. No assembly required: Full-length MHC class I allele discovery by PacBio circular consensus sequencing. *Hum. Immunol.* **2015**, *76*, 891–896. [CrossRef] [PubMed]

68. Gordon, D.; Huddleston, J.; Chaisson, M.J.; Hill, C.M.; Kronenberg, Z.N.; Munson, K.M.; Malig, M.; Raja, A.; Fiddes, I.; Hillier, L.W. Long-read sequence assembly of the gorilla genome. *Science* **2016**, *352*, aae0344. [CrossRef] [PubMed]

69. Warren, W.C.; Hillier, L.W.; Tomlinson, C.; Minx, P.; Kremitzki, M.; Graves, T.; Markovic, C.; Bouk, N.; Pruitt, K.D.; Thibaud-Nissen, F.; et al. A New Chicken Genome Assembly Provides Insight into Avian Genome Structure. *G3: Genes|Genomes|Genet.* **2017**, *7*, 109–117. [CrossRef]

70. Redin, D.; Frick, T.; Aghelpasand, H.; Theland, J.; Käller, M.; Borgström, E.; Olsen, R.-A.; Ahmadian, A. Efficient whole genome haplotyping and high-throughput single molecule phasing with barcode-linked reads. *bioRxiv* **2018**, 356121. [CrossRef]

71. Weisenfeld, N.I.; Kumar, V.; Shah, P.; Church, D.M.; Jaffe, D.B. Direct determination of diploid genome sequences. *Genome Res.* **2017**, *27*, 757–767. [CrossRef]

72. Fuselli, S.; Baptista, R.; Panziera, A.; Magi, A.; Guglielmi, S.; Tonin, R.; Benazzo, A.; Bauzer, L.; Mazzoni, C.; Bertorelle, G. A new hybrid approach for MHC genotyping: High-throughput NGS and long read MinION nanopore sequencing, with application to the non-model vertebrate Alpine chamois (*Rupicapra rupicapra*). *Heredity* **2018**, *121*, 293–303. [CrossRef]

73. Nei, M.; Gu, X.; Sitnikova, T. Evolution by the birth-and-death process in multigene families of the vertebrate immune system. *PNAS* **1997**, *94*, 7799–7806. [CrossRef]

74. Salomonsen, J.; Marston, D.; Avila, D.; Bumstead, N.; Johansson, B.; Juul-Madsen, H.; Olesen, G.D.; Riegert, P.; Skjødt, K.; Vainio, O.; et al. The properties of the single chicken MHC classical class II α chain (B-LA) gene indicate an ancient origin for the DR/E-like isotype of class II molecules. *Immunogenetics* **2003**, *55*, 605–614. [CrossRef]

75. Ren, L.; Yang, Z.; Wang, T.; Sun, Y.; Guo, Y.; Zhang, Z.; Fei, J.; Bao, Y.; Qin, T.; Wang, J.; et al. Characterization of the MHC class II α-chain gene in ducks. *Immunogenetics* **2011**, *63*, 667. [CrossRef]

76. Bollmer, J.L.; Dunn, P.O.; Freeman-Gallant, C.R.; Whittingham, L.A. Social and extra-pair mating in relation to major histocompatibility complex variation in common yellowthroats. *Proc. R. Soc. B Biol. Sci.* **2012**, *279*, 4778–4785. [CrossRef] [PubMed]

77. Zagalska-Neubauer, M.; Babik, W.; Stuglik, M.; Gustafsson, L.; Cichon, M.; Radwan, J. 454 sequencing reveals extreme complexity of the class II Major Histocompatibility Complex in the collared flycatcher. *BMC Evol. Biol.* **2010**, *10*, 395. [CrossRef] [PubMed]

78. Myers, R.M.; Maniatis, T.; Lerman, L.S. Detection and localization of single base changes by denaturing gradient gel electrophoresis. *Methods Enzymol.* **1987**, *155*, 501–527. [PubMed]

79. Orita, M.; Iwahana, H.; Kanazawa, H.; Hayashi, K.; Sekiya, T. Detection of polymorphisms of human DNA by gel electrophoresis as single-strand conformation polymorphisms. *Proc. Natl. Acad. Sci. USA* **1989**, *86*, 2766–2770. [CrossRef] [PubMed]

80. Argüello, J.R.; Little, A.-M.; Bohan, E.; Goldman, J.M.; Marsh, S.G.E.; Madrigal, J.A. High resolution HLA class I typing by reference strand mediated conformation analysis (RSCA). *Tissue Antigens* **1998**, *52*, 57–66. [CrossRef] [PubMed]

81. Vlček, J.; Hoeck, P.E.A.; Keller, L.F.; Wayhart, J.P.; Dolinová, I.; Štefka, J. Balancing selection and genetic drift create unusual patterns of MHCIIβ variation in Galápagos mockingbirds. *Mol. Ecol.* **2016**, *25*, 4757–4772. [CrossRef] [PubMed]

82. Gaigher, A.; Roulin, A.; Gharib, W.H.; Taberlet, P.; Burri, R.; Fumagalli, L. Lack of evidence for selection favouring MHC haplotypes that combine high functional diversity. *Heredity* **2018**, *120*, 396–406. [CrossRef]

83. Worley, K.; Gillingham, M.; Jensen, P.; Kennedy, L.; Pizzari, T.; Kaufman, J.; Richardson, D. Single locus typing of MHC class I and class II B loci in a population of red jungle fowl. *Immunogenetics* **2008**, *60*, 233–247. [CrossRef]

84. Loiseau, C.; Zoorob, R.; Robert, A.; Chastel, O.; Julliard, R.; Sorci, G. Plasmodium relictum infection and MHC diversity in the house sparrow (Passer domesticus). *Proc. R. Soc. B Biol. Sci.* **2010**, *278*, 1264–1272. [CrossRef]

85. Richardson, D.S.; Komdeur, J.; Burke, T.; Von Schantz, T. MHC-based patterns of social and extra-pair mate choice in the Seychelles warbler. *Proc. R. Soc. B Biol. Sci.* **2005**, *272*, 759–767. [CrossRef]

86. Whittingham, L.A.; Dunn, P.O.; Freeman-Gallant, C.R.; Taff, C.C.; Johnson, J.A. Major histocompatibility complex variation and blood parasites in resident and migratory populations of the common yellowthroat. *J. Evol. Biol.* **2018**, *31*, 1544–1557. [CrossRef] [PubMed]

87. Sepil, I.; Lachish, S.; Hinks, A.E.; Sheldon, B.C. Mhc supertypes confer both qualitative and quantitative resistance to avian malaria infections in a wild bird population. *Proc. R. Soc. B Biol. Sci.* **2013**, *280*, 20130134. [CrossRef] [PubMed]

88. O'Connor, E.A.; Strandh, M.; Hasselquist, D.; Nilsson, J.-Å.; Westerdahl, H. The evolution of highly variable immunity genes across a passerine bird radiation. *Mol. Ecol.* **2016**, *25*, 977–989. [CrossRef] [PubMed]

89. Jarvi, S.I.; Bianchi, K.R.; Farias, M.E.; Txakeeyang, A.; McFarland, T.; Belcaid, M.; Asano, A. Characterization of class II β chain major histocompatibility complex genes in a family of Hawaiian honeycreepers: 'amakihi (Hemignathus virens). *Immunogenetics* **2016**, *68*, 461–475. [CrossRef] [PubMed]

90. Roved, J.; Hansson, B.; Tarka, M.; Hasselquist, D.; Westerdahl, H. Evidence for sexual conflict over major histocompatibility complex diversity in a wild songbird. *Proc. R. Soc. B Biol. Sci.* **2018**, *285*, 20180841. [CrossRef] [PubMed]

91. Sepil, I.; Lachish, S.; Sheldon, B.C. Mhc-linked survival and lifetime reproductive success in a wild population of great tits. *Mol. Ecol.* **2013**, *22*, 384–396. [CrossRef] [PubMed]

92. Bollmer, J.L.; Dunn, P.O.; Whittingham, L.A.; Wimpee, C. Extensive MHC Class II B Gene Duplication in a Passerine, the Common Yellowthroat (Geothlypis trichas). *J. Hered.* **2010**, *101*, 448–460. [CrossRef]

93. Alcaide, M.; Liu, M.; Edwards, S.V. Major histocompatibility complex class I evolution in songbirds: Universal primers, rapid evolution and base compositional shifts in exon 3. *PeerJ* **2013**, *1*, e86. [CrossRef]

94. Bonneaud, C.; Sorci, G.; Morin, V.; Westerdahl, H.; Zoorob, R.; Wittzell, H. Diversity of Mhc class I and IIB genes in house sparrows (Passer domesticus). *Immunogenetics* **2004**, *55*, 855–865. [CrossRef]

95. Richardson, D.S.; Westerdahl, H. MHC diversity in two Acrocephalus species: The outbred Great reed warbler and the inbred Seychelles warbler. *Mol. Ecol.* **2003**, *12*, 3523–3529. [CrossRef]

96. Wutzler, R.; Foerster, K.; Kempenaers, B. MHC class I variation in a natural blue tit population (Cyanistes caeruleus). *Genetica* **2012**, *140*, 349–364. [CrossRef] [PubMed]

97. Silva, M.C.; Edwards, S.V. Structure and Evolution of a New Avian MHC Class II B Gene in a Sub-Antarctic Seabird, the Thin-Billed Prion (Procellariiformes: Pachyptila belcheri). *J. Mol. Evol.* **2009**, *68*, 279–291. [CrossRef] [PubMed]

98. Huse, S.M.; Huber, J.A.; Morrison, H.G.; Sogin, M.L.; Welch, D.B. Accuracy and Quality of Massively Parallel DNA Pyrosequencing. *Genome Biol.* **2007**, *8*, R143. [CrossRef] [PubMed]

99. Quince, C.; Lanzen, A.; Davenport, R.J.; Turnbaugh, P.J. Removing noise from pyrosequenced amplicons. *BMC Bioinform.* **2011**, *12*, 38. [CrossRef] [PubMed]

100. Lighten, J.; van Oosterhout, C.; Bentzen, P. Critical review of NGS analyses for *de novo* genotyping multigene families. *Mol. Ecol.* **2014**, *23*, 3957–3972. [CrossRef] [PubMed]

101. Biedrzycka, A.; Sebastian, A.; Migalska, M.; Westerdahl, H.; Radwan, J. Testing genotyping strategies for ultra-deep sequencing of a co-amplifying gene family: MHC class I in a passerine bird. *Mol. Ecol. Resour.* **2017**, *17*, 642–655. [CrossRef] [PubMed]

102. Sommer, S.; Courtiol, A.; Mazzoni, C.J. MHC genotyping of non-model organisms using next-generation sequencing: A new methodology to deal with artefacts and allelic dropout. *BMC Genom.* **2013**, *14*, 542. [CrossRef] [PubMed]

103. Minias, P.; Pikus, E.; Whittingham, L.A.; Dunn, P.O. Evolution of Copy Number at the MHC Varies across the Avian Tree of Life. *Genome Biol. Evol.* **2018**, *11*, 17–28. [CrossRef]

104. Balasubramaniam, S.; Bray, R.D.; Mulder, R.A.; Sunnucks, P.; Pavlova, A.; Melville, J. New data from basal Australian songbird lineages show that complex structure of MHC class II β genes has early evolutionary origins within passerines. *BMC Evol. Biol.* **2016**, *16*, 112. [CrossRef]

105. Biedrzycka, A.; Bielanski, W.; Cmiel, A.; Solarz, W.; Zajac, T.; Migalska, M.; Sebastian, A.; Westerdahl, H.; Radwan, J. Blood parasites shape extreme major histocompatibility complex diversity in a migratory passerine. *Mol. Ecol.* **2018**, *27*, 2594–2603. [CrossRef]

106. Westerdahl, H.; Waldenstrom, J.; Hansson, B.; Hasselquist, D.; von Schantz, T.; Bensch, S. Associations between malaria and MHC genes in a migratory songbird. *Proc. R. Soc. B. Biol. Sci.* **2005**, *272*, 1511–1518. [CrossRef] [PubMed]

107. Westerdahl, H. Passerine MHC: Genetic variation and disease resistance in the wild. *J. Ornithol.* **2007**, *148*, 469–477. [CrossRef]

108. Loiseau, C.; Zoorob, R.; Garnier, S.; Birard, J.; Federici, P.; Julliard, R.; Sorci, G. Antagonistic effects of a Mhc class I allele on malaria-infected house sparrows. *Ecol. Lett.* **2008**, *11*, 258–265. [CrossRef] [PubMed]

109. Radwan, J.; Zagalska-Neubauer, M.; Cichon, M.; Sendecka, J.; Kulma, K.; Gustafsson, L.; Babik, W. MHC diversity, malaria and lifetime reproductive success in collared flycatchers. *Mol. Ecol.* **2012**, *21*, 2469–2479. [CrossRef]

110. Westerdahl, H.; Stjernman, M.; Råberg, L.; Lannefors, M.; Nilsson, J.-Å. MHC-I Affects Infection Intensity but Not Infection Status with a Frequent Avian Malaria Parasite in Blue Tits. *PLoS ONE* **2013**, *8*, e72647. [CrossRef] [PubMed]

111. Dunn, P.O.; Bollmer, J.L.; Freeman-Gallant, C.R.; Whittingham, L.A. MHC variation is related to sexually selected ornament, survival, and parasite resistance in common yellowthroats. *Evol. Int. J. Org. Evol.* **2013**, *67*, 679–687. [CrossRef]

112. Westerdahl, H.; Asghar, M.; Hasselquist, D.; Bensch, S. Quantitative disease resistance: To better understand parasite-mediated selection on major histocompatibility complex. *Proc. R. Soc. B Biol. Sci.* **2011**, *279*, 577–584. [CrossRef] [PubMed]

113. Slade, J.W.G.; Sarquis-Adamson, Y.; Gloor, G.B.; Lachance, M.-A.; MacDougall-Shackleton, E.A. Population differences at MHC do not explain enhanced resistance of song sparrows to local parasites. *J. Hered.* **2016**, *108*, 127–134. [CrossRef]

114. Jarvi, S.I.; Schultz, J.J.; Atkinson, C.T. PCR diagnostics underestimate the prevalence of avian malaria (Plasmodium relictum) in experimentally-infected passerines. *J. Parasitol.* **2002**, *88*, 153–158. [CrossRef]

115. Rivero-de Aguilar, J.; Westerdahl, H.; Martínez-de la Puente, J.; Tomás, G.; Martínez, J.; Merino, S. MHC-I provides both quantitative resistance and susceptibility to blood parasites in blue tits in the wild. *J. Avian Biol.* **2016**, *47*, 669–677. [CrossRef]

116. Bonneaud, C.; Pérez-Tris, J.; Federici, P.; Chastel, O.; Sorci, G. Major histocompatability alleles associated with local resistance to malaria in a passerine. *Evol. Int. J. Org. Evol.* **2006**, *60*, 383–389. [CrossRef]

117. Chappell, P.; Meziane el, K.; Harrison, M.; Magiera, L.; Hermann, C.; Mears, L.; Wrobel, A.G.; Durant, C.; Nielsen, L.L.; Buus, S.; et al. Expression levels of MHC class I molecules are inversely correlated with promiscuity of peptide binding. *eLife* **2015**, *4*, e05345. [CrossRef] [PubMed]

118. Kaufman, J.; Salomonsen, J. The "Minimal Essential MHC" Revisited: Both Peptide-Binding and Cell Surface Expression Level of MHC Molecules are Polymorphisms Selected by Pathogens in Chickens. *Hereditas* **1997**, *127*, 67–73. [CrossRef] [PubMed]

119. Hawley, D.M.; Fleischer, R.C. Contrasting epidemic histories reveal pathogen-mediated balancing selection on class II MHC diversity in a wild songbird. *PLoS ONE* **2012**, *7*, e30222. [CrossRef] [PubMed]

120. Bonneaud, C.; Balenger, S.L.; Russell, A.F.; Zhang, J.W.; Hill, G.E.; Edwards, S.V. Rapid evolution of disease resistance is accompanied by functional changes in gene expression in a wild bird. *PNAS* **2011**, *108*, 7866–7871. [CrossRef]

121. Bonneaud, C.; Balenger, S.L.; Zhang, J.; Edwards, S.V.; Hill, G.E. Innate immunity and the evolution of resistance to an emerging infectious disease in a wild bird. *Mol. Ecol.* **2012**, *21*, 2628–2639. [CrossRef] [PubMed]

122. Potts, W.K.; Wakeland, E.K. Evolution of MHC genetic diversity: A tale of incest, pestilence and sexual preference. *Trends Genet.* **1993**, *9*, 408–412. [CrossRef]

123. Zelano, B.; Edwards, S.V. An Mhc Component to Kin Recognition and Mate Choice in Birds: Predictions, Progress, and Prospects. *Am. Nat.* **2002**, *160*, S225–S237. [CrossRef]

124. Ekblom, R.; Sæther, S.A.; Grahn, M.; Fiske, P.; Kålås, J.A.; Höglund, J. Major histocompatibility complex variation and mate choice in a lekking bird, the great snipe (*Gallinago media*). *Mol. Ecol.* **2004**, *13*, 3821–3828. [CrossRef]

125. Bonneaud, C.; Chastel, O.; Federici, P.; Westerdahl, H.; Sorci, G. Complex *Mhc*-based mate choice in a wild passerine. *Proc. R. Soc. B Biol. Sci.* **2006**, *273*, 1111–1116. [CrossRef]

126. Strandh, M.; Westerdahl, H.; Pontarp, M.; Canbäck, B.; Dubois, M.-P.; Miquel, C.; Taberlet, P.; Bonadonna, F. Major histocompatibility complex class II compatibility, but not class I, predicts mate choice in a bird with highly developed olfaction. *Proc. R. Soc. B Biol. Sci.* **2012**, *279*, 4457–4463. [CrossRef] [PubMed]

127. Løvlie, H.; Gillingham, M.A.F.; Worley, K.; Pizzari, T.; Richardson, D.S. Cryptic female choice favours sperm from major histocompatibility complex-dissimilar males. *Proc. R. Soc. B Biol. Sci.* **2013**, *280*, 20131296. [CrossRef] [PubMed]

128. Griggio, M.; Biard, C.; Penn, D.J.; Hoi, H. Female house sparrows" count on" male genes: Experimental evidence for MHC-dependent mate preference in birds. *BMC Evol. Biol.* **2011**, *11*, 44. [CrossRef] [PubMed]

129. Freeman-Gallant, C.R.; Meguerdichian, M.; Wheelwright, N.T.; Sollecito, S.V. Social pairing and female mating fidelity predicted by restriction fragment length polymorphism similarity at the major histocompatibility complex in a songbird. *Mol. Ecol.* **2003**, *12*, 3077–3083. [CrossRef] [PubMed]

130. Kamiya, T.; O'Dwyer, K.; Westerdahl, H.; Senior, A.; Nakagawa, S. A quantitative review of MHC-based mating preference: The role of diversity and dissimilarity. *Mol. Ecol.* **2014**, *23*, 5151–5163. [CrossRef] [PubMed]

131. Ujvari, B.; Belov, K. Major histocompatibility complex (MHC) markers in conservation biology. *Int. J. Mol. Sci.* **2011**, *12*, 5168–5186. [CrossRef] [PubMed]

132. Radwan, J.; Biedrzycka, A.; Babik, W. Does reduced MHC diversity decrease viability of vertebrate populations? *Biol. Conserv.* **2010**, *143*, 537–544. [CrossRef]

133. Sommer, S. The importance of immune gene variability (MHC) in evolutionary ecology and conservation. *Front. Zool.* **2005**, *2*, 16. [CrossRef]

134. Hughes, A.L. MHC Polymorphism and the Design of Captive Breeding Programs. *Conserv. Biol.* **1991**, *5*, 249–251. [CrossRef]

135. Meyer-Lucht, Y.; Mulder, K.P.; James, M.C.; McMahon, B.J.; Buckley, K.; Piertney, S.B.; Höglund, J. Adaptive and neutral genetic differentiation among Scottish and endangered Irish red grouse (Lagopus lagopus scotica). *Conserv. Genet.* **2016**, *17*, 615–630. [CrossRef]

136. Miller, H.C.; Lambert, D.M. Genetic drift outweighs balancing selection in shaping post-bottleneck major histocompatibility complex variation in New Zealand robins (Petroicidae). *Mol. Ecol.* **2004**, *13*, 3709–3721. [CrossRef] [PubMed]

137. Zhang, B.; Fang, S.-G.; Xi, Y.-M. Major Histocompatibility Complex Variation in the Endangered Crested Ibis Nipponia nippon and Implications for Reintroduction. *Biochem. Genet.* **2006**, *44*, 110–120. [CrossRef] [PubMed]

138. Rodgers, J.R.; Cook, R.G. MHC class Ib molecules bridge innate and acquired immunity. *Nat. Rev. Immunol.* **2005**, *5*, 459–471. [CrossRef] [PubMed]

139. O'Callaghan, C.A.; Bell, J.I. Structure and function of the human MHC class Ib molecules HLA-E, HLA-F and HLA-G. *Immunol. Rev.* **1998**, *16*, 129–138. [CrossRef] [PubMed]

140. Moon, D.A.; Veniamin, S.M.; Parks-Dely, J.A.; Magor, K.E. The MHC of the Duck (*Anas platyrhynchos*) Contains Five Differentially Expressed Class I Genes. *J. Immunol.* **2005**, *175*, 6702–6712. [CrossRef]

141. Shiina, T.; Oka, A.; Imanishi, T.; Hanzawa, K.; Gojobori, T.; Watanabe, S.; Inoko, H. Multiple class I loci expressed by the quail Mhc. *Immunogenetics* **1999**, *49*, 456–460. [CrossRef]

142. Zhang, Q.; Hill, G.E.; Edwards, S.V.; Backström, N. A house finch (*Haemorhous mexicanus*) spleen transcriptome reveals intra- and interspecific patterns of gene expression, alternative splicing and genetic diversity in passerines. *BMC Genom.* **2014**, *15*, 305. [CrossRef]

143. Wang, B.A.; Ekblom, R.; Castoe, T.A.; Jones, E.P.; Kozma, R.; Bongcam-Rudloff, E.; Pollock, D.D.; Hoglund, J. Transcriptome sequencing of black grouse (*Tetrao tetrix*) for immune gene discovery and microsatellite development. *Open Biol.* **2012**, *2*. [CrossRef]

144. Newhouse, D.J.; Hofmeister, E.K.; Balakrishnan, C.N. Transcriptional response to West Nile virus infection in the zebra finch (*Taeniopygia guttata*). *Royal Soc. Open Sci.* **2017**, *4*. [CrossRef]

145. Jax, E.; Wink, M.; Kraus, R.H.S. Avian transcriptomics: Opportunities and challenges. *J. Ornithol.* **2018**, *159*, 599–629. [CrossRef]

146. Pardal, S.; Drews, A.; Alves, J.A.; Ramos, J.A.; Westerdahl, H. Characterization of MHC class I in a long distance migratory wader, the Icelandic black-tailed godwit. *Immunogenetics* **2017**, *69*, 463–478. [CrossRef] [PubMed]

147. Monson, M.S.; Mendoza, K.M.; Velleman, S.G.; Strasburg, G.M.; Reed, K.M. Expression profiles for genes in the turkey major histocompatibility complex B-locus. *Poult. Sci.* **2013**, *92*, 1523–1534. [CrossRef] [PubMed]

148. Koch, M.; Camp, S.; Collen, T.; Avila, D.; Salomonsen, J.; Wallny, H.J.; van Hateren, A.; Hunt, L.; Jacob, J.P.; Johnston, F.; et al. Structures of an MHC class I molecule from B21 chickens illustrate promiscuous peptide binding. *Immunity* **2007**, *27*, 885–899. [CrossRef] [PubMed]

149. Follin, E.; Karlsson, M.; Lundegaard, C.; Nielsen, M.; Wallin, S.; Paulsson, K.; Westerdahl, H. In silico peptide-binding predictions of passerine MHC class I reveal similarities across distantly related species, suggesting convergence on the level of protein function. *Immunogenetics* **2013**, *65*, 299–311. [CrossRef] [PubMed]

150. Pierini, F.; Lenz, T.L. Divergent Allele Advantage at Human MHC Genes: Signatures of Past and Ongoing Selection. *Mol. Biol. Evol.* **2018**, *35*, 2145–2158. [CrossRef] [PubMed]

151. Manczinger, M.; Boross, G.; Kemény, L.; Müller, V.; Lenz, T.L.; Papp, B.; Pál, C. Pathogen diversity drives the evolution of generalist MHC-II alleles in human populations. *PLoS Biol.* **2019**, *17*, e3000131. [CrossRef] [PubMed]

152. Nei, M.; Rooney, A.P. Concerted and birth-and-death evolution of multigene families. *Annu. Rev. Genet.* **2005**, *39*, 121–152. [CrossRef] [PubMed]

153. Kelley, J.; Walter, L.; Trowsdale, J. Comparative genomics of major histocompatibility complexes. *Immunogenetics* **2005**, *56*, 683–695. [CrossRef] [PubMed]

154. Spurgin, L.G.; Van Oosterhout, C.; Illera, J.C.; Bridgett, S.; Gharbi, K.; Emerson, B.C.; Richardson, D.S. Gene conversion rapidly generates major histocompatibility complex diversity in recently founded bird populations. *Mol. Ecol.* **2011**, *20*, 5213–5225. [CrossRef] [PubMed]

155. Ohta, T. Role of diversifying selection and gene conversion in evolution of major histocompatibility complex loci. *PNAS* **1991**, *88*, 6716–6720. [CrossRef] [PubMed]

156. Chen, J.-M.; Cooper, D.N.; Chuzhanova, N.; Ferec, C.; Patrinos, G.P. Gene conversion: Mechanisms, evolution and human disease. *Nat. Rev. Genet.* **2007**, *8*, 762–775. [CrossRef] [PubMed]

157. Alcaide, M.; Edwards, S.; Cadahía, L.; Negro, J. MHC class I genes of birds of prey: Isolation, polymorphism and diversifying selection. *Conserv. Genet.* **2009**, *10*, 1349–1355. [CrossRef]

158. Alcaide, M.; Edwards, S.V.; Negro, J.J. Characterization, polymorphism, and evolution of MHC class IIB genes in birds of prey. *J. Mol. Evol.* **2007**, *65*, 541–554. [CrossRef] [PubMed]

159. Burri, R.; Niculita-Hirzel, H.; Salamin, N.; Roulin, A.; Fumagalli, L. Evolutionary Patterns of MHC Class II B in Owls and Their Implications for the Understanding of Avian MHC Evolution. *Mol. Biol. Evol.* **2008**, *25*, 1180–1191. [CrossRef] [PubMed]

160. Burri, R.; Salamin, N.; Studer, R.A.; Roulin, A.; Fumagalli, L. Adaptive Divergence of Ancient Gene Duplicates in the Avian MHC Class II B. *Mol. Biol. Evol.* **2010**, *27*, 2360–2374. [CrossRef] [PubMed]

161. Gao, L.-Z.; Innan, H. Very low gene duplication rate in the yeast genome. *Science* **2004**, *306*, 1367–1370. [CrossRef]
162. Minias, P.; Whittingham, L.A.; Dunn, P.O. Coloniality and migration are related to selection on MHC genes in birds. *Evol. Int. J. Org. Evol.* **2017**, *71*, 432–441. [CrossRef]
163. O'Connor, E.A.; Cornwallis, C.K.; Hasselquist, D.; Nilsson, J.-Å.; Westerdahl, H. The evolution of immunity in relation to colonization and migration. *Nat. Ecol. Evol.* **2018**, *2*, 841–849. [CrossRef]
164. Berthold, P.; Helbig, A.J.; Mohr, G.; Querner, U. Rapid microevolution of migratory behaviour in a wild bird species. *Nature* **1992**, *360*, 668–670. [CrossRef]
165. Able, K.P.; Belthoff, J.R. Rapid 'evolution' of migratory behaviour in the introduced house finch of eastern North America. *Proc. R. Soc. Lond. Ser. B Biol. Sci.* **1998**, *265*, 2063–2071. [CrossRef]
166. Grayson, P.; Sin, S.Y.; Sackton, T.B.; Edwards, S.V. Comparative genomics as a foundation for evo-devo studies in birds. In *Avian and Reptilian Developmental Biology*; Sheng, G., Ed.; Humana Press: New York, NY, USA, 2017; pp. 11–46.

Review

Major Histocompatibility Complex (MHC) Genes and Disease Resistance in Fish

Takuya Yamaguchi [1] and Johannes M. Dijkstra [2,*]

[1] Laboratory of Fish Immunology, Friedrich-Loeffler-Institute, Boddenblick 5A, 17498 Insel Riems, Germany; Takuya.Yamaguchi@fli.de
[2] Institute for Comprehensive Medical Science, Fujita Health University, Toyoake, Aichi 470-1192, Japan
* Correspondence: dijkstra@fujita-hu.ac.jp; Tel.: +81-562-93-9381; Fax: +81-562-93-8832

Received: 6 February 2019; Accepted: 23 April 2019; Published: 25 April 2019

Abstract: Fascinating about classical major histocompatibility complex (MHC) molecules is their polymorphism. The present study is a review and discussion of the fish MHC situation. The basic pattern of MHC variation in fish is similar to mammals, with MHC class I versus class II, and polymorphic classical versus nonpolymorphic nonclassical. However, in many or all teleost fishes, important differences with mammalian or human MHC were observed: (1) The allelic/haplotype diversification levels of classical MHC class I tend to be much higher than in mammals and involve structural positions within but also outside the peptide binding groove; (2) Teleost fish classical MHC class I and class II loci are not linked. The present article summarizes previous studies that performed quantitative trait loci (QTL) analysis for mapping differences in teleost fish disease resistance, and discusses them from MHC point of view. Overall, those QTL studies suggest the possible importance of genomic regions including classical MHC class II and nonclassical MHC class I genes, whereas similar observations were not made for the genomic regions with the highly diversified classical MHC class I alleles. It must be concluded that despite decades of knowing MHC polymorphism in jawed vertebrate species including fish, firm conclusions (as opposed to appealing hypotheses) on the reasons for MHC polymorphism cannot be made, and that the types of polymorphism observed in fish may not be explained by disease-resistance models alone.

Keywords: fish; MHC; polymorphism; disease resistance; quantitative trait loci (QTL) studies; evolution

---

1. Introduction

1.1. The Polymorphism of MHC Genes

Extensive gene polymorphism (variation between alleles) is unusual because in situations in which an allele is superior to all others or if alleles are neutral, selection or chance occurrence induces allele fixation throughout the population [1,2]. To maintain polymorphism as found for classical MHC, a process called "balancing selection" is necessary, which means there is evolutionary pressure favoring allelic variation within a species [3,4]. Classical MHC genes were reported to be the most polymorphic genes in the human genome [5,6], and extensive classical MHC polymorphism is a common feature among jawed vertebrate species [7]. Long before the responsible genes were identified, phenotypic effects of genetic MHC variation were already known because of rejection of MHC-mismatched ("HLA"-mismatched in human; "H2"-mismatched in mice) allogeneic tissue and cell grafts in humans and mice (reviewed by [8]). Because several of the responsible genes are linked together in a gene-"complex" region, which shows similarity between humans and mice, the genomic region was called the Major Histocompatibility Complex (in this article *Mhc* is used when referring to the region), and the family of genes encoding the targets for allogeneic rejection were called (classical) MHC genes. The later elucidation of the function of MHC molecules in peptide antigen presentation to T cells [9], and the effects of MHC polymorphism on that peptide presentation [10], led many researchers to believe that MHC-mediated allograft rejection was only an artifact, as stated by Michalová et al. in 2000 ([11]; the Jan Klein group): "The allograft reaction is an artifactual manifestation of the true function of the class I and class II loci, which is the presentation of antigenic peptides for recognition by T lymphocytes and thus the initiation of the adaptive immune response." However, although the models explaining MHC polymorphism based on disease resistance variation are intellectually appealing (see below), there has been little experimental support. In 1994, Satta et al. ([12]; the Jan Klein group) stated: " ... it has proved difficult to demonstrate the presence of balancing selection at MHC loci experimentally. Only a few cases of associations between specific MHC alleles and resistance to parasites in natural populations have been reported, and even these are not entirely convincing." Not much has changed since then, despite more than 20 years of research including many more species, with Kelly and Trowsdale stating in 2017 [13]: "It is widely assumed that resistance to infection is driving the extreme MHC variation, although direct evidence for this is limited." The word limited in this citation refers to the fact that only very few cases have been reported that rather convincingly show correlations between MHC sequence polymorphism and pathogen resistance (see below), and that even if those few correlations are true they may be considered "anecdotal evidence" instead of final proof for the "pathogen resistance model". This is not to say that the model is not true, as it has been pointed out that even small advantages that are very hard to capture experimentally can lead to evolutionary selection [12]. Nevertheless, researchers should probably keep an open mind for the possibility of other or additional explanations for MHC polymorphism. An important part of this review considers the fact that in many fish species the observed polymorphism in classical MHC class I shows features which cannot easily be explained by the pathogen resistance model, and that other evolutionary pressures may (additionally) shape fish MHC polymorphism. The present review is a summary of what is currently known about fish MHC, with a focus on polymorphism and disease resistance.

1.2. MHC Variation and Resistance to Infectious Diseases in Tetrapod Species

Classical MHC class I and II molecules present peptide antigens for recognition by T cell receptors (TCR) on $CD8^+$ cytotoxic and $CD4^+$ helper/regulatory T cells, respectively [9,14]. The classical MHC class I and II molecules show extensive allelic polymorphism in residues lining the peptide binding groove, which leads to a presentation of different sets of antigen peptides by different individuals of the same species [10,15]. By comparing synonymous and nonsynonymous nucleotide exchange rates, many of the MHC residues lining the peptide binding groove have been shown to be under evolutionary selection towards sequence variation [16,17]. Within species allelic MHC variation tends to partially predate the most recent speciation events (is inherited from ancestral species), reflected in trans-species allelic sequence lineages, while also new allelic variation keeps being generated by point mutation and/or recombination events supposedly driven by a race of arms with pathogens [18,19]. The allelic MHC variation is commonly believed to increase the protection of a population of species against pathogens that otherwise might more easily evade effective MHC presentation by all individuals in a population (reviewed in [20]). However, differences in resistance to infectious diseases have rarely been strongly linked to *Mhc* haplotypes (see above [12,13,20,21]). Genotype–phenotype linkage studies investigating potential linkage of MHC genotypes with differences in pathogen resistance cannot easily differentiate between effects caused by differences in the MHC peptide binding groove, by other variable features of the MHC molecules (e.g., expression levels), or by variation in linked genes. The most compelling case for explaining differences in resistance to an infectious disease by different MHC alleles presenting different sets of peptides probably has been made for resistance and susceptibility to HIV conferred by certain HLA-B molecules (reviewed in [22]).

1.3. MHC in Fish

Although in teleost fishes (modern bony fishes) allograft rejection and thymus-dependent antibody responses had been known for quite a while (reviewed in [23]), it was only in 1990 that the first fish MHC class I and II genes were found in common carp [24]. Later, MHC class I [25] and class II [26] were found in cartilaginous fish, followed by the detection of MHC class I and II in lobe-finned fishes [27–29]. Jawless fish and invertebrates do not have MHC genes [30]. There now have been a number of reports on genomic organizations of fish MHC genes, on their polymorphism and their expression, and there have also been some reports on teleost MHC at the protein and functional level. Those studies are summarized in the next paragraphs. In short, although the evidence is fragmentary, the functions of classical fish MHC class I and II molecules in fish appear to be similar to those in mammals. Table 1 provides an overview comparing MHC (system) traits in teleost fish and mammals.

Table 1. Summary of MHC (system) traits in teleost fish compared to mammals. *(important references are between brackets).*

MHC molecule types

MHC class I versus class II
Like mammals, most fish possess MHC class I, MHC class IIα, MHC class IIβ and β₂-m molecules. However, some teleost fish lost MHC class II genes in their evolution. [24,31–34]

Classical versus nonclassical MHC class II
As in mammals, in fish both polymorphic classical and nonpolymorphic nonclassical MHC class II are found. The nonclassical MHC class II lineages in teleost fish are not the same as in tetrapods, and, unlike the DM lineage distribution in tetrapods, they are not stably inherited throughout those teleosts that do possess an MHC class II system. [27,35–39]

Classical versus nonclassical MHC class I
Like mammals, all investigated teleost fish possess at least one polymorphic classical MHC class I gene and a number of nonpolymorphic nonclassical ones. The conclassical MHC class I lineages found in tetrapods and fish are not the same. Different from mammals is that throughout ray-finned fish members of an ancient nonclassical MHC class I lineage named "Z" are conserved which show highly conserved features for presumably binding a (modified) peptide ligand, while the Z lineage was lost in tetrapods. [24,40–46]

Genomic organization

Genomic location of classical MHC class II genes
Unlike in mammals, teleost fish classical MHC class II genes are not situated in typical *Mhc* regions and are not linked with the classical MHC class I genes. Even between teleost fish species differences in genomic locations of the classical MHC class II genes can be observed, highlighting the relative plasticity of the teleost fish MHC class II system. [27,47,48]

Genomic location of classical MHC class I genes
As in mammals, fish classical MHC class I genes are linked together with TAP, tapasin, PSMB and other conserved "framework" genes in a typical *Mhc* region. [11,49–53]

PSMB, TAP and tapasin genes in the Mhc region
As in mammals, PSMB, TAP and tapasin genes are located in the teleost fish *Mhc* region, but their number and lineages differ. Unlike in mammals, teleost fish *TAP1* is not linked with the *Mhc* region. Compared to mammals, teleost fish have additional ancient gene copies of the PSMB6/9/12 and PSMB7/10/13 lineages. In some but not all teleost fish, different from humans, considerable allelic or haplotype variation can be found for the *Mhc*-situated PSMB and TAP2 genes. [54–57]

Functions of classical MHC class II

Expression of classical MHC class II
As in mammals, teleost fish classical MHC class II transcripts and molecules appear to be expressed in professional antigen presenting cells like for example B lymphocytes. Furthermore, as in mammals, their expression is upregulated after immune stimulation, which agrees with conservation of promoter motifs similar to those in mammals. [58–65]

Peptide presentation by classical MHC class II
For fish MHC class II, the structure or functions in peptide presentation have not been determined yet. However, fish classical MHC class II molecules have a set of conserved residues which suggest a similar mode of peptide ligand binding as known in mammals. [27]

Allograft rejection
As in mammals, teleost fish classical MHC class II genes have been found linked with allograft rejection. [66]

Indirect evidence for classical MHC class II function
Although fragmentary, there is abundant evidence for helper and regulatory T cell functions in teleost fish similar to as in mammals. Furthermore, teleost fish have CD4, LAG-3 and CD74 molecules that probably all participate in the MHC class II system as in mammals. [31,32,59,60,67–79]

Functions of classical MHC class I

Expression of classical MHC class I
As in mammals, teleost fish classical MHC class I transcripts and molecules are ubiquitously expressed and show the highest expression in epithelial and lymphoid tissues. Furthermore, as in mammals, their expression is upregulated after immune stimulation, which agrees with conservation of promoter motifs similar to those in mammals. Different from mammals is that in teleost fish, which are ectotherm species, the levels of MHC class I can be temperature dependent. [62,80–84]

Peptide presentation
As in mammals, teleost fish classical MHC class I molecules form heterotrimer complexes with β₂-m and peptide ligands of ~9 aa. X-ray crystallography analysis revealed a similar complex structure as known in mammals. [85,86]

Allograft rejection
As in mammals, teleost fish classical MHC class I genes have been found linked with allograft rejection. [87,88]

MHC class I restriction of cell-mediated cytotoxicity by T cells
Although conclusive experiments have not been performed yet, there are several lines of evidence that together suggest that specific cell-mediated cytotoxicity by teleost fish CD8⁺ T cells requires classical MHC class I matching of the target cells as known in mammals. [89–95]

1.4. Associations between MHC Variation and Differences in Disease Resistance in Fish

Some fish species, for example Salmoniformes, can have large numbers of offspring in single broods (sometimes >1000), which is helpful for genotype–phenotype linkage association studies. In regard to linkage analysis of teleost fish genotypes with disease resistance there have been studies specifically dedicated to MHC genes, as well as genome-wide quantitative trait loci (QTL) studies that allow investigation of the impact of linkage groups harboring MHC genes. The present article focuses on the summary of those QTL reports, while largely neglecting the genotype–phenotype studies specifically dedicated to MHC genes. The reason for this is that, generally, we only deem the QTL analyses as sufficiently objective and trustworthy from human behavior and statistical points of view; most genotype–phenotype studies dedicated to fish MHC genes suffer from the statistical modeling weakness that no predictions were made, and we deem it psychologically risky to give (mostly young) researchers tasks that only result in publications if they do find some interesting associations. At a note, in the past the last author (J.M.D.) has been involved in a project partially funded to find differences in anti-virus disease resistance associated with the highly polymorphic classical MHC class I locus in rainbow trout, but could not find such differences (unpublished data). From that project, however, he learned that such negative data tend not to make it into publications (worsening the statistical integrity of the body of published articles) and that there is an enormous career-pressure to present (part of) the data from an (artificial, outcome-selected) angle that would suggest statistical relevance. Luckily, nowadays there have been a number of QTL studies in fish providing the luxury of being able to rely only on those studies for the main conclusions on genotype–phenotype associations. The body of published QTL studies provides no indications that the enormous allelic variation in teleost fish classical MHC class I variation causes differences in disease resistance, but might hint at possible influences of MHC class II allelic variation and of nonclassical MHC class I intact allele versus null-allele variation.

2. Fish Phylogeny

When discussing fish MHC, it is important to realize that this is a phylogenetically hugely diversified group. Fishes constitute approximately half of all vertebrate species [96], and can be divided into multiple clades (Figure 1). The most primitive extant fish are the Agnatha ("no jaws"), which include the extant lamprey and hagfish, and which have an immune system without MHC and TCR [30]. Although the precise timings are a matter of debate, the lineage separation of the Agnatha and the Gnathostomata ("jawed mouth") probably occurred around 565 million years ago (MYA), after which the Chondrichthyes (cartilaginous fish like shark and ray) separated from the Osteichthyes (bony fish) around 465 MYA, followed by the separation between Sarcopterygii (lobe-finned fish and tetrapods) and Actinopterygii (ray-finned fish) around 427 MYA [97,98]. The immune system throughout Gnathostomata appears to be basically similar, with an important role for MHC molecules [30]. The Teleostei (teleosts) constitute the vast majority of the extant Actinopterygii (Figure 1; [96]), and their genomes are characterized by remnants of a whole genome duplication (WGD) event that happened early in the teleost lineage [99]. Because of their abundance and economical importance, the teleosts are the most intensively investigated fish group. Three important teleost clades are the Otocephala (with e.g., herring, catfish, and Cypriniformes like grass carp and zebrafish), the Protacanthopterygii (with e.g., Salmoniformes like rainbow trout and Atlantic salmon) and the Neoteleostei (neoteleosts, e.g., cichlids, pufferfish, medaka, stickleback, cod). In teleost fish, the most comprehensive MHC work has probably been done on Salmoniformes (e.g., rainbow trout and Atlantic salmon) and Cypriniformes (e.g., zebrafish and grass carp), whereas for Neoteleostei, which comprise the biggest group of teleost fish (Figure 1), high quality MHC genetic analyses were performed for medaka.

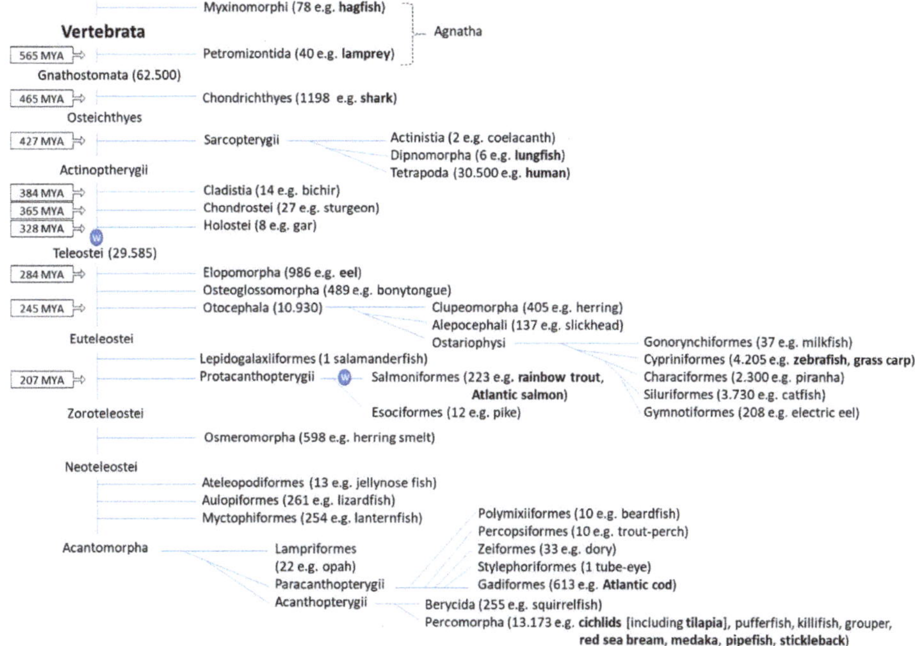

Figure 1. Schematic overview of fish classification. The organization of fish clades, and the estimated number of species which is indicated between brackets for some of them, are based on Nelson et al., 2016 [96]. Also between brackets, simplified popular English names of fish species representative for the clade are given; fish species that are important in the main text are highlighted in bold. For some nodes the time of separation in million years ago (MYA) is given as calculated in [98]. Circles with the letter W refer to whole genome duplication events early in the lineages Teleostei and Salmoniformes. Fish phylogeny and species numbers are under continued discussion, and the figure should be understood as an approximation.

3. Classical and Nonclassical MHC Genes

3.1. Classical and Nonclassical MHC Class II

In mammals, classical MHC class II molecules show extensive allelic polymorphism and are expressed at the surface of professional antigen presenting cells (APCs; for example, B-cells and macrophages) where they have an important immune function in presenting peptides from endocytosed/phagocytosed antigens to TCRαβ$^+$CD4$^+$ T lymphocytes [9,14]. In humans, the classical MHC class II molecules are HLA-DP, -DQ, and -DR, while the non-classical MHC class II molecules are HLA-DM and -DO. The HLA-DM and HLA-DO molecules do not present peptides but have a "peptide-editing" (helping to select for high affinity peptides) function in the peptide loading system of classical MHC class II [100]. Whereas HLA-DO appears to be an evolutionary relatively young diversification from the classical MHC class II lineage, DM lineage genes can already be found in lungfish and DM lineage is stably inherited throughout tetrapod species [27,67]. Despite absence of DM, teleost MHC class II molecules can be distinguished into classical versus nonclassical based on polymorphism, expression patterns, and presence/absence of residues important for peptide ligand binding (e.g., [27,35–38]). The first convincing descriptions of classical MHC class II polymorphism in fish probably were for zebrafish by Ono et al., 1992 [39], and nurse shark by Kasahara et al., 1993 [101]. The old nonclassical MHC class II lineages found in teleost fish were named "DB" and "DE", but unlike the DM lineage in tetrapods these lineages are not stably inherited throughout most teleosts [27].

Whereas indirect evidence indicates the presence of classical MHC class II functions in fish, there are no good clues for allowing speculation on the functions of nonclassical teleost MHC class II. In teleost fish, in many cases, the nonclassical and classical MHC class II genes are not linked with each other [27].

3.2. Classical and Nonclassical MHC Class I

In mammals, classical MHC class I molecules show extensive allelic polymorphism and are expressed at the surface of most nucleated cells where they have an important immune function in presenting peptides from intracellular antigens to TCRαβ$^+$CD8$^+$ T lymphocytes [9,14]. In addition, tetrapod species have a wide variety of nonclassical MHC class I molecules, which are not stably inherited among species clades [40,102], do not show the polymorphism of the classical molecules, and have a wide variety of functions within and outside the immune system [103].

The first description of a fish gene encoding a classical MHC class I molecule, identified by the peptide termini binding residues (reviewed in [41]) and in later studies by extensive allelic polymorphism, was by Grimholt et al., 1993 [42], for Atlantic salmon. Like in tetrapod species, most of the well-investigated teleost species have one, two or three polymorphic classical MHC class I genes per haploid genome that encode proteins with a (predicted) conserved ability to bind peptides, and a variable number of nonclassical genes that are closely related to these classical genes (e.g., [49,50,104,105]); together these teleost molecules have been assigned as the "U lineage" [43]. Besides the U lineage genes, in teleost fish also genes of diverged nonclassical MHC class I lineages now named Z [24,44], S [106], L [45], and P [43] can be found. The S, L and P lineages are not stably inherited throughout teleosts, and their function is unknown [43]. The Z lineage is divided into "typical" and "atypical" molecules, with the typical Z molecules probably representing the more original form which is found in all investigated teleosts but also in spotted gar, bichir and lungfish [29,40,43,44,107]. The "atypical" Z molecules constitute highly differentiated Z forms and are only found in some teleost species [24,43]. The function of typical Z molecules is not known, but comparison with sequence motifs of classical MHC class I suggests that they bind conserved peptides of approximately 8-9 amino acids with a modification of the N-terminus. The conservation among the residues estimated to line the Z binding groove is near-absolute between species as widely divergent as lungfish, gar, bichir and teleosts [40,43], which is unprecedented among MHC molecules. The number of genes for typical Z molecules in different fish species can differ between 1 and ~10 [43,107], and within species considerable differences in their expression patterns and encoded features like cytoplasmic tails are observed [43,44,107]. Like in mammalian genomes, in teleost fish genomes the classical MHC class I genes tend to be linked with each other and in many cases they are also linked with some nonclassical MHC class I genes, whereas other nonclassical MHC class I genes are dispersed over the genome [43,105,108].

4. Allelic Variation in Classical MHC Molecules in Fish

4.1. Allelic Variation in Fish Classical MHC Class II

The important study by Shum et al., 2001 [46], concluded that in investigated fish species, similar as in tetrapods, there has been an evolutionary selection towards within species variation (balancing selection) in those residues of classical MHC class I and class II molecules that line the peptide binding groove and (probably) influence peptide preferences. The types of allelic variation in classical MHC class I and II in mammals and sharks, and classical MHC class II in teleost fish, seem to be relatively similar, and appear to be mostly dedicated to creating variations in the peptide binding groove that (are predicted to) affect the selection of bound peptides (e.g., [46,109,110]). There now have been quite a number of reports on classical MHC class II variation within teleost fish species, but it is not always clear from which gene locus (or how many loci) the variable gene sequences are derived (e.g., see Text S3 in [27]). Convenient for interpretations of classical MHC class II sequence variation is the situation in species like Atlantic salmon and rainbow trout, because these salmonid species have only one classical MHC class II locus with one gene for an alpha chain and one gene for a beta chain [46,109,111]. Interestingly, whereas in tetrapods the classical MHC class II alpha chains tend to show little allelic variation, in Salmoniformes the alpha and beta chain allelic variations appear to be similarly extensive [109].

4.2. Allelic Variation in Fish Classical MHC Class I

The first solid study on allelic polymorphism in fish classical MHC class I was performed in sharks [112], showing a similar level of allelic variation as known in mammals [46]. Although unexpected high levels of diversification among classical MHC class I sequences in a single teleost fish species had been reported before [11,46,51], it was only in 2002 that Aoyagi et al., 2002 [113], showed that widely diversified classical MHC class I sequences found in rainbow trout were alleles. For quantitative comparisons of fish and human levels of classical MHC class I variation see the studies by Shum et al., 2001 [46], and McConnell et al., 2016 [54]. The allelic variation in classical class I molecules in intensively investigated teleost fish species extends far beyond the residues that line the peptide binding groove (Figure 2; [43,113–116]). In a recent article ([117]; see Supplementary File 3 in that article) we summarized the lowest level of amino acid (aa) identity between the deduced $\alpha 1 + \alpha 2$ domains (which are the most important domains for function) of reported allelic classical MHC class I sequences within three well-investigated representative teleost species: Zebrafish, 40%; Atlantic salmon, 47%; Medaka, 53%. In comparison, human classical MHC class I molecule HLA-A2 $\alpha 1 + \alpha 2$ sequence (GenBank P01892 residues 25–206) can be found to share >80% aa identity with HLA-B and HLA-C sequences (e.g., Genbank AAB96790 and AVQ10002), and shares 75% aa identity with murine H-2K (GenBank AAA39553), and 48% aa identity with a grass carp classical MHC class I sequence (GenBank BAD01658); thus, teleost allelic classical MHC class I variation in the peptide binding domains (which are the domains that interact with T cells) can look like variation between widely divergent species. In investigated Cypriniformes and Salmoniformes the allelic classical MHC class I variation is ancient, with some trans-species lineages for $\alpha 1$ domain sequences inherited from before the separation with eel around 284 MYA [43,46]. In contrast, allelic human classical MHC class I lineages were reported to be only shared with big apes from which the human ancestor only separated ~6 MYA [46,97,118,119]. Variation between the three loci *HLA-A, -B* and *-C* can be traced back to the time before the separation between the Apes (Hylobatidae plus Hominidae) and Cercopithecidae (e.g., macaque) ~21 MYA [120,121], but human MHC class I alleles that derived from recombination events that exchanged gene fragments between these loci are rare (e.g., [122]). The most extreme classical MHC class I allelic variation has been described for rainbow trout (*Oncorhynchus mykiss*), with eight highly divergent lineages for the $\alpha 1$ domain, two or three highly divergent lineages for the $\alpha 2$ domain, and even length and sequence variation in the $\alpha 3$ domain, all found in a single gene locus [46,51,80,113,114,123]. The allelic variation in rainbow trout classical MHC class I is further increased by a >10 kb intron between the $\alpha 1$ and $\alpha 2$ domain exons [50], which has been used for allelic recombination events leading to alternative $\alpha 1$-with-$\alpha 2$ combinations (first observed by [46]) as exemplified in Figure 2. The enormous evolutionary pressure necessary to keep this allelic variation appears to be highlighted by the fact that in trout (or other investigated fish) there is no "genomic reservoir" of all these ancient lineages in nonclassical gene or pseudogene copies that could explain the variation in the classical locus by recent interlocus recombination events [43,124]; simply said, the ancient allelic variation appears to be predominantly maintained by selective pressure at the allelic level. Another salmonid fish, Atlantic salmon (*Salmo salar*), also possesses only one polymorphic classical MHC class I gene (named *UBA*) and the allelic variation is reminiscent of that found in trout, with as major difference that most (though not all) of the reported Atlantic salmon sequences have $\alpha 2$ domain sequences of the same lineage [43,114,125,126]. Early research already reported extensive variation among zebrafish classical MHC class I sequences [11,127–129], but only in recent years it was realized how this variation is organized at the genomic level [116]. In a corresponding stretch of haploid genome, zebrafish can have either one, two or three classical MHC class I genes, which can be very different from each other and between individuals. Figure 2 shows an example of the variation between zebrafish classical MHC class I sequences encoded by allelic haplotypes, providing evidence of past recombination events involving the intron 2 between the α1 and α2 domain exons. Although in Cypriniformes evidence for such recombination events creating new combinations of α1 and α2 sequences is not as abundant as in Salmoniformes, also in Cypriniformes this type of recombination appears to have been aided by a large size of intron 2 (e.g., [130]).

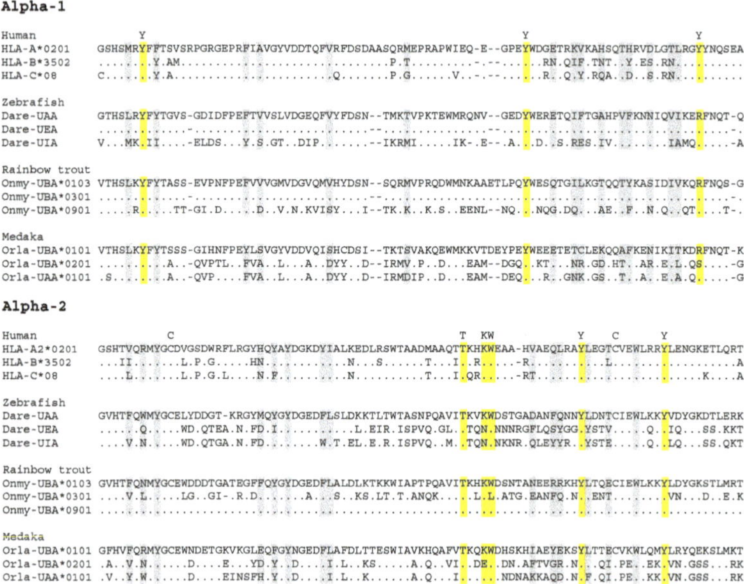

Figure 2. Alignment of deduced classical MHC class I α1 + α2 domain amino acid sequences. This figure is dedicated to showing aspects of within species variation in the most important functional parts of the classical MHC class I molecules, namely the α1 and α2 domains. The three sequences shown each for zebrafish, rainbow trout and medaka were chosen because they reveal past recombination events. In zebrafish, the Dare-UAA and Dare-UEA sequences share an identical α1 sequence, but have very different α2 sequences, indicative of a recombination event involving the intron between the respective exons [11]. On the other hand, Dare-UIA has an α2 sequence which is very similar to Dare-UEA, but has a very different α1 sequence. Among investigated fish species, recombination events involving the intron between the α1 and α2 exons may have been the most abundant in rainbow trout, since many alleles show their recent traces (e.g., [46,113]). This is exemplified here by Onmy-UBA*0103 having an identical versus very different α1 sequence compared to Onmy-UBA*0301 and Onmy-UBA*0901, respectively, whereas the reverse is found for their α2 domains. Investigated medaka haplotypes have two intact classical MHC class I loci, which have consistently been designated *UAA* and *UBA*, although in some haplotypes the "UBA" sequence is quite similar to the UAA sequence (exemplified here by Orla-UBA*0201) and in other haplotypes the UBA sequence is highly divergent from the UAA sequence (exemplified here by Orla-UBA*0101); Nonaka and Nonaka, 2010 [116], explained this situation by interlocus recombination. The levels of divergence among the here depicted zebrafish, rainbow trout and medaka sequences, can be found between individuals of the same species (as allelic or haplotype variation), and are larger than found between three sequences of different human classical MHC class I loci HLA-A, -B and, C. Gray and yellow shading highlight residues that may form part of the peptide binding groove, with yellow shading and a letter indication above the alignment used for conserved residues that bind the peptide ligand termini [41,131]. The letter C above the alignment indicates conserved cysteines. GenBank accessions of the depicted sequences are: HLA-A2, P01892; HLA-B*3502, AAB96790; HLA-C*08, AVQ10002; Dare-UAA, Z46776; Dare-UEA, BC053140; Dare-UIA, KC626502; Onmy-UBA*0103, AF287483 (previous name Onmy-UBA*101); Onmy-UBA*0301, AF287492 (previous name Onmy-UBA*701); Onmy-UBA*0901, AF296366; Orla-UBA*0101, BAD93266; Orla-UBA*0201, AB450999; Orla-UAA*0101, BAD93265.

In Neoteleostei the evolutionary pressure to maintain ancient classical MHC class I variation appears to be less than in Salmoniformes and Cypriniformes. Namely, in most investigated Neoteleostei only classical MHC class I sequences with α1 domain sequences belonging to only one of the eight α1 lineages found in Salmoniformes (named lineage "α1-I") were found, and also in the other domains of the classical MHC class I molecules the trans-species lineage variation was not as ancient as found in Salmoniformes [43]. Interestingly, in salmonid evolution the α1-I lineage seems to have been superior to the other α1 domain lineages (lineages α1-II to α1-VIII) for the establishment of new allelic peptide binding groove variation [43], making it an even more fascinating question why species like Salmoniformes kept all those ancient lineages. The best investigated neoteleost species for classical MHC class I variation is medaka (*Oryzias latipes*), with allelic variation determined for two neighboring classical MHC class I genes *Orla-UAA* and *Orla-UBA*, which to some extent experienced interlocus recombination (example shown in Figure 2) and in some alleles also have a relatively large intron between the α1 and α2 domain exons [49,108,115,132]. Although the level of allelic variation is not as ancient and extensive as found in some Cypriniformes and Salmoniformes, and while all medaka classical MHC class I α1 domain sequences belong to lineage α1-I [43], the level of allelic diversification is still impressive and higher and more ancient than found in humans ([108,115] and see the above calculations). It is unclear in how far the medaka classical MHC class I situation is representative for Neoteleostei. Compared to medaka, in the neoteleost fishes stickleback and Atlantic cod the classical MHC class I genes may be characterized by less and younger diversification, whilst having a higher number of classical genes per haploid genome; however, the extent/quality of research in those species probably does not allow final conclusions on the classical MHC class I situation yet [43,117,133–136].

5. Functional Analyses of Fish Classical MHC Genes and Molecules

5.1. Expression Patterns of Classical MHC Class II

Analyses with Northern dot blots, RT-PCR or polyclonal antisera showed that classical MHC class II genes/molecules in teleost fish are predominantly expressed in B-lymphocytes (e.g., [58,59]) and in polymorphic cells presumably involved in antigen presentation in the thymus [60] and other tissues [61,62]. This expression pattern resembles that of mammalian MHC class II. Furthermore, teleost classical MHC class II expression can be upregulated after immune stimulation (e.g., [63,64,96]), which also is reminiscent of the situation in mammals and which agrees with conserved promoter elements [64,65].

5.2. Expression Patterns of Classical MHC Class I

Northern blot data indicated that classical MHC class I in shark (e.g., [137]) and teleosts (e.g., [80]) are expressed ubiquitously, with highest expression in lymphoid and epithelial tissues. In rainbow trout this was confirmed at the cellular level by an established monoclonal antibody, showing that classical MHC class I molecules were predominantly found in epithelial cells, endothelial cells, and leukocytes [81], as in mammals. By using an established polyclonal antiserum, a similar expression profile was found for classical MHC class I in stickleback [62]. Various studies showed that fish classical MHC class I gene expression can be enhanced after immune stimulation (e.g., [82,85,138,139]), in agreement with conserved promoter elements [50,82,140,141]. Notable is that in common carp the cell-surface expression levels of classical MHC class I were substantially less at lower temperatures, consistent with the lower amounts of β_2-m transcripts found at those temperatures [83]. It has to be realized that fish are ectotherm species, and that the fish adaptive immune system does not work equally well under each naturally encountered temperature [142,143].

5.3. Binding of Peptide Ligands by Classical MHC Class I Molecules

In mammals, both classical MHC class I and MHC class II bind peptide ligands in the groove formed by their membrane-distal domains, but whereas for teleost fish classical MHC class I this has been confirmed, for fish MHC class II this has not been investigated yet. Classical MHC class I heavy chains form complexes with a single immunoglobulin-like domain molecule beta-2-microglobulin (β_2-m), which was also found to be the case in teleosts [84–86]. The binding of β_2-m to the heavy chain is unstable, unless simultaneously a peptide of ~9 aa binds into the groove of the heavy chain (synergistic heterotrimer complex formation), as was also found for teleost fish [85,86]. A recent milestone in fish MHC research was the elucidation of a grass carp heavy chain/β_2-m/peptide heterotrimer structure, which was found to be similar to such structures in tetrapod species [86]. Recently, also interaction was shown between rainbow trout classical MHC class I and tapasin [144], providing additional evidence for similarities in the peptide presenting functions between teleost fish and mammals.

5.4. MHC class I Restriction of Cell-Mediated Cytotoxicity by Lymphocytes

Specific cell-mediated cytotoxicity by lymphocytes in ginbuna crucian carp was found to require syngeneity between the effector cell donor and the target cells [89–92], and in mammals such genetic restriction involves classical MHC class I. Linkage association studies suggested the need of matching classical MHC class I markers for specific cell-mediated cytotoxicity in rainbow trout, but the experimental setups have been too limited to allow firm conclusions on MHC restriction [93,94]. In grouper, a neoteleost fish, direct evidence was provided of specific cell-mediated cytotoxicity against virus-infected autologous cells by CD8$^+$ lymphocytes [95]. In summary, there probably is MHC restriction in fish, but final evidence remains needed.

5.5. Additional, Indirect Indications for Classical MHC Functions in Fish

Except for the above-mentioned, there are also (other) indirect indications for MHC functions in fish similar to as in mammals. For example, clonal expansions of systemic TCR$\alpha\beta$ T cells upon specific immune stimulation have been observed, and there are many observations that support the existence of similar cytokine networks and helper and regulatory T cell functions (e.g., [68,69]; reviewed in [70]). Furthermore, also the T cell education system, concerning the tissue organization of the thymus and the existence of CD4$^-$CD8$^-$, CD4$^+$CD8$^+$, CD4$^+$CD8$^-$ and CD4$^+$CD8$^-$ thymocytes, seems to be similar between teleost fish and mammals [60,71–74]; the shark thymus may be similarly organized, but has been studied less intensively [145]. Teleost fish classical MHC class II molecules possess conserved residues that in mammals can bind to the TCR co-receptor molecule CD4 on helper/regulatory T cells [27] and teleost fish classical MHC class I molecules possess conserved features that in mammals are involved in binding the TCR co-receptor CD8 on cytotoxic T cells [41,113,146]. The CD8 molecule on mammalian cytotoxic T cells is a heterodimer of an alpha and a beta chain, CD8α and CD8β, and fish have genes for both components [147–151]; the cytoplasmic tail of teleost fish CD8α was shown to have LCK kinase binding properties as known in mammals [152]. Unlike mammals, teleost fish have two CD4 molecules, CD4-1 and CD4-2, which mostly are co-expressed by the same T cells and which both have cytoplasmic tails that can bind LCK kinase as known for mammalian CD4 [74–78]; the reason for this is not known. Like mammals, teleost fish also have LAG-3 as a potential receptor molecule for MHC class II complexes [77], but the function of fish LAG-3 has not been studied yet. Similar pathways in fish and mammals for MHC intracellular transport and loading with peptide ligands is suggested by fish possessing a similar set of specialized molecules as known in mammals, such as PSMB, TAP and tapasin molecules for the MHC class I system (see below), and CD74 (aka invariant chain or Ii; [67,79,153]) for the MHC class II system. Peculiarly, teleost fish have two CD74 molecules, CD74a and CD74b, the function of which is not known [59,67,79,154]. A strong indicator for a similar MHC class II functional system as in mammals seems to be that in teleost fish that lost their MHC class II genes also the *CD4-1, CD4-2, LAG-3, CD74a* and *CD74b* genes tend to be lost or to have lost their original function [31,32,117,155].

6. Genomic Organization/Haplotype Variation

In teleost fish, classical MHC class I and II are not linked, and only classical MHC class I genes reside in typical *Mhc* genomic regions. Throughout jawed vertebrate species, despite individual differences, typical *Mhc* genomic regions are found with classical MHC genes plus a conserved set of non-MHC genes amongst which genes for proteins involved in the classical MHC class I peptide loading pathway (reviewed in [156]). After fragmentary reports (e.g., [11,51,55]), Clark et al., 2001 [52], were the first to report a consecutive sequence of a teleost Mhc genomic region (in Fugu). Currently, *Mhc* genomic regions have been analyzed for a considerable number of teleost fish species (e.g., [27,43,50,53]). Distribution in different genomic linkage groups (nonlinkage) of classical MHC class I versus classical MHC class II in the teleost genomes was first reported for zebrafish by Bingulac-Popovic et al., 1997 [47], and later confirmed for other teleost fishes (e.g., [27,48]). Pipefish and, independently, Gadiformes including Atlantic cod (for phylogeny see Figure 1) apparently even lost MHC class II function [31–33,117]. Meanwhile, shark classical MHC class I and II genes were found to be conventionally linked in a typical *Mhc* region as known in tetrapods, concluding that such linkage is the ancestral situation [157–159]. Data suggest that in primitive ray-finned fish, like spotted gar, classical MHC class I and II genes may still be linked together, whereas in teleost fish only the classical MHC class I genes remained in a typical *Mhc* region and the classical MHC class II genes were translocated to other chromosomes [27]. The whole genome duplication early in the teleost lineage resulted in duplications of the *Mhc* region, but one of the duplicated regions, which lost all classical MHC genes and only in some teleost species retained nonclassical MHC class II [27], is usually not discussed as an *Mhc* region and will be neglected in the remaining of this article.

6.1. Teleost Fish Mhc Allelic/Haplotype Sequence Variation in PSMB and TAP2 Genes

At least in mammals, a large cytoplasmic protein complex with peptidase properties called the "proteasome" generates peptides that can be transported through the membrane of the endoplasmic reticulum by heterodimer transporters associated with antigen processing 1 and 2 (TAP1 and TAP2; aka ABCB2 and ABCB3) complexes, and then can be aided/modified/selected by a number of molecules including tapasin (aka TAP binding protein or TAPBP) to bind in the classical MHC class I peptide binding groove [14,160]. During infection, at least in mammals, the proteasome beta (PSMB) subunits PSMB5, PSMB6 and PSMB7 are (partially) exchanged by PSMB8 (aka LMP7), PSMB9 (aka LMP2), PSMB10 (aka MECL1), respectively, creating an "immunoproteasome" with different protease properties, which leads to the generation of different peptides that are presented by classical MHC class I [14,161]. In teleost fish *Mhc* regions, as is probably inherited from a jawed vertebrate ancestor, a gene organization similar (though not identical) to that in many other jawed vertebrates can be found, namely with classical MHC class I genes linked with *PSMB8*, *PSMB9*, *PSMB12*, *PSMB13*, *TAP2* and *tapasin* genes, and in some fish species with *PSMB10* (reviews [56,156]). PSMB12 (previously also named LMP2-like, LMP2/δ, PSMB9L or PSMB9B) and PSMB13 (previously also named PSMB7 or PSMB10) genes are members of the PSMB6/9/12 and PSMB7/10/13 families, respectively, found in teleost fish, and their levels of divergence from the respective other two family members suggest that they may be ancient [51,54,56,57]. The implication of these teleost genes in immune responses was not only suggested by their linkage within the *Mhc* region, but also by the increased expression of teleost *PSMB8*, *PSMB9*, *PSMB10* (named *PSMB7* in [162]), *PSMB12*, *PSMB13*, *TAP2* and *tapasin* after immune stimulation (e.g., [162–165]). In humans, the molecules involved in the classical MHC class I peptide loading pathway show little allelic variation (e.g., [54]), but in rat allelic variations in TAP2 are found which appear to affect the peptides that can be presented by classical MHC class I [166–168] and a similar situation is found for TAP1, TAP2 and tapasin in chicken [169–171]. In the frog Xenopus, not only ancient allelic classical MHC class I variation is observed, but also highly diverged allelic forms of PSMB8, TAP1 and TAP2, and some variation in PSMB9 [54,172,173]. In several investigated teleost fish, like medaka, rainbow trout, Atlantic salmon and zebrafish, very divergent and ancient variation is observed for PSMB8, which is represented by two lineages called PSMB8A and

PSMB8F [49,54,56,174]. The PSMB8A and PSMB8F lineages were already established at the level of cartilaginous fish, their sequences can show >30% amino acid sequence divergence from each other, and they are predicted to confer different protease properties to the immunoproteasome [174–176]; although not necessarily at identical location within the different *Mhc* haplotypes, it was found that *PSMB8A* and *PSMB8F* sequences can segregate in a functionally allelic manner in zebrafish and medaka [176]. Based on inconclusive experiments it was prematurely hypothesized that also in rainbow trout full-length *PSMBA* and *PSMBF* genes segregate as functional alleles [176], whereas there is only evidence for intact trout *PSMB8A* and *PSMB8F* genes being located on different chromosomes without indications for significant allelic variation (see the paragraph below and [56]). Considerable allelic/haplotype sequence divergence can, in some teleost fish species, also be found for *Mhc*-situated *PSMB9*, *PSMB13* and *TAP2*, and zebrafish allelic/haplotype variation for their encoded products can be as high as 14%, 29% and 50% amino acid divergence, respectively ([49,56,176]; calculations by [54]). Zebrafish may also have *Mhc* haplotypes without *PSMB12* (null-allele variation) [54]. In the *Mhc* regions of Salmoniformes, notable levels of polymorphism in the peptide loading pathway genes could not be found, with the exception of *TAP2* [50,56].

The analyses of *Mhc*-situated *PSMB/TAP/tapasin* genes have not been sufficiently exhaustive yet for allowing a definite comparison between teleost fishes on levels of within species allelic/haplotype divergence. In summary, in some teleost fish species, classical MHC class I genes displaying unprecedented levels of allelic/haplotype divergence are closely linked with peptide loading pathway genes that also display considerable allelic/haplotype sequence divergence. Whether the latter has a function in providing the most suitable peptides for binding the classical MHC class I molecules encoded by the respective *Mhc* haplotype, or mainly functions to further increase variation in peptide/MHC complexes between individuals of the same species, remains to be determined.

6.2. Copy Number Differences in MHC Class II Genes

Although in humans the copy number of MHC class I and II genes does not largely vary between individuals, more notable differences in MHC gene numbers between individuals were reported for various other species such as for example rat [177], quail [178], and the frog Xenopus [179]. Also in shark (e.g., [156,180]) and teleost fish, differences in MHC gene number can be observed. Extensive MHC class II B copy number variation in cichlid fishes was concluded [181] and copy number variation was also shown for MHC class II A in the cichlid tilapia [182]. For a summary of observed or suggested MHC class II gene copy number variation in several teleost fishes see Text S3 in reference [27].

6.3. Copy Number Differences in MHC Class I Genes

Southern blot data indicated that rainbow trout individuals differ in their genomic copy number for the MHC class I lineages U, S and L (e.g., [45,106]). Although variation in Z gene sequences between rainbow trout individuals was reported at the cDNA level [124], allelic or copy number variation in salmonid Z genes has not properly been investigated. As mentioned above, zebrafish has a variable number of classical U lineage genes in its *Mhc* region, but it also has a nonclassical U lineage gene situated on another chromosome which displays null-allele (presence/absence of intact gene) variation [183]. Data suggest that zebrafish have copy number variation in genes of the nonclassical MHC class I lineages L and Z, but genes of these lineages are not linked to the zebrafish *Mhc* [105,107]. Some of the zebrafish L lineage genes, however, can be found linked to the classical MHC class II genes [45]. In Neoteleostei, in cichlid fishes and Atlantic cod data suggested copy number variation in U lineage genes [133,134,184,185], and in medaka probable null-allele variation was found for a nonclassical gene of the U lineage situated within the *Mhc* region [49]. In summary, copy number variations in both MHC class I and II genes appear to be common in fish. Null-allele variation in nonclassical MHC class I genes is particularly interesting because in mammals it is known that knockout of such genes can deplete T cell subpopulations that are restricted by their products [186], like for example *MR1* and *CD1d* knockout cause depletions of MAIT cells [187] and NKT cells [188], respectively.

7. The Genomic Organization of the Classical MHC Gene Loci and the Duplicated *Mhc* Regions *Onmy-IA* and *Onmy-IB* in Rainbow Trout

Salmonid fishes probably experienced an additional whole genome duplication event around 60-90 MYA (SGD for salmonid-specific whole genome duplication; [50,189,190]). Figure 3 schematically shows the organization of (i) the rainbow trout *Mhc* region harboring a polymorphic classical MHC class I locus on Chr. 18 (aka linkage group LG-16 or RT-16) which has been called the *Onmy-IA* region, (ii) its SGD-derived duplicated *Mhc* region without classical genes on Chr. 14 (aka LG-3 or RT-3) which has been called the *Onmy-IB* region [50], and (iii) the polymorphic MHC class II locus on Chr. 17 (aka LG-29 or RT-29). Whereas for the single classical MHC class I gene, *UBA* on Chr. 18, extreme levels of allelic diversification were observed (see above; [43,113,114]), this was not the case for four nonclassical genes of the U lineage on Chr. 14, named *UCA*, *UDA*, *UEA* and *UGA* [114,124,191]. Whereas *UCA* and *UDA* are quite similar to each other and even recombined with each other [191], all these nonclassical sequences are considerably different (showing <70% aa identity over the encoded full-length sequence) from the classical *UBA* sequences, and *UEA* and *UGA* are considerably different from each other and from *UCA/UDA* [50,124]. The *UCA* and *UDA* genes display some degree of allelic variation, which unlike in classical sequences is not predominantly dedicated to the membrane-distal domains, while *UEA* and *UGA* seem to be close to monomorphic [124,191]. However, importantly, for *UCA* and *UDA* [124,191], as well for *UEA* [124] and *UGA* ([124]; our unpublished results), also presumable null-allele (pseudogene) sequences or indications for null-alleles were found. The rainbow trout *UCA*, *UDA*, *UEA* and *UGA* genes may all be involved in the immune system, as suggested by their enhanced transcription after viral infection [192]. Among these nonclassical MHC class I molecules, only UGA may bind peptides in a manner very reminiscent of classical MHC class I [124], but in rainbow trout the *UGA* 5′-UTR has additional AUGs plus an inverted repeat (suggesting regulation at the translation level; GenBank accession EU036647) and the UGA cytoplasmic tail has a typical dileucine endosomal targeting motif (GenBank accession AY253140; [193]), which together with the apparent lack of polymorphism argues against classical character. As already concluded previously for Salmoniformes [56], there are no indications for important allelic/haplotype variations for rainbow trout PSMB and tapasin genes situated in the *Mhc* region, but for *TAP2a* (the small font letter "a" refers to being situated in the *Onmy-IA* region) there are indications for 6% allelic amino acid divergence (Figure 3; [50,51]). The investigated *Onmy-IA* and *Onmy-IB* haplotypes each contain three genes of the nonclassical MHC class I lineage Z [56], but there is no information about their possible allelic/haplotype variation.

As mentioned above, rainbow trout have only one classical MHC class II locus with one alpha gene and one beta gene, both which are polymorphic, situated on Chr. 17 (Figure 3). On the same chromosome, at a far distance of ~18 Mb, a gene of the nonclassical MHC class I lineage L, *LDA* [45], is situated, but there is no information suggesting important polymorphism or null-allele variants of that locus.

In short, (1) *Onmy-IA* haplotypes display extreme sequence divergence in the single classical MHC class I gene *UBA* (see the paragraph on allelic polymorphism) and some sequence variation in the associated *TAP2a* genes, (2) the most dramatic variation observed among the *Onmy-IB* haplotypes probably concerns the possible null-allele variation for the nonclassical genes *UCA*, *UDA*, *UEA* and *UGA*, and (3) the trout classical MHC class II locus is characterized by polymorphism of both the single alpha and single beta genes.

In Atlantic salmon, the organization of the *IA*, *IB* and classical MHC class II loci is quite similar, though not identical, to the one shown for rainbow trout in Figure 3 [56,104,194].

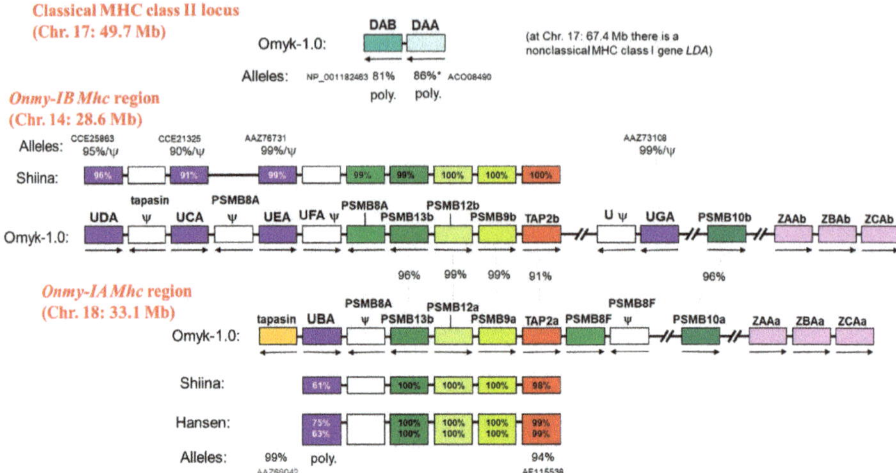

Figure 3. Schematic organization of the rainbow trout *Onmy-IA*, *Onmy-IB* and classical MHC class II loci. MHC, PSMB, TAP2 and tapasin genes are indicated by blocks, and double slashes indicate that there is a short stretch with other genes between them (for those genes see [56]). Blocks representing intact genes are colored based on identity or molecule family, and white blocks indicate pseudogenes (ψ for pseudogene). The indicated chromosome positions are based on the rainbow trout whole-genome dataset Omyk_1.0 accessible at NCBI (https://www.ncbi.nlm.nih.gov/assembly/GCF_002163495.1/). For the *Onmy-IA* and *Onmy-IB* regions, not only the gene organizations found in Omyk_1.0 are shown, but also the shorter stretches reported by Shiina et al., 2005 [50], and Hansen (GenBank HM210571). Numbers indicated within the blocks for the Shiina genes refer to the amino acid identity percentages when comparing with the products of the matching Omyk_1.0 gene. This is done similarly for the Hansen genes, with the first number based on comparison with the matching Omyk_1.0 gene, and the second number based on comparison with the matching Shiina gene. For the PSMB and TAP2 genes, the amino acid identities between the Omyk_1.0 *Onmy-IA* and *Onmy-IB* encoded gene products are also given (the numbers on top of dashed lines). The only classical MHC class I gene is the *UBA* gene situated in the *Onmy-IA* region, and the only classical MHC class II genes are *DAA* (encoding an alpha chain) and *DAB* (encoding a beta chain). They are known to be polymorphic (poly., see the main text). As an approximate measure for levels of variation, if deemed possible and interesting, in the "Alleles" sections for many genes the maximum divergent allelic molecules compared with the Omyk_1.0 encoded gene products are listed with their GenBank accession numbers and the percentages of amino acid identity that they share with the Omyk_1.0 encoded molecules; if in addition for these genes also indications for null-alleles were found (see main text), that is shown by ψ symbol. Arrows indicate gene orientations.

In Figure 3, only the MHC loci in rainbow trout are shown for which extensive variation is well documented. Also on other trout chromosomes MHC genes are located, but they do not overlap with the confidence intervals of the most interesting reported QTL, and will not further be discussed in this article: Chr. 2, nc II (nonclassical MHC class II); Chr. 3, nc II; Chr. 6, nc I (L lineage); Chr. 12, nc II; Chr. 13, nc II; Chr. 22, nc I (U and L lineages); Chr. 24, nc I (S lineage), Chr. 26, nc I (L lineage).

8. Association of Teleost Fish MHC Genes with Disease Resistance

Rainbow trout and Atlantic salmon are the only fish species for which disease resistance related genome-wide QTL studies can readily be linked with whole genome sequence information [190,195] available at the chromosome and linkage group levels (e.g., [196,197]), and for which the locations of the classical MHC class I and II loci are known; for rainbow trout, the *Onmy-IA*, *Onmy-IB* and classical MHC class II loci (see Figure 3) were also physically mapped to chromosomes 18, 14 and 17, respectively [50,198,199]. In Atlantic salmon, the corresponding *Sasa-IA*, *Sasa-IB* and classical MHC class II loci were mapped to chromosomes 27, 14 and 12 in that species [194]. Table 2 summarizes the relevant QTL studies, with underlining highlighting the most important QTL in the respective study. In both rainbow trout and Atlantic salmon, the chromosomes with the *IB* region and the classical MHC class II locus were found among the linkage groups with suggestive or significant QTL. Ozaki et al., 2001 and 2007 [200,201], found that the major QTL in rainbow trout for resistance against IPN virus mapped to a large part of Chr. 14 that includes the *Onmy-IB* region, but finer mapping was not performed. Palti et al., 2015 [202], found a QTL on trout Chr. 14 for resistance against the bacterium *Flavobacterium psychrophilum* (cold water disease), but the most likely region on Chr. 14 for that QTL does not include the *Onmy-IB* region. Moen et al., 2009 [203], mapped a suggestive QTL for resistance against IPN virus in Atlantic salmon to a large region (the region upstream of negative marker BHMS429) of Chr. 14 where the *IB* region in that species (the *Sasa-IB* region) is located, but fine mapping was not performed. Gonen et al., 2015 [204], mapped a suggestive QTL for resistance against SAV virus in Atlantic salmon to Chr. 14, but from their study we are unable to understand the location of the confidence interval of that QTL on the chromosome. Ozaki et al., 2007 [201], mapped a suggestive QTL for resistance against IPN virus in rainbow trout Chr. 17, with the region of most likelihood laying not far from (but not including) the classical MHC class II locus. Khoo et al., 2005 [205], mapped a single QTL for resistance against IHNV virus in rainbow trout to a large region (for location compare [205] with [201]) that includes the classical MHC class II locus, but fine mapping was not performed. In the study by Fraslin et al., 2018 [206], a consistent QTL for resistance against cold water disease in rainbow trout was found on Chr. 17, but the most likely region for that QTL does not include the classical MHC class II locus (see their Table S6 [206]). Verrier et al., 2013 [207], mapped a QTL for resistance against VHS virus in trout to a part of Chr. 17, which does not include the classical MHC class II locus (compare with their Table S1 [207]). In summary, disease resistance related QTL studies in rainbow trout and Atlantic salmon suggest a possible QTL effect of the *IB* and classical MHC class II loci, but much finer mapping is necessary to substantiate those hypotheses. Importantly, however, the salmonid *IA* regions with the classical MHC class I genes are not associated with any of the reported QTL.

Table 2. List of genome-wide QTL studies in rainbow trout and Atlantic salmon that investigated disease resistance or other immune traits. QTL are listed as described in the respective references, with the most important QTK underlined; double underlining is used if among the important QTL one was found especially important. Red font (IB) and green font (II) indications are used to indicate chromosomes with the *IB* region and the classical MHC class II locus, respectively. Indications of (IB) or (II) in bold font refer to the respective locus being located within the confidence interval region mapped to only a part of the chromosome in the respective study, Italic font refers to them being located outside that confidence interval region, and if in normal font we were unable to asses that matter. The *IA* regions with classical MHC class I map to rainbow trout Chr. 18 and to Atlantic salmon Chr. 27, but those chromosomes were not found to harbor QTL in the listed studies.

A) Disease Resistance/Immunity Related QTL Studies in Rainbow Trout		
Resistance against	**QTL encoding linkage group**	**Reference**
Infectious pancreatic necrosis (IPN) virus	Chr. 3, Chr. 7, Chr. 8, Chr. 14 **(IB)**, Chr. 16, Chr-17 *(II)*, Chr. 20, Chr. 24, Chr. 27	Ozaki et al. 2007 [201]
	Chr. 14 **(IB)**, Chr. 16	Ozaki et al. 2001 [200]
Infectious hematopoietic necrosis (IHN) virus	Chr. 17 **(II)**	Khoo et al. 2004 [205]
Viral hemorrhagic septicemia (VHS) virus	Chr. 2, Chr. 3, Chr. 4, Chr. 5, Chr. 17 *(II)*, Chr. 24	Verrier et al. 2013 [207]

Table 2. *Cont.*

Cold water disease *Flavobacterium psychrophilum* (bacterium)	Chr. 2, Chr. 3, Chr. 7, Chr. 10, Chr. 17 (II), Chr. 21, Chr. 24, Chr. 25, Chr. 26, Chr. 29	Fraslin et al. 2018 [206]
	Chr. 3, Chr. 5, Chr. 8, Chr. 10, Chr. 13, Chr. 15, Chr. 25	Vallejo et al. 2017 [208]
	Chr. 8, Chr. 19, Chr. 25	Liu et al. 2015 [209]
	Chr. 1, Chr. 6, Chr. 7, Chr. 8, Chr. 11, Chr. 12, Chr. 14 (IB), Chr. 25	Palti et al. 2015 [202]
	Chr. 2, Chr. 3, Chr. 6, Chr. 8, Chr. 12, Chr. 13, Chr. 20	Vallejo et al. 2014 [210]
	Chr. 5, Chr. 16, Chr. 19	Wiens et al. 2013 [211]
Ceratomyxa shasta (parasite)	Chr. 9, Chr. 16, Chr. 20, Chr. 22, Chr. 29	Nichols et al. 2003 [212]
Whirling disease *Myxobolus cerebralis* (parasite)	Chr. 9	Baerwald et al. 2011 [213]
YAC-1 cells (murine tumor cell line)	Chr. 3	Zimmerman et al. 2004 [214]
B) Disease Resistance/Immunity Related QTL Studies in Atlantic Salmon		
Infectious pancreatic necrosis (IPN) virus	Chr. 26	Houston et al. 2010 [215]
	Chr. 4, Chr. 8, Chr. 26	Houston et al. 2008 [216]
	Chr. 1, Chr. 3, Chr. 4, Chr. 5, Chr. 6, Chr. 7, Chr. 9, Chr.10, Chr. 14 (IB), Chr. 17, Chr. 18, Chr. 19, Chr. 20, Chr. 26	Moen et al 2009 [203]
Pancreas disease Salmonid alphavirus (SAV)	Chr. 2, Chr. 3, Chr. 4, Chr. 7, Chr. 14 (IB), Chr. 26	Gonen et al. 2015 [204]
Infectious Salmon Anaemia (ISA) virus	Chr. 15 (maybe additional weaker QTL, but difficult to interpret)	Moen et al. 2004 [217]
Gyrodactylus salaris (parasite)	Chr. 4, Chr. 5, Chr. 6, Chr. 10, Chr. 13, Chr. 15, Chr. 16, Chr. 17, Chr. 23, Chr. 24	Gilbey et al. 2006 [218]

In other fish species, there also have been a number of genome-wide QTL studies on disease resistance traits, but more information is necessary to interpret them in regard to MHC genes. However, although the location of the classical MHC class I genes is not known in that study, it is interesting that Sawayama et al., 2017 [219], mapped a suggestive QTL for resistance against red sea bream iridoviral disease (RSIVD) in red sea bream to a relatively small confidence interval region that includes classical MHC class II (but finer mapping remains necessary).

Notable are also two studies in Salmoniformes that compared possible effects on differences in disease resistance by the *IA*, *IB* and classical MHC class II linkage groups. Miller et al., 2004 [220], found differences in resistance against IHN virus in Atlantic salmon to be linked with the *Sasa-IB* region, and maybe with the classical MHC class II locus, but not with the *Sasa-IA* region. Johnson et al., 2008 [221], found differences in resistance against the bacterium *F. psychrophilum* in rainbow trout to be linked with *Onmy-IB* and the classical MHC class II locus but not with *Onmy-IA*. Thus, also those two studies suggest that the enormous allelic divergence in the classical MHC class I genes in the *IA* regions has no notable impact on disease resistance.

9. Association of Teleost Fish MHC Genes with Allograft Rejection

In teleosts, both the classical MHC class I and classical MHC class II linkage groups have been associated with allograft rejection. Namely, linkage association studies indicated that MHC class II in gila topminnow is an important marker for scale allograft rejection [66] and that classical MHC class I in rainbow trout is an important marker for erythrocyte allograft rejection ([87] plus http://dx.doi.org/10.1007/s00251-003-0632-3, which the journal originally forgot to print). The erythrocyte allograft rejection in rainbow trout needed previous sensitization or took several weeks and was coincident with CD8 upregulation, thus suggesting T-cell involvement [87,222]. In contrast, studies in channel catfish found association of the classical MHC class I linkage group with immediate spontaneous cell-mediated cytotoxicity against allogeneic cell lines, which is reminiscent of NK-cell activity [88]. The relative importance of NK-cell and T-cell activity in MHC-dependent allograft rejection in fish remains to be determined.

In recent years, it has become accepted that, although rare, mammals can transmit cancer cells by contact over mucosal tissue, as exemplified by a facial tumor in Tasmanian devil and a venereal cancer in dogs [223,224]. Furthermore, spread of transmissible cancer cells among mollusks suggest that these cells can survive transport through water [225]. Because teleost fish live in water and are fully covered by mucosal tissue (they have "living skin" without an outer layer of dead epithelial cells), and, dependent on the species, can live in groups and can be cannibalistic, their intensity of grafting cancer cells and other cells to each other may be more intensive than in mammals. Since allografting has the potential risk of inducing cancer (see above) or graft versus host reaction (GVHR; [226]), teleost fish may need an enhanced ability to kill allografted cells, and we speculate that this at least in part can explain their MHC situation. Namely, the nonlinkage of classical MHC class I and II is expected to enhance the allogeneic variety among the individuals within teleost populations, and the wide allelic/haplotype divergence of the classical MHC class I sequences may provoke a strong immune response by cytotoxic T cells and/or NK cells. Because allelic classical MHC class I molecules in teleosts can have divergence levels which in mammals are only known among xenografts (see above), we speculate that teleost NK cells may play an important role in allograft rejection (for NK cells and xenografts see the review by [227]).

10. Association of Teleost Fish MHC Genes with Partner Selection

There is a line of research which claims an association between MHC variation and preferences for sexual or non-sexual partners in a wide variety of animals including teleost (e.g., reviewed in [228,229]). However, the claims for such associations may not be fully convincing and have been debated (e.g., [230,231]). The most prominent study claiming an MHC-based selection of sexual partners in fish was in stickleback and had Reusch et al., 2001 [232] conclude that female sticklebacks choose their mating partners by MHC class II B gene "counting" in order to assure optimal gene copy numbers in their offspring. However, in our opinion, the setup of this study [232] was lacking in scientific quality, as it was essentially based on comparing the number of MHC class II B fragments amplified by PCR from genomic DNA. Namely, (1) it was not investigated whether the amplified genomic fragments represented intact genes or pseudogenes, (2) it was not determined whether the primers used were exhaustive in amplifying all stickleback MHC class II B alleles present in the investigated population, and (3) the statistics suffered from a lack of prediction. A recent QTL study that investigated partner selection by female sticklebacks only found significant markers on Chr. XIV and Chr. XXI [233], whereas the classical MHC class II loci of stickleback probably are located on Chr. VII [27,234].

Overall, as probably follows from our organization of this paragraph, we are quite skeptical about the research field specifically dedicated to the correlation of MHC with partner choice.

11. Association of Teleost Fish Classical MHC Class I with Behavior and Behavior-Related Growth

For a long time, neurons were thought to be immunoprivileged cells without classical MHC class I expression. However, Corriveau et al. found in 1998 [235] that classical MHC class I in mammals is expressed in some neurons during neuron rearrangement. The authors naturally suggested that the molecules might have a function in such rearrangement. Later studies showed that β_2-m knockout mice, which should be deficient in most MHC class I cell surface expression, have aberrant phenotypes in brain morphology [236] and sexual behavior [237]. Huh et al., 2000 [236], found that in these mice synaptic connections were stabilized by an increase of long-term potentiation and the lack of long-term depression, and Oliveira et al., 2004 [238], found an increased reduction of their synapses after transection of spinal motor neurons. Mice lacking classical MHC class I molecules (KbDb-knockout) have less synapse elimination compared with wild type (WT), and elimination can be restored to WT levels by selectively expressing H2-Db in LGN neurons [239]. Furthermore, pyramidal neurons in KbDb-knockout mice have more extensive cortical connectivity than normal [240]. However, the molecular cascade by which classical MHC class I molecules affect neural plasticity probably is still not understood [240–242]. To our knowledge, the effect of allelic MHC variation on mammalian brain or neurons has not been investigated yet, although in humans a linkage association was found between the *Mhc* region and schizophrenia (reviewed in [243]).

Using a monoclonal antibody established against rainbow trout classical MHC class I, we found staining in some neurons in the brain stem of early rainbow trout fry [60]. We speculated that if classical MHC class I has a function in neural arrangement, as found for mammals, the enormous allelic variation between rainbow trout individuals could lead to differences in their neural systems. A trait influenced by brain development and under balancing selection is behavior [244–246]. We therefore studied the linkage association of rainbow trout classical MHC class I (locus *UBA* in the *Onmy-IA* region) with boldness/aggressiveness versus carefulness/friendliness, because these behavior traits are relatively easy to study as a dimorphism. In a series of experiments independently performed in two different institutes we demonstrated that within the investigated strain of rainbow trout the *UBA**401 allele marker was associated with bold/aggressive behavior and fast growth, whereas the *UBA**4901 allele marker was associated with careful/friendly behavior and comparatively slow growth [247]. The data suggested that the growth differences were caused by differences in feeding and swimming behavior during social competition [247]. However, although we had our initial data confirmed in another institute and the statistics looked convincing [247], currently we are somewhat skeptical about our model because in the bulk of QTL studies on fish behavior a possible linkage between the classical MHC class I locus and behavior was not found (e.g., [248,249]). The only genome-wide QTL report that might support our finding is a study in tilapia that reported linkage association between the classical MHC class I locus and stress-related features [250].

In the future, to obtain final evidence of whether classical MHC class I allelic/haplotype differences can cause differences in the fish nervous system and behavior, it probably will be best to start using transgenic zebrafish or medaka.

12. Discussion

In disease resistance QTL studies in rainbow trout and Atlantic salmon, linkage groups with the highly polymorphic classical MHC class I locus (the *IA* region) were not among the reported QTL regions, whereas QTL were found to be linked with a region including several nonclassical MHC class I genes (the *IB* region) and the region including the classical MHC class II locus (see also [201,251]). Fine mapping and analysis of the respective MHC alleles remains necessary. In case the classical MHC class II locus would be directly responsible for the observed QTL, we speculate that the probable cause concerns allelic differences in peptide binding grooves and the consequential presentation of different peptides, whereas if the *IB* region would be directly responsible, we assume that the probable cause more likely concerns the presence versus absence of certain nonclassical MHC class I genes (null-allele variation) and the consequential absence/presence of T cell subpopulations (although in fish such correlation remains to be determined). We have no hypothesis for why peptide groove variation in MHC class II might cause bigger differences in disease resistance than peptide groove variation in MHC class I. Modern techniques for making transgenic fish, combined with the large number of offspring in single broods of some species like Salmoniformes, would probably allow a final assessment of the theory that different MHC alleles can cause differences in disease resistance against pathogens. Currently, that popular theory is debatable to some extent—even for mammals.

It is puzzling why the allelic classical MHC class I variation in many teleost fish is so divergent and ancient. The maintenance of these ancient allelic lineages probably can't be explained by differences in peptide binding properties, because only some and not all ancient lineages are characterized by unique properties of the peptide binding groove [43] and the variable positions extend far beyond the peptide binding groove (e.g., Figure 2). Especially allelic length and sequence variation in the $\alpha 3$ domain [113] is difficult to explain with a model solely based on pathogen-driven selection for presenting different peptides. The most straightforward hypothesis to explain this extreme classical MHC class I allelic divergence is that it was selected to enhance the vigor of allograft rejection, because that is the process most readily observed in association with allelic MHC variation, and that is the process from which the MHC molecules derived their name. However, for the moment that is only speculation, and more research is needed on MHC-induced allograft rejection in fish.

Classical MHC molecules present a wide variety of peptides, and most pathogens may provide enough different antigens for potentially inducing adaptive T cell responses. Naturally, if immune responses based on subunit or peptide vaccines would be measured, the chances of finding important effects of MHC allelic variation should be higher. The QTL studies listed in this study concerned primary challenges with pathogens, and it would be interesting to perform some QTL studies in fish that would be more dependent on immune memory.

As shown in Figure 3, the MHC situation in rainbow trout provides a unique situation with very different types of allelic/haplotype variation in unlinked classical MHC class I, classical MHC class II, and a fragment with several nonclassical MHC class I. We hope that in the future, more intensive investigation of the influences of these variations on the trout immune system (e.g., on the selected NK and T cell populations) may benefit both fish aquaculture and the general understanding of MHC evolution.

In conclusion, the impressive MHC polymorphism originally found in mice and humans is a common trait among jawed vertebrates including fish. This indicates important evolutionary advantages that likely involve the resistance against pathogens as one of the relevant phenotypes. However, despite decades of knowledge of MHC polymorphism in an increasing number of model species, conclusive evidence for any of the elegant explanatory theories has not been obtained. Classical MHC class I molecules in many teleost fish show allelic variation at two distinct levels, namely the level of relatively young peptide binding groove variation as known in mammals (within lineage variation) and the level of much older variation that also concerns residues outside the binding groove (between lineage variation). For both levels, a strong signature of evolutionary selection is observed. Those two different levels of allelic variation may be driven by different functions, possibly not only involving pathogen resistance. This situation in fish might be fundamentally different from the situation in mammals, but alternatively might represent a more extreme case of a situation that has not yet been recognized in mammals due to its subtlety. Possibly, knowledge of fish MHC may contribute to a better understanding of mammalian MHC.

Funding: This research was funded by German Research Council grant No. FI 604/7-1 for TY.

Conflicts of Interest: The authors declare no conflict of interest.

References

1. Kimura, M.; Ohta, T. The Average Number of Generations until Fixation of a Mutant Gene in a Finite Population. *Genetics* **1969**, *61*, 763–771. [PubMed]
2. Kimura, M. *The Neutral Theory of Molecular Evolution*; Cambridge University Press: Cambridge, MA, USA, 1983.
3. Hedrick, P.W.; Thomson, G. Evidence for balancing selection at HLA. *Genetics* **1983**, *104*, 449–456.
4. Hughes, A.L.; Yeager, M. Natural selection and the evolutionary history of major histocompatibility complex loci. *Front. Biosci.* **1998**, *3*, d509–d516. [CrossRef]
5. Mungall, A.J.; Palmer, S.A.; Sims, S.K.; Edwards, C.A.; Ashurst, J.L.; Wilming, L.; Jones, M.C.; Horton, R.; Hunt, S.E.; Scott, C.E.; et al. The DNA sequence and analysis of human chromosome 6. *Nature* **2003**, *425*, 805–811. [CrossRef]
6. Norman, P.J.; Norberg, S.J.; Guethlein, L.A.; Nemat-Gorgani, N.; Royce, T.; Wroblewski, E.E.; Dunn, T.; Mann, T.; Alicata, C.; Hollenbach, J.A.; et al. Sequences of 95 human MHC haplotypes reveal extreme coding variation in genes other than highly polymorphic HLA class I and II. *Genome Res.* **2017**, *27*, 813–823. [CrossRef]

7. Maccari, G.; Robinson, J.; Ballingall, K.; Guethlein, L.A.; Grimholt, U.; Kaufman, J.; Ho, C.S.; de Groot, N.G.; Flicek, P.; Bontrop, R.E.; et al. IPD-MHC 2.0: An improved inter-species database for the study of the major histocompatibility complex. *Nucleic Acids Res.* **2017**, *45*, D860–D864. [CrossRef] [PubMed]

8. Kelly, A.; Trowsdale, J. Genetics of antigen processing and presentation. *Immunogenetics* **2019**, *71*, 161–170. [CrossRef]

9. Germain, R.N. MHC-dependent antigen processing and peptide presentation: Providing ligands for T lymphocyte activation. *Cell* **1994**, *76*, 287–299. [CrossRef]

10. Rammensee, H.G.; Friede, T.; Stevanoviíc, S. MHC ligands and peptide motifs: First listing. *Immunogenetics* **1995**, *41*, 178–228.

11. Michalova, V.; Murray, B.W.; Sultmann, H.; Klein, J. A contig map of the Mhc class I genomic region in the zebrafish reveals ancient synteny. *J. Immunol.* **2000**, *164*, 5296–5305. [CrossRef] [PubMed]

12. Satta, Y.; O'hUigin, C.; Takahata, N.; Klein, J. Intensity of natural selection at the major histocompatibility complex loci. *Proc. Natl. Acad. Sci. USA* **1994**, *91*, 7184–7188. [CrossRef]

13. Kelly, A.; Trowsdale, J. Introduction: MHC/KIR and governance of specificity. *Immunogenetics* **2017**, *69*, 481–488. [CrossRef] [PubMed]

14. Neefjes, J.; Jongsma, M.L.; Paul, P.; Bakke, O. Towards a systems understanding of MHC class I and MHC class II antigen presentation. *Nat. Rev. Immunol.* **2011**, *11*, 823–836. [CrossRef] [PubMed]

15. Madden, D.R. The three-dimensional structure of peptide-MHC complexes. *Annu. Rev. Immunol.* **1995**, *13*, 587–622. [CrossRef]

16. Hughes, A.L.; Nei, M. Pattern of nucleotide substitution at major histocompatibility complex class I loci reveals overdominant selection. *Nature* **1988**, *335*, 167–170. [CrossRef] [PubMed]

17. Hughes, A.L.; Nei, M. Nucleotide substitution at major histocompatibility complex class II loci: Evidence for overdominant selection. *Proc. Natl. Acad. Sci. USA* **1989**, *86*, 958–962. [CrossRef]

18. Figueroa, F.; Günther, E.; Klein, J. MHC polymorphism pre-dating speciation. *Nature* **1988**, *335*, 265–267. [CrossRef] [PubMed]

19. Parham, P.; Lawlor, D.A.; Lomen, C.E.; Ennis, P.D. Diversity and diversification of HLA-A,B,C alleles. *J. Immunol.* **1989**, *142*, 3937–3950. [PubMed]

20. Kaufman, J. Generalists and Specialists: A New View of How MHC Class I Molecules Fight Infectious Pathogens. *Trends Immunol.* **2018**, *39*, 367–379. [CrossRef]

21. Trowsdale, J.; Knight, J.C. Major histocompatibility complex genomics and human disease. *Annu. Rev. Genomics Hum. Genet.* **2013**, *14*, 301–323. [CrossRef] [PubMed]

22. Goulder, P.J.; Walker, B.D. HIV and HLA class I: An evolving relationship. *Immunity* **2012**, *37*, 426–440. [CrossRef] [PubMed]

23. Iwama, G.; Nakanishi, T. *The Fish Immune System*; Academic Press: San Diego, CA, USA, 1996.

24. Hashimoto, K.; Nakanishi, T.; Kurosawa, Y. Isolation of carp genes encoding major histocompatibility complex antigens. *Proc. Natl. Acad. Sci. USA* **1990**, *87*, 6863–6867. [CrossRef] [PubMed]

25. Hashimoto, K.; Nakanishi, T.; Kurosawa, Y. Identification of a shark sequence resembling the major histocompatibility complex class I alpha 3 domain. *Proc. Natl. Acad. Sci. USA* **1992**, *89*, 2209–2212. [CrossRef] [PubMed]

26. Kasahara, M.; Vazquez, M.; Sato, K.; McKinney, E.C.; Flajnik, M.F. Evolution of the major histocompatibility complex: Isolation of class II A cDNA clones from the cartilaginous fish. *Proc. Natl. Acad. Sci. USA* **1992**, *89*, 6688–6692. [CrossRef] [PubMed]

27. Dijkstra, J.M.; Grimholt, U.; Leong, J.; Koop, B.F.; Hashimoto, K. Comprehensive analysis of MHC class II genes in teleost fish genomes reveals dispensability of the peptide-loading DM system in a large part of vertebrates. *BMC Evol. Biol.* **2013**, *13*, 260. [CrossRef] [PubMed]

28. Betz, U.A.; Mayer, W.E.; Klein, J. Major histocompatibility complex class I genes of the coelacanth *Latimeria chalumnae*. *Proc. Natl. Acad. Sci. USA* **1994**, *91*, 11065–11069. [CrossRef]

29. Sato, A.; Sultmann, H.; Mayer, W.E.; Klein, J. Mhc class I gene of African lungfish. *Immunogenetics* **2000**, *51*, 491–495. [CrossRef]

30. Flajnik, M.F.; Kasahara, M. Origin and evolution of the adaptive immune system: Genetic events and selective pressures. *Nat. Rev. Genet.* **2010**, *11*, 47–59. [CrossRef]

31. Star, B.; Nederbragt, A.J.; Jentoft, S.; Grimholt, U.; Malmstrøm, M.; Gregers, T.F.; Rounge, T.B.; Paulsen, J.; Solbakken, M.H.; Sharma, A.; et al. The genome sequence of Atlantic cod reveals a unique immune system. *Nature* **2011**, *477*, 207–210. [CrossRef]

32. Malmstrøm, M.; Matschiner, M.; Tørresen, O.K.; Star, B.; Snipen, L.G.; Hansen, T.F.; Baalsrud, H.T.; Nederbragt, A.J.; Hanel, R.; Salzburger, W.; et al. Evolution of the immune system influences speciation rates in teleost fishes. *Nat. Genet.* **2016**, *48*, 1204–1210. [CrossRef]

33. Haase, D.; Roth, O.; Kalbe, M.; Schmiedeskamp, G.; Scharsack, J.P.; Rosenstiel, P.; Reusch, T.B. Absence of major histocompatibility complex class II mediated immunity in pipefish, Syngnathus typhle: Evidence from deep transcriptome sequencing. *Biol. Lett.* **2013**, *9*, 20130044. [CrossRef] [PubMed]

34. Ono, H.; Figueroa, F.; O'hUigin, C.; Klein, J. Cloning of the beta 2-microglobulin gene in the zebrafish. *Immunogenetics* **1993**, *38*, 1–10. [CrossRef] [PubMed]

35. Sültmann, H.; Mayer, W.E.; Figueroa, F.; O'Huigin, C.; Klein, J. Organization of Mhc class II B genes in the zebrafish (*Brachydanio rerio*). *Genomics* **1994**, *23*, 1–14. [CrossRef] [PubMed]

36. Harstad, H.; Lukacs, M.F.; Bakke, H.G.; Grimholt, U. Multiple expressed MHC class II loci in salmonids; details of one non-classical region in Atlantic salmon (*Salmo salar*). *BMC Genomics* **2008**, *9*, 193. [CrossRef] [PubMed]

37. Summers, K.; Roney, K.E.; da Silva, J.; Capraro, G.; Cuthbertson, B.J.; Kazianis, S.; Rosenthal, G.G.; Ryan, M.J.; McConnell, T.J. Divergent patterns of selection on the DAB and DXB MHC class II loci in Xiphophorus fishes. *Genetica* **2009**, *135*, 379–390. [CrossRef]

38. Bannai, H.P.; Nonaka, M. Comprehensive analysis of medaka major histocompatibility complex (MHC) class II genes: Implications for evolution in teleosts. *Immunogenetics* **2013**, *65*, 883–895. [CrossRef] [PubMed]

39. Ono, H.; Klein, D.; Vincek, V.; Figueroa, F.; O'hUigin, C.; Tichy, H.; Klein, J. Major histocompatibility complex class II genes of zebrafish. *Proc. Natl. Acad. Sci. USA* **1992**, *89*, 11886–11890. [CrossRef] [PubMed]

40. Dijkstra, J.M.; Yamaguchi, T.; Grimholt, U. Conservation of sequence motifs suggests that the nonclassical MHC class I lineages CD1/PROCR and UT were established before the emergence of tetrapod species. *Immunogenetics* **2018**, *70*, 459–476. [CrossRef]

41. Hashimoto, K.; Okamura, K.; Yamaguchi, H.; Ototake, M.; Nakanishi, T.; Kurosawa, Y. Conservation and diversification of MHC class I and its related molecules in vertebrates. *Immunol. Rev.* **1999**, *167*, 81–100. [CrossRef]

42. Grimholt, U.; Hordvik, I.; Fosse, V.M.; Olsaker, I.; Endresen, C.; Lie, O. Molecular cloning of major histocompatibility complex class I cDNAs from Atlantic salmon (Salmo salar). *Immunogenetics* **1993**, *37*, 469–473. [CrossRef]

43. Grimholt, U.; Tsukamoto, K.; Azuma, T.; Leong, J.; Koop, B.F.; Dijkstra, J.M. A comprehensive analysis of teleost MHC class I sequences. *BMC Evol. Biol.* **2015**, *15*, 32. [CrossRef]

44. Kruiswijk, C.P.; Hermsen, T.T.; Westphal, A.H.; Savelkoul, H.F.; Stet, R.J. A novel functional class I lineage in zebrafish (*Danio rerio*), carp (*Cyprinus carpio*), and large barbus (*Barbus intermedius*) showing an unusual conservation of the peptide binding domains. *J. Immunol.* **2002**, *169*, 1936–1947. [CrossRef]

45. Dijkstra, J.M.; Katagiri, T.; Hosomichi, K.; Yanagiya, K.; Inoko, H.; Ototake, M.; Aoki, T.; Hashimoto, K.; Shiina, T. A third broad lineage of major histocompatibility complex (MHC) class I in teleost fish; MHC class II linkage and processed genes. *Immunogenetics* **2007**, *59*, 305–321. [CrossRef]

46. Shum, B.P.; Guethlein, L.; Flodin, L.R.; Adkison, M.A.; Hedrick, R.P.; Nehring, R.B.; Stet, R.J.; Secombes, C.; Parham, P. Modes of salmonid MHC class I and II evolution differ from the primate paradigm. *J. Immunol.* **2001**, *166*, 3297–3308. [CrossRef] [PubMed]

47. Bingulac-Popovic, J.; Figueroa, F.; Sato, A.; Talbot, W.S.; Johnson, S.L.; Gates, M.; Postlethwait, J.H.; Klein, J. Mapping of Mhc class I and class II regions to different linkage groups in the zebrafish, Danio rerio. *Immunogenetics* **1997**, *46*, 129–134. [CrossRef]

48. Sato, A.; Figueroa, F.; Murray, B.W.; Malaga-Trillo, E.; Zaleska-Rutczynska, Z.; Sultmann, H.; Toyosawa, S.; Wedekind, C.; Steck, N.; Klein, J. Nonlinkage of major histocompatibility complex class I and class II loci in bony fishes. *Immunogenetics* **2000**, *51*, 108–116. [CrossRef] [PubMed]

49. Tsukamoto, K.; Hayashi, S.; Matsuo, M.Y.; Nonaka, M.I.; Kondo, M.; Shima, A.; Asakawa, S.; Shimizu, N.; Nonaka, M. Unprecedented intraspecific diversity of the MHC class I region of a teleost medaka, Oryzias latipes. *Immunogenetics* **2005**, *57*, 420–431. [CrossRef] [PubMed]

50. Shiina, T.; Dijkstra, J.M.; Shimizu, S.; Watanabe, A.; Yanagiya, K.; Kiryu, I.; Fujiwara, A.; Nishida-Umehara, C.; Kaba, Y.; Hirono, I.; et al. Interchromosomal duplication of major histocompatibility complex class I regions in rainbow trout (*Oncorhynchus mykiss*), a species with a presumably recent tetraploid ancestry. *Immunogenetics* **2005**, *56*, 878–893. [CrossRef]

51. Hansen, J.D.; Strassburger, P.; Thorgaard, G.H.; Young, W.P.; Du Pasquier, L. Expression, linkage, and polymorphism of MHC-related genes in rainbow trout, Oncorhynchus mykiss. *J. Immunol.* **1999**, *163*, 774–786.

52. Clark, M.S.; Shaw, L.; Kelly, A.; Snell, P.; Elgar, G. Characterization of the MHC class I region of the Japanese pufferfish (*Fugu rubripes*). *Immunogenetics* **2001**, *52*, 174–185. [CrossRef] [PubMed]

53. Ohashi, K.; Takizawa, F.; Tokumaru, N.; Nakayasu, C.; Toda, H.; Fischer, U.; Moritomo, T.; Hashimoto, K.; Nakanishi, T.; Dijkstra, J.M. A molecule in teleost fish, related with human MHC-encoded G6F, has a cytoplasmic tail with ITAM and marks the surface of thrombocytes and in some fishes also of erythrocytes. *Immunogenetics* **2010**, *62*, 543–559. [CrossRef]

54. McConnell, S.C.; Hernandez, K.M.; Wcisel, D.J.; Kettleborough, R.N.; Stemple, D.L.; Yoder, J.A.; Andrade, J.; de Jong, J.L. Alternative haplotypes of antigen processing genes in zebrafish diverged early in vertebrate evolution. *Proc. Natl. Acad. Sci. USA* **2016**, *113*, E5014–E5023. [CrossRef]

55. Takami, K.; Zaleska-Rutczynska, Z.; Figueroa, F.; Klein, J. Linkage of LMP, TAP, and RING3 with Mhc class I rather than class II genes in the zebrafish. *J. Immunol.* **1997**, *159*, 6052–6060.

56. Grimholt, U. Whole genome duplications have provided teleosts with many roads to peptide loaded MHC class I molecules. *BMC Evol. Biol.* **2018**, *18*, 25. [CrossRef]

57. Murray, B.W.; Sültmann, H.; Klein, J. Analysis of a 26-kb region linked to the Mhc in zebrafish: Genomic organization of the proteasome component beta/transporter associated with antigen processing-2 gene cluster and identification of five new proteasome beta subunit genes. *J. Immunol.* **1999**, *163*, 2657–2666.

58. Rodrigues, P.N.; Hermsen, T.T.; Rombout, J.H.; Egberts, E.; Stet, R.J. Detection of MHC class II transcripts in lymphoid tissues of the common carp (*Cyprinus carpio*, L.). *Dev. Comp. Immunol.* **1995**, *19*, 483–496. [CrossRef]

59. Dijkstra, J.M.; Kiryu, I.; Kollner, B.; Yoshiura, Y.; Ototake, M. MHC class II invariant chain homologues in rainbow trout (*Oncorhynchus mykiss*). *Fish Shellfish Immunol.* **2003**, *15*, 91–105. [CrossRef]

60. Fischer, U.; Dijkstra, J.M.; Kollner, B.; Kiryu, I.; Koppang, E.O.; Hordvik, I.; Sawamoto, Y.; Ototake, M. The ontogeny of MHC class I expression in rainbow trout (*Oncorhynchus mykiss*). *Fish Shellfish Immunol.* **2005**, *18*, 49–60. [CrossRef]

61. Koppang, E.O.; Hordvik, I.; Bjerkås, I.; Torvund, J.; Aune, L.; Thevarajan, J.; Endresen, C. Production of rabbit antisera against recombinant MHC class II beta chain and identification of immunoreactive cells in Atlantic salmon (*Salmo salar*). *Fish Shellfish Immunol.* **2003**, *14*, 115–132. [CrossRef]

62. Scharsack, J.P.; Kalbe, M.; Schaschl, H. Characterization of antisera raised against stickleback (*Gasterosteus aculeatus*) MHC class I and class II molecules. *Fish Shellfish Immunol.* **2007**, *23*, 991–1002. [CrossRef]

63. Boudinot, P.; Blanco, M.; de Kinkelin, P.; Benmansour, A. Combined DNA immunization with the glycoprotein gene of viral hemorrhagic septicemia virus and infectious hematopoietic necrosis virus induces double-specific protective immunity and nonspecific response in rainbow trout. *Virology* **1998**, *249*, 297–306. [CrossRef] [PubMed]

64. Wu, X.M.; Hu, Y.W.; Xue, N.N.; Ren, S.S.; Chen, S.N.; Nie, P.; Chang, M.X. Role of zebrafish NLRC5 in antiviral response and transcriptional regulation of MHC related genes. *Dev. Comp. Immunol.* **2017**, *68*, 58–68. [CrossRef] [PubMed]

65. Syed, M.; Vestrheim, O.; Mikkelsen, B.; Lundin, M. Isolation of the promoters of Atlantic salmon MHCII genes. *Mar. Biotechnol.* **2003**, *5*, 253–260. [CrossRef] [PubMed]

66. Cardwell, T.N.; Sheffer, R.J.; Hedrick, P.W. MHC variation and tissue transplantation in fish. *J. Hered.* **2001**, *92*, 305–308. [CrossRef]

67. Dijkstra, J.M.; Yamaguchi, T. Ancient features of the MHC class II presentation pathway, and a model for the possible origin of MHC molecules. *Immunogenetics* **2019**, *71*, 233–249. [CrossRef]

68. Boudinot, P.; Boubekeur, S.; Benmansour, A. Rhabdovirus infection induces public and private T cell responses in teleost fish. *J. Immunol.* **2001**, *167*, 6202–6209. [CrossRef] [PubMed]

69. Dijkstra, J.M. TH2 and Treg candidate genes in elephant shark. *Nature* **2014**, *511*, E7–E9. [CrossRef] [PubMed]

70. Yamaguchi, T.; Takizawa, F.; Fischer, U.; Dijkstra, J.M. Along the Axis between Type 1 and Type 2 Immunity; Principles Conserved in Evolution from Fish to Mammals. *Biology* **2015**, *17*, 814–859. [CrossRef]

71. Lam, S.H.; Chua, H.L.; Gong, Z.; Wen, Z.; Lam, T.J.; Sin, Y.M. Morphologic transformation of the thymus in developing zebrafish. *Dev. Dyn.* **2002**, *225*, 87–94. [CrossRef]

72. Langenau, D.M.; Zon, L.I. The zebrafish: A new model of T-cell and thymic development. *Nat. Rev. Immunol.* **2005**, *5*, 307–317. [CrossRef]

73. Takizawa, F.; Dijkstra, J.M.; Kotterba, P.; Korytář, T.; Kock, H.; Köllner, B.; Jaureguiberry, B.; Nakanishi, T.; Fischer, U. The expression of CD8α discriminates distinct T cell subsets in teleost fish. *Dev. Comp. Immunol.* **2011**, *35*, 752–763. [CrossRef]

74. Takizawa, F.; Magadan, S.; Parra, D.; Xu, Z.; Korytář, T.; Boudinot, P.; Sunyer, J.O. Novel Teleost CD4-Bearing Cell Populations Provide Insights into the Evolutionary Origins and Primordial Roles of CD4+ Lymphocytes and CD4+ Macrophages. *J. Immunol.* **2016**, *196*, 4522–4535. [CrossRef]

75. Suetake, H.; Araki, K.; Suzuki, Y. Cloning, expression, and characterization of fugu CD4, the first ectothermic animal CD4. *Immunogenetics* **2004**, *56*, 368–374. [CrossRef]

76. Dijkstra, J.M.; Somamoto, T.; Moore, L.; Hordvik, I.; Ototake, M.; Fischer, U. Identification and characterization of a second CD4-like gene in teleost fish. *Mol. Immunol.* **2006**, *43*, 410–419. [CrossRef]

77. Laing, K.J.; Zou, J.J.; Purcell, M.K.; Phillips, R.; Secombes, C.J.; Hansen, J.D. Evolution of the CD4 family: teleost fish possess two divergent forms of CD4 in addition to lymphocyte activation gene-3. *J. Immunol.* **2006**, *177*, 3939–3951. [CrossRef] [PubMed]

78. Taylor, E.B.; Wilson, M.; Bengten, E. The Src tyrosine kinase Lck binds to CD2, CD4–1, and CD4–2 T cell co-receptors in channel catfish, Ictalurus punctatus. *Mol. Immunol.* **2015**, *66*, 126–138. [CrossRef]

79. Yoder, J.A.; Haire, R.N.; Litman, G.W. Cloning of two zebrafish cDNAs that share domains with the MHC class II-associated invariant chain. *Immunogenetics* **1999**, *50*, 84–88. [CrossRef] [PubMed]

80. Hansen, J.D.; Strassburger, P.; Du Pasquier, L. Conservation of an alpha 2 domain within the teleostean world, MHC class I from the rainbow trout Oncorhynchus mykiss. *Dev. Comp. Immunol.* **1996**, *20*, 417–425. [CrossRef]

81. Dijkstra, J.M.; Kollner, B.; Aoyagi, K.; Sawamoto, Y.; Kuroda, A.; Ototake, M.; Nakanishi, T.; Fischer, U. The rainbow trout classical MHC class I molecule Onmy-UBA*501 is expressed in similar cell types as mammalian classical MHC class I molecules. *Fish Shellfish Immunol.* **2003**, *14*, 1–23. [CrossRef] [PubMed]

82. Dijkstra, J.M.; Yoshiura, Y.; Kiryu, I.; Aoyagi, K.; Kollner, B.; Fischer, U.; Nakanishi, T.; Ototake, M. The promoter of the classical MHC class I locus in rainbow trout (*Oncorhynchus mykiss*). *Fish Shellfish Immunol.* **2003**, *14*, 177–185. [CrossRef]

83. Rodrigues, P.N.; Dixon, B.; Roelofs, J.; Rombout, J.H.; Egberts, E.; Pohajdak, B.; Stet, R.J. Expression and temperature-dependent regulation of the beta2-microglobulin (Cyca-B2m) gene in a cold-blooded vertebrate, the common carp (*Cyprinus carpio* L.). *Dev. Immunol.* **1998**, *5*, 263–275. [CrossRef]

84. Antao, A.B.; Chinchar, V.G.; McConnell, T.J.; Miller, N.W.; Clem, L.W.; Wilson, M.R. MHC class I genes of the channel catfish: Sequence analysis and expression. *Immunogenetics* **1999**, *49*, 303–311. [CrossRef]

85. Chen, W.; Jia, Z.; Zhang, T.; Zhang, N.; Lin, C.; Gao, F.; Wang, L.; Li, X.; Jiang, Y.; Li, X.; et al. MHC class I presentation and regulation by IFN in bony fish determined by molecular analysis of the class I locus in grass carp. *J. Immunol.* **2010**, *185*, 2209–2221. [CrossRef] [PubMed]

86. Chen, Z.; Zhang, N.; Qi, J.; Chen, R.; Dijkstra, J.M.; Li, X.; Wang, Z.; Wang, J.; Wu, Y.; Xia, C. The Structure of the MHC Class I Molecule of Bony Fishes Provides Insights into the Conserved Nature of the Antigen-Presenting System. *J. Immunol.* **2017**, *199*, 3668–3678. [CrossRef]

87. Sarder, M.R.; Fischer, U.; Dijkstra, J.M.; Kiryu, I.; Yoshiura, Y.; Azuma, T.; Kollner, B.; Ototake, M. The MHC class I linkage group is a major determinant in the in vivo rejection of allogeneic erythrocytes in rainbow trout (*Oncorhynchus mykiss*). *Immunogenetics* **2003**, *55*, 315–324. [CrossRef]

88. Quiniou, S.M.; Wilson, M.; Bengten, E.; Waldbieser, G.C.; Clem, L.W.; Miller, N.W. MHC RFLP analyses in channel catfish full-sibling families: Identification of the role of MHC molecules in spontaneous allogeneic cytotoxic responses. *Dev. Comp. Immunol.* **2005**, *29*, 457–467. [CrossRef]

89. Somamoto, T.; Nakanishi, T.; Okamoto, N. Specific cell-mediated cytotoxicity against a virus-infected syngeneic cell line in isogeneic ginbuna crucian carp. *Dev. Comp. Immunol.* **2000**, *24*, 633–640. [CrossRef]

90. Somamoto, T.; Nakanishi, T.; Okamoto, N. Role of specific cell-mediated cytotoxicity in protecting fish from viral infections. *Virology* **2002**, *297*, 120–127. [CrossRef]

91. Somamoto, T.; Yoshiura, Y.; Sato, A.; Nakao, M.; Nakanishi, T.; Okamoto, N.; Ototake, M. Expression profiles of TCRbeta and CD8alpha mRNA correlate with virus-specific cell-mediated cytotoxic activity in ginbuna crucian carp. *Virology* **2006**, *348*, 370–377. [CrossRef] [PubMed]

92. Somamoto, T.; Nakanishi, T.; Nakao, M. Identification of anti-viral cytotoxic effector cells in the ginbuna crucian carp, Carassius auratus langsdorfii. *Dev. Comp. Immunol.* **2013**, *39*, 370–377. [CrossRef]

93. Dijkstra, J.M.; Fischer, U.; Sawamoto, Y.; Ototake, M.; Nakanishi, T. Exogenous antigens and the stimulation of MHC class I restricted cell-mediated cytotoxicity: Possible strategies for fish vaccines. *Fish Shellfish Immunol.* **2001**, *11*, 437–458. [CrossRef] [PubMed]

94. Utke, K.; Kock, H.; Schuetze, H.; Bergmann, S.M.; Lorenzen, N.; Einer-Jensen, K.; Köllner, B.; Dalmo, R.A.; Vesely, T.; Ototake, M.; et al. Cell-mediated immune responses in rainbow trout after DNA immunization against the viral hemorrhagic septicemia virus. *Dev. Comp. Immunol.* **2008**, *32*, 239–252. [CrossRef]

95. Chang, Y.T.; Kai, Y.H.; Chi, S.C.; Song, Y.L. Cytotoxic CD8α+ leucocytes have heterogeneous features in antigen recognition and class I MHC restriction in grouper. *Fish Shellfish Immunol.* **2011**, *30*, 1283–1293. [CrossRef]

96. Nelson, J.S.; Grande, T.C.; Wilson, M.V.H. *Fishes of the World*, 5th ed.; John Wiley & Sons: Hoboken, NJ, USA, 2016.

97. Kumar, S.; Hedges, S.B. A molecular timescale for vertebrate evolution. *Nature* **1998**, *392*, 917–920. [CrossRef] [PubMed]

98. Broughton, R.E.; Betancur-R, R.; Li, C.; Arratia, G.; Ortí, G. Multi-locus phylogenetic analysis reveals the pattern and tempo of bony fish evolution. *PLoS Curr.* **2013**, *5*. [CrossRef]

99. Jaillon, O.; Aury, J.M.; Brunet, F.; Petit, J.L.; Stange-Thomann, N.; Mauceli, E.; Bouneau, L.; Fischer, C.; Ozouf-Costaz, C.; Bernot, A.; et al. Genome duplication in the teleost fish Tetraodon nigroviridis reveals the early vertebrate proto-karyotype. *Nature* **2004**, *431*, 946–957. [CrossRef]

100. Painter, C.A.; Stern, L.J. Conformational variation in structures of classical and non-classical MHCII proteins and functional implications. *Immunol. Rev.* **2012**, *250*, 144–157. [CrossRef]

101. Kasahara, M.; McKinney, E.C.; Flajnik, M.F.; Ishibashi, T. The evolutionary origin of the major histocompatibility complex: Polymorphism of class II alpha chain genes in the cartilaginous fish. *Eur. J. Immunol.* **1993**, *23*, 2160–2165. [CrossRef]

102. Hughes, A.L.; Nei, M. Evolution of the major histocompatibility complex: Independent origin of nonclassical class I genes in different groups of mammals. *Mol. Biol. Evol.* **1989**, *6*, 559–579.

103. Adams, E.J.; Luoma, A.M. The adaptable major histocompatibility complex (MHC) fold: Structure and function of nonclassical and MHC class I-like molecules. *Annu. Rev. Immunol.* **2013**, *31*, 529–561. [CrossRef]

104. Lukacs, M.F.; Harstad, H.; Grimholt, U.; Beetz-Sargent, M.; Cooper, G.A.; Reid, L.; Bakke, H.G.; Phillips, R.B.; Miller, K.M.; Davidson, W.S.; et al. Genomic organization of duplicated major histocompatibility complex class I regions in Atlantic salmon (*Salmo salar*). *BMC Genomics* **2007**, *8*, 251. [CrossRef]

105. Dirscherl, H.; McConnell, S.C.; Yoder, J.A.; de Jong, J.L. The MHC class I genes of zebrafish. *Dev. Comp. Immunol.* **2014**, *46*, 11–23. [CrossRef]

106. Shum, B.P.; Rajalingam, R.; Magor, K.E.; Azumi, K.; Carr, W.H.; Dixon, B.; Stet, R.J.; Adkison, M.A.; Hedrick, R.P.; Parham, P. A divergent non-classical class I gene conserved in salmonids. *Immunogenetics* **1999**, *49*, 479–490. [CrossRef] [PubMed]

107. Dirscherl, H.; Yoder, J.A. Characterization of the Z lineage Major histocompatability complex class I genes in zebrafish. *Immunogenetics* **2014**, *66*, 185–198. [CrossRef] [PubMed]

108. Nonaka, M.I.; Aizawa, K.; Mitani, H.; Bannai, H.P.; Nonaka, M. Retained orthologous relationships of the MHC Class I genes during euteleost evolution. *Mol. Biol. Evol.* **2011**, *28*, 3099–3112. [CrossRef] [PubMed]

109. Gómez, D.; Conejeros, P.; Marshall, S.H.; Consuegra, S. MHC evolution in three salmonid species: A comparison between class II alpha and beta genes. *Immunogenetics* **2010**, *62*, 531–542. [CrossRef] [PubMed]

110. Li, W.; Sun, W.X.; Hu, J.F.; Hong, D.W.; Chen, S.L. Molecular characterization, polymorphism and expression analysis of swamp eel major histocompatibility complex class II B gene, after infection by Aeromonas Hydrophilia. *J. Anim. Plant Sci.* **2014**, *24*, 481–491.

111. Stet, R.J.; de Vries, B.; Mudde, K.; Hermsen, T.; van Heerwaarden, J.; Shum, B.P.; Grimholt, U. Unique haplotypes of co-segregating major histocompatibility class II A and class II B alleles in Atlantic salmon (*Salmo salar*) give rise to diverse class II genotypes. *Immunogenetics* **2002**, *54*, 320–331. [CrossRef]

112. Okamura, K.; Ototake, M.; Nakanishi, T.; Kurosawa, Y.; Hashimoto, K. The most primitive vertebrates with jaws possess highly polymorphic MHC class I genes comparable to those of humans. *Immunity* **1997**, *7*, 777–790. [CrossRef]

113. Aoyagi, K.; Dijkstra, J.M.; Xia, C.; Denda, I.; Ototake, M.; Hashimoto, K.; Nakanishi, T. Classical MHC class I genes composed of highly divergent sequence lineages share a single locus in rainbow trout (*Oncorhynchus mykiss*). *J. Immunol.* **2002**, *168*, 260–273. [CrossRef]

114. Kiryu, I.; Dijkstra, J.M.; Sarder, R.I.; Fujiwara, A.; Yoshiura, Y.; Ototake, M. New MHC class Ia domain lineages in rainbow trout (*Oncorhynchus mykiss*) which are shared with other fish species. *Fish Shellfish Immunol.* **2005**, *18*, 243–254. [CrossRef] [PubMed]

115. Nonaka, M.I.; Nonaka, M. Evolutionary analysis of two classical MHC class I loci of the medaka fish, Oryzias latipes: Haplotype-specific genomic diversity, locus-specific polymorphisms, and interlocus homogenization. *Immunogenetics* **2010**, *62*, 319–332. [CrossRef]

116. McConnell, S.C.; Restaino, A.C.; de Jong, J.L. Multiple divergent haplotypes express completely distinct sets of class I MHC genes in zebrafish. *Immunogenetics* **2014**, *66*, 199–213. [CrossRef]

117. Dijkstra, J.M.; Grimholt, U. Major histocompatibility complex (MHC) fragment numbers alone—In Atlantic cod and in general—Do not represent functional variability. Version 2. *F1000Research* **2018**, *7*, 963. [CrossRef]

118. Vogel, T.U.; Evans, D.T.; Urvater, J.A.; O'Connor, D.H.; Hughes, A.L.; Watkins, D.I. Major histocompatibility complex class I genes in primates: Co-evolution with pathogens. *Immunol. Rev.* **1999**, *167*, 327–337. [CrossRef] [PubMed]

119. De Groot, N.G.; Otting, N.; Doxiadis, G.G.; Balla-Jhagjhoorsingh, S.S.; Heeney, J.L.; van Rood, J.J.; Gagneux, P.; Bontrop, R.E. Evidence for an ancient selective sweep in the MHC class I gene repertoire of chimpanzees. *Proc. Natl. Acad. Sci. USA* **2002**, *99*, 11748–11753. [CrossRef]

120. Lafont, B.A.; Buckler-White, A.; Plishka, R.; Buckler, C.; Martin, M.A. Characterization of pig-tailed macaque classical MHC class I genes: Implications for MHC evolution and antigen presentation in macaques. *J. Immunol.* **2003**, *171*, 875–885. [CrossRef] [PubMed]

121. Meredith, R.W.; Janečka, J.E.; Gatesy, J.; Ryder, O.A.; Fisher, C.A.; Teeling, E.C.; Goodbla, A.; Eizirik, E.; Simão, T.L.; Stadler, T.; et al. Impacts of the Cretaceous Terrestrial Revolution and KPg extinction on mammal diversification. *Science* **2011**, *334*, 521–524. [CrossRef]

122. Barber, L.D.; Percival, L.; Valiante, N.M.; Chen, L.; Lee, C.; Gumperz, J.E.; Phillips, J.H.; Lanier, L.L.; Bigge, J.C.; Parekh, R.B.; et al. The inter-locus recombinant HLA-B*4601 has high selectivity in peptide binding and functions characteristic of HLA-C. *J. Exp. Med.* **1996**, *184*, 735–740. [CrossRef] [PubMed]

123. Xia, C.; Kiryu, I.; Dijkstra, J.M.; Azuma, T.; Nakanishi, T.; Ototake, M. Differences in MHC class I genes between strains of rainbow trout (*Oncorhynchus mykiss*). *Fish Shellfish Immunol.* **2002**, *12*, 287–301. [CrossRef] [PubMed]

124. Miller, K.M.; Li, S.; Ming, T.J.; Kaukinen, K.H.; Schulze, A.D. The salmonid MHC class I: More ancient loci uncovered. *Immunogenetics* **2006**, *58*, 571–589. [CrossRef] [PubMed]

125. Grimholt, U.; Drablos, F.; Jorgensen, S.M.; Hoyheim, B.; Stet, R.J. The major histocompatibility class I locus in Atlantic salmon (*Salmo salar*, L.): Polymorphism, linkage analysis and protein modelling. *Immunogenetics* **2002**, *54*, 570–581. [CrossRef]

126. Consuegra, S.; Megens, H.J.; Schaschl, H.; Leon, K.; Stet, R.J.; Jordan, W.C. Rapid evolution of the MH class I locus results in different allelic compositions in recently diverged populations of Atlantic salmon. *Mol. Biol. Evol.* **2005**, *22*, 1095–1106. [CrossRef] [PubMed]

127. Takeuchi, H.; Figueroa, F.; O'Huigin, C.; Klein, J. Cloning and characterization of class I Mhc genes of the zebrafish, Brachydanio rerio. *Immunogenetics* **1995**, *42*, 77–84. [CrossRef]

128. Graser, R.; Vincek, V.; Takami, K.; Klein, J. Analysis of zebrafish Mhc using BAC clones. *Immunogenetics* **1998**, *47*, 318–325. [CrossRef] [PubMed]

129. Sültmann, H.; Murray, B.W.; Klein, J. Identification of seven genes in the major histocompatibility complex class I region of the zebrafish. *Scand. J. Immunol.* **2000**, *51*, 577–585. [CrossRef]

130. Van Erp, S.H.; Dixon, B.; Figueroa, F.; Egberts, E.; Stet, R.J. Identification and characterization of a new major histocompatibility complex class I gene in carp (*Cyprinus carpio*, L.). *Immunogenetics* **1996**, *44*, 49–61. [CrossRef] [PubMed]

131. Matsumura, M.; Fremont, D.H.; Peterson, P.A.; Wilson, I.A. Emerging principles for the recognition of peptide antigens by MHC class I molecules. *Science* **1992**, *257*, 927–934. [CrossRef] [PubMed]

132. Matsuo, M.Y.; Asakawa, S.; Shimizu, N.; Kimura, H.; Nonaka, M. Nucleotide sequence of the MHC class I genomic region of a teleost, the medaka (*Oryzias latipes*). *Immunogenetics* **2002**, *53*, 930–940.

133. Persson, A.C.; Stet, R.J.; Pilström, L. Characterization of MHC class I and beta(2)-microglobulin sequences in Atlantic cod reveals an unusually high number of expressed class I genes. *Immunogenetics* **1999**, *50*, 49–59. [CrossRef]

134. Miller, K.M.; Kaukinen, K.H.; Schulze, A.D. Expansion and contraction of major histocompatibility complex genes: A teleostean example. *Immunogenetics* **2002**, *53*, 941–963.

135. Schaschl, H.; Wegner, K.M. Contrasting mode of evolution between the MHC class I genomic region and class II region in the three-spined stickleback (*Gasterosteus aculeatus*, L.; Gasterosteidae: Teleostei). *Immunogenetics* **2007**, *59*, 295–304. [CrossRef]

136. Tørresen, O.K.; Brieuc, M.S.O.; Solbakken, M.H.; Sørhus, E.; Nederbragt, A.J.; Jakobsen, K.S.; Meier, S.; Edvardsen, R.B.; Jentoft, S. Genomic architecture of haddock (*Melanogrammus aeglefinus*) shows expansions of innate immune genes and short tandem repeats. *BMC Genomics* **2018**, *19*, 240. [CrossRef] [PubMed]

137. Bartl, S.; Baish, M.A.; Flajnik, M.F.; Ohta, Y. Identification of class I genes in cartilaginous fish, the most ancient group of vertebrates displaying an adaptive immune response. *J. Immunol.* **1997**, *159*, 6097–6104.

138. Koppang, E.O.; Dannevig, B.H.; Lie, O.; Ronningen, K.; Press, C.M.L. Expression of MHC class I and II mRNA in a macrophage-like cell line (SHK-1) derived from Atlantic salmon, Salmo salar, L., head kidney. *Fish Shellfish Immunol.* **1999**, *9*, 473–489. [CrossRef]

139. Jørgensen, S.M.; Syvertsen, B.L.; Lukacs, M.; Grimholt, U.; Gjøen, T. Expression of MHC class I pathway genes in response to infectious salmon anaemia virus in Atlantic salmon (*Salmo salar*, L.) cells. *Fish Shellfish Immunol.* **2006**, *21*, 548–560. [CrossRef]

140. Antao, A.B.; Wilson, M.; Wang, J.; Bengten, E.; Miller, N.W.; Clem, L.W.; Chinchar, V.G. Genomic organization and differential expression of channel catfish MHC class I genes. *Dev. Comp. Immunol.* **2001**, *25*, 579–595. [CrossRef]

141. Stet, R.J.; Kruiswijk, C.P.; Dixon, B. Major histocompatibility lineages and immune gene function in teleost fishes: The road not taken. *Crit. Rev. Immunol.* **2003**, *23*, 441–471. [CrossRef]

142. Bly, J.E.; Clem, L.W. Temperature-mediated processes in teleost immunity: In vitro immunosuppression induced by in vivo low temperature in channel catfish. *Vet. Immunol. Immunopathol.* **1991**, *28*, 365–377. [CrossRef]

143. Abram, Q.H.; Dixon, B.; Katzenback, B.A. Impacts of Low Temperature on the Teleost Immune System. *Biology* **2017**, *6*, 39. [CrossRef] [PubMed]

144. Sever, L.; Vo, N.T.K.; Bols, N.C.; Dixon, B. Tapasin's protein interactions in the rainbow trout peptide-loading complex. *Dev. Comp. Immunol.* **2018**, *81*, 262–270. [CrossRef] [PubMed]

145. Criscitiello, M.F.; Ohta, Y.; Saltis, M.; McKinney, E.C.; Flajnik, M.F. Evolutionarily conserved TCR binding sites, identification of T cells in primary lymphoid tissues, and surprising trans-rearrangements in nurse shark. *J. Immunol.* **2010**, *184*, 6950–6960. [CrossRef]

146. Wang, J.; Zhang, N.; Wang, Z.; Yanan, W.; Zhang, L.; Xia, C. Structural insights into the evolution feature of a bony fish CD8αα homodimer. *Mol. Immunol.* **2018**, *97*, 109–116. [CrossRef]

147. Hansen, J.D.; Strassburger, P. Description of an ectothermic TCR coreceptor, CD8 alpha, in rainbow trout. *J. Immunol.* **2000**, *164*, 3132–3139. [CrossRef]

148. Moore, L.J.; Somamoto, T.; Lie, K.K.; Dijkstra, J.M.; Hordvik, I. Characterisation of salmon and trout CD8alpha and CD8beta. *Mol. Immunol.* **2005**, *42*, 1225–1234. [CrossRef]

149. Suetake, H.; Araki, K.; Akatsu, K.; Somamoto, T.; Dijkstra, J.M.; Yoshiura, Y.; Kikuchi, K.; Suzuki, Y. Genomic organization and expression of CD8alpha and CD8beta genes in fugu Takifugu rubripes. *Fish Shellfish Immunol.* **2007**, *23*, 1107–1118. [CrossRef]

150. Hansen, J.D.; Farrugia, T.J.; Woodson, J.; Laing, K.J. Description of an elasmobranch TCR coreceptor: CD8α from Rhinobatos productus. *Dev. Comp. Immunol.* **2011**, *35*, 452–460. [CrossRef]

151. Venkatesh, B.; Lee, A.P.; Ravi, V.; Maurya, A.K.; Lian, M.M.; Swann, J.B.; Ohta, Y.; Flajnik, M.F.; Sutoh, Y.; Kasahara, M.; et al. Elephant shark genome provides unique insights into gnathostome evolution. *Nature* **2014**, *505*, 174–179. [CrossRef]

152. Hayashi, N.; Takeuchi, M.; Nakanishi, T.; Hashimoto, K.; Dijkstra, J.M. Zinc-dependent binding between peptides derived from rainbow trout CD8alpha and LCK. *Fish Shellfish Immunol.* **2010**, *28*, 72–76. [CrossRef]

153. Criscitiello, M.F.; Ohta, Y.; Graham, M.D.; Eubanks, J.O.; Chen, P.L.; Flajnik, M.F. Shark class II invariant chain reveals ancient conserved relationships with cathepsins and MHC class II. *Dev. Comp. Immunol.* **2012**, *36*, 521–533. [CrossRef]

154. Fujiki, K.; Smith, C.M.; Liu, L.; Sundick, R.S.; Dixon, B. Alternate forms of MHC class II-associated invariant chain are not produced by alternative splicing in rainbow trout (*Oncorhynchus mykiss*) but are encoded by separate genes. *Dev. Comp. Immunol.* **2003**, *27*, 377–391. [CrossRef]

155. Wilson, A.B. MHC and adaptive immunity in teleost fishes. *Immunogenetics* **2017**, *69*, 521–528. [CrossRef] [PubMed]

156. Kulski, J.K.; Shiina, T.; Anzai, T.; Kohara, S.; Inoko, H. Comparative genomic analysis of the MHC: The evolution of class I duplication blocks, diversity and complexity from shark to man. *Immunol. Rev.* **2002**, *190*, 95–122. [CrossRef]

157. Ohta, Y.; Okamura, K.; McKinney, E.C.; Bartl, S.; Hashimoto, K.; Flajnik, M.F. Primitive synteny of vertebrate major histocompatibility complex class I and class II genes. *Proc. Natl. Acad. Sci. USA* **2000**, *97*, 4712–4717. [CrossRef] [PubMed]

158. Ohta, Y.; McKinney, E.C.; Criscitiello, M.F.; Flajnik, M.F. Proteasome, transporter associated with antigen processing, and class I genes in the nurse shark Ginglymostoma cirratum: Evidence for a stable class I region and MHC haplotype lineages. *J. Immunol.* **2002**, *168*, 771–781. [CrossRef]

159. Ohta, Y.; Shiina, T.; Lohr, R.L.; Hosomichi, K.; Pollin, T.I.; Heist, E.J.; Suzuki, S.; Inoko, H.; Flajnik, M.F. Primordial linkage of β2-microglobulin to the MHC. *J. Immunol.* **2011**, *186*, 3563–3571. [CrossRef] [PubMed]

160. Blees, A.; Januliene, D.; Hofmann, T.; Koller, N.; Schmidt, C.; Trowitzsch, S.; Moeller, A.; Tampé, R. Structure of the human MHC-I peptide-loading complex. *Nature* **2017**, *551*, 525–528. [CrossRef]

161. Murata, S.; Takahama, Y.; Kasahara, M.; Tanaka, K. The immunoproteasome and thymoproteasome: Functions, evolution and human disease. *Nat. Immunol.* **2018**, *19*, 923–931. [CrossRef] [PubMed]

162. Robledo, D.; Taggart, J.B.; Ireland, J.H.; McAndrew, B.J.; Starkey, W.G.; Haley, C.S.; Hamilton, A.; Guy, D.R.; Mota-Velasco, J.C.; Gheyas, A.A.; et al. Gene expression comparison of resistant and susceptible Atlantic salmon fry challenged with Infectious Pancreatic Necrosis virus reveals a marked contrast in immune response. *BMC Genomics* **2016**, *17*, 279. [CrossRef] [PubMed]

163. Hansen, J.D.; La Patra, S. Induction of the rainbow trout MHC class I pathway during acute IHNV infection. *Immunogenetics* **2002**, *54*, 654–661. [CrossRef]

164. Martin, S.A.; Zou, J.; Houlihan, D.F.; Secombes, C.J. Directional responses following recombinant cytokine stimulation of rainbow trout (*Oncorhynchus mykiss*) RTS-11 macrophage cells as revealed by transcriptome profiling. *BMC Genomics* **2007**, *8*, 150. [CrossRef]

165. Kasthuri, S.R.; Umasuthan, N.; Whang, I.; Lim, B.S.; Jung, H.B.; Oh, M.J.; Jung, S.J.; Yeo, S.Y.; Kim, S.Y.; Lee, J. Molecular characterization and expressional affirmation of the beta proteasome subunit cluster in rock bream immune defense. *Mol. Biol. Rep.* **2014**, *41*, 5413–5427. [CrossRef] [PubMed]

166. Momburg, F.; Roelse, J.; Howard, J.C.; Butcher, G.W.; Hämmerling, G.J.; Neefjes, J.J. Selectivity of MHC-encoded peptide transporters from human, mouse and rat. *Nature* **1994**, *367*, 648–651. [CrossRef]

167. Powis, S.J.; Young, L.L.; Joly, E.; Barker, P.J.; Richardson, L.; Brandt, R.P.; Melief, C.J.; Howard, J.C.; Butcher, G.W. The rat cim effect: TAP allele-dependent changes in a class I MHC anchor motif and evidence against C-terminal trimming of peptides in the ER. *Immunity* **1996**, *4*, 159–165. [CrossRef]

168. Joly, E.; Le Rolle, A.F.; González, A.L.; Mehling, B.; Stevens, J.; Coadwell, W.J.; Hünig, T.; Howard, J.C.; Butcher, G.W. Co-evolution of rat TAP transporters and MHC class I RT1-A molecules. *Curr. Biol.* **1998**, *8*, 169–172. [CrossRef]

169. Walker, B.A.; Hunt, L.G.; Sowa, A.K.; Skjødt, K.; Göbel, T.W.; Lehner, P.J.; Kaufman, J. The dominantly expressed class I molecule of the chicken MHC is explained by coevolution with the polymorphic peptide transporter (TAP) genes. *Proc. Natl. Acad. Sci. USA* **2011**, *108*, 8396–8401. [CrossRef] [PubMed]

170. Van Hateren, A.; Carter, R.; Bailey, A.; Kontouli, N.; Williams, A.P.; Kaufman, J.; Elliott, T. A mechanistic basis for the co-evolution of chicken tapasin and major histocompatibility complex class I (MHC I) proteins. *J. Biol. Chem.* **2013**, *288*, 32797–32808. [CrossRef]

171. Tregaskes, C.A.; Harrison, M.; Sowa, A.K.; van Hateren, A.; Hunt, L.G.; Vainio, O.; Kaufman, J. Surface expression, peptide repertoire, and thermostability of chicken class I molecules correlate with peptide transporter specificity. *Proc. Natl. Acad. Sci. USA* **2016**, *113*, 692–697. [CrossRef] [PubMed]

172. Namikawa, C.; Salter-Cid, L.; Flajnik, M.F.; Kato, Y.; Nonaka, M.; Sasaki, M. Isolation of Xenopus LMP-7 homologues. Striking allelic diversity and linkage to MHC. *J. Immunol.* **1995**, *155*, 1964–1971.

173. Ohta, Y.; Powis, S.J.; Lohr, R.L.; Nonaka, M.; Pasquier, L.D.; Flajnik, M.F. Two highly divergent ancient allelic lineages of the transporter associated with antigen processing (TAP) gene in Xenopus: Further evidence for co-evolution among MHC class I region genes. *Eur. J. Immunol.* **2003**, *33*, 3017–3027. [CrossRef]

174. Noro, M.; Nonaka, M. Evolution of dimorphisms of the proteasome subunit beta type 8 gene (PSMB8) in basal ray-finned fish. *Immunogenetics* **2014**, *66*, 325–334. [CrossRef]

175. Kandil, E.; Namikawa, C.; Nonaka, M.; Greenberg, A.S.; Flajnik, M.F.; Ishibashi, T.; Kasahara, M. Isolation of low molecular mass polypeptide complementary DNA clones from primitive vertebrates. Implications for the origin of MHC class I-restricted antigen presentation. *J. Immunol.* **1996**, *156*, 4245–4253. [PubMed]

176. Tsukamoto, K.; Miura, F.; Fujito, N.T.; Yoshizaki, G.; Nonaka, M. Long-lived dichotomous lineages of the proteasome subunit beta type 8 (PSMB8) gene surviving more than 500 million years as alleles or paralogs. *Mol. Biol. Evol.* **2012**, *29*, 3071–3079. [CrossRef] [PubMed]

177. Roos, C.; Walter, L. Considerable haplotypic diversity in the RT1-CE class I gene region of the rat major histocompatibility complex. *Immunogenetics* **2005**, *56*, 773–777. [CrossRef]

178. Hosomichi, K.; Shiina, T.; Suzuki, S.; Tanaka, M.; Shimizu, S.; Iwamoto, S.; Hara, H.; Yoshida, Y.; Kulski, J.K.; Inoko, H.; et al. The major histocompatibility complex (Mhc) class IIB region has greater genomic structural flexibility and diversity in the quail than the chicken. *BMC Genomics* **2006**, *7*, 322. [CrossRef] [PubMed]

179. Flajnik, M.F.; Kasahara, M.; Shum, B.P.; Salter-Cid, L.; Taylor, E.; Du Pasquier, L. A novel type of class I gene organization in vertebrates: A large family of non-MHC-linked class I genes is expressed at the RNA level in the amphibian Xenopus. *EMBO J.* **1993**, *12*, 4385–4396. [CrossRef] [PubMed]

180. Bartl, S.; Weissman, I.L. Isolation and characterization of major histocompatibility complex class IIB genes from the nurse shark. *Proc. Natl. Acad. Sci. USA* **1994**, *91*, 262–266. [CrossRef] [PubMed]

181. Málaga-Trillo, E.; Zaleska-Rutczynska, Z.; McAndrew, B.; Vincek, V.; Figueroa, F.; Sültmann, H.; Klein, J. Linkage relationships and haplotype polymorphism among cichlid Mhc class II B loci. *Genetics* **1998**, *149*, 1527–1537. [PubMed]

182. Murray, B.W.; Shintani, S.; Sültmann, H.; Klein, J. Major histocompatibility complex class II A genes in cichlid fishes: Identification, expression, linkage relationships, and haplotype variation. *Immunogenetics* **2000**, *51*, 576–586. [CrossRef]

183. Dirscherl, H.; Yoder, J.A. A nonclassical MHC class I U lineage locus in zebrafish with a null haplotypic variant. *Immunogenetics* **2015**, *67*, 501–513. [CrossRef]

184. Sato, A.; Klein, D.; Sültmann, H.; Figueroa, F.; O'hUigin, C.; Klein, J. Class I mhc genes of cichlid fishes: Identification, expression, and polymorphism. *Immunogenetics* **1997**, *46*, 63–72. [CrossRef] [PubMed]

185. Murray, B.W.; Nilsson, P.; Zaleska-Rutczynska, Z.; Sültmann, H.; Klein, J. Linkage Relationships and Haplotype Variation of the Major Histocompatibility Complex Class I A Genes in the Cichlid Fish Oreochromis niloticus. *Mar. Biotechnol* **2000**, *2*, 437–448. [PubMed]

186. Anderson, C.K.; Brossay, L. The role of MHC class Ib-restricted T cells during infection. *Immunogenetics* **2016**, *68*, 677–691. [CrossRef]

187. Treiner, E.; Duban, L.; Bahram, S.; Radosavljevic, M.; Wanner, V.; Tilloy, F.; Affaticati, P.; Gilfillan, S.; Lantz, O. Selection of evolutionarily conserved mucosal-associated invariant T cells by MR1. *Nature* **2003**, *422*, 164–169. [CrossRef] [PubMed]

188. Chen, Y.H.; Chiu, N.M.; Mandal, M.; Wang, N.; Wang, C.R. Impaired NK1+ T cell development and early IL-4 production in CD1-deficient mice. *Immunity* **1997**, *6*, 459–467. [CrossRef]

189. Allendorf, F.W.; Thorgaard, G.H. Tetraploidy and the evolution of salmonid fishes. In *Evolutionary Genetics of Fishes*; Turner, B.J., Ed.; Plenum Press: New York, NY, USA, 1984; pp. 1–53.

190. Berthelot, C.; Brunet, F.; Chalopin, D.; Juanchich, A.; Bernard, M.; Noël, B.; Bento, P.; Da Silva, C.; Labadie, K.; Alberti, A.; et al. The rainbow trout genome provides novel insights into evolution after whole-genome duplication in vertebrates. *Nat. Commun.* **2014**, *5*, 3657. [CrossRef] [PubMed]

191. Dijkstra, J.M.; Kiryu, I.; Yoshiura, Y.; Kumanovics, A.; Kohara, M.; Hayashi, N.; Ototake, M. Polymorphism of two very similar MHC class Ib loci in rainbow trout (*Oncorhynchus mykiss*). *Immunogenetics* **2006**, *58*, 152–167. [CrossRef]

192. Landis, E.D.; Purcell, M.K.; Thorgaard, G.H.; Wheeler, P.A.; Hansen, J.D. Transcriptional profiling of MHC class I genes in rainbow trout infected with infectious hematopoietic necrosis virus. *Mol. Immunol.* **2008**, *45*, 1646–1657. [CrossRef]

193. Pond, L.; Kuhn, L.A.; Teyton, L.; Schutze, M.P.; Tainer, J.A.; Jackson, M.R.; Peterson, P.A. A role for acidic residues in di-leucine motif-based targeting to the endocytic pathway. *J. Biol. Chem.* **1995**, *270*, 19989–19997. [CrossRef]

194. Grimholt, U. MHC and Evolution in Teleosts. *Biology* **2016**, *5*, 6. [CrossRef]

195. Lien, S.; Koop, B.F.; Sandve, S.R.; Miller, J.R.; Kent, M.P.; Nome, T.; Hvidsten, T.R.; Leong, J.S.; Minkley, D.R.; Zimin, A.; et al. The Atlantic salmon genome provides insights into rediploidization. *Nature* **2016**, *533*, 200–205. [CrossRef] [PubMed]

196. Phillips, R.B.; Nichols, K.M.; DeKoning, J.J.; Morasch, M.R.; Keatley, K.A.; Rexroad, C.E., 3rd; Gahr, S.A.; Danzmann, R.G.; Drew, R.E.; Thorgaard, G.H. Assignment of rainbow trout linkage groups to specific chromosomes. *Genetics* **2006**, *174*, 1661–1670. [CrossRef]

197. Palti, Y.; Genet, C.; Luo, M.C.; Charlet, A.; Gao, G.; Hu, Y.; Castaño-Sánchez, C.; Tabet-Canale, K.; Krieg, F.; Yao, J.; et al. A first generation integrated map of the rainbow trout genome. *BMC Genomics* **2011**, *12*, 180. [CrossRef] [PubMed]

198. Fujiwara, A.; Kiryu, I.; Dijkstra, J.M.; Yoshiura, Y.; Nishida-Umehara, C.; Ototake, M. Chromosome mapping of MHC class I in rainbow trout (*Oncorhynchus mykiss*). *Fish Shellfish Immunol.* **2003**, *14*, 171–175. [CrossRef] [PubMed]

199. Phillips, R.B.; Zimmerman, A.; Noakes, M.A.; Palti, Y.; Morasch, M.R.; Eiben, L.; Ristow, S.S.; Thorgaard, G.H.; Hansen, J.D. Physical and genetic mapping of the rainbow trout major histocompatibility regions: evidence for duplication of the class I region. *Immunogenetics* **2003**, *55*, 561–569. [CrossRef] [PubMed]

200. Ozaki, A.; Sakamoto, T.; Khoo, S.; Nakamura, K.; Coimbra, M.R.; Akutsu, T.; Okamoto, N. Quantitative trait loci (QTLs) associated with resistance/susceptibility to infectious pancreatic necrosis virus (IPNV) in rainbow trout (*Oncorhynchus mykiss*). *Mol. Genet. Genomics* **2001**, *265*, 23–31.

201. Ozaki, A.; Khoo, S.; Yoshiura, Y.; Ototake, M.; Sakamoto, T.; Dijkstra, J.M.; Okamoto, N. Identification of Additional Quantitative Trait Loci (QTL) Responsible for Susceptibility to Infectious Pancreatic Necrosis Virus in Rainbow Trout. *Fish Pathol.* **2007**, *42*, 131–140. [CrossRef]

202. Palti, Y.; Vallejo, R.L.; Gao, G.; Liu, S.; Hernandez, A.G.; Rexroad, C.E., 3rd; Wiens, G.D. Detection and Validation of QTL Affecting Bacterial Cold Water Disease Resistance in Rainbow Trout Using Restriction-Site Associated DNA Sequencing. *PLoS ONE* **2015**, *10*, e0138435. [CrossRef]

203. Moen, T.; Baranski, M.; Sonesson, A.K.; Kjøglum, S. Confirmation and fine-mapping of a major QTL for resistance to infectious pancreatic necrosis in Atlantic salmon (*Salmo salar*): Population-level associations between markers and trait. *BMC Genomics* **2009**, *10*, 368. [CrossRef]

204. Gonen, S.; Baranski, M.; Thorland, I.; Norris, A.; Grove, H.; Arnesen, P.; Bakke, H.; Lien, S.; Bishop, S.C.; Houston, R.D. Mapping and validation of a major QTL affecting resistance to pancreas disease (salmonid alphavirus) in Atlantic salmon (*Salmo salar*). *Heredity* **2015**, *115*, 405–414. [CrossRef]

205. Khoo, S.K.; Ozaki, A.; Nakamura, F.; Arakawa, T.; Ichimoto, S.; Nickolov, R.; Sakamoto, T.; Akutsu, T.; Mochizuki, M.; Denda, I.; et al. Identification of a novel chromosomal region associated with infectious hematopoietic necrosis (IHN) resistance in rainbow trout Oncorhynchus mykiss. *Fish Pathol.* **2004**, *39*, 95–101. [CrossRef]

206. Fraslin, C.; Dechamp, N.; Bernard, M.; Krieg, F.; Hervet, C.; Guyomard, R.; Esquerré, D.; Barbieri, J.; Kuchly, C.; Duchaud, E.; et al. Quantitative trait loci for resistance to Flavobacterium psychrophilum in rainbow trout: effect of the mode of infection and evidence of epistatic interactions. *Genet. Sel. Evol.* **2018**, *50*, 60. [CrossRef]

207. Verrier, E.R.; Dorson, M.; Mauger, S.; Torhy, C.; Ciobotaru, C.; Hervet, C.; Dechamp, N.; Genet, C.; Boudinot, P.; Quillet, E. Resistance to a rhabdovirus (VHSV) in rainbow trout: Identification of a major QTL related to innate mechanisms. *PLoS ONE* **2013**, *8*, e55302. [CrossRef]

208. Vallejo, R.L.; Liu, S.; Gao, G.; Fragomeni, B.O.; Hernandez, A.G.; Leeds, T.D.; Parsons, J.E.; Martin, K.E.; Evenhuis, J.P.; Welch, T.J.; et al. Similar Genetic Architecture with Shared and Unique Quantitative Trait Loci for Bacterial Cold Water Disease Resistance in Two Rainbow Trout Breeding Populations. *Front. Genet.* **2017**, *8*, 156. [CrossRef]

209. Liu, S.; Vallejo, R.L.; Palti, Y.; Gao, G.; Marancik, D.P.; Hernandez, A.G.; Wiens, G.D. Identification of single nucleotide polymorphism markers associated with bacterial cold water disease resistance and spleen size in rainbow trout. *Front. Genet.* **2015**, *6*, 298. [CrossRef] [PubMed]

210. Vallejo, R.L.; Palti, Y.; Liu, S.; Evenhuis, J.P.; Gao, G.; Rexroad, C.E., 3rd; Wiens, G.D. Detection of QTL in rainbow trout affecting survival when challenged with Flavobacterium psychrophilum. *Mar. Biotechnol.* **2014**, *16*, 349–360. [CrossRef] [PubMed]

211. Wiens, G.D.; Vallejo, R.L.; Leeds, T.D.; Palti, Y.; Hadidi, S.; Liu, S.; Evenhuis, J.P.; Welch, T.J.; Rexroad, C.E., 3rd. Assessment of genetic correlation between bacterial cold water disease resistance and spleen index in a domesticated population of rainbow trout: identification of QTL on chromosome Omy19. *PLoS ONE* **2013**, *8*, e75749. [CrossRef]

212. Nichols, K.M.; Bartholomew, J.; Thorgaard, G.H. Mapping multiple genetic loci associated with Ceratomyxa shasta resistance in Oncorhynchus mykiss. *Dis. Aquat. Organ.* **2003**, *56*, 145–154. [CrossRef]

213. Baerwald, M.R.; Petersen, J.L.; Hedrick, R.P.; Schisler, G.J.; May, B. A major effect quantitative trait locus for whirling disease resistance identified in rainbow trout (*Oncorhynchus mykiss*). *Heredity* **2011**, *106*, 920–926. [CrossRef] [PubMed]

214. Zimmerman, A.M.; Evenhuis, J.P.; Thorgaard, G.H.; Ristow, S.S. A single major chromosomal region controls natural killer cell-like activity in rainbow trout. *Immunogenetics* **2004**, *55*, 825–835.

215. Houston, R.D.; Haley, C.S.; Hamilton, A.; Guy, D.R.; Mota-Velasco, J.C.; Gheyas, A.A.; Tinch, A.E.; Taggart, J.B.; Bron, J.E.; Starkey, W.G.; et al. The susceptibility of Atlantic salmon fry to freshwater infectious pancreatic necrosis is largely explained by a major QTL. *Heredity* **2010**, *105*, 318–327. [CrossRef]

216. Houston, R.D.; Haley, C.S.; Hamilton, A.; Guy, D.R.; Tinch, A.E.; Taggart, J.B.; McAndrew, B.J.; Bishop, S.C. Major quantitative trait loci affect resistance to infectious pancreatic necrosis in Atlantic salmon (*Salmo salar*). *Genetics* **2008**, *178*, 1109–1115. [CrossRef] [PubMed]

217. Moen, T.; Fjalestad, K.T.; Munck, H.; Gomez-Raya, L. A multistage testing strategy for detection of quantitative trait Loci affecting disease resistance in Atlantic salmon. *Genetics.* **2004**, *167*, 851–858. [CrossRef]

218. Gilbey, J.; Verspoor, E.; Mo, T.A.; Sterud, E.; Olstad, K.; Hytterød, S.; Jones, C.; Noble, L. Identification of genetic markers associated with Gyrodactylus salaris resistance in Atlantic salmon Salmo salar. *Dis. Aquat. Organ.* **2006**, *25*, 119–129. [CrossRef] [PubMed]

219. Sawayama, E.; Tanizawa, S.; Kitamura, S.I.; Nakayama, K.; Ohta, K.; Ozaki, A.; Takagi, M. Identification of Quantitative Trait Loci for Resistance to RSIVD in Red Sea Bream (*Pagrus major*). *Mar. Biotechnol.* **2017**, *19*, 601–613. [CrossRef]

220. Miller, K.M.; Winton, J.R.; Schulze, A.D.; Purcell, M.K.; Ming, T.J. Major histocompatibility complex loci are associated with susceptibility of Atlantic salmon to infectious hematopoietic necrosis virus. *Environ. Biol. Fishes* **2004**, *69*, 307–316. [CrossRef]

221. Johnson, N.A.; Vallejo, R.L.; Silverstein, J.T.; Welch, T.J.; Wiens, G.D.; Hallerman, E.M.; Palti, Y. Suggestive association of major histocompatibility IB genetic markers with resistance to bacterial cold water disease in rainbow trout (*Oncorhynchus mykiss*). *Mar. Biotechnol.* **2008**, *10*, 429–437. [CrossRef]

222. Fischer, U.; Utke, K.; Ototake, M.; Dijkstra, J.M.; Kollner, B. Adaptive cell-mediated cytotoxicity against allogeneic targets by CD8-positive lymphocytes of rainbow trout (*Oncorhynchus mykiss*). *Dev. Comp. Immunol.* **2003**, *27*, 323–337. [CrossRef]

223. Pearse, A.M.; Swift, K. Allograft theory: Transmission of devil facial-tumour disease. *Nature* **2006**, *439*, 549. [CrossRef]

224. Murgia, C.; Pritchard, J.K.; Kim, S.Y.; Fassati, A.; Weiss, R.A. Clonal origin and evolution of a transmissible cancer. *Cell* **2006**, *126*, 477–487. [CrossRef] [PubMed]

225. Metzger, M.J.; Villalba, A.; Carballal, M.J.; Iglesias, D.; Sherry, J.; Reinisch, C.; Muttray, A.F.; Baldwin, S.A.; Goff, S.P. Widespread transmission of independent cancer lineages within multiple bivalve species. *Nature* **2016**, *534*, 705–709. [CrossRef]

226. Qin, Q.W.; Ototake, M.; Nagoya, H.; Nakanishi, T. Graft-versus-host reaction (GVHR) in clonal amago salmon, *Oncorhynchus rhodurus*. *Vet. Immunol. Immunopathol.* **2002**, *89*, 83–89. [CrossRef]

227. Griesemer, A.; Yamada, K.; Sykes, M. Xenotransplantation: Immunological hurdles and progress toward tolerance. *Immunol. Rev.* **2014**, *258*, 241–258. [CrossRef]

228. Singh, P.B. Chemosensation and genetic individuality. *Reproduction* **2001**, *121*, 529–539. [CrossRef]

229. Bernatchez, L.; Landry, C. MHC studies in nonmodel vertebrates: What have we learned about natural selection in 15 years? *J. Evol. Biol.* **2003**, *16*, 363–377. [CrossRef] [PubMed]

230. Overath, P.; Sturm, T.; Rammensee, H.G. Of volatiles and peptides: In search for MHC-dependent olfactory signals in social communication. *Cell Mol. Life Sci.* **2014**, *71*, 2429–2442. [CrossRef]

231. Lobmaier, J.S.; Fischbacher, U.; Probst, F.; Wirthmüller, U.; Knoch, D. Accumulating evidence suggests that men do not find body odours of human leucocyte antigen-dissimilar women more attractive. *Proc. Biol. Sci.* **2018**, *285*, 20180566. [CrossRef]

232. Reusch, T.B.; Haberli, M.A.; Aeschlimann, P.B.; Milinski, M. Female sticklebacks count alleles in a strategy of sexual selection explaining MHC polymorphism. *Nature* **2001**, *414*, 300–302. [CrossRef]

233. Bay, R.A.; Arnegard, M.E.; Conte, G.L.; Best, J.; Bedford, N.L.; McCann, S.R.; Dubin, M.E.; Chan, Y.F.; Jones, F.C.; Kingsley, D.M.; et al. Genetic Coupling of Female Mate Choice with Polygenic Ecological Divergence Facilitates Stickleback Speciation. *Curr. Biol.* **2017**, *27*, 3344–3349. [CrossRef]

234. Glazer, A.M.; Killingbeck, E.E.; Mitros, T.; Rokhsar, D.S.; Miller, C.T. Genome Assembly Improvement and Mapping Convergently Evolved Skeletal Traits in Sticklebacks with Genotyping-by-Sequencing. *G3* **2015**, *5*, 1463–1472. [CrossRef]

235. Corriveau, R.A.; Huh, G.S.; Shatz, C.J. Regulation of class I MHC expression in the developing and mature CNS by Neural activity. *Neuron* **1998**, *21*, 505–520. [CrossRef]

236. Huh, G.S.; Boulanger, L.M.; Du, H.; Riquelme, P.A.; Brotz, T.M.; Shatz, C.J. Functional requirement for class I MHC in CNS development and plasticity. *Science* **2000**, *290*, 2155–2159. [CrossRef]

237. Loconto, J.; Papes, F.; Chang, E.; Stowers, L.; Jones, E.P.; Takada, T.; Kumanovics, A.; Fischer Lindahl, K.; Dulac, C. Functional expression of murine V2R pheromone receptors involves selective association with the M10 and M1 families of MHC class Ib molecules. *Cell* **2003**, *112*, 607–618. [CrossRef]

238. Oliveira, A.L.; Thams, S.; Lidman, O.; Piehl, F.; Hokfelt, T.; Karre, K.; Linda, H.; Cullheim, S. A role for MHC class I molecules in synaptic plasticity and regeneration of neurons after axotomy. *Proc. Natl. Acad. Sci. USA* **2004**, *101*, 17843–17848. [CrossRef] [PubMed]

239. Lee, H.; Brott, B.K.; Kirkby, L.A.; Adelson, J.D.; Cheng, S.; Feller, M.B.; Datwani, A.; Shatz, C.J. Synapse elimination and learning rules co-regulated by MHC class I H2-Db. *Nature* **2014**, *509*, 195–200. [CrossRef] [PubMed]

240. Adelson, J.D.; Sapp, R.W.; Brott, B.K.; Lee, H.; Miyamichi, K.; Luo, L.; Cheng, S.; Djurisic, M.; Shatz, C.J. Developmental Sculpting of Intracortical Circuits by MHC Class I H2-Db and H2-Kb. *Cereb. Cortex* **2016**, *26*, 1453–1463. [CrossRef] [PubMed]

241. Shatz, C.J. MHC class I: An unexpected role in neuronal plasticity. *Neuron* **2009**, *64*, 40–45. [CrossRef]

242. Elmer, B.M.; McAllister, A.K. Major histocompatibility complex class I proteins in brain development and plasticity. *Trends Neurosci.* **2012**, *35*, 660–670. [CrossRef]

243. Misra, M.K.; Damotte, V.; Hollenbach, J.A. The Immunogenetics of neurological disease. *Immunology* **2018**, *153*, 399–414. [CrossRef]

244. Benus, R.F.; Bohus, B.; Koolhaas, J.M.; van Oortmerssen, G.A. Heritable variation for aggression as a reflection of individual coping strategies. *Experientia* **1991**, *47*, 1008–1019. [CrossRef] [PubMed]

245. Sokolowski, M.B.; Pereira, H.S.; Hughes, K. Evolution of foraging behavior in Drosophila by density-dependent selection. *Proc. Natl. Acad. Sci. USA* **1997**, *94*, 7373–7377. [CrossRef]

246. Campbell, J.M.; Carter, P.A.; Wheeler, P.A.; Thorgaard, G.H. Aggressive behavior.; brain size and domestication in clonal rainbow trout lines. *Behav. Genet.* **2015**, *45*, 245–254. [CrossRef]

247. Azuma, T.; Dijkstra, J.M.; Kiryu, I.; Sekiguchi, T.; Terada, Y.; Asahina, K.; Fischer, U.; Ototake, M. Growth and behavioral traits in Donaldson rainbow trout (*Oncorhynchus mykiss*) cosegregate with classical major histocompatibility complex (MHC) class I genotype. *Behav. Genet.* **2005**, *35*, 463–478. [CrossRef]

248. Wright, D.; Nakamichi, R.; Krause, J.; Butlin, R.K. QTL analysis of behavioral and morphological differentiation between wild and laboratory zebrafish (*Danio rerio*). *Behav. Genet.* **2006**, *36*, 271–284. [CrossRef]

249. Liu, S.; Vallejo, R.L.; Gao, G.; Palti, Y.; Weber, G.M.; Hernandez, A.; Rexroad, C.E., 3rd. Identification of single-nucleotide polymorphism markers associated with cortisol response to crowding in rainbow trout. *Mar. Biotechnol.* **2015**, *17*, 328–337. [CrossRef]

250. Cnaani, A.; Zilberman, N.; Tinman, S.; Hulata, G.; Ron, M. Genome-scan analysis for quantitative trait loci in an F2 tilapia hybrid. *Mol. Genet. Genomics* **2004**, *272*, 162–172. [CrossRef]

251. Phillips, R.B.; Ventura, A.B.; Dekoning, J.J.; Nichols, K.M. Mapping rainbow trout immune genes involved in inflammation reveals conserved blocks of immune genes in teleosts. *Anim. Genet.* **2013**, *44*, 107–113. [CrossRef]

 *cells*

Review

An RNA Metabolism and Surveillance Quartet in the Major Histocompatibility Complex

Danlei Zhou [1,2,3,*], **Michalea Lai** [2,3], **Aiqin Luo** [1,*] **and Chack-Yung Yu** [2,3,*]

[1] School of Life Science, Beijing Institute of Technology, Beijing 100081, China
[2] The Abigail Wexner Research Institute at Nationwide Children's Hospital, Columbus, OH 43205, USA
[3] Department of Pediatrics, The Ohio State University, Columbus, OH 43205, USA
* Correspondence: Danlei.Zhou@nationwidechildrens.org (D.Z.); Bitluo@bit.edu.cn (A.L.);
   Chack-Yung.Yu@nationwidechildrens.org (C.-Y.Y.)

Received: 19 July 2019; Accepted: 29 August 2019; Published: 30 August 2019

Abstract: At the central region of the mammalian major histocompatibility complex (MHC) is a complement gene cluster that codes for constituents of complement C3 convertases (C2, factor B and C4). Complement activation drives the humoral effector functions for immune response. Sandwiched between the genes for serine proteinase factor B and anchor protein C4 are four less known but critically important genes coding for essential functions related to metabolism and surveillance of RNA during the transcriptional and translational processes of gene expression. These four genes are NELF-E (RD), SKIV2L (SKI2W), DXO (DOM3Z) and STK19 (RP1 or G11) and dubbed as NSDK. NELF-E is the subunit E of negative elongation factor responsible for promoter proximal pause of transcription. SKIV2L is the RNA helicase for cytoplasmic exosomes responsible for degradation of de-polyadenylated mRNA and viral RNA. DXO is a powerful enzyme with pyro-phosphohydrolase activity towards 5′ triphosphorylated RNA, decapping and exoribonuclease activities of faulty nuclear RNA molecules. STK19 is a nuclear kinase that phosphorylates RNA-binding proteins during transcription. STK19 is also involved in DNA repair during active transcription and in nuclear signal transduction. The genetic, biochemical and functional properties for NSDK in the MHC largely stay as a secret for many immunologists. Here we briefly review the roles of (a) NELF-E on transcriptional pausing; (b) SKIV2L on turnover of deadenylated or expired RNA 3′→5′ through the Ski-exosome complex, and modulation of inflammatory response initiated by retinoic acid-inducible gene 1-like receptor (RLR) sensing of viral infections; (c) DXO on quality control of RNA integrity through recognition of 5′ caps and destruction of faulty adducts in 5′→3′ fashion; and (d) STK19 on nuclear protein phosphorylations. There is compelling evidence that a dysregulation or a deficiency of a NSDK gene would cause a malignant, immunologic or digestive disease.

Keywords: DXO; DOM3Z; NELF-E; RD; SKIV2L; SKI2W; STK19; RP1; NSDK; RLR; miR1236; SVA; RNA quality control; 5′→3′ RNA decay; 3′→5′ mRNA turnover; antiviral immunity; interferon β; promoter-proximal transcriptional pause; exosomes; nuclear kinase; hepatocellular carcinoma; Ski complex; trichohepatoenteric syndrome; melanoma

---

1. Introduction

In 1984, Professor Rodney Porter's team in Oxford reported the cloning and confirmed the physical linkage of the genes for human complement C4A and C4B, factor B and C2 through overlapping cosmids [1]. These genes are located at the central region of the human major histocompatibility complex (MHC), a region that is associated with numerous complex autoimmune diseases [2–7]. They code for the essential constituents of the classical, lectin and alternative pathways C3 convertases. The complement system is a major effector arm for both innate and adaptive immune systems [8]. While C2 and factor B are only separated by an intergenic distance of 421 bp [9], it was a bit puzzling that the genes for C4 and factor B are separated by a large gap of 30 kb. Thus, it was satisfying when a genomic DNA probe upstream of the C4 promoter region successfully fished out an 1.1 kb cDNA clone that was dubbed as RP (to memorialize the late Professor Rodney Porter) [10,11]. Working from the other end, Meo's team reported the cloning of RD that is up to 149 bp tail-to-tail with factor B [12]. The protein sequence of RD is fascinating because of its 23 consecutive copies of positively charged arginine (R) and negatively charged aspartic acid (D), not to mention its leucine zipper and RNA binding motif. Between RD and RP, we found two more genes, SKI2W and DOM3Z [13–15].

Through biochemical and molecular biologic research to decipher their functions, these four "novel" genes turn out to be extremely old in evolution. Except for RP, structurally and functionally related genes and proteins can be traced all the way back to yeast or metazoans. The modern names for these four genes are NELF-E, SKIV2L, DXO and STK19 (NSDK), respectively. Unlike the Class I and Class II and complement that have distinct immunologic functions involving proteins, NSDK seems to play fundamental life functions in eukaryotes and actively engaged in the surveillance of RNA integrity including the 5′ cap and 3′ tail. In the process of cell metabolism, these four genes play critical roles on transcriptional control, RNA turnover and signaling. They may also fine-tune the innate immune response or type I interferon-mediated inflammation elicited by infections of RNA viruses. Here we give a brief account on the molecular genetics and biology of this RNA metabolism and surveillance "quartet" in the MHC (Figure 1, Table 1).

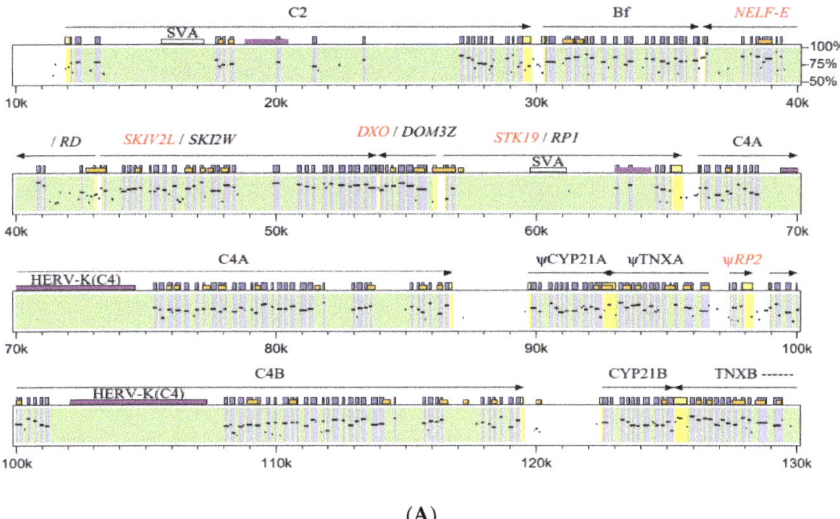

(A)

Figure 1. *Cont.*

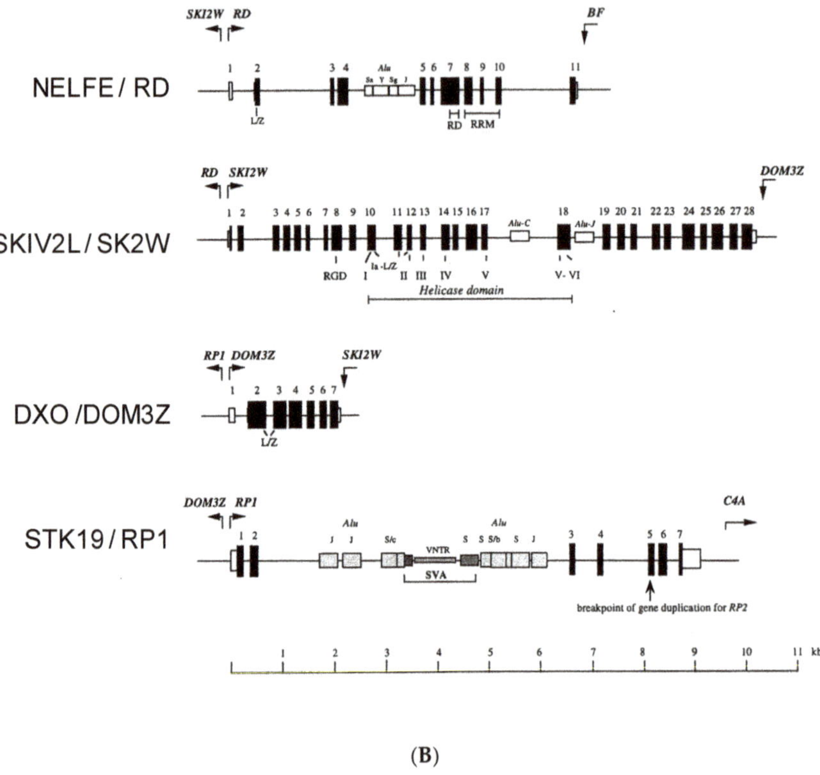

(B)

Figure 1. Map and gene structures of the NSDK (NELF-E, SKIV2L, DOX and STK19) quartet in the major histocompatibility complex (MHC) complement gene cluster. (**A**) A molecular map comparing the gene organization of human and mouse MHC complement gene cluster with RNA surveillance quartet consisting of NELF-E, SKIV2L, DXO and STK19 at the intergenic region between complement factor B and complement C4A gene. Original gene names are shown as alternatives. The mouse sequences with identities from 50–100% are plotted below the human sequences. Exons for human genes are shown as solid, purple boxes. The genomic region spanned for each gene is shaded green. The 5′ and 3′ untranslated sequences are in yellow boxes. The conserved coding sequences between human and mouse genes are highlighted in purple. Locations of CpG rich dinucleotides are shown as orange boxes. Large, repetitive retroelements SVA and endogenous retrovirus HERV-K(C4) are shown in burgundy. Arrows with solid lines represent configurations of structural genes. *RP2* and *TNXA* are partially duplicated gene fragments. *CYP21A* is a pseudogene in human. Pseudogenes and gene fragments are labeled with ψ in front of gene names. Genomic DNA sequences are obtained from the following accession numbers: Human, U89335-U89337, AF019413, M59815, M59816, L26260-L26263, U07856, AF059675 and AF077974; mouse, AF030001, AF049850 and AF109906. Numberings below box represents length in human genomic DNA in kb (2k = 2 kb). The 5′ region of complement C2 gene is not well defined. Located approx. 26.4 kb upstream of the gene for C2 isoform 5 and in opposite orientation is the zinc finger and BTB domain containing protein 12 (ZBTB12). Present at the 3′ end of the C2 gene and in opposite orientation is a genetic element for long non-coding RNA (lncRNA C2-AS1, not shown). (**B**) Exon-intron structures of human NELF-E, SKIV2L, DXO and STK19 (modified from refs [15,16]).

Table 1. Characteristics of the NSDK genes for RNA metabolism and surveillance quartet in the class III region of the human major histocompatibility complex.

Official Symbol	Original Name; Other Earlier Names	Gene, Transcripts and Proteins (Location, cDNA, aa)	Expression	Specific Features	Function	Website
NELF-E	**RD** RDP; RDBP; D6S45	11 exons 6.7 kb 380 aa	Ubiquitous; highest in testis	CpG rich, leucine zipper motif, RD repeats, RNA recognition motif (RRM); miR-1236 in intron 3	A subunit of the negative RNA elongation factor that represses transcriptional elongation by RNA polymerase II	https://www.ncbi.nlm.nih.gov/gene/7936
SKIV2L	**SKI2W** HLP, SKI2, 170A, DDX13, DDX49, SKIV2, SKIV2L1, THES2	28 exons 11 kb 1246 aa	Ubiquitous; highest in spleen	CpG rich, leucine zipper motif, helicase domain, RGD and Alu element	Unwinds RNA secondary structures, RNA turnover, antiviral defense, modulates type 1 interferon expression	https://www.ncbi.nlm.nih.gov/gene/6499
DXO	**DOM3Z** DOM3L, NG6, RAI1	7 exons 2.2 kb 396 aa	Ubiquitous; highest in testis and adrenal gland	CpG rich; leucine zipper motif	RNA quality control, decapping and 5′→3′ RNA decay	https://www.ncbi.nlm.nih.gov/gene/1797
STK19	**RP1** and **RP2** HLA-RP1, G11, D6S60, D6S60E	9 exons 11 kb 364 aa or 7 exons 9.1 kb 254 aa	Ubiquitous; highest in adrenal gland	Partial gene duplication in the RCCX modules; SVA (SINE - 21 CpG rich VNTRs-Alu) in intron; 8 Alu elements	Nuclear Serine/Threonine kinase	https://www.ncbi.nlm.nih.gov/gene/8859

2. NELF-E or RD

To Start a Long Journey, Get Prepared, Do a Test Run, Pause to Check Things Out, Modify and Let Go:
The Amazing RD or Negative Elongation Factor Subunit E (NELF-E) in Gene Transcription

The human RD gene, also known as NELF-E and D6S45, is located between complement factor
B (Bf) and the SKIV2L gene. This gene was discovered by Meo and colleagues in 1988 [12]. RD is
organized in a tail-to-tail configuration with Bf (Figure 1). There are 205 nucleotides between the 3′ ends
of RD and Bf. The 5′ regulatory region of the RD gene is rich in CpG sequences [17]. Transcripts for
RD are ubiquitously expressed but the highest expression level is found in the testis.

The RD gene is 6.7 kb in size and consists of 11 exons. It encodes 380 amino acids. The calculated
molecular weight of the RD protein is 46 kDa [17,18]. The RD protein contains a leucine-zipper motif
at the N terminal region, which facilitates interactions with other proteins [12,17,19]. A striking feature
of RD, however, is a tract of tandem repeats with positively and negatively charged amino acids,
which are predominantly Arg-Asp (RD), from residues 184 to 243, which are encoded by exon 7. Close
to the carboxyl terminal region is an RNA recognition motif (RRM), which is located between residues
264 and 327 and encoded by exons 8–10. Thus, RD belongs to ribonucleoprotein family. The high
degree of sequence similarities between human and murine RD both at DNA and amino acid levels
suggested functional importance [12,15,17,20].

One feature for the regulation of gene expression in metazoans is the promoter-proximal
transcriptional pausing of RNA polymerase II. This phenomenon is carried out by two complexes known
as the negative elongation factor (NELF) and DRB (5,6-dichloro-1-β-ᴅ-ribofuranosyl-benzimidazole)
sensitivity inducing factor (DSIF). NELF acts cooperatively with DSIF to strongly represses
transcriptional elongation by RNA polymerase II (RNAPII), about 25–50 nucleotides downstream
of transcriptional initiation, particularly in the absence of P-TEFb (positive transcription elongation
factor b) [21]. Promoter proximal pausing of transcription is a universal process for RNA polymerase
II transcripts that allows modifications of nascent transcripts with 5′ m^7G caps, proper transcript
processing and elongation checkpoint control. In an active transcription process, the cyclin dependent
kinase 9 (Cdk9) in P-TEFb may phosphorylate the carboxyl domain of the RNAPII, DSIF and NFLF-E at
sites next to its RNA recognition motif, and release the paused transcripts for processive elongation [22]
(Figure 2).

NELF consists of four subunits, A, B, C or D, and E. The C and D subunits of NELF are alternatively
spliced products of the same gene [23]. Remarkably, the RD protein is identified as the E-subunit
or the smallest subunit of the NELF complex [21]. NELF-E is important for NELF function because
it has a functional RNA recognition and binding (RRM) domain. Mutations of the RRM domain in
NELF-E impair transcription repression without affecting protein-protein interactions with other NELF
subunits [23].

The biochemical activities for various NELF subunits were determined mainly by Hiroshi Handa's
team in Yokohama, Japan. They reconstituted NELF-like complex with epitope-tagged NELF-A,
NELF-B, NELF-D, and NELF-E expressed in insect Sf9 cells and investigated their interactions [24].
It was shown that NELF binds to DSIF-RNA Polymerase II (RNAPII) complex [21,23]. The leucine
zipper of NELF-E interacts directly with NELF-B but not with NELF-A or NELF-D. NELF-D interacts
directly with NELF-A and NELF-B. However, NELF-B does not interact directly with NELF-A.
Consistent with the model that NELF-B and NELF-D (or NELF-C) that bring NELF-A and NELF-E
together via 3 protein-protein interactions, and NELF-A binds to RNAPII directly. The NELF complex
needs interactions of 4 subunits, A–(D or C)–B–E. NELF-C and NELF-D are present in distinct NELF
complexes [24].

In the nucleus, NELF interacts with the nuclear cap-binding complex (CBC) and participates in
the 3′ end processing of mRNA. Both cap-binding proteins CBP-80 and CBP-20 of CBC bind to NELF-E.
The region from amino acids 244 to 380 of NELF-E containing an RNA-recognition motif is necessary
and sufficient for the interaction between NELF and CBC-RNA. Knockdowns of NELF-E and CBP80 in

HeLa cells rendered the free NELF subunits and CBP20 unstable, which are subjected to degradation rapidly, and abolished functions of the NELF holocomplex and the CBC [25].

On the transcription and processing of small nuclear RNA (snRNAs) that are essential for the assembly and functions of spliceosomes, and small nucleolar RNAs (snoRNAs) that are integral to the biosynthesis of ribosomes by RNAPII, NELF-E and ARS2 (arsenite resistance protein 2) play their respective roles progressively through binding to the cap-binding complex CBC. The binding of NELF-E allows promoter proximal pausing, and the subsequent replacement of NELF-E by ARS2 to the same binding site on CBC plus concurrent binding by PHAX (phosphorylated adaptor for RNA export) enable a continuous process to produce these distinct groups of RNA and prepare them for further processing in the cytoplasm or the nucleus [26].

The mRNAs for histones are special as they may not contain polyadenylated 3' ends. The RRM of NELF-E may bind to the stem-loop structure in the 3' end of the "replication-dependent" histone mRNAs and enhance their stability. Notably, NELF appears to physically associate with histone gene loci and forms the so-called "NELF-bodies", which are probably engaged in the processing of histone mRNA. Knockdowns of NELF-E and CBP-80 in HeLa cells led to increased accumulation of the aberrant, polyadenylated form of histone mRNAs [25].

NELF-E was studied on the regulation of gene expression for human immunodeficiency virus HIV. Recombinant protein with the RNA-recognition motif (RRM) of NELF-E was shown to bind the stem of viral RNA with TAR (transactivation response element) [27]. Such binding (between NELF-E and TAR) inhibits the human immunodeficiency virus type 1 basal transcription from the long terminal repeat (HIV-LTR) [28]. Cyclin-dependent kinase 9 (Cdk9) in P-TEFb can phosphorylate NELF-E at sites next to its RNA recognition motif, and thereby making NELF-E no longer able to bind to TAR nor to repress HIV transcription [22].

Poly-(ADP-ribose) polymerase 1 (PARP1) can covalently transfer of ADP-ribose from NAD^+ onto substrate proteins, consequently modulates diverse biological processes of ribosylated proteins. In a human cell line, Gibson et al. reported that PARP1 interacted with NELF and mediated ADP ribosylation of NELF-E [29,30]. In a manner similar to Cdk9 of P-TEFb-mediated phosphorylation, ADP-ribosylation near the RRM of NELF-E ablates its ability to bind RNA, and therefore releases RNAPII from NELF-dependent pausing. ADP-ribosylation of NELF-E is dependent on phosphorylation by Cdk9/P-TEFb. PARP1-dependent ADP-ribosylation of NELF-E reinforces P-TEFb-mediated RNAPII pause release. Depletion or inhibition of PARP1, or mutagenesis of the four negatively charged glutamate residues in NELF-E (E122, E151, E172, and E374) to glutamines (Q), resulted in a substantial reduction in NELF-E modification by PARP-1. Thus, it was concluded that PARP1-dependent ADP ribosylation of phosphorylated NELF-E (and NELF-A) is necessary for efficient release of RNA polymerase II into productive elongation [29,30].

In a study of double-stranded (ds) DNA break and DNA repair, NELF-E is found preferentially recruited to nearby transcriptionally active genes at sites of DNA ds-breaks to underpin transcriptional repression in a PARP-1 dependent manner. The transcriptional pause allows the machinery for DNA repair to carry out its function and avoid a potential collision of the two systems (DNA repair and transcription) at the damage site. A slight decrease in ribosylation of NELF-E (by PARP-1) at the site of DNA breaks also ensures the pause of transcription imposed by NELF [31].

In a different study of UV-light induced DNA damage (but not ds-breaks) and changes of RNA metabolism, it was found that phosphorylation of NELF-E at serine 115 by MK2 (MAPK activated protein kinase II) promotes the recruitment of 14-3-3 dimers and rapid dissociation of the NELF complex from chromatin. This process is accompanied by elongation of transcription through RNA polymerase II [32].

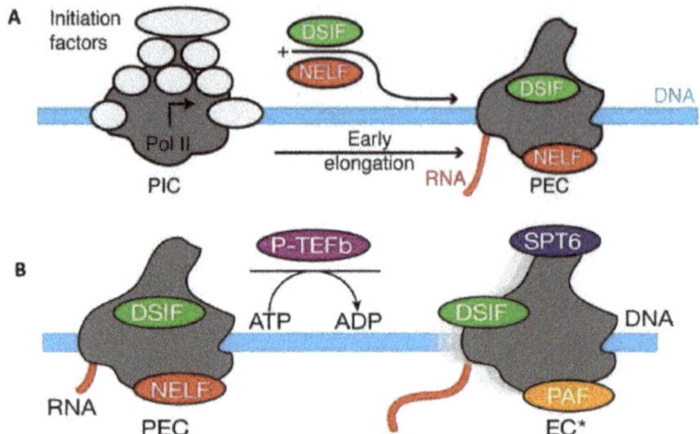

Figure 2. Pause and go of RNA polymerase II during transcription. (**A**) Schematic of conversion of the RNAPII pre-initiation complex (PIC) to a promoter proximally paused RNAPII-DSIF-NELF elongation complex (PEC) [33]. (**B**) Schematic showing conversion of the paused PEC to the activated RNA Pol II-DSIF-PAF-SPT6 elongation complex (EC*). SPT6, homolog of yeast *suppressor of Ty6* (adapted from references [33,34], with permission from Dr. Patrick Cramer, Max Planck Institute for Biophysical Chemistry, Germany).

A cryo-electron microscopy structure with paused transcription complex consisting of RNAPII purified from porcine, human DSIF expressed in bacteria, and human NELF complex expressed in insect cells has been presented with 3.2 Å resolution (Figure 2A). This paused RNAPII–DSIF–NELF elongation complex structure contains a tilted DNA–RNA hybrid that is incompatible with binding of nucleoside triphosphate substrates. The NELF complex binds to the funnel structure of the RNAPII and restricts its mobility. During this paused state, through its two tentacle-like structures, NELF has flexible contacts with the DSIF and the exiting RNA. NELF also prevents the binding of the anti-pausing transcription elongation factor IIS [33].

The same research team continued to show cryo-EM structures of the elongation complex with an active transcription process. To re-activate the paused complex, P-TEFb phosphorylates RNAPII, DSIF and NELF. NELF is displaced by Pol II associated factor PAF. Homolog of suppressor for transposon Ty insertion 6 (SPT6) binds to RNAPII and then set off the transcriptional elongation process (Figure 2B) [34].

Considering the fundamental relevance of NELF-E on the cellular process such as gene expression, dysregulated production or aberrant function of NELF-E could have detrimental effects on life processes. Reports on disease associations of NELF-E are relatively scarce but there are recent articles related to NELF-E in carcinogenesis. NELF-E was preferentially overexpressed in hepatocellular carcinoma (HCC), especially in HCC with portal vein invasion (PVI) [35]. NELF-E via somatic copy number alterations appeared to enhance oncogene MYC signaling and activation, which promoted HCC progression [36]. NELF-E protein levels could be an independent risk factor for early intra-hepatic recurrence of HCC within two years of surgery [35].

A Gene Within a Gene: miR-1236 in Intron 3 of NELF-E

While overexpression of NELF-E may be associated with carcinogenesis, a genetic element embedded in the NELF-E gene serves as a tumor suppressor. Between nucleotides 2104 and 2205 of the human NELF-E gene is the precursor for a micro-RNA, miR-1236. This is a splice-dependent mirtronic miRNA that co-expresses with NELF-E [37,38]. miR-1236 is transcribed by RNA polymerase II. The promoter region of miR-1236 is the same as the NELF-E promoter.

miR-1236 can inhibit tumors migration, proliferation and invasion activity by binding to target mRNAs and down-regulating their protein synthesis. It can also induce target gene expression by binding to gene promoter in the nucleus. Binding of miR-1236 to target mRNAs can lead to abolition of hypoxia-induced epithelial–mesenchymal transition (EMT) and inhibition of migration and invasion of tumor cells. miR-1236-3p can suppress ovarian cancer metastasis by binding to the 3′ untranslated region of zinc-finger E-box binding homeobox 1 (ZEB1) mRNA, and downregulate ZEB1 gene expression [39]. ZEB1 is an EMT inducers. miR-1236 can inhibit the invasion and metastasis of gastric cancer possibly through decreasing phosphorylation by Ak transforming (AKT) kinase and binding to metastasis-associated protein MTA2 [40]. miR-1236 may bind to the promoter of p21 and induces nuclear p21 expression. p21 is a cyclin dependent kinase inhibitor and a regulator of cell cycle progression. miR-1236 and p21 expression are significantly decreased in renal cell carcinoma. Targeted activation of p21 by miR-1236 can inhibit cell proliferation, may have therapeutic potential in the treatment of renal cell carcinoma [41].

3. SKIV2L or SKI2W

Quality Controls I—Clearing Broken, Bad, Used and Viral RNA Products in the Cell Requires An Engine to Unwind Helical structures of RNA for Degradation—The Helicase for Cytoplasmic Exosomes SKIV2L (Ski2w or DDX13)

The Ski complex is a heterotetrameric complex consisting of Ski2, Ski3 and Ski8 in 1:1:2 stoichiometry that is associated with the cytoplasmic RNA exosome for 3′→5′ mRNA decay [42]. Early investigations revealed that yeast Ski2p is engaged in the inhibition of translation from polyadenylation-negative RNA, and has antiviral activities as yeast viruses are RNA structures without poly-A tails [43]. Mutations in these Ski genes in yeasts gave rise to the super-killing phenotype with viral infections [44,45].

Exosomes are highly conserved complex structures responsible for mRNA turnover and therefore fundamental for regulation of gene expression and cell differentiation in eukaryotes. Each exosome consists of 9 or 10 structural proteins that forms a channel structure by which mRNA molecules thread through and get degraded 3′→5′ by an exoribonuclease [46]. The Ski complex serves to unwind the secondary or helical structures of single stranded mRNA or double stranded RNA from some viruses. The Ski3 and Ski8 proteins form the scaffold structure for the Ski complex by which the Ski2 RNA helicase performs its unwinding activity to prepare RNA to thread through the exosome for degradation in the cytoplasm [47,48]. In the nucleus, a similar exosome structure exists but the corresponding RNA helicase is the Mtr4 or KIAA0052 [49,50].

The functional human SKI2W gene in the HLA was discovered in 1995 [13]. It also documented the existence of a SKI2W-related gene, KIAA0052, in the human genome. Because of the striking sequence similarities with the yeast antiviral gene SKI2, SKI2W was renamed to SKIV2L, and KIAA0052 (MTR4) was renamed to SKIV2L2. The human SKIV2L gene is located between RD (NELF-E) and DOM3Z (DXO) at the MHC class III region. It spans 11 kb and consists of 28 exons that codes for a 4 kb transcript. SKIV2L transcripts are ubiquitously expressed in various tissues, especially in the spleen, lymph nodes, duodenum, appendix and gall bladder. The intergenic region between NELF-E and SKIV2L is close to minimum and it was shown that a 176 bp DNA fragment from this region had bidirectional promoter activity for both NELF-E and SKIV2L [51].

The Skiv2l protein has a molecular weight ~140 kDa that consists of 1,246 amino acids [13,17,52]. The Skiv2l protein has structural motifs characteristic of an RNA helicase, which include motifs for binding and hydrolysis of nucleotide triphosphates such as ATP, RNA binding and a DEVH-box [13,53]. Thus, another name for Skiv2l is DDX13 [54]. Skiv2l uses ATP as an energy source to unwind helical or secondary structures of RNA molecules. In addition, Skiv2l contains a leucine zipper motif for protein-protein interaction [13,54], an RGD motif that may be a ligand for adhesion molecules, and an acidic motif with KDLL sequence for endoplasmic reticulum targeting (possibly for unfold protein response). The Skiv2l protein also consists of four potential N-linked glycosylation sites and two potential sulfation sites [13,17].

Fusion proteins with human Ski2w immobilized on polyvinylidene fluoride (PVDF) membrane were able to hydrolyze γ-^{32}P-labeled ATP, which would be an energy source for helicase activities [13].

Immunocytochemistry experiments using specific rabbit antiserum showed that Skiv2l proteins in HeLa cells are localized to the nucleolus and the cytoplasm, a phenomenon that was independently confirmed in transfectants expressing tagged-Skiv2l protein (Figure 3) [52]. Co-sedimentation experiments for ribosome profile through sucrose gradient centrifugation and fractionations revealed that Ski2w is associated with the 40S subunit of ribosomes [52].

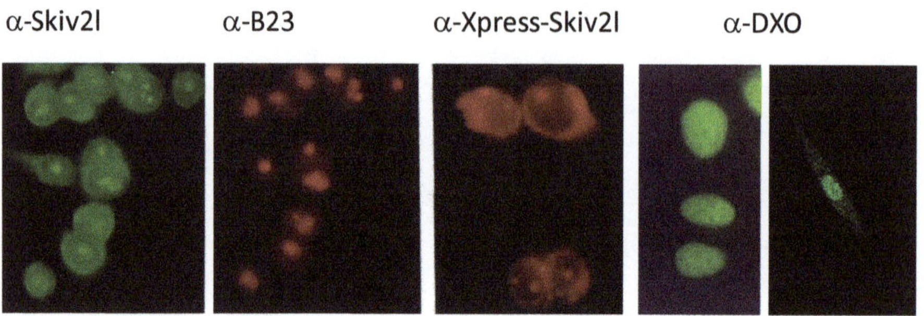

Figure 3. Cellular localization of Skiv2l and DXO in HeLa cells under immunofluorescent microscopy. Human HeLa cells were stained by antibodies against human Skiv2l (Ski2w), nucleoli-specific B23 (nucleophosmin), anti-Xpress for HeLa transfectants with Xpress-Skiv2l fusion protein, and antibodies against DXO (Dom3z). The first three panels illustrate the presence of Skiv2l in the nucleoli and cytoplasm. The fourth and fifth panels show the dominant location of DXO (Dom3z) in the nuclei of HeLa cells. Photographs were originally taken from UV-microscope with 10×40 magnifications (modified from [15,52,55]).

The roles of human Skiv2l on the fine regulation of antiviral defense and RNA turnover was examined recently [55]. Discrimination of foreign RNA such as viral RNA with 5′ triphosphate instead of a m^{7}G cap for host mRNA, or double-stranded RNA that seldom exists in a eukaryotic cell, are achieved by RNA helicases such as retinoic acid activating gene 1 (RIG1) or melanoma differentiation-associated protein 5 (MDA5) [56,57]. Stimulation with RIG1/MDA5 ligands [such as thapsigargin or poly(I:C)] on primary macrophages with *knockdown* of Skiv2l, but not Xrn1 or Ski3, led to substantial increase of transcription for interferon-β. Thus, Skiv2l specifically suppressed or regulated the production of type I interferon from macrophages in response to ligands that mimic viral infections. This is because Xrn1 knockdown or Ski3 (TTC37) knockdown in similar experiments had no effects on IFN-β expression [56]. Skiv2l probably fine-tunes the antiviral response by RIG1/MDA5 to prevent the over inflammation that can cause tissue injuries.

During a viral infection or an inflammation, the host cells are under stress and endoplasmic reticula in cells are subjected to excessive protein synthesis that overwhelms the posttranslational machinery to properly fold and modify protein molecules. Such stress triggers the unfolded protein response (UPR) through inositol-requiring enzyme 1 (IRE-1) for endonuclease cleavage of specific cellular mRNAs. One of them is a premature mRNA, which retains an unspliced intron, for the X-box *binding protein* 1 (XBP-1). Splicing out the intronic sequences from the premature mRNA of XBP-1 in the cytoplasm generates ligands for endogenous retinoic acid-inducible gene-like receptors (RLRs). In the absence of Skiv2l, such UPR would trigger elevated production of type I interferon (IFN) production. Therefore, SKIV2L is an important negative regulator of the RLR response to exogenous RNA ligands including RNA viruses [56]. Under such a scenario, the absence or partial functional deficiency of Skiv2l could generate unchecked or dysregulated production of ligands or sensors for RLR. The results would be autoinflammation in the absence of a viral infection, or increased susceptibility to systemic autoimmune disease. Table 2 lists results of recent attempts to investigate the potential roles of Skiv2l mutations in human diseases.

It was mind-boggling to find that a deficiency in Skiv2l or another Ski complex protein Ski3 or TTC37 causes the syndromic intractable diarrhea known as tricho-hepato-enteric syndrome or THES [58,59]. This extremely rare syndrome is characterized by (1) the presence of woolly and easily breakable scalp hair, (2) facial dysmorphism and hypertelorism of paired organs such as eyes, (3) intrauterine restriction in growth with low birth weights, (4) dysfunctional liver with hematochromatosis and coarse appearance, (5) intractable diarrhea that may initiate soon after birth or at neonatal stage, persists despite prolonged bowl rest. Many baby patients with THES require parenteral nutrition, become dehydrated and marasmic, and incapable to thrive [60–63].

Table 2. A list of human diseases associated with mutations of SKIV2L.

Disease	SNP	Relationship to Disease	Remarks
SLE (Systemic Lupus Erythematosus)	T allele of rs419788 in intron 6	Confer disease susceptibility in additive pattern	One copy confers a low risk of disease and two copies result in greater susceptibility [64]
AMD (Age-related macular degeneration)	R151Q (rs438999)	May exert a functional effect in AMD	Strong LD (linkage disequilibrium) with Bf R32Q (rs641153) [65]
	Intronic SNP (rs429608)	AMD genetic protective factors	Protective effect for AMD [66] in Han Chinese and Japanese populations [67,68]
	rs429608 and rs453821	Significantly associated with neovascular AMD	Not associated significantly with PCV [69]
PCV (Polypoidal Choroidal Vasculopathy)	3'UTR (rs2075702)	Significant association with PCV	Decreased risk of developing PCV [70]
SD/THES (Syndromic diarrhea/tricho-hepato-enteric syndrome)	c.1635_1636insA (p.Gly546Argfs*35) c.2266C>T (p.Arg756*) c.2442G>A (p.Trp814*) c.848G>A (p.Trp283*) c.1022T>G (p.Val341Gly) c.2572del (p.Val858*) c.2662_2663del (p.Arg888Glyfs*12) c.1434del (p.Ser479Alafs*3)	Deleterious mutations detected in six individuals with typical SD/THES	Molecular defects in SKIV2L cause SD/THES [58,71]
	c.1891G>A p.Gly631Ser c.3187C>T p.Arg1063*	Two new mutations found in a trichohepatoenteric syndrome patient	The patient was a Malaysian child [72]
	(c.1891G>A) (c.1120C>T)	Two variants identified in THES patients	These two mutations can cause THES [73]
	c.1420G>T (p.Q474*) c.3262G>T (p.E1088*)	Novel compound heterozygous nonsense mutations were identified in THES patient	Decreased levels of SKIV2L protein expression in blood mononuclear cells [74]
	25 exons (p.Glu1038 fs*7 (c.3112_3140del))	A rare mutation in THES patient	This patient died at three year old [75]
IBD (Inflammatory Bowel Disease)	c.354+5G>A	Identified a novel splicing mutation in a patient with IBD (ulcerative colitis)	This mutation was related to Inflammatory Bowel Disease [76]

Further analyses revealed that most THES patients had defects in the immune system with low levels of switched immunoglobulins, highly variable responses to vaccination including unresponsiveness of specific antigens, impaired T cell and NK cell functions such as low production of γ-interferon and abnormal T cell proliferation, and severe Epstein Barr Virus (EBV) infection [77]. It appears that 60% of the THES patients had a genetic deficiency of TTC37 [59] and 40% had a deficiency of SKIV2L [78], and many of those homozygous deficiencies were the results of consanguineous marriage. Since the discovery of THES, a spectrum of disease phenotypes and severity has emerged, and the disease has been reported in multiple racial groups [74,75,79–81]. It has been noticed that THES patients with deficiencies of SKIV2L experienced more severe disease in terms of liver damage and impairment of prenatal growth than those with deficiencies of TTC37 [78]. THES patients with deficiency of Skiv2l but not deficiency of TTC37 exhibited a remarkable type I interferon stimulated gene expression profile [56].

It is not clear for the mechanism between the loss of function for SKIV2L or TTC37 and the detrimental phenotypes of THES, which include the failure to thrive, inability in food digestion, poor quality in hair texture, and impairments of liver and immune functions.

Intriguingly, elevated expression levels of type I interferon stimulated genes in the absence of apparent viral infections is a marked phenomenon of systemic autoimmune diseases including systemic lupus erythematosus and juvenile dermatomyositis [82–84]. Thus, it would be appropriate to examine if there were genetic or acquired deficiency, dysregulation or impaired functions of Skiv2l in autoimmune, inflammatory and infectious diseases [56].

4. DXO or DOM3Z

Quality Controls II—Destroy Disqualified Products—DXO (DOM3Z) to Degrade Miscapped or Uncapped RNA from the Start to the End

The discovery of human DOM3Z (DXO) was reported by Yang and colleagues in 1998 [14]. This gene is arranged head-to-head orientation with STK19 (RP1) and tail-to-tail with SKIV2L (SKI2W) [15,17]. The 5′ ends of DOM3Z and STK19 overlap but they are in opposite configurations. The exon 1 of STK19 is present at the intron 2 of DOM3Z genomic region. The 3′ ends of DOM3Z and SKI2W are separated by 59 base pairs.

The DOM3Z gene has CpG-rich sequences at its 5′ region, a feature conserved in housekeeping genes in simple and complex eukaryotes. Human DOM3Z is ubiquitously expressed, but the highest expression levels are found in the adrenals and the testes. The first exon (123 bp) of DOM3Z is noncoding because there is an in-frame TGA stop codon at nucleotides 19–21. The full length DOM3Z cDNA is 1386 bp and its genomic DNA is 2.2 kb in size, which consists of 7 exons encoding 396 amino acids. Between residues 103 to 124 of the Dom3z protein is a leucine zipper motif for protein-protein interactions [14]. Yang and colleagues also found that human DOM3Z has homologs in round worm *Caenorhabditis elegans* (DOM-3), baker's yeast, fission yeast and possibly flowering plant Arabidopsis [14,15,17]. However, no other known protein motifs were identified at that time.

Most progress in studying the biologic functions of DOM3Z homologs occurred in yeasts, which is named **Rai1**. Through genetic deletion and complementation analyses, Xue and coworkers showed in year 2,000 that Rai1 interacts Rat1 (also named Xrn2). The yeast Rat1 is a nuclear 5′→3′ exoribonuclease that is important in RNA processing and RNA degradation [85]. Rai1 stabilizes and activates the exonuclease activities of Rat1. Genetic mutants of Rai1 affects the proper (5′ and 3′) processing of the 5.8S ribosomal RNA. Deletion of Rai1 gene leads to defective biogenesis of the 60S ribosome but that can be suppressed by high copy of RAT1. Intriguingly, double mutants of Rai1 and putative RNA helicase gene Ski2 in yeast were synthetically lethal [85], suggesting Rai1 and Ski2 proteins have overlapping functions that cannot be both dysfunctional.

Xiang, Jiao, Tong, Kiledjian and colleagues in 2009 reported crystal structures of *S. pombe* Rat1 in complex with Rai1 in 2.2 Å resolution [86], and Rai1 and its murine homolog Dom3z in 2.0 Å resolution. It was shown that Rai1 can stabilize Rat1 secondary structure to facilitate Rat1 to degrade RNAs more effectively. They identified a large pocket in Rai1 and in Dom3z that contains highly conserved residues, including three acidic side chains that coordinates a divalent cation such as Mg^{+2} or Mn^{+2}. They demonstrated that Rai1 has pyrophosphohydrolase activity towards 5′ triphosphorylated RNA [86]. This team of investigators continued to demonstrate that Rai1 possesses decapping and endonuclease activities, which can remove the entire cap structure dinucleotide from an mRNA in vitro and in yeast cells, especially towards mRNAs with unmethylated, faulty or aberrant caps [87]. Thus, it is proposed that Rai1 is involved in quality control process to ensure mRNA 5′-end integrity.

The same research team also found a related protein for Rai1 in yeast known as Ydr370C, which is mainly expressed in the cytoplasm. The crystal structure of the Ydr370C protein from *Kluyveromyces lactis* (named as DXO1) was solved and its biochemical activity investigated [88]. Ydr370C has robust decapping activity for incompletely capped mRNA and 5′→3′ distributive exoribonuclease activity, but no activity for pyrophosphohydrolase. The cytoplasmic Ydr370C was renamed Dxo1 because of its decapping and exoribonuclease activities for mRNA. If both Dxo1 and Rai1 in yeast are disrupted, mRNAs with incomplete caps are produced and readily detectable under normal growth conditions. Therefore, yeast possesses two partially redundant proteins that detect and degrade incompletely capped mRNAs [88]. In mammals including humans, there is only one homolog for Rai1 and Dxo1, Dom3z. Dom3z possesses all three functional activities of Rai1 and Dxo1, which are pyrophosphohydrolase, decapping, and 5′ to 3′ exoribonuclease activities. Thus, Dom3z is given an alternative name as Dxo. Dom3z preferentially degrades defectively capped and inefficiently spliced pre-mRNAs (Figure 4A) [89,90]. Collectively, Rai1, Dxo1 and Dxo identify a novel class of proteins that functions in a quality control mechanism to ensure proper mRNA 5′-end fidelity or integrity [89].

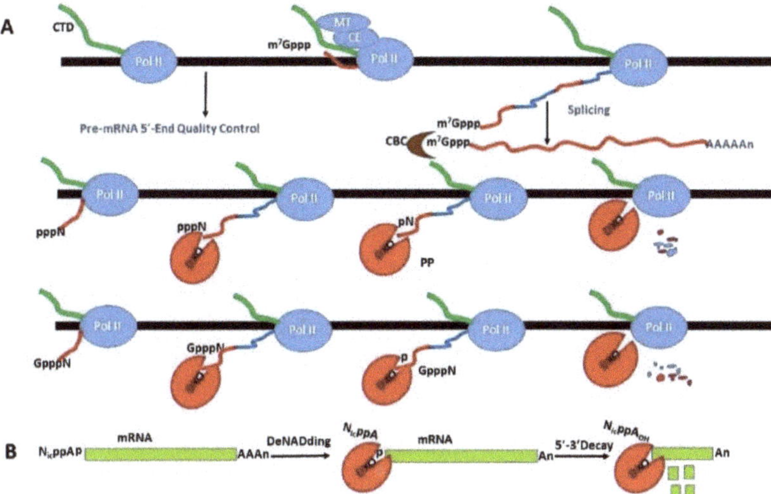

Figure 4. DXO decapping and DXO-mediated decay of NAD$^+$-capped mRNA. (**A**) DXO can remove the diphosphate from a triphosphorylated cap of mRNA, the entire cap structure of an unmethylated guanosine-capped mRNA, or aberrant pre-mRNAs. (**B**) DXO hydrolyzes the phosphodiester linkage 3′ to the adenosine to remove the entire NAD$^+$ moiety and generates an RNA molecule with 5′ monophosphate. Modified and redrawn from ref [89] for panel A, and ref [91] for panel B (with permission from Dr. Mike Kiledjian, Rutgers University, NJ).

If the first (^{m7}GpppN$_m$pRNA) or the second nucleotide (^{m7}GpppN$_m$pN$_m$pRNA) of the mRNA were methylated at the ribose 2′-O position, it would be protected from Dxo-mediated decapping and degradation [92].

Over the past forty years, it has been firmly established that mRNA in eukaryotes are predominantly modified by a 7-methyl-guanosine cap at the 5′ end during the initiation of gene transcription. This m^{7}G cap helps the processes for (a) RNA splicing to remove intronic sequences, (b) the nucleus-cytoplasm transport of mature mRNA for translation, and (c) protection of mature mRNA from degradation. It has been presumed that such capping process in eukaryotes was constitutive and efficient with little errors until yeast Rai1 is shown to play an important role on the quality control of mRNA capping and stability. Moreover, it has been found recently that some bacteria RNA can be capped at the 5′ end with nicotinamide adenine dinucleotide (NAD$^+$) to increase their stability. Unexpectedly, such NAD$^+$

caps are also present in 5–10% eukaryotic mRNA and some small nucleolar RNA (snoRNA). Notably, NAD⁺-capped mRNAs are *not* translatable. They are also less stable and subject to Dxo-mediated decapping and 5′ →3′ decay, a process known as deNADding (Figure 4B) [91,93].

5. STK19 (RP1/G11)

An On-and-Off Switch—The Nuclear Kinase STK19 (RP1 or G11) Whose Malfunction Can Cause Melanoma

While the 5′ regulatory regions of the duplicated C4 genes were characterized, a canonical polyadenylation signal (AATAAA) was found in the promoter region of each C4 gene, suggesting the presence of a novel gene. Thus, cDNA clones were isolated and genomic DNA sequences for this novel gene were determined [10,11,94]. This gene was originally named RP to memorialize the late Professor Rodney Porter, in his research unit the human MHC complement genes were cloned and characterized.

In human, an intact RP1 gene is located 611 base pairs upstream of the first C4 gene, and a partially duplicated RP2 gene is present upstream of each of the succeeding C4 gene (Figure 1) [10,17]. Immediately upstream of RP1 is DOM3Z. RP1 is in head-to-head configuration with DOM3Z, and tail-to-head with C4 [17]. Upstream of RP2 is another partially duplicated gene fragment for tenascin X, TNXA. Human RP is part of a four-gene duplication complex in the HLA termed the RCCX module, which stands for R̲P, C̲4, C̲YP21 and TNX̲ (Figure 1) [95,96].

RP1 is expressed in all tissues and its highest expression is in the adrenals and the testes of human and rodents. Homologs of RP1 are predominantly present among vertebrates but an exception is found in a sea anemone (*Exaiptasia pallida*, LOC110250916/XP_020913230.1). The RP1 protein contains a bipartite nuclear localization signal and was shown to be a nuclear protein [10]. Subsequently, recombinant protein of this gene, which was also named G11 by Duncan Campbell and colleagues [97], was shown to be a s̲erine/t̲hreonine k̲inase [98]. This gene was renamed to STK19, which uncoupled publications for some of the original observations [10,95,99]. Other names shown at the NCBI website for this gene include D6S60; D6S60E; HLA-RP1.

STK19 gene contains 7 or 9 exons. The 9-exon STK19 gene is 11 kb in size and codes for 364 amino acids with a calculated molecular weight of 41.5 kDa. This long form of STK19 gene overlaps to the 5′ region of the DOM3Z gene. There is a variant coding for 368 amino acids that is due to an alternative splicing, which results in an extension of 12 nucleotides from the 3′ end of exon 5 to the intron 6, leading to the addition of four more amino acids after R221 (VCDC). The 7-exon version of STK19 is 9.1 kb in size that has a 267 bp intergenic distance from DOM3Z. It codes for 254 amino acids with a calculated molecular weight of 30 kDa. The first 110 amino acid residues encoded by the first two exons are absent in the shorter RP1 transcript. Murine STK19 has the small version with 254 amino acids.

Scattered in the introns of the human RP1 gene are 8 copies of s̲hort i̲nterspersed n̲uclear e̲lements named *Alu*. One of them is an integral component of a mobile retroelement SVA that consists of a distinct S̲INE, 21 copies of GC-rich v̲ariable n̲umber t̲andem r̲epeats (V̲NTRs), and an A̲lu element [10,15,17]. The SVA element in RP1 was flanked by short direct repeating sequences of 13 nucleotides, which was indicative of insertion by a discrete composite genetic unit (Figure 1B). The SVA in RP1 was the first identification and documentation of such distinct and discrete composite retroelements into a human genome [10]. The highly repetitive nature of the SVA sequences render specific biologic studies challenging. It was demonstrated certain member(s) of SVA has the capability to "retrotranspose" with help of the reverse transcriptase coded by active long interspersed element LINE-1 [100–102].

In terms of gene function, STK19 encodes a nuclear protein [10] that is a kind of serine/threonine kinase. The STK19 protein can bind ATP at its kinase domain, phosphorylate certain proteins at Ser/Thr residues, such as α-casein and histones [98]. The actual physiological properties and function, the substrates of the kinase activity for STK19, and the relationship with disease for STK19 are being gradually revealed [103]. Results of the interactome analyses revealed that nine proteins bind to STK19 (https://www.ncbi.nlm.nih.gov/gene/8859, [104]). These proteins are DNA polymerase subunit E or POLE, RNA polymerase II subunit G or POLR2G, transcription factor SP3, mRNA splicing factor SF3BF4, basic helix-loop-helix protein e40 or BHLHE40 that controls circadian rhythm and cell differentiation, inosine monophosphate dehydrogenase 1 or IMPDH1 that is an enzyme essential for guanine nucleotide biosynthesis through catalysis of synthesis for xanthine monophosphate from inosine 5-monophosphate, membrane transporter for nucleosides or solute carrier 29 or SLC29A1, preferentially expressed antigen in melanoma PRAME that acts as a repressor of retinoic acid receptor and may promote growth of cancer cells, and tripartite motif protein TRIM23 that may be a ADP ribosylation factor for guanine nucleotide binding protein and plays a role in the formation of intracellular transport vesicles and for their movement from one compartment to another in the cell. These interacting proteins may be substrates or cofactors of STK19 kinase, implicating that STK19 could be engaged in DNA replication, RNA transcription, RNA splicing, nucleoside/nucleotide synthesis and transport, and control of gene expression for development and differential cellular functions.

While STK19 knockdown in cells revealed little effects on transcription, it is required for recovery of transcription after DNA damage and therefore important for the transcription-related DNA damage response [105]. STK19 interacts with Cockayne syndrome B protein (CSB, also named ERCC6) after DNA damage. Cells lacking STK19 were UV-sensitive.

By analysis of large-scale exome data from melanoma and uninvolved tissues, Eran Hodis and colleagues showed STK19 is a novel melanoma gene for its somatic hotspot of point mutations. While the D89N mutation of SKT19 has a 5% frequency in a melanoma [106], no SKT19 mutations were detectable in nevus associated-melanomas [107]. There are 10% gene mutations in STK19 in basal cell carcinoma (BCC). Strikingly, all of those mutations resulted in D89N substitution [108]. Recently, Yin and colleagues reported STK19 can phosphorylate NRAS at Ser-89 and activate oncogenic NRAS to promote melanoma genesis. When STK19 has the D89N change, it is a gain of function mutation that phosphorylates or interacts more efficiently with NRAS and enhances melanocyte transformation [109].

In addition, by next-generation sequencing and bioinformatics analysis, STK19 was found to have novel T/C missense single-nucleotide variants in Chr6:31947203 in induced pluripotent stem cell (iPSC) lines, which can be used to trace the genotype of original cells [110].

Through meta-analyses of published GWAS data, STK19, SKIV2L and NELF-E appeared to be potential pleiotropic genes for metabolic syndrome and inflammation [111]. Moreover, Amare et al. reported that genetic variant of STK19 gene correlated with schizophrenia by bivariate GWAS [112]. Other functions of STK19 remain to be demonstrated.

6. The Physical Linkage of NELF-E and SKIV2L (RD and SKI2W), DXO and STK19 (DOM3Z and RP1) and the Emergence of NSDK Quartet in the MHC

We ask when did the genes for NELF-E, SKIV2L, DXO and STK19 start to link together physically as a group in the animal kingdom, and their relationships with the class I and class II genes in the emergence of the MHC in the vertebrates. We attempted to obtain answers to these questions through analyzing data deposited in the National Center for Biotechnology Information (NCBI/PubMed: https://www.ncbi.nlm.nih.gov/home/ or https://www.ncbi.nlm.nih.gov/pubmed/).

The evolution history of NELF-E is intriguing because the presence of a leucine zipper and the RRM, which is a characteristic of a NELF-E like molecule, can be found in *Drosophila* but there are no RD repeats (https://www.ncbi.nlm.nih.gov/protein/NP_648241.1, [113]). The presence of the RD (Arg-Asp) and/or RE (Arg-Glu) repeats is definitive in zebrafish NELF-E (https://www.ncbi.nlm.nih. gov/protein/XP_005170084.1, [114]). Related gene or protein function for SKIV2L can be found in yeast, and for DXO in *C. elegans* [115]. STK19 homologs with high sequence identities are predominantly present among vertebrates. There is no evidence for a physical linkage for any two of the NSDK quartet prior to vertebrates.

For the data deposited in the NCBI, most chromosomal segments are relatively short, and their linkage status is not yet known. The neighboring gene linkages of NELF-E and SKIV2L in one chromosomal segment, and DXO and STK19 in another chromosomal segment are observable in shark, zebrafish, and coelacanth. Genes in each of these pairs are organized in "head-to-head" configuration and therefore they share the same 5′ regulatory sequences in opposite orientations (Figure 1B). The physical linkage of all four genes (i.e., NELF-E, SKIV2L, DXO and STK19) together as a group seems to be definitive in frogs [116]. Again, we should emphasize that data available for analyses are limited.

In egg-laying mammal platypus, MHC class I and class II genes are present on chromosome X3, but most (if not all) of the class III genes form a linkage unit on chromosome X5. The assembly of the class III genes (including NSDK with complement convertases), class I genes and class II genes as the MHC locus appears intact in placental mammals.

While the NSDK quartet genes express ubiquitously, it is not known whether there are concerted controls of gene expression (such as genes in operons in prokaryotes), or differential responses to regulation during growth and development.

7. Conclusions

NELF-E-SKIV2L-DXO-STK19 is a complex genetic unit located between complement factor B and component C4 in the class III region of the major histocompatibility complex (MHC) [98,117]. NELF-E, SKIV2L and DXO make important roles in transcription, processing and degradation of a variety of RNAs and antiviral immunity [24,56]. As such, NELF-E or RD is a subunit of the negative elongation factor (NELF). It is an RNA-binding protein that imposes a promoter proximal pause in transcription by the RNA polymerase II. The purpose is to allow capping and quality checking prior to full strength elongation of RNA for gene expression [24,30]. SKI2W or SKIV2L RNA helicase, an engine to unwind helical structures of RNA molecules and propel them through the tunnel formed by cytoplasmic exosomes for 3′→5′ degradation. SKIV2L is an important negative regulator of the RIG-I-like receptors (RLR) mediated antiviral response and an important modulator of the type I interferon response in the cytoplasm [56]. The RLR [56,57], toll-like receptors, and complement systems are innate immune response effectors that engage in pattern recognitions of differentially modified biomolecules to differentiate self and non-self [118,119].

Genetic studies revealed that homozygous deficiency of human SKIV2L contributes to the pathogenesis of syndromic diarrhea or trichohepatoenteric syndrome (THES). Regulation of gene expression is critical in determining cell identity, development and responses to the environment. DXO or Dom3z possesses pyrophosphohydrolase, decapping and 5′-3′ exoribonuclease activities that can preferentially degrade incompletely 7-methylguanosine (m^{7}G) capped, or NAD$^+$ capped mRNAs from the 5′ end [89]. STK19 is a Serine/Threonine nuclear protein kinase. STK19 is important for the transcription-related DNA damage response [105,109]. STK19 phosphorylates N-RAS at Ser-89. Dysregulated phosphorylation of NRAS at S89 activates and promotes melanoma genesis [109]. It is necessary to point out that the biochemical and functional properties of STK19 are still largely unexplored. It is also not clear whether and how do the NSDK quartet proteins interact with each other, and the advantage for these four genes being tightly linked together.

Thus, NSDK seems to engage in the surveillance of host RNA integrity at the "head" and at the "tail" [120], in the destruction and turnover of faulty or expired RNA molecules or RNA viruses, and in the fine-tuning of innate immunity. The roles of NSDK in carcinogenesis, infectious and autoimmune diseases are only starting to emerge. We look forward to further studies on disease associations, functional mechanisms and therapeutics of these very old but non-classical MHC genes.

Author Contributions: All authors contribute to literature search and writing of the manuscript.

Funding: This research was partially funded by grants from the National Institute of Arthritis, Musculoskeletal and Skin Diseases (NIAMS) of National Institutes of Health R01 AR073311 and R21 AR070509, and the Cure JM Foundation (C.-Y.Y.).

Acknowledgments: We are indebted to Ms Bi Zhou and colleagues of our research teams for contributions.

Conflicts of Interest: The authors declare no conflicts of interest.

Abbreviations

ARS2	arsenite resistance protein 2;
CBC	cap-binding complex;
CDK9	cyclin dependent kinase 9;
CSB	Cockayne syndrome B protein (also named ERCC6);
DRB	5,6-dichloro-1-β-D-ribofuranosylbenzimidazole;
DSIF	DRB sensitivity inducing factor;
DXO	decapping exoribonuclease;
EC *	activated elongation complex;
EMT	epithelial–mesenchymal transition;
HCC	hepatocellular carcinoma;
IRE-1	Inositol-requiring enzyme 1;
MDA5	melanoma differentiation-associated protein 5;
MK2	MAPK-activated protein kinase II;
NSDK	NELF-E (RD), SKIV2L (SKI2W), DXO (DOM3Z) and STK19;
NELF	negative elongation factor;
P-TEFb	positive transcription elongation factor b;
PAF	Pol II associated factor;
PARP-1	poly-ADP-ribose polymerase 1;
PEC	paused elongation complex;
PHAX	phosphorylated adaptor for RNA export;
RD	a gene with Arg-Asp repeats, also known as NELF-E;
RIG-I	retinoic acid-inducible gene 1;
RLRS	RIG-I-like receptors;
RNAPII	RNA polymerase II;
RP	the gene upstream of complement C4, also known as STK19;
RRM	RNA recognition and binding domain;
SKI	yeast genes by which mutations lead to superkilling phenotypes after viral infection;
snRNA	small nuclear RNA;
snoRNA	small nucleolar RNA;
SPT6	yeast homolog of suppressor for transposon Ty insertion, 6;
UPR	unfolded protein response;
TAR	transactivation response element in HIV;
ZEB1	zinc-finger E-box binding homeobox 1

References and Note

1. Carroll, M.C.; Campbell, R.D.; Bentley, D.R.; Porter, R.R. A molecular map of the human major histocompatibility complex class III region linking complement genes C4, C2 and factor B. *Nature* **1984**, *307*, 237–241. [CrossRef] [PubMed]

2. Horton, R.; Wilming, L.; Rand, V.; Lovering, R.C.; Bruford, E.A.; Khodiyar, V.K.; Lush, M.J.; Povey, S.; Talbot, C.C., Jr.; Wright, M.W.; et al. Gene map of the extended human MHC. *Nat. Rev. Genet.* **2004**, *5*, 889–899. [CrossRef] [PubMed]

3. Miller, F.W.; Cooper, R.G.; Vencovsky, J.; Rider, L.G.; Danko, K.; Wedderburn, L.R.; Lundberg, I.E.; Pachman, L.M.; Reed, A.M.; Ytterberg, S.R.; et al. Genome-wide association study of dermatomyositis reveals genetic overlap with other autoimmune disorders. *Arthritis Rheum.* **2013**, *65*, 3239–3247. [CrossRef] [PubMed]

4. Dawkins, R.; Leelayuwat, C.; Gaudieri, S.; Tay, G.; Hui, J.; Cattley, S.; Martinez, P.; Kulski, J. Genomics of the major histocompatibility complex: haplotypes, duplication, retroviruses and disease. *Immunol. Rev.* **1999**, *167*, 275–304. [CrossRef] [PubMed]

5. Hauptmann, G.; Bahram, S. Genetics of the central MHC. *Curr. Opin. Immunol.* **2004**, *16*, 668–672. [CrossRef] [PubMed]

6. Lintner, K.E.; Wu, Y.L.; Yang, Y.; Spencer, C.H.; Hauptmann, G.; Hebert, L.A.; Atkinson, J.P.; Yu, C.Y. Early Components of the Complement Classical Activation Pathway in Human Systemic Autoimmune Diseases. *Front. Immunol.* **2016**, *7*, 36. [CrossRef] [PubMed]

7. Savelli, S.L.; Roubey, R.A.S.; Kitzmiller, K.J.; Zhou, D.; Nagaraja, H.N.; Mulvihill, E.; Barbar-Smiley, F.; Ardoin, S.P.; Wu, Y.L.; Yu, C.Y. Opposite Profiles of Complement in Antiphospholipid Syndrome (APS) and Systemic Lupus Erythematosus (SLE) Among Patients With Antiphospholipid Antibodies (aPL). *Front. Immunol.* **2019**, *10*, 885. [CrossRef]

8. Ricklin, D.; Hajishengallis, G.; Yang, K.; Lambris, J.D. Complement: a key system for immune surveillance and homeostasis. *Nat. Immunol.* **2010**, *11*, 785–797. [CrossRef]

9. Wu, L.C.; Morley, B.J.; Campbell, R.D. Cell-specific expression of the human complement protein factor B gene: evidence for the role of two distinct 5′-flanking elements. *Cell* **1987**, *48*, 331–342. [CrossRef]

10. Shen, L.; Wu, L.C.; Sanlioglu, S.; Chen, R.; Mendoza, A.R.; Dangel, A.W.; Carroll, M.C.; Zipf, W.B.; Yu, C.Y. Structure and genetics of the partially duplicated gene RP located immediately upstream of the complement C4A and the C4B genes in the HLA class III region. Molecular cloning, exon-intron structure, composite retroposon, and breakpoint of gene duplication. *J. Biol. Chem.* **1994**, *269*, 8466–8476.

11. Yu, C.Y. The complete exon-intron structure of a human complement component C4A gene. DNA sequences, polymorphism, and linkage to the 21-hydroxylase gene. *J. Immunol.* **1991**, *146*, 1057–1066. [PubMed]

12. Levi-Strauss, M.; Carroll, M.C.; Steinmetz, M.; Meo, T. A previously undetected MHC gene with an unusual periodic structure. *Science* **1988**, *240*, 201–204. [CrossRef] [PubMed]

13. Dangel, A.W.; Shen, L.; Mendoza, A.R.; Wu, L.C.; Yu, C.Y. Human helicase gene SKI2W in the HLA class III region exhibits striking structural similarities to the yeast antiviral gene SKI2 and to the human gene KIAA0052: emergence of a new gene family. *Nucleic Acids Res.* **1995**, *23*, 2120–2126. [CrossRef] [PubMed]

14. Yang, Z.; Shen, L.; Dangel, A.W.; Wu, L.C.; Yu, C.Y. Four ubiquitously expressed genes, RD (D6S45)-SKI2W (SKIV2L)-DOM3Z-RP1 (D6S60E), are present between complement component genes factor B and C4 in the class III region of the HLA. *Genomics* **1998**, *53*, 338–347. [CrossRef] [PubMed]

15. Yang, Z.; Yu, C.Y. Organizations and gene duplications of the human and mouse MHC complement gene clusters. *Exp. Clin Immunogenet.* **2000**, *17*, 1–17. [CrossRef]

16. Yu, C.Y.; Yang, Z.; Blanchong, C.A.; Miller, W. The human and mouse MHC class III region: a parade of 21 genes at the centromeric segment. *Immunol. Today* **2000**, *21*, 320–328.

17. Yang, Z.; Qu, X.; Yu, C.Y. Features of the two gene pairs RD-SKI2W and DOM3Z-RP1 located between complement component genes factor B and C4 at the MHC class III region. *Front. Biosci.* **2001**, *6*, D927–D935. [CrossRef]

18. Cheng, J.; Macon, K.J.; Volanakis, J.E. cDNA cloning and characterization of the protein encoded by RD, a gene located in the class III region of the human major histocompatibility complex. *Biochem. J.* **1993**, *294*, 589–593. [CrossRef]

19. Surowy, C.S.; Hoganson, G.; Gosink, J.; Strunk, K.; Spritz, R.A. The human RD protein is closely related to nuclear RNA-binding proteins and has been highly conserved. *Gene* **1990**, *90*, 299–302. [CrossRef]

20. Speiser, P.W.; White, P.C. Structure of the human RD gene: A highly conserved gene in the class III region of the major histocompatibility complex. *DNA* **1989**, *8*, 745–751. [CrossRef]

21. Yamaguchi, Y.; Takagi, T.; Wada, T.; Yano, K.; Furuya, A.; Sugimoto, S.; Hasegawa, J.; Handa, H. NELF, a multisubunit complex containing RD, cooperates with DSIF to repress RNA polymerase II elongation. *Cell* **1999**, *97*, 41–51. [CrossRef]

22. Fujinaga, K.; Irwin, D.; Huang, Y.; Taube, R.; Kurosu, T.; Peterlin, B.M. Dynamics of human immunodeficiency virus transcription: P-TEFb phosphorylates RD and dissociates negative effectors from the transactivation response element. *Mol. Cell Biol.* **2004**, *24*, 787–795. [CrossRef] [PubMed]

23. Yamaguchi, Y.; Inukai, N.; Narita, T.; Wada, T.; Handa, H. Evidence that negative elongation factor represses transcription elongation through binding to a DRB sensitivity-inducing factor/RNA polymerase II complex and RNA. *Mol. Cell Biol.* **2002**, *22*, 2918–2927. [CrossRef] [PubMed]

24. Narita, T.; Yamaguchi, Y.; Yano, K.; Sugimoto, S.; Chanarat, S.; Wada, T.; Kim, D.K.; Hasegawa, J.; Omori, M.; Inukai, N.; et al. Human transcription elongation factor NELF: identification of novel subunits and reconstitution of the functionally active complex. *Mol. Cell Biol.* **2003**, *23*, 1863–1873. [CrossRef] [PubMed]

25. Narita, T.; Yung, T.M.; Yamamoto, J.; Tsuboi, Y.; Tanabe, H.; Tanaka, K.; Yamaguchi, Y.; Handa, H. NELF interacts with CBC and participates in 3′ end processing of replication-dependent histone mRNAs. *Mol. Cell* **2007**, *26*, 349–365. [CrossRef] [PubMed]

26. Schulze, W.M.; Cusack, S. Structural basis for mutually exclusive co-transcriptional nuclear cap-binding complexes with either NELF-E or ARS2. *Nat. Commun.* **2017**, *8*, 1302. [CrossRef] [PubMed]

27. Rao, J.N.; Neumann, L.; Wenzel, S.; Schweimer, K.; Rosch, P.; Wohrl, B.M. Structural studies on the RNA-recognition motif of NELF E, a cellular negative transcription elongation factor involved in the regulation of HIV transcription. *Biochem. J.* **2006**, *400*, 449–456. [CrossRef] [PubMed]

28. Pagano, J.M.; Kwak, H.; Waters, C.T.; Sprouse, R.O.; White, B.S.; Ozer, A.; Szeto, K.; Shalloway, D.; Craighead, H.G.; Lis, J.T. Defining NELF-E RNA binding in HIV-1 and promoter-proximal pause regions. *PLoS Genet.* **2014**, *10*, e1004090. [CrossRef] [PubMed]

29. Gibson, B.A.; Zhang, Y.; Jiang, H.; Hussey, K.M.; Shrimp, J.H.; Lin, H.; Schwede, F.; Yu, Y.; Kraus, W.L. Chemical genetic discovery of PARP targets reveals a role for PARP-1 in transcription elongation. *Science* **2016**, *353*, 45–50. [CrossRef]

30. Guo, J.; Price, D.H. RNA polymerase II transcription elongation control. *Chem. Rev.* **2013**, *113*, 8583–8603. [CrossRef]

31. Awwad, S.W.; Abu-Zhayia, E.R.; Guttmann-Raviv, N.; Ayoub, N. NELF-E is recruited to DNA double-strand break sites to promote transcriptional repression and repair. *EMBO Rep.* **2017**, *18*, 745–764. [CrossRef] [PubMed]

32. Borisova, M.E.; Voigt, A.; Tollenaere, M.A.X.; Sahu, S.K.; Juretschke, T.; Kreim, N.; Mailand, N.; Choudhary, C.; Bekker-Jensen, S.; Akutsu, M.; et al. p38-MK2 signaling axis regulates RNA metabolism after UV-light-induced DNA damage. *Nat. Commun.* **2018**, *9*, 1017. [CrossRef]

33. Vos, S.M.; Farnung, L.; Urlaub, H.; Cramer, P. Structure of paused transcription complex Pol II-DSIF-NELF. *Nature* **2018**, *560*, 601–606. [CrossRef] [PubMed]

34. Vos, S.M.; Farnung, L.; Boehning, M.; Wigge, C.; Linden, A.; Urlaub, H.; Cramer, P. Structure of activated transcription complex Pol II-DSIF-PAF-SPT6. *Nature* **2018**, *560*, 607–612. [CrossRef] [PubMed]

35. Iida, M.; Iizuka, N.; Tsunedomi, R.; Tsutsui, M.; Yoshida, S.; Maeda, Y.; Tokuhisa, Y.; Sakamoto, K.; Yoshimura, K.; Tamesa, T.; et al. Overexpression of the RD RNA binding protein in hepatitis C virus-related hepatocellular carcinoma. *Oncol. Rep.* **2012**, *28*, 728–734. [CrossRef] [PubMed]

36. Dang, H.; Takai, A.; Forgues, M.; Pomyen, Y.; Mou, H.; Xue, W.; Ray, D.; Ha, K.C.H.; Morris, Q.D.; Hughes, T.R.; et al. Oncogenic Activation of the RNA Binding Protein NELFE and MYC Signaling in Hepatocellular Carcinoma. *Cancer Cell* **2017**, *32*, 101–114.e8. [CrossRef] [PubMed]

37. Chen, S.Y.; Teng, S.C.; Cheng, T.H.; Wu, K.J. miR-1236 regulates hypoxia-induced epithelial-mesenchymal transition and cell migration/invasion through repressing SENP1 and HDAC3. *Cancer Lett.* **2016**, *378*, 59–67. [CrossRef]

38. Butkyte, S.; Ciupas, L.; Jakubauskiene, E.; Vilys, L.; Mocevicius, P.; Kanopka, A.; Vilkaitis, G. Splicing-dependent expression of microRNAs of mirtron origin in human digestive and excretory system cancer cells. *Clin. Epigenetics* **2016**, *8*, 33. [CrossRef]

39. Wang, Y.; Yan, S.; Liu, X.; Zhang, W.; Li, Y.; Dong, R.; Zhang, Q.; Yang, Q.; Yuan, C.; Shen, K.; et al. miR-1236-3p represses the cell migration and invasion abilities by targeting ZEB1 in high-grade serous ovarian carcinoma. *Oncol. Rep.* **2014**, *31*, 1905–1910. [CrossRef]

40. An, J.X.; Ma, M.H.; Zhang, C.D.; Shao, S.; Zhou, N.M.; Dai, D.Q. miR-1236-3p inhibits invasion and metastasis in gastric cancer by targeting MTA2. *Cancer Cell Int.* **2018**, *18*, 66. [CrossRef]

41. Wang, C.; Tang, K.; Li, Z.; Chen, Z.; Xu, H.; Ye, Z. Targeted p21(WAF1/CIP1) activation by miR-1236 inhibits cell proliferation and correlates with favorable survival in renal cell carcinoma. *Urol. Oncol.* **2016**, *34*, 59.e23–59.e34. [CrossRef] [PubMed]

42. Synowsky, S.A.; Heck, A.J. The yeast Ski complex is a hetero-tetramer. *Protein Sci.* **2008**, *17*, 119–125. [CrossRef] [PubMed]

43. Widner, W.R.; Wickner, R.B. Evidence that the SKI antiviral system of Saccharomyces cerevisiae acts by blocking expression of viral mRNA. *Mol. Cell Biol.* **1993**, *13*, 4331–4341. [CrossRef] [PubMed]

44. Ridley, S.P.; Sommer, S.S.; Wickner, R.B. Superkiller mutations in Saccharomyces cerevisiae suppress exclusion of M2 double-stranded RNA by L-A-HN and confer cold sensitivity in the presence of M and L-A-HN. *Mol. Cell Biol.* **1984**, *4*, 761–770. [CrossRef] [PubMed]

45. Brown, J.T.; Bai, X.; Johnson, A.W. The yeast antiviral proteins Ski2p, Ski3p, and Ski8p exist as a complex in vivo. *RNA* **2000**, *6*, 449–457. [CrossRef]

46. Schneider, C.; Tollervey, D. Threading the barrel of the RNA exosome. *Trends Biochem. Sci.* **2013**, *38*, 485–493. [CrossRef]

47. Halbach, F.; Reichelt, P.; Rode, M.; Conti, E. The yeast ski complex: crystal structure and RNA channeling to the exosome complex. *Cell* **2013**, *154*, 814–826. [CrossRef]

48. Anderson, J.S.; Parker, R.P. The 3′ to 5′ degradation of yeast mRNAs is a general mechanism for mRNA turnover that requires the SKI2 DEVH box protein and 3′ to 5′ exonucleases of the exosome complex. *EMBO J.* **1998**, *17*, 1497–1506. [CrossRef]

49. Weick, E.M.; Puno, M.R.; Januszyk, K.; Zinder, J.C.; DiMattia, M.A.; Lima, C.D. Helicase-Dependent RNA Decay Illuminated by a Cryo-EM Structure of a Human Nuclear RNA Exosome-MTR4 Complex. *Cell* **2018**, *173*, 1663–1677.e21. [CrossRef]

50. Thoms, M.; Thomson, E.; Bassler, J.; Gnadig, M.; Griesel, S.; Hurt, E. The Exosome Is Recruited to RNA Substrates through Specific Adaptor Proteins. *Cell* **2015**, *162*, 1029–1038. [CrossRef]

51. Lee, S.G.; Song, K. Identification and characterization of a bidirectional promoter from the intergenic region between the human DDX13 and RD genes. *Mol. Cells* **2000**, *10*, 47–53. [CrossRef] [PubMed]

52. Qu, X.; Yang, Z.; Zhang, S.; Shen, L.; Dangel, A.W.; Hughes, J.H.; Redman, K.L.; Wu, L.C.; Yu, C.Y. The human DEVH-box protein Ski2w from the HLA is localized in nucleoli and ribosomes. *Nucleic Acids Res.* **1998**, *26*, 4068–4077. [CrossRef] [PubMed]

53. Fuller-Pace, F.V. RNA helicases: modulators of RNA structure. *Trends Cell Biol.* **1994**, *4*, 271–274. [CrossRef]

54. Lee, S.G.; Song, K. Genomic organization of the human DDX13 gene located between RD and RP1 in the class III MHC complex. *Mol. Cells* **1997**, *7*, 414–418. [PubMed]

55. Yang, Z. Molecular genetic and biochemical studies of the human and mouse MHC complement gene clusters. PhD Dissertation, The Ohio State University, Columbus, OH, USA, 1999.

56. Eckard, S.C.; Rice, G.I.; Fabre, A.; Badens, C.; Gray, E.E.; Hartley, J.L.; Crow, Y.J.; Stetson, D.B. The SKIV2L RNA exosome limits activation of the RIG-I-like receptors. *Nat. Immunol.* **2014**, *15*, 839–845. [CrossRef] [PubMed]

57. Brisse, M.; Ly, H. Comparative Structure and Function Analysis of the RIG-I-Like Receptors: RIG-I and MDA5. *Front. Immunol.* **2019**, *10*, 1586. [CrossRef] [PubMed]

58. Fabre, A.; Charroux, B.; Martinez-Vinson, C.; Roquelaure, B.; Odul, E.; Sayar, E.; Smith, H.; Colomb, V.; Andre, N.; Hugot, J.P.; et al. SKIV2L mutations cause syndromic diarrhea, or trichohepatoenteric syndrome. *Am. J. Hum. Genet.* **2012**, *90*, 689–692. [CrossRef]

59. Hartley, J.L.; Zachos, N.C.; Dawood, B.; Donowitz, M.; Forman, J.; Pollitt, R.J.; Morgan, N.V.; Tee, L.; Gissen, P.; Kahr, W.H.; et al. Mutations in TTC37 cause trichohepatoenteric syndrome (phenotypic diarrhea of infancy). *Gastroenterology* **2010**, *138*, 2388–2398. [CrossRef]

60. Avery, G.B.; Villavicencio, O.; Lilly, J.R.; Randolph, J.G. Intractable diarrhea in early infancy. *Pediatrics* **1968**, *41*, 712–722.

61. Stankler, L.; Lloyd, D.; Pollitt, R.J.; Gray, E.S.; Thom, H.; Russell, G. Unexplained diarrhoea and failure to thrive in 2 siblings with unusual facies and abnormal scalp hair shafts: a new syndrome. *Arch. Dis. Child.* **1982**, *57*, 212–216. [CrossRef]

62. Girault, D.; Goulet, O.; Le Deist, F.; Brousse, N.; Colomb, V.; Cesarini, J.P.; de Potter, S.; Canioni, D.; Griscelli, C.; Fischer, A.; et al. Intractable infant diarrhea associated with phenotypic abnormalities and immunodeficiency. *J. Pediatr.* **1994**, *125*, 36–42. [CrossRef]

63. Dweikat, I.; Sultan, M.; Maraqa, N.; Hindi, T.; Abu-Rmeileh, S.; Abu-Libdeh, B. Tricho-hepato-enteric syndrome: a case of hemochromatosis with intractable diarrhea, dysmorphic features, and hair abnormality. *Am. J. Med. Genet. A* **2007**, *143A*, 581–583. [CrossRef] [PubMed]

64. Fernando, M.M.; Stevens, C.R.; Sabeti, P.C.; Walsh, E.C.; McWhinnie, A.J.; Shah, A.; Green, T.; Rioux, J.D.; Vyse, T.J. Identification of two independent risk factors for lupus within the MHC in United Kingdom families. *PLoS Genet.* **2007**, *3*, e192. [CrossRef] [PubMed]

65. McKay, G.J.; Silvestri, G.; Patterson, C.C.; Hogg, R.E.; Chakravarthy, U.; Hughes, A.E. Further assessment of the complement component 2 and factor B region associated with age-related macular degeneration. *Invest. Ophthalmol. Vis. Sci.* **2009**, *50*, 533–539. [CrossRef] [PubMed]

66. Kopplin, L.J.; Igo, R.P., Jr.; Wang, Y.; Sivakumaran, T.A.; Hagstrom, S.A.; Peachey, N.S.; Francis, P.J.; Klein, M.L.; SanGiovanni, J.P.; Chew, E.Y.; et al. Genome-wide association identifies SKIV2L and MYRIP as protective factors for age-related macular degeneration. *Genes Immun.* **2010**, *11*, 609–621. [CrossRef] [PubMed]

67. Lu, F.; Shi, Y.; Qu, C.; Zhao, P.; Liu, X.; Gong, B.; Ma, S.; Zhou, Y.; Zhang, Q.; Fei, P.; et al. A genetic variant in the SKIV2L gene is significantly associated with age-related macular degeneration in a Han Chinese population. *Invest. Ophthalmol. Vis. Sci.* **2013**, *54*, 2911–2917. [CrossRef] [PubMed]

68. Yoneyama, S.; Sakurada, Y.; Mabuchi, F.; Sugiyama, A.; Kubota, T.; Iijima, H. Genetic variants in the SKIV2L gene in exudative age-related macular degeneration in the Japanese population. *Ophthalmic Genet.* **2014**, *35*, 151–155. [CrossRef] [PubMed]

69. Liu, K.; Chen, L.J.; Tam, P.O.; Shi, Y.; Lai, T.Y.; Liu, D.T.; Chiang, S.W.; Yang, M.; Yang, Z.; Pang, C.P. Associations of the C2-CFB-RDBP-SKIV2L locus with age-related macular degeneration and polypoidal choroidal vasculopathy. *Ophthalmology* **2013**, *120*, 837–843. [CrossRef] [PubMed]

70. Kondo, N.; Honda, S.; Kuno, S.; Negi, A. Role of RDBP and SKIV2L variants in the major histocompatibility complex class III region in polypoidal choroidal vasculopathy etiology. *Ophthalmology* **2009**, *116*, 1502–1509. [CrossRef] [PubMed]

71. Fabre, A.; Martinez-Vinson, C.; Goulet, O.; Badens, C. Syndromic diarrhea/Tricho-hepato-enteric syndrome. *Orphanet. J. Rare Dis.* **2013**, *8*, 5. [CrossRef] [PubMed]

72. Lee, W.S.; Teo, K.M.; Ng, R.T.; Chong, S.Y.; Kee, B.P.; Chua, K.H. Novel mutations in SKIV2L and TTC37 genes in Malaysian children with trichohepatoenteric syndrome. *Gene* **2016**, *586*, 1–6. [CrossRef] [PubMed]

73. Zheng, B.; Pan, J.; Jin, Y.; Wang, C.; Liu, Z. Targeted next-generation sequencing identification of a novel missense mutation of the SKIV2L gene in a patient with trichohepatoenteric syndrome. *Mol. Med. Rep.* **2016**, *14*, 2107–2110. [CrossRef] [PubMed]

74. Hiejima, E.; Yasumi, T.; Nakase, H.; Matsuura, M.; Honzawa, Y.; Higuchi, H.; Okafuji, I.; Yorifuji, T.; Tanaka, T.; Izawa, K.; et al. Tricho-hepato-enteric syndrome with novel SKIV2L gene mutations: A case report. *Medicine (Baltimore)* **2017**, *96*, e8601. [CrossRef] [PubMed]

75. Xinias, I.; Mavroudi, A.; Mouselimis, D.; Tsarouchas, A.; Vasilaki, K.; Roilides, I.; Lacaille, F.; Giouleme, O. Trichohepatoenteric syndrome: A rare mutation in SKIV2L gene in the first Balkan reported case. *SAGE Open Med. Case Rep.* **2018**, *6*, 2050313X18807795. [CrossRef] [PubMed]

76. Ashton, J.J.; Andreoletti, G.; Coelho, T.; Haggarty, R.; Batra, A.; Afzal, N.A.; Beattie, R.M.; Ennis, S. Identification of Variants in Genes Associated with Single-gene Inflammatory Bowel Disease by Whole-exome Sequencing. *Inflamm. Bowel Dis.* **2016**, *22*, 2317–2327. [CrossRef]

77. Vely, F.; Barlogis, V.; Marinier, E.; Coste, M.E.; Dubern, B.; Dugelay, E.; Lemale, J.; Martinez-Vinson, C.; Peretti, N.; Perry, A.; et al. Combined Immunodeficiency in Patients With Trichohepatoenteric Syndrome. *Front. Immunol.* **2018**, *9*, 1036. [CrossRef] [PubMed]

78. Bourgeois, P.; Esteve, C.; Chaix, C.; Beroud, C.; Levy, N.; consortium, T.c.; Fabre, A.; Badens, C. Tricho-Hepato-Enteric Syndrome mutation update: Mutations spectrum of TTC37 and SKIV2L, clinical analysis and future prospects. *Hum. Mutat.* **2018**, *39*, 774–789. [CrossRef]

79. Verloes, A.; Lombet, J.; Lambert, Y.; Hubert, A.F.; Deprez, M.; Fridman, V.; Gosseye, S.; Rigo, J.; Sokal, E. Tricho-hepato-enteric syndrome: Further delineation of a distinct syndrome with neonatal hemochromatosis phenotype, intractable diarrhea, and hair anomalies. *Am. J. Med. Genet.* **1997**, *68*, 391–395. [CrossRef]

80. Poulton, C.; Pathak, G.; Mina, K.; Lassman, T.; Azmanov, D.N.; McCormack, E.; Broley, S.; Dreyer, L.; Gration, D.; Taylor, E.; et al. Tricho-hepatic-enteric syndrome (THES) without intractable diarrhoea. *Gene* **2019**, *699*, 110–114. [CrossRef]

81. Monies, D.M.; Rahbeeni, Z.; Abouelhoda, M.; Naim, E.A.; Al-Younes, B.; Meyer, B.F.; Al-Mehaidib, A. Expanding phenotypic and allelic heterogeneity of tricho-hepato-enteric syndrome. *J. Pediatr. Gastroenterol. Nutr.* **2015**, *60*, 352–356. [CrossRef]

82. Lintner, K.E.; Patwardhan, A.; Rider, L.G.; Abdul-Aziz, R.; Wu, Y.L.; Lundstrom, E.; Padyukov, L.; Zhou, B.; Alhomosh, A.; Newsom, D.; et al. Gene copy-number variations (CNVs) of complement C4 and C4A deficiency in genetic risk and pathogenesis of juvenile dermatomyositis. *Ann. Rheum. Dis.* **2016**, *75*, 1599–1606. [CrossRef] [PubMed]

83. Baechler, E.C.; Batliwalla, F.M.; Karypis, G.; Gaffney, P.M.; Ortmann, W.A.; Espe, K.J.; Shark, K.B.; Grande, W.J.; Hughes, K.M.; Kapur, V.; et al. Interferon-inducible gene expression signature in peripheral blood cells of patients with severe lupus. *Proc. Natl. Acad. Sci. USA* **2003**, *100*, 2610–2615. [CrossRef] [PubMed]

84. Ronnblom, L.; Pascual, V. The innate immune system in SLE: type I interferons and dendritic cells. *Lupus* **2008**, *17*, 394–399. [CrossRef] [PubMed]

85. Xue, Y.; Bai, X.; Lee, I.; Kallstrom, G.; Ho, J.; Brown, J.; Stevens, A.; Johnson, A.W. Saccharomyces cerevisiae RAI1 (YGL246c) is homologous to human DOM3Z and encodes a protein that binds the nuclear exoribonuclease Rat1p. *Mol. Cell Biol.* **2000**, *20*, 4006–4015. [CrossRef] [PubMed]

86. Xiang, S.; Cooper-Morgan, A.; Jiao, X.; Kiledjian, M.; Manley, J.L.; Tong, L. Structure and function of the 5'->3' exoribonuclease Rat1 and its activating partner Rai1. *Nature* **2009**, *458*, 784–788. [CrossRef] [PubMed]

87. Jiao, X.; Xiang, S.; Oh, C.; Martin, C.E.; Tong, L.; Kiledjian, M. Identification of a quality-control mechanism for mRNA 5'-end capping. *Nature* **2010**, *467*, 608–611. [CrossRef] [PubMed]

88. Chang, J.H.; Jiao, X.; Chiba, K.; Oh, C.; Martin, C.E.; Kiledjian, M.; Tong, L. Dxo1 is a new type of eukaryotic enzyme with both decapping and 5'-3' exoribonuclease activity. *Nat. Struct. Mol. Biol.* **2012**, *19*, 1011–1017. [CrossRef] [PubMed]

89. Jiao, X.; Chang, J.H.; Kilic, T.; Tong, L.; Kiledjian, M. A mammalian pre-mRNA 5' end capping quality control mechanism and an unexpected link of capping to pre-mRNA processing. *Mol. Cell* **2013**, *50*, 104–115. [CrossRef] [PubMed]

90. Zhai, L.T.; Xiang, S. mRNA quality control at the 5' end. *J. Zhejiang Univ. Sci. B* **2014**, *15*, 438–443. [CrossRef]

91. Kiledjian, M. Eukaryotic RNA 5'-End NAD(+) Capping and DeNADding. *Trends Cell Biol* **2018**, *28*, 454–464. [CrossRef]

92. Picard-Jean, F.; Brand, C.; Tremblay-Letourneau, M.; Allaire, A.; Beaudoin, M.C.; Boudreault, S.; Duval, C.; Rainville-Sirois, J.; Robert, F.; Pelletier, J.; et al. Correction: 2'-O-methylation of the mRNA cap protects RNAs from decapping and degradation by DXO. *PLoS ONE* **2018**, *13*, e0202308. [CrossRef] [PubMed]

93. Jiao, X.; Doamekpor, S.K.; Bird, J.G.; Nickels, B.E.; Tong, L.; Hart, R.P.; Kiledjian, M. 5' End Nicotinamide Adenine Dinucleotide Cap in Human Cells Promotes RNA Decay through DXO-Mediated deNADding. *Cell* **2017**, *168*, 1015–1027.e10. [CrossRef] [PubMed]

94. Belt, K.T.; Yu, C.Y.; Carroll, M.C.; Porter, R.R. Polymorphism of human complement component C4. *Immunogenetics* **1985**, *21*, 173–180. [CrossRef] [PubMed]

95. Yang, Z.; Mendoza, A.R.; Welch, T.R.; Zipf, W.B.; Yu, C.Y. Modular variations of the human major histocompatibility complex class III genes for serine/threonine kinase RP, complement component C4, steroid 21-hydroxylase CYP21, and tenascin TNX (the RCCX module). A mechanism for gene deletions and disease associations. *J. Biol. Chem.* **1999**, *274*, 12147–12156. [PubMed]

96. Yu, C.Y.; Chung, E.K.; Yang, Y.; Blanchong, C.A.; Jacobsen, N.; Saxena, K.; Yang, Z.; Miller, W.; Varga, L.; Fust, G. Dancing with complement C4 and the RP-C4-CYP21-TNX (RCCX) modules of the major histocompatibility complex. *Prog. Nucleic Acid Res. Mol. Biol.* **2003**, *75*, 217–292. [PubMed]

97. Sargent, C.A.; Anderson, M.J.; Hsieh, S.L.; Kendall, E.; Gomez-Escobar, N.; Campbell, R.D. Characterisation of the novel gene G11 lying adjacent to the complement C4A gene in the human major histocompatibility complex. *Hum. Mol. Genet.* **1994**, *3*, 481–488. [CrossRef] [PubMed]

98. Gomez-Escobar, N.; Chou, C.F.; Lin, W.W.; Hsieh, S.L.; Campbell, R.D. The G11 gene located in the major histocompatibility complex encodes a novel nuclear serine/threonine protein kinase. *J. Biol. Chem.* **1998**, *273*, 30954–30960. [CrossRef] [PubMed]

99. Blanchong, C.A.; Zhou, B.; Rupert, K.L.; Chung, E.K.; Jones, K.N.; Sotos, J.F.; Zipf, W.B.; Rennebohm, R.M.; Yu, C.Y. Deficiencies of human complement component C4A and C4B and heterozygosity in length variants of RP-C4-CYP21-TNX (RCCX) modules in caucasians. The load of RCCX genetic diversity on major histocompatibility complex-associated disease. *J. Exp. Med.* **2000**, *191*, 2183–2196. [CrossRef] [PubMed]

100. Ostertag, E.M.; Goodier, J.L.; Zhang, Y.; Kazazian, H.H., Jr. SVA elements are nonautonomous retrotransposons that cause disease in humans. *Am. J. Hum. Genet.* **2003**, *73*, 1444–1451. [CrossRef]

101. Hancks, D.C.; Kazazian, H.H., Jr. SVA retrotransposons: Evolution and genetic instability. *Semin Cancer Biol* **2010**, *20*, 234–245. [CrossRef]

102. Savage, A.L.; Bubb, V.J.; Breen, G.; Quinn, J.P. Characterisation of the potential function of SVA retrotransposons to modulate gene expression patterns. *BMC Evol. Biol.* **2013**, *13*, 101. [CrossRef] [PubMed]

103. Lehner, B.; Semple, J.I.; Brown, S.E.; Counsell, D.; Campbell, R.D.; Sanderson, C.M. Analysis of a high-throughput yeast two-hybrid system and its use to predict the function of intracellular proteins encoded within the human MHC class III region. *Genomics* **2004**, *83*, 153–167. [CrossRef]

104. STK19 serine/threonine kinase 19 [Homo sapiens (human)]. Available online: https://www.ncbi.nlm.nih.gov/gene/8859 (accessed on 1 June 2019).

105. Boeing, S.; Williamson, L.; Encheva, V.; Gori, I.; Saunders, R.E.; Instrell, R.; Aygun, O.; Rodriguez-Martinez, M.; Weems, J.C.; Kelly, G.P.; et al. Multiomic Analysis of the UV-Induced DNA Damage Response. *Cell Rep.* **2016**, *15*, 1597–1610. [CrossRef] [PubMed]

106. Hodis, E.; Watson, I.R.; Kryukov, G.V.; Arold, S.T.; Imielinski, M.; Theurillat, J.P.; Nickerson, E.; Auclair, D.; Li, L.; Place, C.; et al. A landscape of driver mutations in melanoma. *Cell* **2012**, *150*, 251–263. [CrossRef] [PubMed]

107. Shitara, D.; Tell-Marti, G.; Badenas, C.; Enokihara, M.M.; Alos, L.; Larque, A.B.; Michalany, N.; Puig-Butille, J.A.; Carrera, C.; Malvehy, J.; et al. Mutational status of naevus-associated melanomas. *Br. J. Dermatol.* **2015**, *173*, 671–680. [CrossRef] [PubMed]

108. Bonilla, X.; Parmentier, L.; King, B.; Bezrukov, F.; Kaya, G.; Zoete, V.; Seplyarskiy, V.B.; Sharpe, H.J.; McKee, T.; Letourneau, A.; et al. Genomic analysis identifies new drivers and progression pathways in skin basal cell carcinoma. *Nat. Genet.* **2016**, *48*, 398–406. [CrossRef] [PubMed]

109. Yin, C.; Zhu, B.; Zhang, T.; Liu, T.; Chen, S.; Liu, Y.; Li, X.; Miao, X.; Li, S.; Mi, X.; et al. Pharmacological Targeting of STK19 Inhibits Oncogenic NRAS-Driven Melanomagenesis. *Cell* **2019**, *176*, 1113–1127.e16. [CrossRef] [PubMed]

110. Ishikawa, T. Next-generation sequencing traces human induced pluripotent stem cell lines clonally generated from heterogeneous cancer tissue. *World J. Stem Cells* **2017**, *9*, 77–88. [CrossRef]

111. Kraja, A.T.; Chasman, D.I.; North, K.E.; Reiner, A.P.; Yanek, L.R.; Kilpelainen, T.O.; Smith, J.A.; Dehghan, A.; Dupuis, J.; Johnson, A.D.; et al. Pleiotropic genes for metabolic syndrome and inflammation. *Mol. Genet. Metab.* **2014**, *112*, 317–338. [CrossRef]

112. Amare, A.T.; Vaez, A.; Hsu, Y.H.; Direk, N.; Kamali, Z.; Howard, D.M.; McIntosh, A.M.; Tiemeier, H.; Bultmann, U.; Snieder, H.; et al. Bivariate genome-wide association analyses of the broad depression phenotype combined with major depressive disorder, bipolar disorder or schizophrenia reveal eight novel genetic loci for depression. *Mol. Psychiatry* **2019**. [CrossRef]

113. Negative elongation factor E [Drosophila melanogaster]. Available online: https://www.ncbi.nlm.nih.gov/protein/NP_648241.1 (accessed on 1 June 2019).

114. Negative elongation factor E isoform X2 [Danio rerio]. Available online: https://www.ncbi.nlm.nih.gov/protein/XP_005170084.1 (accessed on 1 June 2019).

115. Paulsen, J.E.; Capowski, E.E.; Strome, S. Phenotypic and molecular analysis of mes-3, a maternal-effect gene required for proliferation and viability of the germ line in C. elegans. *Genetics* **1995**, *141*, 1383–1398. [PubMed]

116. Session, A.M.; Uno, Y.; Kwon, T.; Chapman, J.A.; Toyoda, A.; Takahashi, S.; Fukui, A.; Hikosaka, A.; Suzuki, A.; Kondo, M.; et al. Genome evolution in the allotetraploid frog Xenopus laevis. *Nature* **2016**, *538*, 336–343. [CrossRef] [PubMed]

117. Milner, C.M.; Campbell, R.D. Genetic organization of the human MHC class III region. *Front. Biosci.* **2001**, *6*, D914–D926. [CrossRef] [PubMed]

118. Janeway, C.A., Jr.; Medzhitov, R. Innate immune recognition. *Annu. Rev. Immunol.* **2002**, *20*, 197–216. [CrossRef] [PubMed]

119. Medzhitov, R.; Janeway, C., Jr. Innate immunity. *N. Engl. J. Med.* **2000**, *343*, 338–344. [CrossRef] [PubMed]

120. Note added at update: A recent review that is highly relevant and provides comprehensive background knowledge to this article: Cramer, P. Organization and regulation of gene transcription. *Nature* **2019**, *573*, 45–54.

Review

Long Noncoding RNA *HCP5*, a Hybrid HLA Class I Endogenous Retroviral Gene: Structure, Expression, and Disease Associations

Jerzy K. Kulski [1,2]

[1] Faculty of Health and Medical Sciences, UWA Medical School, The University of Western Australia, Crawley, WA 6009, Australia; kulski@me.com or yurek.kulski@uwa.edu.au; Tel.: +61-047-799-9943

[2] Department of Molecular Life Science, Division of Basic Medical Science and Molecular Medicine, Tokai University School of Medicine, Isehara 259-1193, Japan

Received: 5 May 2019; Accepted: 17 May 2019; Published: 20 May 2019

Abstract: The *HCP5* RNA gene (NCBI ID: 10866) is located centromeric of the *HLA-B* gene and between the *MICA* and *MICB* genes within the major histocompatibility complex (MHC) class I region. It is a human species-specific gene that codes for a long noncoding RNA (lncRNA), composed mostly of an ancient ancestral endogenous antisense 3′ long terminal repeat (LTR, and part of the internal *pol* antisense sequence of endogenous retrovirus (ERV) type 16 linked to a human leukocyte antigen (HLA) class I promoter and leader sequence at the 5′-end. Since its discovery in 1993, many disease association and gene expression studies have shown that *HCP5* is a regulatory lncRNA involved in adaptive and innate immune responses and associated with the promotion of some autoimmune diseases and cancers. The gene sequence acts as a genomic anchor point for binding transcription factors, enhancers, and chromatin remodeling enzymes in the regulation of transcription and chromatin folding. The *HCP5* antisense retroviral transcript also interacts with regulatory microRNA and immune and cellular checkpoints in cancers suggesting its potential as a drug target for novel antitumor therapeutics.

Keywords: HCP5; lncRNA; MHC; HLA; human endogenous retrovirus (HERV); cancer; autoimmune diseases; competing endogenous RNA (ceRNA); human immunodeficiency virus (HIV); human papillomavirus (HPV)

1. Introduction

The human major histocompatibility complex (MHC), also known as the human leukocyte antigen (HLA), covers 0.13% of the human genome and spans ~4 Mbp on the short arm of chromosome six at position 6p21 within a region that contains more than 250 annotated genes and pseudogenes [1,2]. The classical class I and class II regions within the MHC have extensive patterns of linkage disequilibrium (LD), and a high degree of single nucleotide polymorphisms (SNPs) at the HLA genes can differentiate worldwide populations [1,3–5]. HLA polymorphisms are a crucial determinant of the adaptive immune response to infectious agents, allograft success, or rejection and self/nonself immune recognition that can contribute to more autoimmune diseases than any other region of the genome [1,2,6–8]. Apart from the adaptive immune response, MHC class I molecules have a role in brain development, synaptic plasticity, axonal regeneration, and immune-mediated neurodegeneration [9–12]. At least half of the molecules encoded by this highly polymorphic locus are involved in antigen processing and presentation, inflammation regulation, the complement system, and the innate and adaptive immune responses, highlighting the importance of the MHC in immune-mediated autoimmune and infectious diseases [1,2]. Polymorphisms expressed by the MHC genomic region influence many critical biological traits and individuals' susceptibility to the development of chronic autoimmune diseases such as type I diabetes, rheumatoid arthritis, celiac disease, psoriasis, ankylosing spondylitis, multiple sclerosis, Graves' disease, schizophrenia, bipolar disorder, inflammatory bowel disease, and dermatomyositis [2,6–8]. Furthermore, different viral infections and cancers are associated strongly with the suppression of MHC genomic expression activity, particularly in the region of the MHC class I and class II loci [6,13–15].

There are tens of thousands of genomic loci that express microRNA (miRNA) [16] and lncRNA [17–20], but only about 50 have been investigated in any great detail with respect to their role in the regulation of the immune system and disease [21–24]. Although there are many miRNA and lncRNA loci within the MHC genomic region, they have been ignored largely in favor of studies on polymorphisms of the HLA class I and class II gene loci in health, disease, and transplantation cell/tissue/organ typing [25,26]. This review focuses on the structure and function of only one of these HLA lncRNA, the *HCP5* lncRNA, which is located between the *MICA* and *MICB* genes and ~105 kb centromeric of the *HLA-B* gene.

In 1993, Vernet et al. [27] discovered a novel coding sequence belonging to a new multicopy pseudogene family P5 that they mapped within the HLA class I region and named *P5-1* (alias for *HCP5*). They found that it expressed a 2.5-kb transcript in human B-cells, phytohemagglutinin-activated lymphocytes, a natural killer-like cell line, normal spleen, hepatocellular carcinoma, neuroblastoma, and other non-lymphoid tissue, but not in T-cells. *HCP5 (P5-1)* appeared to be a hybrid sequence created by nonhomologous recombination between two pseudogenes or nonmobile genetic elements that possibly produced a protein comprising 219 amino acids (aa's) [28]. A few years later, the *HCP5 (P5-1)* gene was mapped precisely to a region between the *MICA* and *MICB* genes and downstream, at the centromeric end, of the two classical HLA class I genes *HLA-B* and *HLA-C* (Figure 1) [29]. In 1999, Kulski and Dawkins [30] used the computer programs Censor and RepeatMasker and dot-plot DNA and RNA sequence analyses to demonstrate that the *HCP5* gene sequence and its transcripts were composed mainly of the 3'LTR and *pol* sequences of an ancient *HERV16* insertion, which was a member of the HERVL or class III category of endogenous retroviruses (ERVs) in the human and mammalian genomes [31,32].

Because *HCP5* expressed an antisense transcript that was complementary to retrovirus *pol* mRNA sequences and a 3'LTR, Kulski and Dawkins [30] suggested that it might have a role in immunity to retrovirus infection. They considered that the lncRNA of *HCP5* might hybridize with retroviral sense mRNA sequences to suppress viral transcription, translation, and transport. Eight years later, a single-nucleotide polymorphism (rs2395029) in the *HCP5* gene was associated with *HLA-B*57:01* and correlated with a lower HIV-1 viral set point [33], indicating that these two alleles within a particular haplotype may have a role in viral control [34]. However, when Yoon et al. [35] tested the antisense/antiviral hypothesis for *HCP5* by infecting TZM-bl cells in vitro with HIV-1 and plasmids expressing high levels of *HCP5* transcripts, they observed no restriction with infectivity throughout the viral life cycle. They concluded from their findings that the *HCP5* gene had no direct antiviral effect, and that the association of an *HCP5* variant with viral control most likely was due to an *HLA-B*57:01*-related effect or other functional variants in the haplotype or both. In fact, it appears that the role of *HCP5* in immunity and human disease is far more complex than previously envisioned, and that its antiviral affects might occur by way of some secondary mechanisms such as the possible involvement of miRNA inhibition rather than by hybridization of the *HCP5* transcript with the complementary viral *pol* transcripts.

During the last two decades, *HCP5* SNPs have been associated with many different diseases in genome-wide association studies (GWASs), gene expression studies, and cancer studies investigating tissue and cellular biomarkers of tumor progression and inhibition. To better understand the genetics, molecular biology, and functions of *HCP5*, this paper reviewed the available data and literature on the genomic organization, structure, and function of the *HCP5* gene (HLA complex P5 (non-protein coding), HGNC:21659) in health and disease (MIM:604676), particularly its association with autoimmune diseases, cancer, and infections by way of its endogenous interactions with miRNA and various gene targets. Table S1 lists the online databases and repositories that were searched and interrogated to find the available data on *HCP5*. Table S2 lists a summary of the downloaded functional associations in all of the datasets linked with Harmonizome, which is an integrated knowledge base connecting big data with a collection of information about genes and proteins from 114 datasets provided by 66 online resources (Table S1).

2. *HCP5* Genomic Organization, Gene Structure, and the *HERV16* Antisense Transcript

2.1. Human MHC Class I Genomic Structure as Duplication Blocks and HCP5 Location within an HERV16 Duplicated Sequence

The HLA class I genomic region within chromosome six is composed of highly polymorphic frozen blocks [36] that were generated by a series of duplication, deletion, and various genomic rearrangement events during mammalian and primate evolution. Numerous crossover events of a tripartite gene combination (duplicon) of *MIC*, *HERV16*, and *HLA class I* generated the duplication blocks and the HLA class I structural organization of the HLA class I and MIC genes within the beta and alpha blocks of the MHC genomic region in an unidentified ancestor [37]. The ancient endogenous retroviral sequence *HERV16* is repeated at least twelve times, along with the HLA class I coding and noncoding sequences within the alpha and beta block of the MHC class I region, and it appears, along with other retroelements, to have been a recombination site for many of the duplication events involving unequal crossovers [37,38]. The beta-block of approximately 362 kb between the *POU5F1* and *MCCD1* genes harbors the *MICA*, *MICB*, *HLA-B*, and the *HLA-C* genes as well as the *HCP5* gene that is centromeric of a short, fragmented *HLA-X* pseudogene (645 bp) and telomeric of the *HCG26* long noncoding RNA gene (Figure 1). The *HCP5* gene is located between the *MICA* and *MICB* genes at the centromeric end of the HLA beta block, and it also has a neighboring 3538 bp *HERV16* fragmented duplication (alias *P5.8*) at chr6:31383746-31387283 between the *HLA-S* (alias *HLA-17*) pseudogene (919 bp) and the *MICA-AS1* RNA gene [30,38]. The *HCP5* gene is different from all the other duplicated *HERV16* sequences in the MHC in that the ancestral endogenous antisense 3′LTR and part of the internal *pol* antisense sequence of *HERV16* (exon 2) is linked to an HLA class I promoter and leader sequence at the 5′-end (exon 1) that regulates its expression (see Section 2.5).

There are at least four *HCP5* gene sequence variants in the genomic databases to consider: *HCP5-1* with 2630 bp (or 2432 to 2658 bp); *HCP5-2* of 23 kb; *HCP5-3* of 575 bp; and *HCP5-4* of 465 bp ((HGNC: 21659, Entrez Gene: 10866, Ensembl: ENSG00000206337, OMIM: 604676, UniProtKB: Q6MZN7, Vega: OTTHUMG00000031282, and GCID: GC06P031400). The original *P5.1* sequence (GenBank L06175) reported by Vernet et al. [27] was 2535 bp. According to the online Ensembl database (February 2019), *HCP5* has four transcript or splice variants and one gene allele that is associated with 64 phenotypes. In this review, the *HCP5* gene of 2,630 bp is defined by the NCBI Gene ID: 10,866 (updated 31 January 2019) and by its location on the GRCh38.p12 assembly (annotation release 109 in March 2018) at chr6:31,463,180–31,465,809 (NC_000006.12) with a transcript length of 2547 bp, including eight adenines of the RNA polyA tail (NCBI ncRNA reference sequence NR_040662.1). It spans 2630 bp of a 5153 bp (nt positions from 717 to 3346) extended *HCP5* gene reference with 5′ and 3′ noncoding regions (GenBank AB088109.1) and the chromosome six reference sequence NC_000006.12. DNA sequence alignment of NR_040662.1 and AB088109.1 or NC_000006.12 revealed that the *HCP5* gene is composed of two exons: a 100 bp exon 1 and a 91 bp intron and 2355 bp exon 2. The HERV16 sequence begins at nucleotide position 150 within exon 2 of *HCP5* and occupies most of exon 2 of the HCP5 RNA sequence NR_040662.1.

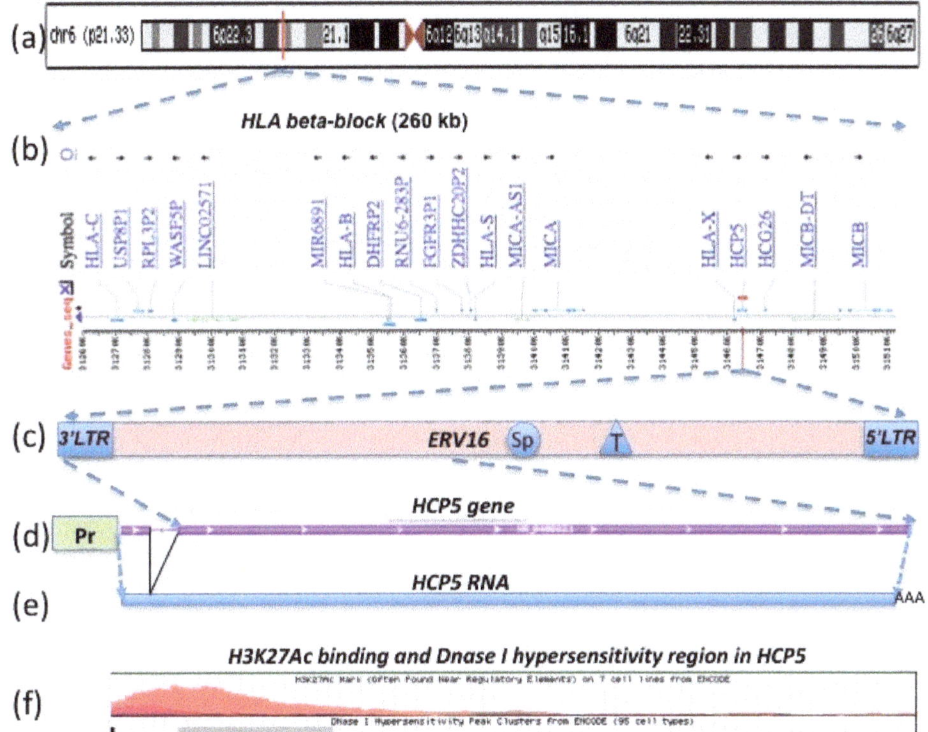

Figure 1. Location of the *HCP5* gene (**d**) within the *ERV16* element (**c**) and the human leukocyte antigen (HLA) class I region of the beta-block (**b**) on chromosome 6 at 6p21.33 (**a**). The HLA class I promoter region (the green rectangle labeled as Pr) for the 2630 bp *HCP5* gene (**d**) initiates transcription of the 2547 bp lnc *HCP5* RNA (**e**). The 91 bp intron in the *HCP5* gene is represented by the thin grey line between the violet rectangular lines (**d**). The *AluSp*, THE1B, and *LTR162B* insertions within the 6173 bp *ERV16* sequence [30] (**c**) are indicated by the labeled circle, triangle, and rectangles, respectively (see Table 1 for more details). The H3K27Ac binding (orange curve) and Dnase I hypersensitivity region (grey horizontal rectangle) associated with the *HCP5* gene sequence (**d**) and sourced from the University of California, Santa Cruz (UCSC) genomic browser (Table S1) are shown on line (**f**).

Instead of simply annotating the *HCP5* gene at positions 31,463,180–31,465,809 on chromosome six, the Ensembl human genome assembly GRCh38.p10 that is linked to various other gene and genomic databases (OMIM, UniProtKB, HGNC, Vega, and GCID) complicated the genomic location by placing the variant (Ensembl: ENSG00000206337) at the extended genomic coordinate of 6:31,400,702 to 31,477,506 (76,805 nt). This extended genomic coordinate for the *HCP5* large variant is confusing because the 5′-end is within the *MICA* gene sequence ~76,000 bp from the transcription start site and overlaps *MICA* by 60% rather than within the 5′-UTR promoter region closer to the transcription start site of the 2630 bp *HCP5* variant (Figure 1, Table 1). Furthermore, these extended genomic coordinates for the large *HCP5* variant can be misleading with respect to the location of SNPs within the *HCP5* gene and the transcription factor regulatory elements within the 5′-UTR of the *HCP5* gene. For example, the *HCP5* SNP rs6938467 (T/G) that is associated significantly with Kawasaki disease in Korean children [39] is positioned at chr6:31,440,139 (GRCh38.p12) well outside the *HCP5* RNA reference sequence (NR_040662.1) and upstream of the *LINC01149* locus starting within the *MICA* gene in the genomic location of chr6:31,400,702–31,465,809 and ENSG00000206337. Therefore, this expanded version of the large *HCP5* variant sequence (ENSG00000206337) with a genomic length of 65,108 to 76,000 bp has been excluded from this review except for occasionally noting whether or not the published *HCP5* SNPs in disease association studies lie inside or outside the 2630 bp version of the *HCP5* gene sequence.

2.2. HCP5 RNA, a Complementary Sequence of ERV16 (an ERV Class III (ERVL))

The gene structure of *HCP5* is unusual because it is a hybrid sequence of an *HLA* class I gene fragment in exon 1 and the fragmented portions of the 3′LTR and internal sequence of *ERV16* in exon 2 [30], annotated by RepeatMasker as *3′LTR16B* and complementary (minus strand) *ERV3-16A3* (Table 1). The entire *ERV16* sequence from *5′LTR16B* to *3′LTR16B* with the internal *ERV3-16A3* is located on the reverse strand between chr6:31,469,592 to 31,463,420 (6173 bp), and it is 2.4 times longer than the *HCP5* gene sequence of 2630 bp (Figure 1). The internal *ERV3-16A3* sequence contained between the flanking LTRs has about 60% similarity to *HERVL*, mainly within the *pol* gene region and with only a slight identity to parts of the *gag* gene. An assumed primer binding sequence (5′-TGGCTTCAGGAGTGGTCC-3′) for leucine-tRNA is present two nucleotides downstream from the 3′ end of the *5′-LTR16B* sequence, which has identity with 15 of 18 nucleotides of *HERVL* [30]. Thus, the *HCP5* 2547 bp transcript of the reference sequence NR_040662.1 terminates about a third of the way from the *3′LTR16B* to the *5′LTR16B2* of the *ERV16* sequence, with the two *LTR16B* sequences possibly acting as promoters and/or enhancers for *HCP5* and various other expressed sequences within the vicinity [30]. The *ERV16* sequence downstream of *HCP5* is also interrupted with insertions from other retroelements including *AluSp*, *THE1B*, and *AluSx* (Table 1), indicating that this *ERV16* sequence is at least as old as the wave of *AluSp/x* insertions that occurred in primates about 37 million years ago [40].

Table 1. The chromosomal location of *HCP5* and the *ERV16* (*LTR16B and ERV3-16A3*) retroelement between the pseudogene *HLA-X* and the *HCG26* lncRNA gene within the HLA class I region [30].

Sequence Name	Location on Chr6 *	Length bp	Orient.	Feature
MERC21	31,461,669–31,462,057	389	+	LTR fragment
HLA-X	31,461,846–31,462,490	645	+	Silent pseudogene
MERC21	31,462,544–31,462,625	82	+	LTR fragment
HLA-UTR	31,462,625–31,463,419	794	+	HLA-5′UTR + exon 1
HCP5-5′ncr	31,462,464–31,463,413	950	+	HLA class I promoter homologs
DNase region	31,463,001–31,463,190	190	+	100% reliability score
HCP5	31,463,180–31,465,809	2630	+	lncRNA gene
3′LTR16B	31,463,420–31,463,715	296	−	3′LTR (nt 150–446 in HCP5 RNA)
ERV3-16A3	31,463,716–31,464,933	1218	−	Internal (nt 447–2547 in HCP5 RNA)
AluSp	31,465,920–31,466,225	306	−	Insertion within ERV3-16A3
ERV3-16A3	31,466,237–31,467,707	1471	−	Internal
THE1B	31,467,708–31,468,050	343	+	Fragmented insertion within ERV3-16A3
ERV3-16A3	31,468,051–31,468,201	151	−	Internal
ERV3-16A3	31,468,362–31,468,644	283	−	Internal
5′LTR16B2	31,468,881–31,469,043	163	−	5′LTR
AluSx	31,469,044–31,469,282	239	−	Insertion within LTR16B
5′LTR16B2	31,469,283–31,469,592	310	−	5′LTR

Table 1. *Cont.*

Sequence Name	Location on Chr6 *	Length bp	Orient.	Feature
L2	31,469,593–31,470,361	769	–	L2 LINE
MLT1E2	31,470,734–31,471,317	584	–	ERL-MaLR, LTR fragment
L2	31,471,325–31,472,047	723	–	L2 LINE
HCG26	31,471,229–31,472,408	1180	+	LncRNA

* Chr6:31461669-31472408, GRCh38.p12 assembly (annotation release 109 in March 2018). 5′ncr is 5′ noncoding region. Orientation of the sequence is on the +ve DNA strand or the −ve (complementary) DNA strand.

2.3. HCP5 and the Human Papillomavirus (HPV) Minor Structural Protein Interacting Protein (PMSP) Gene

A surprising finding during a BLAST search of *HCP5* sequences in GenBank was that a portion of the *HCP5* RNA (NR_040662.1, 1844–2011 bp) shares 99% identity with the 186 bp sequence of the papillomavirus minor structural protein interacting protein *(PMSP)* gene sequence (AJ437509.2) (Figure 2) that is only 7% of the length of the entire *HCP5* sequence. *PMSP* codes for a 61 aa peptide (UniProtKB-Q8TCT4) that interacts with the capsid L2 protein of HPV1, HPV11, and HPV16 in the cytoplasm, and both proteins are transported into cellular nuclear dots for viral assembly [41]. In this regard, the 7% of *HCP5* appears to be the solitary genomic locus for *PMSP* gene expression. Alternatively, the *PMSP* gene might be found at a different genomic locus that is still to be identified. The start of the *PMSP* gene (1 to 22 bp) has 100% identity with *HCP5* (NR_040662.1) at nt positions 37 to 58, and the remainder of the *PMSP* gene has 98% identity with 164 bp of *HCP5* (NR_040662.1) at nt positions 1847 to 2012 located in the *ERV16* portion of *HCP5* (Figure 2). Thus, both the *PMSP* gene and the *HCP5* gene of the human genome share the same part of the *ERV16* sequence, possibly generated as a hybrid at some time before or early in primate history. The 164 bp portion of the *PMSP* sequence within *HCP5* is also present in macaque monkeys with 91% identity, but not in the chimpanzee or gorilla where the *HCP5* sequence between the two *MIC* genes has been deleted in available sequences [42]. The *PMSP* gene also has about 88% identity for 141 bp within the multicopy *Homo sapiens* zinc ribbon domain containing one antisense pseudogene (*ZNRD1ASP*) transcript variant 1, noncoding RNA (2450 bp) at three locations within the class I region of the MHC on chr6:31,465,114–31,465,281 in *MICB*; 30,001,688–30,001,549 between *HLA-J* and *ZNRD1*; and 29,713,247–29,713,129 between *HLA-F* and the *IFITM4P* noncoding RNA gene. The *PMSP* portion of the *HCP5* transcript might be upregulated during warts infections in humans; plantar warts are associated significantly with the SNP rs9267257 that is located at 6:31488485 within the *LOC102725068*, an intergenetic variant position between *HCP5* and *MICB* [43]. Conversely, Karim et al. [44] showed that episomal copies of *HPV16* and *HPV18* in undifferentiated keratinocytes downregulated the expression of *HCP5* and the *HLA-A, -B, -C,* and *-G* gene products that were involved with the antigen presenting pathway in order to allow persistence of latent infections. However, expression of the *HCP5* gene that was downregulated substantially by human papillomavirus (HPV) infection was again upregulated 2.2-fold in HPV-infected keratinocytes after 24 h of treatment with polyI:C compared to greater than three-fold upregulation in HPV-uninfected cells during the same period [44]. Further studies are needed to confirm the possible association between *HCP5*, *PMSP*, and the assembly of the HPV capsids in various papillomaviral infections.

2.4. HLA Class I Leader Sequence and Cytomegalovirus (CMV) Signal Peptides

The *HCP5* RNA sequence has an open reading frame (ORF) that may code for a peptide of 132 aa (Q6M2N7.1) with a domain resembling HLA class I signal peptides [28,30]. It has another ORF with the first 24 of 52 aa of the putative HCP5 peptide starting from methionine and sharing identity with at least 18 of 24 aa of the leader peptide of HLA-A, -B, and -C, whereas the remaining 28 aa have no sequence similarity to any known protein [30]. The first 23 aa, RVMESRTLLLLFSGAVALIOTWA, of the putative 52 aa HCP5 peptide sequence shares at least 65% (15/23 aa) identity and up to 78.3% (18/23 aa) identity with the leader peptides of HLA-A, -B, and -C. However, it is only the MESRTLL portion of the sequence that is found at the start of the 61 aa peptide that is encoded by the *PMSP* gene (Figure 2). Also, the HCP5 peptide sequence VMESRTLLL shares 74% identity with the VMAPRTLIL sequence within the signal peptides of HLA-A and the UL40 protein of the human cytomegalovirus (HCMV), which binds and stabilizes cell surface expression of both the endogenous HLA-E and the HCMV-encoded MHC-I homologue gpUL18 to regulate differentially two distinct natural killer cell evasion pathways [45,46]. HLA-E and gpUL18 bind a restricted subset of peptides derived from the leader peptides of other class I molecules, but whether they bind the putative HCP5 leader peptide or the PMSP leader peptide is not known.

```
PMSP   1                                              ATGGAGTCCCGAACCCTCCTCC--  22
                                                      ||||||||||||||||||||||
HCP5   1     GACTCAGATTCTCCCCAGACGCCAAGGTTGCGGGTCATGGAGTCCCGAACCCTCCTCCTG  60

PMSP         ------------------------------------------------------------
HCP5   61    CTGTTCTCGGGAGCCGTGGCCCTGATCCAGACCTGGGCAGATTACAATTACAATCAAGGC  120


PMSP   23    -------------------------------------------------CAACATATTTCTC  35
                                                              |||||||||||||
HCP5   1801  GCATTCTATCTTATCAGGGCCTACAGTAACATGCCAAAAGTTGCTTCCAACATATTTCTC  1860

PMSP   36    TGCTTTGGATGGGGCATATTTCTGTGCTGTGGATGACATGGCCTTACTCCAGAATCCCAG  95
             ||||||||||||||||||||||||||||||||||||||||||||||||||||||||||||
HCP5   1861  TGCTTTGGATGGGGCATATTTCTGTGCTGTGGATGACATGGCCTTACTCCAGAATCCCAG  1920

PMSP   96    GCCCTCCACTGTGACTCTCCTACTGGTGCTTGGTTCAGCTCCACCCCAAATCTTACCCCA  155
             ||||||||||||||||||||||||||||||||||||||||||||||||||||||||||||
HCP5   1921  GCCCTCCACTGTGACTCTCCTACTGGTGCTTGGTTCAGCTCCACCCCAAATCTTACCCCA  1980

PMSP   156   CCACTGGCACTTTCAGCACCAGGGGGTCTGA                               186
             ||||||||||||||||||||||||||||||
HCP5   1981  CCACTGGCACTTTCAGCACCAGGGGGTCTGAAGGATGGTGACTGCGCCATGGCCTGGATC  2040
```

Figure 2. DNA nucleotide alignment between the papillomavirus minor structural protein (PMSP) (AJ437509.2) and *HCP5* RNA sequence (NR_040662.1).

2.5. HLA-Class I Gene Promoters and Enhancers within HCP5

The *HCP5* genomic sequence has a putative HLA-class I gene promoter for transcriptional activation of the *P5-1* mRNA (Figure 1 and Table S3) that is linked to a 159 bp sequence homologous to exon 1 (signal peptide) and part of intron 1 of classical class I genes [28,30]. Table S3 shows the Clustal W alignment of 360 bp of sequence, including the promoter regions of the *HCP5* and *HLA-A* genes, approximately 120 bp of the start of the *HCP5* RNA sequence, and 159 bp of the start of the *HLA-A* mRNA sequence. The 5′ nonretroviral promoter region of the *HCP5* gene has at least 88% identity with the promoter regions of various class I HLA genes including *HLA-A, -B, -C, -E, -F,* and *-G* [44] (Table S3). There is also 84% to 87% identity between the *HCP5* sequence and the first ~159 bp of the HLA class I transcripts.

The *HCP5* gene promoter and the associated exon and intron 1 of MHC class I genes are located within a 1075 bp genomic region between the *MER4C* and *HERV* sequence (Table 1). The putative gene promoter and the peptide-coding region are linked to the *3′LTR16B* fragment and are in reverse orientation to the *HLA-X* pseudogene (Figure 1) that is positioned between *MER4C* and *MER21B* fragment within the *HCP5* genomic sequence [30]. The HLA-class I gene promoter of *HCP5* has an intact TATA box, CAAT box, and regulatory response sequences for interferon, kappa B1, and kappa B2 that have been described for the regulatory complex region of class I promoters [28,30,47]. Numerous copies of a motif (AGAAA) for heat shock transcription factor are distributed in forward and reverse orientation within the *HCP5* gene promoter region, and a potential cap signal sequence (5′-GACTCAGATTC-3′) for transcription initiation is located 26 nt downstream from the TATA box [30].

Although few experiments have been performed directly on the *HCP5* promoter, the high sequence similarity with the HLA class I promoters suggests that it is regulated and expressed similarly to *HLA-A, -B, -C, -G,* and *-F* but with differences dependent on polymorphisms and mutations providing sequence variants [47] (Table S3) and tissue-specific effects on expression [48]. Regulatory elements and transcription factors associated with the MHC class I promoter region were reviewed recently [47–50] and highlighted that the MHC class I promoter region with the MHC class I NF-κB enhanceosome, together with the RFX-complex, ATF1/CREB, and the NFY-complex (binding the SXY-module), activated MHC class I transcription. Additionally, type II interferons (IFN-γ) can activate MHC class I transcription by upregulation of *IRF1*, which bind to the interferon-sensitive response elements (ISREs) of the MHC class I promoter [51,52]. Gene Expression Omnibus (GEO) expression profiles reveal that *HCP5* gene expression is regulated by IFN-γ (Table 2) along with other MHC class I genes [53–55]. Recently, TGFbeta was shown to induce over-expression of the *HCP5* gene and activity of the S-mothers against decapentaplegic homolog 3 (SMAD3) protein complex in adenocarcinomas [56]. Also, the ubiquitous transcription factor SP1, which is known to regulate MHC class I gene expression both constitutively and in a tissue-specific manner [48], was shown to upregulate HCP5 expression and induce the development of osteosarcoma [57]. Moreover, the osteosarcoma study also found that the transcription factors STAT3, E2F1, and NF-κB1 had no effect on *HCP5* expression. In contrast, NF-κB binding to Enhancer A is necessary for both constitutive and induced MHC class I expression [58], and NF-κB-induced MHC class I expression is most prominent for the *HLA-A* locus, which contains two NF-κB binding sites in its Enhancer A region [49,59]. Figure 3 shows the comparison and strong correlation between *HCP5* and *HLA-B* expression in the same 27 tissues that were the results of the NCBI BioProject PREJEB4337. In general, *HLA-B* transcription is expressed at much higher levels with a range of 1 to 590 reads per kilobase of transcript per million mapped reads (RPKM) than *HCP5* transcription with a range of 1 to 39 RPKM.

Table 2. Effect of interferons and cytokines on HCP5 RNA expression (nd, no difference).

Cytokine	Cell Type	No: Controls/Tests	Up or Down	Geoprofile (206082_at)	Reference PMID
IL10	Peripheral blood mononuclear cells	4/4	down	GDS4551	23449998
KSHV-IRF4	lymphoma	2/2	up	GDS4956	24335298
IFN-alpha	natural killer cells	1/3	up	GDS4163	20334827
IFN-gamma	keratinocytes	5/4	up	GDS1846	14760888
IFN-alpha	hepatocytes (24 h)	3/3	up	GDS4390	22248663
IL28B	hepatocytes (24 h)	3/3	up	GDS4390	22248663
IFN-II	lung epithelium (24 h	4/4	up	GDS2341	16800785
IFN-gamma	bronchial epithelium (24 h)	5/5	up	GDS1256	15985639
IFNg + DXM	bronchial epithelium (24 h)	5/5	up	GDS1256	15985639
IFN-gamma	macrophages	6/6	up	GDS4232	22140520
IFN-gamma	microglia (24 h)	4/4	up	GDS1036	16163375
TNF	endothelial	4/4	nd	GDS1542	16617158
TNF	keratinocytes	4/4	up	GDS1289	15722350
IL1	macrophages	5/5	nd	GDS3005	18498781
IL6	macrophages	5/5	nd	GDS3005	18498781
IL6	macrophages	3/3	nd	GDS3290	18511485
Cigarette smoke	small airway cells	12/12	down	GDS2486	19106307

DXM is dexamethasone.

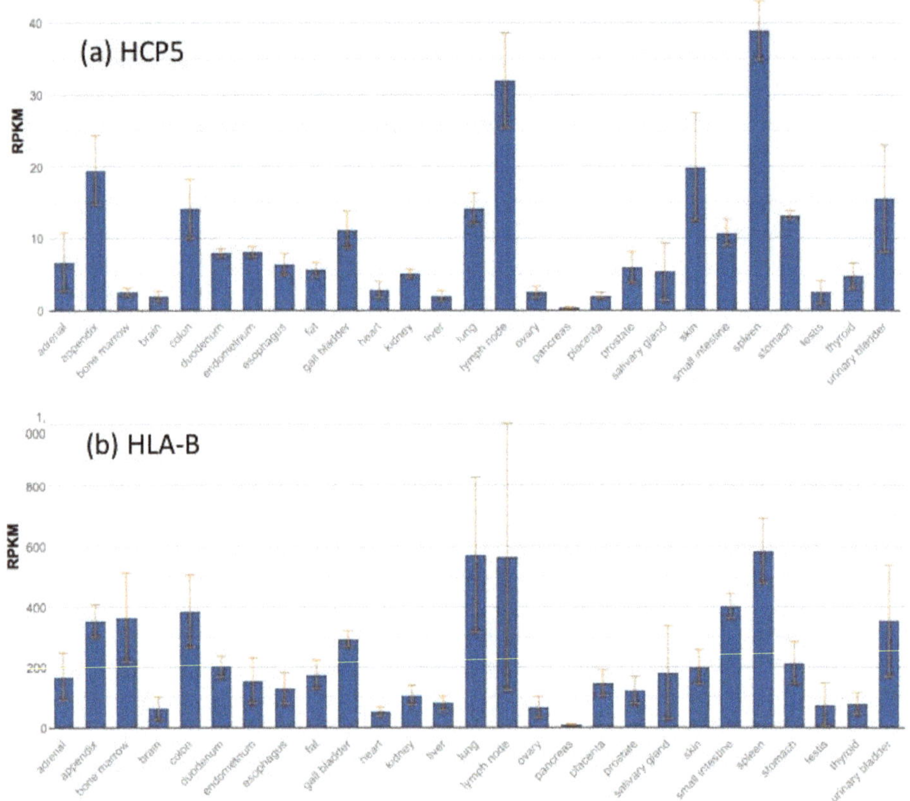

Figure 3. *HCP5* (**a**) and *HLA-B* (**b**) RNA sequences in 27 different normal tissues from 95 human individuals. NCBI BioProject PREJEB4337. RPKM (reads per kilobase of transcript per million mapped reads) on the *y*-axis is a normalized unit of transcript expression [60].

The class I transactivator (CITA) NLRC5 (alias NOD4, NOD27, and CLR16.1) induces genes involved in the MHC class I-dependent antigen processing and presentation pathway; it activates the promoters of MHC class I and related genes by co-binding within the SXY module of MHC class I to form a CITA enhanceosome and induces the expression of classical MHC class I, nonclassical class I, the beta-2 microglobulin, immunoproteasome component (*PSMB9*), peptide transporter (*TAP1*) [61–63], and *HCP5* [64].

The *HCP5* RNA also contains the 3′*LTR16B2* and a substantial portion of the internal *ERV16* sequence (Table 1). It should be noted however that there are ~860 *ERV16* copies of *LTR16B2* with the *ERV16* internal regions and ~18,100 copies of the solitary *LTR16B* within the human genome [31]. These ERVs provide an enormous reservoir of autonomous gene regulatory modules, some of which play important roles in normal regulation of genes and gene networks and influence the transcription of host genes. At least one of these 5′*LTR16B* sequences is known to act as a promoter for the activation of the isoform anaplastic lymphoma kinase (ALK) of the receptor tyrosine kinase (RTK) in ~11% of skin cutaneous melanomas [65]. The regulatory elements of HERV/LTRs tend to locate near to the genes involved in immune responses in order to interact with them, indicating that these regulatory elements play an important role in controlling the immune regulatory network [66]. Therefore, the possibility exists that the 3′*LTR16B2* antisense sequence within the lncRNA of *HCP5* might interact with one or other *LTR16B2* sense sequences distributed across the human genome to regulate the expression of *LTR16B2*-regulated genes outside the MHC genomic region.

2.6. HCP5-CCCTC-Binding Factor (CTCF) Connection and Chromatin Structural Regulation

There are two high-probability CCCTC-binding factor (CTCF) binding sites within the body of the *HCP5* genomic sequence that demarcate the likely position of open and closed chromatin (Figure 4). The *CTCF* gene (ID: 10664) encodes the transcriptional repressor CTCF, also known as 11-zinc protein or CCCTC-binding factor, which is involved in many cellular processes including transcriptional regulation, insulator activity, regulation of interferon-gamma (IFNg) induction of MHC class I and class II gene expression [49,66,67], V(D)J recombination in B and T cells [68], and regulation of chromatin architecture [69]. The primary role of CTCF is in the regulation of the 3D structure of chromatin by binding together strands of DNA to form chromatin loops and anchor DNA to the nuclear lamina [70], but it also has an essential role in blocking the interaction between enhancers and promoters to prevent oncogenesis [71]. This nuclear protein is able to use different combinations of the zinc finger domains to bind different DNA target sequences and proteins, and depending upon the context of the site, the protein can interact with a histone acetyltransferase (HAT)-containing complex as a transcriptional activator or a histone deacetylase (HDAC)-containing complex as a transcriptional repressor. If the protein is bound to a transcriptional insulator element, it can block communication between enhancers and upstream promoters, thereby regulating imprinted expression.

The insulator factor CTCF controls MHC class I and II gene expression and is required for the formation of long-distance chromatin interactions [49,66,67]. A search of the ENCODE database revealed that the proximal end of the *HCP5* sequence had two CTCF binding sites in addition to peak clusters of DNase I hypersensitivity, H3K4me3, and H3K27ac with Z scores >1.8 (Figure 4). In this regard, the first 100 nucleotides of *HCP5* exon 1 has a DNase I hypersensitivity region of CpG island DNA (Figure 1f, Table 1) that has 100% sequence homology with CpG island DNA on chr 22 (NCBI, AJ236677.1) [72]. A number of published studies have shown that there are differential methylation sites such as CTCF within the *HCP5* gene, and that hypomethylation is associated strongly with the regulation of a number of phenotypes and diseases (see Section 4, *HCP5* methylomics). The two experimentally unverified CTCF binding sites in the Ensembl database were at the *HCP5* core positions ENSR0000078786 and ENSR00000195544. Of the 123 cells studied for position ENSR0000078786, 2 were active, 28 inactive, 3 poised, 6 repressed, and the regulatory activity was unknown for 84 cell types. The cells in the active or poised positions for ENSR0000078786 were DND-41 (a human leukemic T-cell line with a p53 mutation) and MCF7, and those for ENSR00000787587 were A549 (an epithelial lung carcinoma), a keratinocyte, and MCF7. The presence of CTCF binding sites at the *HCP5* locus (Figure 4) is noteworthy given the importance of the CTCF transcription factor in diverse genomic regulatory functions including regulating chromatin structure, gene expression, and development of various acute and chronic diseases. IFNg probably activates the CTCF binding sites at the *HCP5* locus [51–55], but this requires experimental confirmation.

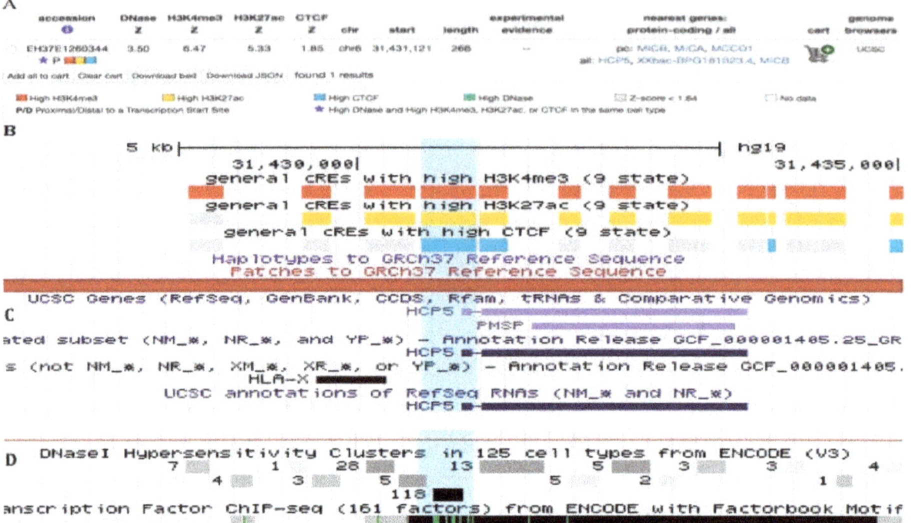

Figure 4. CTCF-binding sites proximal to and within the *HCP5* gene. (**A**) ENCODE accession number (EH37E1260344) and statistical scores for presence of binding sites for CTCF, H3K4me3, and H3K27ac and DNaseI hypersensitivity clusters. (**B**) The ENCODE regulation tracking details from the UCSC genome browser (Table S1) show the CTCF locations (blue horizontal blocks on line 3) relative to (**C**) the *HCP5* and *HLA-X* gene positions below the horizontal red line. In (**B**) the cREs with a high probability for binding H3K4me3 are the red blocks on line 1, and those with high probability for binding H3K27ac are the yellow blocks on line 2. (**D**) The location of the DNaseI hypersensitivity clusters and 161 transcription factor binding sites from ChIP-seq data analysis archived in ENCODE are shown below the *HLA-X* and *HCP5* genes.

2.7. Structural Deletion of HCP5 and MICA Genes

Examination of the *HCP5* locus in the Database of Genomic Variants (Table S1) revealed that this gene region is the site of copy number variants (CNVs) with multiple reported deletions (losses) and duplications (gains). The deletion of this locus is associated mostly with a null haplotype for the *MICA* and *MICB* genes and LD with *HLA-B48:01*—reported initially to occur in 3.7% of East Asian individuals [42,73]. Subsequently, the *HLA-B48:01* allele was genotyped at a frequency of 26% or higher in some ethnic groups such as the Ami of Taiwan [74] or the Angaite Amerindian community in Paraguay [75]. In a cohort of 2026 disease-free Caucasians and African Americans, 38 deletions and four duplications of the *HCP5* locus were detected [76]—29 of the deletions had the same apparent breakpoints and seven had different breakpoints. For one of the reported sample sets of 65% Caucasians and 34% African Americans, deletions were detected in 31 African Americans and only seven Caucasians [76]. In another study of 1854 samples that represented 790 Europeans and 1064 individuals from 51 different populations, *HCP5* locus deletions were detected in 34 ethnically diverse individuals (e.g., Chinese, Japanese, Palestinian, Pakistani, and Yoruban) and in only three Europeans [77]. The unforeseen deletion of *HCP5* might interfere with the *HCP5* genotyping assay for abacavir hypersensitivity [78]. However, an alternative and simple proxy assay for *HLA-B*57* and *HLA-B*48:01* could be the structurally polymorphic *AluMICB*, which is a lineage indel within intron 1 of the *MICB* gene [79].

The *HCP5* genomic region has been deleted in gorillas and chimpanzees because of recombination and fusion between the *MICB* and *MICA* genes [80–82]. On the other hand, the transcriptional activity and function of the *P5.1.1/HCP5*-like ortholog in the Macaque monkey is not known, although its structure has diverged substantially from that of *HCP5* in the human. The human lineages studied to date carry two functional MIC genes, *MICA* and *MICB*, and a number of unprocessed pseudogenes (*MICC, D, E, F,* and *G*) [1,2]. *MICA* and *MICB* appear to have diverged from each other 33–44 million years ago [83,84], before the divergence of chimpanzee and human. Nevertheless, deletion of *HCP5* and some forms of the *MIC* genes from humans, gorillas, and chimpanzees highlights the plasticity of this intergenic genomic region in different primate species.

In Japan, the *HLA-B*48:01* haplotype was estimated to have a 3.2% frequency with 62% of these haplotypes carrying the *MIC/HCP5* null configuration. Based on the current size of the Japanese population and a homozygote rate of 0.1025%, more than 80,000 individuals could be expected to have the homozygous *HLA-B*48:01* haplotype [85]. Apparently, the *MIC/HCP5*-deficient individuals have no overt clinical symptoms of endangered health, unlike those without MHC class I or class II molecules who suffer from mild to severe immune deficiency syndromes and diseases. In an unpublished study, homozygous *HLA-B*48:01* individuals were reported to have no immunological defect in T cell or B cell populations, no change in T cell or B cell receptor repertoire, no obvious change in immunological signals, and no known susceptibility to any diseases [85]. However, ethnic Thais with the *HLA-B48* haplotypes might be susceptible to developing secondary dengue hemorrhagic fever [86], and deletion between the two MIC genes was associated with nasopharyngeal carcinoma in Malaysian Chinese [87] but not Southern Chinese Han [88].

3. *HCP5* Single Nucleotide Variant (SNV) Associations with Disease

Many reviews are available that outline the enormous advancements made during the last twenty years to improve our understanding of the MHC locus and genetic susceptibility to autoimmune and infectious diseases because of the availability of dense genotyping platforms and hybridization chips such the custom-made Illumina Infinium SNP chip (Immunochip) [8,89]. The online Genome-Wide Association Studies (GWAS) Catalog [90] revealed that *HCP5* has 131 associations in 86 studies (86 variants) with most severe consequences associated with SNP in the regulatory and transcript region from chromosomal position 31,462,917 to 31,466,072. Some of the different diseases and phenotypes that were associated with the internal and 5′ or 3′ UTR single nucleotide variants (SNVs) of the *HCP5* gene, using dense genotyping platforms such as Illumina's HumanHap550 BeadChips or candidate genotyping by PCR, are listed in Table 3. However, many of these associations need to be considered with caution because they are likely to be part of extended haplotypes or in LD with other identified or unidentified genes in the MHC. In most of these association studies, the *HCP5* SNV was in strong LD with SNV from other MHC gene loci. This is not surprising given the relatively large genetic diversity within the HLA class I genomic blocks observed between different HLA haplotypes [5], which has resulted in the formation of ancestral haplotypes with relatively frozen polymorphic blocks at the HLA class I and class II loci [36].

Table 3. Diseases and phenotypes associated with *HCP5* SNV and retroelement (RE).

SNV ID	Chr Position	RE or Pseudogene	HCP5 Position	Ref	Disease or Phenotype
rs3094228	6:31462150	HLA-X	-	[91]	Thyroid peroxidase antibody positivity
rs115846244	6:31462283	HLA-X	-	[43]	Tonsillectomy
rs4360170	6:31462582	MERC21	-	[43]	Cold sores
rs3094605	6:31462917	HLA promoter	5′UTR	[92]	Lung cancer
rs3099840	6:31462944	HLA promoter	5′UTR	[92]	Lung cancer
rs3132090	6:31462975	HLA promoter	5′UTR	[92]	Lung cancer
rs2255221	6:31463914	ERV3-16A3	internal	[93]	HIV-1 control
rs2395029	6:31464003	ERV3-16A3	internal	[94]	Flucloxacillin drug liver injury
rs2395029	6:31464003	ERV3-16A3	internal	[33,34]	HIV-1 control

Table 3. *Cont.*

SNV ID	Chr Position	RE or Pseudogene	HCP5 Position	Ref	Disease or Phenotype
rs2395029	6:31464003	ERV3-16A3	internal	[95]	AIDS and HIV-1 control
rs2395029	6:31464003	ERV3-16A3	internal	[96]	Psoriasis and psoriatic arthritis
rs2395029	6:31464003	ERV3-16A3	internal	[93]	HIV-1 control
rs2395029	6:31464003	ERV3-16A3	internal	[97]	Abacavir-induced hypersensitivity
rs3130907	6:31464036	ERV3-16A3	internal	[92]	Lung cancer
rs2284178	6:31464348	ERV3-16A3	internal	[98]	MICB measurement
rs3094014	6:31465781	ERV3-16A3	3′UTR	[92]	Lung cancer
rs75640364	6:31465789	ERV3-16A3	3′UTR	[99]	Herpes zoster
rs3128986	6:31465916	ERV3-16A3	3′UTR	[92]	Lung cancer
rs79022003	6:31465935	AluSp	3′UTR	[100]	Eosinophil counts
rs3131620	6:31466054	AluSp	3′UTR	[92]	Lung cancer
rs3094604	6:31466334	ERV3-16A3	3′UTR	[101]	Squamous cell lung carcinoma
rs3094604	6:31466334	ERV3-16A3	3′UTR	[92]	Lung cancer
rs3094604	6:31466334	ERV3-16A3	3′UTR	[102]	MMR vaccine-related febrile seizures
rs3131619	6:31466554	ERV3-16A3	3′UTR	[103]	Myositis
rs3094013	6:31466589	ERV3-16A3	3′UTR	[104]	Idiopathic inflammatory myopathies
rs3131618	6:31466589	ERV3-16A3	3′UTR	[103]	Myositis
rs3094013	6:31466844	ERV3-16A3	3′UTR	[43]	Tonsillectomy
rs2523675	6:31468255	ERV3-16A3	3′UTR	[105]	Relapse after cord blood transplantation
rs2518028	6:31468270	ERV3-16A3	3′UTR	[105]	Relapse after cord blood transplantation
rs2523673	6:31469218	LTR16B/AluSx	3′UTR	[100]	Platelet count

3.1. HIV and AIDS

Fellay et al. [33,34] were the first to find a significant association of the *HCP5* gene variant rs2395029 (T>G) with disease nonprogression in a cohort of HIV-infected individuals. The association was particularly strong with 10% of the variation among individuals during the viral asymptomatic set point (stabilized viral load) period of infection with the minor C/G-allele being associated with a lower viral load. Their hypothesis was that the association observed for the variation in HIV-1 set point was due mainly to *HLA-B*57:01*, although they acknowledged the possible effect of the *HCP5* variation because it was an endogenous retroviral element (ERV) with sequence homology to retroviral *pol* genes, and that its transcripts were known to be expressed in lymphocytes with the possible production of predicted short protein antigens with an amino acid substitution at the SNP rs2395029 [28,30]. A model in which a combined haplotypic effect of *HCP5*, *HLA-B*57:01*, and the SNV rs9264942 located in the 5′ region of the *HLA-C* gene, 35 kb away from transcription initiation and 156 kb telomeric of the *HCP5* gene on the HIV-1 set point, was consistent with the observation that suppression of viremia can be maintained in *B*57:01* patients with undetectable viral load, even if HIV-1 undergoes mutations that allow escape from cytotoxic T lymphocyte (CTL)–mediated restriction. Therefore, *HCP5* seemed to be a good candidate to interact with HIV. Furthermore, Fellay et al. [33] found that the strongest association with progression to AIDS included a set of seven polymorphisms located in and near to the ring finger protein 39 (*RNF39*) and zinc ribbon domain-containing 1 (*ZNRD1*) genes in the MHC regions.

The association of the *HCP5* polymorphism at rs2395029-G with *HLA-B*57:01* and HIV nonprogression to AIDS was confirmed in follow-up studies [93,95]. Limou et al. [95] used their own Genomics of Resistance to Immunodeficiency Virus (GRIV) cohort and the AIDS GWAS of the Euro-CHAVI (Center for HIV/AIDS Vaccine Immunology) cohort and replicated the results of Fellay et al. [33]. As expected, they obtained the strongest association with the *HCP5* rs2395029 minor SNP and the second strongest association with the *C6orf48* rs9368699 SNP, which was in LD with *HCP5* and with several SNPs located in the MHC class I and class III region including *HLA-B, MICB, PSORS1C1* (class I region), *TNXB, TNF, LTB, BAT1, BAT2, BAT3,* and *RDBP* (class III region). Their study suggested an independent role for the *ZNRD1* gene in disease progression. Of the 50 best signals found in their meta-analysis, 46 originated from the HLA locus, emphasizing the critical role of the MHC in the control of HIV-1 replication and delayed disease progression.

A GWAS of a multiethnic cohort of HIV-1 controllers and progressors [93] obtained similar results as in the previous two studies [33,95] but with the added novelty that (1) the nature of the HLA-viral peptide interaction was the major factor modulating durable control of HIV infection, (2) *HCP5* rs2395029-G was a proxy not only for *HLA-B*57:01* but also for many protective and risk HLA alleles (predominantly at *HLA-B*), and (3) with an independent effect on *HLA-C* gene expression that together differentially affected the response to HIV and delayed progression to AIDS [106,107]. However, *HCP5* rs2395029 was not associated with viral load at set point in African populations [108], instead, the viral load was associated with the *HLA-B*57:03* allele [92] suggesting a difference in individuals of African and European ancestry. The rs2395029-G polymorphism is missing from the Yoruban population of the Niger-Congo, and they have the *HLA-B*57:03* allele instead of the *HLA-B*57:01* of Europeans [108,109].

The Catano et al. [110] study of mainly European Americans was noteworthy because they showed in their univariate analysis that the *HCP5* minor allele was associated with a slow disease course and lower viral loads, whereas in the multivariate models, after partitioning out the protective effects of *HLA-B*57*, the *HCP5* minor allele was associated with disease acceleration and enhanced viral replication. These contradictory associations for *HCP5* are generally obscured, possibly because of the very strong LD between this allele and a subset of protective *HLA-B*57* alleles. Furthermore, they found that *HCP5* and *HLA-C* alleles stratified the *HLA-B*57*-containing genotypes into those that associated with either disease retardation or progression, "providing one explanation for the long-standing conundrum of why some *HLA-B*57*-carrying individuals are long-term non-progressors, whereas others exhibit progressive disease." Their study highlighted the strong dependence of genotype–phenotype relationships upon cohort design, phenotype selection, LD patterns, and populations studied. However, in a more recent study [109], the minor alleles of *HCP5* rs2395029, *HLA-C* rs9264942, and *ZNRD1* rs3689068 were associated strongly with lower viral load among antiretroviral-naïve individuals who had a shorter time to first viral load of less than 51 copies/mL during combination antiretroviral therapy, even after adjustment for viral load before combination antiretroviral therapy. The authors of a 2016 study [109] concluded that more studies are needed to elucidate the function and mechanisms of these SNVs in relation to HIV disease progression and disease course as well as to clarify whether these are functional SNV or whether they simply reflect a strong LD with other SNVs that are actually functional but, as yet, unidentified.

In 2010, Yoon et al. [35] infected TZM-bl cells in vitro with HIV-1 and plasmids expressing high levels of *HCP5* transcripts; they showed that the *HCP5* gene had no direct antiviral effect. This implied that the association of an *HCP5* variant with viral control more likely was due to various other complex interactions and epistatic factors. Thus, *HCP5* rs2395029 was deemed more as a useful genetic marker or proxy for *HLA-B*57:01* in Europeans; and it was used as such by other researchers to show that hypersensitivity in HIV patients to abacavir treatment was associated with *HLA-B*57:01* and HIV [97,110,111]. Abacavir is a potent nucleoside reverse transcriptase inhibitor, and it is an integral part of antiretroviral therapy in combination with other antivirals with good efficacy and a favorable long-term toxicity profile in the treatment of HIV. However, serious hypersensitivity to abcavir is recognized as a prohibitive and life-threatening treatment in approximately 8% of Caucasians and

3% of African Americans. Rodriguez-Novoa et al. [97] concluded from their analysis of 245 HIV patients that "the use of HCP5 rs2395029 testing could be as useful as HLA-B*57:01 typing to prevent the abacavir hypersensitivity reaction. Given that HCP5 testing is cheaper, less time-consuming and easier to perform than HLA typing, it may confidently replace the latter in clinical settings." However, in 2012, when Melis et al. [78] genotyped both the *HCP5* SNP and *HLA-B*57:01* in a set of 1888 samples, they found a good correlation, but they also found that one *HLA-B*57:01*-positive sample tested negative for the *HCP5* SNP, and that *HCP5* could not be amplified in two samples as a consequence of a homozygous deletion of the *HCP5* gene. They concluded that copy number variation and incomplete LD interfered with the *HCP5* genotyping assay for abacavir hypersensitivity, and that ethnicities should be considered when using the *HCP5* SNP as a surrogate marker for *HLA-B*57:01*. Discovering the absence of *HCP5* SNPs in two individuals was not a surprising result because the *HCP5* gene was known to be deleted along with the *MICA* gene in a number of Asian HLA haplotypes including the *HLA-B*48* haplotype in Japanese [42,73]. Despite the occasional deletion or structural mutation, genotyping for the *HCP5* rs2395029 minor allele is still a quick and practical method for assessing the possibility of abacavir hypersensitivity associated with *HLA-B*57:01* in order to avoid potentially fatal consequences [112]. Also, the structurally polymorphic *AluMICB*, which is a lineage indel within intron 1 of the *MICB* gene, possibly could be used as an alternative and simple proxy assay for *HLA-B*57* and *HLA-B*48:01* [79]. Ultimately, genotyping for *HLA-B*57* is essential because the most widely accepted hypothesis proposes that abacavir alters the repertoire of peptides that are able to bind to MHC, allowing for presentation of novel self-peptides, which in the absence of abacavir would not bind to *HLA-B*57:01* [113]. These altered peptides are perceived as foreign by T cells, and a cytotoxic response is triggered that can result in the abacavir hypersensitivity reaction [114].

3.2. Herpes Zoster

Crosslin et al. [99] reported that MHC genomic regions of *HCP5*, especially at SNPs rs116062713 (now rs75640364) and rs114864815 (now rs77349273), were strongly associated with susceptibility to herpes zoster (shingles) caused by the varicella zoster virus. The *HCP5* SNP marker rs75640364 is in the 3'UTR of *HCP5* at chr6:31,465,789 and SNP rs77349273 is a 2 kb upstream variant of *HCP5* located between *MICA* and *HLA-X*.

3.3. Autoimmune Disease and Drug Hypersensitivity

Apart from HIV infection, AIDS, and abacavir hypersensitivity, the *HCP5* [rs2395029] SNVs also were associated with other adverse drug reactions, autoimmune diseases, and abnormal phenotypes. The same *HCP5* and *HLA-B* genotypes were associated with psoriasis (PS) and psoriatic arthritis (PSA) [96,115] and found to be a major determinant of flucloxacillin-induced liver injury [94]. Liu et al. [96] found that the *HCP5* G2V polymorphism at rs2395029 had the highest odds ratio with both psoriasis (PS) and psoriatic arthritis (PSA) for 223 Caucasian individuals, and its effect was independent (not in significant LD) of the most highly associated SNP rs10484554 that was 34.7 kb upstream of *HLA-C*. The MHC, in particular the HLA class I region, is the only genomic region that has been shown to be consistently associated with PS. The nine top-ranking SNPs in the Liu et al. [96] study were from the MHC, and seven were significant, even when statistically adjusted for multiple testing. The authors also noted that the same significant *HCP5* SNV in psoriasis and HIV infections is not surprising since psoriasis can be triggered by infection with HIV and other viruses. Hence, it is possible that *HCP5-C* carriers mount a strong immune reaction to viral infection, and when psoriasis is associated with other genes such as *corneodesmosin*, *POU5F1*, *MICA*, and *HLA-C* in the MHC and genes outside the MHC in genetically susceptible individuals [116], then this reaction might lead to excessive inflammation in skin and joints.

The antimicrobial agent flucloxacillin is a common cause of drug-induced liver injury (DILI), but the genetic basis for susceptibility remains unclear. A study of 51 cases of flucloxacillin DILI, along with a replication study of another 23 cases, found that rs2395029(G) carriers were at highly increased risk (odds ratio 80, p = 8.7×10^{-33}) [94]. The rare homozygote rs2395029(G;G) increases the odds of flucloxacillin -induced liver injury by 45×.

Many associations have been attributed to *HCP5* SNVs outside the gene locus, either upstream in the *HLA-X* pseudogene or beyond the *HCP5* 3′UTR, positioned almost up to the locus for the *HCG* lncRNA. For example, SNP rs3099844 was associated with nevirapine-induced Steven–Johnson Syndrome (SJS), toxic epidermal necrolysis (TEN) drug reactions [117], systemic lupus erythematosus (SLE), anti-Ro/SSA [118], Sjogren Syndrome, leukopenia, and lymphoma [119]. Thus, SNV was referred to incorrectly as being in the *HCP5* gene when it was actually positioned in *LOC102725068* (chr6:31,479,918–31,494,794 of GRCh38/hg38), which was the *MICB-DT* gene that coded for the 14,877 bp *MICB* divergent transcript. The *MICB-DT* lncRNA is interesting in its own right because it immediately neighbors the *MICB* gene, has many transcription factor binding sites, and is associated with various diseases or phenotypes such as myositis, asthma, mumps, plantar warts, blood protein measures, MICB protein levels, and lymphocyte and monocyte counts [90].

In a recent genome-wide association analysis of a total of 915 children with Kawasaki disease and 4553 controls in the Korean population, the susceptibility locus for the disease was identified by Kim et al. [39] to be *NMNAT2* on chromosome 1q25.3 (rs2078087) and the human leukocyte antigen (HLA) region on chromosome 6p21.3 (*HLA-C*, *HLA-B*, *MICA*, and *HCP5* with rs9380242, rs9378199, rs9266669, and rs6938467, respectively). The *HCP5* rs6938467 (T/G) is positioned at chr6:31,440,139 (GRCh38.p12) well outside the *HCP5* RNA reference sequence (NR_040662.1) and upstream of the *LINC01149* locus. However, this is within the alternative genomic location of the old Hugo Gene Nomenclature Committee (HGNC definition) of *HCP5* (HGNC ID:21659) using Gencode Gene ENSG00000206337.10 and the Gencode Transcript ENST00000414046.2 starting within the *MICA* gene in the genomic location of chr6:31,400,702–31,465,809, which provided a transcript, including UTRs, of 65,108 bp in sequence length.

3.4. Transplantation

HCP5 upstream and downstream SNVs were associated with disease relapse in 53 patients with unrelated cord blood transplantation (UCBT) when compared to HLA-matched unrelated donors [105]. The diagnosed diseases before transplantation were transfusion-dependent thalassemia (19), genetic diseases (10), anemias (8), leukemias (11), and other neoplastic diseases (5). Of the 58 SNVs that were analyzed by genotyping, seven SNVs were associated with the risk of relapse, and two of these SNVs, rs2523675 and rs2518028, were located ~2.5 kb downstream of *HCP5* but still within the tailing *ERV16* (*ERV3-16A3* and *5′LTR16B2*) sequence of ~4.6 kb. The other SNVs were in the *MICD* gene and the *HLA-DOA* gene. Hematological disease relapses after UCBT were defined as recurrence of malignancy and/or relapse of nonmalignant hematological disorders defined by conversion to partial or complete nonresponse. Although the two *HCP5* SNVs resulted in 2.75 and 4.52 times greater risk of relapse for the recipients than the donors, a possible molecular mechanism for the relapse was not provided.

3.5. Cancer

A number of different SNVs located in the *MICA* and *MICB* genes and in the genomic region between them have been associated with cancer. In a recent analysis of eight GWAS datasets with 17,153 cases and 239,337 controls by Yuan et al. [92], at least six *HCP5* SNVs (including rs3130907 within the HCP5 sequence) were associated significantly with lung cancer susceptibility along with the novel risk SNV rs114020893 in the *lncRNA NEXN-AS1* region at 1p31.1. The authors noted that lung cancer risk-related loci (6p21 and 15q25) were enriched in lncRNAs, such as *HCP5*, *RP11-650L12.2*, *XXbac-BPG27H4.8*, and *HCG17*, and that using noncoding regions in GWAS and gene-based and pathway-based analyses should be complementary to protein coding-related approaches [92].

4. *HCP5* Methylomics

The study of differentially methylated sites, differential gene expression, and epigenetic mechanisms represent a complementary method to genetic association studies for the identification of molecular and biological pathways that contribute to good health during a normal life cycle and to clinical heterogeneity of autoimmune and chronic diseases and cancer [120]. Recent studies have identified differentially methylated sites within, or neighboring, the *HCP5* gene sequence associated with epigenetic regulation of various disease phenotypes (obesity, SLE) and in response to fetal development, aging, HIV infection, and vaccination (Table 4).

Table 4. Differentially methylated positions of HCP5' in six recent studies.

Disease/Phenotype	CpG Site	Position Genecode	Feature	Delta	*p*-Value	Reference
Age-related expression	cg25843003	31431312	LTR16B2	−0.23	2.3×10^{-13}	[121]
in monocytes and T cells	cg01082299	31431969	ERV3-16	−0.001	6.86×10^{-6}	
SLE: Anti-dsDNA'	cg25843003	31431312	LTR16B2	−0.044	5.2×10^{-10}	[122]
	cg00218406	31431407	LTR16B2	−0.067	8.7×10^{-9}	
	cg01082299	31431969	ERV3-16	−0.04	1.2×10^{-7}	
	cg18808777	31431503	ERV3-16	−0.051	3.0×10^{-10}	
HIV associated marker	cg00218406	31431407	LTR16B2			[123]
Influenza vaccination	CpGsite	31428956	MER21B	−0.301	1.31×10^{-4}	[124]
		31431456	LTR16B2			
		31431457	LTR16B2	−0.401	2.0×10^{-7}	
		31433586	3'UTR			
Endometrium	cg25843003	31431312	LTR16B2	−0.064		[125]
Pre to receptive	cg00218406	31431407	LTR16B2	−0.058		
	cg21684411	31431573	ERV3-16	−0.071		
BMI	cg00218406	31431407	LTR16B2	0.005	8.03×10^{-7}	[64]
BMI	cg25843003	31431312	LTR16B2	0.002	1.51×10^{-6}	
Obesity	cg00218406	31431407	LTR16B2	0.048	4.34×10^{-5}	
Obesity	cg25843003	31431312	LTR16B2	0.020	5.98×10^{-5}	

The negative delta values reflect hypomethylation and gene upregulation and the positive delta values reflect hypermethylation and gene downregulation. SLE is Systematic Lupus Erythematosus. The p-values were mostly False Discovery Rate (FDR) adjusted.

Hypomethylation of *HCP5* was associated with autoantibody production against dsDNA, Sjogren's syndrome-related antigen A (SSA), Smith (Sm) antigen, and ribonucleoprotein (RNP) in SLE [122,126], with gene expression and humoral immune response to influenza [124], with hypomethylated PSORS1C1-associated allopurinol-induced severe cutaneous adverse reactions in Han Chinese [127], accelerated aging in chronic HIV infection [123], endometrial receptivity [125], sexual bias in the human placental sexome [128,129], age-related monocyte and T cell gene expression [121], lung adenocarcinoma [130], and with hypomethylated *POU5F1*-associated ankylosing spondylitis [131]. In contrast to the more common observation of hypomethylation, HCP5 hypermethylation was associated with obesity and BMI in an epigenome-wide association study of adiposity in Ghanaian African migrants using whole blood to measure DNA methylation [64].

Age-related HCP5 DNA methylation was associated with gene expression in human monocytes and T cells [121], and the expressed genes that linked to potentially functional age-related methylation sites were enriched with antigen processing and presentation MHC class I and class II genes that were implicated in 'parainflammation' and the development of age-related chronic inflammatory diseases and autoimmune diseases. The total effects of age on gene expression (which increased with age) were significant (FDR < 0.05) for seven MHC genes—*HLA-B, -E, -DPA1, -DPB1, TAP2, TAPBP,* and *HCP5*—with hypomethylation within and/or near to all of those genes. On the other hand, Gross et al. [123] found an HCP5 CpG DNA methylation signature in blood cells of patients with chronic, well-controlled HIV infection that correlated with accelerated aging, and that it also was independently associated with HLA expression and corresponding HIV control. The level of methylation at HCP5 was correlated with a patient's CD4$^+$/CD8$^+$ T cell ratio to provide further evidence that the observed changes were functional [123]. Chronic HIV infection, even when viral loads were kept below the level of detection, is associated with early onset of diseases linked to aging, including cardiovascular disease, kidney disease, cancer, and premature death. Highly active antiretroviral therapy (HAART) controls the burden of HIV, without curing the infection, enabling HIV-infected patients to live for many decades, provided they continue their medications. The increased methylation changes in HIV-infected patients found beyond their chronological age suggested about a five-year increase in aging compared to healthy controls [123].

Systemic lupus erythematosus (SLE) is a chronic inflammatory autoimmune disease of unknown etiology that can affect most organs and is characterized by the development of autoantibodies associated with specific clinical manifestations implicated in the pathogenesis of lupus nephritis and decreased survival [122,126]. The genetic risk factors suggested for SLE include alleles in *IRF5, STAT4, BLK, TNFAIP3, TNIP1, FCGR2B,* and other genes [132]. Genome-wide DNA methylation analysis of SLE revealed persistent hypomethylation of interferon genes and compositional changes to CD4$^+$ T cell populations. For example, Chung et al. [122] characterized the methylation status of 467,314 CpG sites in 326 women with SLE DNA methylation profiling, performed using the Infinium HumanMethylation450 BeadChip (Illumina), and they identified and replicated significant associations between anti-dsDNA autoantibody production and the methylation status of 16 CpG sites in 11 genes. Differential methylation for these CpG sites was also associated with anti-SSA, anti-Sm, and anti-RNP autoantibody production. Overall, associated CpG sites were hypomethylated in autoantibody-positive samples compared to autoantibody-negative cases. In the discovery/replication analysis, associations with hypomethylated CpG sites were within genes (*IFIT1, IFI44L, MX1, RSAD2, OAS1,* and *EIF2AK2*) that were either induced by type 1 interferon or that regulated type 1 interferon signaling (NLRC5). Except for hypomethylation at *HCP5* and the *PSMB8* gene in the class III region, differential methylation of CpG sites within the MHC was not strongly associated with autoantibody production. Thus, hypomethylation of CpG sites within *HCP5* and other genes from different pathways that could not be explained by DNA sequence variation were associated strongly with anti-dsDNA, anti-SSA, anti-Sm, and anti-RNP production in SLE.

In a system-wide association study between DNA methylation, gene expression, and humoral immune response to influenza [124], a cohort of 158 individuals who were 50 to 70 years old showed that *HCP5* along with *HLA-B* and *HLA-DQB2* had an important role in methylation expression, particularly when the humoral immune response to influenza was measured by a hemagglutination inhibition assay (HAI). Only two genes showed association in all three independent analyses: *ADARB2*, an inhibitor of adenosine deaminase activity (RNA editing), and *SPEG*, a kinase with a known function in myocyte development. The small number of genes, including *HCP5*, that overlapped across two or more methods and at multiple time points were *HLA-B, HLA-DQB2,* the histone deacetylase *HDAC4, RWDD2B, PTPRN2* that (de)phosphorylates phosphoinositols in an insulin regulatory role, *DNAH2, HCP5, FAM24B, LOC399815,* and the transcription factor genes *PAX7* and *PAX9*. Many genes (~640 genes) that were identified in one of these analyses had direct protein–protein interacting genes identified in the other analyses, revealing that the impact of methylation on humoral immunity is complex and highly dependent upon the immune outcome. Zimmermann et al. [124] reported that methylation levels of a CpG within the gene body of *HLA-B* (hypermethylation) were strongly associated with HAI and had an opposite trend to that of *HCP5* (hypomethylation).

Human and animal studies have identified that the placenta expresses select transcripts in a sexually dimorphic manner [129]. A microarray-based study identified sex-dependent differences in the placental transcriptomic profile in males and females (sexome) with isolated cells derived from human placental villi [128]. The four cell types examined included cytotrophoblasts, synctiotrophoblasts, and arterial and venous endothelial cells. For sex-dependent differences, the males demonstrated enrichment of signaling pathways previously reported to mediate graft versus host disease and transcripts involved in immune function and inflammation such as *HLA-DQB1* (syncytiotrophoblast), *HLA-DQA1* (syncytiotrophoblast and cytotrophoblast), *HCP5* (cytotrophoblast), *NOS1* (cytotrophoblast), *FSTL3, PAPPA, SPARCL, FCGR2C* (trophoblast epithelium), *CD34* (cytotrophoblast), *HLA-F* (cytotrophoblast), and *BCL2* (syncytiotrophoblast). Males demonstrated a greater in utero vulnerability, and the findings of Cvitic et al. [128] suggested that these effects may partially be due to reduced maternal–fetal compatibility for males who were then required to up-regulate immune-associated transcripts in an attempt to combat an attack by the maternal immune system.

In a genome-wide methylome analysis of endometrial biopsies collected from 17 healthy fertile-aged women from prereceptive and receptive phases of a menstrual cycle, Kukushkina et al. [125] found that extracellular matrix organization and immune response were the pathways most affected by methylation changes during the transition from prereceptive to receptive phase. The overall methylome remained relatively stable during the two time points of the menstrual cycle with small-scale changes affecting 5% of the studied CpG sites (22,272 out of 437,022 CpGs, FDR < 0.05). The study confirmed that the differential methylation of *KRTAP17-1, CASP8, RANBP3L, WT1, MPP7, PTPRN2,* and *HCP5* between the early and midsecretory phases were similar to those observed in the previous studies [133,134]. The differential methylation of *PTPRN2* and *HCP5* in the endometrium is an interesting connection, given that they both were differentially methylated in a system-wide association study between DNA methylation, gene expression, and humoral immune response to influenza [134]. The *PTPRN2* (NCBI gene: 5799) gene product (de)phosphorylates phosphoinositols, has an insulin regulatory role, and it may be an autoantigen in insulin-dependent diabetes mellitus, but its actual function as a methylation site in the endometrium or human monocytes and T cells is not known.

In an epigenome-wide association study using whole blood measures of adiposity in 547 Ghanaian African migrants, Meeks et al. [64] found that obesity and body mass index (BMI) were related to *HCP5* hypermethylation and 18 differentially methylated positions (DMPs) for BMI, 23 for waist circumference, and three for obesity. Fourteen DMP overlapped between BMI and waist circumference. The two epigenome-wide loci that were significantly hypomethylated for both general adiposity and abdominal adiposity were *CPT1A* (carnitine palmitoyltransferase 1A) and *BCAT1* (branched chain amino acid transaminase 1), whereas *NLRC5* ((NLR family CARD domain containing 5) and most other DMPs including those for six HLA genes—*HCP5, HLA-B, TAP1, TAP2, PSMB8,* and *HLA-E*—were hypermethylated. The hypermethylation of *NLRC5* was highly significant, and this gene is known to regulate the expression of MHC class I genes and to limit the activation of inflammatory pathways [54–56]. Thus, the results of Meeks et al. [64] suggested that obesity might suppress the adaptive immune response and induce inflammation that could also result in insulin resistance.

Coit et al. [131] identified a total of 68 differentially methylated sites between ankylosing spondylitis (AS) patients and osteoarthritis controls. *HCP5* and *POU5F1* were both hypomethylated in *HLA-B*27*-positive compared to *HLA-B*27*-negative AS patients. They suggested that *HLA-B*27* might play a role in AS in part through epigenetic linkage disequilibrium-inducing epigenetic dysregulation. The *POU5F1* gene (alias *OCT4*) is located at the telomeric end of the MHC beta-block, ~98 kb upstream of the *HLA-C* gene, and its role in methylation is well described [120].

HCP5 is known to be involved in lung cancer [130], and Yuan et al. [92] described at least six *HCP5* SNVs, including rs3130907, that were associated significantly with lung cancer susceptibility (Table 3). Previously, Orvis et al. [135] showed that inactivation of the *BRG1* gene, also known as *SMARCA4*, which encodes the ATPase subunits of the SW1/SNF chromatin remodeling complex, contributed to non-small cell lung cancer aggressiveness by altering nucleosome positioning in a wide range of genes as well as by downregulating the expression of *HCP5* and all of the classical and nonclassical HLA class I genes.

Presumably, hypomethylation of *HCP5* leads to added interactions and connectivity with proteins and other RNA sequences, especially with miRNA regulators. Studies on hypermethylation of *HCP5* are still lacking, and such studies might provide a more contrasting view of the action of methylation on the function of *HCP5* in health and disease. However, the overall absence of hypermethylation data for *HCP5* might be related to the DNA methylation paradox, whereby methylation of the transcribed region and the region of transcription initiation have opposite effects on gene expression [120]. Although methylation can affect gene expression in both directions depending on the genomic region, there are more negative correlations in the 5′ UTR, while positive correlations are more common in the gene body region. While this was the case for *HLA-B*, the reverse was observed for *HCP5* in humoral immune response to influenza [124].

5. *HCP5*, Gene Targets, and Transcription Factors in Interaction Networks

Many hundreds of different transcription factors (TFs) are believed to target the *HCP5* sequence and regulate its expression. Although only a few experiments have examined the relationship between particular transcription factors and the expression of *HCP5* and its neighboring genes in the region between *MICA* and *MICB*, a variety of datasets have predicted connections between *HCP5* and many known TFs. Table S4 shows five Internet databases sourced from Harmonizome and GeneCards (Table S1) that associated TFs with *HCP5*. For example, the dataset of JASPAR Predicted Transcription Factor Targets predicted that 57 transcription factors were associated with regulating the expression of *HCP5*, whereas the MotifMap Predicted Transcription Factor Targets dataset predicted only seven associations: alpha-CP1, E2A, ETS2, MAFA, NF-kB, NF-Y, and TEF-1. On the other hand, the TRANSFAC curated dataset only listed PTF1A as a TF, interacting with the *HCP5* gene in low- or high-throughput transcription factor functional studies. PTF1A is the pancreas-specific transcription factor 1a, with a role in mammalian pancreatic development and in determining whether cells allocated to the pancreatic buds continue towards pancreatic organogenesis or revert back to duodenal fates. Also, *HCP5* is only one of 233 target genes for the PTF1A transcription factor. In contrast, the TRANSFAC dataset predicted that 13 TFs regulated the expression of the *HCP5* gene: ATF2, ELF3, ETS1, HINFP, JDP2, LEF1, LTF, MYC, NFE2L2, RUNX1, SMAD4, SMARCA2, and SPI1. Most of these predicted transcription factors also targeted many other genes as part of interaction networks, cascades, or divergent pathways.

In a study of interactions between *HCP5* and transcription factors, Warner et al. [136] showed that *HCP5*, together with *NOD2* and *IL-8*, was associated strongly with decreased viability of cells in a study of the inactivation of the *NF-kB1* gene by knockout. In their datafile, the other MHC genes that were strongly associated with decreased viability were *HLA-A, -DQA1, -DRB1, -DRB4, -DOA,* and *-DOB* but not *HLA-B, -E, -G, -F, -DRB5, -DMA, -DRA –DQA2, -DQB1, -DRB3, -DMB, -DPB1, -DPA1, MICA, MICB, TNF, LTA, LTB, C4A,* and *C4B* in the HEK293 cell line. Thus, NF-kB1 (located on chr 4) regulates *HCP5* and the gene expression of some other MHC genes as well as a wide variety of biological functions, including inappropriate activation associated with inflammatory diseases, inappropriate immune cell development, and delayed cell growth. This study [136] also demonstrated that although *HCP5* was often in LD with many genes in the MHC, it could be activated or suppressed independently from most of them.

Coit et al. [131] identified a total of 68 differentially methylated sites in a study of ankylosing spondylitis (AS) patients and osteoarthritis controls; *HCP5* and *POU5F1* were both hypomethylated in *HLA-B*27*-positive compared to *HLA-B*27*-negative AS patients. This predicted cis relationship between the transcription factor *POU5F1* and *HCP5* is interesting, given that both genes are located in the beta block of the HLA class I region [1,2]. Also, Meeks et al. [64] in an epigenome-wide association study of measures of adiposity among Ghanaians showed that *NLRC5* and *HCP5, HLA-B, TAP1,*

TAP2, *PSMB8*, and *HLA-E* were all significantly hypermethylated for both general adiposity and abdominal adiposity.

The interactive website Pathwaynet [137] predicts both the presence of a functional association and the most likely interaction type among human genes or their protein products on a whole-genome scale. It is based on a large compendium of refined regulatory interactions within 77 tissues, with their curated pathways taken from primary experimental datasets such as 690 ChIP-Seq datasets, numerous mass spectrometry of metabolites, protein–protein interactions, disease samples, etc., in order to capture the interaction networks. Figure 5 shows the top 15 genes that interacted with *HCP5*, as predicted by Pathwaynet [137], with a high relationship confidence of between 0.9247 (*BTN3A3* and *IRF9*) and 0.9523 (*HLA-F*). All of the genes were within the MHC region except for *UBE2L6* (chr11q12.1), *TRIM22* (chr11p15.4), *IRF1* (chr5q31.1), *CASP1* (chr11q22.3), and *IRF9* (chr14q12). Pathwaynet did not specify the type of functional relationships between *HCP5* and these 15 genes or gene products; therefore, the predicted gene interactions should be considered with considerable caution. However, it is evident that the non-HCP5 genes have roles in antigen processing and presentation, the proteasome, graft versus host disease, allograft rejection, autoimmune disease, response to type 1 interferon or interferon gamma, regulation of viral reproduction, IL-6- and IL-12-mediated and NOD-like signaling pathways, and signal transduction by the p53 class I mediator. This connection is supported to a large degree by the top 52 genes that were positively associated with *HCP5* gene expression in the Comparative Toxicogenomics Database (CTD) datasets (Table 5). The *HCP5* interaction and regulation is probably by way of the methylome and the competitive endogenous RNA regulatory networks, although this premise needs to be investigated further in both in vivo and in vitro experiments and association studies.

Table 5. Top 52 *HCP5* gene interactions associated with Comparative Toxicogenomics Database (CTD) studies.

Gene Symbol	Gene ID	Interaction Count	Gene Symbol	Gene ID	Interaction Count
PTGS2	5743	71	ICAM1	3383	9
TNF	7124	55	NOS3	4846	9
IL1B	3553	52	PON1	5444	9
PTGS1	5742	42	SCARB1	949	9
CASP3	836	28	AHR	196	8
BCL2	596	21	APOA1	335	8
AGT	183	20	TXN1	22166	8
CTNNB1	1499	20	VEGFA	7422	8
CCND1	595	18	ABCA1	19	7
NFKB1	4790	18	ALB	213	7
RELA	5970	17	CCL2	6347	7
TNFSF10	8743	17	IKBKB	3551	7
NOS2	4843	14	ITGA2B	3674	7
PARP1	142	14	MAPK1	5594	7
NFKBIA	4792	13	MAPK3	5595	7
BAX	581	12	MYC	4609	7
CDKN1A	1026	12	PPARG	5468	7
IL4	3565	12	SELP	6403	7
IL6	3569	11	TCF4	6925	7
ITGB3	3690	11	ABCB1	5243	6
MMP9	4318	11	AKT1	207	6
CASP9	842	10	CYSLTR1	10800	6
IL1A	3552	10	JUN	3725	6
PPARA	5465	10	MMP2	4313	6
BIRC5	332	9	PLAUR	5329	6
CASP8	841	9	RB1	5925	6

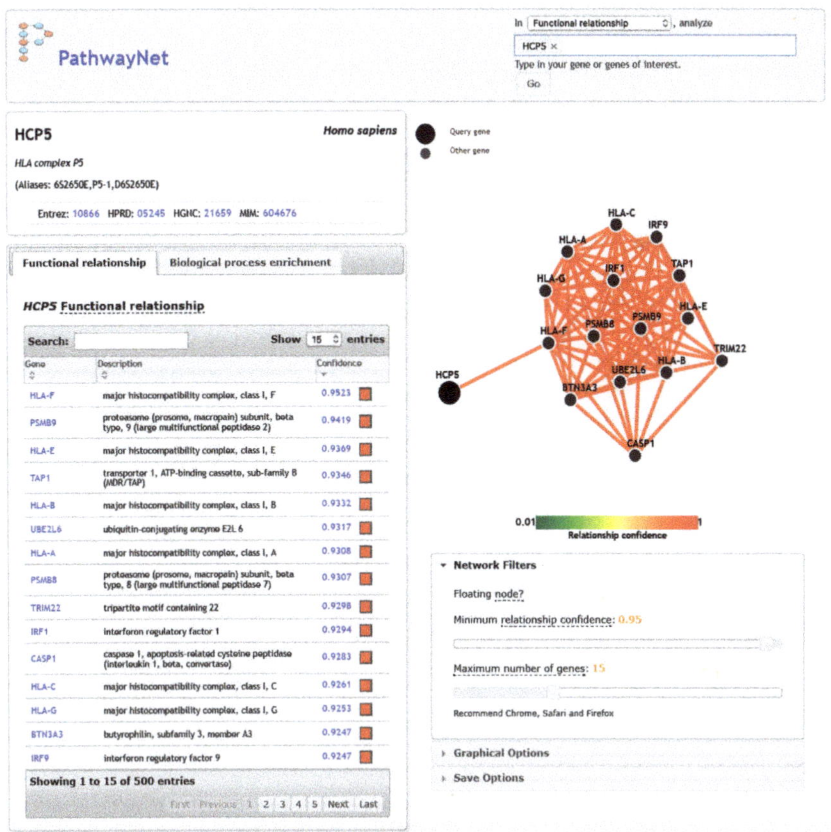

Figure 5. *HCP5* functional relationships in PathwayNet (Table S1) [137].

6. *HCP5* Gene Expression and Gene Interactions

In 1993, Vernet et al. [27] originally reported that *HCP5* expressed a 2.5 kb transcript in human B cells, phytohemagglutinin-activated lymphocytes, a natural killer-like cell line, normal spleen, hepatocellular carcinoma, neuroblastoma, and other nonlymphoid tissue but not in T cells. Since then, numerous studies of genome-wide gene expression using dense Affymetrix expression arrays were published, but the findings rarely reported directly on the expression of *HCP5*. To identify the expression profile of *HCP5* in various scenarios, the databanks needed to be investigated and interrogated separately. In this way, it was possible to find the particular pattern of *HCP5* expression especially in comparison to the other class I and class II genes. *HCP5* is widely expressed at low levels, but it is primarily expressed at higher levels in cells of the immune system such as spleen, blood, and thymus (http://smd-www.stanford.edu/), consistent with potential roles in autoimmunity and cancer.

Harmonizome (Table S1) was a good starting point to find data about *HCP5* expression under various experimental conditions [138]. The gene expression results of *HCP5* in 53 tissues from 8555 samples (570 donors) were sourced from GTEx RNA-seq using the University of California, Santa Cruz (UCSC) online browser (Table S1). Also, interrogation of the online NCBI Gene Expression Omnibus (GEO) with the keyword "HCP5" produced 1771 results to review. Eighty-nine results were related to up and down differential expression, 6 results to the keyword 'immunity', 375 results to 'cancer', 28 to 'HIV', 50 to 'virus', 36 to 'interferon', 42 to 'host defense', and 4 to 'MHC'. These results were browsed with a visual profile of the effects of treatments and experiments on the gene expression of *HCP5* and/or other genes of investigative choice. Nine studies in GEO confirmed that IFN and IL28B upregulated *HCP5* RNA in some cell types, whereas IL10 downregulated *HCP5* RNA in peripheral blood mononuclear cells (PBMCs) (Table 2). There were little or no significant data for other cytokine-positive or -negative effects on *HCP5*.

In comparison, Table S5 shows the ~37 drugs and chemicals that induce or suppress *HCP5* expression with effects on inferred diseases, and that were identified in the Comparative Toxicogenomics Database (CTD) [139]. Based on the data in Table S5, *HCP5* gene expression seems to be decreased by various immunosuppressants and neurotoxins. This includes the immunosuppressive aflatoxin B1 that increased methylation of the *HCP5* gene [140]. In addition, the CTD database [139] revealed that *HCP5* reacted with 714 different genes in various gene expression studies. The top 52 genes that *HCP5* interacted with most often included *PTGS2*, *TNF*, *IL1B*, *PTGS1*, and *CASP3* (Table 5). The summary information provided by NCBI RefSeq for each of these five genes was the following: PTGS1 and PTGS2 are prostaglandin-endoperoxide synthases or cyclooxygenases, key enzymes in the biosynthesis of prostaglandin that are regulated by specific stimulatory events involved in inflammation and mitogenesis. The tumor necrosis factor (*TNF*) gene that is located downstream of *HCP5* in the MHC class III region encodes a multifunctional, proinflammatory cytokine that is a member of the tumor necrosis factor (TNF) superfamily. This cytokine is mainly secreted by macrophages. It is involved in the regulation of a wide spectrum of biological processes including cell proliferation, differentiation, apoptosis, lipid metabolism, and coagulation. It has been implicated in a variety of diseases including autoimmune diseases, insulin resistance, and cancer. Knockout studies in mice suggested that TNF also has a neuroprotective function. IL1B is a cytokine expressed on chromosome two and produced by activated macrophages as a proprotein, which is proteolytically processed to an active form by caspase 1 (CASP1/ICE). This cytokine is an important mediator of the inflammatory response and is involved in a variety of cellular activities including cell proliferation, differentiation, and apoptosis. IL1B induces PTGS2/COX2 in the central nervous system and contributes to inflammatory pain hypersensitivity. CASP3 or caspase 3 is a protease with a central role in the execution phase of cell apoptosis. It inactivates poly(ADP-ribose) polymerase, while it cleaves and activates sterol regulatory element binding proteins as well as caspases 6, 7, and 9. Also, it is the predominant caspase involved in the cleavage of amyloid-beta 4A precursor protein, which is associated with neuronal death in Alzheimer's disease. Therefore, it is evident that *HCP5* RNA is strongly associated with the inflammatory innate immune response as well as adaptive immune responses as indicated by its coexpression with various class I genes in expression studies in the databases (e.g., GEO) and the published literature.

6.1. HCP5 Expression in HIV-Infected Cells

Given that *HCP5* has been associated strongly with viral suppression in HIV-infected cells in GWAS (Table 3), it is surprising that so few papers have specifically addressed the correlation between *HCP5* gene expression levels and HIV levels or response to HIV infection [35]. However, there are a few studies in the GEO database to suggest that *HCP5* expression in response to HIV is induced, suppressed, or unaffected in some cell types. For example, *HCP5* RNA was significantly lower in the three HIV-negative controls than in three samples of jejunal mucosal cells from HIV patients on highly active antiviral therapy [141]. Also, *HCP5* transcription activity was high in three mononuclear cell samples, but it was low or absent in three T cell samples and three fibroblast samples with HIV DNA integration sites [142]. Similarly, *HCP5* RNA was higher in the brains of 10 of 26 patients receiving antiretroviral therapy for HIV-associated neurocognitive disorder than in nine uninfected controls [143]. However, in some studies, there was no significant effect of HIV on *HCP5* RNA levels. For example, there was no difference in the *HCP5* RNA levels of 23 infected and 12 noninfected peripheral blood mononuclear cell samples [144], little difference between eight infected and eight uninfected macrophage samples [145], and little or no difference between five uninfected and 15 HIV-infected CD4$^+$ samples and 15 CD8+ T cell samples [146]. Alternatively, *HCP5* RNA was low or absent in three T cell samples infected with HIV-based vector or three samples treated with TNF-alpha, but it was relatively higher in the three untreated T cell samples [147]. Unfortunately, in the one study on RUNX1 in the regulation of HIV, no data were provided about *HCP5* RNA levels. Thus, based on these limited analyses, the role of *HCP5* in HIV infection and AIDS remains unclear.

6.2. HCP5 Expression in Cancer

HCP5 has been found upregulated or downregulated in a number of different cancers. The interactions between *HCP5* and three transcription factors with potential antioncogenic functions are noteworthy. *HCP5* was confirmed as one of the *KAT8* (alias hMOF) downregulated genes by qPCR and ChIP in the hMOF siRNA knockdown HeLa cells and 20 of 28 clinically diagnosed ovarian cancer tissues [148]. KAT8 (lysine acetyltransferase 8) encodes a member of the MYST histone acetylase protein family involved with the p53 pathway and chromatin organization as well as with the suppression of epithelial to mesenchymal transition and tumor progression [149]. In contrast, Teng et al. [150] demonstrated that the *HCP5* transcribed sequence interacted with an miRNA sequence and the runt-related transcriptional regulator RUNX1 in a feedback loop to regulate the malignant behavior of glioma cells of the brain. Another noteworthy interaction was between *HCP5* and SATB1 (special AT-rich sequence binding protein 1), which is a nuclear matrix-associated DNA binding protein that functions as a chromatin organizer. SATB1 is highly expressed in aggressive breast cancer cells and promotes growth and metastasis by reprogramming gene expression [151]. It also enhanced *HCP5* epigenetically and suppressed the oncogenic long noncoding RNA urothelial carcinoma-associated 1 (*UCA1*) in breast cancer cells. Recently, Zhao and Li [57] showed that transcription factor SP1 induced upregulation of *HCP5*, which in turn promoted the development of osteosarcoma, whereas inhibition of *HCP5* expression reversed cell invasion and epithelial–mesenchymal transition. In addition, *HCP5* is overexpressed in tumor tissues of patients with lung adenocarcinoma, and it is positively correlated with poor prognosis specifically in patients who are smokers with *EGFR* and *KRAS* mutations [56]. *HCP5* also was overexpressed in lymph node metastasis of small cell lung cancer [130,152], glioma tissue [150], colorectal cancerous tissue [153], and cancers of the colon [154], thyroid [155], cervix [156], and breast [151,157].

Interrogation of the TCNG Cancer Network Galaxy Database (Table S1) produced 206 networks for genes regulating or regulated by the *HCP5* gene as estimated from publicly available cancer gene expression data. There are ~1010 genes that were predicted to interact with *HCP5* either as a child node (regulated gene) or parent node (regulating gene). In about 590 interactions, *HCP5* was the parent or regulatory node, and in the remainder (420 nodes) *HCP5* was the child or regulated node. For example, *HLA-A, -B, -G, -H, and –J* were ranked as child nodes downstream of the *HCP5* parent node in four experimental arrays on breast cancer and one experiment on colon cancer. That is, *HCP5* was predicted to regulate the HLA-class I genes in those experiments. On the other hand, *HCP5* was predicted to be the regulated or child node with *NLRC5*, a member of the NOD-like receptor family that acts as a transcriptional activator of MHC class I genes [61–63], that was the parent or regulatory node in some gene expression experiments such as between adenocarcinoma and squamous cell carcinoma in non-small-cell lung carcinoma, breast cancer cell line profiles, non-Hodgkins lymphoma cell lines, complex genetic sarcomas, meningiomas, prostate cancer, and uveal melanoma primary tumors. Since there were far too many data to review here, further information on the activation or suppression of *HCP5* gene expression in many different cancers can be obtained by interrogating the TCNG Cancer Network Galaxy Database with 'HCP5' as the search query and following the links including those to the expression arrays at GEO. A more detailed account of *HCP5* RNA interaction in micro RNA regulatory networks in cancer is provided in the following section.

7. *HCP5* lncRNA Interactions with Regulatory miRNA in Cancer

In recent years, an increasing number of lncRNA, including *HCP5*, were found to have potential functions in cancer [158–160]. The oncogenic lncRNA appear to regulate the transcription and translation of neighboring and distant genes by cis and trans-regulatory functions in a series of biological steps involving dosage compensation, genomic imprinting, and cell cycle dysregulation leading to cancer and its progression [158–160]. A particularly important mechanism to emerge from many of these studies is the role of lncRNAs to bind with regulatory miRNA that control antioncogenic or oncogenic pathways. This three-way binding interaction between lncRNA, miRNA, and regulatory protein coding genes, such as those coding for regulatory transcription factors, has become known as the competing endogenous RNA (ceRNA) mechanism/network [161,162]. In this regard, association studies (Table 3), expression data analysis, and knockdown experiments (Tables 2 and 4–6, Tables S4 and S5) have shown that *HCP5* can promote or suppress cancers depending on the *HCP5* allelic form and the cancer type. Since 2016, at least ten different cancer types were found to occur and/or progress by way of the *HCP5*–miRNA–gene regulator interactions or the ceRNA mechanism (Table 6).

Table 6. *HCP5*–miRNA–protein coding gene regulator interactions.

Antioncogenic miRNA	HCP5 Action	Gene Symbol and Up- or Downregulated	Cancer Type	Reference
miR-203	sponge	TGFb/SMAD3	Lung	[56]
miR-106b-5p	network	CTSS/FGL2	Lung	[130]
miR-17-5b	network	PDCD1LG2/PDL2	Lung	[130]
miR-126		GSR/ASCL1 ↑	Lung metastasis	[152]
		MET/GRM8/DACH1 ↓	Lung metastasis	[152]
miR-139	absorption	RUNX1 ↑	Glioma	[150]
miR-139-5p		ZEB1	Colorectal	[153]
		APIG1 ↓	Colon	[154]
miR-22-3p	sponge	ST6GAL2 ↑	Thyroid	[155]
mi-186-5p	sponge	ST6GAL2 ↑	Thyroid	[155]
miR-216a-5p	sponge	ST6GAL2 ↑	Thyroid	[155]
miR-15a	adsorption	MACC1 ↑	Cervical	[156]
miR-155	sponge	Complement genes	Breast	[157]
miR-128	sponge	MACC1 ↑	Gastric	[163]
miR-101	sponge	NDUFB6 ↓	Gastric	[163]
miR-103a	sponge	NDUFB6 ↓	Gastric	[163]
24 different mIR	ceRNA	TAP1 ↑	Lymphoma	[164]
	ceRNA	PSMB9 ↑	Lymphoma	[164]
	ceRNA	KLF2 ↑	Lymphoma	[164]
	ceRNA	GIPC1 ↑	Lymphoma	[164]
	ceRNA	ETN3A3 ↑	Lymphoma	[164]
	ceRNA	ETN3A1 ↑	Lymphoma	[164]
	ceRNA	CD47 ↑	Lymphoma	[164]
	ceRNA	CCDC50 ↑	Lymphoma	[164]
	ceRNA	LMBR1L ↑	Lymphoma	[164]
	ceRNA	HERC6 ↑	Lymphoma	[164]

↑ is upregulation and ↓ is downregulation.

In an integrated analysis on the dosage effect of lncRNAs in lung adenocarcinoma, Wei et al. [130] found that the protein coding genes *CTSS*, *FGL2*, and *PDCD1LG2* (alias *PDL2*) competed with *HCP5* (ENSG00000206337) and formed a single regulatory subnet with the miRNAs miR-106b-5p and miR-17-5b. This in part confirmed a previous study that found that *HCP5* was involved in the process of lung cancer by competing with *PDL2*, an immune checkpoint gene, and *FGL2*, a therapeutic target to suppress carcinogenesis [92]. This also was consistent with their finding that at least six *HCP5* SNPs, including rs3130907, were associated significantly with lung cancer susceptibility along with the novel risk SNP rs114020893 in the *lncRNA NEXN-AS1* region at 1p31.1.

HCP5 expression was positively correlated with the oncogenesis of a pathological grade of glioma tissues, and knockdown of *HCP5* exerted tumor-suppressive effects in human glioma cells by allowing an increase in expression of the miRNA tumor suppressor miR-139 [150]. The malignant behavior of glioma cells in the brain appears to be regulated by an *HCP5–miR-139–RUNX1* feedback loop, whereby *RUNX1* increased the promoter activities and expression of *HCP5* that binds to the tumor suppressor miRNA-139 (miR-139) and, therefore, acts as an oncogene. *HCP5* absorbed the tumor suppressor miR-139 to downregulate its expression. Upregulated *RUNX1* also inhibited apoptosis in glioma cell-lines U87 and U251. *RUNX1* down-regulation by knockdown using miR-139 as an inhibitor had the opposite effect on apoptosis and exerted tumor-suppressive effects in human glioma cells. MiR-139 inhibited *RUNX1* expression by targeting the 3′-UTR, and *HCP5* knockdown suppressed *RUNX1* expression by allowing miR-139 overexpression [150]. Thus, it was concluded that *HCP5* promotes cell proliferation, cell migration, and invasion (motility) and inhibits apoptosis, as do many other lncRNAs participating in glioma phenotypes [165].

RUNX1 also is known to suppress HIV reactivation in T cells, resulting in a negative correlation between *RUNX1* expression and viral load. The pharmacologic inhibition of *RUNX1* by a small molecule inhibitor, Ro5-3335, synergized with the histone deacetylase (HDAC) inhibitor SAHA (Vorinostat), enhanced the activation of latent HIV-1 in cell lines and PBMCs from patients [166]. However, the effect of *HCP5* on *RUNX1* expression in HIV-infected cells is not known, although this effect might occur by way of epigenetic or transcriptomic regulation.

Liang et al. [155] found that lncRNA *HCP5* promoted follicular thyroid carcinoma (FTC) progression as a competing endogenous RNA (ceRNA) sponge for miR-22-3p, mi-186-5p, and miR-216a-5p and activated alpha-2,6-sialyltransferase 2 (*ST6GAL2*). Functional experiments showed that *HCP5* promoted *ST6GAL2*, which in turn mediated the proliferation, migration, invasiveness, and angiogenic ability of FTC cells. In comparison, Yu et al. [156] showed that overexpressed *HCP5* promoted the development of cervical cancer by absorbing miRNA-15a to promote expression of the MET transcription regulator MACC1 (*MACC1*). MiRNA-15a overexpression in vitro inhibited *MACC1* expression and suppressed the proliferation of cervical cancer cells. In contrast, miRNA-15a knockdown experiments or absorption by *HCP5* allowed increased *MACC1* expression and the proliferation of cervical cancer cells. Furthermore, *HCP5* and *MACC1* were overexpressed in cervical cancer tissues compared to paracancerous tissue, and the survival rate of patients with cervical cancer was negatively correlated to *HCP5* expression and positively correlated to miRNA-15a. Luciferase reporter gene assay also showed that miRNA-15a bound directly to either *HCP5* or *MACC1*. Because HPV16 and HPV18 are associated strongly with cervical cancer progression, the question arises whether the *PMSP* gene [38] of *HCP5* is also expressed and what role it might have with the increased proliferation of cervical cancer cells.

Both *MACC1* and *HCP5* are expressed in gastric cancer, and their expression may be regulated by miRNAs. In a metabolic network analysis, Mo et al. [163] found that *HCP5* was coexpressed with 34 metabolic-related protein-coding genes and five lncRNAs, and regulated by three miRNAs, miR-128, miR-101, and miR-103a, that were downregulated in gastric cancer. The *TOPORS-AS1* lncRNA and the NADH ubiquinone oxireductase subunit B6 (*NDUFB6*) coding gene that were associated with *HCP5* in the network analysis were downregulated in gastric cancer samples [163]. In some cancers, TGFbeta might induce *HCP5* transcription via the activity of SMAD3 [56], and increased levels of *HCP5* RNA might either directly or indirectly affect *GSR, ASCL1, MET, GRM8, DACHI* [152], *ETN3A1, ETN3A3, CCDC50, HERC6, TAP1,* and *PSMB9* [164] as well as many other genes (Table 6).

Based on a data expression analysis, Olgun et al. [157] found that *HCP5* was one of seven lncRNA that interacted with miRNA in breast cancer. *HCP5* bound to miR-155 at the hub of the ceRNA network of interactions in the basal subtype of breast cancer. This basal subtype in breast cancer was characterized by a positive correlation between immune cell infiltration and aggressiveness with a key role for interferon signaling and the induction of cell proliferation by the complement cascade. Olgun et al. [157] detected *C2, C3, C3AR1, C4A,* and *C7* complement genes in the basal ceRNA interactions, suggesting that the complement cascade pathway may be significant for progression of the basal subtype.

In an analysis of differentially expressed profiles of lncRNAs and mRNA in ceRNA networks during transformation of diffuse large B cell lymphoma (DLBCL), Tian et al. [164] identified *HCP5* as a key regulator interacting with many miRNA and protein coding genes associated with transcription (*KLF2*), cell adhesion and proliferation (*CD47*), lipid metabolism (*BTN3A1*), and the adaptive immune response (*TAP1, PSMB9*). However, they concluded that the molecular function of *HCP5* remained unknown.

HCP5 has been associated with various other cancers including cutaneous melanoma [167], HPV-infected head and neck squamous cell carcinoma [168], squamous cell carcinoma cells [169], HCV-induced liver cancer [170], and upper tract urothelial carcinoma [171] in which ceRNA cross-talk has yet to be tested. The ceRNA network and cross-talk mechanism also might have a role in autoimmune diseases such as idiopathic thrombocytopenia [172], viral infections, and *HCP5*-associated phenotypes (Tables 3, 4 and 6) that, as yet, have not been examined for interactions with miRNA and with the other lncRNA regulators and protein coding genes.

8. Perspective

Since the discovery of *HCP5* in 1993 [27], a large amount of data has been gathered about its expression, function, and disease associations. Much of this data, however, is buried in large datasets that require considerable effort to locate, analyze, validate, and interpret [89,90,137–139] (Table S1). Nevertheless, sufficient amounts of published and unpublished data that were retrieved from online public databanks for this review revealed that *HCP5* had important roles in health and disease, particularly with respect to its role as a ceRNA regulator and biomarker in autoimmune diseases and cancer. GWAS indicated that the *HCP5* SNV rs2395029 was a potential marker for abacavir-induced hypersensitivity, a marker for *HLA-B*57:01* in populations of mainly Caucasian or Hispanic descent, as well as flucloxacillin drug liver injury, HIV control, and psoriasis and psoriatic arthritis (Table 3). Other SNVs within the *HCP5* gene or within 2.5 kb of the 5′ or 3′ UTR region are useful association markers for various diseases including myositis, herpes infection, cancer, and risk of relapse after transplantation (Table 3). Methylomic studies have associated *HCP5* strongly with HIV progression, SLE, ankylosing spondylitis (AS) and obesity. Various expression studies have shown that *HCP5* is a useful biomarker for interferon and IL28-related inflammatory response, monocyte response to influenza A infection, and various cancers (Table 4, Table 6, and Table S5). Many of the *HCP5*-associated diseases such as SLE, AS, psoriasis and psoriatic arthritis, myositis, obesity, and cancer are associated also with accelerated aging, morbidity, and mortality.

The *HCP5* gene within the MHC class I genomic region has evolved by exaptation from an ancient endogenous retroviral and gained a new regulatory function by sequestering the MHC promoter and enhancer region from a fragmented/deleted ancient HLA class I gene. It appears to have generated the *PMSP* sequence that codes for a 61 aa peptide that binds to the HPV L2 capsid protein for viral assembly [41]. The function of the PMSP peptide in different cell types other than its interaction with HPV in keratinocytes still needs to be determined and whether the *PMSP* gene is regulated or continuously expressed along with the *HCP5* transcript. The *HCP5* promoter also contains a 22 nt *RUNX1* sequence (Table S3) that might contribute to the interaction between the *HCP5* transcript and the Runx transcription factor in glioma and in the monocytes of HIV patients. Because the *HCP5* promoter has many of the canonical TF binding sites of the HLA class I promoter (Table S3), *HCP5* is often expressed concomitantly with classical and nonclassical class I genes, unless they are differentially separated from each other by epigenomic or ceRNA regulatory processes. For example, in some instances, *HCP5* might be upregulated, whereas class I HLA genes are downregulated as exemplified in the study of humoral immune response to influenza [124]. While it is notable that *HCP5* is often in LD with the HLA class I genes and that their expression is often coordinated in response to various stimulators or suppressors, functional interactions between the products of these genes suggest that *HCP5* has an associated role in antigen processing and presentation, the proteasome, graft versus host disease, allograft rejection, autoimmune disease, response to type 1 interferon or interferon gamma, and in the regulation of viral reproduction, especially in HIV restriction [6,14,24,50,52,107]. The mechanisms by which HLA class I genes and the lncRNA genes *HCP5* and *ZNRD1* might interact with HIV also are worth investigating further because these lncRNAs could be exploited therapeutically using small RNA inhibitors [109,110]. Similarly, the interaction of *HCP5*, microRNA, and protein coding genes in cancer (Table 6) suggests that *HCP5* could be targeted for knockdown or knockout in antitumor therapeutics.

It is evident from the results and reports presented in this review that *HCP5* contributes to regulating viral and autoimmune diseases and cancer, and it can be upregulated or downregulated depending on its response to various exogenous or endogenous stimulators or suppressors. One shared feature of the investigated cancerous and noncancerous diseases is hypomethylation of the *HCP5* gene and upregulation of its transcript, which highlights the enormous potential of this lncRNA as a diagnostic biomarker in these pathologies. The *HCP5* gene sequence is a differentially methylated site associated with the epigenetic regulation of some disease phenotypes, such as obesity [64] and SLE [132], and it may also act in response to fetal development [128], aging [121], HIV [123] or influenza infection, and vaccination [124]. Presumably, hypomethylation of *HCP5* leads to added interactions and connectivity with protein and other RNA sequences, especially with microRNA regulators. Although increased *HCP5* levels seem to be a common event upon viral infection and in response to interferon stimulation and some cancers, the consequences of *HCP5* upregulation are diverse. The mode of *HCP5* action, whether acting as a promoter or suppressor, probably depends on the occurring downstream events. From the mechanistic point of view, *HCP5* either acts as a differentially methylated site involved in epigenetic regulation, or it is involved in transcriptional regulation of protein coding genes by sequestering miRNA as reported for a variety of different cancers (Table 6). Although transcriptional regulation by sequestration of transcription factors or regulatory miRNA seems to be the predominant mode of action by *HCP5* in cancerous diseases, it might also act as a specific or nonspecific 'sponge' for miRNAs in noncancerous diseases. The involvement of *HCP5* lncRNA in various human diseases and cancers underscores the importance of understanding its functions in the ceRNA networks as an important step towards future drug development. Perturbations in cellular regulatory functions due to interactions between *HCP5* and transcription factors probably contribute to some malignancies. For instance, recent studies of the interaction between *HCP5* and Runx family proteins suggests that they may play key roles in stem cell biology, particularly in regulating apoptosis and the G0/G1 transition by way of the ceRNA networks [150]. Runx1 increases the promoter activities and expression of *HCP5*, and it is also known to suppress HIV reactivation in T cells perhaps by way of *HCP5* mediation [166]. In many of the recent studies, the function of the *HCP5* gene is described as a defense-response gene often acting in unison with the inflammatory, innate, and adaptive immune response systems. However, *HCP5* also has disease progression and oncogenic effects that suggest that it may act like a double-edged sword to defend and to attack depending on other endogenous and exogenous regulators in the pathway.

The absence of the *HCP5* gene from the MHC genomic region of chimpanzees and gorillas and some human haplotypes that carry the *HLA-B*48:01* allele appears not to be deleterious or life-threatening [85]. Great apes and humans without the *HCP5* gene seem to live in relatively good health, which suggests that the deletion, if not harmful, might even confer some selective advantage [173]. Nevertheless, it is evident from an exploration of the recent literature and accumulating public databases that *HCP5* has many regulatory or associated functions at the genetic, molecular, and physiological levels. It appears to interact with numerous other genes and/or their products by way of intermediaries such as transcription factors and miRNA, *LTR16B2* genomic repeats, and as an enhancer or superenhancer in the regulation of other genes and in chromatid structural changes. There is still much to learn about the ceRNA actions of *HCP5*, and many of the future surprises about this hybrid MHC class I endogenous retroviral gene undoubtedly will arise from knockdown, knockout, and knockin gene expression studies. The human homologous *HLA-B*48:01* haplotype is a naturally occurring *HCP5-MICB* deletion or knockout haplotype in the human population that warrants genetic and epidemiological analysis to help elucidate the importance of these genes for most humans who still have the intact functional versions. Although the large databanks, datasets, and available publications have provided important insights into the functions of *HCP5*, much work still remains in order to elucidate the actual mechanisms and role of this intriguing MHC class I hybrid retroelement in immunity, health, and disease. More detailed studies are needed, particularly applying knockdown, knockout, and knockup studies, to find out more about its function as a regulator of various molecular and biological processes related to health and disease and the MHC.

9. Conclusions

HCP5 is a unique human-specific gene within the MHC class I genomic region that encodes a hybrid HLA class I endogenous retroviral lncRNA with peptide coding potential. It has functional relationships with many other genes within or outside the MHC genomic region that are involved with antigen processing and presentation, the interferon regulatory pathway, and epigenomic and ceRNA networks; however, many of these functional interactions between multiple genetic variants are still poorly understood. *HCP5* gene SNVs and neighboring upstream and downstream SNVs have been associated with HIV viral load, HPV infection, autoimmune diseases, disease relapse after transplantation, and various cancers. Much still needs to be determined about the defensive and pathological functions of *HCP5* and its structural and functional role in RNA editing and signaling to enable epigenetic plasticity and immune response pathways. Judging from the recent new findings about the possible oncogenic role of *HCP5* as a 'sponge' for sequestering regulatory miRNA in cancer, new insights about its diverse mechanisms and functions in health and disease undoubtedly will continue to emerge and surprise in the near future.

Supplementary Materials: The following are available online at http://www.mdpi.com/2073-4409/8/5/480/s1. Table S1, Databases and Repositories (docx). Table S2, At least 1899 *HCP5* functional associations with biological entities found within the Harmonizome datasets (xls). Table S3, Clustal W(1.83) alignments of the transcription factor binding motifs and proximal promoter region of *HCP5*, *HLA-A*, *-B*, *-C*, *-G*, and *-F* (xls). Table S4, Internet databases with predicted *HCP5*-associated transcription factors (TFs) (docx). Table S5, Chemical and drug effects on *HCP5* expression and inferred diseases (xls).

Funding: This research received no external funding.

Acknowledgments: I thank the staff of Cells' editorial office, especially Daniela Wang for her assistance with the Special Issue on the MHC in Health and Disease. I also thank my assistant guest editors Hans Djikstra and Takashi Shiina for their invaluable contribution and all the authors who submitted and published their papers in this Special Issue. It was hard and relentless work, but ultimately satisfying and rewarding with an erudite outcome.

Conflicts of Interest: The author declares no conflict of interest.

References

1. Shiina, T.; Inoko, H.; Kulski, J.K. An update of the HLA genomic region, locus information and disease associations: 2004. *Tissue Antigens* **2004**, *64*, 631–649. [CrossRef]
2. Shiina, T.; Hosomichi, K.; Inoko, H.; Kulski, J.K. The HLA genomic loci map: Expression, interaction, diversity and disease. *J. Hum. Genet.* **2009**, *54*, 15–39. [CrossRef]
3. Alter, I.; Gragert, L.; Fingerson, S.; Maiers, M.; Louzoun, Y. HLA class I haplotype diversity is consistent with selection for frequent existing haplotypes. *PLoS Comput. Biol.* **2017**, *13*, e1005693. [CrossRef]
4. Meyer, D.; Single, R.M.; Mack, S.J.; Erlich, H.A.; Thomson, G. Signatures of Demographic History and Natural Selection in the Human Major Histocompatibility Complex Loci. *Genetics* **2006**, *173*, 2121–2142. [CrossRef]

5. Shiina, T.; Ota, M.; Shimizu, S.; Katsuyama, Y.; Hashimoto, N.; Takasu, M.; Anzai, T.; Kulski, J.K.; Kikkawa, E.; Naruse, T.; et al. Rapid Evolution of Major Histocompatibility Complex Class I Genes in Primates Generates New Disease Alleles in Humans via Hitchhiking Diversity. *Genetics* **2006**, *173*, 1555–1570. [CrossRef]

6. Trowsdale, J.; Knight, J.C. Major Histocompatibility Complex Genomics and Human Disease. *Annu. Rev. Genom. Hum. Genet.* **2013**, *14*, 301–323. [CrossRef] [PubMed]

7. Mosaad, Y.M. Clinical Role of Human Leukocyte Antigen in Health and Disease. *Scand. J. Immunol.* **2015**, *82*, 283–306. [CrossRef]

8. Matzaraki, V.; Kumar, V.; Wijmenga, C.; Zhernakova, A. The MHC locus and genetic susceptibility to autoimmune and infectious diseases. *Genome Biol.* **2017**, *18*, 76–92. [CrossRef] [PubMed]

9. Cabarrocas, J.; Bauer, J.; Piaggio, E.; Liblau, R.; Lassmann, H. Effective and selective immune surveillance of the brain by MHC class I-restricted cytotoxic T lymphocytes. *Eur. J. Immunol.* **2003**, *33*, 1174–1182. [CrossRef]

10. Lazarczyk, M.J.; Kemmler, J.E.; Eyford, B.A.; Short, J.A.; Varghese, M.; Sowa, A.; Dickstein, D.R.; Yuk, F.J.; Puri, R.; Biron, K.E.; et al. Major Histocompatibility Complex class I proteins are critical for maintaining neuronal structural complexity in the aging brain. *Sci. Rep.* **2016**, *6*, 1–13. [CrossRef]

11. Nelson, P.A.; Sage, J.R.; Wood, S.C.; Davenport, C.M.; Anagnostaras, S.G.; Boulanger, L.M. MHC class I immune proteins are critical for hippocampus-dependent memory and gate NMDAR-dependent hippocampal long-term depression. *Learn. Mem.* **2013**, *20*, 505–517. [CrossRef]

12. Ohtsuka, M.; Inoko, H.; Kulski, J.K.; Yoshimura, S. Major histocompatibility complex (Mhc) class Ib gene duplications, organization and expression patterns in mouse strain C57BL/6. *BMC Genom.* **2008**, *9*, 178. [CrossRef] [PubMed]

13. Prugnolle, F.; Manica, A.; Charpentier, M.; Guégan, J.F.; Guernier, V.; Balloux, F. Pathogen-Driven Selection and Worldwide HLA Class I Diversity. *Curr. Biol.* **2005**, *15*, 1022–1027. [CrossRef]

14. Garrido, F.; Perea, F.; Bernal, M.; Sánchez-Palencia, A.; Aptsiauri, N.; Ruiz-Cabello, F. The Escape of Cancer from T Cell-Mediated Immune Surveillance: HLA Class I Loss and Tumor Tissue Architecture. *Vaccines* **2017**, *5*, 7. [CrossRef] [PubMed]

15. Morrison, B.J.; Steel, J.C.; Morris, J.C. Reduction of MHC-I expression limits T-lymphocyte-mediated killing of Cancer-initiating cells. *BMC Cancer* **2018**, *18*, 469. [CrossRef] [PubMed]

16. Borchert, G.M.; Holton, N.W.; Williams, J.D.; Hernan, W.L.; Bishop, I.P.; Dembosky, J.A.; Elste, J.E.; Gregoire, N.S.; Kim, J.-A.; Koehler, W.W.; et al. Comprehensive analysis of microRNA genomic loci identifies pervasive repetitive-element origins. *Mob. Genet. Elem.* **2011**, *1*, 8–17. [CrossRef]

17. Sun, M.; Kraus, W.L. From Discovery to Function: The Expanding Roles of Long NonCoding RNAs in Physiology and Disease. *Endocr. Rev.* **2015**, *36*, 25–64. [CrossRef] [PubMed]

18. Mercer, T.R.; Mattick, J.S. Structure and function of long noncoding RNAs in epigenetic regulation. *Nat. Struct. Mol. Biol.* **2013**, *20*, 300–307. [CrossRef]

19. Mattick, J.S. The State of Long Non-Coding RNA Biology. *Non-Coding RNA* **2018**, *4*, 17. [CrossRef]

20. Liu, S.J.; Horlbeck, M.A.; Cho, S.W.; Birk, H.S.; Malatesta, M.; He, D.; Attenello, F.J.; Villalta, J.E.; Cho, M.Y.; Chen, Y.; et al. CRISPRi-based genome-scale identification of functional long noncoding RNA loci in human cells. *Science* **2017**, *355*, eaah7111. [CrossRef]

21. Sigdel, K.R.; Cheng, A.; Wang, Y.; Duan, L.; Zhang, Y. The Emerging Functions of Long Noncoding RNA in Immune Cells: Autoimmune Diseases. *J. Immunol. Res.* **2015**, *2015*, 1–9. [CrossRef]

22. Geng, H.; Tan, X.-D. Functional diversity of long non-coding RNAs in immune regulation. *Genes Dis.* **2016**, *3*, 72–81. [CrossRef]

23. Spurlock, C.F.; Crooke, P.S.; Aune, T.M. Biogenesis and Transcriptional Regulation of Long Noncoding RNAs in the Human Immune System. *J. Immunol.* **2016**, *197*, 4509–4517. [CrossRef] [PubMed]

24. Chen, Y.G.; Satpathy, A.T.; Chang, H.Y. Gene regulation in the immune system by long noncoding RNAs. *Nat. Immunol.* **2017**, *18*, 962–972. [CrossRef] [PubMed]

25. Clark, P.M.; Chitnis, N.; Shieh, M.; Kamoun, M.; Johnson, F.B.; Monos, D. Novel and Haplotype Specific MicroRNAs Encoded by the Major Histocompatibility Complex. *Sci. Rep.* **2018**, *8*, 3832. [CrossRef] [PubMed]

26. Gensterblum-Miller, E.; Wu, W.; Sawalha, A.H. Novel Transcriptional Activity and Extensive Allelic Imbalance in the Human MHC Region. *J. Immunol.* **2018**, *200*, 1496–1503. [CrossRef] [PubMed]

27. Vernet, C.; Ribouchon, M.-T.; Chimini, G.; Jouanolle, A.-M.; Sidibe, I.; Pontarotti, P. A novel coding sequence belonging to a new multicopy gene family mapping within the human MHC class I region. *Immunogenetics* **1993**, *38*, 47–53. [CrossRef] [PubMed]

28. Avoustin, P.; Ribouchon, M.T.; Vernet, C.; N'Guyen, B.; Crouau-Roy, B.; Pontarotti, P. Non-homologous recombination within the Major Histocompatibility Complex creates a transcribed hybrid sequence. *Mamm. Genome* **1994**, *5*, 771–776. [CrossRef] [PubMed]

29. Shiina, T.; Tamiya, G.; Oka, A.; Yamagata, T.; Yamagata, N.; Kikkawa, E.; Goto, K.; Mizuki, N.; Watanabe, K.; Fukuzumi, Y.; et al. Nucleotide Sequencing Analysis of the 146-Kilobase Segment around the IkBL and MICA Genes at the Centromeric End of the HLA Class I Region. *Genomics* **1998**, *47*, 372–382. [CrossRef]

30. Kulski, J.K.; Dawkins, R.L. The P5 multicopy gene family in the MHC is related in sequence to human endogenous retroviruses HERV-L and HERV-16. *Immunogenetics* **1999**, *49*, 404–412. [CrossRef]

31. Babaian, A.; Mager, D.L. Endogenous retroviral promoter exaptation in human cancer. *Mob. DNA* **2016**, *7*, 24. [CrossRef] [PubMed]

32. Ito, J.; Sugimoto, R.; Nakaoka, H.; Yamada, S.; Kimura, T.; Hayano, T.; Inoue, I. Systematic identification and characterization of regulatory elements derived from human endogenous retroviruses. *PLos Genet.* **2017**, *13*, e1006883. [CrossRef]

33. Fellay, J.; Shianna, K.V.; Ge, D.; Colombo, S.; Ledergerber, B.; Weale, M.; Zhang, K.; Gumbs, C.; Castagna, A.; Cossarizza, A.; et al. A Whole-Genome Association Study of Major Determinants for Host Control of HIV-1. *Science* **2007**, *317*, 944–947. [CrossRef]

34. Fellay, J.; Ge, D.; Shianna, K.V.; Colombo, S.; Ledergerber, B.; Cirulli, E.T.; Urban, T.J.; Zhang, K.; Gumbs, C.E.; Smith, J.P.; et al. Common Genetic Variation and the Control of HIV-1 in Humans. *PLoS Genet.* **2009**, *5*, e1000791. [CrossRef]

35. Yoon, W.; Ma, B.-J.; Fellay, J.; Huang, W.; Xia, S.-M.; Zhang, R.; Shianna, K.V.; Liao, H.-X.; Haynes, B.F.; Goldstein, D.B. A polymorphism in the HCP5 gene associated with HLA-B*5701 does not restrict HIV-1 in vitro. *AIDS* **2010**, *24*, 155–157. [CrossRef]

36. Dawkins, R.; Leelayuwat, C.; Gaudieri, S.; Tay, G.; Hui, J.; Cattley, S.; Martinez, P.; Kulski, J. Genomics of the major histocompatibility complex: Haplotypes, duplication, retroviruses and disease. *Immunol. Rev.* **1999**, *167*, 275–304. [CrossRef] [PubMed]

37. Kulski, J.K.; Gaudieri, S.; Martin, A.; Dawkins, R.L. Coevolution of PERB11 (MIC) and HLA Class I Genes with HERV-16 and Retroelements by Extended Genomic Duplication. *J. Mol. Evol.* **1999**, *49*, 84–97. [CrossRef]

38. Kulski, J.K.; Gaudieri, S.; Inoko, H.; Dawkins, R.L. Comparison Between Two Human Endogenous Retrovirus (HERV)-Rich Regions Within the Major Histocompatibility Complex. *J. Mol. Evol.* **1999**, *48*, 675–683. [CrossRef]

39. Kim, J.-J.; Yun, S.W.; Yu, J.J.; Yoon, K.L.; Lee, K.-Y.; Kil, H.-R.; Kim, G.B.; Han, M.-K.; Song, M.S.; Lee, H.D.; et al. A genome-wide association analysis identifies NMNAT2 and HCP5 as susceptibility loci for Kawasaki disease. *J. Hum. Genet.* **2017**, *62*, 1023–1029. [CrossRef] [PubMed]

40. Kapitonov, V.; Jurka, J. The age of Alu subfamilies. *J. Mol. Evol.* **1996**, *42*, 59–65. [CrossRef]

41. Görnemann, J.; Hofmann, T.G.; Will, H.; Müller, M. Interaction of Human Papillomavirus Type 16 L2 with Cellular Proteins: Identification of Novel Nuclear Body-Associated Proteins. *Virology* **2002**, *303*, 69–78. [CrossRef]

42. Komatsu-Wakui, M.; Tokunaga, K.; Ishikawa, Y.; Kashiwase, K.; Moriyama, S.; Tsuchiya, N.; Ando, H.; Shiina, T.; Geraghty, D.E.; Inoko, H.; et al. MIC-A polymorphism in Japanese and a MIC-A-MIC-B null haplotype. *Immunogenetics* **1999**, *49*, 620–628. [CrossRef]

43. Tian, C.; Hinds, D.A.; Hromatka, B.S.; Kiefer, A.K.; Eriksson, N.; Tung, J.Y. Genome-wide association and HLA region fine-mapping studies identify susceptibility loci for multiple common infections. *BioRxiv* **2016**, *8*, 599. [CrossRef] [PubMed]

44. Karim, R.; Meyers, C.; Backendorf, C.; Ludigs, K.; Offringa, R.; van Ommen, G.-J.B.; Melief, C.J.M.; van der Burg, S.H.; Boer, J.M. Human Papillomavirus Deregulates the Response of a Cellular Network Comprising of Chemotactic and Proinflammatory Genes. *PLoS ONE* **2011**, *6*, e17848. [CrossRef] [PubMed]

45. Prod'homme, V.; Tomasec, P.; Cunningham, C.; Lemberg, M.K.; Stanton, R.J.; McSharry, B.P.; Wang, E.C.Y.; Cuff, S.; Martoglio, B.; Davison, A.J.; et al. Human Cytomegalovirus UL40 Signal Peptide Regulates Cell Surface Expression of the NK Cell Ligands HLA-E and gpUL18. *J. Immunol.* **2012**, *188*, 2794–2804. [CrossRef] [PubMed]

46. Halenius, A.; Gerke, C.; Hengel, H. Classical and non-classical MHC I molecule manipulation by human cytomegalovirus: So many targets—But how many arrows in the quiver? *Cell. Mol. Immunol.* **2015**, *12*, 139–153. [CrossRef] [PubMed]

47. Ramsuran, V.; Hernández-Sanchez, P.G.; O'hUigin, C.; Sharma, G.; Spence, N.; Augusto, D.G.; Gao, X.; García-Sepúlveda, C.A.; Kaur, G.; Mehra, N.K.; et al. Sequence and Phylogenetic Analysis of the Untranslated Promoter Regions for *HLA* Class I Genes. *J. Immunol.* **2017**, *198*, 2320–2329. [CrossRef]

48. Barbash, Z.S.; Weissman, J.D.; Campbell, J.A.; Mu, J.; Singer, D.S. Major Histocompatibility Complex Class I Core Promoter Elements Are Not Essential for Transcription in vivo. *Mol. Cell. Biol.* **2013**, *33*, 4395–4407. [CrossRef]

49. van den Elsen, P.J. Expression Regulation of Major Histocompatibility Complex Class I and Class II Encoding Genes. *Front. Immunol.* **2011**, *2*, 48. [CrossRef]

50. Jongsma, M.L.M.; Guarda, G.; Spaapen, R.M. The regulatory network behind MHC class I expression. *Mol. Immunol.* **2017**. [CrossRef]

51. Gobin, S.J.P.; Peijnenburg, A.; Keijsers, V.; van den Elsen, P.J. Site alpha Is Crucial for Two Routes of IFNgamma-Induced MHC Class I Transactivation: The ISRE-Mediated Route and a Novel Pathway Involving CIITA. *Immunity* **1997**, *6*, 601–61111. [CrossRef]

52. Zhou, F. Molecular Mechanisms of IFN-γ to Up-Regulate MHC Class I Antigen Processing and Presentation. *Int. Rev. Immunol.* **2009**, *28*, 239–260. [CrossRef]

53. Indraccolo, S.; Pfeffer, U.; Minuzzo, S.; Esposito, G.; Roni, V.; Mandruzzato, S.; Ferrari, N.; Anfosso, L.; Dell'Eva, R.; Noonan, D.M.; et al. Identification of genes selectively regulated by IFNs in endothelial cells. *J. Immunol.* **2007**, *178*, 1122–1135. [CrossRef]

54. Johnson-Huang, L.M.; Suárez-Fariñas, M.; Pierson, K.C.; Fuentes-Duculan, J.; Cueto, I.; Lentini, T.; Sullivan-Whalen, M.; Gilleaudeau, P.; Krueger, J.G.; Haider, A.S.; et al. A Single Intradermal Injection of IFN-γ Induces an Inflammatory State in Both Non-Lesional Psoriatic and Healthy Skin. *J. Invest. Dermatol.* **2012**, *132*, 1177–1187. [CrossRef]

55. Rock, R.B.; Hu, S.; Deshpande, A.; Munir, S.; May, B.J.; Baker, C.A.; Peterson, P.K.; Kapur, V. Transcriptional response of human microglial cells to interferon-γ. *Genes Immun.* **2005**, *6*, 712–719. [CrossRef]

56. Jiang, L.; Wang, R.; Fang, L.; Ge, X.; Chen, L.; Zhou, M.; Zhou, Y.; Xiong, W.; Hu, Y.; Tang, X.; et al. HCP5 is a SMAD3-responsive long non-coding RNA that promotes lung adenocarcinoma metastasis via miR-203/SNAI axis. *Theranostics* **2019**, *9*, 2460–2474. [CrossRef]

57. Zhao, W.; Li, L. SP1-induced upregulation of long non-coding RNA HCP5 promotes the development of osteosarcoma. *Pathol. Res. Pract.* **2019**, *215*, 439–445. [CrossRef]

58. Tilburgs, T.; Meissner, T.B.; Ferreira, L.M.R.; Mulder, A.; Musunuru, K.; Ye, J.; Strominger, J.L. NLRP2 is a suppressor of NF-κB signaling and HLA-C expression in human trophoblasts. *Biol. Reprod.* **2017**, *96*, 831–842. [CrossRef]

59. Gobin, S.J.P.; Keijsers, V.; van Zutphen, M.; van den Elsen, P.J. The Role of Enhancer A in the Locus-Specific Transactivation of Classical and Nonclassical HLA Class I Genes by Nuclear Factor κB. *J. Immunol.* **1998**, *161*, 2276–2283.

60. Fagerberg, L.; Hallström, B.M.; Oksvold, P.; Kampf, C.; Djureinovic, D.; Odeberg, J.; Habuka, M.; Tahmasebpoor, S.; Danielsson, A.; Edlund, K.; et al. Analysis of the Human Tissue-specific Expression by Genome-wide Integration of Transcriptomics and Antibody-based Proteomics. *Mol. Cell. Proteom.* **2014**, *13*, 397–406. [CrossRef]

61. Downs, I.; Vijayan, S.; Sidiq, T.; Kobayashi, K.S. CITA/NLRC5: A critical transcriptional regulator of MHC class I gene expression: CITA/NLRC5. *BioFactors* **2016**, *42*, 349–357. [CrossRef]

62. Kobayashi, K.S.; van den Elsen, P.J. NLRC5: A key regulator of MHC class I-dependent immune responses. *Nat. Rev. Immunol.* **2012**, *12*, 813–820. [CrossRef]

63. Vijayan, S.; Sidiq, T.; Yousuf, S.; van den Elsen, P.J.; Kobayashi, K.S. Class I transactivator, NLRC5: A central player in the MHC class I pathway and cancer immune surveillance. *Immunogenetics* **2019**, *71*, 273–282. [CrossRef]

64. Meeks, K.A.C.; Henneman, P.; Venema, A.; Burr, T.; Galbete, C.; Danquah, I.; Schulze, M.B.; Mockenhaupt, F.P.; Owusu-Dabo, E.; Rotimi, C.N.; et al. An epigenome-wide association study in whole blood of measures of adiposity among Ghanaians: The RODAM study. *Clin. Epigenetics* **2017**, *9*, 103. [CrossRef]

65. Wiesner, T.; Lee, W.; Obenauf, A.C.; Ran, L.; Murali, R.; Zhang, Q.F.; Wong, E.W.P.; Hu, W.; Scott, S.N.; Shah, R.H.; et al. Alternative transcription initiation leads to expression of a novel ALK isoform in cancer. *Nature* **2015**, *526*, 453–457. [CrossRef]

66. Majumder, P.; Boss, J.M. CTCF Controls Expression and Chromatin Architecture of the Human Major Histocompatibility Complex Class II Locus. *Mol. Cell. Biol.* **2010**, *30*, 4211–4223. [CrossRef]

67. Ottaviani, D.; Lever, E.; Mao, S.; Christova, R.; Ogunkolade, B.W.; Jones, T.A.; Szary, J.; Aarum, J.; Mumin, M.A.; Pieri, C.A.; et al. CTCF binds to sites in the major histocompatibility complex that are rapidly reconfigured in response to interferon-gamma. *Nucleic Acids Res.* **2012**, *40*, 5262–5270. [CrossRef]

68. Chaumeil, J.; Skok, J.A. The role of CTCF in regulating V(D)J recombination. *Curr. Opin. Immunol.* **2012**, *24*, 153–159. [CrossRef]

69. Phillips, J.E.; Corces, V.G. CTCF: Master Weaver of the Genome. *Cell* **2009**, *137*, 1194–1211. [CrossRef]

70. Guelen, L.; Pagie, L.; Brasset, E.; Meuleman, W.; Faza, M.B.; Talhout, W.; Eussen, B.H.; de Klein, A.; Wessels, L.; de Laat, W.; et al. Domain organization of human chromosomes revealed by mapping of nuclear lamina interactions. *Nature* **2008**, *453*, 948–951. [CrossRef]

71. Bailey, C.; Metierre, C.; Feng, Y.; Baidya, K.; Filippova, G.; Loukinov, D.; Lobanenkov, V.; Semaan, C.; Rasko, J. CTCF Expression is Essential for Somatic Cell Viability and Protection Against Cancer. *Int. J. Mol. Sci.* **2018**, *19*, 3832. [CrossRef]

72. Cross, S.H.; Clark, V.H.; Simmen, M.W.; Bickmore, W.A.; Maroon, H.; Langford, C.F.; Carter, N.P.; Bird, A.P. CpG island libraries from human chromosomes 18 and 22: Landmarks for novel genes. *Mamm. Genome* **2000**, *11*, 373–383. [CrossRef]

73. Ota, M.; Bahram, S.; Katsuyama, Y.; Saito, S.; Nose, Y.; Sada, M.; Ando, H.; Inoko, H. On the MICA deleted-MICB null, HLA-B*4801 haplotype. *Tissue Antigens* **2000**, *56*, 268–271. [CrossRef] [PubMed]

74. Middleton, D.; Menchaca, L.; Rood, H.; Komerofsky, R. New allele frequency database: http://www. allelefrequencies.net. *Tissue Antigens* **2003**, *61*, 403–407. [CrossRef] [PubMed]

75. Aida, K.; Russomando, G.; Kikuchi, M.; Candia, N.; Franco, L.; Almiron, M.; Ubalee, R.; Hirayama, K. High frequency of MIC null haplotype (HLA-B48-MICA-del-MICB*0107 N) in the Angaite Amerindian community in Paraguay. *Immunogenetics* **2002**, *54*, 439–441. [CrossRef]

76. Shaikh, T.H.; Gai, X.; Perin, J.C.; Glessner, J.T.; Xie, H.; Murphy, K.; O'Hara, R.; Casalunovo, T.; Conlin, L.K.; D'Arcy, M.; et al. High-resolution mapping and analysis of copy number variations in the human genome: A data resource for clinical and research applications. *Genome Res.* **2009**, *19*, 1682–1690. [CrossRef]

77. Itsara, A.; Cooper, G.M.; Baker, C.; Girirajan, S.; Li, J.; Absher, D.; Krauss, R.M.; Myers, R.M.; Ridker, P.M.; Chasman, D.I.; et al. Population Analysis of Large Copy Number Variants and Hotspots of Human Genetic Disease. *Am. J. Hum. Genet.* **2009**, *84*, 148–161. [CrossRef] [PubMed]

78. Melis, R.; Lewis, T.; Millson, A.; Lyon, E.; McMillin, G.A.; Slev, P.R.; Swensen, J. Copy Number Variation and Incomplete Linkage Disequilibrium Interfere with the *HCP5* Genotyping Assay for Abacavir Hypersensitivity. *Genet. Test. Mol. Biomark.* **2012**, *16*, 1111–1114. [CrossRef]

79. Dunn, D.; Ota, M.; Inoko, H.; Kulski, J.K. Association of MHC dimorphic Alu insertions with HLA class I and MIC genes in Japanese HLA-B48 haplotypes. *Tissue Antigens* **2003**, *62*, 259–262. [CrossRef]

80. Anzai, T.; Shiina, T.; Kimura, N.; Yanagiya, K.; Kohara, S.; Shigenari, A.; Yamagata, T.; Kulski, J.K.; Naruse, T.K.; Fujimori, Y.; et al. Comparative sequencing of human and chimpanzee MHC class I regions unveils insertions/deletions as the major path to genomic divergence. *Proc. Natl. Acad. Sci.* **2003**, *100*, 7708–7713. [CrossRef] [PubMed]

81. Kulski, J.K.; Shiina, T.; Anzai, T.; Kohara, S.; Inoko, H. Comparative genomic analysis of the MHC: The evolution of class I duplication blocks, diversity and complexity from shark to man. *Immunol. Rev.* **2002**, *190*, 95–122. [CrossRef] [PubMed]

82. Wilming, L.G.; Hart, E.A.; Coggill, P.C.; Horton, R.; Gilbert, J.G.R.; Clee, C.; Jones, M.; Lloyd, C.; Palmer, S.; Sims, S.; et al. Sequencing and comparative analysis of the gorilla MHC genomic sequence. *Database* **2013**, *2013*, bat011. [CrossRef] [PubMed]

83. Gaudieri, S.; Giles, K.M.; Kulski, J.K.; Dawkins, R.L. Duplication and polymorphism in the MHC: Alu generated diversity and polymorphism within the PERB11 gene family. *Hereditas* **1997**, *127*, 37–46. [CrossRef]

84. Hughes, A.L.; Yeager, M.; Ten Elshof, A.E.; Chorney, M.J. A new taxonomy of mammalian MHC class I molecules. *Immunol. Today* **1999**, *20*, 22–26. [CrossRef]

85. Bahram, S. MIC genes: From genetics to biology. *Adv. Immunol.* **2000**, *76*, 1–60. [PubMed]

86. Vejbaesya, S.; Luangtrakool, P.; Luangtrakool, K.; Kalayanarooj, S.; Vaughn, D.W.; Endy, T.P.; Mammen, M.P.; Green, S.; Libraty, D.H.; Ennis, F.A.; et al. TNF and LTA Gene, Allele, and Extended HLA Haplotype

Associations with Severe Dengue Virus Infection in Ethnic Thais. *J. Infect. Dis.* **2009**, *199*, 1442–1448. [CrossRef]

87. Low, J.S.Y.; Chin, Y.M.; Mushiroda, T.; Kubo, M.; Govindasamy, G.K.; Pua, K.C.; Yap, Y.Y.; Yap, L.F.; Subramaniam, S.K.; Ong, C.A.; et al. A Genome Wide Study of Copy Number Variation Associated with Nasopharyngeal Carcinoma in Malaysian Chinese Identifies CNVs at 11q14.3 and 6p21.3 as Candidate Loci. *PLoS ONE* **2016**, *11*, e0145774. [CrossRef] [PubMed]

88. Wang, W.; Tian, W.; Zhu, F.; Li, L.; Cai, J.; Wang, F.; Liu, K.; Jin, H.; Wang, J. MICA Gene Deletion in 3411 DNA Samples from Five Distinct Populations in Mainland China and Lack of Association with Nasopharyngeal Carcinoma (NPC) in a Southern Chinese Han population. *Ann. Hum. Genet.* **2016**, *80*, 319–326. [CrossRef] [PubMed]

89. Caillat-Zucman, S. New insights into the understanding of MHC associations with immune-mediated disorders: MHC-associated immune diseases. *HLA* **2017**, *89*, 3–13. [CrossRef] [PubMed]

90. Buniello, A.; MacArthur, J.A.L.; Cerezo, M.; Harris, L.W.; Hayhurst, J.; Malangone, C.; McMahon, A.; Morales, J.; Mountjoy, E.; Sollis, E.; et al. The NHGRI-EBI GWAS Catalog of published genome-wide association studies, targeted arrays and summary statistics 2019. *Nucleic Acids Res.* **2019**, *47*, D1005–D1012. [CrossRef]

91. Medici, M.; Porcu, E.; Pistis, G.; Teumer, A.; Brown, S.J.; Jensen, R.A.; Rawal, R.; Roef, G.L.; Plantinga, T.S.; Vermeulen, S.H.; et al. Identification of Novel Genetic Loci Associated with Thyroid Peroxidase Antibodies and Clinical Thyroid Disease. *PLoS Genet.* **2014**, *10*, e1004123. [CrossRef] [PubMed]

92. Yuan, H.; Liu, H.; Liu, Z.; Owzar, K.; Han, Y.; Su, L.; Wei, Y.; Hung, R.J.; McLaughlin, J.; Brhane, Y.; et al. A Novel Genetic Variant in Long Non-coding RNA Gene NEXN-AS1 is Associated with Risk of Lung Cancer. *Sci. Rep.* **2016**, *6*, 34234. [CrossRef]

93. The International HIV Controllers Study. The Major Genetic Determinants of HIV-1 Control Affect HLA Class I Peptide Presentation. *Science* **2010**, *330*, 1551–1557. [CrossRef] [PubMed]

94. Daly, A.K.; Donaldson, P.T.; Bhatnagar, P.; Shen, Y.; Pe'er, I.; Floratos, A.; Daly, M.J.; Goldstein, D.B.; John, S.; Nelson, M.R.; et al. HLA-B*5701 genotype is a major determinant of drug-induced liver injury due to flucloxacillin. *Nat. Genet.* **2009**, *41*, 816–819. [CrossRef]

95. Limou, S.; Le Clerc, S.; Coulonges, C.; Carpentier, W.; Dina, C.; Delaneau, O.; Labib, T.; Taing, L.; Sladek, R.; ANRS Genomic Group; et al. Genomewide Association Study of an AIDS-Nonprogression Cohort Emphasizes the Role Played by *HLA* Genes (ANRS Genomewide Association Study 02). *J. Infect. Dis.* **2009**, *199*, 419–426. [CrossRef] [PubMed]

96. Liu, Y.; Helms, C.; Liao, W.; Zaba, L.C.; Duan, S.; Gardner, J.; Wise, C.; Miner, A.; Malloy, M.J.; Pullinger, C.R.; et al. A Genome-Wide Association Study of Psoriasis and Psoriatic Arthritis Identifies New Disease Loci. *PLoS Genet.* **2008**, *4*, e1000041. [CrossRef]

97. Rodriguez-Novoa, S.; Cuenca, L.; Morello, J.; Cordoba, M.; Blanco, F.; Jimenez-Nacher, I.; Soriano, V. Use of the HCP5 single nucleotide polymorphism to predict hypersensitivity reactions to abacavir: Correlation with HLA-B*5701. *J. Antimicrob. Chemother.* **2010**, *65*, 1567–1569. [CrossRef]

98. Suhre, K.; Arnold, M.; Bhagwat, A.M.; Cotton, R.J.; Engelke, R.; Raffler, J.; Sarwath, H.; Thareja, G.; Wahl, A.; DeLisle, R.K.; et al. Connecting genetic risk to disease end points through the human blood plasma proteome. *Nat. Commun.* **2017**, *8*, 14357. [CrossRef]

99. Crosslin, D.R.; Carrell, D.S.; Burt, A.; Kim, D.S.; Underwood, J.G.; Hanna, D.S.; Comstock, B.A.; Baldwin, E.; de Andrade, M.; Kullo, I.J.; et al. Genetic variation in the HLA region is associated with susceptibility to herpes zoster. *Genes Immun.* **2015**, *16*, 1–7. [CrossRef]

100. Astle, W.J.; Elding, H.; Jiang, T.; Allen, D.; Ruklisa, D.; Mann, A.L.; Mead, D.; Bouman, H.; Riveros-Mckay, F.; Kostadima, M.A.; et al. The Allelic Landscape of Human Blood Cell Trait Variation and Links to Common Complex Disease. *Cell* **2016**, *167*, 1415–1429. [CrossRef] [PubMed]

101. McKay, J.D.; Hung, R.J.; Han, Y.; Zong, X.; Carreras-Torres, R.; Christiani, D.C.; Caporaso, N.E.; Johansson, M.; Xiao, X.; Li, Y.; et al. Large-scale association analysis identifies new lung cancer susceptibility loci and heterogeneity in genetic susceptibility across histological subtypes. *Nat. Genet.* **2017**, *49*, 1126–1132. [CrossRef]

102. Feenstra, B.; Pasternak, B.; Geller, F.; Carstensen, L.; Wang, T.; Huang, F.; Eitson, J.L.; Hollegaard, M.V.; Svanström, H.; Vestergaard, M.; et al. Common variants associated with general and MMR vaccine–related febrile seizures. *Nat. Genet.* **2014**, *46*, 1274–1282. [CrossRef] [PubMed]

103. Miller, F.W.; Chen, W.; O'Hanlon, T.P.; Cooper, R.G.; Vencovsky, J.; Rider, L.G.; Danko, K.; Wedderburn, L.R.; Lundberg, I.E.; Pachman, L.M.; et al. Genome-wide association study identifies HLA 8.1 ancestral haplotype alleles as major genetic risk factors for myositis phenotypes. *Genes Immun.* **2015**, *16*, 470–480. [CrossRef]

104. Rothwell, S.; Cooper, R.G.; Lundberg, I.E.; Miller, F.W.; Gregersen, P.K.; Bowes, J.; Vencovsky, J.; Danko, K.; Limaye, V.; Selva-O'Callaghan, A.; et al. Dense genotyping of immune-related loci in idiopathic inflammatory myopathies confirms HLA alleles as the strongest genetic risk factor and suggests different genetic background for major clinical subgroups. *Ann. Rheum. Dis.* **2016**, *75*, 1558–1566. [CrossRef] [PubMed]

105. Chen, D.-P.; Chang, S.-W.; Jaing, T.-H.; Wang, W.-T.; Hus, F.-P.; Tseng, C.-P. Single nucleotide polymorphisms within HLA region are associated with disease relapse for patients with unrelated cord blood transplantation. *PeerJ* **2018**, *6*, e5228. [CrossRef]

106. van Manen, D.; Kootstra, N.A.; Boeser-Nunnink, B.; Handulle, M.A.; van't Wout, A.B.; Schuitemaker, H. Association of HLA-C and HCP5 gene regions with the clinical course of HIV-1 infection. *AIDS* **2009**, *23*, 19–28. [CrossRef]

107. van Manen, D.; van 't Wout, A.B.; Schuitemaker, H. Genome-wide association studies on HIV susceptibility, pathogenesis and pharmacogenomics. *Retrovirology* **2012**, *9*, 70. [CrossRef]

108. Shrestha, S.; Aissani, B.; Song, W.; Wilson, C.M.; Kaslow, R.A.; Tang, J. Host genetics and HIV-1 viral load set-point in African–Americans. *AIDS* **2009**, *23*, 673–677. [CrossRef]

109. Thørner, L.W.; Erikstrup, C.; Harritshøj, L.H.; Larsen, M.H.; Kronborg, G.; Pedersen, C.; Larsen, C.S.; Pedersen, G.; Gerstoft, J.; Obel, N.; et al. Impact of polymorphisms in the HCP5 and HLA-C, and ZNRD1 genes on HIV viral load. *Infect. Genet. Evol.* **2016**, *41*, 185–190. [CrossRef]

110. Catano, G.; Kulkarni, H.; He, W.; Marconi, V.C.; Agan, B.K.; Landrum, M.; Anderson, S.; Delmar, J.; Telles, V.; Song, L.; et al. HIV-1 Disease-Influencing Effects Associated with ZNRD1, HCP5 and HLA-C Alleles Are Attributable Mainly to Either HLA-A10 or HLA-B*57 Alleles. *PLoS ONE* **2008**, *3*, e3636. [CrossRef]

111. Colombo, S.; Rauch, A.; Rotger, M.; Fellay, J.; Martinez, R.; Fux, C.; Thurnheer, C.; Günthard, H.F.; Goldstein, D.B.; Furrer, H.; et al. The *HCP5* Single-Nucleotide Polymorphism: A Simple Screening Tool for Prediction of Hypersensitivity Reaction to Abacavir. *J. Infect. Dis.* **2008**, *198*, 864–867. [CrossRef] [PubMed]

112. Sousa-Pinto, B.; Correia, C.; Gomes, L.; Gil-Mata, S.; Araújo, L.; Correia, O.; Delgado, L. HLA and Delayed Drug-Induced Hypersensitivity. *Int. Arch. Allergy Immunol.* **2016**, *170*, 163–179. [CrossRef] [PubMed]

113. Illing, P.T.; Vivian, J.P.; Dudek, N.L.; Kostenko, L.; Chen, Z.; Bharadwaj, M.; Miles, J.J.; Kjer-Nielsen, L.; Gras, S.; Williamson, N.A.; et al. Immune self-reactivity triggered by drug-modified HLA-peptide repertoire. *Nature* **2012**, *486*, 554–558. [CrossRef] [PubMed]

114. Ostrov, D.A.; Grant, B.J.; Pompeu, Y.A.; Sidney, J.; Harndahl, M.; Southwood, S.; Oseroff, C.; Lu, S.; Jakoncic, J.; de Oliveira, C.A.F.; et al. Drug hypersensitivity caused by alteration of the MHC-presented self-peptide repertoire. *Proc. Natl. Acad. Sci.* **2012**, *109*, 9959–9964. [CrossRef]

115. Denny, J.C.; Bastarache, L.; Ritchie, M.D.; Carroll, R.J.; Zink, R.; Mosley, J.D.; Field, J.R.; Pulley, J.M.; Ramirez, A.H.; Bowton, E.; et al. Systematic comparison of phenome-wide association study of electronic medical record data and genome-wide association study data. *Nat. Biotechnol.* **2013**, *31*, 1102–1111. [CrossRef]

116. Kisiel, B.; Kisiel, K.; Szymański, K.; Mackiewicz, W.; Biało-Wójcicka, E.; Uczniak, S.; Fogtman, A.; Iwanicka-Nowicka, R.; Koblowska, M.; Kossowska, H.; et al. The association between 38 previously reported polymorphisms and psoriasis in a Polish population: High predicative accuracy of a genetic risk score combining 16 loci. *PLoS ONE* **2017**, *12*, e0179348. [CrossRef] [PubMed]

117. Borgiani, P.; Di Fusco, D.; Erba, F.; Marazzi, M.C.; Mancinelli, S.; Novelli, G.; Palombi, L.; Ciccacci, C. *HCP5* genetic variant (RS3099844) contributes to Nevirapine-induced Stevens Johnsons Syndrome/Toxic Epidermal Necrolysis susceptibility in a population from Mozambique. *Eur. J. Clin. Pharmacol.* **2014**, *70*, 275–278. [CrossRef] [PubMed]

118. Ceccarelli, F.; Perricone, C.; Borgiani, P.; Ciccacci, C.; Rufini, S.; Cipriano, E.; Alessandri, C.; Spinelli, F.R.; Sili Scavalli, A.; Novelli, G.; et al. Genetic Factors in Systemic Lupus Erythematosus: Contribution to Disease Phenotype. *J. Immunol. Res.* **2015**, *2015*, 1–11. [CrossRef]

119. Colafrancesco, S.; Ciccacci, C.; Priori, R.; Latini, A.; Picarelli, G.; Arienzo, F.; Novelli, G.; Valesini, G.; Perricone, C.; Borgiani, P. *STAT4, TRAF3IP2, IL10,* and *HCP5* Polymorphisms in Sjögren's Syndrome: Association with Disease Susceptibility and Clinical Aspects. *J. Immunol. Res.* **2019**, *2019*, 1–8. [CrossRef] [PubMed]

120. Jones, P.A. Functions of DNA methylation: Islands, start sites, gene bodies and beyond. *Nat. Rev. Genet.* **2012**, *13*, 484–492. [CrossRef]

121. Reynolds, L.M.; Taylor, J.R.; Ding, J.; Lohman, K.; Johnson, C.; Siscovick, D.; Burke, G.; Post, W.; Shea, S.; Jacobs, D.R.; et al. Age-related variations in the methylome associated with gene expression in human monocytes and T cells. *Nat. Commun.* **2014**, *5*, 5366. [CrossRef] [PubMed]

122. Chung, S.A.; Nititham, J.; Elboudwarej, E.; Quach, H.L.; Taylor, K.E.; Barcellos, L.F.; Criswell, L.A. Genome-Wide Assessment of Differential DNA Methylation Associated with Autoantibody Production in Systemic Lupus Erythematosus. *PLoS ONE* **2015**, *10*, e0129813. [CrossRef] [PubMed]

123. Gross, A.M.; Jaeger, P.A.; Kreisberg, J.F.; Licon, K.; Jepsen, K.L.; Khosroheidari, M.; Morsey, B.M.; Swindells, S.; Shen, H.; Ng, C.T.; et al. Methylome-wide Analysis of Chronic HIV Infection Reveals Five-Year Increase in Biological Age and Epigenetic Targeting of HLA. *Mol. Cell* **2016**, *62*, 157–168. [CrossRef]

124. Zimmermann, M.T.; Oberg, A.L.; Grill, D.E.; Ovsyannikova, I.G.; Haralambieva, I.H.; Kennedy, R.B.; Poland, G.A. System-Wide Associations between DNA-Methylation, Gene Expression, and Humoral Immune Response to Influenza Vaccination. *PLoS ONE* **2016**, *11*, e0152034. [CrossRef] [PubMed]

125. Kukushkina, V.; Modhukur, V.; Suhorutšenko, M.; Peters, M.; Mägi, R.; Rahmioglu, N.; Velthut-Meikas, A.; Altmäe, S.; Esteban, F.J.; Vilo, J.; et al. DNA methylation changes in endometrium and correlation with gene expression during the transition from pre-receptive to receptive phase. *Sci. Rep.* **2017**, *7*, 3916. [CrossRef] [PubMed]

126. Lin, H.; Sui, W.; Tan, Q.; Chen, J.; Zhang, Y.; Ou, M.; Xue, W.; Li, F.; Cao, C.; Sun, Y.; et al. Integrated analyses of a major histocompatibility complex, methylation and transcribed ultra-conserved regions in systemic lupus erythematosus. *Int. J. Mol. Med.* **2016**, *37*, 139–148. [CrossRef]

127. Cheng, L.; Sun, B.; Xiong, Y.; Hu, L.; Gao, L.; Lv, Q.; Zhou, M.; Li, J.; Chen, X.; Zhang, W.; et al. The minor alleles *HCP5* rs3099844 A and *PSORS1C1* rs3131003 G are associated with allopurinol-induced severe cutaneous adverse reactions in Han Chinese: A multicentre retrospective case-control clinical study. *Br. J. Dermatol.* **2018**, *178*, e191–e193. [CrossRef]

128. Cvitic, S.; Longtine, M.S.; Hackl, H.; Wagner, K.; Nelson, M.D.; Desoye, G.; Hiden, U. The Human Placental Sexome Differs between Trophoblast Epithelium and Villous Vessel Endothelium. *PLoS ONE* **2013**, *8*, 12. [CrossRef]

129. Rosenfeld, C.S. Sex-Specific Placental Responses in Fetal Development. *Endocrinology* **2015**, *156*, 3422–3434. [CrossRef]

130. Wei, Y.; Yan, Z.; Wu, C.; Zhang, Q.; Zhu, Y.; Li, K.; Xu, Y. Integrated analysis of dosage effect lncRNAs in lung adenocarcinoma based on comprehensive network. *Oncotarget* **2017**, *8*, 71430–71446. [CrossRef]

131. Coit, P.; Kaushik, P.; Kerr, G.; Walsh, J.; Dubreuil, M.; Reimold, A.; Sawalha, A. Genome-Wide DNA Methylation Analysis in Ankylosing Spondylitis [abstract]. *Arthritis Rheumatol.* **2018**, *70*. abstract number 2061.

132. Koga, M.; Kawasaki, A.; Ito, I.; Furuya, T.; Ohashi, J.; Kyogoku, C.; Ito, S.; Hayashi, T.; Matsumoto, I.; Kusaoi, M.; et al. Cumulative association of eight susceptibility genes with systemic lupus erythematosus in a Japanese female population. *J. Hum. Genet.* **2011**, *56*, 503–507. [CrossRef] [PubMed]

133. Houshdaran, S.; Zelenko, Z.; Irwin, J.C.; Giudice, L.C. Human endometrial DNA methylome is cycle-dependent and is associated with gene expression regulation. *Mol. Endocrinol.* **2014**, *28*, 1118–1135. [CrossRef] [PubMed]

134. Saare, M.; Modhukur, V.; Suhorutshenko, M.; Rajashekar, B.; Rekker, K.; Sõritsa, D.; Karro, H.; Soplepmann, P.; Sõritsa, A.; Lindgren, C.M.; et al. The influence of menstrual cycle and endometriosis on endometrial methylome. *Clin. Epigenetics* **2016**, *8*, 2. [CrossRef] [PubMed]

135. Orvis, T.; Hepperla, A.; Walter, V.; Song, S.; Simon, J.; Parker, J.; Wilkerson, M.D.; Desai, N.; Major, M.B.; Hayes, D.N.; et al. BRG1/SMARCA4 Inactivation Promotes Non-Small Cell Lung Cancer Aggressiveness by Altering Chromatin Organization. *Cancer Res.* **2014**, *74*, 6486–6498. [CrossRef]

136. Warner, N.; Burberry, A.; Pliakas, M.; McDonald, C.; Núñez, G. A Genome-wide Small Interfering RNA (siRNA) Screen Reveals Nuclear Factor-κB (NF-κB)-independent Regulators of NOD2-induced Interleukin-8 (IL-8) Secretion. *J. Biol. Chem.* **2014**, *289*, 28213–28224. [CrossRef] [PubMed]

137. Park, C.Y.; Krishnan, A.; Zhu, Q.; Wong, A.K.; Lee, Y.-S.; Troyanskaya, O.G. Tissue-aware data integration approach for the inference of pathway interactions in metazoan organisms. *Bioinformatics* **2015**, *31*, 1093–1101. [CrossRef] [PubMed]

138. Rouillard, A.D.; Gundersen, G.W.; Fernandez, N.F.; Wang, Z.; Monteiro, C.D.; McDermott, M.G.; Ma'ayan, A. The harmonizome: A collection of processed datasets gathered to serve and mine knowledge about genes and proteins. *Database* **2016**, *2016*, baw100. [CrossRef]

139. Davis, A.P.; Grondin, C.J.; Johnson, R.J.; Sciaky, D.; McMorran, R.; Wiegers, J.; Wiegers, T.C.; Mattingly, C.J. The Comparative Toxicogenomics Database: Update 2019. *Nucleic Acids Res.* **2019**, *47*, D948–D954. [CrossRef]

140. Rieswijk, L.; Claessen, S.M.H.; Bekers, O.; van Herwijnen, M.; Theunissen, D.H.J.; Jennen, D.G.J.; de Kok, T.M.C.M.; Kleinjans, J.C.S.; van Breda, S.G.J. Aflatoxin B1 induces persistent epigenomic effects in primary human hepatocytes associated with hepatocellular carcinoma. *Toxicology* **2016**, *350–352*, 31–39. [CrossRef]

141. Lerner, P.; Guadalupe, M.; Donovan, R.; Hung, J.; Flamm, J.; Prindiville, T.; Sankaran-Walters, S.; Syvanen, M.; Wong, J.K.; George, M.D.; et al. The Gut Mucosal Viral Reservoir in HIV-Infected Patients Is Not the Major Source of Rebound Plasma Viremia following Interruption of Highly Active Antiretroviral Therapy. *J. Virol.* **2011**, *85*, 4772–4782. [CrossRef]

142. Mitchell, R.S.; Beitzel, B.F.; Schroder, A.R.W.; Shinn, P.; Chen, H.; Berry, C.C.; Ecker, J.R.; Bushman, F.D. Retroviral DNA Integration: ASLV, HIV, and MLV Show Distinct Target Site Preferences. *PLoS Biol.* **2004**, *2*, e234. [CrossRef]

143. Borjabad, A.; Morgello, S.; Chao, W.; Kim, S.-Y.; Brooks, A.I.; Murray, J.; Potash, M.J.; Volsky, D.J. Significant Effects of Antiretroviral Therapy on Global Gene Expression in Brain Tissues of Patients with HIV-1-Associated Neurocognitive Disorders. *Plos Pathog.* **2011**, *7*, e1002213. [CrossRef]

144. Ockenhouse, C.F.; Bernstein, W.B.; Wang, Z.; Vahey, M.T. Functional Genomic Relationships in HIV-1 Disease Revealed by Gene-Expression Profiling of Primary Human Peripheral Blood Mononuclear Cells. *J. Infect. Dis.* **2005**, *191*, 2064–2074. [CrossRef] [PubMed]

145. Brown, J.N.; Kohler, J.J.; Coberley, C.R.; Sleasman, J.W.; Goodenow, M.M. HIV-1 Activates Macrophages Independent of Toll-Like Receptors. *PLoS ONE* **2008**, *3*, e3664. [CrossRef]

146. Hyrcza, M.D.; Kovacs, C.; Loutfy, M.; Halpenny, R.; Heisler, L.; Yang, S.; Wilkins, O.; Ostrowski, M.; Der, S.D. Distinct Transcriptional Profiles in Ex Vivo CD4$^+$ and CD8$^+$ T Cells Are Established Early in Human Immunodeficiency Virus Type 1 Infection and Are Characterized by a Chronic Interferon Response as Well as Extensive Transcriptional Changes in CD8+ T Cells. *J. Virol.* **2007**, *81*, 3477–3486. [CrossRef]

147. Lewinski, M.K.; Bisgrove, D.; Shinn, P.; Chen, H.; Hoffmann, C.; Hannenhalli, S.; Verdin, E.; Berry, C.C.; Ecker, J.R.; Bushman, F.D. Genome-Wide Analysis of Chromosomal Features Repressing Human Immunodeficiency Virus Transcription. *J. Virol.* **2005**, *79*, 6610–6619. [CrossRef] [PubMed]

148. Liu, N.; Zhang, R.; Zhao, X.; Su, J.; Bian, X.; Ni, J.; Yue, Y.; Cai, Y.; Jin, J. A potential diagnostic marker for ovarian cancer: Involvement of the histone acetyltransferase, human males absent on the first. *Oncol. Lett.* **2013**, *6*, 393–400. [CrossRef]

149. Luo, H.; Shenoy, A.K.; Li, X.; Jin, Y.; Jin, L.; Cai, Q.; Tang, M.; Liu, Y.; Chen, H.; Reisman, D.; et al. MOF Acetylates the Histone Demethylase LSD1 to Suppress Epithelial-to-Mesenchymal Transition. *Cell Rep.* **2016**, *15*, 2665–2678. [CrossRef] [PubMed]

150. Teng, H.; Wang, P.; Xue, Y.; Liu, X.; Ma, J.; Cai, H.; Xi, Z.; Li, Z.; Liu, Y. Role of HCP5-miR-139-RUNX1 Feedback Loop in Regulating Malignant Behavior of Glioma Cells. *Mol. Ther.* **2016**, *24*, 1806–1822. [CrossRef]

151. Lee, J.-J.; Kim, M.; Kim, H.-P. Epigenetic regulation of long noncoding RNA UCA1 by SATB1 in breast cancer. *Bmb Rep.* **2016**, *49*, 578–583. [CrossRef]

152. Wang, Z.; Lu, B.; Sun, L.; Yan, X.; Xu, J. Identification of candidate genes or microRNAs associated with the lymph node metastasis of SCLC. *Cancer Cell Int.* **2018**, *18*, 161–171. [CrossRef]

153. Yang, C.; Sun, J.; Liu, W.; Yang, Y.; Chu, Z.; Yang, T.; Gui, Y.; Wang, D. Long noncoding RNA HCP5 contributes to epithelial-mesenchymal transition in colorectal cancer through ZEB1 activation and interacting with miR-139-5p. 11. *Am. J. Transl. Res.* **2019**, *11*, 953–963.

154. Yun, W.-K.; Hu, Y.-M.; Zhao, C.-B.; Yu, D.-Y.; Tang, J.-B. HCP5 promotes colon cancer development by activating AP1G1 via PI3K/AKT pathway. *Eur. Rev. Med Pharmacol. Sci.* **2019**, *23*, 2786–2793.

155. Liang, L.; Xu, J.; Wang, M.; Xu, G.; Zhang, N.; Wang, G.; Zhao, Y. LncRNA HCP5 promotes follicular thyroid carcinoma progression via miRNAs sponge. *Cell Death Dis.* **2018**, *9*, 372. [CrossRef] [PubMed]

156. Yu, Y.; Shen, H.; Fang, D.; Meng, Q.; Xin, Y. LncRNA HCP5 promotes the development of cervical cancer. *Eur. Rev. Med. Pharmacol. Sci.* **2018**, *22*, 4812–4819. [PubMed]

157. Olgun, G.; Sahin, O.; Tastan, O. Discovering lncRNA mediated sponge interactions in breast cancer molecular subtypes. *BMC Genomics* **2018**, *19*, 650. [CrossRef] [PubMed]

158. Jiang, Q.; Ma, R.; Wang, J.; Wu, X.; Jin, S.; Peng, J.; Tan, R.; Zhang, T.; Li, Y.; Wang, Y. LncRNA2Function: A comprehensive resource for functional investigation of human lncRNAs based on RNA-seq data. *BMC Genom.* **2015**, *16*, S2. [CrossRef]

159. Yan, X.; Hu, Z.; Feng, Y.; Hu, X.; Yuan, J.; Zhao, S.D.; Zhang, Y.; Yang, L.; Shan, W.; He, Q.; et al. Comprehensive Genomic Characterization of Long Non-coding RNAs across Human Cancers. *Cancer Cell* **2015**, *28*, 529–540. [CrossRef]

160. Lanzós, A.; Carlevaro-Fita, J.; Mularoni, L.; Reverter, F.; Palumbo, E.; Guigó, R.; Johnson, R. Discovery of Cancer Driver Long Noncoding RNAs across 1112 Tumour Genomes: New Candidates and Distinguishing Features. *Sci. Rep.* **2017**, *7*, 41544. [CrossRef]

161. De Giorgio, A.; Krell, J.; Harding, V.; Stebbing, J.; Castellano, L. Emerging Roles of Competing Endogenous RNAs in Cancer: Insights from the Regulation of PTEN. *Mol. Cell. Biol.* **2013**, *33*, 3976–3982. [CrossRef] [PubMed]

162. Salmena, L.; Poliseno, L.; Tay, Y.; Kats, L.; Pandolfi, P.P. A ceRNA Hypothesis: The Rosetta Stone of a Hidden RNA Language? *Cell* **2011**, *146*, 353–358. [CrossRef]

163. Mo, X.; Li, T.; Xie, Y.; Zhu, L.; Xiao, B.; Liao, Q.; Guo, J. Identification and functional annotation of metabolism-associated lncRNAs and their related protein-coding genes in gastric cancer. *Mol. Genet. Genom. Med.* **2018**, *6*, 728–738. [CrossRef]

164. Tian, L.; He, Y.; Zhang, H.; Wu, Z.; Li, D.; Zheng, C. Comprehensive analysis of differentially expressed profiles of lncRNAs and mRNAs reveals ceRNA networks in the transformation of diffuse large B-cell lymphoma. *Oncol. Lett.* **2018**, *16*, 882–890. [CrossRef] [PubMed]

165. Li, J.; Zhu, Y.; Wang, H.; Ji, X. Targeting Long Noncoding RNA in Glioma: A Pathway Perspective. *Mol. Ther. Nucleic Acids* **2018**, *13*, 431–441. [CrossRef]

166. Klase, Z.; Yedavalli, V.S.R.K.; Houzet, L.; Perkins, M.; Maldarelli, F.; Brenchley, J.; Strebel, K.; Liu, P.; Jeang, K.-T. Activation of HIV-1 from Latent Infection via Synergy of RUNX1 Inhibitor Ro5-3335 and SAHA. *PLoS Pathog.* **2014**, *10*, e1003997. [CrossRef] [PubMed]

167. Ma, X.; He, Z.; Li, L.; Yang, D.; Liu, G. Expression profiles analysis of long non-coding RNAs identified novel lncRNA biomarkers with predictive value in outcome of cutaneous melanoma. *Oncotarget* **2017**, *8*, 1–10. [CrossRef]

168. Nohata, N.; Abba, M.C.; Gutkind, J.S. Unraveling the oral cancer lncRNAome: Identification of novel lncRNAs associated with malignant progression and HPV infection. *Oral Oncol.* **2016**, *59*, 58–66. [CrossRef]

169. Gallant-Behm, C.; Espinosa, J. ΔNp63α utilizes multiple mechanisms to repress transcription in squamous cell carcinoma cells. *Cell Cycle* **2013**, *12*, 409–416. [CrossRef]

170. Goossens, N.; Hoshida, Y. Hepatitis C virus-induced hepatocellular carcinoma. *Clin. Mol. Hepatol.* **2015**, *21*, 105. [CrossRef]

171. Sanford, T.; Porten, S.; Meng, M.V. Molecular Analysis of Upper Tract and Bladder Urothelial Carcinoma: Results from a Microarray Comparison. *PLoS ONE* **2015**, *10*, e0137141. [CrossRef] [PubMed]

172. Sood, R.; Wong, W.; Jeng, M.; Zehnder, J.L. Gene expression profile of idiopathic thrombocytopenic purpura (ITP). *Pediatr. Blood Cancer* **2006**, *47*, 675–677. [CrossRef] [PubMed]

173. Gaudieri, S.; Kulski, J.K.; Dawkins, R.L. MHC and Disease Associations in Nonhuman Primates. In *Molecular Biology and Evolution of Blood Group and MHC Antigens in Primates*; Blancher, A., Klein, J., Socha, W.W., Eds.; Springer: Berlin, Germany, 1997; pp. 464–490.

Article

The Major Histocompatibility Complex of Old World Camels—A Synopsis

Martin Plasil [1,2], Sofia Wijkmark [1], Jean Pierre Elbers [3], Jan Oppelt [4,], Pamela Anna Burger [3] and Petr Horin [1,2,*]

1   Department of Animal Genetics, Veterinary and Pharmaceutical University, Palackeho trida 1, 612 42 Brno, Czech Republic; plasilma@vfu.cz (M.P.); wijkmark.sofia@gmail.com (S.W.)
2   Ceitec VFU, RG Animal Immunogenomics, Palackeho trida 1, 612 42 Brno, Czech Republic
3   Research Institute of Wildlife Ecology, Department of Integrative Biology and Evolution, Vetmeduni Vienna, Savoyenstraße 1, 1160 Wien, Austria; jeanpierre.elbers@vetmeduni.ac.at (J.P.E.); pamela.burger@vetmeduni.ac.at (P.A.B.)
4   Ceitec MU, Masaryk University, Kamenice 753/5, 625 00 Brno, Czech Republic; jan.oppelt@mail.muni.cz
5   Faculty of Science, National Centre for Biomolecular Research, Masaryk University, Kamenice 753/5, 625 00 Brno, Czech Republic
*   Correspondence: horin@dior.ics.muni.cz; Tel.: +420-541-562-292

Received: 13 September 2019; Accepted: 3 October 2019; Published: 5 October 2019

Abstract: This study brings new information on major histocompatibility complex (MHC) class III sub-region genes in Old World camels and integrates current knowledge of the MHC region into a comprehensive overview for Old World camels. Out of the MHC class III genes characterized, *TNFA* and the LY6 gene family showed high levels of conservation, characteristic for MHC class III loci in general. For comparison, an MHC class II gene *TAP1*, not coding for antigen presenting molecules but functionally related to MHC antigen presenting functions was studied. *TAP1* had many SNPs, even higher than the MHC class I and II genes encoding antigen presenting molecules. Based on this knowledge and using new camel genomic resources, we constructed an improved genomic map of the entire MHC region of Old World camels. The MHC class III sub-region shows a standard organization similar to that of pig or cattle. The overall genomic structure of the camel MHC is more similar to pig MHC than to cattle MHC. This conclusion is supported by differences in the organization of the MHC class II sub-region, absence of functional DY genes, different organization of MIC genes in the MHC class I sub-region, and generally closer evolutionary relationships of camel and porcine MHC gene sequences analyzed so far.

Keywords: MHC; major histocompatibility complex; Old World camels; camels; dromedary; Bactrian camel; SNP

---

1. Introduction

The major histocompatibility complex (MHC) is a genomic region of critical importance for vertebrate immune functions. Class I and II MHC molecules are responsible for the presentation of peptides to T cells of intracellular and extracellular origin, respectively [1]. Different MHC molecules encoded by different alleles of multiple MHC genes can bind and present different groups of structurally similar antigenic oligopeptides. Compared to highly specific T- and B-cell receptors, MHC molecules are less specific. This phenomenon allows MHC molecules to cope with the immense diversity of protein antigens based on inherited genetic variation. Different antigen presenting molecules are encoded by different allelic variants of MHC class I and class II genes. The MHC region is characterized by many loci with immune as well as with non-immune functions. Of these, multiple loci code for antigen presenting molecules that possess many alleles and have higher heterozygosity typically than predicted by neutrality [2]. As a result, the MHC is one of the most complex and most polymorphic regions of the mammalian genome [3]. Both parasite-mediated and sexual selection shape functional MHC polymorphism in mammalian populations [4]. However, different taxonomic families, and even different species within a family, have evolved various strategies allowing them to successfully cope with the selection pressure of extremely variable pathogens and with other environmental challenges [5–8]. Therefore, further studies of mammalian species and families are needed for obtaining a comprehensive and unbiased picture of MHC function and diversity [9].

The first camelids appeared in North America, 45–40 Mya [10]. While the ancestors of New World camelids migrated to South America, the ancestors of Old World camelids migrated into Asia (around 6–7 Mya) and continued to spread further west. It is now generally accepted that the divergence between one-humped dromedary (*Camelus dromedarius*) and two-humped Bactrian (*Camelus bactrianus*) camel occurred sometime between 4 and 7 Mya [11,12]. Due to the advancements in camel evolutionary studies, it is well established that extant Old World camelids are composed of three species – *Camelus dromedarius*, *Camelus bactrianus* and *Camelus ferus* [11–14]. The other recent part of the *Camelidae* family is represented by New World camelids, which include alpaca (*Vicugna pacos*), vicugna (*Vicugna vicugna*), llama (*Lama glama*) and guanaco (*Lama guanicoe*).

Camels are an economically and socially important species for a large part of the world. Since ancient times, they have been used as animals of burden, for milk and meat production, as well as for riding and racing. They represent a sustainable livestock species with several specific features [15]. In addition to their adaptation to the arid and dusty environment of deserts, such as specialized water and fat metabolism, stress response to high temperatures, and adaptation to intense ultraviolet radiation, they are more resistant to infectious diseases when compared to other species living in the same geographical area, and they also produce milk with a high content of antimicrobial agents. [16–18].

While genome and transcriptome analyses reveal genes underlying the adaptations of camels to the harsh environment [11,19], another specific feature of camels, their immunogenome, has been only poorly studied so far [11,15]. Camelids are the only mammals producing heavy-chain homodimeric antibodies without a light chain and with the antigen-binding fragment reduced to a single heavy chain variable domain [20]. Similarly, camel T-cell receptors and their genes are quite unique in terms of their structure and function, especially the TR γ/δ loci. In dromedary, (*Camelus dromedarius*), γ/δ rearranged genes, somatic hypermutation increases repertoire diversity, very much similar to the process of affinity maturation of immunoglobulin genes [21].

In this context, it is surprising that only little attention has been paid to MHC studies not only in Old World camels but in all camelids. The first information on the existence of MHC in dromedary and alpaca genomes was provided by Antczak [22], using cytogenetic techniques. In Old and New World camelids, the MHC is located on the long arm of chromosome 20 [23,24], and the first characterization of dromedary and Bactrian MHC class II and class I genes was provided by Plasil et al. [24,25]. The most common structure of the mammalian MHC region, represented by the order MHC class II–MHC class III–MHC class I genes, was identified in camels as well [24]. The organization of the MHC genes within these regions also follows the general pattern observed in other mammalian species [25]. A low level of polymorphism of the class II genes *DRA*, *DRB* and *DQA* genes was observed in all three Old World camel species, as well as for class I genes [24,25]. Surprisingly, MHC class I related *MR1* and *MICA* loci were found to be more polymorphic than a classical MHC class I locus, *B-67* [25].

Considering the gaps in the so far studied general organization of the camel MHC, and based on a new dromedary genome assembly [26], the aim of this work was to characterize selected MHC class III genes, and to integrate the new and previously published knowledge into an improved overview of the MHC region with a special focus on *Camelus dromedarius*. For this purpose, MHC-encoded molecules with different functions were selected. *TNFA* was chosen as an example of a typical, well conserved MHC class III gene, while the *LY6G6* genes were selected as a family of genes with a common origin and located in close physical proximity. *TAP1* classified as an MHC class II gene, was selected to determine variation of gene encoding non-antigen presenting molecules, and to compare it with variation observed for the studied MHC class II [24] and MHC class III genes.

2. Materials and Methods

2.1. Analysis of Selected Non-Antigen Presenting MHC Genes

The camel DNA samples used in this study come from collections at the Research Institute of Wildlife Ecology, Vetmeduni Vienna and UPVS Brno. Bactrian camel samples were collected at three different locations in Mongolia and one sample was collected from a breeder in Austria, while dromedary samples were from Jordan, Saudi Arabia, UAE, Qatar, Sudan, Kenya, Kazakhstan and Nigeria. Numbers of analyzed camels of both species are presented in Table 1. All samples were collected commensally during veterinary procedures for previous research projects (GACR 523/09/1972; PI: P. Horin; FWF P1084-B17 and P24706-B25; PI: P. Burger).

Table 1. List of analyzed genes and numbers of individuals of *C. dromedarius* and *C. bactrianus*.

Locus	*Camelus bactrianus* (n)	*Camelus dromedarius* (n)
TNFA	21	22
TAP1	10	8
Ly6G6C	10	10
Ly6G6D	10	10
Ly6G6E	10	10
Ly6G6F	10	10

An MHC class II gene *TAP1*, and MHC class III *TNFA*, and the *LY6G6* gene family were selected for this study. All these genes were already annotated. We amplified and sequenced them as described previously for MHC class I genes [25]. Briefly, individual gene sequences were extracted from sequence resources at NCBI, as seen in Tables S1–S6. Primers were designed using either the Primer3 Webtool or NCBI PrimerBLAST in Table 2. PCR was performed according to standard protocol using either KAPA 2G Robust HS or KAPA LongRange HS (KapaBiosystems, Wilmington, DE, USA). Sequencing was performed on Illumina MiSeq platform (San Diego, CA, USA) using 500 cycles PE chemistry. Data analysis was performed as previously described in Plasil et al. [25].

Table 2. List of primers used in amplification of selected major histocompatibility complex (MHC) genes in camels.

Name	Sequence 5′ → 3′	Locus	Tm (°C)	Product Length
TAP1-1-F	CATTACCCCAGTGTGGACTTCT	*TAP1*	64.4	8268
TAP1-1-R	GCAACCAAAGGAAATTGAAAAC			
LY6G6C-1-F	GGAGGGACCGTTGGAATTAT	*LY6G6C*	68.0	3491
LY6G6C-1-R	GGCGGCTTTTCTGTCAATAG			
LY6G6D-1-F	CCTCCCCTTTTATTGCCCTA	*LY6G6D*	68.0	2479
LY6G6D-1-R	CCCCATATCACTCCTTCAGC			
LY6G6E-1-F	CACAAGTGGTCACGGTCTCT	*LY6G6E*	68.0	4703
LY6G6E-1-R	GAGTGCTACTTCCCAGTCCAG			
LY6G6F-1-F	GCGTTTATCTGGGCTCTGTT	*LY6G6F*	65.0	3854
LY6G6F-1-R	CCACCCTGTCACTGGCTACT			
TNF-1-F	TGCCTGAGTGTCTGAAAGTCC	*TNFA*	66.0	3667
TNF-1-R	CCCACATACACAGCAGGAACT			

All SNP positions are numbered according to their respective reference sequences; for all loci, their accession numbers can be found in supplementary files.

2.2. New Annotation of the MHC Region and Phylogenetic Analyses

We used the newly available resource of the *C. dromedarius* genome assembly CamDro3 [26], which improves upon work published by Elbers et al. [27], for refining our previous characterization of the MHC region to produce a detailed map of all three MHC class regions [24,25]. The CamDro3 assembly is a result of upgrading the CamDro2 assembly, similarly to Elbers et al. [27], where CamDro1 assembly was upgraded to CamDro2 [28]. The CamDro2 assembly was re-scaffolded using the original Dovetail Chicago and Hi-C reads with the HiRise pipeline [29] in an attempt to fix local misassemblies. We then filled in gaps using our PacBio long-reads (SRA accession: SRP050586) [27] with PBJelly v. 15.8.24 [30] twice instead of one time, which was done for CamDro2. Instead of polishing the assembly with Pilon [31], we used a standard variant calling workflow, which increased the RNA-Seq mapping rates relative to the Pilon-polished assembly. Briefly, we first mapped, trimmed and error-corrected Illumina short-insert sequences (SRA accession: SRR2002493) [28], using BBMap v. 38.12 (https://sourceforge.net/projects/bbmap/) with the vslow and usejni settings to the PBJelly assembly. We then sorted and indexed the resulting BAM file with Sambamba v. 0.6.7 [32], and called variants with CallVariants v. 38.12 (https://sourceforge.net/projects/bbmap/). We finally used BCFtools v. 1.2 (http://samtools.github.io/bcftools/) to generate a consensus sequence, for which we filled in gaps using ABYSS Sealer v. 2.1.0 [33], using default settings except for a bloom filter size of 40 GB, and multiple K values from 90 to 20 in increments of 10.

For phylogenetic analyses of the MHC genes studied here, we included annotated sequences from other mammalian species available at NCBI, as seen in Tables S1–S6. Phylogenetic trees were constructed in MEGA7 using the maximum-likelihood (ML) approach with 1000 bootstraps and the best-fitting model was tested according to the Bayesian information criterion (BIC) [34]. I.e., the best evolutionary model for *TAP1*, *TNFA*, *LY6G6E* and *LY6G6F* followed the Tamura 3-parameter with gamma distribution (5 categories with parameters 0.5098, 0.1968, 0.5569 and 0.8816, respectively) [35]. The *LY6G6C* and *LY6G6D* trees were based on the Jukes-Cantor and Kimura 2-parameter [36,37], respectively.

We have constructed a schematic overview of the MHC regions of selected species to compare their general similarities as well as to summarize evolutionary relationships between individual genes. For this purpose, we used the above-mentioned dromedary assembly CamDro3 as well as assemblies of the following species: Cattle (ARS-UCD1.2 (GCF_002263795.1)), pig (Sscrofa11.1 (GCF_000003025.6)), dog (CanFam3.1 (GCF_000002285.3)), human (GRCh38.p12 (GCF_000001405.38)), and data from our previous studies [24,25,38].

We have also prepared a synteny plot to directly compare the MHC regions of camelids. We lifted over NCBI annotations (release 101) for the *V. pacos* assembly GCF_000164845.2 to the DNA Zoo *V. pacos* assembly (https://www.dnazoo.org/assemblies/Vicugna_pacos) with flo (https://github.com/wurmlab/flo). We then extracted the MHC region (sequences and annotations between genes *RXRB* and *TRIM27*) for the DNA Zoo *V. pacos* assembly and the MHC regions for *C. dromedarius* (CamDro3 assembly), *C. bactrianus* (CamBac2 assembly, GCF_000767855.1 scaffolded with CamDro3), and *C. ferus* genomes (CamFer2 assembly, GCF_000311805.1 scaffolded with CamDro3) [26]. Annotations for CamDro3, CamBac2, and CamFer2 were generated with MAKER v. 2.31.10 following an approach similar to Elbers et al. [27,39,40]. We generated a synteny plot of the MHC regions for *V. pacos* (DNA Zoo V. pacos), *C. bactrianus* (CamBac2), *C. dromedarius* (CamDro3), and *C. ferus* (CamFer2) with Synima [41].

3. Results

3.1. The MHC Class II Gene TAP 1

Compared to the available annotated coding sequences (CDSs) of Old World camels (XM_010994582.1:25-2265), one substitution was found in the CDS of the *TAP1* gene. Out of the total of 19 SNPs within the CDS, 12 were non-synonymous and 7 were shared between *Camelus bactrianus* and *Camelus dromedaries*, as seen in Table S7, Figure S1. We detected a discrepancy between the here developed *TAP1* CDS and the sequences available at NCBI, as seen in Table S1. While the alignment of the published dromedary and Bactrian camel sequences showed significant differences between sequence positions 520-585 of the CDS, we could not confirm them in our sequences as all analysed sequences bear close similarity to the sequence of *C. dromedaries*, as seen in Figure S2. A potential deletion between positions 265 and 306 bp was identified in the CDS of *C. ferus* obtained from NCBI, as seen in Table S1.

A phylogenetic analysis of the camel *TAP1* sequences showed high level of conservation within *Camelidae*. Despite its lower bootstrap support, it suggested a relatively close evolutionary relationship with the domestic pig, *Sus scrofa*, as seen in Figure 1. Additional information regarding the sequence similarity of the analyzed *TAP1* CDSs is available in the Table S1.

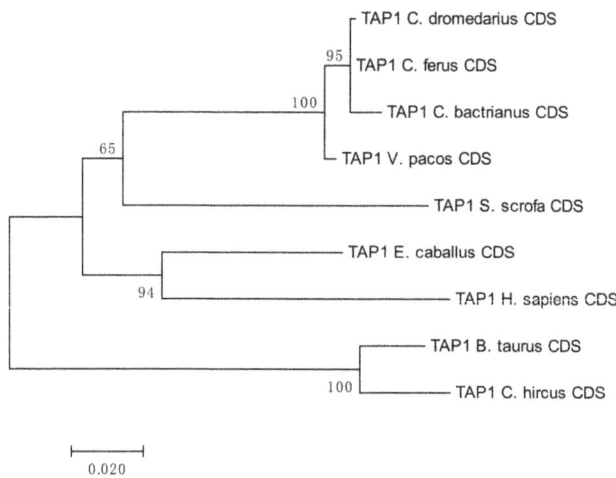

Figure 1. Phylogenetic tree of selected *TAP1* coding sequences (CDSs). The tree is based on the maximum-likelihood method using Tamura 3-parameter model with gamma distribution (five categories, parameter = 0.5098). Number of bootstraps = 1000.

3.2. MHC Class III Genes

3.2.1. The TNFA Gene

The three SNPs found within the *TNFA* CDS in exon 2 (1 non-synonymous) and exon 3 (1 synonymous, 1 non-synonymous) were shared between *C. bactrianus* and *C. dromedaries*, as seen in Figure S3. The haplotypes were conserved across the entire CDS, i.e., the three SNPs were either in homozygous or heterozygous constitution, and the frequency of heterozygote individuals was 0.48 and 0.27 in dromedaries and Bactrian camels, respectively. A comparison with other publicly available camelid sequences did not reveal any additional potential SNPs, as seen in Figure S4. The phylogenetic tree showed an overall high level of conservation within *Camelidae*, and closer evolutionary relationships with sequences from cattle and goat, as seen in Figure 2. Additional information regarding sequence similarity of the analyzed *TNFA* CDSs is available in the Table S2.

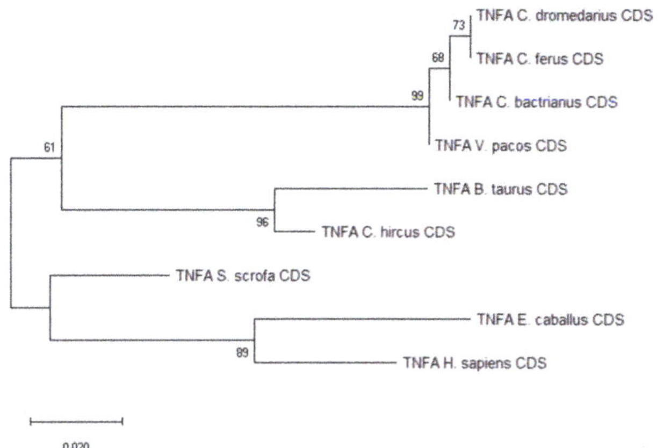

Figure 2. Phylogenetic tree of the *TNFA* CDSs from selected species. The maximum-likelihood method was used based on the Tamura 3-parameter model with gamma distribution (five categories, parameter = 0.1968). The number of bootstraps was 1000.

3.2.2. The LY6G6 Gene Family

Four genes of this family, *LY6G6C*, *LY6G6D*, *LY6G6E*, and *LY6G6F* were analyzed and have been annotated in the three species of Old World camelids. While we discovered only one non-synonymous SNP within the *LY6G6C* CDS at position 218 (G/A) present in *C. dromedaries*, as seen in Figure S5, the sequences retrieved from NCBI showed another non-synonymous SNP at position 235 (C/A) as seen in Figure S6. However, this SNP was not confirmed by our dataset. The frequency of heterozygotes for the *LY6G6C* SNP 218 in *C. dromedarius* was 0.2. The CDS of the *LY6G6D* contained one synonymous SNP (G/T at position 405) observed only in *C. bactrianus*, as seen in Figure S7. The frequency of heterozygotes for this SNP was 0.2 in *C. bactrianus*. Comparison with other available sequences from NCBI did not reveal any additional SNPs in Old World camelids, as seen in Figure S8.

In the family *Camelidae*, the *LY6G6E* gene has been annotated as *sperm acrosome membrane-associated protein 4-like*. Figure S9 shows our analyses, as well as sequences retrieved from NCBI, that the *LY6G6E* CDSs were monomorphic in the subfamily *Camelini*. However, a comparison with the CDS of *Vicugna pacos* (*Lamini*) revealed two SNP positions (319 T/C and 358 G/A), where the latter was non-synonymous. Comparison with the CDS of *Bos taurus* showed that the CDSs compared were not of the same length.

One synonymous (A/G, at position 150) and one non-synonymous SNP (C/T, at position 659) were identified in the CDS of *LY6G6F*, both shared between *C. bactrianus* and *C. dromedaries*, as seen in Figure S10. The frequencies of heterozygotes in *C. bactrianus* and *C. dromedarius*, for both SNPs were 0.3 and 0.5, respectively. Individuals of both species were either heterozygous or homozygous in both SNPs, but never homozygous for one and heterozygous for the other. Comparison of the sequences available on NCBI confirmed both SNPs, as seen in Figure S11.

The results of the phylogenetic analyses summarized in Figures 3–6 document a high level of conservation within the *LY6G6* family CDSs in the *Camelidae*. *LY6G6C* showed the closest evolutionary relationship with the sequence of *Sus scrofa*, while *LY6G6D* was closer to *Bos taurus* and *Capra hircus*. *LY6G6E* was close to sequences of all other included Cetartyodactyla *Sus scrofa*, *Bos taurus* and *Capra hircus*, and *LY6G6F* was also closest to *Sus scrofa*. In humans, *LY6G6E* has a status of pseudogene. Sequence similarities of CDSs used for constructing the phylogenetic trees are provided in Tables S3–S6.

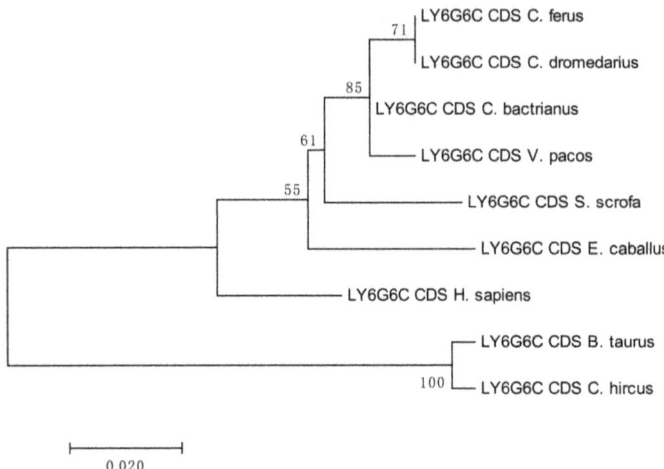

Figure 3. Phylogenetic tree of the *LY6G6C* CDSs from selected species. The maximum-likelihood method was used based on the Jukes-Cantor model. The number of bootstraps was 1000.

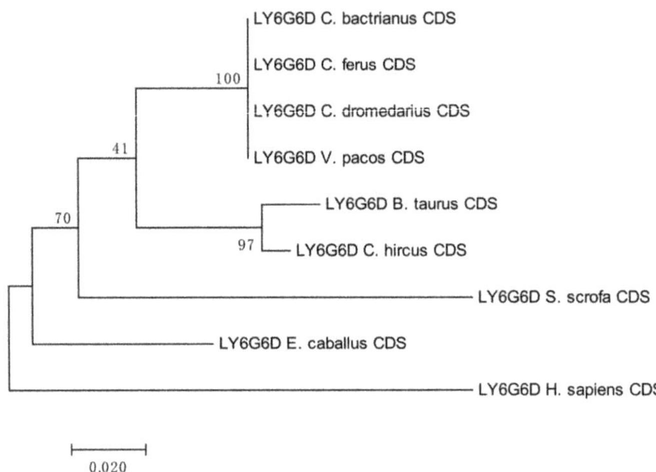

Figure 4. Phylogenetic tree of the *LY6G6D* CDSs from selected species. The maximum-likelihood method was used based on the Kimura 2-parameter model. The number of bootstraps was 1000.

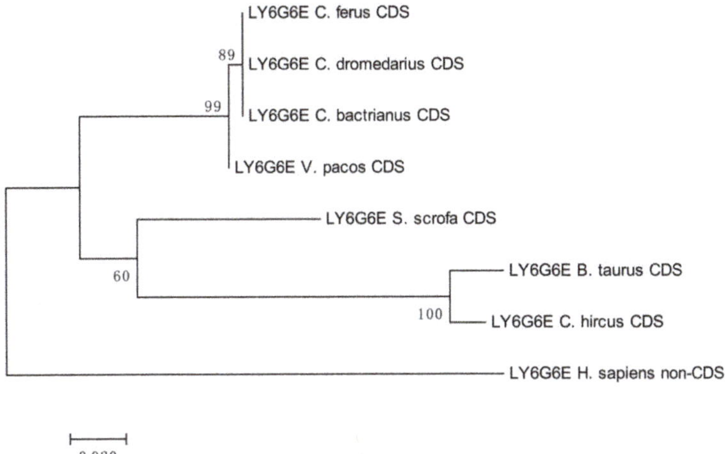

Figure 5. Phylogenetic tree of the *LY6G6E* CDSs from selected species. The maximum-likelihood method was used based on the Tamura 3-parameter model with gamma distribution (five categories, parameter = 0.5569). The number of bootstraps was 1000.

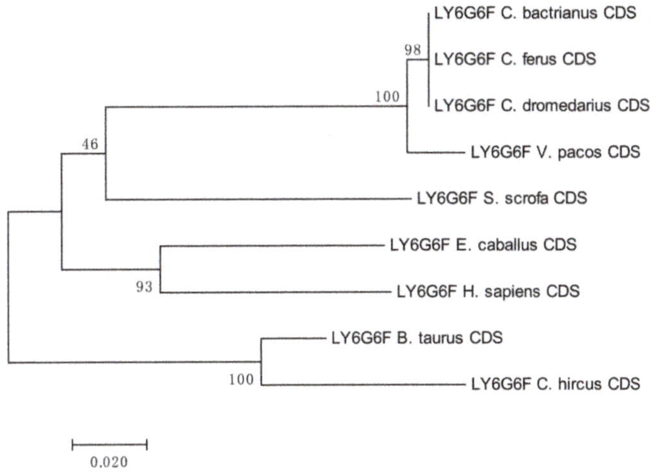

Figure 6. Phylogenetic tree of the *LY6G6F* CDSs from selected species. The maximum-likelihood method was used based on the Tamura 3-parameter model with gamma distribution (five categories, parameter = 0.8816). The number of bootstraps was 1000.

3.3. New Annotation of the MHC Region in the Dromedary

A new version of the organization of the MHC region in the dromedary is shown in Figure 7. A complete annotation record will be available in Lado et al. [26]. A graphical summary of the evolutionary relationships for the genes studied is presented in Figure 8. The synteny plot for available camelid MHC regions is shown in Figure 9.

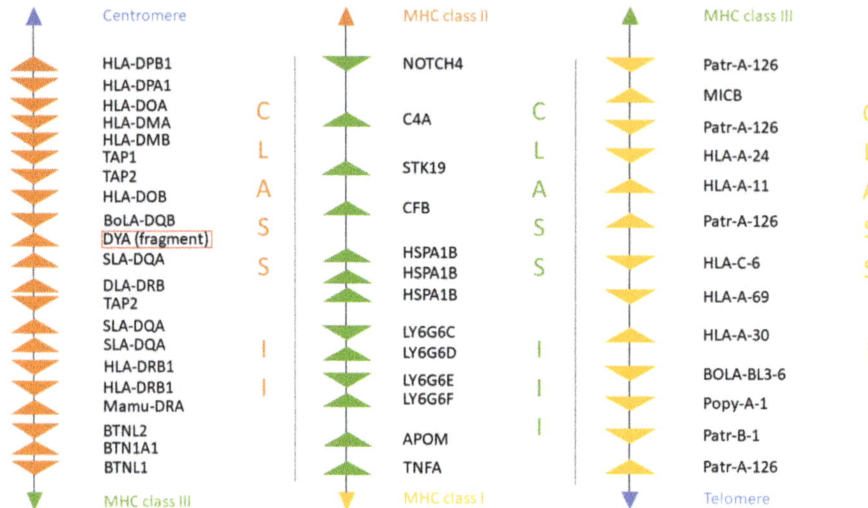

Figure 7. The organization of the MHC region in *C. dromedarius*. MHC class II genes are shown in orange. MHC class III genes are shown in green. MHC class I genes are shown in yellow. Spaces between genes do not represent real distances. Details of the annotation are available in Table S8.

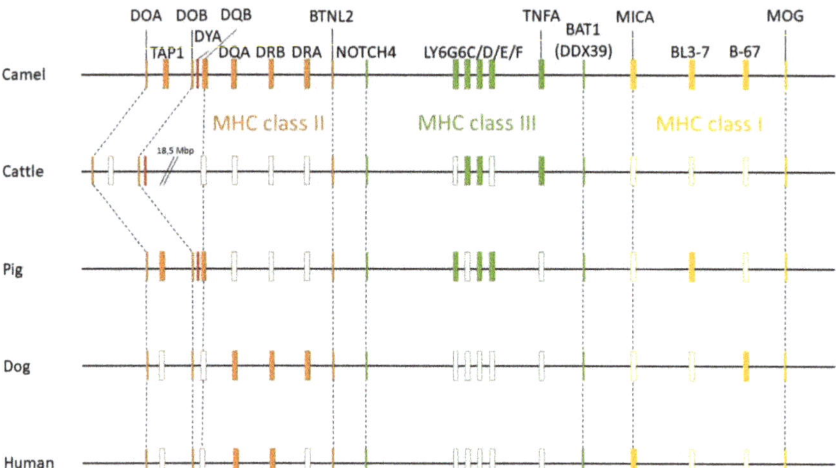

Figure 8. Summary of evolutionary relationships of MHC genes of dromedary, cattle, pig, dog and human. Relative gene positions are presented as rectangles. The filled rectangles highlight the closest relationships observed between the dromedary and the respective species for a given locus within the groups studied.

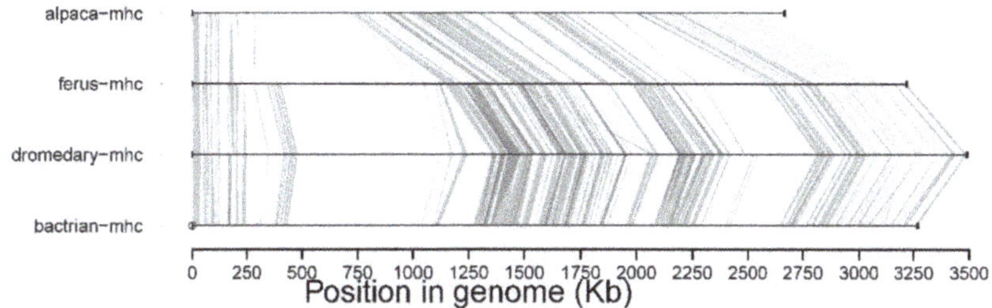

Figure 9. The synteny plot of the camelid MHC region showing a general structural conservation among all camelids. Assemblies of the dromedary (CamDro3) [26]; Bactrian (CamBac2, GCF_000767855.1 scaffolded with CamDro3 as a reference); wild Bactrian (CamFer2, GCF_000311805.1 scaffolded with CamDro3 as a reference); and alpaca (DNA Zoo *V. pacos*) were used.

4. Discussion

This study brings new information on MHC class III sub-region genes in Old World camels. These data, along with new camel genomic resources, allowed us to construct an improved genomic map of the entire MHC region of Old World camels.

The MHC class III region, where no genes coding for antigen presenting molecules are located, represented a gap in our knowledge of the camel MHC. For bridging this gap in the map, we selected different types of class III genes with diverse immunological functions. Based on this, we could make a comparison of their variability with so far known genes from other MHC regions previously studied in the same groups of dromedary and Bactrian camels [24,25]. The data then primarily served for allowing us an overall characterization of the camelid MHC. As such, they do not represent a detailed analysis of the MHC class III region, which still is our pending task.

The *TNFA* (tumor necrosis factor alpha) is a cytokine playing a major role in the activation of both acute and chronical inflammatory processes, and contributing to the activation cascade of mother cytokines and chemokines connected with inflammation [42]. Besides this role, the *TNFA* also contributes to the regulation of apoptosis and carcinogenesis [43,44]. While there are no reports on the role of TNF-alpha and/or of the *TNFA* gene in diseases of camels, in cattle, a closely-related species, *TNFA* polymorphisms have been associated with multiple infectious diseases and with fertility [45–48]. The SNPs identified in this report thus represent potentially useful markers for analyzing various camel diseases. *TNFA*, subject to strong purifying selection [42], may be also used as a marker of a conserved part of the MHC in close physical proximity of MHC genes under positive selection pressure [48].

The LY6G6 (lymphocyte antigen 6 family, member G6) gene family is located within the MHC class III region, spanning approximately 20 kb. In humans (assembly GRCh38.p12) and mice (assembly GRCm38.p4), the family contains four individual genes, *LY6G6C*, *LY6G6D*, *LY6G6E*, and *LY6G6F*. LY6 proteins containing cysteine-rich domain are attached to the cell surface by a GPI anchor, which is involved in signal transduction. LY6 proteins might play a role in the hematopoietic differentiation [49].

This is a first report on the existence of this group of MHC class III genes in camels. Although little is currently known about their functions, they represent suitable anchor and marker loci for the MHC class III region. Sequences of all four selected genes showed high levels of conservation within Old World camels. SNPs were rather rare and were often observed in only one camel species; *LY6G6E* was even completely monomorphic in both species. Interestingly, while the *LY6G6E* sequence has the status of pseudogene in humans, we did not find any features of non-functional *LY6G6E* sequences in the two camel species. These findings are in agreement with those of other closely related mammals, such as *Bos taurus* and *Sus scrofa*, suggesting that this gene is probably functionally important for the Cetartiodactyla.

Having characterized selected MHC class III genes in Old World camels, we were able to build upon our previous studies of MHC class I and MHC class II regions and to update the general organization of the MHC region, describe its similarities with New World camelids, as well as with relevant mammalian species, and to point out its unique features.

The MHC class III genes studied here do not encode antigen presenting molecules and are not involved in the mechanisms of antigen presentation. For comparison, we chose *TAP1*, a gene not coding for antigen presenting molecules, but playing an important role in the process of antigen loading of the MHC class I molecules, located in the MHC class II region [50]. Certain *TAP1* variants may have a large impact on MHC class I expression and interaction with pathogens, at least in some species [51,52]. While *TNFA* and the gene family LY6 showed high levels of conservation characteristic for MHC class III loci in general, the camel *TAP1* was not only polymorphic, but the numbers of SNPs detected were higher than in the MHC class I and class II genes examined so far [24,25].

4.1. MHC Region Organization and Diversity in Old World Camels

The presented MHC organization is primarily based on dromedary sequences, but due to the high conservation and similarity of this region between dromedaries and Bactrian camels, we can take firm conclusions about the latter's MHC. The overall nucleotide and amino acid sequence similarities between the two species ranged across the MHC region (delimited by the genes *TRIM27* and *RXRB*) on chromosome 20 of the new dromedary assembly CamDro3 [26], with no major differences between the three sub-regions, as seen in Figure 9. Extensive trans-species polymorphism (TSP), i.e., allele sharing, is also a typical feature of the MHC of camelids, including llamas [24,25].

The diversity of the MHC class I, class I–related and class II genes is generally lower than expected, based on a comparative analysis of multiple vertebrate species [53]. This finding is in agreement with a lower genome-wide diversity in dromedaries than in the wild and domestic Bactrian camels [28]. We were unable to make a reliable estimate of polymorphisms of class II *DQB* genes. Their exon 2 encoded amino acid sequences containing the antigen-binding site are by four amino-acids longer due to a 12 bp insertion in both camel species. They contain multiple SNPs, but we were unable to identify individual loci [24]. It is possible that certain specificities in *DQB* sequences, such as GC rich regions and/or long repeats cause technical problems as described, e.g., for the horse *DQB* region [54]. Interestingly, some of the genes encoding other than antigen presenting molecules and located in the MHC class I and II regions, such as *MICA* and *TAP1*, were found to have higher polymorphism than classical MHC class I and II genes. On the other hand, the class II *DRA* gene, monomorphic in most mammalian species, is polymorphic both in dromedaries and Bactrian camels [24]. Recently identified MHC-linked microsatellite loci also showed unusually low diversity both in dromedaries and bactrians, often with only two or three alleles [55]. Bottlenecks experienced in the evolutionary history of the three species, and also in recent times due to domestication, might be the cause of the reduced genome-wide variability in dromedaries. However, it is not clear to what extent these processes contributed to the low diversity of the immunogenome. Additionally, the reasons for higher numbers of SNPs observed in genes coding for other than antigen presenting MHC molecules remains especially puzzling.

4.2. Cross-Species Comparisons with New World Camelids and Other Mammals

A direct comparison with alpaca sequences presented here, as well as in our previous work [24,25], showed that the studied loci are generally highly conserved in camelids. Based on the data available so far, as seen in Figure 9, it seems that the organization of the MHC genomic region in *Vicugna pacos* is highly similar if not nearly identical to Old World camels.

When compared to other species, the physical location and the overall organization of the MHC in camels follows the common general structure of the mammalian MHC region [24]. However, phylogenetic relationships of camel MHC genes do not always follow relationships based on other (neutral) nuclear genes [24,25,38,56], suggesting that different MHC sub-regions might have followed different evolutionary pathways. In addition, so far, we have been unable to identify a *DYA* locus, specific for the *Bovidae*, the phylogenetically closest family. Although *DYA*-like sequences were previously annotated in camelid genomes, as seen in Figure 7, a more focused analysis showed that they are more similar to *DQA* rather than to *DYA* sequences, and that it was impossible to identify this region within the camel MHC physical map [38]. Like in camels, the porcine MHC class II region contains a non-functional fragment of DY gene (GenBank Gene ID: 100135048), although it is annotated as *DYB*, while the camel sequence is annotated as *DYA*. Moreover, in cattle, the physical position of the *TAP1* gene is different from other mammalian species, including camels, as seen in Figure 8. Overall, the organization of the camel MHC class II region is more similar to *Sus scrofa* rather than to *Bos taurus*.

Nucleotide sequences of genes located in the MHC class II sub-region are generally closer to dog and human, with the exception of *DQB* and *TAP1*, which are closely related to their porcine counterparts, as seen in Figure 7 [38]. On the other hand, the MHC class III genes and the LY6G6 gene family, are more related to the corresponding porcine sequences, with the exception of *LY6G6D* and *TNFA*, which are closer to the bovine sequences than to its porcine counterpart. Despite a probably common origin and close physical proximity of the LY6G6 gene family, the *LY6G6D* does not behave identically in phylogenetic trees, as seen in Figures 3–6 [57]. It thus seems that different MHC sub-regions have not always the same evolutionary history, reflecting probably the fact that the selection pressures exerted on the immune system are different between camels and cattle.

The evolution of the MHC class I sub-region appears to be even more complex, as shown in our previous study [25]. The MHC class I locus *B-67* is closely related to the canine locus *DLA-88*, the locus *BL3-7* is related to the porcine *SLA-11*, while *MICA* is closest to the relevant human sequences. While the camel MIC genes follow the pattern of the human *MICA/B* genes, cattle have at least three functional MIC genes with unique sequences [25,58]. In contrast to these differences between camel and cattle, high similarities with the pig MHC class I genes were often observed. Especially, the MHC class I gene *BL3-7*, a locus of unclear status, highly similar to the annotated sequence *BL3-6* in alpacas, is also closely related to the locus *SLA-11* in pig, and is one MHC locus with unknown function and unusual structure [25,59]. Interestingly, we have also found close similarities between camels and pigs for two other complex immunogenomic regions, the natural killer complex (NKC) and the leukocyte receptor complex (LRC), encoding natural killer cell receptor molecules. Both regions differ from their cattle counterparts, resembling more the pig NKC and LRC, respectively [60]. However, in this context, it is important to note that the annotation of MHC class I loci from CamDro3, as presented in Figure 7, is based on calculations of their sequence similarities with other species. Taking into consideration that sequence similarities alone are often not informative for a correct assignment of their classical or non-classical status in non-model organisms, we do not have enough information for definitive conclusions about this status for class I genes in camels either.

5. Conclusions

This first characterization of specific MHC class III genes added to the general structure and variability of the MHC in Old World camels, allowing a synopsis of the current knowledge of the MHC region of Old World camelids. Novel phylogenetic analyses between camelids and other domestic mammals showed high conservation of this region among Old and New World camels, and a closer evolutionary relationship with pigs rather than with cattle. The knowledge of the MHCs structure, function and evolution in camels is important for understanding their immune response to diseases in their specific and extreme environments.

Supplementary Materials: The supplementary materials are available online at http://www.mdpi.com/2073-4409/8/10/1200/s1.

Author Contributions: M.P.: investigation, methodology, formal analysis, validation, visualization, writing – original draft, writing – review & editing; S.W.: investigation, formal analysis; J.P.E.: data curation, formal analysis, visualization, writing – review & editing; J.O.: formal analysis; P.A.B.: conceptualization, funding acquisition, project administration, resources, writing – review & editing; P.H.: conceptualization, funding acquisition, project administration, supervision, writing – original draft, writing – review & editing.

Funding: This work was supported by the Austrian Science Foundation (FWF) project P29623-B25 and by CEITEC - Central European Institute of Technology with research infrastructure supported by the project

CZ.1.05/1.1.00/02.0068 financed from European Regional Development Fund. J.P. Elbers is funded by the FWF project P29623-B25 granted to P.A. Burger.

Acknowledgments: We thank all camel owners and veterinarian colleagues for their kind cooperation during sample collection, O. Abdelhadi, A. Abdussamad, H. Burgsteiner, B. Faye, G. Gassner, J. Juhasz, G. Konuspayeva, D. Modry, P. Nagy, A. and J. Perret, M. Qablan and R. Saleh. Open Access Funding by the Austrian Science Fund (FWF).

Conflicts of Interest: The authors declare no conflict of interest.

References

1. Janeway, C.A.; Travers, P.; Walport, M.; Shlomchik, M.J. *Immunobiology: The Immune System in Health and Disease*, 6th ed.; Taylor & Francis Group, Garland Science: New York, NY, USA, 2005.
2. Hedrick, P.W.; Whittam, T.S.; Parham, P. Heterozygosity at individual amino acid sites: Extremely high levels for HLA-A and-B genes. *Proc. Natl. Acad. Sci. USA* **1991**, *88*, 5897–5901. [CrossRef] [PubMed]
3. Kumánovics, A.; Takada, T.; Lindahl, K.F. Genomic organization of the mammalian MHC. *Annu. Rev. Immunol.* **2003**, *21*, 629–657. [CrossRef] [PubMed]
4. Winternitz, J.C.; Minchey, S.G.; Garamszegi, L.Z.; Huang, S.; Stephens, P.R.; Altizer, S. Sexual selection explains more functional variation in the mammalian major histocompatibility complex than parasitism. *Proc. Biol. Sci.* **2013**, *280*, 20131605. [CrossRef] [PubMed]
5. Rocha, R.G.; Magalhães, V.; López-Bao, J.V.; van der Loo, W.; Llaneza, L.; Alvares, F.; Esteves, P.J.; Godinho, R. Alternated selection mechanisms maintain adaptive diversity in different demographic scenarios of a large carnivore. *BMC Evol. Biol.* **2019**, *19*, 90. [CrossRef] [PubMed]
6. Aguilar, A.; Roemer, G.; Debenham, S.; Binns, M.; Garcelon, D.; Wayne, R.K. High MHC diversity maintained by balancing selection in an otherwise genetically monomorphic mammal. *Proc. Natl. Acad. Sci. USA* **2004**, *101*, 3490–3494. [CrossRef]
7. Mikko, S.; Røed, K.; Schmutz, S.; Andersson, L. Monomorphism and polymorphism at Mhc DRB loci in domestic and wild ruminants. *Immunol. Rev.* **1999**, *167*, 169–178. [CrossRef]
8. Doxiadis, G.G.; Otting, N.; de Groot, N.G.; Bontrop, R.E. Differential evolutionary MHC class II strategies in humans and rhesus macaques: Relevance for biomedical studies. *Immunol. Rev.* **2001**, *183*, 76–85. [CrossRef]
9. Bernatchez, L.; Landry, C. MHC studies in nonmodel vertebrates: What have we learned about natural selection in 15 years? *J. Evol. Biol.* **2003**, *16*, 363–377. [CrossRef]
10. Burger, P.A.; Ciani, E.; Faye, B. Old World camels in a modern world—A balancing act between conservation and genetic improvement. *Anim. Genet.* **2019**, in press. [CrossRef]
11. Wu, H.; Guang, X.; Al-Fageeh, M.B.; Cao, J.; Pan, S.; Zhou, H.; Zhang, L.; Abutarboush, M.H.; Xing, Y.; Xie, Z. Camelid genomes reveal evolution and adaptation to desert environments. *Nat. Commun.* **2014**, *5*, 5188. [CrossRef]
12. Ji, R.; Cui, P.; Ding, F.; Geng, J.; Gao, H.; Zhang, H.; Yu, J.; Hu, S.; Meng, H. Monophyletic origin of domestic bactrian camel (Camelus bactrianus) and its evolutionary relationship with the extant wild camel (Camelus bactrianus ferus). *Anim. Genet.* **2009**, *40*, 377–382. [CrossRef] [PubMed]
13. Silbermayr, K.; Orozco-terWengel, P.; Charruau, P.; Enkhbileg, D.; Walzer, C.; Vogl, C.; Schwarzenberger, F.; Kaczensky, P.; Burger, P.A. High mitochondrial differentiation levels between wild and domestic Bactrian camels: A basis for rapid detection of maternal hybridization. *Anim. Genet.* **2010**, *41*, 315–318. [CrossRef] [PubMed]
14. Sequencing, T.B.C.G.; Analysis Consortium. Genome sequences of wild and domestic bactrian camels. *Nat. Commun.* **2012**, *3*, 1202. [CrossRef]
15. Burger, P.A. The history of Old World camelids in the light of molecular genetics. *Trop. Anim. Health Pro.* **2016**, *48*, 905–913. [CrossRef] [PubMed]
16. Wernery, U.; Kinne, J. Foot and mouth disease and similar virus infections in camelids: A review. *Rev. Sci. Tech. Oie* **2012**, *31*, 907–918. [CrossRef]
17. Dirie, M.F.; Abdurahman, O. Observations on little known diseases of camels (Camelus dromedarius) in the Horn of Africa. *Rev. Sci. Tech. Oie* **2003**, *22*, 1043–1050. [CrossRef]
18. Al Kanhal, H.A. Compositional, technological and nutritional aspects of dromedary camel milk. *Int. Dairy J.* **2010**, *20*, 811–821.

19. Ali, A.; Baby, B.; Vijayan, R. Camel Genome-from Desert to Medicine. *Front. Genet.* **2019**, *10*, 17. [CrossRef]

20. Muyldermans, S. Single domain camel antibodies: Current status. *Rev. Mol. Biotech.* **2001**, *74*, 277–302. [CrossRef]

21. Ciccarese, S.M.; Burger, P.; Ciani, E.; Castelli, V.; Linguiti, G.; Plasil, M.; Massari, S.; Horin, P.; Antonacci, R. The camel adaptive immune receptors repertoire as a singular example of structural and functional genomics. *Front. Genet.* **2019**. Under review. [CrossRef]

22. Antczak, D. Major histocompatibility complex genes of the dromedary camel. In *Proceedings of the Qatar Foundation Annual Research Conference, Doha, Qatar, 24–25, November, 2013*; Hamad bin Khalifa University Press (HBKU Press): Doha, Qatar, 2013; p. BIOP015. [CrossRef]

23. Avila, F.; Baily, M.P.; Perelman, P.; Das, P.J.; Pontius, J.; Chowdhary, R.; Owens, E.; Johnson, W.E.; Merriwether, D.A.; Raudsepp, T. A comprehensive whole-genome integrated cytogenetic map for the alpaca (Lama pacos). *Cytogenet. Genome Res.* **2014**, *144*, 196–207. [CrossRef] [PubMed]

24. Plasil, M.; Mohandesan, E.; Fitak, R.R.; Musilova, P.; Kubickova, S.; Burger, P.A.; Horin, P. The major histocompatibility complex in Old World camelids and low polymorphism of its class II genes. *BMC Genomics* **2016**, *17*, 167. [CrossRef] [PubMed]

25. Plasil, M.; Wijkmark, S.; Elbers, J.P.; Oppelt, J.; Burger, P.; Horin, P. The major histocompatibility complex of Old World camelids: Class I and class I-related genes. *HLA* **2019**, *93*, 203–215. [CrossRef] [PubMed]

26. Lado, S.; Elbers, J.P.; Rogers, M.F.; Perelman, P.L.; Proskuryakova, A.A.; Serdyukova, N.A.; Johnson, W.E.; Horin, P.; Corander, J.; Murphy, D.; et al. Reference-guided assembly of two Old World camel genomes and genomic diversity of Old World camelid immune response genes. Manuscript in preparation.

27. Elbers, J.P.; Rogers, M.F.; Perelman, P.L.; Proskuryakova, A.A.; Serdyukova, N.A.; Johnson, W.E.; Horin, P.; Corander, J.; Murphy, D.; Burger, P.A. Improving Illumina assemblies with Hi-C and long reads: An example with the North African dromedary. *Mol. Ecol. Resour.* **2019**, *19*, 1015–1026. [CrossRef]

28. Fitak, R.R.; Mohandesan, E.; Corander, J.; Burger, P.A. The de novo genome assembly and annotation of a female domestic dromedary of North African origin. *Mol. Ecol. Resour.* **2016**, *16*, 314–324. [CrossRef]

29. Putnam, N.H.; O'Connell, B.L.; Stites, J.C.; Rice, B.J.; Blanchette, M.; Calef, R.; Troll, C.J.; Fields, A.; Hartley, P.D.; Sugnet, C.W. Chromosome-scale shotgun assembly using an in vitro method for long-range linkage. *Genome Res.* **2016**, *26*, 342–350. [CrossRef]

30. English, A.C.; Richards, S.; Han, Y.; Wang, M.; Vee, V.; Qu, J.; Qin, X.; Muzny, D.M.; Reid, J.G.; Worley, K.C. Mind the gap: Upgrading genomes with Pacific Biosciences RS long-read sequencing technology. *PLoS ONE* **2012**, *7*, e47768. [CrossRef]

31. Walker, B.J.; Abeel, T.; Shea, T.; Priest, M.; Abouelliel, A.; Sakthikumar, S.; Cuomo, C.A.; Zeng, Q.; Wortman, J.; Young, S.K. Pilon: An integrated tool for comprehensive microbial variant detection and genome assembly improvement. *PLoS ONE* **2014**, *9*, e112963. [CrossRef]

32. Tarasov, A.; Vilella, A.J.; Cuppen, E.; Nijman, I.J.; Prins, P. Sambamba: Fast processing of NGS alignment formats. *Bioinformatics* **2015**, *31*, 2032–2034. [CrossRef]

33. Jackman, S.D.; Vandervalk, B.P.; Mohamadi, H.; Chu, J.; Yeo, S.; Hammond, S.A.; Jahesh, G.; Khan, H.; Coombe, L.; Warren, R.L. ABySS 2.0: Resource-efficient assembly of large genomes using a Bloom filter. *Genome Res.* **2017**, *27*, 768–777. [CrossRef] [PubMed]

34. Kumar, S.; Stecher, G.; Tamura, K. MEGA7: Molecular evolutionary genetics analysis version 7.0 for bigger datasets. *Mol. Biol. Evol.* **2016**, *33*, 1870–1874. [CrossRef] [PubMed]

35. Tamura, K. Estimation of the number of nucleotide substitutions when there are strong transition-transversion and G+ C-content biases. *Mol. Biol. Evol.* **1992**, *9*, 678–687. [PubMed]

36. Jukes, T.H.; Cantor, C.R. Evolution of protein molecules. *Mammal. Prot. Metab.* **1969**, *3*, 132.

37. Kimura, M. A simple method for estimating evolutionary rates of base substitutions through comparative studies of nucleotide sequences. *J Mol. Evol.* **1980**, *16*, 111–120. [CrossRef]

38. Plasil, M. Comparative genomics of the major histocompatibility complex MHC. Ph.D. Thesis, Masaryk University, Brno, Czech Republic, 26 October 2018.

39. Cantarel, B.L.; Korf, I.; Robb, S.M.; Parra, G.; Ross, E.; Moore, B.; Holt, C.; Alvarado, A.S.; Yandell, M. MAKER: An easy-to-use annotation pipeline designed for emerging model organism genomes. *Genome Res.* **2008**, *18*, 188–196. [CrossRef]

40. Holt, C.; Yandell, M. MAKER2: An annotation pipeline and genome-database management tool for second-generation genome projects. *BMC Bioinformatics* **2011**, *12*, 491. [CrossRef]

41. Farrer, R.A. Synima: A Synteny imaging tool for annotated genome assemblies. *BMC Bioinform.* **2017**, *18*, 507. [CrossRef]

42. Chu, W.M. Tumor necrosis factor. *Cancer Lett.* **2013**, *328*, 222–225. [CrossRef]

43. Pan, S.; An, P.; Zhang, R.; He, X.; Yin, G.; Min, W. Etk/Bmx as a tumor necrosis factor receptor type 2-specific kinase: Role in endothelial cell migration and angiogenesis. *Mol. Cell. Biol.* **2002**, *22*, 7512–7523. [CrossRef]

44. Odbileg, R.; Konnai, S.; Ohashi, K.; Onuma, M. Molecular cloning and phylogenetic analysis of inflammatory cytokines of Camelidae (llama and camel). *J. Vet. Med. Sci.* **2005**, *67*, 921–925. [CrossRef] [PubMed]

45. Ranjan, S.; Bhushan, B.; Panigrahi, M.; Kumar, A.; Deb, R.; Kumar, P.; Sharma, D. Association and expression analysis of single nucleotide polymorphisms of partial tumor necrosis factor alpha gene with mastitis in crossbred cattle. *Anim. Biotechnol.* **2015**, *26*, 98–104. [CrossRef] [PubMed]

46. Lendez, P.A.; Passucci, J.A.; Poli, M.A.; Gutierrez, S.E.; Dolcini, G.L.; Ceriani, M.C. Association of TNF-α gene promoter region polymorphisms in bovine leukemia virus (BLV)-infected cattle with different proviral loads. *Arch. Virol.* **2015**, *160*, 2001–2007. [CrossRef] [PubMed]

47. Kawasaki, Y.; Aoki, Y.; Magata, F.; Miyamoto, A.; Kawashima, C.; Hojo, T.; Okuda, K.; Shirasuna, K.; Shimizu, T. The effect of single nucleotide polymorphisms in the tumor necrosis factor-α gene on reproductive performance and immune function in dairy cattle. *J. Reprod. Develop.* **2014**, *60*, 173–178. [CrossRef]

48. Seitzer, U.; Gerdes, J.; Müller-Quernheim, J. Genotyping in the MHC locus: Potential for defining predictive markers in sarcoidosis. *Resp. Res.* **2001**, *3*, 6. [CrossRef]

49. Mallya, M.; Campbell, R.D.; Aguado, B. Characterization of the five novel Ly-6 superfamily members encoded in the MHC, and detection of cells expressing their potential ligands. *Protein. Sci.* **2006**, *15*, 2244–2256. [CrossRef]

50. Trowsdale, J.; Hanson, I.; Mockridge, I.; Beck, S.; Townsendt, A.; Kelly, A. Sequences encoded in the class II region of the MHC related to the 'ABC' superfamily of transporters. *Nature* **1990**, *348*, 741. [CrossRef]

51. Kaufman, J. Co-evolution with chicken class I genes. *Immunol. Rev.* **2015**, *267*, 56–71. [CrossRef]

52. Praest, P.; Luteijn, R.D.; Brak-Boer, I.G.J.; Lanfermeijer, J.; Hoelen, H.; Ijgosse, L.; Costa, A.I.; Gorham, R.D.; Lebbink, R.J.; Wiertz, E. The influence of TAP1 and TAP2 gene polymorphisms on TAP function and its inhibition by viral immune evasion proteins. *Mol. Immunol.* **2018**, *101*, 55–64. [CrossRef]

53. Kulski, J.K.; Shiina, T.; Anzai, T.; Kohara, S.; Inoko, H. Comparative genomic analysis of the MHC: The evolution of class I duplication blocks, diversity and complexity from shark to man. *Immunol. Rev.* **2002**, *190*, 95–122. [CrossRef]

54. Viļuma, A.; Mikko, S.; Hahn, D.; Skow, L.; Andersson, G.; Bergström, T.F. Genomic structure of the horse major histocompatibility complex class II region resolved using PacBio long-read sequencing technology. *Sci. Rep.* **2017**, *7*, 45518. [CrossRef] [PubMed]

55. Wijacki, J.; (Department of Animal Morphology, Physiology and Genetics, Mendel University, Brno, Czech Republic). Personal communication, 2019.

56. Wang, Q.; Yang, C. The phylogeny of the Cetartiodactyla based on complete mitochondrial genomes. *Int. J. Biol.* **2013**, *5*, 30. [CrossRef]

57. Mallya, M.; Campbell, R.D.; Aguado, B. Transcriptional analysis of a novel cluster of LY-6 family members in the human and mouse major histocompatibility complex: Five genes with many splice forms. *Genomics* **2002**, *80*, 113–123. [CrossRef] [PubMed]

58. Birch, J.; Sanjuan, C.D.J.; Guzman, E.; Ellis, S.A. Genomic location and characterisation of MIC genes in cattle. *Immunogenetics* **2008**, *60*, 477–483. [CrossRef]

59. Renard, C.; Vaiman, M.; Chiannilkulchai, N.; Cattolico, L.; Robert, C.; Chardon, P. Sequence of the pig major histocompatibility region containing the classical class I genes. *Immunogenetics* **2001**, *53*, 490–500. [CrossRef]

60. Futas, J.; Oppelt, J.; Jelinek, A.; Elbers, J.P.; Wijacki, J.; Knoll, A.; Burger, P.A.; Horin, P. Natural killer cell receptor genes in camels: Another mammalian model. *Front. Genet.* **2019**, *10*, 620. [CrossRef]

Article

Intramolecular Domain Movements of Free and Bound pMHC and TCR Proteins: A Molecular Dynamics Simulation Study

Rudolf Karch [1], Claudia Stocsits [1], Nevena Ilieva [2,3] and Wolfgang Schreiner [1,*

[1] Section of Biosimulation and Bioinformatics, Center for Medical Statistics, Informatics and Intelligent Systems (CeMSIIS), Medical University of Vienna, Spitalgasse 23, A-1090 Vienna, Austria

[2] Institute of Information and Communication Technologies (IICT), Bulgarian Academy of Sciences, Acad. G. Bonchev Str., Block 25A, 1113 Sofia, Bulgaria

[3] CERN-TH, Esplanade des Particules 1, 1211 Geneva, Switzerland

* Correspondence: wolfgang.schreiner@meduniwien.ac.at; Tel.: +43-1-40400-66790

Received: 31 May 2019; Accepted: 12 July 2019; Published: 13 July 2019

Abstract: The interaction of antigenic peptides (p) and major histocompatibility complexes (pMHC) with T-cell receptors (TCR) is one of the most important steps during the immune response. Here we present a molecular dynamics simulation study of bound and unbound TCR and pMHC proteins of the LC13-HLA-B*44:05-pEEYLQAFTY complex to monitor differences in relative orientations and movements of domains between bound and unbound states of TCR-pMHC. We generated local coordinate systems for MHC α1- and MHC α2-helices and the variable T-cell receptor regions TCR V_α and TCR V_β and monitored changes in the distances and mutual orientations of these domains. In comparison to unbound states, we found decreased inter-domain movements in the simulations of bound states. Moreover, increased conformational flexibility was observed for the MHC α2-helix, the peptide, and for the complementary determining regions of the TCR in TCR-unbound states as compared to TCR-bound states.

Keywords: molecular dynamics simulation; major histocompatibility complex; antigen; T-cell receptor; domain movements

---

1. Introduction

The interaction between T-cell receptors (TCR) and antigenic peptides (p) presenting major histocompatibility complexes (pMHC) is a crucial step in adaptive immune response [1]. It triggers the generation of cell-mediated immunity to pathogens and other antigens. The response is driven by TCRs specifically recognizing antigenic peptides bound to and presented by MHC (pMHC) of infected or transformed cells. MHC class I molecules are membrane bound proteins on the surface of antigen presenting cells consisting of an α-chain and a β2-microglobulin. Short fragments of antigens are presented in a pocket formed by the α1 and α2 domains. The MHC-bound peptides are presented to the TCR for examination. The TCR itself is a heterodimer on the surface of T-cells and is formed by an α-chain and a β-chain (each chain containing a variable and a constant region) [2]. The binding site of the T-cell receptor is constituted by the two variable regions TCR V_α and TCR V_β, which specifically recognize the α-helices of the MHC peptide-binding groove as well as parts of the antigen [3]. The interaction of TCR and pMHC is weak but highly specific, therefore being able to distinguish self-antigens from foreign peptides [1]. Additional molecules and co-receptors, e.g., the CD8, a T-cell surface glycoprotein, help establish a stable, specific and sensitive interaction. Correct binding of the TCR to pMHC induces a kinase mediated signaling cascade and activation of the T-cell immune response.

The TCR approaches the MHC in a diagonal orientation relative to the pMHC-binding groove, driven by long-range electrostatic interactions and low affinity binding events [4,5]. During the binding process the complementarity determining regions (CDRs) of the TCR V_α (CDR1α, CDR2α, CDR3α) and TCR V_β domains (CDR1β, CDR2β CDR3β) are positioned over the N- and C-terminus of the peptide and form contacts [6,7]. Local conformational readjustment occurs after binding of the TCR to the pMHC [8–10]. Complexation of TCR- and pMHC-molecules takes place via hydrogen bonds, salt links, and van der Waals interactions with different relative contributions [4].

Protein-protein interfaces within crystal structures of TCR-pMHC complexes have been evaluated in numerous studies and provided insight into the molecular mechanism behind the interaction and recognition mechanism of TCR with the pMHC [6,11,12] but lack information about the intermolecular approach and reorientation. Details of the binding and spatial rearrangements of the TCR-pMHC interaction on an atomistic scale remain to be uncovered.

Molecular dynamics (MD) is a powerful tool to investigate dynamic molecular processes and movements by solving Newton's equations of motion and has long been an adequate device to complement experimental crystallographic and biophysical studies on proteins. The first MD studies focused on pMHC complexes and were of comparably short duration due to the limitations in computational resources. Early MD simulations on a very short timescale ≤ 0.6 ns by Toh et al. [13] suggested that antagonist peptides bound to HLA-DR4, an MHC class II, display greater flexibility in the putative TCR binding regions compared to agonist peptides. Similar results on epitope flexibility, albeit with MD simulations of 100 ns duration, were reported by Reboul et al. [14]. Zacharias and Springer [12] studied the conformational flexibility of the MHC Class I peptide-binding region (α1- α2 domain) in peptide bound and free states with several independent MD simulations of 26 ns each and reported an increased conformational heterogeneity in the absence of a peptide ligand. As to TCR-pMHC interactions, Wolfson et al. [15] performed a series of 15 ns MD simulations studying peptide and TCR mutations and found that the footprint of the TCR on the pMHC is insensitive to mutations of the TCR peptide-binding loops, but peptide mutations can make multiple local changes to TCR-pMHC interface dynamics and interactions. A similar trend was observed in 20 ns long MD simulations of the LC13 T-cell receptor in complex with HLA-B*0801 and the Epstein-Barr virus (EBV) derived peptide FLRGRAYGL [16]. The results of this study revealed a high mobility of the CDR-loops of the TCR as well as the presence of two conformational clusters in the peptide's structure, underlying the backbone flexibility of the peptide. Very recently, Tedeschi et al. [17] used MD simulations to complement their experimental studies on an unusual placement of an EBV epitope in the binding groove of an ankylosing spondylitis (AS)-associated HLA-B27 allele that allowed CD8+ T cell activation and inferred from computational analysis that the strongest risk factor for AS, i.e., B*2705, is able to elicit anti-viral T cell immune-responses even when the binding groove might be partially occupied by the epitope. In the present study, we performed MD simulations of bound and unbound TCR-pMHC configurations of a LC13-HLA-B*44:05-pEEYLQAFTY complex in order to evaluate if and how intramolecular domain movements and relative orientations (namely between TCR V_α and TCR V_β, and MHC α1 and MHC α2) change due to complexation of TCR with pMHC molecules.

2. Methods and Materials

2.1. System and Start Configurations

We used a set of three different start configurations: In the first set MD simulations were started from the crystal structure of a complex consisting of a LC13 TCR, a self-peptide (EEYLQAFTY) from the ATP binding cassette protein ABCD3, and a MHC of the HLA-B*44:05 type (PDB-ID: 3KPS, see Figure 1). This system was chosen because it is available at a high resolution of 2.7 Å. Moreover, Macdonald et al. [18] determined binding characteristics and immunogenicity of three MHC alleles (HLA-B*44:02, HLA-B*44:03, and HLA-B*44:05) in complex with the ABCD3 nine-mer peptide and the LC13 TCR and Ferber et al. [19] analyzed TCR binding orientations over pMHC. In previous papers we used the 3KPS system for geometric analyses of alloreactive HLA α-helices [11,20] as well as for finding semi-rigid domains in MD simulations [21]. In the present study we consider differences in the dynamics of intramolecular domains between bound and unbound states of the 3KPS system. In the second and third set of configurations, simulations were started from isolated, i.e., unbound TCR- and pMHC-molecules as derived from the original 3KPS crystal structure.

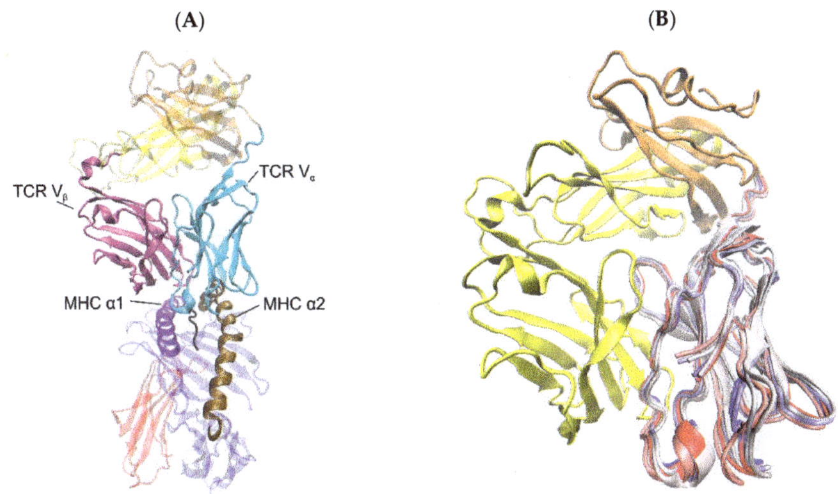

Figure 1. Cartoon representations of TCR and pMHC molecules. (**A**) TCR-pMHC complex from the crystal structure 3KPS with highlighted domains TCR V$_\alpha$ (cyan), TCR V$_\beta$ (purple), MHC α1 (violet), and MHC α2 (ochre), TCR constant domain α (orange), TCR constant domain β (yellow), MHC α3 (blue), MHC β-microglobulin (red), peptide (black). (**B**) TCR alone with overlaid snapshots of domain TCR V$_\alpha$ at every 8 ns from a 40 ns simulation run of the TCR-pMHC complex to illustrate the extent of domain movements. The beginning of the trajectory appears in red, the middle in white, and the end in blue. The figures were produced using VMD [22].

2.2. Molecular Dynamics Simulations

MD simulations were run in an explicit water solvent using GROMACS 5.1.1 [23]. We used the amber99sb-ildn force field [24] to simulate the protein placed in a rhombic dodecahedral box with 2 nm minimum distance between the protein atoms and the box boundary. The whole system contains approximately 397,000 atoms, consisting of 826 protein residues and approximately 128,000 solvent molecules. Based on the SPC water model [25] the protein was solvated and neutralized by replacing solvent molecules with Na^+ and Cl^- ions to reach a salt concentration of 0.15 mol/L.

After solvation the whole complex was energetically minimized using the steepest-descent approach with 10,000 steps. For equilibration we set the temperature of the system to 310 K and the equations of motion were followed for 100 ps using position restraint MD and a Berendsen-thermostat with a time constant of 0.1 ps. This equilibration under NVT conditions (constant number of particles, volume and temperature) was followed by a 100 ps equilibration in an NPT ensemble (constant number of particles, pressure and temperature) to control the pressure by a Berendsen-barostat set to 1 bar with a time constant of 1.0 ps.

The MD production runs were carried out with a time-step of 4 fs using the LINCS constraint algorithm for all bonds and virtual sites for hydrogen atoms. For the Van der Walls and Coulomb interactions a cut-off radius of 1.4 nm was applied. Long-range Coulomb interactions were computed using the particle-mesh Ewald method with a maximum grid spacing of 0.12 nm for the FFT and an interpolation order of 4. Velocity-rescaling temperature coupling was set to 310 K and Berendsen isotropic pressure coupling was set to 1 bar. Coordinates were written to the trajectory file on hard disk every 40 ps, resulting in 1000 frames for each run. This procedure was repeated for each run in the set of initial configurations. We performed 30 independent MD simulation runs with different initial velocities and with a length of 40 ns each: 10 runs for the TCR-pMHC complex (bound configurations), 10 runs for a free pMHC molecule, and 10 runs for a free TCR molecule (unbound configurations). For reasons of comparison, two additional runs of 400 ns in length (one for the free pMHC- and the other for the free TCR-molecule) and one 144 ns run for the TCR-pMHC complex were performed. Prior to the analysis of the trajectories, translational and rotational motions of the protein relative to the equilibrated structure were removed.

2.3. Distance between Domains

One commonly used method among many others [26–28] to characterize relative movements of intramolecular domains is the calculation of distances between groups of atoms (domains) as a function of simulation time. We evaluated the distances d between the geometric centers of domains, based on the C_α atoms in the protein backbone.

2.4. Relative Orientation of Two Intramolecular Domains

Based on the methods described previously [29] we monitored the relative orientations of intramolecular domains during the time-course of the simulations. For the definition of domains we used the C_α atoms of the protein backbone. As a first pair of domains we chose the two epitope binding domains TCR V_α (104 residues) and TCR V_β (117 residues), both involved in the formation of the interface with pMHC. As a second pair we considered the antigen presenting sites of MHC, namely MHC α1 (25 residues) and MHC α2 (31 residues).

During an MD simulation the atoms within each domain move according to the equations of motion and thus their relative orientation changes from time step to time step. Principle component analysis (PCA) is commonly used to analyze the motions of domains within the conformations obtained during MD simulations [26]. PCA as such yields eigenvectors unique in directions but ambiguous in orientation. To arrive at standardized eigenvectors allowing for comparison between different MD-runs we chose the crystal structure as a reference frame and re-oriented (if necessary) the resulting eigenvectors after performing the PCA [29]. The three orthonormal eigenvectors from the PCA were used to define an eigenvector matrix to serve as a local coordinate system, moving together with the domain, see Figure 2. For each frame of an MD simulation we monitored the relative orientation between local coordinate systems of the domains.

To actually characterize the relative orientations of domains and their changes over time, we computed the cosines between the corresponding eigenvectors pairs $(\mathbf{v}_1, \mathbf{w}_1)$, $(\mathbf{v}_2, \mathbf{w}_2)$ and $(\mathbf{v}_3, \mathbf{w}_3)$ of the respective local coordinate systems.

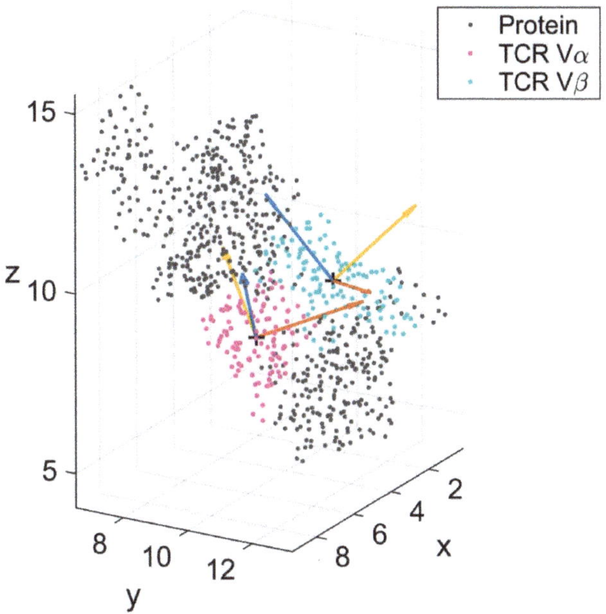

Figure 2. Schematic representation of TCR V$_\alpha$ (magenta) and TCR V$_\beta$ (cyan) with local eigenvectors. The remaining parts of the complex (labelled Protein) are shown in gray. Eigenvectors ($\mathbf{v}_1$, $\mathbf{v}_2$, $\mathbf{v}_3$ and $\mathbf{w}_1$, $\mathbf{w}_2$, $\mathbf{w}_3$) are displayed for a single frame of the trajectory and colored blue ($\mathbf{v}_1$, $\mathbf{w}_1$), orange ($\mathbf{v}_2$, $\mathbf{w}_2$) and yellow ($\mathbf{v}_3$, $\mathbf{w}_3$). Units of the coordinate axes are in nm.

2.5. Domain Deformations and Fluctuations

To quantify intra-domain deformations we computed the RMSD (root-mean-square deviation) of each frame with respect to the first frame of each trajectory, considering only the C$_\alpha$ atoms within the respective domain. The root mean square deviation RMSD(t) of a group of atoms in a molecule at time t with respect to a given reference structure is a measure of deformations (deviations in shape), i.e., the square root of averaged squared distances between the atom positions at time t and the positions in the reference structure. Here, the starting structure of the production runs (i.e., the first frame of the trajectory) was chosen as the reference structure. As such, RMSD(t) is a statistical measure of the distance of a group of atoms at a particular time t with respect to the same group in the reference structure and was calculated as:

$$\text{RMSD}(t) = \sqrt{\frac{1}{N} \sum_{i=1}^{N} \left| \mathbf{r}_i(t) - \mathbf{r}_i^{\text{ref}} \right|^2} \tag{1}$$

where $\mathbf{r}_i(t)$ is the position of atom i at time t after least square fitting to the reference structure, N is the total number of atoms in the group of atoms considered, and $\mathbf{r}_i^{\text{ref}}$ is the position of atom i in the reference structure. On the other hand, the root mean square fluctuation RMSF(i) is a statistical measure of the deviation between the position of atom i (or a group of atoms, e.g., a residue) and some reference position $\mathbf{r}_i^{\text{ref}}$:

$$\text{RMSF}(i) = \sqrt{\frac{1}{T} \sum_{t_j=1}^{T} \left| \mathbf{r}_i(t_j) - \mathbf{r}_i^{\text{ref}} \right|^2} \tag{2}$$

where T is the time interval over which the average is taken and t_j is the time of frame j during the simulation of the respective trajectory. Here, we have chosen the time-averaged position of atom i as the reference position, i.e., $\mathbf{r}_i^{\text{ref}} = <\mathbf{r}_i>$. Whereas the RMSD($t$) is an average taken over all atoms within a group for a specific time t, the RMSF(i) is an average over time for a specific atom i.

3. Results

3.1. Relative Movements between TCR V_α and TCR V_β

The variable regions TCR V_α and TCR V_β interact with the pMHC surface and are therefore important during TCR-pMHC recognition. Interestingly, if TCR and pMHC are free (i.e., unbound), the distance d between the domains TCR V_α and TCR V_β is not distinctively different from the bound configurations (results from ten 40 ns runs): $< d_{\text{free}} > = 2.523$ nm, $< d_{\text{bound}} > = 2.537$ nm, $\Delta < d > = 0.014$ nm (difference of mean distances $< d >$ between bound and unbound configurations), $\sigma_{\text{free}} = 0.005$ nm (standard deviation of distances for unbound configurations), and $\sigma_{\text{bound}} = 0.002$ nm (standard deviation of distances for bound configurations), see Figure 3.

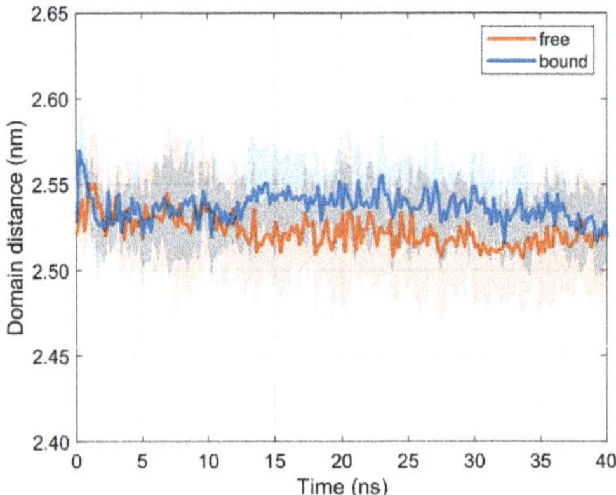

Figure 3. Time course of inter-domain distance d between domains TCR Vα and TCR Vβ. Mean ± SD for unbound (orange) and bound (blue) configurations from ten 40 ns runs with distance evaluation every 0.2 ns.

Deformations within each domain can be characterized by RMSD-values. Figure 4 shows boxplots of RMSD-values computed with respect to the first frame of each trajectory for the unbound (runs 1 to 10) and bound states (runs 11 to 20). Deformations within TCR V_α and TCR V_β are virtually unaffected by the binding of the pMHC und TCR molecules (see Figure 4, Tables 1 and 2): For TCR V_α the overall mean ± SD of RMSD-values from ten 40 ns runs were 0.115 ± 0.002 nm for the unbound states and 0.101 ± 0.002 nm for the bound states. For TCR V_β the respective values were 0.096 ± 0.003 nm (unbound) and 0.101 ± 0.002 nm (bound).

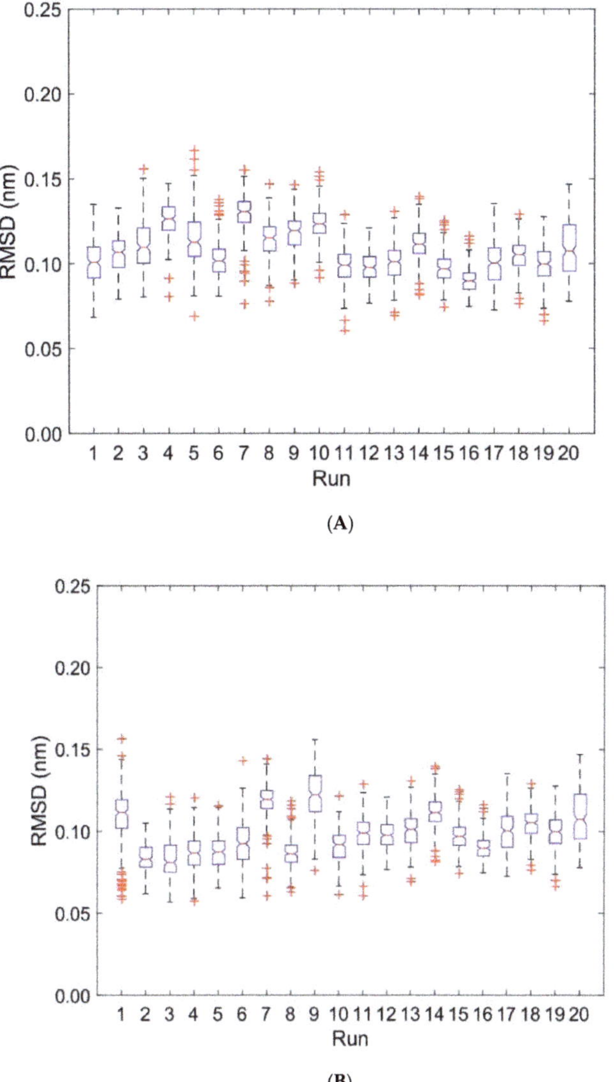

(A)

(B)

Figure 4. Boxplots of RMSD for TCR V_α and TCR V_β. RMSD-values are computed for every 0.2 ns with respect to the first frame of each trajectory. Run numbers 1 to 10 correspond to unbound states, 11 to 20 to bound states. (**A**) RMSD for TCR V_α domain. (**B**) RMSD for TCR V_β domain.

Table 1. RMSD (mean ± SD) in nm for various domains of the unbound molecules TCR and pMHC as calculated from 20 (10 for free TCR and 10 for free MHC) 40 ns MD runs (1–10) and from two (free TCR and free MHC) 400 ns MD runs (11).

Run	TCR V_α	TCR V_β	MHC $\alpha1$	MHC $\alpha2$	MHC Peptide
1	0.101 ± 0.013	0.109 ± 0.017	0.087 ± 0.019	0.132 ± 0.021	0.157 ± 0.015
2	0.106 ± 0.011	0.084 ± 0.009	0.059 ± 0.013	0.111 ± 0.014	0.061 ± 0.017
3	0.112 ± 0.015	0.083 ± 0.011	0.075 ± 0.011	0.079 ± 0.011	0.123 ± 0.012
4	0.126 ± 0.010	0.087 ± 0.011	0.061 ± 0.010	0.117 ± 0.019	0.117 ± 0.034
5	0.114 ± 0.016	0.087 ± 0.009	0.066 ± 0.009	0.079 ± 0.019	0.085 ± 0.017
6	0.102 ± 0.011	0.093 ± 0.014	0.083 ± 0.011	0.153 ± 0.039	0.096 ± 0.020
7	0.129 ± 0.011	0.119 ± 0.011	0.061 ± 0.010	0.097 ± 0.013	0.085 ± 0.014
8	0.114 ± 0.011	0.087 ± 0.010	0.084 ± 0.016	0.073 ± 0.013	0.104 ± 0.020
9	0.118 ± 0.011	0.122 ± 0.016	0.076 ± 0.012	0.118 ± 0.031	0.064 ± 0.018
10	0.124 ± 0.009	0.091 ± 0.010	0.080 ± 0.015	0.107 ± 0.016	0.094 ± 0.014
11	0.108 ± 0.016	0.085 ± 0.013	0.077 ± 0.013	0.067 ± 0.015	0.085 ± 0.011

Table 2. RMSD (mean ± SD) in nm for various domains of the bound complex TCR-pMHC as calculated from ten 40 ns MD runs (1–10) and from one 144 ns MD run (11).

Run	TCR V_α	TCR V_β	MHC $\alpha1$	MHC $\alpha2$	MHC Peptide
1	0.099 ± 0.010	0.071 ± 0.007	0.074 ± 0.009	0.076 ± 0.010	0.052 ± 0.010
2	0.098 ± 0.008	0.069 ± 0.008	0.070 ± 0.009	0.094 ± 0.011	0.056 ± 0.007
3	0.101 ± 0.011	0.087 ± 0.018	0.093 ± 0.014	0.095 ± 0.022	0.048 ± 0.007
4	0.111 ± 0.010	0.069 ± 0.007	0.070 ± 0.015	0.127 ± 0.024	0.082 ± 0.015
5	0.098 ± 0.009	0.080 ± 0.012	0.076 ± 0.007	0.072 ± 0.009	0.046 ± 0.006
6	0.090 ± 0.008	0.064 ± 0.007	0.079 ± 0.011	0.120 ± 0.022	0.047 ± 0.006
7	0.100 ± 0.013	0.068 ± 0.007	0.084 ± 0.013	0.117 ± 0.028	0.052 ± 0.011
8	0.105 ± 0.009	0.067 ± 0.008	0.084 ± 0.014	0.134 ± 0.021	0.057 ± 0.011
9	0.100 ± 0.011	0.070 ± 0.008	0.076 ± 0.011	0.079 ± 0.014	0.057 ± 0.016
10	0.109 ± 0.016	0.066 ± 0.006	0.067 ± 0.016	0.105 ± 0.029	0.083 ± 0.011
11	0.097 ± 0.011	0.073 ± 0.010	0.073 ± 0.013	0.081 ± 0.012	0.066 ± 0.008

We characterized the relative orientation of the domains TCR V_α and TCR V_β by the directional cosines between eigenvectors attached to each domain and sampled over time for all trajectories as outlined in the methods section and in our previous paper [29], see Figure 5. The eigenvectors $\mathbf{v}_1$ and $\mathbf{w}_1$ correspond to the main extensions of the respective domain. To quantify the differences between bound and unbound states in the frequency distributions of the respective angles α between $\mathbf{v}_1$ and $\mathbf{w}_1$ we used the test statistic D of the Kolmogorov Smirnov test. For $\mathbf{v}_1$ and $\mathbf{w}_1$ the results from ten 40 ns runs were $D = 0.215$, $< \alpha(\mathbf{v}_1,\mathbf{w}_1) >_{free} = 58.96°$, $\sigma_{free} = 2.15°$, $< \alpha(\mathbf{v}_1,\mathbf{w}_1) >_{bound} = 59.88°$, $\sigma_{bound} = 1.75°$. Larger differences were observed for the angles β between $\mathbf{v}_2$ and $\mathbf{w}_2$: $D = 0.643$, $< \beta(\mathbf{v}_2,\mathbf{w}_2) >_{free} = 57.38°$, $\sigma_{free} = 2.36°$, $< \beta(\mathbf{v}_2,\mathbf{w}_2) >_{bound} = 62.78°$, $\sigma_{bound} = 3.78°$.

These results may be summarized as follows: The domains TCR V_α and TCR V_β remain relatively unchanged in their main orientation after complexation of TCR and pMHC, as seen by the small changes of the angles between $\mathbf{v}_1$ and $\mathbf{w}_1$ (58.96° on average for the unbound state versus 59.88° in the bound state), see Figure 5A. An additional change in orientation is due to a rotation around these major axes as reflected in the variation of the angles β between $\mathbf{v}_2$ and $\mathbf{w}_2$. As a result of complexation their mean values change from 57.38° for the unbound states to 62.78° for the bound states, see Figure 5B.

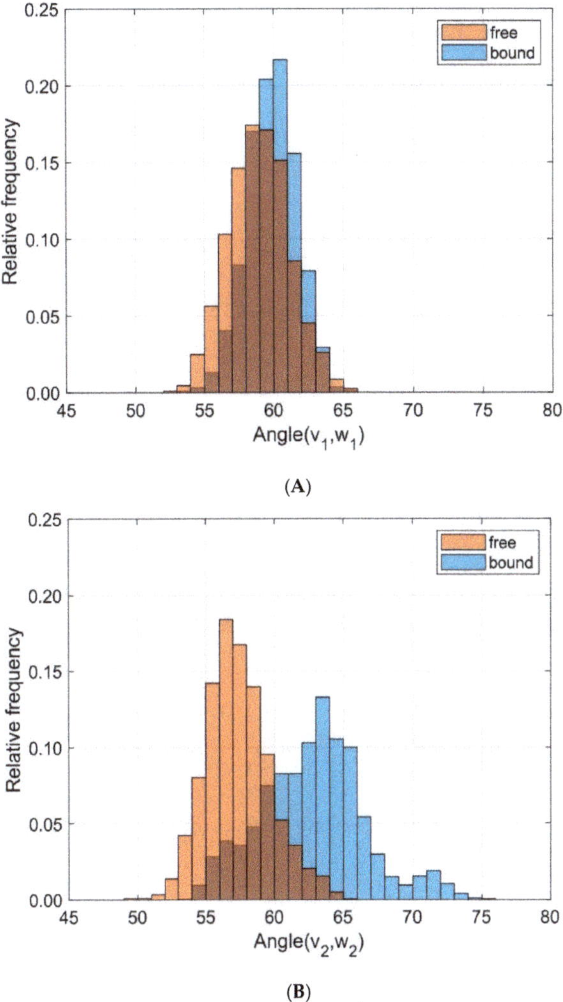

(A)

(B)

Figure 5. Frequency distribution of angles between corresponding principal component eigenvectors of domains TCR V_α and TCR V_β for bound (blue) and unbound states (orange). (**A**): Angles between eigenvectors $\mathbf{v}_1$, $\mathbf{w}_1$ corresponding to the main extensions ($D = 0.215$); (**B**): Angles between $\mathbf{v}_2$, $\mathbf{w}_2$ ($D = 0.643$).

3.2. Relative Movements between MHC α1 and MHC α2

The two α-helices of the MHC, together with a β-floor, form the peptide binding cleft that presents antigen fragments to be examined by the TCR. Together with the variable domains of the TCR, TCR V_α and TCR V_β, they are most important for the formation of the TCR-pMHC complex and for a successful initiation of the immune response. We thus also monitored intramolecular movements of the two α-helices in the free (i.e., unbound) and bound states. If TCR and pMHC are unbound, the distance d between domains MHC α1 and MHC α2 practically does not change with respect to the bound state (results from ten 40 ns runs): $<d_{\text{free}}> = 1.676$ nm, $<d_{\text{bound}}> = 1.649$ nm, $\Delta<d> = 0.028$ nm (difference of mean distances $<d>$ between unbound and bound configurations), $\sigma_{\text{free}} = 0.004$ nm (standard deviation of distances for unbound configurations), and $\sigma_{bound} = 0.004$ nm (standard deviation of distances for bound configurations), see Figure 6.

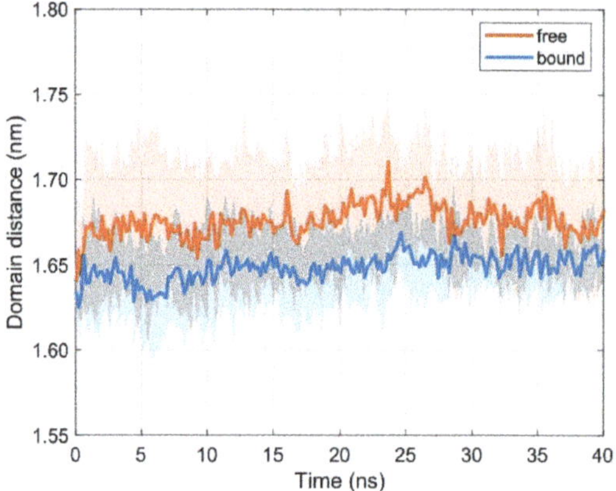

Figure 6. Time course of inter-domain distance d between domains MHC α1 and MHC α2. Mean ± SD for unbound (orange) and bound (blue) configurations from ten 40 ns runs with distance evaluation every 0.2 ns.

We computed RMSD-values of each domain MHC α1 and MHC α2 with respect to the first frame of each trajectory for the unbound (runs 1 to 10) and bound states (runs 11 to 20), see Figure 7, Tables 1 and 2: For the MHC α1 domain, the boxplots of RMSD do not show any systematic dependence on the binding state of pMHC and TCR (the overall mean ± SD of RMSD-values from ten 40 ns runs were 0.073 ± 0.003 nm for the unbound states and 0.077 ± 0.003 nm for the bound states). In contrast to the MHC α1 domain, the RMSD-values of the MHC α2 domain exhibited a pronounced difference between unbound and bound configurations: The unbound state not only shows larger RMSD-values than the bound state (mean 0.107 nm versus 0.077 nm), but also a three-fold variability in RMSD-values (SD 0.009 nm versus 0.003 nm).

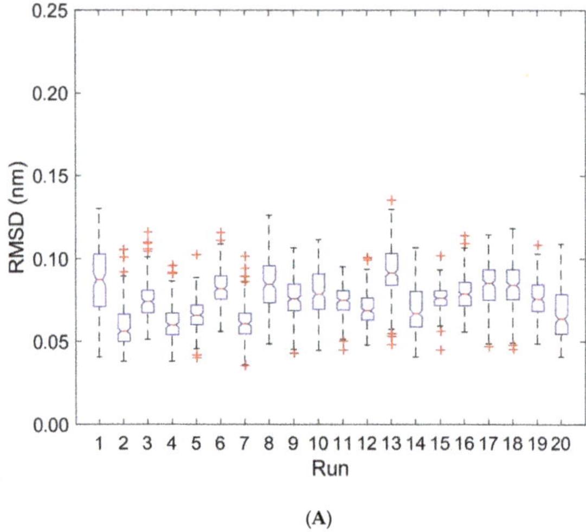

(A)

Figure 7. *Cont.*

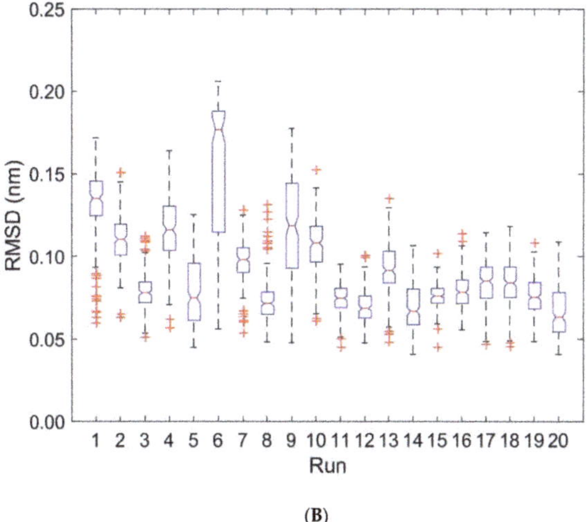

(B)

Figure 7. Boxplots of RMSD for MHC α1 and MHC α2. RMSD values are computed every 0.2 ns with respect to the first frame of each trajectory. Run numbers 1 to 10 correspond to unbound states, 11 to 20 to bound states. (**A**): RMSD for the MHC α1 domain. (**B**): RMSD for the MHC α2 domain.

To monitor the relative orientation between the helical domains MHC α1 and MHC α2 we calculated directional cosines between corresponding eigenvectors and examined their distributions over time, see Figure 8. The angles α between the eigenvectors $\mathbf{v}_1$ and $\mathbf{w}_1$ (corresponding to the main extensions of the domains) were slightly different between unbound and bound states (results from ten 40 ns runs): $D = 0.255$, $< \alpha(\mathbf{v}_1,\mathbf{w}_1) >_{\text{free}} = 20.75°$, $\sigma_{\text{free}} = 2.74°$, $< \alpha(\mathbf{v}_1,\mathbf{w}_1) >_{\text{bound}} = 22.50°$, $\sigma_{\text{bound}} = 1.85°$. More pronounced differences and larger fluctuations were observed for the angles β between $\mathbf{v}_2$ and $\mathbf{w}_2$: $D = 0.290$, $< \beta(\mathbf{v}_2,\mathbf{w}_2) >_{\text{free}} = 86.57°$, $\sigma_{\text{free}} = 12.26°$, $< \beta(\mathbf{v}_2,\mathbf{w}_2) >_{\text{bound}} = 81.16°$, $\sigma_{\text{bound}} = 11.06°$. These eigenvectors $\mathbf{v}_2$ and $\mathbf{w}_2$ point away from the axis of the helices at right angles and indicate a larger variation in the relative orientation between MHC α1 and MHC α2 in the unbound state as compared to the bound state.

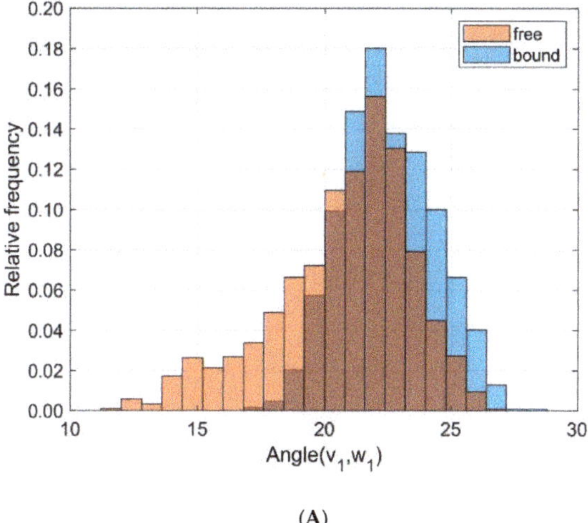

(A)

Figure 8. *Cont.*

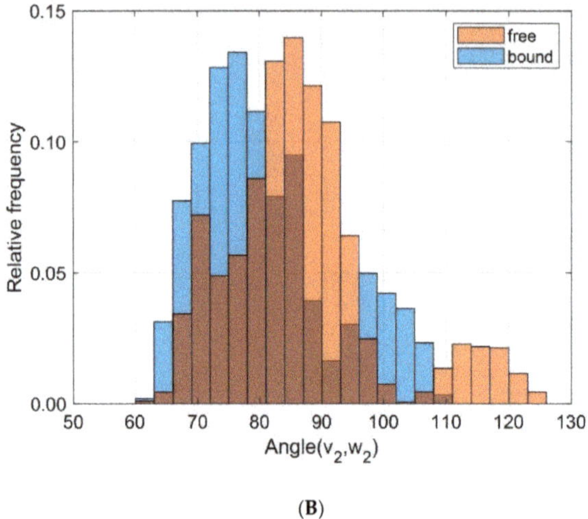

(**B**)

Figure 8. Frequency distribution of angles between corresponding principal component eigenvectors of domains MHC α1 and MHC α2 for bound (blue) and unbound states (orange). (**A**): Angles between eigenvectors $\mathbf{v}_1$, $\mathbf{w}_1$ corresponding to the main extensions ($D = 0.255$); (**B**): Angles between $\mathbf{v}_2$, $\mathbf{w}_2$ ($D = 0.290$).

4. Discussion

Crystallographic studies of TCR-pMHC complexes have delivered important insights regarding the structural components involved in stabilizing the TCR-pMHC complex. Despite the huge success and the valuable contributions of these studies, they provide only a limited static view. On the other hand, molecular dynamics simulations allow for following the dynamics of single atoms or groups of atoms with a time-resolution in the order of pico-seconds. Although the interaction between a TCR and a pMHC is weak (10^3–10^5 M^{-1} s^{-1} [30]), it represents an important initial step in T-cell activation. Detailed studies on protein-protein complexes suggest that the binding process of protein molecules can be divided into three phases [31]: A first and short phase of diffusion (initial contacts form within 2 ns) is followed by an intermediate phase where most native contacts are established (between 50–200 ns). In the last phase the final stereospecific complex is formed (this takes hundreds of nanoseconds). Various studies have reported conformational changes at the TCR-pMHC binding interface in the course of complex formation, in particular at the CDRs and peptides and less at the MHC [32–34]. In the current work, we present molecular dynamics simulations to study the differences in intramolecular domain movements between bound and unbound states of a specific TCR-pMHC complex.

For the analyses of domain orientations, we attached local coordinate systems (PCA-eigenvectors) to the domains and calculated the cosines between corresponding eigenvectors. Based on three sets of independent simulations with different initial configurations (TCR-pMHC bound, TCR unbound, pMHC unbound) we could show that binding of TCR to pMHC decreases the flexibility in the inter-domain movements, in particular the dynamics of TCR V$_\alpha$ relative to TCR V$_\beta$ and of MHC α1 relative to MHC α2, see the distribution of angles between corresponding principal component eigenvectors in Figures 5 and 8.

Further examination of domain RMSD-values revealed that MHC α2-dynamics clearly depends on the binding states (i.e., bound versus unbound) of TCR and pMHC molecules, see Figure 9. Conformational flexibility, as characterized by the RMSD-values of MHC α2, was distinctively larger in the unbound states as compared to the bound states. On the contrary, differences between bound and unbound states in RMSD-values of the TCR domains V$_\alpha$ and V$_\beta$ were considerably less pronounced, see Figure 10. The decrease of domain RMSD-values of MHC α2 is consistent with the dynamics of the peptide placed in the binding groove between the two MHC α-helices: Root mean square fluctuations (RMSF) of individual residues of the peptide during the course of the simulations show distinctively larger values in the unbound as compared to the bound states (Figure 11). In particular, central amino acid residues at positions 4–7 within the peptide (Leu, Gln, Ala, Phe) display reduced dynamic flexibility upon binding of the pMHC with the TCR, in agreement with previous studies of various peptide/MHC-I complexes and TCRs [35,36], see [6,37–39] for reviews. On the other hand, the N- and C-termini of positions 1 and 9 in the peptide are much less influenced by TCR-binding due to the network of hydrogen bonds from the β-sheet and from the lateral α-helices which stabilize the peptide binding domain [12]. Macdonald et al. [18] emphasized the importance of the C-terminal region (p6–p8) of the peptide for TCR recognition: In particular, the small p6 (Ala) and p8 (Thr) residues enabled the aromatic p7 (Phe) residue to protrude within a central pocket of the TCR, sandwiched between Tyr31α and Tyr100β; the p6-Ala made interactions with the CDR3α and the CDR3β loop, while p8-Thr and p7-Phe formed contacts with the CDR3β loop.

Residues in the variable domains V_α and V_β of the T-cell receptor show less pronounced differences in RMSF-values between free and bound states, with the exception of residues 30 (Thr), 31 (Tyr), 58 (Arg), 96–98 (Gly, Gly, Thr), 100 (Tyr), and 102 (Gly) in the TCR V_α domain and residues 50–52 (Gln, Asn, Glu), 71–73 (Glu, Gly, Ser), 98–100 (Gln, Ala, Tyr), and 105 (Glu) in the TCR V_β domain (Figure 12). Apart from residue 58 in the V_α domain and residues 71–73 in the V_β domain, all the other residues with a distinctively larger RMSF-value in the unbound states are part of the CDR-loops CDR1α (30, 31), CDR3α (96–98, 100, 102), CDR2β (50–52), and CDR3β (98–100, 105). These results are in agreement with several studies that have found a reduced mobility of the TCR [6,40] and conformational changes in the CDR-loops [4,7,41] upon formation of the TCR-pMHC complex. In particular, we observed a structural change between the free and bound states in the CDR3α loop of residues 95–97 (Ala, Gly, Gly) from a 3_{10}-helix (free) to a coil (bound), where Gly96α and Gly97α form interactions with Arg62 and Ile66 of the MHC α1 helix, respectively [18] (see supplementary Figure S1). As to the type of interactions between CDR-loops and pMHC, Macdonald et al. [18] reported predominantly van der Waals interactions and H-bonds, including one salt bridge between Glu52β of CDRβ2 and Arg79 of the α1 helix. These crystallographic studies have also shown that CDR1α and CDR2α loops make contacts with the α2 helix of pMHC, whereas contacts to the α1 helix are established by the CDR2β and CDR3α loops (CDR3α also makes interactions with the α2 helix and the peptide). While the CDR1β loop participates only minimally in the TCR-pMHC interactions, the contacts with the peptide are dominated by the CDR3β loop, which sits centrally above the peptide-binding cleft, adjoining the CDR3α loop (see also Figure 12C).

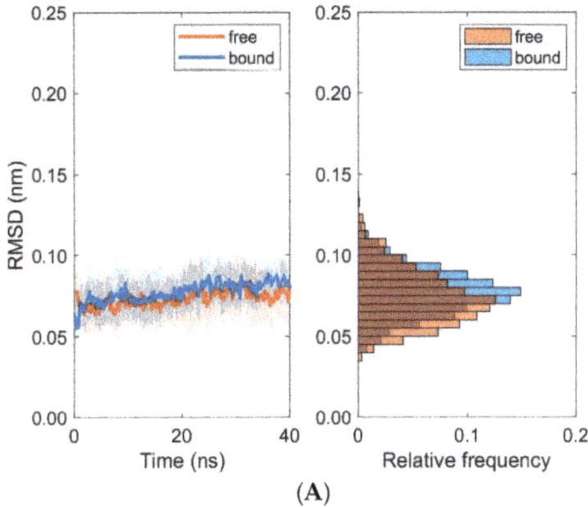

Figure 9. *Cont.*

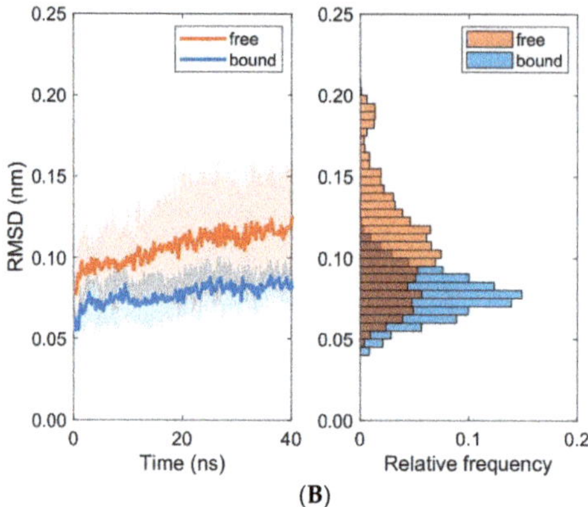

Figure 9. Time course and distribution of RMSD-values for domains MHC $\alpha1$ (**A**) and MHC $\alpha2$ (**B**). Time-series (left panels, mean ± SD) and histograms (right panels) display pooled RMSD-values from ten 40 ns MD simulations each for bound and unbound (free) states.

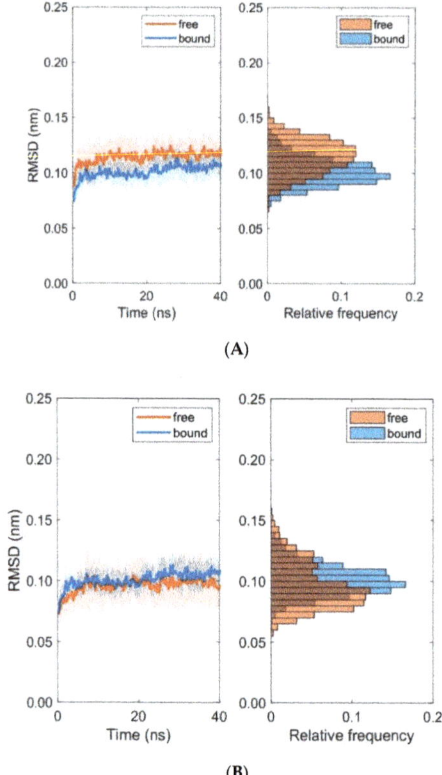

Figure 10. Time course and distribution of RMSD-values for domains TCR V_α (**A**) and TCR V_β (**B**). Time-series (left panels, mean ± SD) and histograms (right panels) display pooled RMSD-values from ten 40 ns MD simulations each for bound and unbound (free) states.

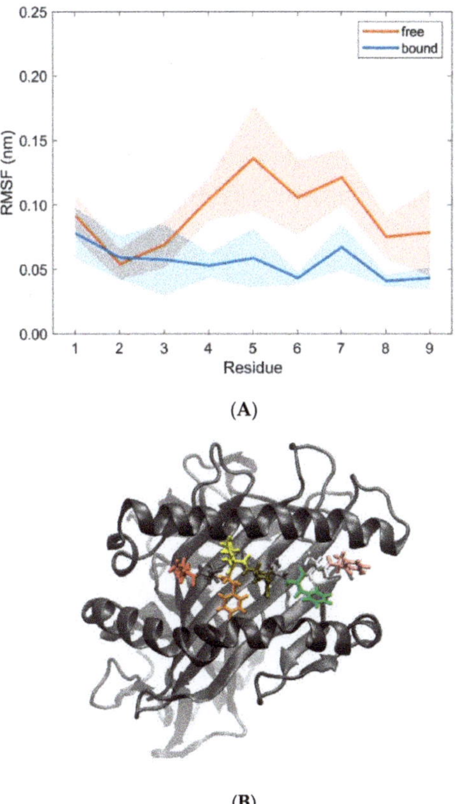

(A)

(B)

Figure 11. (**A**) RMSF-values (mean ± SD) of MHC peptide residues in free (orange) and bound (blue) states calculated from 10 pooled 40 ns MD simulations each for unbound (free) and bound states. (**B**) Cartoon representation of the 3KPS MHC binding groove showing the lateral α1- and α2-helices and the β-sheet (grey) together with the peptide (displayed as licorice) consisting of nine residues (position 1: Glu, red; position 9: Tyr, pink) as obtained from a snapshot of a 40 ns MD simulation of a TCR-free pMHC. The figure in panel B was produced using VMD [22].

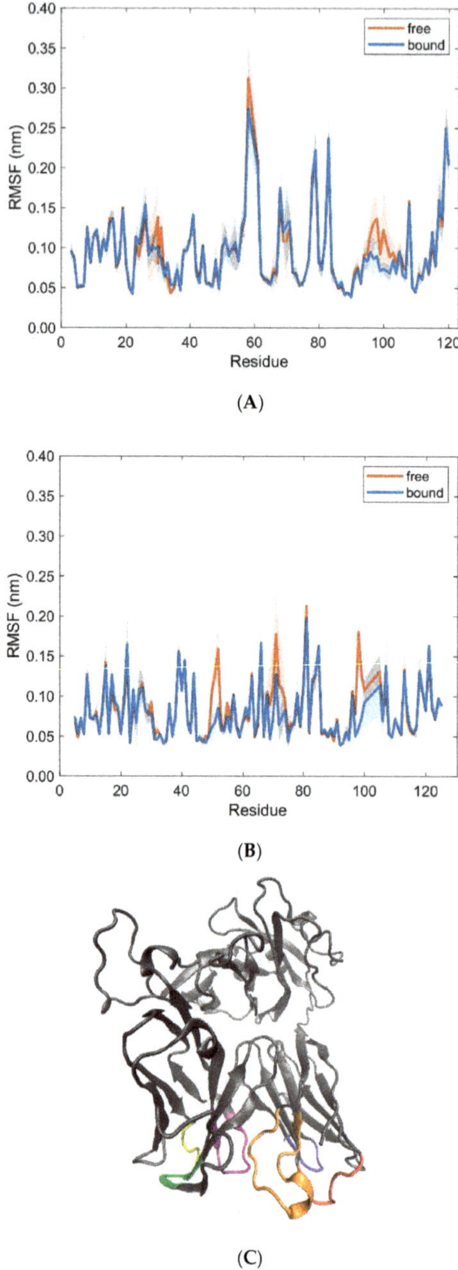

Figure 12. RMSF-values (mean ± SD) of TCR residues in the variable domains V_α (**A**) and V_β (**B**) in free (orange) and bound (blue) states calculated from 10 pooled 40 ns MD simulations each for unbound (free) and bound states. (**C**) Cartoon representation of the 3KPS TCR (grey) together with colored CDR loops CDR1α (24–31, red), CDR2α (48–55, blue), CDR3α (93–104, orange), CDR1β (26–31, yellow), CDR2β (48–55, green), CDR3β (95–107, purple) as obtained from the crystal structure of the 3KPS system. The figure in panel C was produced using VMD [22].

Only minor changes were observed in the inter-domain distance between TCR V_α relative to TCR V_β and MHC $\alpha1$ relative to MHC $\alpha2$ as a consequence of binding. Fluctuations within domains (TCR V_α, TCR V_β, MHC $\alpha1$, and MHC $\alpha2$) are virtually unaffected by the binding state, see Figures 4 and 7. In addition, angles between corresponding eigenvectors of MHC $\alpha1$ and MHC $\alpha2$ show larger variation in the unbound states, indicating a wider range of the groove between the two α-helices, see Figures 5 and 8.

A study based on static crystallographic data [42] demonstrated that the angle between TCR V_α and TCR V_β domains varies depending on the presence of peptides, whereby domains show larger variations in the unbound state and smaller variations in the bound state of the TCR. These findings are consistent with our MD-based results regarding the angles between vectors $\mathbf{v}_1$ and $\mathbf{w}_1$.

As to the design of our MD experiments, we performed 10 independent 40 ns runs for each of the 3 configurations studied (pMHC free, TCR free, TCR-pMHC complexed). This relatively short simulation length was chosen because it is more efficient (in terms of sampling the conformational space) to run various short independent MD simulations than a single long one [43], see, e.g., the variations of RMSD-values between the runs in the boxplots of Figure 7. On the other hand, RMSD mean values as calculated from ten pooled 40 ns runs were consistent with the respective values obtained from one 144 ns run (TCR-pMHC complex) und two 400 ns runs (TCR and pMHC free), see run 11 in Tables 1 and 2.

There are several limitations in the present study, in particular the influence of the lipid bilayer on TCR-pMHC interactions. Since the MHC and TCR are membrane-embedded proteins, the influence of the membrane environment itself is an essential ingredient for TCR-pMHC interaction and T cell response [44]. Various advanced MD studies have been performed to evaluate the role of the lipid bilayer for TCR-pMHCII systems: In a ground-breaking study, Wan et al. [45] performed MD simulations of a membrane-embedded TCR-pMHC-CD4 complex. Although the duration of the simulation was limited to 10 ns due to the size of the system (more than 300.000 atoms), the computed structural and thermodynamic properties, such as binding free energies of TCR-pMHC and pMHC-CD4, were in fair agreement with experimental data. Bello et al. [46] studied the influence of the membrane on the dynamics and energetics of peptide-bound and peptide-free MHC class II molecules in aqueous and membrane-bound environments by 100 and 150 ns MD simulations and found that the presence of the membrane might restrict the conformational flexibility of the peptide-binding cleft. More recently, Bello et al. [47] explored the energetic and dynamic behavior of a pMHCII-TCR complex embedded in two opposing lipid membranes using three independent 300 ns-long unbiased MD simulations. This study provided evidence of the main contributors to the pMHCII-TCR complex formation and identified key residues involved in this molecular recognition process.

The formation of the immunological synapse incorporates co-receptors, integrins, and various signaling proteins that trigger signaling cascades resulting in T-cell activation. Therefore, future MD studies of the 3KPS system should not only include the lipid membrane, but also co-receptors (CD8, CD3 complex) and other co-stimulatory molecules to assess their influence on the system [6].

Another limitation of the present study relates to the simulation time of 40 ns, which is way too short to reliably cover all processes involved, e.g., in TCR-signaling or even dissociation of the TCR-pMHC complex. Several authors have used coarse-grained simulations and elastic network models for modeling protein flexibility and interactions, see the recent reviews by Kmiecik et al. [48,49]. Moreover, steered MD [50] has been applied to extend simulation times to more realistic regimes.

5. Conclusions

We have performed MD simulations with different starting configurations to monitor differences in intramolecular domain movements between bound and unbound TCR-pMHC molecules of the LC13-HLA-B*44:05-pEEYLQAFTY complex. The results provide some insights into alterations of the dynamics as a consequence of complex formation: For the domains MHC $\alpha1$ and MHC $\alpha2$ the angles between corresponding eigenvectors show larger variation in the unbound state, indicating a wider range of the groove between the two α-helices. For the TCR V_α and TCR V_β domains, angles between corresponding eigenvectors indicate little movements in their main orientation but significant rotations around these major axes. The MHC $\alpha2$-helix showed larger RMSD flexibility for unbound states. The observed structural changes in the CDR3α loop upon binding together with the large RMSF-values of the same region in the unbound state (see Figure 12A) are consistent with the view that structural changes involved in protein binding correlate with intrinsic motions of proteins in the unbound state [51]. Moreover, the structural transition from a 3_{10}-helix to a coil in the CDR3α loop upon complex formation supports the hypothesis that (at least for the CDR3α loop) TCR-pMHC binding can be characterized by the induced fit model, whereby structural rearrangements of the TCR CDRs ensure to achieve the most stable complex [3]. Differences in conformational flexibility (as characterized by RMSD- and RMSF-values) between free and bound forms both for the MHC α-helices and the peptide are consistent with the view that the TCR perceives the peptide and the MHC as a single, composite ligand, making only little distinction between them [7]. Although even our 30 relatively short MD simulations of 40 ns each revealed interesting dynamical features not available from crystallographic data alone, the length of simulations should be extended in subsequent studies to cover processes with longer characteristic time scales.

Supplementary Materials: The following are available online at http://www.mdpi.com/2073-4409/8/7/720/s1, Figure S1: Cartoon representations of CDR3α loops (Ile91, Leu92, Pro93, Leu94, Ala95, Gly96, Gly97, Thr98, Ser99, Tyr100, Gly102, Lys103, Leu104, Thr105) in free (A) and bound (B) states. Residues Ala95 (blue), Gly96 (grey), and Gly97 (grey) exhibit a structural change from a 3_{10}-helix (free, panel A) to a coil (bound, panel B). VMD [22] was used to identify structural changes and to produce the figures. Video S1: Movie of one 144 ns run of the TCR-pMHC complex started from the crystal structure 3KPS.

Author Contributions: R.K. performed MD simulations, implemented key parts of the algorithms, performed evaluations and wrote parts of the manuscript. C.S. performed MD simulations, evaluations and wrote parts of the manuscript. N.I. performed initial MD simulations and contributed concepts in preparation of the project. W.S. contributed concepts for evaluation and coordinated the study.

Data Availability: For this study data have been downloaded from the publically available protein data bank (https://www.rcsb.org/).

Funding: Work was carried out by staff of the Medical University of Vienna, no external funding was used.

Acknowledgments: Parts of the MD simulations were performed using resources of the Vienna Scientific Cluster 3 (VSC3). We thank Michael Cibena (CeMSIIS) for his help with running the simulations and preparing the manuscript. The authors gratefully acknowledge the helpful suggestions of the unknown reviewers.

Conflicts of Interest: The authors declare that there were no conflict of interest.

References

1. Rudolph, M.G.; Wilson, I.A. The specificity of TCR/pMHC interaction. *Curr. Opin. Immunol.* **2002**, *14*, 52–65. [CrossRef]
2. Garcia, K.C.; Degano, M.; Stanfield, R.L.; Brunmark, A.; Jackson, M.R.; Peterson, P.A.; Teyton, L.; Wilson, I.A. An alphabeta T cell receptor structure at 2.5 A and its orientation in the TCR-MHC complex. *Science* **1996**, *274*, 209–219. [CrossRef]
3. Armstrong, K.M.; Piepenbrink, K.H.; Baker, B.M. Conformational changes and flexibility in T-cell receptor recognition of peptide-MHC complexes. *Biochem. J.* **2008**, *415*, 183–196. [CrossRef]
4. Rudolph, M.G.; Stanfield, R.L.; Wilson, I.A. How TCRs bind MHCs, peptides, and coreceptors. *Annu. Rev. Immunol.* **2006**, *24*, 419–466. [CrossRef]
5. Bjorkman, P.J. MHC restriction in three dimensions: A view of T cell receptor/ligand interactions. *Cell* **1997**, *89*, 167–170. [CrossRef]
6. Kass, I.; Buckle, A.M.; Borg, N.A. Understanding the structural dynamics of TCR-pMHC complex interactions. *Trends Immunol.* **2014**, *35*, 604–612. [CrossRef]
7. Baker, B.M.; Scott, D.R.; Blevins, S.J.; Hawse, W.F. Structural and dynamic control of T-cell receptor specificity, cross-reactivity, and binding mechanism. *Immunol. Rev.* **2012**, *250*, 10–31. [CrossRef]

8. Reiser, J.B.; Gregoire, C.; Darnault, C.; Mosser, T.; Guimezanes, A.; Schmitt-Verhulst, A.M.; Fontecilla-Camps, J.C.; Mazza, G.; Malissen, B.; Housset, D. A T cell receptor CDR3beta loop undergoes conformational changes of unprecedented magnitude upon binding to a peptide/MHC class I complex. *Immunity* **2002**, *16*, 345–354. [CrossRef]

9. Ma, Z.; Janmey, P.A.; Finkel, T.H. The receptor deformation model of TCR triggering. *FASEB J.* **2008**, *22*, 1002–1008. [CrossRef]

10. Choudhuri, K.; Van Der Merwe, P.A. Molecular mechanisms involved in T cell receptor triggering. *Semin. Immunol.* **2007**, *19*, 255–261. [CrossRef]

11. Ribarics, R.; Kenn, M.; Karch, R.; Ilieva, N.; Schreiner, W. Geometry Dynamics of Alpha- Helices in Different Class I Major Histocompatibility Complexes. *J. Immunol. Res.* **2015**, *2015*, 20. [CrossRef]

12. Zacharias, M.; Springer, S. Conformational flexibility of the MHC class I alpha1-alpha2 domain in peptide bound and free states: A molecular dynamics simulation study. *Biophys. J.* **2004**, *87*, 2203–2214. [CrossRef]

13. Toh, H.; Kamikawaji, N.; Tana, T.; Sasazuki, T.; Kuhara, S. Molecular dynamics simulations of HLA-DR4 (DRB1*0405) complexed with analogue peptide: Conformational changes in the putative T-cell receptor binding regions. *Protein Eng.* **1998**, *11*, 1027–1032. [CrossRef]

14. Reboul, C.F.; Meyer, G.R.; Porebski, B.T.; Borg, N.A.; Buckle, A.M. Epitope flexibility and dynamic footprint revealed by molecular dynamics of a pMHC-TCR complex. *PLoS Comput. Biol.* **2012**, *8*, e100240. [CrossRef]

15. Wolfson, M.Y.; Nam, K.; Chakraborty, A.K. The effect of mutations on the alloreactive T cell receptor/peptide-MHC interface structure: A molecular dynamics study. *J. Phys. Chem. B* **2011**, *115*, 8317–8327. [CrossRef]

16. Stavrakoudis, A. Insights into the structure of the LC13 TCR/HLA-B8-EBV peptide complex with molecular dynamics simulations. *Cell Biochem. Biophys.* **2011**, *60*, 283–295. [CrossRef]

17. Tedeschi, V.; Alba, J.; Paladini, F.; Paroli, M.; Cauli, A.; Mathieu, A.; Sorrentino, R.; D'Abramo, M.; Fiorillo, T.M. Unusual Placement of an EBV Epitope into the Groove of the Ankylosing Spondylitis-Associated HLA-B27 Allele Allows CD8+ T Cell Activation. *Cells* **2019**, *8*, 572. [CrossRef]

18. Macdonald, W.A.; Chen, Z.; Gras, S.; Archbold, J.K.; Tynan, F.E.; Clements, C.S.; Bharadwaj, M.; Kjer-Nielsen, L.; Saunders, P.M.; Wilce, M.C.; et al. T Cell Allorecognition via Molecular Mimicry. *Immunity* **2009**, *31*, 897–908. [CrossRef]

19. Ferber, M.; Zoete, V.; Michielin, O. T-cell receptors binding orientation over peptide/MHC class I is driven by long-range interactions. *PLoS ONE* **2012**, *7*, e51943. [CrossRef]

20. Ribarics, R.; Karch, R.; Ilieva, N.; Schreiner, W. Geometric analysis of alloreactive HLA α-helices. *Biomed Res. Int.* **2014**, *2014*, 943186. [CrossRef]

21. Kenn, M.; Ribarics, R.; Ilieva, N.; Schreiner, W. Finding semirigid domains in biomolecules by clustering pair-distance variations. *Biomed Res. Int.* **2014**, *2014*, 731325. [CrossRef]

22. Humphrey, W.; Dalke, A.; Schulten, K. VMD: Visual molecular dynamics. *J. Mol. Graph.* **1996**, *14*, 33–38. [CrossRef]

23. Hess, B.; Kutzner, C.; van der Spoel, D.; Lindahl, E. GROMACS 4: Algorithms for Highly Efficient, Load-Balanced, and Scalable Molecular Simulation. *J. Chem. Theory Comput.* **2008**, *4*, 435–447. [CrossRef]

24. Lindorff-Larsen, K.; Piana, S.; Palmo, K.; Maragakis, P.; Klepeis, J.L.; Dror, R.O.; Shaw, D.E. Improved side-chain torsion potentials for the Amber ff99SB protein force field. *Proteins* **2010**, *78*, 1950–1958. [CrossRef]

25. Berendsen, H.J.C.; Postma, J.P.M.; Van Gunsteren, W.F.; Hermans, J. Interaction models for water in relation to protein hydration. In *Intermolecular Forces*; Springer: Dordrecht, The Netherlands, 1981; pp. 331–342.

26. Amadei, A.; Linssen, A.B.M.; Berendsen, H.J. Essential Dynamics of Proteins. *Proteins* **1993**, *17*, 412–425. [CrossRef]

27. Kim, H.J.; Choi, M.Y.; Kim, H.J.; Llinás, M. Conformational Dynamics and Ligand Binding in the Multi-Domain Protein PDC109. *PLoS ONE* **2010**, *5*, e9180. [CrossRef]

28. Hub, J.S.; de Groot, B.L. Detection of functional modes in protein dynamics. *PLoS Comput. Biol.* **2009**, *5*, e1000480. [CrossRef]

29. Schreiner, W.; Karch, R.; Ribarics, R.; Cibena, M.; Ilieva, N. Relative Movements of Domains in Large Molecules of the Immune System. *J. Immunol. Res.* **2015**, *2015*, 210675. [CrossRef]

30. Gakamsky, D.M.; Luescher, I.F.; Pecht, I. T cell receptor-ligand interactions: A conformational preequilibrium or an induced fit. *Proc. Natl. Acad. Sci. USA* **2004**, *101*, 9063. [CrossRef]

31. Ahmad, M.; Gu, W.; Geyer, T.; Helms, V. Adhesive water networks facilitate binding of protein interfaces. *Nat. Commun.* **2011**, *2*, 261. [CrossRef]

32. Lee, J.K.; Stewart-Jones, G.; Dong, T.; Harlos, K.; Di Gleria, K.; Dorrell, L.; Douek, D.C.; van der Merwe, P.A.; Jones, E.Y.; McMichael, A.J. T cell cross-reactivity and conformational changes during TCR engagement. *J. Exp. Med.* **2004**, *200*, 1455–1466. [CrossRef]

33. Borbulevych, O.Y.; Piepenbrink, K.H.; Gloor, B.E.; Scott, D.R.; Sommese, R.F.; Cole, D.K.; Sewell, A.K.; Baker, B.M. T cell receptor cross-reactivity directed by antigen-dependent tuning of peptide-MHC molecular flexibility. *Immunity* **2009**, *31*, 885–896. [CrossRef]

34. Borbulevych, O.Y.; Piepenbrink, K.H.; Baker, B.M. Conformational melding permits a conserved binding geometry in TCR recognition of foreign and self molecular mimics. *J. Immunol.* **2011**, *186*, 2950–2958. [CrossRef]

35. Madden, D.R.; Garboczi, D.N.; Wiley, D.C. The antigenic identity of peptide-MHC complexes: A comparison of the conformations of five viral peptides presented by HLA-A2. *Cell* **1993**, *75*, 693–708. [CrossRef]

36. Tynan, F.E.; Burrows, S.R.; Buckle, A.M.; Clements, C.S.; Borg, N.A.; Miles, J.J.; Beddoe, T.; Whisstock, J.C.; Wilce, M.C.; Silins, S.L.; et al. T cell receptor recognition of a 'super-bulged' major histocompatibility complex class I-bound peptide. *Nat. Immunol.* **2005**, *6*, 1114–1122. [CrossRef]

37. Wieczorek, M.; Abualrous, E.T.; Sticht, J.; Alvaro-Benito, M.; Stolzenberg, S.; Noé, F.; Freund, C. Major Histocompatibility Complex (MHC) Class I and MHC Class II Proteins: Conformational Plasticity in Antigen Presentation. *Front. Immunol.* **2017**, *8*, 292. [CrossRef]

38. Ayres, C.M.; Corcelli, S.A.; Baker, B.M. Peptide and Peptide-Dependent Motions in MHC Proteins: Immunological Implications and Biophysical Underpinnings. *Front. Immunol.* **2017**, *8*, 935. [CrossRef]

39. Natarajan, K.; Jiang, J.; May, N.A.; Mage, M.G.; Boyd, L.F.; McShan, A.C.; Sgourakis, N.G.; Bax, A.; Margulies, D.H. The Role of Molecular Flexibility in Antigen Presentation and T Cell Receptor-Mediated Signaling. *Front. Immunol.* **2018**, *9*, 1657. [CrossRef]

40. Hawse, W.F.; Champion, M.M.; Joyce, M.V.; Hellman, L.M.; Hossain, M.; Ryan, V.; Pierce, B.G.; Weng, Z.; Baker, B.M. Cutting edge: Evidence for a dynamically driven T cell signaling mechanism. *J. Immunol.* **2012**, *188*, 5819–5823. [CrossRef]

41. Rossjohn, J.; Gras, S.; Miles, J.J.; Turner, S.J.; Godfrey, D.I.; McCluskey, J. T cell antigen receptor recognition of antigen-presenting molecules. *Annu. Rev. Immunol.* **2015**, *33*, 169–200. [CrossRef]

42. Hoffmann, T.; Krackhardt, A.M.; Antes, I. Quantitative Analysis of the Association Angle between T-cell Receptor Vα/Vβ Domains Reveals Important Features for Epitope Recognition. *PLoS Comput. Biol.* **2015**, *11*, e1004244. [CrossRef]

43. Adler, M.; Beroza, P. Improved Ligand Binding Energies Derived from Molecular Dynamics: Replicate Sampling Enhances the Search of Conformational Space. *J. Chem. Inf. Model.* **2013**, *53*, 2065–2072. [CrossRef]

44. He, H.T.; Bongrand, P. Membrane dynamics shape TCR-generated signaling. *Front. Immunol.* **2012**, *3*, 90. [CrossRef]

45. Wan, S.; Flower, D.R.; Coveney, P.V. Toward an atomistic understanding of the immune synapse: Large-scale molecular dynamics simulation of a membrane-embedded TCR-pMHC-CD4 complex. *Mol. Immunol.* **2008**, *45*, 1221–1230. [CrossRef]

46. Bello, M.; Correa-Basurto, J. Molecular dynamics simulations to provide insights into epitopes coupled to the soluble and membrane-bound MHC-II complexes. *PLoS ONE* **2013**, *8*, e72575. [CrossRef]

47. Bello, M.; Correa-Basurto, J. Energetic and flexibility properties captured by long molecular dynamics simulations of a membrane-embedded pMHCII-TCR complex. *Mol. Biosyst.* **2016**, *12*, 1350–1366. [CrossRef]

48. Kmiecik, S.; Kouza, M.A.O.; Badaczewska-Dawid, A.E.; Kloczkowski, A.; Kolinski, A.A.O. Modeling of Protein Structural Flexibility and Large-Scale Dynamics: Coarse-Grained Simulations and Elastic Network Models. *Int. J. Mol. Sci.* **2018**, *19*, 3496. [CrossRef]

49. Kmiecik, S.; Gront, D.; Kolinski, M.; Wieteska, L.; Dawid, A.E.; Kolinski, A. Coarse-Grained Protein Models and Their Applications. *Chem. Rev.* **2016**, *116*, 7898–7936. [CrossRef]

50. Cuendet, M.A.; Michielin, O. Protein-protein interaction investigated by steered molecular dynamics: The TCR-pMHC complex. *Biophys. J.* **2008**, *95*, 3575–3590. [CrossRef]

51. Tobi, D.; Bahar, I. Structural changes involved in protein binding correlate with intrinsic motions of proteins in the unbound state. *Proc. Natl. Acad. Sci. USA* **2005**, *102*, 18908–18913. [CrossRef]

Article

Unusual Placement of an EBV Epitope into the Groove of the Ankylosing Spondylitis-Associated HLA-B27 Allele Allows CD8+ T Cell Activation

Valentina Tedeschi [1,†], **Josephine Alba** [2,†], **Fabiana Paladini** [1], **Marino Paroli** [3], **Alberto Cauli** [4], **Alessandro Mathieu** [4], **Rosa Sorrentino** [1], **Marco D'Abramo** [2,*] and **Maria Teresa Fiorillo** [1,*]

[1] Department of Biology and Biotechnology 'Charles Darwin', Sapienza University of Rome, 00185 Rome, Italy; valentina.tedeschi@uniroma1.it (V.T.); fabiana.paladini@uniroma1.it (F.P.); rosa.sorrentino@uniroma1.it (R.S.)
[2] Department of Chemistry, Sapienza University of Rome, 00185 Rome, Italy; josephine.alba@uniroma1.it
[3] Division of Clinical Immunology and Rheumatology, Department of Biotechnology and Medical Surgical Sciences, Sapienza University of Rome, 00185 Rome, Italy; marino.paroli@uniroma1.it
[4] Rheumatology Unit, Department of Medical Sciences and Public Health, University and AOU of Cagliari, Monserrato, 09042 Cagliari, Italy; cauli@unica.it (A.C.); amath@unica.it (A.M.)
* Correspondence: marco.dabramo@uniroma1.it (M.D.); mariateresa.fiorillo@uniroma1.it (M.T.F.); Tel.: +39-0649693263 (M.D.); +39-0649917708 (M.T.F.)
† These authors contributed equally to this work.

Received: 2 April 2019; Accepted: 8 June 2019; Published: 11 June 2019

Abstract: The human leukocyte antigen HLA-B27 is a strong risk factor for Ankylosing Spondylitis (AS), an immune-mediated disorder affecting axial skeleton and sacroiliac joints. Additionally, evidence exists sustaining a strong protective role for HLA-B27 in viral infections. These two aspects could stem from common molecular mechanisms. Recently, we have found that the HLA-B*2705 presents an EBV epitope (pEBNA3A-RPPIFIRRL), lacking the canonical B27 binding motif but known as immunodominant in the HLA-B7 context of presentation. Notably, 69% of B*2705 carriers, mostly patients with AS, possess B*2705-restricted, pEBNA3A-specific CD8+ T cells. Contrarily, the non-AS-associated B*2709 allele, distinguished from the B*2705 by the single His116Asp polymorphism, is unable to display this peptide and, accordingly, B*2709 healthy subjects do not unleash specific T cell responses. Herein, we investigated whether the reactivity towards pEBNA3A could be a side effect of the recognition of the natural longer peptide (pKEBNA3A) having the classical B27 consensus (KRPPIFIRRL). The stimulation of PBMC from B*2705 positive patients with AS in parallel with both pEBNA3A and pKEBNA3A did not allow to reach an unambiguous conclusion since the differences in the magnitude of the response measured as percentage of IFNγ-producing CD8+ T cells were not statistically significant. Interestingly, computational analysis suggested a structural shift of pEBNA3A as well as of pKEBNA3A into the B27 grooves, leaving the A pocket partially unfilled. To our knowledge this is the first report of a viral peptide: HLA-B27 complex recognized by TCRs in spite of a partially empty groove. This implies a rethinking of the actual B27 immunopeptidome crucial for viral immune-surveillance and autoimmunity.

Keywords: HLA-B27; viral peptides; computational analysis; ankylosing spondylitis

---

1. Introduction

The antigen processing and presentation pathway enables MHC class I molecules, in humans named Human Leukocyte antigens (HLAs), to alert CD8+ T lymphocytes as part of the immune-surveillance program [1]. To this aim, HLA class I molecules, expressed on the surface of almost all cells, display short peptides, usually nine residues in length, originating from endogenously synthesized microbial or cellular proteins, which are recognized by CD8+ T cells through their T cell receptors (TCRs) [2,3].

These peptide fragments are generated by proteasome degradation in the cytosol and the portion able to escape the shredding by cytosolic aminopeptidases, is translocated into the endoplasmic reticulum (ER) through the transporter associated with antigen processing (TAP) proteins [2]. Inside the ER, two resident aminopeptidases, ERAP1 and ERAP2, further shorten the peptides at the N-terminally extended end prior to loading into the HLA class I alleles [4,5].

The HLA class I molecules are heterotrimeric proteins of a polymorphic heavy chain assembling with the invariant beta2-microglobulin whose stability is strongly dependent on the presence of a bound peptide allowing the complex to reach a closed conformation. Hence, the bound peptide sits into a groove made by (A–F) pockets of the heavy chain where a network of interactions via hydrogen bonds and salt bridges ensures that the N- and C-termini and the main chain of the peptide fragment are anchored [6,7]. There is considerable variability in the contacts between the side chains of the peptide residues and the polymorphic pockets among all different HLA class I molecules whereas the N- and C-terminal interactions are well conserved [7,8]. Accordingly, thermodynamic data suggest that these interactions could be crucial for the formation and stabilization of the peptide/HLA class I complexes [9].

Among HLA class I alleles, the HLA-B27 is the focus of intense investigation given its involvement in some autoimmune disorders such as Ankylosing Spondylitis, the prototype of a group of chronic rheumatic inflammatory diseases named Spondyloarthropathies and, in the immune protection against some viral infections such as HIV, HCV, EBV, and Influenza virus [10–13]. Nevertheless, the molecular mechanisms linking the HLA-B27 to autoimmunity and to antiviral defence are still largely undefined. Such a duality could stem from overlapping pathways related directly or indirectly to the quality of its peptide repertoire [14]. Further evidence in this direction comes from genome wide association studies (GWAS) pointing out that both Endoplasmic Reticulum aminopeptidase 1 (ERAP1), which is in an epistatic relationship with the HLA-B27, and ERAP2, are risk factors for Ankylosing Spondylitis (AS) [15–17]. It is well acknowledged that both ER resident enzymes have a central role in shaping the immunopeptidome of HLA-B27 molecules [4,5]. Notably, ERAP1 allelic variants conferring a higher enzymatic activity as well as the presence of ERAP2 have been shown to be predisposing factors for AS [4,14,16,17].

The HLA-B27 immunopeptidome is characterized by the preferred binding of 9-mer peptides with an arginine at second position (P2) and hydrophobic or basic residues at their C-termini [13,18,19]. However, our and other groups have shown an expansion of the B27 repertoire to peptides with glutamine or lysine at the primary anchor position at P2 [20,21]. Moreover, several studies have highlighted the high flexibility of the HLA-B27 groove which can even accommodate longer peptides, either N- or C-terminally extended, whose availability is expected to be influenced by the allelic specificity and presence/absence of ERAP enzymes [22,23].

At present, more than 170 different HLA-B27 subtypes have been identified. The HLA-B*2705 is the ancestral allele which shows a worldwide association with AS [24]. Interestingly, some other HLA-B27 alleles, such as HLA-B*2709, relatively frequent in Sardinia (20% of all B27 alleles), do not predispose to the disease [25–27]. It is relevant that the HLA-B*2705 and B*2709 alleles differ by the unique polymorphism at aa 116 (Asp to His) which is positioned in the F pocket of the peptide binding groove [28]. This single substitution is critical for the peptide consensus sequence motifs, for the structural and dynamic features as well as for the T cell repertoire distinguishing the two B27 alleles [28–30].

Very recently, we have described in a high number of HLA-B*2705 subjects, mostly patients with AS, the presence of a HLA-B27-restricted CD8+ T cell reactivity against an EBV epitope from EBNA3A (pEBNA3A; RPPIFIRRL) which does not have a proper B27 binding consensus motif. Notably, the B*2709 allele is not a presenting molecule for such a peptide and consequently, in B*2709 positive individuals this immune response does not occur [31].

Herein, we investigate whether the reactivity towards pEBNA3A in the HLA-B*2705 context of presentation is a consequence of the recognition of the longer peptide KRPPIFIRRL (pKEBNA3A) endowed with a canonical B27 binding motif. Moreover, by computational approaches, we model the structural and dynamic behaviour of these complexes and discuss how the pEBNA3A and pKEBNA3A peptides fit into the groove of the B*2705 and the B*2709 alleles.

2. Materials and Methods

2.1. Study Subjects

Eighty-five HLA-B*2705-positive subjects (74 AS-patients and 11 HD) and nine HLA-B*2709-positive controls were enrolled in this study. Diagnosis of AS has been made according to modified New York criteria. The HLA-B*27 subtype was determined by serological analysis using ME1 mAb (specificity: HLA-B*27; -B*07; -B*42; -B*67; -B*73 and -B*w22) and genomic analysis. The study received the approval by the Ethics Committee of the University of Cagliari (365/09/CE). All subjects provided written informed consent prior to enrolment.

2.2. Synthetic Peptides

pEBNA3A (RPPIFIRRL; aa 379–387), an immunodominant EBV epitope restricted by HLA-B*07 [32] and its N-terminal-extended version pKEBNA3A (KRPPIFIRRL) have been used in this study. For the binding assay, a self-peptide restricted by HLA-B*27 (TIS, RRLPIFSRL; aa 325–333) [31] and an immunodominant EBV-epitope HLA-B*35-restricted (YPLHEQHGM; aa 458–466) were included. Peptides (purity > 95%) were purchased by Aurogene (Rome, Italy).

2.3. Cell Lines

TAP-defective CEM 174.T2 cells (ATCC number: CRL-1992™) and HMy2.C1R cells (ATCC number: CRL-1993™), stably expressing B*2705 or B*2709 were used. Cells were cultured in heat-inactivated 10% fetal bovine serum (FBS; Euroclone Spa, Pero, Milan, Italy)/RPMI 1640 medium (Euroclone Spa, Pero, Milan, Italy) supplemented with 2 mmol/L L-glutamine, 100 U/mL penicillin, 100 µg/mL streptomycin. T2 and C1R stable transfectants were maintained in medium added with G418 (800 µg/mL) or hygromycin B (200 µg/mL), respectively.

2.4. PBMC Stimulation

PBMC (4×10^6) separated from whole blood by density gradient separation (Cedarlane Laboratories Ltd., Burlington, ON, Canada) were incubated with the indicated peptides (20 µM) and cultured for 12 days at 2×10^6 cells/mL in 5% FBS/RPMI complete medium, and 10 U/mL of human recombinant interleukin 2 (IL2) (Roche Applied Science, Penzberg, Germany). On day 3 and 9, fresh medium containing 20 U/mL of IL2 was replenished. After 12 days of antigen pulsing, PBMC were used for intracellular IFNγ staining.

2.5. Intracellular IFNγ Staining

Briefly, C1R transfectants were incubated overnight with pEBNA3A or pKEBNA3A (30 µM) or with medium alone and thoroughly washed before to be plated with antigen-stimulated PBMC at 5:1 PBMC/APC ratio. After 1 h, the cells were treated with brefeldin A (10 µg/mL) at 37 °C for 16 h. Cells were stained with the anti-CD8-FITC mAb (BioLegend, San Diego, CA, USA), 20′ on ice, fixed with 4% paraformaldehyde 20′ on ice, permeabilized with 1% BSA/0.1% saponin in PBS 1X for 5′ at RT and then resuspended in 2% paraformaldehyde/1% BSA. Finally, cells were stained by an anti-IFNγ-APC mAb (BioLegend, San Diego, CA, USA) for 20′ at RT. The samples were acquired by a FACSCalibur flow cytometer (Becton Dickinson, Franklin Lakes, NJ, USA) and analysed by FlowJo software (Tree Star Inc, Ashland, OR, USA).

2.6. T2.B27 Stabilization Assay

The capability of the selected peptides to stabilize the B*2705 and B*2709 molecules on the cell surface of T2 stable transfectants was assessed according to the procedure previously described by Tedeschi et al. [31]. The surface amount of B27 molecules on T2 transfectants was determined by using ME1 mAb and a rabbit anti-mouse IgG-FITC (Jackson ImmunoResearch Europe, Suffolk, UK) as secondary antibody. An antibody of IgG isotype was also employed as control. The results are shown as the mean ± SD of three independent experiments.

2.7. Statistics

Data obtained by IFNγ-production assays from pEBNA3A- or pKEBNA3A-stimulated PBMCs from B*2705-positive carriers were compared by Mann Whitney U-test; the pEBNA3A response in B*2705- or B*2709-positive subjects was compared by Fisher's two-tailed exact test. A p value < 0.05 was considered statistically significant.

2.8. Modelling of the Complexes

The EBV-derived epitope (pEBNA3A; RPPIFIRRL) was modelled by means of the mutagenesis tool included in the Pymol software package (The PyMOL Molecular Graphics System. Version 1.7.0.2, Schrödinger, LLC, New York, NY, USA, www.pymol.org), starting from the crystallographic structure of pVIR peptide (RRKWRRWHL) [33]. A lysine residue was added to the pEBNA3A peptide to model the 10-mer peptide (pKEBNA3A; KRPPIFIRRL).

Molecular docking calculations have been performed to verify the peptide binding-groove insertion. The best pose was refined using the Galaxy Refine Complex tool [34] in order to remove unfavorable interactions.

2.9. Computational Methods

We performed a molecular dynamics simulation for each of the HLA-B27 subtypes in complex with the corresponding peptide and an additional ligand-free simulation as a control for each of the HLA-B27 subtypes. The simulations have been performed by means of the Gromacs software package (version 5.0.7) [35] using the OPLS-AA force field [36]. All the simulations, lasting ~200 ns each, were performed in a cubic box with the SPC/E water model [37], imposing a distance of 1.2 nm between the protein and the box. The systems were neutralized and simulated at physiological concentration of Na^+ and Cl^- (0.15 M). The temperature was kept constant at 310 K by coupling the system to an external bath [38] with a coupling time constant $\tau T = 0.1$ ps. The systems have been coupled to a pressure bath (1 bar) with a coupling time constant $\tau P = 1.0$ ps.

The crystallographic structures (PDB 1OGT for B*2705, 1OF2 for B*2709; 1WOV for TIS: B*2705 and 1WOW for TIS:B*2709) were used to start the ligand-free and the TIS-complex simulations. To simulate the EBV-epitope and the complexes formed by the synthetic peptides, a homology modelling procedure has been applied using the corresponding crystal structures (see above).

To evaluate the different subtype-dependent peptide mobility, their entropies (S) have been calculated by means of the Schlitter method [39]. The differences in the peptide entropies between the two HLA-B27 subtypes were computed according to: $T\Delta S = T\ (S_{pep/B*2709} - S_{pep/B*2705})$.

The principal components (Essential Dynamics, ED) analysis has been performed on the MD trajectories to describe the concerted motions associated to the largest collective atomic fluctuations [40].

3. Results

3.1. HLA-B27 Allele-Specific Presentation of a Suboptimal EBV Derived Peptide

In previous work, we have described in HLA-B*2705 carriers the presence of HLA-B27-restricted CD8+ T lymphocytes specific for an EBV antigen (pEBNA3A; RPPIFIRRL) known as immunodominant in the context of HLA-B7 restriction and suboptimal for HLA-B27. Noteworthy, this occurs in a B27 allele-dependent manner since the same CD8+ T cell reactivity could not be observed in HLA-B*2709 positive healthy donors [31]. To date, we have screened 85 HLA-B*2705 positive individuals (74 patients with AS and 11 healthy controls) and found that 69% of them display pEBNA3A-specific, B*2705-mediated CD8+ T cell responses monitored through the IFNγ production. To this aim, PBMC from these individuals have been cultured in the presence of pEBNA3A (20 μM). After 12 days, they were exposed to pEBNA3A-pulsed C1R.B27 transfectants and the IFNγ production detected in the CD8+ T cell subset (Figure 1A).

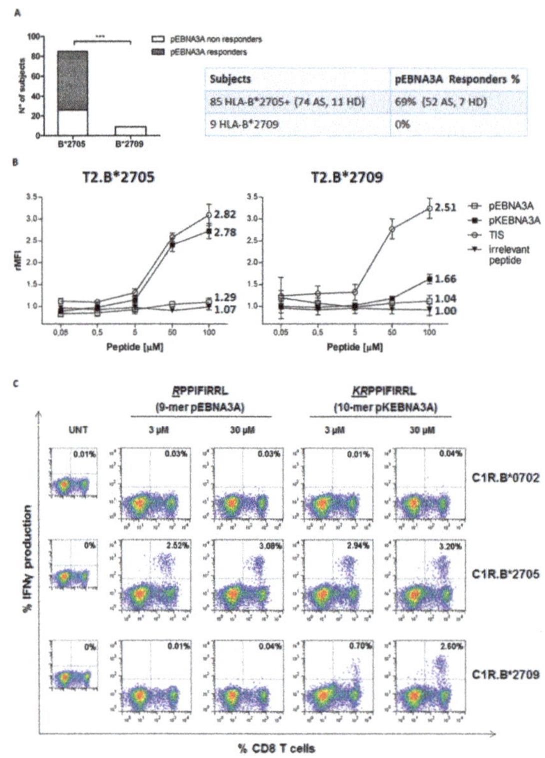

Figure 1. B*2705 but not B*2709-positive carriers mount a response to the 9-mer pEBNA3A epitope. (**A**) CD8+ T cell responses to pEBNA3A detected by IFNγ production analyzed in 85 B*2705 carriers (74 patients with Ankylosing Spondylitis, indicated as AS, and 11 healthy donors, indicated as HD) and in 9 B*2709 healthy donors. 69% of B*2705 carriers but none of B*2709 individuals show reactivity against pEBNA3A antigen (p value < 0.0001 calculated by Fisher's two-tailed exact test). (**B**) Staining of B*2705 and B*2709 molecules expressed on the cell surface of T2 transfectants performed by using ME1 mAb after 16 h of cell incubation with pEBNA3A, pKEBNA3A, or TIS (RRLPIFSRL) used as positive reference as well as with an irrelevant peptide at the indicated concentrations. Results are expressed as relative mean fluorescence intensity (rMFI) obtained with peptide-treated compared to untreated cells. Values represent the mean ± SEM of three independent experiments. The numbers on the right indicate the fold increase of rMFI obtained with 100 μM compared to 0.5 μM of each peptide. (**C**) Flow cytometry plots reporting IFNγ production by PBMCs from a representative B*2705 positive patient with AS expanded in the presence of pEBNA3A (12 days) and exposed to indicated C1R transfectants pulsed with pEBNA3A or pKEBNA3A (3 or 30 μM) before intracellular staining. Numbers indicate the percentage of IFNγ-secreting CD8+ T cells.

As expected, none of the nine B*2709 healthy subjects, analysed in the previous and in the current study, responded to the 9-mer pEBNA3A epitope (Figure 1A).

Next, we asked whether the immune-reactivity against pEBNA3A could be a side-effect of the cross-recognition with the longer natural peptide (pKEBNA3A; KRPPIFIRRL) extended by a Lys at the N-terminus and, therefore, due to the Arg at P2, restoring the canonical B27 consensus motif. At first, we performed cell surface stabilization assays on T2.B*2705 and T2.B*2709 transfectants incubated with increasing concentrations of pKEBNA3A (10-mer), pEBNA3A (9-mer) or TIS used as positive reference. The results in Figure 1B confirmed the inability of pEBNA3A to stabilize either HLA-B*2705 or HLA-B*2709 molecules on B27 transfectants as previously shown in Tedeschi et al. [31]. pKEBNA3A showed instead a dose-response curve nearly overlapping to that of the positive reference TIS peptide on T2.B*2705 cells. When T2.B*2709 transfectants were used, the 10-mer displayed a detectable binding only at the highest dose (Figure 1B). Therefore, pKEBNA3A could be considered as a strong binder for B*2705 and as an intermediate/poor ligand for the B*2709.

PBMC from 15 B*2705 subjects (13 AS patients and two healthy individuals), known to be pEBNA3Aresponders, were then cultured in medium containing pEBNA3A, for 12 days, before re-stimulation with C1R.B*2705, C1R.B*2709 or C1R.B7 cell transfectants pulsed with both pEBNA3A and pKEBNA3A epitopes at 3 or 30 μM. As expected, pEBNA3A triggered IFNγ secretion by CD8+ T cells only when presented by C1R.B*2705 but not by C1R.B*2709 cells (Figure 1C) [31]. Very interestingly, pKEBNA3A was able to activate CD8+ T cell effector functions when displayed in complex with either B*2705 or B*2709 molecules (Figures 1C and S1). This is consistent with the peptide binding data showing a strong binding of pKEBNA3A to B*2705 and a moderate but reproducible stabilization of B*2709 allele. Accordingly, as shown in Figure 1C (right panels), the re-stimulation of PBMC with the 10-mer triggered a slightly higher number of IFNγ producing CD8+ T cells when compared to the shorter pEBNA3A epitope complexed with the B*2705 molecules. This difference, although not statistically significant ($p > 0.05$), is found in 13 out of 15 samples analyzed (Figure S1).

3.2. Little Evidence for pKEBNA3A Production in Vivo

The results illustrated above allowed us to speculate that the in vivo antigen processing mechanism could produce pKEBNA3A besides pEBNA3A and that in B27 positive individuals the former antigen is dominant. To better address this point, we asked whether in subjects typed as B*2709 positive, a specific CD8+ T cell response could arise upon stimulation with pKEBNA3A. To this aim, PBMC from 3 B*2709 positive individuals have been analysed for reactivity to the 10-mer pKEBNA3A epitope but in none of them a detectable response was observed (data not shown).

A comparative analysis in B*2705 carriers (seven AS patients and two healthy subjects) by a concurrent stimulation of PBMC with pEBNA3A and pKEBNA3A epitopes was then performed. Figure 2 shows a trend of higher magnitude of CD8+ T cell response, monitored as IFNγ producing cells, when the initial antigenic stimulation was done with the shorter peptide. Interestingly, and in agreement with the data shown in Figure 1C, a more effective T cell activation was achieved when the first stimulation was carried out with pEBNA3A and the re-stimulation with pKEBNA3A. Re-stimulation with pKEBNA3A presented by C1R.B*2709 yielded a lower production of IFNγ (Figures 2 and S1). Overall, these results do not support an immunodominance of pKEBNA3A over pEBNA3A.

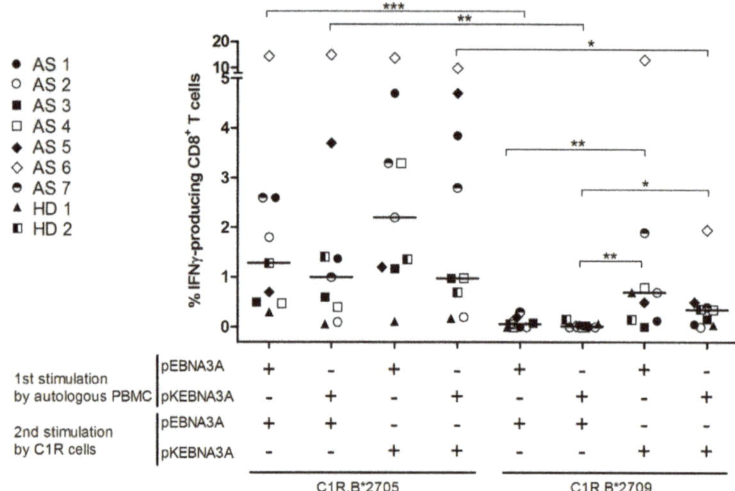

Figure 2. Usage of pEBNA3A in the first stimulation followed by pKEBNA3A as boost leads to the most effective CD8+ T response. IFNγ production by CD8+ T cells was assessed in nine B*2705 positive carriers (seven patients with AS indicated as AS and two healthy subjects indicated as HD). PBMCs were first stimulated with pEBNA3A or pKEBNA3A (12 days) and then re-stimulated with C1R expressing B*2705 or B*2709 molecules, pre-pulsed with the 9-mer or the N-extended version. The percentages of IFNγ-producing CD8+ T cells were compared by Mann Whitney test; *** *p* value < 0.001, ** *p* value < 0.01, * *p* value < 0.05.

3.3. Peptide Residues Distance from Binding Groove Pockets

It is known that the HLA peptide is inserted in specific pockets of the binding groove [41]. In particular, the first two amino acid residues are important to anchor the peptide in the A and B pockets of the binding groove. Therefore, to study the conformational behaviour of pEBNA3A and pKEBNA3A in complex with the B*2705 and B*2709 molecules, MD simulations were performed (Figure 3). To understand the peptide positioning in these pockets, the distance of the P1 residue from the central amino acid of the A and B pockets (residues 5 and 45 were chosen, respectively) has been calculated in the case of both peptides in complex with B*2705 and B*2709 molecules. The results shown in Table 1 clearly indicate that, along the MD trajectory, the first residue P1 interacts with the B pocket in both the 9-mer and 10-mer epitopes.

Interestingly, the analysis of the P9/P10 peptide residue interactions shows that the P10 of pKEBNA3A in complex with the B*2709 almost completely filled the F pocket. In all the other systems, the pocket F was mostly unfilled (Table 1).

Moreover, the estimated entropy differences shown in Table 2 revealed higher entropies of the peptides in complex with the B*2705 subtype with respect to the peptides bound to B*2709. This indicates that both peptides, pEBNA3A and pKEBNA3A, are entropically favoured when bound to the HLA-B*2705 with respect to the HLA-B*2709. As a control, we successfully checked that our entropy estimates for the TIS peptide bound to both the B*2709 and the B*2705, were in agreement with a previous computational study [42].

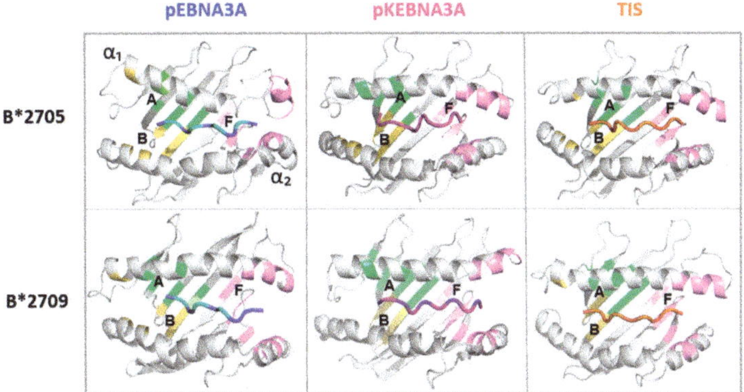

Figure 3. Representative snapshots as obtained by MD simulations of the peptide:HLA-B27 complexes. The upper panels show HLA-B*2705 in complex with pEBNA3A (blue), pKEBNA3A (magenta), TIS (orange) peptides. The bottom panels show the peptide:HLA-B*2709 complexes. A (green), B (yellow) and F (pink) pockets are indicated.

Table 1. Peptide position in the binding groove.

Peptide: HLA-B27 Complex	P1 Interaction with A Pocket	P1 Interaction with B Pocket	P9 or P10 Interaction with F Pocket
pEBNA3A: B*2705	No	Yes	No §
pEBNA3A: B*2709	No	Yes	No §
pKEBNA3A: B*2705	No §	Yes	No
pKEBNA3A: B*2709	No	Yes	Yes
TIS: B*2705 = TIS: B*2709	Yes	No	Yes
	§ high fluctuation		

Peptide interactions in the A and B pockets. P1 and P9/P10 represent the 1st and the last residue, respectively. A distance of 0.4 nm was chosen as interaction cut-off (i.e., "yes" indicates a mean residue distance < 0.4 nm and "no" a mean residue distance > 0.4 nm).

Table 2. Peptide entropy differences.

Peptides	TΔS kJ/mol
pEBNA3A: HLA-B*2709-05	-77.6 ± 10.5
pKEBNA3A: HLA-B*2709-05	-47.4 ± 14.7
TIS: HLA-B*2709-05	41.4 ± 1.3

The ΔS is the difference between the entropies of the peptide bound to the B*2709 and that bound to the B*2705.

The epitope recognition by TCR might also be influenced by the peptide exposure. However, no relevant differences emerged when the solvent peptide exposure areas of pEBNA3A and pKEBNA3A in complex with the two B27 alleles were compared. As shown in Table 3, the differences between the two subtypes are quite small (within 2 nm^2), indicating a similar degree of solvent exposure.

Table 3. Peptide solvent exposure.

Peptides	Solvent Exposure (nm^2)
pEBNA3A: HLA-B*2705	16.7 ± 0.1
pEBNA3A: HLA-B*2709	16.7 ± 0.1
pKEBNA3A: HLA-B*2705	18.2 ± 0.1
pKEBNA3A: HLA-B*2709	18.9 ± 0.1

3.4. Binding Grooves Dynamics

To study the influence of the peptides on the binding groove conformational behaviour, we applied the essential dynamics (ED) analysis to the groove motions.

Starting from the concatenated trajectories of the binding groove residues (residues from one to 175), the ED analysis showed that their motions were described for the 68.5% by the first five eigenvectors.

From the projection of the groove alpha carbons on the first 2 eigenvectors, we obtained the common essential subspace explored by the eight binding grooves, corresponding to the six peptides: HLA-B27 complexes displayed in Figure 4 and the empty B*2705 and B*2709 molecules, used as references.

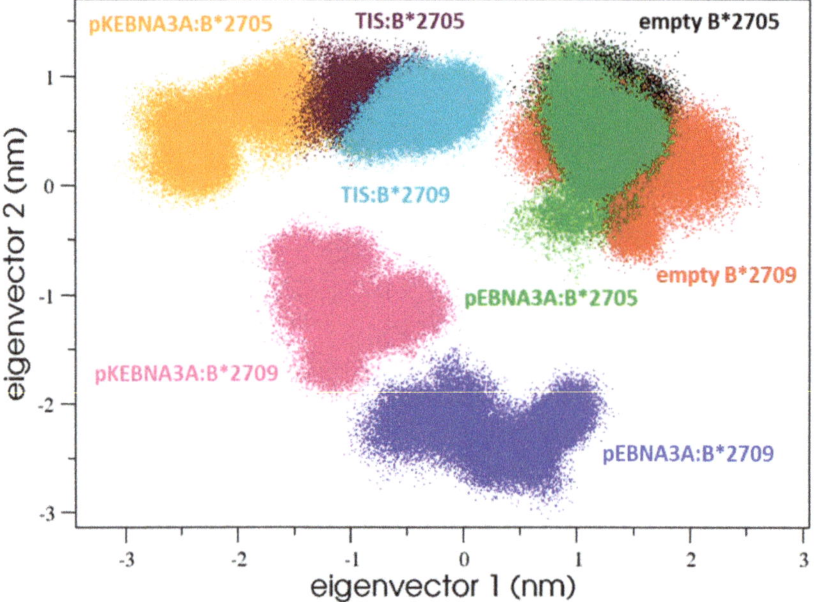

Figure 4. C-alpha binding-groove projections on their common essential subspace, highlighting four main regions featured by: (1) the empty HLA-B*2705/09 and pEBNA3A:B*2705; (2) pKEBNA3A:B*2705 and TIS:B*2705/09; (3) pEBNA3A:HLA-B*2709; (4) pKEBNA3A:HLA-B*2709 complexes. Each point represents a structure sampled by MD simulation.

Interestingly, four main conformational regions sampled by these systems can be identified. One of them characterized the two empty HLA-B27 subtypes as well as the pEBNA3A:HLA-B*2705 complex. Thus, the presence of the 9-mer peptide in the HLA-B*2705 subtype did not seem to modify the groove conformation, being its projection overlapped to the ligand-free one.

In line with our findings, previous computational [42] and experimental [43,44] studies highlighted a subtype-dependent conformational flexibility. In particular, by means of isotope-edited IR spectroscopy [44], observed larger flexibilities and/or an opening of the binding groove for the B*2705 either for peptide-filled or peptide-devoid molecules [43,44]. These observations seem in accordance with the enhanced structural variability observed in the pEBNA3A:HLA-B*2705 complex and reported in Figure 3 (top left panel), where a lack of secondary structure is observed (along the MD trajectory) in a few residues of one helix defining the F pocket.

The second region on such a subspace characterizes the pKEBNA3A:B*2705 and the TIS: B*2705/09 complexes, which are the complexes displaying the highest stability in the binding assay (Figure 4).

The two remaining regions describing the peptide:B*2709 complexes fell in distinct areas of the plane, indicating that both pEBNA3A and pKEBNA3A peptides affected the B*2709 conformational dynamics in different ways (Figure 4).

Additional analysis of the local contacts showed that the P1 residue of the pEBNA3A makes an H-bond with the Glu45 of B*2705 (Table S1). The same behaviour is also observed in the TIS:B27 complexes but not in the pEBNA3A:B*2709 (Table S1).

In summary, the analysis of the binding groove dynamics showed a similar behaviour of TIS: B*2705 and pKEBNA3A:B*2705 complexes indicating similar structural-dynamical effects induced by these peptides on the B*2705. This is consistent with the functional binding data showing that high rMFIs were reached upon binding of both TIS and pKEBNA3A with B*2705 molecules (Figure 1C).

Our data also confirm previous findings suggesting the subtype-dependent conformational dynamics as an intrinsic feature that distinguishes the two HLA-B27 subtypes [42,44]. That is, our computational procedure provides a physically sound explanation of the different binding modes of three peptides bound to either HLA-B27 subtype. However, it should be noted that a generalization of these results as well as the possibility to predict the binding modes by "in-silico" approaches requires several additional MD simulations of different peptides bound to the HLAs, which are beyond the scope of our work and computationally unfeasible.

4. Discussion

To our knowledge, this is the first piece of evidence that the HLA-B*2705 is able to unleash a CD8+ T cell response even though presenting a suboptimal viral peptide which possibly occupies the binding groove leaving out the A pocket. This is a relevant observation for an HLA-B allele which predisposes to Spondyloarthritis, particularly to AS, but which is also protective in several viral infections [10–13]. At first glance, this finding suggests that the actual B*2705 peptidome can indeed be more extended than that detected by high throughput technologies.

The pEBNA3A epitope (RPPIFIRRL), herein investigated, but already described by us [31] as an uncanonical HLA-B*2705 restricted antigen, evokes a CD8+ T cell reactivity in 69% of B*2705 carriers, either patients with AS or healthy controls. This high percentage of pEBNA3A responders substantiates that this peptide can act as an effective antigen in the B*2705 context of presentation although lacking a suitable B27 consensus sequence. As a matter of fact, both mathematical predictive algorithms and experimental binding data have assigned it as a putative non-B27-ligand (Figure 1B) [31]. Therefore, the propensity to load and present suboptimal viral/microbial antigen would be a plausible trait of the HLA-B*2705 by which to broaden the spectrum of triggered immune responses giving a simple immunological explanation for the superior viral immune-surveillance [12–14]. It must also be considered that this same feature, through the yield of unstable peptide:HLA complexes, would fuel the tendency of the HLA-B*2705 to produce aberrant forms implicated in the pathogenetic mechanisms of the associated Spondyloarthritis [45].

It is noteworthy that the non-AS-associated HLA-B*2709 allele, which differs from the B*2705 for a single amino acid change in the F pocket [13,25,27], is unable to present this suboptimal epitope [31]. Accordingly, no pEBNA3A-driven CD8+ T cell responses have been found in the nine B*2709 healthy donors screened so far (Figure 1A).

These data prompted us to check whether the CD8+ T cell reactivity towards pEBNA3A in the B*2705 carriers could be a side-effect of the recognition of the longer natural 10-mer (pKEBNA3A). Of note, the N-terminal extension by Lys in pKEBNA3A peptide (KRPPIFIRRL) re-established a typical B27 consensus motif with an Arg in P2. Correspondingly, peptide binding assays on T2.B27 cells pointed out that this N-terminally extended peptide acquired the features of a high and moderate binder for the B*2705 and 09 molecules, respectively (Figure 1B). Nonetheless, functional CD8+ T cell data obtained by stimulating PBMC from B*2705 subjects with the 9-mer and 10-mer in a comparative way and re-stimulating with the same peptides in all four combinations (Figure 2), did not support our hypothesis according to which the latter peptide is immunodominant over the former. In fact, more robust T cell responses, monitored as % of IFNγ-producing CD8+ T cells, have been reached when pEBNA3A was used in the first stimulation. Of note, the best functional outcome was obtained using pEBNA3A in the first followed by pKEBNA3A in the second stimulation. This is most probably due to the higher stability of pKEBNA3A:B27 complexes compared to pEBNA3A:B27 complexes on the C1R transfectants (Figure 2). Overall, these data suggest that the CD8+ T cell repertoire is prevalently pEBNA3A-oriented, most likely because there could be in vivo a sharp prevalence of pEBNA3A:B27 vs. pKEBNA3A:B27 complexes on cell surface. This can be due to the activity of ERAP2, or even ERAP1, which can trim the N-terminal lysine from pKEBNA3A while sparing pEBNA3A from further degradation since the proline at position 2, and the residue immediately upstream of it (arginine), cannot be cut by either aminopeptidase [46] and, therefore, albeit less stable, the pEBNA3A:B*2705 complex results more abundant. The same cannot be seen in the HLA-B*2709 positive subjects, probably due to the lack of flexibility that would be necessary to accommodate the shorter peptide. In principle, the amount of pEBNA3A available to HLA-B27 loading should depend on level of expression [47] and enzymatic activity of ERAP and increases in parallel to it. Notably, AS associates with ERAP1 allelic variants endowed with high enzymatic activity as well as with the presence of ERAP2 [15–17]. Hence, we predict that the available quantity of pEBNA3A would be higher in the HLA-B27 positive patients with AS than in B27 healthy subjects. In this regard, it would be informative to look for both pEBNA3A and pKEBNA3A in the B*2705 and B*2709 peptidomes from EBV-B lymphoblastoid cell lines.

Interestingly, we have not detected CD8+ T cell reactivity to either pEBNA3A or pKEBNA3A in the B*2709 healthy donors although the latter peptide can bind and be presented by B*2709 molecules. This has been proved by binding assays (Figure 1B) and by IFNγ production from CD8+ T cells of B*2705 subjects re-stimulated with C1R-B*2709 cells pulsed with pKEBNA3A (Figures 2 and S1). The reason of the lack of such T cell reactivity in B*2709-positive individuals is hard to explain and presumably is a sum of several factors among which the different nature (flexibility/plasticity) of B*2709 versus the B*2705, the availability of these viral peptides influenced by the different ERAP backgrounds (polymorphisms/enzymatic activity) and the mechanisms that shaped the B*2709 vs. the B*2705 restricted T cell repertoire [4,5,14,43,48,49]. Ultimately, this evidence supports the possibility that in vivo pKEBNA3A is absent.

By means of computational analysis, we tried to gain insights into the binding mode of pEBNA3A within the B27 grooves looking for conformational/dynamic peculiarities underlying the different behaviour of the B*2705 and B*2709 pair of alleles in respect to this uncanonical antigen presentation.

As already speculated in the previous study [31] and supported by functional experiments with analogue peptides, MD simulations infer that the pArg1 acts as a principal anchor in both pEBNA3A:HLA-B*2705 and pEBNA3A:HLA-B*2709 complexes making contacts exclusively with the B pocket into the two grooves. As a consequence of this shifting, the peptide is not stabilized into the A pocket which remains empty. Additional MD simulations of different peptide:HLA complexes could be very useful to confirm such a behaviour. However, this peptide conformation has been already disclosed, by the X-ray crystallography, for Tax8 an octameric peptide from human T cell lymphotropic virus-1 when complexed with the HLA-A2 molecule and, more recently, for a truncated 7-mer EBV peptide (LMP2 343-349) in the HLA-A*11:01 antigen-binding cleft [6,50]. Moreover, the peptidome analysis of the B*5101, a HLA class I allele associated with Behçet's disease, displayed the presence of 10-mers with Pro or Ala at P2 together with their N-terminally truncated forms allowing to speculate about a non-canonical binding mode for these shorter peptides which leaves aside the A pocket [51].

Another important point is that in both pEBNA3A:B*2705 and pEBNA3A:B*2709 complexes, the F pocket is only intermittently filled by the pLeu9 as indicated by large fluctuations observed along the MD trajectories. This implies a remarkable solvent exposition of the C-terminal peptide moiety in both complexes which could favour the interaction with the TCR but it is also suggestive of a rather weak interaction of the peptide into the groove. Considering the T-cell activation data, it is likely that this conformational flexibility at the peptide C-terminus is compensated in the case of the B*2705 through favourable interactions involving the N-terminal moiety of the peptide. Accordingly, pArg1 of pEBNA3A makes a hydrogen-bond with the Glu45 in the B*2705 but not in the B*2709.

Really unexpected is the pKEBNA3A binding mode in the two B27 grooves. Indeed, MD trajectories highlighted a common peptide shift with respect to the shorter pEBNA3A antigen. This means that pLys1, but not pArg2, fits into the B pocket in either B*2705 or B*2709 molecules. Notably, this result is consistent with our recent data showing that also Lys, beside Arg and Gln, could serve as principal B27 anchor residue interacting with the B pocket [21].

The analysis of pKEBNA3A residue interactions showed that the P10, exclusively in the B*2709 binding groove, almost completely fills the F pocket. This further interaction gives a clue to explain why the 10-mer peptide acquires the capability to bind to the B*2709 and be presented to CD8+ T cells inducing their activation. Therefore, the reason of the lack of reactivity in HLA-B*2709 individuals is most probably due to an in vivo shortage of the 10-mer compared to the 9-mer peptide.

Interestingly, the analysis of the groove motions points out that the peptides affect the conformational space accessible to the B27 grooves and, remarkably, a common conformational region is sampled by the three complexes (Figure 4) showing the highest stability induced by the peptide binding (Figure 1B).

Some limitations of this study prevented us to give a definite answer to our hypothesis. First of all, our data neither prove nor disprove the immunodominance of the longer and "canonical" B27-binder peptide (pKEBNA3A) over the shorter and "uncanonical" pEBNA3A antigen. The identification of both or even one of the two peptides in the B*2705 and B*2709 peptidomes from EBV-B lymphoblastoid cell lines would support our hypothesis. Nevertheless, the lack of specific memory T cells in the B*2709 individuals towards the longer peptide which can nonetheless be accommodated in the groove, argues against the *in vivo* availability of this epitope. A second point concerns the intrinsic limitations of the computational methods used to gain insights into the conformations of the two epitopes in complex with both B27 molecules. Unfortunately, a first attempt to make crystals of pEBNA3A and pKEBNA3A in complex with the HLA-B*2705 has been unsuccessful (Dr. Bernhard Loll, personal communication).

In conclusion, this study shows for the first time that the strongest risk factor for AS, i.e., B*2705, is able to elicit anti-viral T cell immune-responses even when the binding groove might be partially occupied by the epitope as inferred from computational analysis. This feature is not shared by the very close B*2709, which is not a risk factor for AS. This evidence hints a deeper definition of the B27 immunopeptidome to gain insights into its pathogenic role in Spondyloarthritis as well as in the immune-surveillance.

Supplementary Materials: The following are available online at http://www.mdpi.com/2073-4409/8/6/572/s1, Figure S1: B*2709 allele is able to present pKEBNA3A, Table S1. Hydrogen bonds between the peptides and the HLA-B27 binding groove.

Author Contributions: V.T. and F.P. performed the cellular experiments and analysed data. J.A. and M.D. performed M.D. simulations and analysis. M.P., A.C. and A.M. collected the donor material and contributed to analysis of data. M.T.F., R.S. and M.D. designed and supervised the study and wrote the manuscript. All authors contributed to revise and finalize the manuscript.

Funding: This study was supported by Fondazione Ceschina to M.T.F. and R.S. and by Sapienza Università di Roma through Progetti di Ateneo to M.T.F., R.S. and M.D.

Acknowledgments: The authors wish to thank Silvana Caristi and Federica Lucantoni for the excellent technical assistance and all subjects participating in the study. They also thank CINECA for computational support to M.D. and J.A. They are grateful to Bernhard Loll (Freie Universität Berlin, Berlin, Germany) for sharing unpublished data and for fruitful discussion.

Conflicts of Interest: The authors declare no conflict of interest.

References

1. Neefjes, J.; Jongsma, M.L.; Paul, P.; Bakke, O. Towards a systems understanding of MHC class I and MHC class II antigen presentation. *Nat. Rev. Immunol.* **2011**, *11*, 823–836. [CrossRef] [PubMed]
2. Saunders, P.M.; van Endert, P. Running the gauntlet: From peptide generation to antigen presentation by MHC class I. *Tissue Antigens* **2011**, *78*, 161–170. [CrossRef] [PubMed]
3. Gras, S.; Burrows, S.R.; Turner, S.J.; Sewell, A.K.; McCluskey, J.; Rossjohn, J. A structural voyage toward an understanding of the MHC-I-restricted immune response: Lessons learned and much to be learned. *Immunol. Rev.* **2012**, *250*, 61–81. [CrossRef] [PubMed]
4. López de Castro, J.A. How ERAP1 and ERAP2 Shape the Peptidomes of Disease-Associated MHC-I Proteins. *Front. Immunol.* **2018**, *9*, 2463. [CrossRef] [PubMed]
5. Evnouchidou, I.; van Endert, P. Peptide trimming by endoplasmic reticulum aminopeptidases: Role of MHC class I binding and ERAP dimerization. *Hum. Immunol.* **2019**. [CrossRef] [PubMed]
6. Khan, A.R.; Baker, B.M.; Ghosh, P.; Biddison, W.E.; Wiley, D.C. The structure and stability of an HLA-A*0201/octameric tax peptide complex with an empty conserved peptide-N-terminal binding site. *J. Immunol.* **2000**, *164*, 6398–6405. [CrossRef] [PubMed]
7. Madden, D.R.; Gorga, J.C.; Strominger, J.L.; Wiley, D.C. The structure of HLAB27 reveals nonomer self-peptides bound in an extended conformation. *Nature* **1991**, *353*, 321–325. [CrossRef] [PubMed]
8. Josephs, T.M.; Grant, E.J.; Gras, S. Molecular challenges imposed by MHC-I restricted long epitopes on T cell immunity. *Biol. Chem.* **2017**, *398*, 1027–1036. [CrossRef] [PubMed]
9. Davies, M.N.; Hattotuwagama, C.K.; Moss, D.S.; Drew, M.G.; Flower, D.R. Statistical deconvolution of enthalpic energetic contributions to MHC-peptide binding affinity. *BMC Struct. Biol.* **2006**, *6*, 5. [CrossRef] [PubMed]
10. Brown, M.A.; Kenna, T.; Wordsworth, B.P. Genetics of ankylosing spondylitis—Insights into pathogenesis. *Nat. Rev. Rheumatol.* **2016**, *12*, 81–91. [CrossRef] [PubMed]
11. Ranganathan, V.; Gracey, E.; Brown, M.A.; Inman, R.D.; Haroon, N. Pathogenesis of ankylosing spondylitis—Recent advances and future directions. *Nat. Rev. Rheumatol.* **2017**, *13*, 359–367. [CrossRef] [PubMed]
12. Neumann-Haefelin, C. HLA-B27-mediated protection in HIV and hepatitis C virus infection and pathogenesis in spondyloarthritis: Two sides of the same coin? *Curr. Opin. Rheumatol.* **2013**, *25*, 426–433. [CrossRef] [PubMed]
13. Sorrentino, R.; Böckmann, R.A.; Fiorillo, M.T. HLA-B27 and antigen presentation: At the crossroads between immune defense and autoimmunity. *Mol. Immunol.* **2014**, *57*, 22–27. [CrossRef] [PubMed]
14. Vitulano, C.; Tedeschi, V.; Paladini, F.; Sorrentino, R.; Fiorillo, M.T. The interplay between HLA-B27 and ERAP1/ERAP2 aminopeptidases: From anti-viral protection to spondyloarthritis. *Clin. Exp. Immunol.* **2017**, *190*, 281–290. [CrossRef] [PubMed]
15. Burton, P.R.; Clayton, D.G.; Cardon, L.R.; Craddock, N.; Deloukas, P.; Duncanson, A.; Ward, M.M.; Learch, T.L.; Weisman, M.H.; Brown, M.; et al. Association scan of 14,500 nonsynonymous SNPs in four diseases identifies autoimmunity variants. *Nat. Genet.* **2007**, *39*, 1329–1337. [CrossRef] [PubMed]
16. Evans, D.M.; Spencer, C.C.; Pointon, J.J.; Su, Z.; Harvey, D.; Kochan, G.; Oppermann, U.; Dilthey, A.; Pirinen, M.; Stone, M.A.; et al. Interaction between ERAP1 and HLA-B27 in ankylosing spondylitis implicates peptide handling in the mechanism for HLA-B27 in disease susceptibility. *Nat. Genet.* **2011**, *43*, 761–767. [CrossRef] [PubMed]
17. Robinson, P.C.; Costello, M.E.; Leo, P.; Bradbury, L.A.; Hollis, K.; Cortes, A.; Lee, S.; Joo, K.B.; Shim, S.C.; Weisman, M.; et al. ERAP2 is associated with ankylosing spondylitis in HLA-B27-positive and HLA-B27-negative patients. *Ann. Rheum. Dis.* **2015**, *74*, 1627–1629. [CrossRef] [PubMed]
18. Marcilla, M.; López de Castro, J.A. Peptides: The cornerstone of HLA-B27 biology and pathogenetic role in spondyloarthritis. *Tissue Antigens* **2008**, *71*, 495–506. [CrossRef] [PubMed]
19. de Castro, J.A. HLA-B27-bound peptide repertoires: Their nature, origin and pathogenetic relevance. *Adv. Exp. Med. Biol.* **2009**, *649*, 196–209. [PubMed]
20. Infantes, S.; Lorente, E.; Barnea, E.; Beer, I.; Barriga, A.; Lasala, F.; Jiménez, M.; Admon, A.; López, D. Natural HLA-B*2705 protein ligands with glutamine as anchor motif: Implications for HLA-B27 association with spondyloarthropathy. *J. Biol. Chem.* **2013**, *288*, 10882–10889. [CrossRef] [PubMed]

21. Yair-Sabag, S.; Tedeschi, V.; Vitulano, C.; Barnea, E.; Glaser, F.; Melamed Kadosh, D.; Taurog, J.D.; Fiorillo, M.T.; Sorrentino, R.; Admon, A. The Peptide Repertoire of HLA-B27 may include Ligands with Lysine at P2 Anchor Position. *Proteomics* **2018**, *18*, e1700249. [CrossRef] [PubMed]

22. Urban, R.G.; Chicz, R.M.; Lane, W.S.; Strominger, J.L.; Rehm, A.; Kenter, M.J.; UytdeHaag, F.G.; Ploegh, H.; Uchanska-Ziegler, B.; Ziegler, A. A subset of HLA-B27 molecules contains peptides much longer than nonamers. *Proc. Natl. Acad. Sci. USA* **1994**, *91*, 1534–1538. [CrossRef] [PubMed]

23. Chen, L.; Fischer, R.; Peng, Y.; Reeves, E.; McHugh, K.; Ternette, N.; Hanke, T.; Dong, T.; Elliott, T.; Shastri, N.; et al. Critical role of endoplasmic reticulum aminopeptidase 1 in determining the length and sequence of peptides bound and presented by HLA-B27. *Arthritis Rheumatol.* **2014**, *66*, 284–294. [CrossRef] [PubMed]

24. Khan, M.A. An Update on the Genetic Polymorphism of HLA-B*27 With 213 Alleles Encompassing 160 Subtypes (and Still Counting). *Curr. Rheumatol. Rep.* **2017**, *19*, 9. [CrossRef] [PubMed]

25. D'Amato, M.; Fiorillo, M.T.; Carcassi, C.; Mathieu, A.; Zuccarelli, A.; Bitti, P.P.; Tosi, R.; Sorrentino, R. Relevance of residue 116 of HLA-B27 in determining susceptibility to ankylosing spondylitis. *Eur. J. Immunol.* **1995**, *25*, 3199–3201. [CrossRef] [PubMed]

26. Paladini, F.; Taccari, E.; Fiorillo, M.T.; Cauli, A.; Passiu, G.; Mathieu, A.; Punzi, L.; Lapadula, G.; Scarpa, R.; Sorrentino, R. Distribution of HLA-B27 subtypes in Sardinia and continental Italy and their association with spondylarthropathies. *Arthritis Rheum.* **2005**, *52*, 3319–3321. [CrossRef] [PubMed]

27. Paladini, F.; Fiorillo, M.T.; Tedeschi, V.; Cauli, A.; Mathieu, A.; Sorrentino, R. Ankylosing Spondylitis: A Trade Off of HLA-B27, ERAP, and Pathogen Interconnections? Focus on Sardinia. *Front. Immunol.* **2019**, *10*, 35. [CrossRef] [PubMed]

28. Fiorillo, M.T.; Sorrentino, R. T cell responses against viral and self-epitopes and HLA-B27 subtypes differently associated with Ankylosing Spondylitis. *Adv. Exp. Med. Biol.* **2009**, *649*, 255–262.

29. Fiorillo, M.T.; Meadows, L.; D'Amato, M.; Shabanowitz, J.; Hunt, D.F.; Appella, E.; Sorrentino, R. Susceptibility to ankylosing spondylitis correlates with the C-terminal residue of peptides presented by various HLA-B27 subtypes. *Eur. J. Immunol.* **1997**, *27*, 368–373. [CrossRef] [PubMed]

30. Uchanska-Ziegler, B.; Ziegler, A.; Schmieder, P. Structural and dynamic features of HLA-B27 subtypes. *Curr. Opin. Rheumatol.* **2013**, *25*, 411–418. [CrossRef] [PubMed]

31. Tedeschi, V.; Vitulano, C.; Cauli, A.; Paladini, F.; Piga, M.; Mathieu, A.; Sorrentino, R.; Fiorillo, M.T. The Ankylosing Spondylitis-associated HLA-B*2705 presents a B*0702-restricted EBV epitope and sustains the clonal amplification of cytotoxic T cells in patients. *Mol. Med.* **2016**, *22*, 215–223. [CrossRef] [PubMed]

32. Hill, A.; Worth, A.; Elliott, T.; Rowland-Jones, S.; Brooks, J.; Rickinson, A.; McMichael, A. Characterization of two Epstein-Barr virus epitopes restricted by HLA-B7. *Eur. J. Immunol.* **1995**, *25*, 18–24. [CrossRef] [PubMed]

33. Hulsmeyer, M.; Fiorillo, M.T.; Bettosini, F.; Sorrentino, R.; Saenger, W.; Ziegler, A.; Uchanska-Ziegler, B. Dual, HLA-B27 subtype-dependent conformation of a self-peptide. *J. Exp. Med.* **2004**, *199*, 271–281. [CrossRef] [PubMed]

34. Ko, J.; Park, H.; Heo, L.; Seok, C. GalaxyWEB server for protein structure prediction and refinement. *Nucleic Acids Res.* **2012**, *40*, W294–W297. [CrossRef] [PubMed]

35. Abraham, M.J.; Murtola, T.; Schulz, R.; Páll, S.; Smith, J.C.; Hess, B.; Lindahl, E. GROMACS: High performance molecular simulations through multi-level parallelism from laptops to supercomputers. *SoftwareX* **2015**, *1–2*, 19–25. [CrossRef]

36. Kaminski, G.A.; Friesner, R.A.; Tirado-Rives, J.; Jorgensen, W.L. Evaluation and Reparametrization of the OPLS-AA Force Field for Proteins via Comparison with Accurate Quantum Chemical Calculations on Peptides. *J. Phys. Chem. B* **2001**, *105*, 6474–6487. [CrossRef]

37. Pekka, M.; Lennart, N. Structure and Dynamics of the TIP3P, SPC, and SPC/E Water Models at 298 K. *J. Phys. Chem. A* **2001**, *105*, 9954–9960. [CrossRef]

38. Berendsen, H.J.C.; Postma, J.P.M.; van Gunsteren, W.F.; Di Nola, A.; Haak, J.R. Molecular dynamics with coupling to an external bath. *J. Chem. Phys.* **1984**, *81*, 3684–3690. [CrossRef]

39. Schlitter, J. Estimation of absolute and relative entropies of macromolecules using the covariance matrix. *Chem. Phys. Lett.* **1993**, *215*, 617–621. [CrossRef]

40. Amadei, A.; Linssen, A.B.; Berendsen, H.J.C. Essential dynamics of proteins. *Proteins* **1993**, *17*, 412–425. [CrossRef]

41. van Deutekom, H.W.; Keşmir, C. Zooming into the binding groove of HLA molecules: Which positions and which substitutions change peptide binding most? *Immunogenetics* **2015**, *67*, 425–436. [CrossRef] [PubMed]

42. Narzi, D.; Becker, C.M.; Fiorillo, M.T.; Uchanska-Ziegler, B.; Ziegler, A.; Böckmann, R.A. Dynamical characterization of two differentially disease associated MHC class I proteins in complex with viral and self-peptides. *J. Mol. Biol.* **2012**, *415*, 429–442. [CrossRef] [PubMed]

43. Fabian, H.; Huser, H.; Loll, B.; Ziegler, A.; Naumann, D.; Uchanska-Ziegler, B. HLA-B27 heavy chains distinguished by a micropolymorphism exhibit differential flexibility. *Arthritis Rheum.* **2010**, *62*, 978–987. [CrossRef] [PubMed]

44. Fabian, H.; Loll, B.; Huser, H.; Naumann, D.; Uchanska-Ziegler, B.; Ziegler, A. Influence of inflammation-related changes on conformational characteristics of HLA-B27 subtypes as detected by IR spectroscopy. *FEBS J.* **2011**, *278*, 1713–1727. [CrossRef] [PubMed]

45. Colbert, R.A.; Tran, T.M.; Layh-Schmitt, G. HLA-B27 misfolding and ankylosing spondylitis. *Mol. Immunol.* **2014**, *57*, 44–51. [CrossRef] [PubMed]

46. Fruci, D.; Romania, P.; D'Alicandro, V.; Locatelli, F. Endoplasmic reticulum aminopeptidase 1 function and its pathogenic role in regulating innate and adaptive immunity in cancer and major histocompatibility complex class I-associated autoimmune diseases. *Tissue Antigens* **2014**, *84*, 177–186. [CrossRef]

47. Paladini, F.; Fiorillo, M.T.; Vitulano, C.; Tedeschi, V.; Piga, M.; Cauli, A.; Mathieu, A.; Sorrentino, R. An allelic variant in the intergenic region between ERAP1 and ERAP2 correlates with an inverse expression of the two genes. *Sci. Rep.* **2018**, *8*, 10398. [CrossRef]

48. Fabian, H.; Huser, H.; Narzi, D.; Misselwitz, R.; Loll, B.; Ziegler, A.; Böckmann, R.A.; Uchanska-Ziegler, B.; Naumann, D. HLA-B27 subtypes differentially associated with disease exhibit conformational differences in solution. *J. Mol. Biol.* **2008**, *376*, 798–810. [CrossRef]

49. Kosmrlj, A.; Read, E.L.; Qi, Y.; Allen, T.M.; Altfeld, M.; Deeks, S.G.; Pereyra, F.; Carrington, M.; Walker, B.D.; Chakraborty, A.K. Effects of thymic selection of the T-cell repertoire on HLA class I-associated control of HIV infection. *Nature* **2010**, *465*, 350–354. [CrossRef]

50. Xiao, Z.; Ye, Z.; Tadwal, V.S.; Shen, M.; Ren, E.C. Dual non-contiguous peptide occupancy of HLA class I evoke antiviral human CD8 T cell response and form neo-epitopes with self-antigens. *Sci. Rep.* **2017**, *7*, 5072. [CrossRef]

51. Guasp, P.; Alvarez-Navarro, C.; Gomez-Molina, P.; Martín-Esteban, A.; Marcilla, M.; Barnea, E.; Admon, A.; López de Castro, J.A. The Peptidome of Behçet's Disease-Associated HLA-B*51:01 Includes Two Subpeptidomes Differentially Shaped by Endoplasmic Reticulum Aminopeptidase 1. *Arthritis Rheumatol.* **2016**, *68*, 505–515. [CrossRef] [PubMed]

Article

HLA-DQA1 and HLA-DQB1 Alleles, Conferring Susceptibility to Celiac Disease and Type 1 Diabetes, Are More Expressed Than Non-Predisposing Alleles and Are Coordinately Regulated

Federica Farina [1,†], Stefania Picascia [2,†], Laura Pisapia [1], Pasquale Barba [1], Serena Vitale [2], Adriana Franzese [3], Enza Mozzillo [3], Carmen Gianfrani [2,‡] and Giovanna Del Pozzo G [1,*,‡]

1   Institute of Genetics and Biophysics "Adriano Buzzati Traverso"-CNR, 80131 Naples, Italy
2   Institute of Biochemistry and Cell Biology-CNR, 80131 Naples, Italy
3   Department of Translational Medical Science (DISMET), Section of Pediatrics, University of Naples Federico II, 80131 Naples, Italy
*   Correspondence: giovanna.delpozzo@igb.cnr.it
†   Co-first authors.
‡   Co-last authors.

Received: 25 June 2019; Accepted: 16 July 2019; Published: 19 July 2019

Abstract: HLA DQA1*05 and DQB1*02 alleles encoding the DQ2.5 molecule and HLA DQA1*03 and DQB1*03 alleles encoding DQ8 molecules are strongly associated with celiac disease (CD) and type 1 diabetes (T1D), two common autoimmune diseases (AD). We previously demonstrated that DQ2.5 genes showed a higher expression with respect to non-CD associated alleles in heterozygous DQ2.5 positive (HLA DR1/DR3) antigen presenting cells (APC) of CD patients. This differential expression affected the level of the encoded DQ2.5 molecules on the APC surface and established the strength of gluten-specific CD4$^+$ T cells response. Here, we expanded the expression analysis of risk alleles in patients affected by T1D or by T1D and CD comorbidity. In agreement with previous findings, we found that DQ2.5 and DQ8 risk alleles are more expressed than non-associated alleles also in T1D patients and favor the self-antigen presentation. To investigate the mechanism causing the high expression of risk alleles, we focused on HLA DQA1*05 and DQB1*02 alleles and, by ectopic expression of a single mRNA, we modified the quantitative equilibrium among the two transcripts. After transfection of DR7/DR14 B-LCL with HLA-DQA1*05 cDNA, we observed an overexpression of the endogenous DQB1*02 allele. The DQ2.5 heterodimer synthesized was functional and able to present gluten antigens to cognate CD4$^+$ T cells. Our results indicated that the high expression of alpha and beta transcripts, encoding for the DQ2.5 heterodimeric molecules, was strictly coordinated by a mechanism acting at a transcriptional level. These findings suggested that, in addition to the predisposing HLA-DQ genotype, also the expression of risk alleles contributed to the establishment of autoimmunity.

Keywords: autoimmunity; risk genes; expression; regulation

---

1. Introduction

The human leukocyte antigen (HLA) class II heterodimeric molecules, composed by the alpha and beta chains, are encoded by many different alleles, which generate the high polymorphism characteristic of this locus. The genes encoding DR and DQ isotypes are the main risk factors associated with several autoimmune diseases. Type 1 diabetes (T1D) and celiac disease (CD) are autoimmune disorders, affecting between 0.5% and 1% of the general population. Very frequently, these two diseases co-occur in families, and approximately 4–9% of patients with T1D also have CD, while patients with

CD are at increased risk of developing T1D [1]. CD and T1D share immunopathogenic mechanisms, although the autoreactive T cells and autoantibodies are directed against different autoantigens, such as insulin, GADA65, and IA-2 in T1D [2] and tissue transglutaminase in CD [3]. The HLA-DRB1, HLA-DQA1, and HLA-DQB1 genes display a major component of familial clustering in both T1D and CD. Subjects at high risk to develop T1D carry either DQ2.5 haplotype, encoded by DQA1*05:01 and DQB1*02:01 or DQ8 haplotype, encoded by DQA1*03:01 and DQB1*03:02 [4]., whereas the most prominent association of CD is with HLA-DQ2.5 molecules [5]. The alleles encoding DQ2.5 molecule are in linkage disequilibrium (LD) with HLA-DRB1*03, while the DQ8 alleles are in LD with HLA-DRB1*04. More specifically, considering the subtype specificity, the DRB1*03:01 is found coupled to DQA1*05:01 and DQB1*02:01 almost exclusively, thus creating the haplotype called "DR3", while DRB1*04:01 is found in LD with DQA1*03:01 but can have either DQB1*03:01 or DQB1*03:02 included in the haplotype. Both are referred to as "DR4" haplotypes, although that carrying the DQB1*03:02 allele (DQ8 haplotype) predisposes to disease and that with DQB1*03:01 allele (DQ7 haplotype) is protective [4,5]. T1D risk depends not only on the haplotypic context but also on the genotypic assets of the risk alleles. In fact, the highest risk is conferred by DR3/DR4 heterozygous genotype. The contribution of DR3/DR4 is higher than the sum of the individual DR3 and DR4 haplotypes. This group comprised T1D subjects carrying DRB1*03:01-DQA1*05:01-DQB1*02 on one chromosome and DRB1*04:01-DQA1*03:01-DQB1*03:02 on the other chromosome. One hypothesis for the increased risk is the putative presence of DQ heterodimers encoded by alleles in *trans*, in addition to the DQ molecules encoded by alleles in *cis*, on the cell surface of immune cells [6,7].

We have previously demonstrated [8] that APCs from celiac patients, carrying the DR3-DQ2.5 haplotype in homozygosis (DR3/DR3) or in heterozygosis (DR1/DR3) have a comparable ability to stimulate the CD4$^+$ T cell activation and proliferation when challenged with an equivalent amount of gluten antigen. This finding is a consequence of the differential expression of DQA1*05 and DQB1*02 alleles compared to non-CD-associated ones, affecting the DQα1*05 and DQβ1*02 chain amount and the HLA-DQ2.5 surface heterodimers density. In the present work, we measured the expression of DQ2.5 and DQ8 risk genes in APC from patients affected by T1D and from patients affected by T1D and CD comorbidity, that could be relevant in the activation of antigen specific CD4 lymphocytes.

Previous papers from our lab have demonstrated the stoichiometric balance of two messengers encoding alpha and beta protein chains, by a mechanism coordinating the transcription and the mRNA processing. A ribonucleoprotein complex, binding the 5′UTR and 3′UTR of HLA class II transcripts, regulates the nucleus-cytoplasm export and degradation of these mRNA [9,10]. Moreover, it has been demonstrated that the ectopic over-expression of the DRB1 gene determines an increase of the endogenous DRA mRNA in non-professional APC, such as melanoma M14 cell line. In order to investigate the mechanism determining the high and the coordinated expression of DQA1*05 and DQB1*02 risk alleles, we have set up experiments aimed to perturb the stoichiometric balance of two mRNA and to assess the functionality of resulting heterodimers in the antigen presentation.

2. Materials and Methods

2.1. Patients Enrollment and Selection of Antigen Presenting Cells

Type 1 diabetes patients were enrolled at the Department of Pediatrics, University Hospital "Federico II" of Naples, where they are regularly followed up. Patients or their parents, in the case of children under 12 years old, were informed about the objective of the study and provided written informed consent in accordance with the ethical standards of the institutional, the national research committee, and in accordance with the 1964 Helsinki declaration and its later ethical amendments.

Peripheral blood mononuclear cells (PBMC) were isolated from blood samples of patients and healthy donors by Ficoll-Paque gradient separation. PBMC were used to prepare genomic DNA and total RNA. B lymphoblastoid cell lines (B-LCL), used as APCs, were obtained, as previously reported. All subjects enrolled were genotyped for DQA1 and DQB1 by using AllSet Gold SSP

HLA-DQ Low Res kit (Thermo Fisher Scientific, Monza, Italy) and the genotypes are indicated in Table 1. In the functional test with T cells and in the nucleofection experiments we used DR1/DR3 positive B-LCL (B-LCL#5) from CD patient carrying DQA1*05-DQA1*01/DQB1*02-DQB1*05 genotype that was previously described [8] and DR7/DR14 positive B-LCL from a healthy donor carrying DQA1*02-DQA1*01/DQB1*02-DQB1*05 genotype.

Table 1. HLA-DQ and HLA-DR genotype and phenotype of T1D patients enrolled in the study.

Code	Cells [a]	Diagnosis	DQA1 Genotype	DQB1 Genotype	DQ Phenotype	DR Phenotype
§1	B-LCL	T1D	*03/*05	*02/*03	DQ2.5/DQ8	DR3/DR4
§2	B-LCL	T1D	*03/*05	*02/*02	DQ2.5/DQ2.3	DR3/DR9
§3	B-LCL	T1D	*05/*05	*02/*02	DQ2.5/DQ2.5	DR3/DR3
§4	B-LCL	T1D with CD	*03/*05	*02/*03	DQ2.5/DQ8	DR3/DR4
§5	B-LCL	T1D with CD	*01/*03	*05/*03	DQ8/DQ5	DR4/DR16
§6	B-LCL	T1D with CD	*01/*05	*02/*05	DQ2.5/DQ5	DR1/DR3
§7	B-LCL	T1D with CD	*02/*05	*02/*02	DQ2.5/DQ2.2	DR3/DR7
§8	B-LCL	T1D with CD	*05/*05	*02/*02	DQ2.5/DQ2.5	DR3/DR3
§9	PBMC	T1D	*01/*05	*02/*05	DQ2.5/DQ5	DR1/DR3
§10	PBMC	T1D	*01/*05	*02/*06	DQ2.5/DQ6	DR3/DR15
§11	PBMC	T1D	*01/*05	*02/*05	DQ2.5/DQ5	DR1/DR3
§12	PBMC	T1D	*01/*03	*03/*05	DQ8/DQ5	DR4/DR1
§13	PBMC	T1D	*03/*05	*02/*03	DQ2.5/DQ8	DR3/DR4
§14	PBMC	T1D	*03/*05	*02/*03	DQ2.5/DQ8	DR3/DR4
§15	PBMC	T1D	*01/*03	*03/*05	DQ8/DQ5	DR4/DR1
§16	PBMC	T1D	*05/*05	*02/*03	DQ2.5/DQ7	DR3/DR5
§17	PBMC	T1D	*01/*05	*02/*05	DQ2.5/DQ5	DR3/DR1
§18	PBMC	T1D	*03/*03	*03/*03	DQ8/DQ8	DR4/DR4
§19	PBMC	T1D	*01/*03	*03/*05	DQ8/DQ5	DR4/DR1
§20	PBMC	T1D	*03/*03	*03/*03	DQ8/DQ8	DR4/DR4
§21	PBMC	T1D	*01/*03	*03/*05	DQ8/DQ5	DR4/DR1

[a] the immune cell source for HLA genotyping.

2.2. Monoclonal Antibodies and Flow Cytometry Analysis

B-LCLs were harvested when they were at sub-confluence and were suspended at 10^6 cells/mL in ice-cold PBS, 10% FCS and 1% NaN_3. Then, 100 µl of cell suspension was plated in a 96 V-bottom plate and labeled with 10 µg/mL of primary or isotypes control monoclonal antibodies, or with 10µl of the hybridoma supernatant, previously titrated. The cells were incubated at 4 °C in the dark for 30 min, washed and thereafter labeled with secondary antibodies at a final concentration of 1 µg/mL in 3% BSA/PBS for an additional 30 min at 4 °C in the dark. The primary monoclonal antibodies used to reveal the cell surface HLA DQ expression were: SFR20-DQa5, a rat anti-HLADQA1*05 purified from a hybridoma supernatant, kindly provided by Prof Radka [11]; 2.12E11, a murine anti-HLA-DQB1*02, kindly provided by Prof L. Sollid [12]. Fluorochrome conjugated anti-rat IgG(-PE) and anti-mouse IgG (FITC) were used as secondary antibodies. All phenotypes were analyzed with FACSCanto II system and elaborated using the DIVA software (BD Biosciences, Milan, Italy).

2.3. cDNA Cloning and B-LCL Nucleofection

The full-length DQA1*05 and DQB1*02 cDNA were prepared using retrotranscribed RNA from B-LCL of homozygous DR3/DR3 positive CD patient (B-LCL#1), previously described [8]. By specific primers, we cloned both cDNA in pcDNA3 vector and named the constructs pDQA105 and pDQB102. Using pDQA105 as a template, we cloned the cDNA deleted of 3′UTR (pDQA105Δ). The correct cloning was assessed by sequencing (Eurofins, Munchen, Germany). We transfected DR1/DR3 B-LCL#5 with three constructs and DR7/DR14 B-LCL with pDQA105 by nucleofection using Amaxa Cell line Nucleofector kit V (Lonza, Euroclone, Milan Italy). Cells were harvested after 24 h for the RNA preparation and functional test and after 48 h for flow cytometry experiments.

2.4. RNA Quantization

Total RNA was prepared with the Aurum™ Total RNA kit (BIORAD, Milan, Italy), and 0.5 μg of RNA was used for reverse transcriptase reactions, performed using an iScript™ cDNA Synthesis kit (BIORAD). The number of specific transcripts was measured by qPCR using the Quanti Tect SYBR Green PCR Kit (BIORAD) through the DNA Engine Opticon Real-Time PCR Detection System (BIORAD). Each reaction was run in triplicates in the presence of 0.2 mM primers synthesized by Eurofins, and each experiment was performed four times. The primer sequences are reported in Table 2. The relative amount of specific transcripts was calculated by the comparative cycle threshold method [13] and β-actin transcript was used for normalization.

Table 2. Primers used for qPCR.

Gene	Primers	Sequences 5′ → 3′
β-Actin	ACT-F ACT-R	TCATGAAGTGTGACGTTGACA CCTAGAAGCATTTGCGGTGCAC
HLA-DQA1*01	DQA1*01-F DQA1*R	CGGTGGCCTGAGTTCAGCAA GGAGACTTGGAAAACACTGTGACC
HLA-DQA1*02	DQA1*02-F DQA1*R	AAGTTGCCTCTGTTCCACAGAC GGAGACTTGGAAAACACTGTGACC
HLA-DQA1*03	DQA1*03-F DQA1*R	CTCTGTTCCGCAGATTTAGAAGA GGAGACTTGGAAAACACTGTGACC
HLA-DQA1*05	DQA1*05-F DQA1*R	CTCTGTTCCGCAGATTTAGAAGA GGAGACTTGGAAAACACTGTGACC
HLA-DQB1*02	DQB1*02-F DQB1*R	TCTTGTGAGCAGAAGCATCT CAGGATCTGGAAGGTCCAGT
HLA-DQB1*03	DQB1*03-F DQB1*R	CGGAGTTGGACACGGTGTGC CAGGATCTGGAAGGTCCAGT
HLA-DQB1*05	DQB1*05 DQB1*R	ACAACTACGAGGTGGCGTACC CAGGATCTGGAAGGTCCAGT
HLA-DQB1*06	DQB1*06-F DQB1*R	CAGATCAAAGTCCGGTGGTTTC CAGGATCTGGAAGGTCCAGT

The newly synthesized RNA transcripts were captured by Click-iT Nascent RNA Capture Kit (ThermoFisher), according to manufacturer's instructions. Briefly, B-LCLs, seeded at 50% confluency, were labeled with 0.2 mM ethynyl uridine (EU) and incubated at 37 °C for 16 h. Total RNA was prepared with TRIzol reagent (Life Technologies, Thermo Fisher Scientific, Monza, Italy). The EU-labeled RNAs were biotinylated with 0.5 mM biotin azide in Click-iT reaction buffer (Thermo Fisher Scientific, Monza, Italy). The biotinylated RNAs were precipitated and resuspended in distilled water. Purified RNA (0.5 μg) was bound to 25 μL of Dynabeads MyOne Streptavidin T1 magnetic beads in Click-iT RNA binding buffer. The RNA captured on the beads was used as template for cDNA synthesis. Reverse transcription was performed using the SuperScript VILO cDNA Synthesis Kit (Invitrogen, Thermo Fisher Scientific, Monza, Italy) following the manufacturer's instructions. The number of specific transcripts was measured by qPCR using SsoAdvanced SYBR Green PCR Kit (BIORAD). The apparatus and primers used were the same as described above.

2.5. T Cell Functional Assay

A gliadin-reactive T cell line (TCL) was previously established from the jejunal biopsies of a DR3/DR3 homozygous CD patient in disease remission. The TCL was obtained by stimulating intestinal cells with irradiated autologous PBMC (1.5×10^6) and gliadin peptide and expanded by cyclic stimulations with allogenic PBMC and phytohemagglutinin (PHA, 0.5 μg/mL) in complete medium (X-Vivo 15 medium supplemented with 5% AB-pooled human serum and antibiotics, Lonza). Cells were fed by adding IL-2 (50 UI/mL, R&D System) every 3 days. This TCL was reactive to DQ2.5-glia-γ1 peptide (PQQPQQSFPQQQQPA) [14]. When in resting phase, T cells (3×10^4) were co-incubated with indicated B-LCL (1×10^5) pulsed with DQ2.5-glia-γ1 peptide at concentration 0.1 and 1 μM. In some experiments B-LCL transfected with a pDQA105 construct or with pcDNA (empty vector) were used. Cell supernatants (50 μL) were collected after 48 h for the evaluation of INF-γ production, using standard sandwich ELISA procedure. The DQ2.5-glia-γ1 peptide was provided by CASLO ApS, (Kongens Lyngby, Denmark).

2.6. Statistical Analysis

All results are shown as the mean of at least three independent experiments. Statistical analysis was performed using the unpaired Student's t-test with a two-tailed distribution. A p-value less than 0.05 was considered significant.

3. Results

3.1. The DQA1*03 and DQB1*03 mRNA Associated with T1D Risk Were More Abundant than the mRNA of Non-T1D Related Alleles

We genotyped the DNA of B-LCL and PBMC from patients affected by T1D alone or by T1D and CD comorbidity in order to identify the HLA-DQ genes (Table 1). The HLA-DR phenotype was designed by LD. We measured, by qPCR, the amount of DQA1*03, DQA1*05, DQB1*03, and DQB1*02 mRNA with respect to transcripts of alleles not associated with diseases. A value of 100% of mRNA amount was assigned when DQA1*05 (patients §3, §8), DQA1*03 (patients §18 and §20), DQB1*02 (patients §2, §3, §7, §8) and DQB1*03 (patients §18, §20) alleles were in homozygosis. For patients DQA1 and DQB1 heterozygous, the mRNA amount was expressed as a percentage of the total messenger of the gene.

Our results demonstrated that in heterozygous B-LCL, the amount of DQA1*05 (patients §6, §7) and DQA1*03 (patient §5) mRNA was higher respect to DQA1*01 (patients §5, §6) and DQA1*02 (patient §7) mRNA, as showed in Figure 1A. Similar results were obtained in heterozygous PBMC, in which either DQA1*05 (patients §9, §10, §11, §17) and DQA1*03 (patients §12, §15, §19, §21) mRNA are more abundant than DQA1*01 transcript (Figure 1C). Moreover, either B-LCL (patients §1, §4, Figure 1A) and PBMC (patients §13, §14, Figure 1C) carrying DQ2.5/DQ8 genotypes showed a comparable amount of DQA1*03 and DQA1*05 mRNA, with exception of B-LCL §2 in which DQA1*05 was more abundant than DQA1*03 (Figure 1A). Similarly, we demonstrated that DQB1 risk alleles were more expressed with respect to the alleles non-T1D associated. The heterozygous B-LCL of patients §5 and §6, carrying DQB1*03 or DQB1*02, respectively, showed mRNA amount higher than DQB1*05 (Figure 1B), as well as the heterozygous PBMC, in which DQB1*02 allele (patients §9, §10, §11, §17) displayed a quantity of transcript greater than non-predisposing alleles (DQB1*05 and DQB1*06, Figure 1D). Finally, PBMC §12, §15, §19, and §21 (Figure 1D) expressed a higher level of DQB1*03 mRNA with respect to DQB1*05 mRNA, with the exception of PBMC §16 (Figure 1D) that showed a lower amount of DQB1*03 mRNA, in comparison to DQB1*02. Finally, the B-LCL of patients §1 and §4 (Figure 1B) and PBMC of patients §13 and §14, characterized by DQ2.5/DQ8 genotypes, showed a comparable amount of DQB1*03 and DQB1*02 mRNA (Figure 1D). We calculated the significance among the expression values of two alleles for each patient and for the group carrying the same alleles (Figure 1). The fold change (FC) of the expression value for each patient were more clearly reported in Table 3. Overall, these results demonstrated that disease predisposing alleles are, surprisingly, expressed at higher levels than non-associated alleles in all heterozygous APC. When DQA1*05 and DQB1*03 risk alleles, as well as DQB1*02 and DQB1*03, are in the same DQ2.5/DQ8 genotype, the cells expressed equal amounts of mRNA in all determinations done.

Table 3. DQA1 and DQB1 mRNA fold change in different APC.

Cells	Patients	mRNA Alleles	Fold Change	Cells	Patients	mRNA Alleles	Fold Change
B-LCL	§1 §4	DQA1*05 = DQA1*03	1	B-LCL	§5	DQB1*03 > DQB1*05	3.2
B-LCL	§2	DQA1*05 > DQA1*03	2.7	B-LCL	§6	DQB1*02 > DQB1*05	4.6
B-LCL	§5	DQA1*03 > DQA1*01	3.2	B-LCL	§1 §4	DQB1*02 = DQB1*03	1
B-LCL	§6	DQA1*05 > DQA1*01	3.0	PBMC	§9 §11 §17	DQB1*02 > DQB1*05	5.6
B-LCL	§7	DQA1*05 > DQA1*02	2.7	PBMC	§10	DQB1*02 > DQB1*06	5.9
PBMC	§9 §10 §11 §17	DQA1*05 > DQA1*01	4.8	PBMC	§19 §21 §12 §15	DQB1*03 > DQB1*05	3.1
PBMC	§12 §15 §19 §21	DQA1*03 > DQA1*01	2.9	PBMC	§13 §14	DQB1*02 = DQB1*03	1
PBMC	§13 §14	DQA1*05 = DQA1*03	1	PBMC	§16	DQB1*02 > DQB1*03	3.3

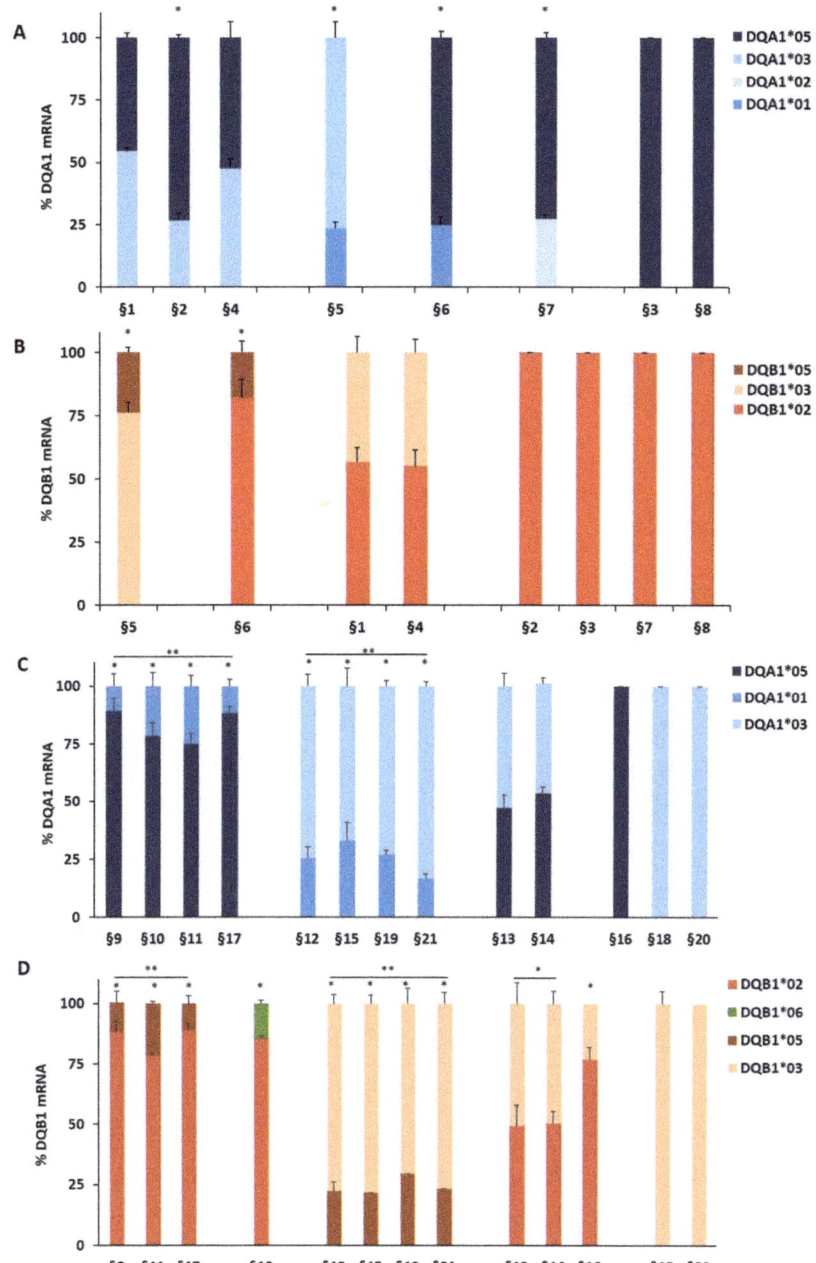

Figure 1. Expression of DQB1 and DQA1 genes. The gene expression is showed as percentages of the total DQA1 and DQB1 transcripts for each APC. Panel **A** and **C** show the expression of DQA1 alleles in B-LCL and PBMC, respectively; panel **B** and **D** show the expression of DQB1 alleles in B-LCL and PBMC, respectively. The APC are grouped by their genotypes; the significance, among the expression values of two alleles, was calculated for each patient and for the group with at least two samples (* $p < 0.05$, ** $p < 0.005$).

3.2. Analysis of Co-Regulated Expression of DQA1*05 and DQB1*02 mRNA

We investigated on the mechanism causing the differential expression of HLA genes associated with T1D and CD risk by nucleofection of the full-length DQA1*05 or DQB1*02 cDNA (pDQA105 and pDQB102) and the DQA1*05 deleted of 3'UTR (pDQA105Δ) cDNA in DR1/DR3 B-LCL#5. The aim was to verify if the ectopic expression of each alpha or beta mRNA affected the modulation of the other mRNA. 48 h after nucleofection with pDQA105 we observed an increased surface expression of DQα1*05 chain (Figure 2A), as expected, and an unexpected increment of DQ1β*02 chain (2,1 fold change of MFI) expressed by endogenous gene (Figure 2B). Similarly, when we transfected the B-LCL#5 with pDQB102, we observed a 2.3-fold increase of DQ1β*02 MFI (Figure 2B) and, 1.7-fold increment of DQα1*05 chain (Figure 2A). No variation in the MFI of DQα1*05 and DQ1β*02 was observed when we transfected pDQA105Δ (Figure 2A,B).

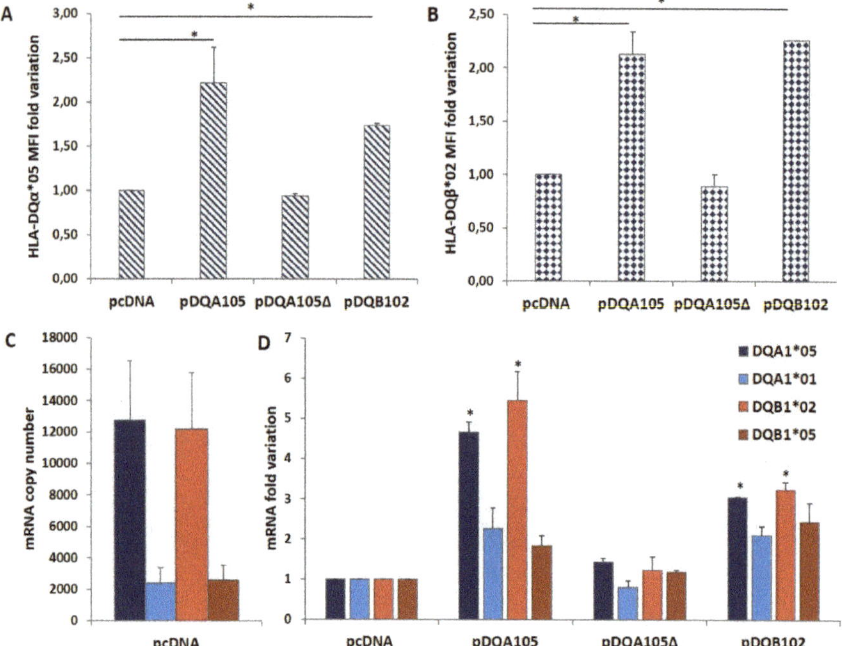

Figure 2. Coregulated expression of DQA1*05 and DQB1*02 CD risk alleles in DR1/DR3 B-LCL#5. This cell line was transfected with pcDNA, pDQA105, pDQA105Δ, and pDQB102 constructs. The surface expression of DQα1*05 (panel **A**) and DQ1β*02 (panel **B**) was measured by flow cytometry and reported as mean fluorescence intensity (MFI). The amount of DQA1*05 and DQB1*02 mRNA related to diseases, as well as the DQA1*01 and DQB1*05 mRNA, non-associated to pathologies, was reported as copy number in panel **C** and as fold variation in panel **D**. The *p*-value was calculated respect to cells transfected with empty vectors (* $p < 0.05$).

To assess if the overexpression of surface molecules corresponded to the increment of transcripts, we performed an absolute quantification of DQA1 and DQB1 mRNA by qPCR, using allele-specific primers (Table 2) at 48 h after nucleofection. In Figure 2C, we showed the copy number of transcripts expressed by non-transfected DR1/DR3 B-LCL#5. As already reported in [8], at a steady-state, the amount of endogenous DQA1*05 and DQB1*02 CD-associated transcripts was 5-fold higher, in comparison to the amount of non-CD associated DQA1*01 and DQB1*05 mRNA. Following the nucleofection of DR1/DR3 B-LCL#5 with pDQA105, we observed a 4.7-fold increase of DQA1*05 mRNA, including ectopic and endogenous transcripts, and 5.5-fold increase of DQB1*02 endogenous mRNA (Figure 2D). A slight increase has been quantified also for non–T1D associated transcripts, DQA1*01 (2,3 fold) and DQB1*05 (1,8 fold) mRNA, since they are not significant (Figure 2D). When we transfected pDQA105Δ, the cDNA deleted of 3'UTR, we did not observe any mRNA increase. Analogously, following the transfection of pDQB102, we observed an approximate 3-fold increase in DQA1*05 mRNA and endogenous DQB1*02 mRNA (Figure 2D), while the other transcripts showed only a low increase.

To confirm this phenomenon, we repeated the same transfection in DR7/DR14 B-LCL lacking the DQA1*05 allele. The ectopic expression of the pDQA105 construct induced the synthesis of DQα1*05 surface molecule (Figure 3A) and DQA1*05 mRNA (Figure 3B). Notably, we appreciated a 19.5-fold increase in the endogenous DQB1*02 mRNA (Figure 3B). Although in a reduced amount, the DQA1*01, DQA1*02, and DQB1*05 transcripts increased (5-, 4.9-, and 7.4-fold, respectively, Figure 3B). These results demonstrated that the ectopic expression of a full-length DQA1*05 cDNA induced a marked rise of DQB1*02 mRNA expression.

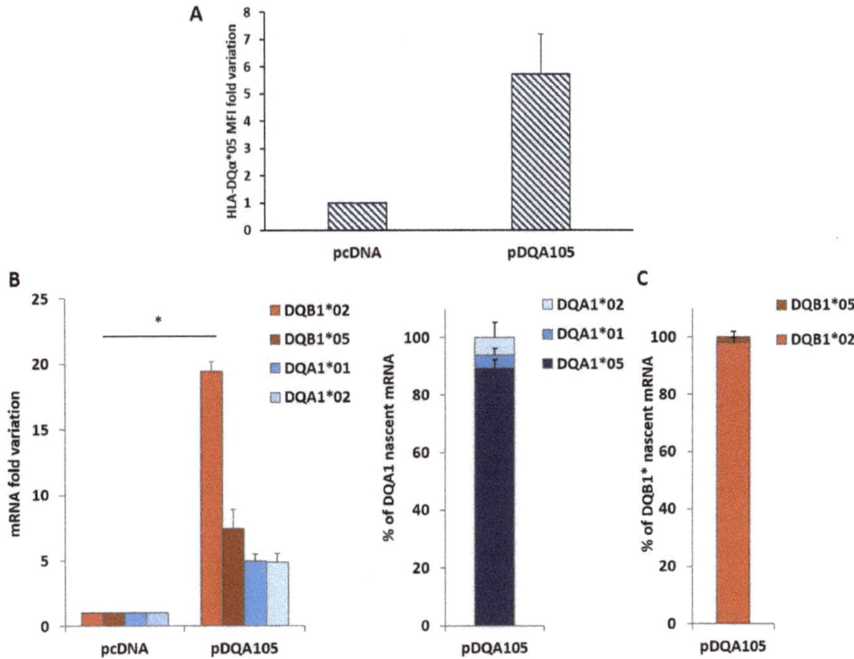

Figure 3. Coregulated expression of DQA1*05 and DQB1*02 CD risk alleles in DR7/DR14 B-LCL and nascent RNA. This cell line was transfected with pcDNA and pDQA105 constructs. In panel **A** we show the MFI assessed by flow cytometry. Panel **B** reports the fold variation of endogenous DQA1*01, DQA1*02, DQB1*02 and DQB1*05 mRNA. P value was calculated respect to cells transfected with empty vector (* $p < 0.05$). Panel **C** shows the quantity of nascent endogenous mRNA following pDQA105 transfection. The left histogram shows the ectopic DQA1*05 mRNA in addition to the endogenous DQA1*01 and DQA1*02 mRNA, as percentage of total DQA1 transcript. The right histogram reported the amount of nascent endogenous DQB1*02 and DQB1*05 mRNA as percentage of total DQB1 transcript.

We investigated if the overexpression of endogenous mRNA, following the nucleofection is determined by *de novo* transcription. We treated DR7/DR14 B-LCL with the 5-ethynyluridine (EU) that is incorporated into the nascent RNA. New synthetized EU-labeled RNA, is quantified by cDNA preparation and qPCR. The results are reported in Figure 3C as the percentage of each transcript on the 100% total mRNA. We observed that the high expression of DQA1*05 mRNA with respect to DQA1*01 and DQA1*02 is mainly represented by newly synthetized mRNA (left histogram) and this transcript induced *de novo* transcription of endogenous DQB1*02 (Figure 3C, right histogram). The amount of DQA1*01, DQA1*02, and DQB1*05 mRNA was unaffected.

3.3. DR7/DR14 B-LCLs Transfected with pDQA105 Efficiently Stimulated Gluten-Specific CD4$^+$ T Cells

We investigated whether the new synthesized DQ2.5 heterodimer, expressed following APC transfection, was able to stimulate a CD4$^+$ T-cell line, previously established from the intestinal biopsies of a homozygous DR3/DR3 CD patient [14]. We stimulated T cells with autologous DR3/DR3 B-LCL#1 or with DR1/DR3 B-LCL#5 transfected with pcDNA or pDQA105 constructs, pulsed with γ-peptide. After 48 h, we measured the INF-γ production on supernatant by ELISA. We found that CD4 T cells, in response to DR1/DR3 B-LCL#5 transfected with pDQA105, produced a similar amount of INF-γ (2340 pg/mL), with respect to cells transfected by pcDNA and compared to autologous DR3/DR3 B-LCL#1 (mean INF-γ: 2125 pg/mL and 1864 pg/mL, respectively, Figure 4A). Overall, these results demonstrated that heterozygous APC, overexpressing full-length DQA1*05 cDNA, retained the same ability to present gluten peptides and stimulate a T-cell response than homozygous APC. We next evaluated the antigen presenting properties of DR7/DR14 B-LCL after transfection with DQA1*05 gene lacking in that cell line and we assessed the activation of the CD4$^+$ T cells specific for DQ2.5-γ-1 peptide at 0.1 and 1 μM suboptimal concentrations. We found that the INF-γ induced by DR7/DR14 B-LCL transfected with pDQA105 (2939.7 pg/mL at 0.1, and 3859.3 pg/mL at 1 μM) are not dissimilar to those produced in response to homozygous DR3/DR3 B-LCL #1 (2718.3 pg/mL at 0.1 and 4063.5 pg/mL at 1 μM) at both concentrations (Figure 4B). Similar results have been obtained when T cells were stimulated with DR1/DR3 B-LCL#1 (3089.5 pg/mL at 0.1 μM, and 4482,75 pg/mL at 1 μM). Conversely, a low cytokine production has been detected after stimulation of T cells with DR7/DR14 B-LCL transfected with empty vector (1556.9 pg/mL at 0.1 μM and 3233.98 pg/mL 1 μM), since the DQ2.2 molecule, expressed by this APC, is able to present the gliadin antigenic peptide with low affinity [14,15]. Overall, these data demonstrated that APC lacking the DQ2.5 molecule, presented antigenic gliadin peptide and became capable of stimulating the gliadin-specific CD4 T cells if an ectopic expression of DQA1*05 alleles is induced.

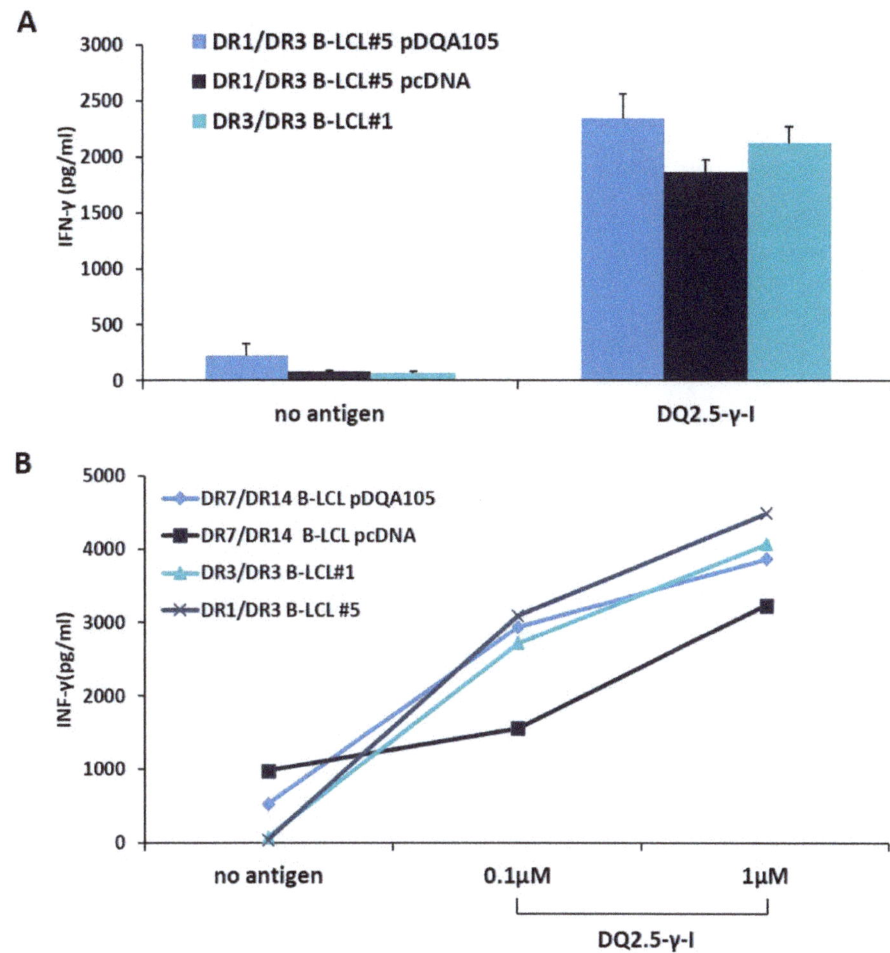

Figure 4. The antigen presenting properties of B-LCLs transfected with pDQA105. The antigen presenting capability of B-LCL pulsed with immunodominant DQ2.5-γ1 gliadin peptide were measured by assessing the activation of intestinal CD4$^+$ T cell line. Panel **A** shows the IFN-γ production by CD4 stimulated by DR3/DR3 B-LCL#1 and DR1/DR3 B-LCL#5 transfected with pcDNA or pDQA105 constructs. Panel **B** shows the IFN-γ production by CD4 stimulated with DR3/DR3 B-LCL#1, DR1/DR3 B-LCL#5, and DR7/DR14 B-LCL transfected with pcDNA or pDQA105 constructs.

4. Discussion

Although the autoimmune diseases, such as celiac disease and type 1 diabetes, are polygenic disorders, the main genetic risk factor is represented by HLA class II genes. The encoded HLA molecules expressed on the surface of APC have a key role in presenting the self-antigens to inflammatory CD4$^+$ T lymphocytes. A number of experimental evidences demonstrated that the expression of these molecules, irrespective of the nature of the antigen, influences the achievement of the HLA-peptide complex threshold, needed for activation and proliferation of autoreactive CD4$^+$ T cells [16,17]. We have previously demonstrated that DQA1*05 and DQB1*02 alleles, predisposing to CD, when carried by heterozygous APC, are differentially expressed. This marked expression of CD associated alleles in the APC of heterozygous patients resulted in the production of high concentration of DQ2.5 molecules, comparable to those produced in DQ2.5 homozygous cells. As a consequence, the strength of the gliadin-specific CD4$^+$ T cells response is mainly dependent on the antigen dose [8]. In the present work, we investigated HLA expression in B-LCL and PBMC from patients affected by T1D and in patients with co-occurrence of T1D and CD. Most heterozygous subjects carried DQA1*03 and DQB1*03 alleles encoding DQ8 molecules, a smaller group presented DQA1*05 and DQB1*02 alleles translating in DQ2.5 heterodimer and the third group displayed all four alleles encoding DQ2.5/DQ8 molecules (Table 1). The results of mRNA quantization confirmed in T1D the phenomenon already demonstrated in CD. Either DQA1*03 and DQA1*05 risk alleles show an expression higher than 50% when in heterozygous with another DQA1 allele non-associated to T1D while, if the two risk alleles are present in the same genotype, we observed a comparable expression. The high expression was revealed also for DQB1*03 allele in LD with DQA1*03 and for DQB1*02 in LD with DQA1*05, in terms of relative percentage.

Notably, in DQ2.5/DQ8 positive APC that carry the two risk haplotypes, we found a comparable amount of DQA1*03 and DQA1*05, as well as that of DQB1*03 and DQB1*02 mRNA. These findings indicate that among the risk alleles there is not an expression hierarchy and their protein products are expressed at same level. The consistency of our observation is fulfilled when DQ2.5 haplotype is not associated to DQ8, as we found that DQA1*05 is more abundant than DQA1*03, and DQB1*02 is more abundant than DQB1*03. In conclusion, our findings suggested that the allele-specific transcripts have to reach a sort of "threshold level" to mediate the self-antigen presentation, and that a precise hierarchy of expression exists between the DQ alleles predisposing, or not, to autoimmunity. In addition, we found that all T1D risk alleles are regulated accordingly to a mechanism responsible for increased gene expression, that guarantees a coordinated mRNA synthesis of DQA1 and DQB1 genes mapped on the same chromosome.

To investigate this mechanism, we used a known T cells model system that allows us to explore also the functional aspect. Following transfection of DQA1*05 cDNA in a B-LCL expressing the DQ2.5 molecule, we observed an increased endogenous amount of DQB1*02 mRNA, reaching the same quantity of DQA1*05 transcripts. Differently, by transfecting pDQA105Δ we did not observe variations, confirming the important role of 3′UTR in the modulation of transcript stability [10]. Following transfection of pDQA105 in the B-LCL not carrying DQA1*05 allele, we observed a rise in the expression of the endogenous DQB1*02 allele (Figure 2). In conclusion, we demonstrated that the expression of the two genes encoding the surface molecule DQ2.5 is coordinated. To investigate whether the DQB1*02 synthesis was due to the de novo transcription, we monitored and quantified the incorporation of EU, the uridine analog labeling the nascent RNA (Figure 3B) and demonstrated that the stoichiometric balance between the two transcripts of risk alleles in LD is determined by a coordinated transcription. When we modified their equilibrium by ectopic DQA1*05 expression, we stimulated the transcription of endogenous DQB1*02 mRNA.

To explain the mRNA coordinate transcription we suggested that the chromatin, through the formation of loop structures, arranged the promoter of alpha allele close to the promoter of beta allele ensuring the coordination of transcription [18]. Moreover, in our previous papers, we demonstrated that the balance of two transcripts was determined by a coordination not only of the transcription but also of the RNA processing. We proposed that UTRs of alpha and beta transcripts are associated, just after their synthesis, with a protein complex that guarantees a coordinate export from the nucleus and a coordinate turnover [10].

Finally, the antigen presentation experiments confirmed that the heterozygous DR1/DR3 B cells retains the same ability to present the gliadin antigen than homozygous DR3/DR3 B cells and that the overexpression of DQA1*05 did not further increase the density of DQ2.5 heterodimer. This result clearly indicated that DQ2.5 expression reaches a "plateau" and the immune response is dependent on the dose of antigen. However in DR7/DR14 B cells expressing DQ2.2 heterodimer the ectopic expression of DQA1*05 gene induces the new synthesis of DQ2.5 heterodimer, which is able to present gluten antigen and to stimulate gliadin-specific CD4$^+$ T cells.

In conclusion, we demonstrated that the two alleles in LD, DQA1*05, and DQB1*02, encoding the DQ2.5 molecule, and DQA1*03 and DQB1*03, encoding the DQ8 molecule, showed a balanced expression. A coordinated transcription is probably ensured by the chromatin structure while the binding of the protein complex with the UTRs regulated their balanced turnover [8].

Our overall findings demonstrated that, in addition to predisposing genotype, the high expression of risk alleles is relevant in the establishment of autoimmunity. In fact, the magnitude of the T cell activation and proliferation is related not only to the nature of epitopes but also to the amount of the antigen-HLA class II complexes expressed on the surface of APC.

Author Contributions: Methodology, F.F. and S.P.; formal analysis, L.P.; software, P.B; data curation, S.V.; resources, A.F and E.M.; supervision, funding acquisition, writing, review and editing, C.G. and G.D.P.G.

Funding: This work was supported by CNR-DSB Progetto Bandiera "InterOmics" 2017 and by PO FESR 2014-2020 "SATIN" from Regione Campania.

Acknowledgments: We acknowledge the Italian Foundation of Celiac Disease for having supported the research cited in the reference Pisapia et al, 2016, by Grant N. FC 014/2014. The IGB-IBBC FACS facility is also acknowledged.

Conflicts of Interest: The authors have no conflicts of interest to declare.

References

1. Romanos, J.; Bakker, S.F.; Onengut-gumuscu, S.; Simsek, S.; Rewers, M.; Mulder, C.J.; Liu, E.; Rich, S.S. Contrasting the Genetic Background of Type 1 Diabetes and Celiac Disease Autoimmunity. *Diab. Care* **2015**, *38*, 37–44.

2. Van Lummel, M.; Duinkerken, G.; Van Veelen, P.A.; De Ru, A.; Cordfunke, R.; Zaldumbide, A.; Gomez-Touriño, I.; Arif, S.; Peakman, M.; Drijfhout, J.W.; et al. Posttranslational modification of HLA-DQ binding islet autoantigens in type 1 diabetes. *Diabetes* **2014**, *63*, 237–247. [CrossRef] [PubMed]

3. Abadie, V.; Sollid, L.M.; Barreiro, L.B.; Jabri, B. Integration of Genetic and Immunological Insights into a Model of Celiac Disease Pathogenesis. *Annu. Rev. Immunol.* **2011**, *29*, 493–525. [CrossRef] [PubMed]

4. Noble, J.A.; Valdes, A.M. Genetics of the HLA region in the prediction of type 1 diabetes. *Curr. Diab. Rep.* **2011**, *11*, 533–542. [CrossRef] [PubMed]

5. Meresse, B.; Malamut, G.; Cerf-Bensussan, N. Celiac Disease: An Immunological Jigsaw. *Immunity* **2012**, *36*, 907–919. [CrossRef] [PubMed]

6. Erlich, H.; Valdes, A.M.; Noble, J.; Carlson, J.A.; Varney, M.; Concannon, P.; Mychaleckyj, J.C.; Todd, J.A.; Bonella, P.; Fear, A.L.; et al. HLA DR-DQ Haplotypes and Genotypes and Type 1 Diabetes Risk. *Diabetes* **2008**, *57*, 1084–1092. [CrossRef] [PubMed]

7. Noble, J.A. Immunogenetics of type 1 diabetes: A comprehensive review. *J. Autoimmun.* **2015**, *64*, 101–112. [CrossRef] [PubMed]

8. Pisapia, L.; Camarca, A.; Picascia, S.; Bassi, V.; Barba, P.; Del Pozzo, G.; Gianfrani, C. HLA-DQ2.5 genes associated with celiac disease risk are preferentially expressed with respect to non-predisposing HLA genes: Implication for anti-gluten T cell response. *J. Autoimmun.* **2016**, *70*, 63–72. [CrossRef] [PubMed]

9. Corso, C.; Pisapia, L.; Citro, A.; Cicatiello, V.; Barba, P.; Cigliano, L.; Abrescia, P.; Maffei, A.; Manco, G.; Del Pozzo, G. EBP1 and DRBP76/NF90 binding proteins are included in the major histocompatibility complex class II RNA operon. *Nucleic Acids Res.* **2011**. [CrossRef] [PubMed]

10. Pisapia, L.; Cicatiello, V.; Barba, P.; Malanga, D.; Maffei, A.; Hamilton, R.S.; Del Pozzo, G. Co-regulated expression of alpha and beta mRNAs encoding HLA-DR surface heterodimers is mediated by the MHCII RNA operon. *Nucleic Acids Res.* **2013**, *41*, 3772–3786. [CrossRef] [PubMed]

11. Amar, A.; Radka, S.F.; Holbeck, S.L.; Kim, S.J.; Nepom, B.S.; Nelson, K.; Nepom, G.T. Characterization of specific HLA-DQ alpha allospecificities by genomic, biochemical, and serologic analysis. *J. Immunol.* **1987**, *138*, 3986–3990. [PubMed]

12. Viken, H.D.; Paulsen, G.; Sollid, L.M.; Lundin, K.E.A.; Tjønnfjord, G.E.; Thorsby, E.; Gaudernack, G. Characterization of an HLA-DQ2-specific monoclonal antibody. Influence of amino acid substitutions in DQβ 1*0202. *Hum. Immunol.* **1995**, *42*, 319–327. [CrossRef]

13. Livak, K.J.; Schmittgen, T.D. Analysis of relative gene expression data using real-time quantitative PCR and the 2-ΔΔCT method. *Methods* **2001**, *25*, 402–408. [CrossRef] [PubMed]

14. Camarca, A.; Anderson, R.P.; Mamone, G.; Fierro, O.; Facchiano, A.; Costantini, S.; Zanzi, D.; Sidney, J.; Auricchio, S.; Sette, A.; et al. Intestinal T cell responses to gluten peptides are largely heterogeneous: Implications for a peptide-based therapy in celiac disease. *J.Imunol.* **2009**, *182*, 4158–4166. [CrossRef] [PubMed]
15. Vader, W.; Stepniak, D.; Kooy, Y.; Mearin, L.; Thompson, A.; van Rood, J.J.; Spaenij, L.; Koning, F. The HLA-DQ2 gene dose effect in celiac disease is directly related to the magnitude and breadth of gluten-specific T cell responses. *Proc. Natl. Acad. Sci. USA* **2003**, *100*, 12390–12395. [CrossRef] [PubMed]
16. Cavalli, G.; Hayashi, M.; Jin, Y.; Yorgov, D.; Santorico, S.A.; Holcomb, C.; Spritz, R.A.; Dinarello, C.A. MHC class II super-enhancer increases surface expression of HLA-DR and HLA-DQ and affects cytokine production in autoimmune vitiligo. *Proc. Natl. Acad. Sci. USA* **2015**. [CrossRef] [PubMed]
17. Raj, P.; Rai, E.; Song, R.; Khan, S.; Wakeland, B.E.; Viswanathan, K.; Arana, C.; Liang, C.; Zhang, B.; Dozmorov, I.; et al. Regulatory polymorphisms modulate the expression of HLA class II molecules and promote autoimmunity. *eLife* **2016**, *5*, e12089. [CrossRef] [PubMed]
18. Gianfrani, C.; Pisapia, L.; Picascia, S.; Strazzullo, M.; Del Pozzo, G. Expression level of risk genes of MHC class II is a susceptibility factor for autoimmunity: New insights. *J. Autoimmun.* **2018**, *89*, 1–10. [CrossRef] [PubMed]

Article

A New Pedigree-Based SNP Haplotype Method for Genomic Polymorphism and Genetic Studies

Zareen Vadva [1,†], Charles E. Larsen [1,2,*,†], Bennett E. Propp [1], Michael R. Trautwein [1], Dennis R. Alford [1,‡] and Chester A. Alper [1,2,*]

1   Program in Cellular and Molecular Medicine, Boston Children's Hospital, Boston, MA 02115, USA
2   Department of Pediatrics, Harvard Medical School, Boston, MA 02115, USA
*   Correspondence: charles.larsen@childrens.harvard.edu (C.E.L.);
    chester.alper@childrens.harvard.edu (C.A.A.); Tel.: +1-617-713-8855 (C.E.L.)
†   These authors contributed equally to this work.
‡   Deceased.

Received: 28 June 2019; Accepted: 31 July 2019; Published: 5 August 2019

Abstract: Single nucleotide polymorphisms (SNPs) are usually the most frequent genomic variants. Directly pedigree-phased multi-SNP haplotypes provide a more accurate view of polymorphic population genomic structure than individual SNPs. The former are, therefore, more useful in genetic correlation with subject phenotype. We describe a new pedigree-based methodology for generating non-ambiguous SNP haplotypes for genetic study. SNP data for haplotype analysis were extracted from a larger Type 1 Diabetes Genetics Consortium SNP dataset based on minor allele frequency variation and redundancy, coverage rate (the frequency of phased haplotypes in which each SNP is defined) and genomic location. Redundant SNPs were eliminated, overall haplotype polymorphism was optimized and the number of undefined haplotypes was minimized. These edited SNP haplotypes from a region containing *HLA-DRB1* (DR) and *HLA-DQB1* (DQ) both correlated well with HLA-typed DR,DQ haplotypes and differentiated HLA-DR,DQ fragments shared by three pairs of previously identified megabase-length conserved extended haplotypes. In a pedigree-based genetic association assay for type 1 diabetes, edited SNP haplotypes and HLA-typed HLA-DR,DQ haplotypes from the same families generated essentially identical qualitative and quantitative results. Therefore, this edited SNP haplotype method is useful for both genomic polymorphic architecture and genetic association evaluation using SNP markers with diverse minor allele frequencies.

Keywords: disease association; haplotype; HLA polymorphism; major histocompatibility complex (MHC); pedigree; phase; protocol; single nucleotide polymorphism (SNP); T1DGC; type 1 diabetes (T1D)

---

1. Introduction

Evidence that specific markers in or near candidate susceptibility genes mark susceptibility to type 1 diabetes (T1D) was first obtained by association studies, wherein positivity rates of major histocompatibility complex (MHC) alleles in patients were compared with those in an "ethnically-matched" control population (so-called standard "patient vs. control" association studies) [1,2]. Variations on such patient vs. control association studies are still widely favored [3,4] for studying this complex genetic disease [5–7]. However, results from patient vs. control association studies can be confounded by population stratification [8–10]. Ethnic matching of patients and control subjects helps to reduce confusion of a purely subpopulation genetic marker (that could be increased in populations at elevated risk for disease), with a genetic marker for a susceptibility gene, that is often also a subpopulation marker when disease incidence differs considerably among ethnic subpopulations.

Thirty-five years ago, we developed a method that minimizes genetic association study population stratification using a family-based haplotyping approach to determine the frequencies of alleles and haplotypes in T1D-affected pedigrees [11]. The "disease vs. family control haplotype" method yielded sets of T1D (DIS; occurring in patients) and family control (FC; not found in any patient in the family) haplotypes for comparison. The underlying haplotyping method was originally implemented using the HLA and MHC complement gene ("complotype") typing to identify megabase (Mb)-length haplotypes fixed (i.e., at relatively high frequency) in a population (i.e., ancestral (AHs) or conserved extended haplotypes (CEHs)) and their regional haplotypic fragments [12–14]. The population-level existence of AHs/CEHs and their regional MHC fragments has been validated repeatedly using pedigree-based haplotyping methods, but CEHs are often undetectable using maximum likelihood techniques based on underlying data from unrelated subjects [15,16].

Here, we adapted that pedigree-based method to create a modern version based only on single nucleotide polymorphism (SNP) data. A validated method of this type should be useful in future studies both within and outside the human MHC to study both short- and long-range population-level haplotype sequence fixity and as a source for genetic association assays. We chose to validate the method using MHC data because: (a) of the availability of overlapping HLA and SNP typing data from two earlier studies; and, (b) of the vast prior information available from this region including its significant population-level genetic polymorphism and the long-range haplotype sequence fixity in many populations (including among families with members that are affected by T1D).

We used both the Type 1 Diabetes Genetics Consortium (T1DGC) MHC Fine Mapping study (containing biallelic dense SNP [17,18] and polymorphic HLA allele [19,20] genotypes) and the T1DGC ImmunoChip study (containing primarily biallelic dense SNP genotypes [21]) databases. Both databases provided data collected from T1D-affected subjects, their siblings and their parents. A subset of pedigrees overlapped in the two databases. We used the ImmunoChip study database to generate pedigree-phased dense SNP haplotypes that were assigned DIS or FC status by the original methodology within a 240 kb region showing the strongest genetic association to T1D within the human genome [3,4,22] containing the genes *HLA-DRB1* (DR), *HLA-DQA1* and *HLA-DQB1* (DQ) (together, the HLA-DR/DQ region). We optimized the selection of SNP data for haplotype analysis from a larger SNP dataset based on minor allele frequency (MAF) variation and redundancy, coverage rate and genomic location. Finally, we compared those edited SNP haplotype variants with pedigree-analyzed classically-typed HLA-DR,DQ haplotypes from the same families that were available from the earlier MHC Fine Mapping study, in order to test their relative ability to detect genetic association with T1D.

2. Materials and Methods

Our goal was to design a method to convert SNP genotype data obtained in families (pedigrees) into phased haplotypes edited to remove redundant and less informative SNPs to produce an optimized final set of unambiguous fully pedigree-phased edited SNP haplotypes useful for a variety of genetic and genomic assays. The new core method of this process (Section 2.3) is based on optimization, namely which SNPs to remove ("triage") and which to maintain in the finalized edited haplotypes. We describe a step-wise process for the creation of these edited SNP haplotypes. We then present an alternative method. As a direct test of the efficacy of the method, we test the extent to which the edited 27-SNP haplotypes correlated with the specific classically-defined 4-digit HLA pedigree-phased haplotypes. The final section describes one application of these haplotypes: a previously described family-based genetic association assay for T1D, using either edited SNP or classically-defined *HLA-DRB1*, *-DQA1*, *-DQB1* haplotypes for the same region of the MHC.

2.1. T1DGC Datasets

Two different T1DGC datasets were analyzed in this study. Both studies contain mostly families with multiple-affected children from several geographical cohorts. The MHC Fine Mapping dataset (June, 2009 (final) data freeze) consisted of 2298 families from nine geographical cohorts: Asia-Pacific

(AP), British Diabetic Association, Danish, Europe (EUR), Human Biological Data Interchange, Joslin, North America (NA), United Kingdom (UK) and Sardinia. The MHC Fine Mapping study provided both 4-digit HLA and SNP genotyping data. The T1DGC ImmunoChip dataset (dbGaP Study Accession: phs000911.v1.p1) consisted of 2708 families from four of the same geographical cohorts: AP, EUR, NA and UK, and it provided dense SNP typing data alone. Only 2609 of those families were affected sib pair families having at least two children with T1D. The dataset also included 19 families with only one affected child and 35 families with no T1D-affected member. A total of 1067 families were shared between both T1DGC datasets.

2.2. Genotype Extraction and Pedigree-Phased Haplotypes

PLINK [23] extracted and combined family demographic, phenotypic and genotypic data from the T1DGC ImmunoChip database in a genomic region stretching from *HLA-DRA* to *MTCO3P1* (Figure 1) to create a standard pedigree file. Unless stated otherwise, all SNP position (pos) data are for human chromosome 6 from dbSNP build GRCh38.p12. The boundary SNPs were rs14004 (pos: 32439932) and rs3104402 (pos: 32713899). Thus, the region length was nearly 274 kb. PLINK determined the total genotyping rate to be 0.998885, with all 217 SNPs and 10791 subjects passing filters and quality-control measures. Separate analyses of the same final region, but using data in which the region extracted in PLINK and later phased was significantly larger, gave essentially identical results in downstream studies (data not shown).

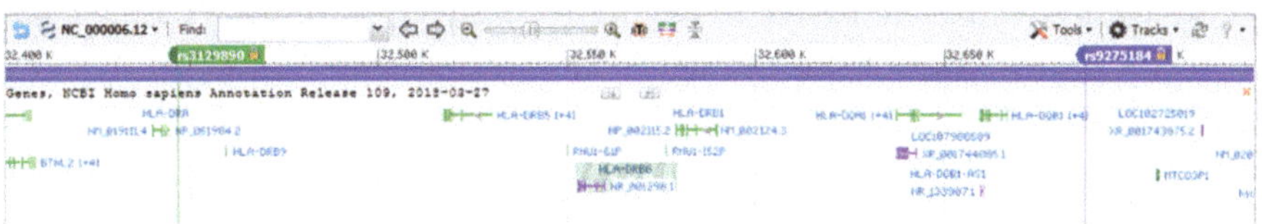

Figure 1. Genomic map of HLA-DR/DQ region in the human major histocompatibility complex (MHC) reference sequence. The map shows a slightly larger region than that phased in MERLIN. The two marked single nucleotide polymorphisms (SNPs) represent the boundaries of the phased 101 SNP haplotypes from which SNPs were "pre-triaged" for redundancy to create the initial 37-SNP haplotypes for further editing.

Family genotype data and 1383 non-genotyped (missing) founder placeholders were phased from the pedigree file into haplotypes using MERLIN (version 1.1.2) [24]. We used the "best" haplotype estimation mode in MERLIN to provide us with haplotypes that correspond to the most likely pattern of gene flow. We then analyzed the phased haplotypes of a sub-region containing 101 contiguous SNPs in the HLA-DR/DQ region. The SNPs ranged from rs3129890 to rs9275184 (Figure 1). Haplotype crossovers for each family were assessed by determining instances in which a haplotype changed from the first to last SNP in the 101 SNP HLA-DR/DQ region, and families in which crossovers occurred were subsequently removed from further analysis. We removed the few families with such apparent crossovers for two reasons: (a) apparent de novo haplotype crossovers (i.e., within the families studied) occasionally are inaccurate and can occur due to de novo mutations or rare SNP typing or MERLIN phasing errors, and we wished to minimize such complexities; and, (b) the method described here is not intended to identify de novo haplotype crossovers. Although our method is directed at identifying population-level common and rare SNP haplotypes, the output of the method would be useful for comparison with output from those rare families with apparent crossovers for detecting and/or validating de novo haplotype mutations or crossovers. Finally, we note that genotyping errors (considered extremely infrequent in these two datasets) would likely have only minor effects on the results for the common and minor variants we studied (as genotyping errors would result in either unphased or singleton variants).

2.3. Creating Finalized Founder SNP Haplotypes in the HLA-DR/DQ Region

Using the phased founder (i.e., parental) SNP haplotypes in our dataset, we designed a work flow to remove ("pre-triage") SNPs to increase the number of fully-phased haplotypes. A "fully-phased" haplotype is a haplotype defined at every SNP (i.e., assigned a phased nucleotide at every SNP position). Phased coverage at every SNP was first quantified for the entire set of founder haplotypes. "Coverage" is the percentage of all haplotypes that contain an assigned (i.e., phased) nucleotide at any given SNP. Separately, SNP MAF was provided by T1DGC for every SNP (all of which were biallelic). Six SNP MAF categories were used to create separate SNP groups for pre-triage. We arbitrarily decided to set a preliminary goal of retaining only 36–37% of all the SNPs within the region to optimize resultant haplotype diversity, coverage and SNP spatial distribution. We chose a bell-shaped distribution of MAFs for the initial pre-triage such that a higher percentage of the final SNPs would be in the three categories between 11% and 40% MAFs and fewer in the 1–10% and 41–50% categories. Within each MAF category, SNPs with higher coverage rates were retained unless the resulting spatial distribution within the region would be grossly asymmetric. Thus, priority was given to higher coverage. Supplementary Table S1 shows the 37 SNPs chosen for the original analysis and the 10 SNPs edited out in the following step. The MAF distribution of these SNPs was four in the 1–5% range, three in the 6–10% range, six in the 11–20% range, 12 in the 21–30% range, seven in the 31–40% range and five in the 41–50% range.

After the pre-triage step, we sorted haplotype sequences to isolate the fully-phased haplotypes. The overall coverage rate for all SNPs ranged from 79.1 to 90.9%. We then sorted the fully-phased SNP haplotype variants from highest to lowest frequency and tested for SNP redundancy among the haplotype variants. We then determined, for each SNP in a given MAF range, the haplotype at which it had a different allele from the first haplotype. If there was a SNP that changed alone in any variant among the group of haplotypes comprising the top 90%, then it was kept. For SNP allele pairs (or higher groupings) that changed together among the SNP haplotypes, we determined whether the SNPs were biallelic as a unit (i.e., whether they existed as only two variants among all haplotypes). If SNP pairs (or larger groupings) were biallelic among the top 95% of all haplotypes (i.e., were "redundant"), then the SNP(s) with the lower coverage was/were eliminated. Re-sorting founder haplotypes based on each new set of SNPs, sorting the haplotypes by highest to lowest frequency, and checking for additional SNP redundancy was repeated until the SNP redundancy was eliminated.

2.4. An Alternative Triaging Method

We tested an alternative SNP-editing haplotype method in which there was no pre-triaging of SNPs to determine whether the numbers and polymorphic complexity of the resultant edited haplotypes differed significantly. Thus, the method began in the last paragraph of Section 2.3 beginning with all 101 SNPs from the HLA-DR,DQ region (instead of only the 37 shown in Supplementary Table S1), and the triaging process in the last step resulted in a final number of 39 SNPs in edited haplotypes (data not shown). Several parallel studies were conducted with these haplotypes for comparison with our main 27-SNP edited haplotype results, and the downstream results were similar.

2.5. Identifying SNP Haplotype Variants for MHC CEHs and Identifying CEHs from SNP Haplotype Variants Based on the T1DGC MHC Fine Mapping Study

HLA (at the four-digit level) and SNP genotyping data from the MHC Fine Mapping study were provided by T1DGC. As described previously [25], the T1DGC HLA typing methodology did not target all polymorphic sites. Some alleles were not distinguished. For example [25], *HLA-DQB1*02:01*, found on DR3 haplotypes, and *HLA-DQB1*02:02*, found on DR7 haplotypes, were both assigned the **02:01* allele in the T1DGC data. Here, we maintain that assignment when referring to T1DGC data, but we provide the appropriate alleles [13,14,26] in named CEHs or their HLA-DR,DQ fragments. All genotyping data for the MHC region were phased together in MERLIN.

Two CEHs ([HLA-B8,SC01,DR3] and [HLA-B18,F1C30,DR3]) are at particularly high frequency among European Caucasian families affected by T1D, and we had prior knowledge that these two CEHs differed in or near the HLA-DR/DQ region [14,27]. To enhance our ability to differentiate the HLA-DR,DQ variants of these two CEHs in haplotypes lacking or unphased for either of the two HLA-C,B fragment variants distinguishing them, we analyzed 524 B8,DR3 and 214 B18,DR3 haplotypes fully defined at *HLA-C, -B, -DRB1, -DQA1*, and *-DQB1* to identify five SNPs in the MHC Fine Mapping study useful as SNP haplotype surrogates (Table 1). These SNPs are located both telomeric to and within the genomic region used from the T1DGC ImmunoChip data in the main results presented here. Although each of the two CEHs were represented by some minor 5-SNP haplotype variants (Table 1), none of the B8,DR3 haplotypes had the dominant B18,DR3 5-SNP haplotype and none of the B18, DR3 haplotypes had the dominant B8,DR3 5-SNP haplotype.

Table 1. T1DGC MHC Fine Mapping SNPs to distinguish B8,DR3 and B18,DR3 CEHs [1].

dbSNP Variants CEH	rs2076536	rs3117103	rs3135363	rs6901541	rs4999342	Cell Line Sequence		
B8,DR3	T	T	G	C	C	COX		
B18,DR3	C	A	A	T	T	QBL		
T1DGC Variants CEH	rs2076536	rs3117103	rs3135363	rs6901541	rs4999342	% Dominant Sequence	% Other Sequences	% Unphased
B8,DR3	1	1	3	2	2	95.0	0.8	4.2
B18,DR3	3	4	1	4	4	79.4	10.7	9.8

[1] Seq = Sequence. Shown are the reference sequence (rs) SNP alleles for two different MHC conserved extended haplotypes (CEHs). dbSNP data were provided by NCBI (https://www.ncbi.nlm.nih.gov/snp/).

In several other cases, we performed a reverse analysis using the MHC Fine Mapping data. When a specific HLA-DR,DQ haplotype had a relatively high-frequency 27-SNP haplotype identified from the T1DGC ImmunoChip data, we analyzed the *HLA-C* and *HLA-B* alleles of both the dominant and most frequent minor 27-SNP haplotypes using the HLA typing data from the MHC Fine Mapping study for the 1067 families overlapping between the studies.

2.6. Correlating Edited SNP Haplotypes and HLA Haplotypes Overlapping in the Two T1DGC Datasets

We used two methods to correlate the dominant HLA-DR,DQ haplotypes determined in the MHC Fine Mapping study with the major edited 27-SNP haplotypes determined from the ImmunoChip dataset using the 1067 families shared between the two T1DGC datasets. We determined first the dominant HLA-DR,DQ haplotype corresponding to each major edited 27-SNP haplotype. Separately, we quantified the percentages of the two most frequent edited 27-SNP haplotypes along with the percentage of unphased (at even a single SNP) 27-SNP haplotypes corresponding to each of the major classically-defined HLA-DR,DQ haplotypes.

Finally, we compared the 27-SNP haplotypes and the HLA-typed DR,DQ haplotypes in these shared families for statistical results in the T1D gene association assay both in terms of ranking of and relative numbers of haplotypes distributed between DIS and FC designations. To perform the gene association assay based on HLA-DR,DQ haplotypes in the MHC Fine Mapping study, we categorized the haplotypes based on their 4-digit alleles and then combined them into haplotype groups based on a nomenclature presented previously [3]. We categorized only those haplotypes (>97% of all haplotypes) that correlated with the major edited 27-SNP haplotypes determined from the ImmunoChip dataset (Section 2.6). We calculated a DIS/FC haplotype ratio of the HLA-DR,DQ haplotypes and compared it with the *HLA-DRB1-DQB1* patient/control (P/C) ratio for T1D susceptibility presented previously [3]. Both ratios were also compared based on the relative rank of the haplotypes. We defined a haplotype with a DIS/FC ratio >1 as a susceptibility haplotype and a haplotype with a DIS/FC ratio <0.5 as a protective haplotype, with neutral haplotypes falling within a DIS/FC ratio between 0.5 and 1.0.

2.7. Assigning Disease and Family Control Status to Haplotypes for a Genetic Association Assay

Using the final set of all fully-phased edited 27-SNP haplotypes, we assigned DIS and FC status to founder haplotypes. A DIS haplotype was defined as any parental haplotype in a patient with T1D. A FC haplotype was defined as any parental haplotype only in unaffected members of the same family. Subjects assigned unknown disease status were treated as unaffected members of the pedigree. Finally, to equalize the number of DIS and FC haplotypes based on their parental contribution, we removed any founder lacking either a DIS or FC haplotype. Thus, only haplotypes from founders who had one DIS and one FC haplotype were retained. This was designed to maximize ethnic identity distribution between DIS and FC haplotypes.

2.8. Statistical Analyses

DIS and FC SNP haplotypes were ranked separately based on their frequencies within each of the two categories. Pearson's chi-squared (χ^2) test was performed to determine whether there was a statistical difference between the raw number (n) distribution of identical DIS vs. FC SNP haplotypes if DIS and FC haplotypes were each observed at $n \geq 5$. Significance was set at $p < 0.05$. A Bonferroni correction was applied to adjust for significance for multiple comparison tests.

3. Results

3.1. Identifying Fully-Defined Edited SNP Haplotypes from the T1DGC ImmunoChip Study

The MERLIN output for the T1DGC ImmunoChip dataset was 10790 founder haplotypes containing 101 SNPs in the region (Figure 1). Of those haplotypes, 913 were undefined at all positions and many haplotypes were either partially undefined or unphaseable due to missing pedigree genotype data. Due to MERLIN-assigned haplotype crossovers within the studied region, 114 families (4.2% of all families in the dataset) were removed from further analysis.

The pre-triage method used to select SNPs resulted in 6194 fully-defined (at every SNP) 37-SNP haplotypes (57% of the original haplotypes). Upon further removal of 10 redundant SNPs (Supplementary Table S1), the number of fully-defined haplotypes increased to 6309 27-SNP haplotypes (58% of the original haplotypes). Among the 6309 haplotypes were 94 unique haplotype variants. Of these, 15 variants each existed above 1% (Table 2) and 41 variants were single examples (<1% of all haplotypes).

As compared with the pre-triage results, the non-pre-triage method resulted in fewer fully-defined 39-SNP haplotypes ($n = 5695$). Most of the results presented in the rest of this report, therefore, focus on the fully-defined 27-SNP haplotypes resulting from the pre-triage method.

3.2. Comparison of Overlapping Families in T1DGC Studies: Testing SNP Haplotype Method vs. HLA Typing

Of the 6309 27-SNP haplotypes from the entire T1DGC ImmunoChip dataset (Table 2), 2561 (41%) were from families also in the T1DGC MHC Fine Mapping database. Of those 27-SNP haplotypes shared by the two studies, 2466 (96.3%) were among the top 19 variants: Table 3 shows the total numbers of 27-SNP haplotypes for each of those major variants along with the 4-digit alleles or 2-digit specificities of the major HLA-DR,DQ haplotypes, groups or fragments that dominated them. Each variant was dominated by a particular HLA-DR,DQ haplotype or haplotype group. For example, the most common variant among the group, variant 1, was the HLA-DR4,DQ8 haplotype (a group specificity composed of haplotypes containing a wide variety of DR4 alleles (e.g., *HLA-DRB1*04:01, *04:02, *04:03*) in addition to *HLA-DQA1*03:01* and *HLA-DQB1*03:02*). In contrast, the second most common variant, variant 2, was predominantly *HLA-DRB1*03:01, -DQB1*02:02* (DR3,DQ2) fragments of the [HLA-B8,SC01,DR3] CEH, but the DR3,DQ2 fragment of the [HLA-B18,F1C30,DR3] CEH was variant 4. Two DR,DQ haplotypes (*HLA-DRB1*13:02, -DQB1*06:04* and *HLA-DRB1*13:01, -DQB1*06:03*) each dominated two other 27-SNP haplotype variant groups (variants 10 and 19 and variants 11 and 13, respectively).

Table 2. The most frequent T1DGC ImmunoChip study edited 27-SNP haplotypes [1].

SNP Variant Name	Total (*n*)	Percentage	SNP Variant Name	Total (*n*)	Percentage
Variant 1	1786	28.3	Variant 19	45	0.7
Variant 2	1064	16.9	Variant 20	26	0.4
Variant 3	517	8.2	Variant 21	22	0.3
Variant 4	467	7.4	Variant 22	22	0.3
Variant 5	400	6.3	Variant 23	21	0.3
Variant 6	308	4.9	Variant 24	10	0.2
Variant 7	296	4.7	Variant 25	8	0.1
Variant 8	285	4.5	Variant 26	8	0.1
Variant 9	154	2.4	Variant 27	8	0.1
Variant 10	134	2.1	Variant 28	8	0.1
Variant 11	112	1.8	Variant 29	8	0.1
Variant 12	96	1.5	Variant 30	7	0.1
Variant 13	79	1.3	Variant 31	6	0.1
Variant 14	77	1.2	Variant 32	4	0.1
Variant 15	71	1.1	Variant 33	4	0.1
Variant 16	59	0.9	Variant 34	4	0.1
Variant 17	56	0.9	Variant 68	1	0.0
Variant 18	55	0.9			

[1] These edited SNP haplotype variants are those that existed at $n \geq 4$ in the entire T1DGC ImmunoChip study or were otherwise named in the main text (Variant 68). The SNP haplotype sequences for all of the edited SNP haplotypes named here are given in Supplementary Table S2.

Table 3. Major edited 27-SNP haplotypes shared by both T1DGC studies.

Edited SNP Haplo Rank	Variant Name	SNP Haplo Total (*n*)	% Defined SNP Haplos	Dominant HLA-DR,DQ Haplotype			
				HLA-DRB1	*HLA-DQA1*	*HLA-DQB1*	HLA Abbrev.
1	Variant 1	729	28.5	04:xx	03:01	03:02	DR4,DQ8
2	Variant 2	447	17.5	03:01	05:01	02:01	B8,DR3,DQ2
3	Variant 4	205	8.0	03:01	05:01	02:01	B18,DR3,DQ2
4	Variant 3	202	7.9	01:01	01:01	05:01	DR0101,DQ5
5	Variant 5	166	6.5	07:01	02:01	02:02	DR7,DQ2
6	Variant 6	124	4.8	04:xx	03:01	03:01/03:04	DR4,DQ7
7	Variant 7	116	4.5	15:01	01:02	06:02	DR15,DQ6
8	Variant 8	106	4.1	11:xx	05:01	03:01	DR11,DQ3
9	Variant 9	62	2.4	08:01	04:01	04:02	DR8,DQ4
10	Variant 10	53	2.1	13:02	01:02	06:04	DR1302,DQ6 var1
11	Variant 11	40	1.6	13:01	01:03	06:03	DR1301,DQ6 var1
11	Variant 12	40	1.6	16:01	01:02	05:02	DR16,DQ5
13	Variant 14	34	1.3	07:01	02:01	03:03	DR7,DQ3
14	Variant 13	30	1.2	13:01	01:03	06:03	DR1301,DQ6 var2
15	Variant 15	29	1.1	09:01	03:01	03:03	DR9,DQ3
16	Variant 17	23	0.9	12:01	05:01	03:01	DR12,DQ3
17	Variant 19	22	0.9	13:02	01:02	06:04	DR1302,DQ6 var2
18	Variant 18	21	0.8	14:01/14:04	01:01	05:03	DR14,DQ5
19	Variant 16	17	0.7	01:02	01:01	05:01	DR0102,DQ5
	TOTAL	2466	96.3				

Table 4 and Supplementary Table S3 show the opposite information to that of Table 3: the degree to which a particular dominant 27-SNP haplotype from Table 3 represented the entire group of HLA-DR,DQ haplotypes (as defined by fully-phased *HLA-DRB1*, *-DQA1*, *-DQB1* alleles) was remarkably high. Except for the three DR,DQ haplotypes mentioned above that were found in two different dominant 27-SNP haplotypes, few to none of the most frequent HLA-DR,DQ haplotypes contained a secondary 27-SNP haplotype variant (Table S3). Most of the differences between the total numbers of specific HLA-DR,DQ haplotypes and the total numbers of the dominant 27-SNP haplotype representing those DR,DQ haplotypes were caused by the failure of full phasing among the 27 SNPs (Table S3). Thus, the major 27-SNP haplotypes correlated directly with the major HLA-DR,DQ haplotypes.

Table 4. Major HLA-DR,DQ haplotypes shared in T1DGC studies: their dominant 27-SNP haplotype and their percentages [1].

DR,DQ Haplo Rank	HLA Haplo Abbrev.	DR,DQ Total (*n*)	% all DR,DQ Defined Haplos	Dominant SNP Haplotype	1st Total (*n*)	% of This DR,DQ Haplotype Group	% of Fully-Defined in This Group
1	DR4,DQ8	1024	27.4	Variant 1	722	70.5%	99.2%
2	All DR3,DQ2	950	25.4	Variant 2	441	46.4%	67.7%
3	DR0101,DQ5	290	7.8	Variant 3	185	63.8%	99.5%
4	DR7,DQ2	230	6.2	Variant 5	165	71.7%	100.0%
5	DR15,DQ6	188	5.0	Variant 7	110	58.5%	99.1%
6	DR11,DQ3	182	4.9	Variant 8	102	56.0%	98.1%
7	DR4,DQ7	155	4.1	Variant 6	106	68.4%	98.1%
8	DR1301,DQ6	108	2.9	Variant 11	40	37.0%	58.8%
9	DR1302,DQ6	104	2.8	Variant 10	51	49.0%	65.4%
10	DR8,DQ4	89	2.4	Variant 9	58	65.2%	90.6%
11	DR16,DQ5	62	1.7	Variant 12	40	64.5%	100.0%
11	DR7,DQ3	54	1.4	Variant 14	34	63.0%	94.4%
13	DR9,DQ3	45	1.2	Variant 15	29	64.4%	100.0%
14	DR14,DQ5	36	1.0	Variant 18	21	58.3%	87.5%
15	DR12,DQ3	30	0.8	Variant 17	22	73.3%	95.7%
16	DR0102,DQ5	27	0.7	Variant 16	17	63.0%	100.0%
	TOTAL	3574	95.7	TOTAL	2143		

[1] The HLA haplotype abbreviations used here are those from Table 3 with minor exceptions. Here, the test haplotype is the HLA-DR,DQ (DR,DQ) haplotype. Therefore, for example, the entire DR3,DQ2 group is analyzed. The last column gives the percentage of each DR,DQ haplotype group represented by the dominant 27-SNP haplotype among all fully-defined 27-SNP haplotypes. The second most frequent 27-SNP haplotype and their percentages of each DR,DQ haplotype group as well as the total untyped or unphased 27-SNP haplotypes for each DR,DQ group are given in Supplementary Table S3.

3.3. Summary of Edited SNP Haplotypes Distinguishing DR,DQ Haplotypes and Specific CEHs that Share HLA-DR,DQ Alleles

Table 5 shows, by direct comparison, the high degree to which the major 27-SNP variants directly correlated with specific HLA-DR,DQ haplotypes or haplotypic groups. For 15 of the 19 most common 27-SNP variants, 95% or more of all individual haplotypes in the group were part of a single HLA-DR,DQ haplotype and four (variants 1, 6, 8 and 18) comprised a haplotypic group, and all 19 of the top 27-SNP variants reach the 85% or higher level of this metric.

Table 5. Dominant MHC CEHs in major 27-SNP edited haplotypes of the DR,DQ region [1].

SNP Haplo Var. Name	Dom. DR,DQ Haplo (DRB1,DQA1,DQB1)	SNP Haplo Total (*n*)	Dom. DR,DQ Haplo Total (*n*)	Dom. CEH of DR,DQ Var.	Dom. DR,DQ CEH Total (*n; %*)
Variant 1	*04:xx,03:01,03:02*	729	722	None	**
Variant 2	*03:01,05:01,02:01*	447	441	[HLA-C7,B8,SC01,DR3]	**
Variant 3	*01:01,01:01,05:01*	202	185	*	**
Variant 4	*03:01,05:01,02:01*	205	202	[HLA-C5,B18,F1C30,DR3]	**
Variant 5	*07:01,02:01,02:02*	166	165	*	**
Variant 6	*04:xx,03:01,03:01/03:04*	124	106	*	**
Variant 7	*15:01,01:02,06:02*	116	110	[HLA-C7,B7,SC31,DR15]	31; 52% ***
Variant 8	*11:xx,05:01,03:01*	106	102	None	**
Variant 9	*08:01,04:01,04:02*	62	58	[HLA-C7,B39,unk,DR8]	11; 19%
Var. 10	*13:02,01:02,06:04*	53	51	[HLA-C3,B40,SC02,DR13]	26; 51%
Var. 11	*13:01,01:03,06:03*	40	40	[HLA-C12,B38,SC21,DR13]	10; 25%
Var. 12	*16:01,01:02,05:02*	40	40	[HLA-C12,B39,unk,DR16]	9; 23%
Var. 13	*13:01,01:03,06:03*	30	27	[HLA-C3,B15,unk,DR13]	8; 30%
Var. 14	*07:01,02:01,03:03*	34	34	[HLA-C6,B57,SC61,DR7]	20; 59%
Var. 15	*09:01,03:01,03:03*	29	29	[HLA-C7,B7,unk,DR9]	5; 17%
Var. 16	*01:02,01:01,05:01*	17	17	[HLA-C8,B14,SC2(1,2),DR1]	11; 65%
Var. 17	*12:01,05:01,03:01*	23	22	[HLA-C5,B44,unk,DR12]	4; 18%
Var. 18	*14:01/14:04,01:01,05:03*	21	21	[HLA-C4,B35,unk,DR14]	6; 29%
Var. 19	*13:02,01:02,06:04*	22	22	[HLA-C7,B15,unk,DR13]	6; 27%
	TOTAL	2466	2394		

[1] Dom. = Dominant; Haplo = Haplotype; Var. = Variant. * The dominant CEH of this group was not determined; ** The totals for these CEHs were not determined; *** Only 60 of 110 haplotypes were evaluated.

As with HLA-DR,DQ 4-digit allelic haplotypes, there is a dominant long-range CEH specific for most 27-SNP haplotype variants (Table 5). Furthermore, two major 27-SNP variants distinguish different CEH fragments of three HLA-DR,DQ haplotypes: HLA-DR3,DQ2 by SNP variants 2 and 4; HLA-DR1302,DQ0604 by SNP variants 10 and 19; and HLA-DR1301,DQ0603 by SNP variants 11 and 13. The CEHs represented by variants 2 and 4 are well known, and variant 10 is well characterized [13,14]: the class I fragment alleles are (*HLA-C*03:04,B*40:01*) and its complotype is SC02. The putative CEH represented by variant 19 (Table 5) has not been previously characterized. SNP variant 19's class I fragment alleles are (*HLA-C*07:01,B*15:17*)—a rare centromeric class I haplotype. The CEHs represented by variants 11 and 13 are also less well characterized. The variant 11 CEH ([HLA-C12,B38,SC21,DR1301,DQ0603]) is a class II variant of the well-known Ashkenazi CEH [HLA-C12,B38,SC21,DR0402,DQ0302] (unpublished observations), but they appear to differ elsewhere in class I as well: the DR4,DQ8 CEH is dominated by *HLA-A*26:01* [13,14], whereas six of the ten variant 11 DR13, DQ6 CEH examples (Table 5) bear *HLA-A*02:01* (two others bear *HLA-A*26:01*). The variant 13 putative CEH ([HLA-C0303,B1501,unk,DR1301,DQ0603] may be a class II variant of either of two previously identified DR4 AHs [13].

Thus, 27-SNP haplotype variants may be useful in identifying previously unidentified or only partially characterized AHs/CEHs. Other than the ones mentioned above, another putative CEH [HLA-C7,B39,unk,DR8], represented by variant 9, has not been, to our knowledge, previously characterized. Of the 11 examples we found of this haplotype group, nine had the (*HLA-C*07:02,B*39:06*) fragment (six with *HLA-A*24:02*) and the other two contained the class I haplotype (*HLA-A*02:01,C*07:02,B*39:01*). As another example, the putative CEH [HLA-C12,B39,unk,DR16] of variant 12 has also not been described previously. Of the nine examples of this putative CEH, seven had the (*HLA-C*12:03,B*39:01*) centromeric class I fragment and the other two contained the class I haplotype (*HLA-A*02:01,C*12:03,B*39:06*).

Some edited 27-SNP haplotypes contain a secondary CEH or putative CEH. For example, the variant 7 SNP haplotype has a secondary well-characterized CEH ([HLA-C12,B18,S042,DR15,DQ6]; $n = 8$ (13% of 60 total DR15, DQ6 haplotypes evaluated)). A second variant 12 haplotype ($n = 7$; 18% of all variant 12's defined HLA-DR,DQ haplotypes) may be a CEH: [HLA-C7,B44,unk,DR16] with the centromeric class I fragment (*HLA-C*07:04,B*44:02*). Finally, three other previously unreported putative CEHs contain, at 4-digit resolution, the following class I fragments: (a) SNP variant 15: (*HLA-C*07:02,B*07:02*) (although this is a common Caucasian HLA-C,B fragment); (b) SNP variant 17: (*HLA-C*05:01,B*44:02*); and, (c); SNP variant 18: (*HLA-C*04:01,B*35:01*). Further work (e.g., sequence analysis in class III) is required in order to confirm the CEH status for each of these apparently fixed long-range haplotypes.

3.4. Establishing and Analyzing the Designated DIS and FC SNP Haplotypes in the T1DGC ImmunoChip Study and Analyzing SNP Haplotypes for Genetic Association with T1D

Of the 6309 fully-defined 27-SNP haplotypes, 4272 were DIS and 2037 were FC haplotypes, comprised of 94 different haplotype variants ($n = 62$ DIS and $n = 69$ FC variants). We removed 603 founders who had no FC haplotype. With 27-SNPs, 5364 fully-phased SNP-haplotypes ($n = 3360$ DIS and 2004 FC haplotypes) remained. There were 87 different haplotype variants ($n = 61$ DIS and $n = 61$ FC variants), including 45 singleton haplotypes (<1% of all haplotypes). The most common DIS haplotype was variant 1 ($n = 1244$, 37% of all DIS haplotypes), and the most common FC haplotype was variant 7 ($n = 254$, 13% of all FC haplotypes).

We then equalized the number of DIS and FC haplotypes, using only founders with one DIS and one FC haplotype, which resulted in 2004 DIS and 2004 FC haplotypes. There were 75 different haplotype variants ($n = 43$ DIS and $n = 61$ FC variants). The most common DIS haplotype was variant 1 ($n = 916$, 46% of all DIS haplotypes), and the most common FC haplotype remained as variant 7 (Table 6). Among the haplotypes shown in Table 6 (where $n \geq 5$), Pearson's chi-squared test showed a statistically significant difference between DIS and FC SNP haplotype frequencies (χ^2 = 1198.15, df = 14, p = 4.34 × 10^{-247}; p-adjusted = 6.51 × 10^{-246}).

Table 6. Analysis of equalized fully-phased 27-SNP edited disease (DIS) and family control (FC) haplotypes from the ImmunoChip study [1].

SNP Haplo Var. Name	DIS Haplo (*n*)	FC Haplo (*n*)	Total (*n*)	DIS/FC Haplo Ratio	DIS Haplo Rank	FC Haplo Rank	χ^2 *
Variant 1	916	238	1154	3.85	1	2	398.34
Variant 2	416	204	620	2.04	2	3	72.49
Variant 3	108	190	298	0.57	4	5	22.56
Variant 4	249	46	295	5.41	3	12	139.69
Variant 7	5	254	259	0.02	14	1	239.39
Variant 5	47	190	237	0.25	6	5	86.28
Variant 8	18	195	213	0.09	10	4	147.08
Variant 6	73	120	193	0.61	5	7	11.45
Variant 11	8	71	79	0.11	13	8	50.24
Variant 9	45	32	77	1.41	7	15	2.19
Variant 14	2	68	70	0.03	19	9	–
Variant 10	25	43	68	0.58	8	13	4.76
Variant 12	20	39	59	0.51	9	14	6.12
Variant 13	4	52	56	0.08	17	10	–
Variant 18	1	49	50	0.02	22	11	–
Variant 15	18	24	42	0.75	10	17	0.86
Variant 17	5	26	31	0.19	14	16	14.23
Variant 16	9	17	26	0.53	12	19	2.46
Variant 19	4	16	20	0.25	17	20	–
Variant 20	1	19	20	0.05	22	18	–
Variant 21	1	10	11	0.10	22	21	–
Variant 23	1	10	11	0.10	22	21	–
Variant 25	5	1	6	5.00	14	29	–
Variant 29	1	5	6	0.20	22	23	–
Variant 28	2	3	5	0.67	19	24	–
Variant 33	2	2	4	1.00	19	27	–
Variant 31	1	3	4	0.33	22	24	–
Variant 27	1	3	4	0.33	22	24	–
Variant 32	1	2	3	0.50	22	27	–
Misc. Haplos	15	72	87				
TOTAL	2004	2004	4008				1198.15

[1] Misc. = Miscellaneous; Haplo = Haplotype; Var. = Variant. * Chi-squared statistic of DIS and FC SNP haplotypes each ≥ 5 in frequency.

3.4.1. Analyzing Genetic Association with T1D among Families Overlapping in the Two T1DGC Studies Using the Designated DIS and FC SNP Haplotypes from the ImmunoChip Study

Using only overlapping families from both the MHC Fine Mapping and ImmunoChip datasets, we observed 2561 fully-phased 27-SNP haplotypes, including 1747 DIS and 814 FC haplotypes. We then equalized the number of DIS and FC haplotypes, keeping haplotypes based on parental contribution to the patients, which resulted in 808 DIS and 808 FC haplotypes (Table 7). This group of 27-SNP haplotypes was composed of 48 different variants (n = 26 DIS and n = 42 FC variants). The most common DIS variant was variant 1 (n = 373, 46% of all DIS haplotypes), and the most common FC variant was variant 7 (n = 103, 13% of all FC haplotypes).

Table 7. Analysis of equalized DIS and FC SNP haplotypes each ≥ 5 in frequency among overlapping families in both T1DGC studies [1].

SNP Haplo Var. Name	DIS Haplo (n)	FC Haplo (n)	Total (n)	DIS/FC Haplo Ratio	DIS Haplo Rank	FC Haplo Rank	χ^2 *
Variant 1	373	102	475	3.66	1	2	154.61
Variant 2	169	84	253	2.01	2	3	28.56
Variant 4	105	24	129	4.38	3	9	50.86
Variant 3	44	72	116	0.61	4	5	6.76
Variant 7	1	103	104	0.01	17	1	–
Variant 5	21	78	99	0.27	6	4	32.82
Variant 8	11	70	81	0.16	8	6	42.98
Variant 6	26	47	73	0.55	5	7	6.04
Variant 9	17	16	33	1.06	7	13	0.03
Variant 11	2	30	32	0.07	13	8	–
Variant 10	10	16	26	0.63	9	13	1.38
Variant 12	6	18	24	0.33	10	12	6.00
Variant 13	2	20	22	0.10	13	10	–
Variant 18	1	19	20	0.05	17	11	–
Variant 15	3	12	15	0.25	12	15	–
Variant 17	1	11	12	0.09	17	16	–
Variant 19	4	5	9	0.80	11	18	–
Variant 16	2	6	8	0.33	13	17	–
Variant 28	2	1	3	2.00	13	19	–
Variant 27	1	1	2	1.00	17	19	–
Misc. Haplos	7	73	80				
TOTAL	808	808	1616				330.04

[1] Misc.: Miscellaneous; Haplo: Haplotype; Var: Variant. * Chi-squared statistic of DIS and FC SNP haplotypes each ≥5 in frequency.

Pearson's chi-squared test showed a statistically significant difference between DIS and FC SNP haplotype frequencies (χ^2 = 330.04, df = 9, p = 1.09 × 10^{-65}; p-adjusted = 1.09 × 10^{-64}). The results of these tests performed in overlapping families between the two T1DGC datasets largely mirror the results of the SNP haplotype method and genetic association assay performed on the entire ImmunoChip dataset (see Section 3.4).

3.4.2. Analyzing Genetic Association with T1D among Families Overlapping in the Two T1DGC Studies Using the Designated DIS and FC HLA-DR,DQ Haplotypes from the MHC Fine Mapping Study

To compare the statistical results of our genetic association assay based on the edited 27-SNP haplotypes in the overlapping families from both T1DGC datasets, we performed a genetic association analysis using the same families using only the HLA-DR,DQ typing to determine the haplotype identities from the MHC Fine Mapping dataset. We initially identified 3735 HLA-DR,DQ haplotypes (n = 2564 DIS and n = 1171 FC haplotypes). We then equalized the number of DIS and FC haplotypes, keeping haplotypes based on parental contribution to the patients, which resulted in 1171 DIS and 1171 FC haplotypes (Table 8). The most common DIS haplotype group was DR4,DQ8 (n = 508, 43% of all DIS haplotypes), and the most common FC haplotype was DR15,DQ0602 (n = 166, 14% of all FC haplotypes).

Pearson's chi-squared test showed a statistically significant difference between DIS and FC SNP haplotype frequencies (χ^2 = 693.71, df = 10, p = 1.40 × 10^{-142}; p-adjusted = 1.54 × 10^{-141}) when DIS and FC haplotypes were each greater than or equal to five in frequency (Table 8). The results of this genetic association assay give qualitatively similar results to the genetic association assay performed on the same overlapping families using the edited 27-SNP haplotype method (see Section 3.4.1).

Table 8. Analysis of equalized DIS and FC HLA-DR,DQ haplotypes each ≥ 5 in frequency among overlapping families in both T1DGC studies [1].

DR,DQ Haplo Var. Name	DIS Haplo (n)	FC Haplo (n)	Total (n)	DIS/FC Haplo Ratio	DIS Haplo Rank	FC Haplo Rank	χ^2 *
DR4,DQ8	508	87	595	5.84	1	6	297.88
DR3,DQ2	416	133	549	3.13	2	2	145.88
DR0405,DQ2	7	4	11	1.75	12	16	–
DR8,DQ4	30	26	56	1.15	5	12	0.29
DR13,DQ0604	21	30	51	0.70	7	10	1.59
DR0901,DQ0303	11	19	30	0.58	10	14	2.13
DR1,DQ0501	72	132	204	0.55	3	3	17.65
DR16,DQ0502	13	25	38	0.52	8	13	3.79
DR4,DQ7	27	68	95	0.40	6	8	17.69
DR7,DQ2	31	118	149	0.26	4	5	50.80
DR13,DQ0603	8	77	85	0.10	11	7	56.01
DR11,DQ0301	12	132	144	0.09	9	3	100.00
DR12,DQ0301	1	17	18	0.06	13	15	–
DR14,DQ0503	1	30	31	0.03	13	10	–
DR15,DQ0602	1	166	167	0.01	13	1	–
DR0701,DQ0303	0	49	49	0.00	16	9	–
Misc. Haplos	12	58	70				
TOTAL	1171	1171	2342				693.71

[1] Misc.: Miscellaneous; Haplo: Haplotype; Var: Variant. *Chi-square statistic of DIS and FC HLA-DR,DQ haplotypes each ≥ 5 in frequency.

4. Discussion

The T1DGC MHC databases used in this study are two of the largest family-based dense SNP datasets available for allele and genetic evaluation. Furthermore, many of the genotyped pedigrees in these datasets include both parents and multiple children. These datasets thus provide a rich resource for direct observational pedigree-based haplotype phasing. The datasets have the added benefit of having a significant portion of the genotype data within a part of the human genome (the MHC) that is (a) highly gene-dense; (b) highly polymorphic; (c) the most completely characterized (on a population-based level) Mb region of the human genome; and, (d) linked to and/or associated with a wide variety of phenotypes-including the one (T1D) for which the datasets were designed.

Additionally, one of the two T1DGC datasets (the MHC Fine Mapping study) has HLA genotype data at 4-digit resolution, and there are many overlapping families with the other dataset (the ImmunoChip study). These facts, and prior knowledge of both the polymorphic nature and population-level polymorphic genomic architecture of the region, allowed us to correlate directly observed pedigree-phased edited SNP haplotypes with HLA haplotypes and longer-ranged CEHs. This provided a means of testing the degree to which the edited SNP haplotypes generated using our SNP editing process were representative of the same haplotypes defined by classical HLA-DR,DQ typing.

A major obstacle for pedigree-based observational definition of SNP haplotypes is the significant ambiguity in haplotype assignment, especially for biallelic SNPs, inherently created by relatively small pedigrees [28]. Furthermore, due to 1383 missing parents ("founders"), a large percentage of family genotype data was lacking in the ImmunoChip database we used. Nevertheless, genotype data can often be phased into defined haplotypes at many markers even using only haploidentical siblings. Using our strategy of prioritizing inclusion of high "coverage" SNPs (Supplementary Table S1; SNPs at which the percentage of fully-defined (-phased) haplotypes is maximal), we were able to define 58% of the haplotypes containing 27 SNPs.

Separately, we optimized the polymorphic nature of the resultant SNP haplotypes by choosing SNPs with a wide array of MAFs and removing any SNPs within any given MAF range that appeared to be "redundant." Redundant SNPs are those that form only a biallelic SNP haplotype with any other SNP. That is, a SNP is redundant to another SNP if essentially all (≥95%) of the resultant independent fully-defined haplotypes contain only two of the theoretically four possible SNP haplotype combinations of the two tested SNPs. We maximized the final haplotype definition by removing the redundant SNP with the lower coverage.

Our results show that the method provides results remarkably similar in polymorphic detail as compared with classical four-digit HLA typing at three loci (*HLA-DRB1, -DQA1* and *-DQB1*) in the same genomic region. Indeed, the edited 27-SNP haplotypes could, for at least three separate HLA-DR,DQ haplotypes, distinguish pairs of different haplotype variants among identical 4-digit HLA-DR,DQ variants. For the pair of HLA-DR3,DQ2 variants (representing variants of the CEHs [HLA-B8,SC01,DR3] and [HLA-B18,F1C30,DR3]), this was not surprising. It was already known that these two DR3,DQ2 variants, while nearly identical in a 106 kb region overlapping with the 240 kb region we analyzed [27], have different alleles at another locus (*HLA-DRB3*) within the region we studied [13,14].

Conversely, the edited 27-SNP haplotypes could not distinguish variants of the HLA-DR4, DQ8 haplotype group (those that differ, at the third and fourth digits, in classical *HLA-DRB1* typing, but share the *HLA-DQA1*03:01* and *HLA-DQB1*03:02* alleles). This is also not particularly surprising. The T1DGC ImmunoChip dataset only included three SNPs within the *HLA-DRB1* locus itself. Although we used all three *HLA-DRB1* SNPs among our starting 37 SNPs (and triaged out one of them due to redundancy for the 27-SNP analysis), these SNPs were clearly insufficient to distinguish alleles that differ within a particular *HLA-DRB1* exonic sequence. The results suggest, however, that the DR4,DQ8 haplotypes may share a highly similar sequence throughout the HLA-DR,DQ region (other than at *HLA-DRB1*) in a way similar (although not within the same boundaries) to that of the DR3,DQ2 haplotype group [27].

Long-range (>1 Mb) human MHC haplotypes of highly fixed sequence markers existing at relatively high frequency among many geoethnic populations have been identified as CEHs [12–15,26]. Several MHC CEH pair variants and groups share sequence identity (e.g., Class III or HLA-DR,DQ blocks) surrounded both telomerically and centromerically by regions in which the CEHs differ significantly [12–14,26,27]. In this report, although we did not analyze all of the dominant CEHs in every edited SNP haplotype variant, our results clearly demonstrate that the SNP haplotype variants we evaluated in the 240 kb region are strongly genetically linked to and can act as surrogate markers of these long-range differences. As our results demonstrate (Table 5), the 27-SNP haplotypes in this single HLA class II region can be exploited to identify previously unreported putative CEHs.

T1D genetic association results, based on HLA-DR,DQ alleles and haplotype variants as well as individual SNP alleles and SNP haplotype variants, have been previously reported, in some cases using underlying data overlapping with those we analyzed. Our HLA-DR,DQ genetic association results (Table 8) among the 1067 families shared between the two T1DGC datasets largely parallel results from the two largest previously published analyses of T1D genetic association with HLA-DR,DQ haplotypes [3,25]. The earlier of the two publications was a 2007 meta-analysis of HLA-DR,DQ haplotype variant risk effects on T1D summarizing results from 38 studies conducted worldwide [3]. The relative distribution and ranks of T1D susceptibility haplotype variants as determined by DIS to FC ratios in our study essentially are identical to those found based on the 2007 study's summary "patient to control" (P/C) ratios. Although we grouped all HLA-DR4,DQ8 haplotypes together, and there are a few specific HLA-DR4,DQ8 haplotypes in the 2007 meta-analysis that are not among the highest susceptibility group, the latter composed only 2% of the T1DGC dataset we used. The remaining 98% of HLA-DR4,DQ8 haplotypes we studied were all HLA-DR,DQ variants within the top 10 group of P/C ratios in the 2007 study [3].

In a 2008 report of HLA-DR,DQ haplotypes in a different subset of the T1DGC MHC Fine Mapping study [25], a family-based patient to control genetic association analysis showed a statistically different DR,DQ haplotype distribution among patients and controls in Caucasian (largely of European origin) subjects ($p = 5 \times 10^{-124}$) that parallels our study results. The study also found a rank hierarchy of haplotype risk for T1D (based on odds ratios) that was similar to both the 2007 meta-analysis [3] and the results we present here. In summary, the HLA-DR,DQ haplotype analysis we present here, with which we compare our edited 27-SNP haplotype analyses, is consistent with prior results.

The results of the analysis of HLA-DR,DQ haplotypes in overlapping families from the MHC Fine Mapping dataset (Table 8) were largely in parallel with the results of the analysis of the edited 27-SNP haplotypes (Table 7) in overlapping families from the ImmunoChip dataset. Thus, for both structural variant analysis and T1D genetic association analysis, the core method of edited SNP haplotypes provided a data source that was essentially as useful as 4-digit HLA-DR,DQ typing. Both methods of HLA class II variant designation captured variant 1 (DR4,DQ8) as the most frequent haplotype among all haplotypes and the most frequent DIS haplotype. Both methods also showed that variant 7 (DR15,DQ6) was one of the most protective haplotypes among all haplotypes and the most frequent haplotype among FC haplotypes. The genetic association assays based on HLA-DR,DQ haplotypes and edited 27-SNP haplotypes among overlapping families also gave qualitatively (and to a large extent, quantitatively) similar results. Variant 1 (among SNP haplotypes) and DR4,DQ8 (among HLA-DR,DQ haplotypes) both showed the highest ratio of DIS to FC haplotype frequency.

Finally, the edited 27-SNP haplotype variants among the overlapping families of the two T1DGC studies were representative of the edited 27-SNP haplotypes from the entire ImmunoChip dataset. Among the 34 most frequent fully-defined 27-SNP haplotype variants in the entire ImmunoChip dataset (Table 2), only 14 haplotype variants (3.9% of the total haplotypes) were not represented among SNP haplotypes in the overlapping families (Table 7). Among the 29 27-SNP haplotype variants existing at least once each as a DIS and as a FC haplotype in the entire ImmunoChip dataset (Table 6), only nine were not similarly represented among the 27-SNP haplotypes in the overlapping families (Table 7). Thus, the genetic association assays based on the edited 27-SNP haplotypes from the overlapping family subset showed qualitatively similar results to those edited 27-SNP haplotypes from the entire ImmunoChip dataset. Variants 1 and 7, both among overlapping families and in the entire ImmunoChip dataset, showed the largest differences in frequency ratios of DIS to FC haplotypes.

We did not compare our pedigree-phased and -defined structural haplotypes with those that might be generated from the same underlying genotype data using any of the numerous maximum likelihood statistical methods available to "impute" SNP haplotypes using unrelated individuals. However, for those investigators interested in testing or comparing the accuracy of various haplotype imputation methodologies (either with each other or with the directly phased and defined haplotypes produced herein), these two T1DGC datasets would seem to be ideal resources with which to conduct such studies. It would be a useful validation procedure for proponents of haplotype imputation, and any future designers of haplotype imputation methodologies, to use databases such as these two T1DGC MHC databases. Our prediction is that imputed haplotypes guessed at by maximum likelihood statistical methods using the same source genotypes used in this study would show quantitative inaccuracy as compared with the direct observational results presented here [15]. However, at the very least, such comparisons might lead to improved methodologies for haplotype imputation in those (unfortunately) common situations in which geneticists must use databases containing only genotype data from unrelated subjects.

In conclusion, we believe the method developed here to optimize SNP haplotype analysis may prove useful as a tool for a wide variety of end uses. The underlying method clearly provides structural information that parallels that of HLA typing and is, therefore, validated in the most intensively studied region of the human genome. The method can be used to analyze genetic association with a genetic phenotype, as we have presented here for the complex autoimmune disease T1D. The method can also be used to evaluate both regional and longer-range population-level genomic architecture. This opens up the entire human genome to the study of long-range AH/CEH structures that have thus far been limited almost entirely to the MHC.

Supplementary Materials: The following are available online at http://www.mdpi.com/2073-4409/8/8/835/s1. Table S1: SNPs used for edited haplotypes in HLA-DR/DQ region, Table S2: SNP sequences for the most frequent T1DGC ImmunoChip study edited 27-SNP haplotypes, Table S3: Major HLA-DR,DQ haplotypes shared in T1DGC studies: their secondary and unphased 27-SNP haplotypes and their percentages.

Author Contributions: The following were the individual author contributions to this study: conceptualization, C.A.A., C.E.L., M.R.T. and Z.V.; methodology, D.R.A., C.A.A., C.E.L., B.E.P., M.R.T. and Z.V.; software, M.R.T. and Z.V.; validation, D.R.A., C.E.L. and Z.V.; formal analysis, C.E.L., M.R.T. and Z.V.; investigation, D.R.A., C.E.L., B.E.P., M.R.T. and Z.V.; resources, M.R.T. and Z.V.; data curation, D.R.A., C.E.L., M.R.T. and Z.V.; writing—original draft preparation, C.E.L. and Z.V.; writing—review and editing, C.A.A., C.E.L., B.E.P., M.R.T. and Z.V.; visualization, C.E.L. and Z.V.; supervision, C.A.A. and C.E.L.; project administration, C.A.A. and C.E.L.; funding acquisition, C.A.A.

Funding: This research was funded by Juvenile Diabetes Research Foundation grant 1-2008-472 and institutional funds from the Program in Cellular and Molecular Medicine, Boston Children's Hospital.

Acknowledgments: This research was performed under the auspices of the Type 1 Diabetes Genetics Consortium (T1DGC), a collaborative clinical study sponsored by the National Institute of Diabetes and Digestive and Kidney Diseases (NIDDK), National Institute of Allergy and Infectious Diseases (NIAID), National Human Genome Research Institute (NHGRI), National Institute of Child Health and Human Development (NICHD), and Juvenile Diabetes Research Foundation International (JDRF) and supported by grant U01 DK062418 from the National Institutes of Health. Genotyping was performed by the Sanger Institute (Hinxton, UK) which is supported by The Wellcome Trust. The T1DGC genotyping and phenotyping was conducted by the T1DGC Investigators and supported by the National Institute of Diabetes and Digestive and Kidney Diseases (NIDDK). The data from the T1DGC MHC Fine Mapping study reported here were supplied by the NIDDK Central Repositories. The data from the T1DGC ImmunoChip study were supplied by dbGaP. This manuscript was not prepared in collaboration with Investigators of the T1DGC study and does not necessarily reflect the opinions or views of the T1DGC study, the NIDDK Central Repositories, or the NIDDK.

Conflicts of Interest: The authors declare no conflict of interest. The funders had no role in the design of the study; in the collection, analyses, or interpretation of data; in the writing of the manuscript, or in the decision to publish the results.

References

1. Cudworth, A.G.; Woodrow, J.C. Genetic susceptibility in diabetes mellitus: Analysis of the HLA association. *Br. Med. J.* **1976**, *2*, 846–848. [CrossRef] [PubMed]

2. Platz, P.; Jackobsen, B.K.; Morlin, N.; Ryder, L.P.; Svejgaard, A.; Thomsen, M.; Christy, M.; Kromann, H.; Benn, J.; Nerup, J.; et al. HLA-D and -DR antigens in genetic analysis of insulin dependent diabetes mellitus. *Diabetologia* **1981**, *21*, 108–115. [CrossRef] [PubMed]

3. Thomson, G.; Valdes, A.M.; Noble, J.A.; Kockum, I.; Grote, M.N.; Najman, J.; Erlich, H.A.; Cucca, F.; Pugliese, A.; Steenkiste, A.; et al. Relative predispositional effects of HLA class II DRB1-DQB1 haplotypes and genotypes on type 1 diabetes: A meta-analysis. *Tissue Antigens* **2007**, *21*, 110–127. [CrossRef] [PubMed]

4. Hu, X.; Deutsch, A.J.; Lenz, T.L.; Onengut-Gumuscu, S.; Han, B.; Chen, W.M.; Howson, J.M.M.; Todd, J.A.; Bakker, P.I.W.; Rich, S.S.; et al. Additive and interaction effects at three amino acid positions in HLA-DQ and HLA-DR molecules drive type 1 diabetes risk. *Nat. Genet.* **2015**, *21*, 898–905. [CrossRef] [PubMed]

5. Steck, A.K.; Rewers, M.J. Genetics of type 1 diabetes. *Clin. Chem.* **2011**, *57*, 176–185. [CrossRef]

6. Katsarou, A.; Gudbjörnsdottir, S.; Rawshani, A.; Dabelea, D.; Bonifacio, E.; Anderson, B.J.; Jacobsen, L.M.; Schatz, D.A.; Lernmark, Å. Type 1 diabetes mellitus. *Nat. Rev. Dis. Primers* **2017**, *3*, 17016. [CrossRef]

7. Alper, C.A.; Larsen, C.E.; Trautwein, M.R.; Alford, D.R. A stochastic epigenetic Mendelian oligogenic disease model for type 1 diabetes. *J. Autoimmun.* **2019**, *96*, 123–133. [CrossRef]

8. Balding, D.J. A tutorial on statistical methods for population association studies. *Nat. Rev. Genet.* **2006**, *7*, 781–791. [CrossRef]

9. Liu, N.; Zhang, K.; Zhao, H. Haplotype-association analysis. In *Genetic Dissection of Complex Traits*, 2nd ed.; Rao, D.C., Gu, C.C., Eds.; Academic Press: San Diego, CA, USA, 2008; pp. 335–405.

10. Alper, C.A.; Larsen, C.E. *Major Histocompatibility Complex: Disease Associations*; In eLS; John Wiley Sons, Ltd.: Chichester, UK, 2015.

11. Raum, D.; Awdeh, Z.; Yunis, E.J.; Alper, C.A.; Gabbay, K.H. Extended major histocompatibility complex haplotypes in type 1 diabetes mellitus. *J. Clin. Investig.* **1984**, *74*, 449–454. [CrossRef]

12. Awdeh, Z.L.; Raum, D.; Yunis, E.J.; Alper, C.A. Extended HLA/complement allele haplotypes: Evidence for T/t-like complex in man. *Proc. Natl. Acad. Sci. USA* **1983**, *80*, 259–263. [CrossRef]

13. Dawkins, R.; Leelayuwat, C.; Gaudieri, S.; Tay, G.; Hui, J.; Cattley, S.; Martinez, P.; Kulski, J. Genomics of the major histocompatibility complex: Haplotypes, duplication, retroviruses and disease. *Immunol. Rev.* **1999**, *167*, 275–304. [CrossRef] [PubMed]

14. Yunis, E.J.; Larsen, C.E.; Fernandez-Viña, M.; Awdeh, Z.L.; Romero, T.; Hansen, J.A.; Alper, C.A. Inheritable variable sizes of DNA stretches in the human MHC: Conserved extended haplotypes and their fragments or blocks. *Tisssue Antigens* **2003**, *62*, 1–20. [CrossRef]

15. Alper, C.A.; Larsen, C.E.; Dubey, D.P.; Awdeh, Z.L.; Fici, D.A.; Yunis, E.J. The haplotype structure of the human major histocompatibility complex. *Hum. Immunol.* **2006**, *67*, 73–84. [CrossRef] [PubMed]

16. Walsh, E.C.; Mather, K.A.; Schaffner, S.F.; Farwell, L.; Daly, M.J.; Patterson, N.; Cullen, M.; Carrington, M.; Bugawan, T.L.; Erlich, H.; et al. An integrated haplotype map of the human major histocompatibility complex. *Am. J. Hum. Genet.* **2003**, *73*, 580–590. [CrossRef] [PubMed]

17. Brown, W.M.; Pierce, J.; Hilner, J.E.; Perdue, L.H.; Lohman, K.; Li, L.; Venkatesh, R.B.; Hunt, S.; Mychaleckyj, J.C.; Deloukas, P. Type 1 Diabetes Genetics Consortium. Overview of the MHC fine mapping data. *Diab. Obes. Metab.* **2009**, *11*, 2–7. [CrossRef] [PubMed]

18. Rich, S.S.; Akolkar, B.; Concannon, P.; Erlich, H.; Hilner, J.E.; Julier, C.; Morahan, G.; Nerup, J.; Nierras, C.; Pociot, F.; et al. Overview of the Type 1 Diabetes Genetics Consortium. *Genes Immun.* **2009**, *10*, S1–S4. [CrossRef] [PubMed]

19. Mychaleckyj, J.C.; Noble, J.A.; Moonsamy, P.V.; Carlson, J.A.; Varney, M.D.; Post, J.; Helmberg, W.; Pierce, J.J.; Bonella, P.; Fear, A.L.; et al. HLA genotyping in the international Type 1 Diabetes Genetics Consortium. *Clin. Trials* **2010**, *7*, S75–S87. [CrossRef]

20. Noble, J.A.; Valdes, A.M.; Varney, M.D.; Carlson, J.A.; Moonsamy, P.; Fear, A.L.; Lane, J.A.; Lavant, E.; Rappner, R.; Louey, A.; et al. HLA class I and genetic susceptibility to type 1 diabetes. Results from the Type 1 Diabetes Genetics Consortium. *Diabetes* **2010**, *59*, 2972–2979. [CrossRef]

21. Morahan, G.; Mehta, M.; James, I.; Chen, W.M.; Akolkar, B.; Erlich, H.A.; Hilner, J.E.; Julier, C.; Nerup, J.; Nierras, C.; et al. Tests for genetic interactions in type 1 diabetes. Linkage and stratification analyses of 4422 affected sib-pairs. *Diabetes* **2011**, *60*, 1030–1040. [CrossRef]

22. He, C.; Hamon, S.; Li, D.; Barral-Rodriguez, S.; Ott, J. Type 1 Diabetes Genetics Consortium. MHC fine mapping of human type 1 diabetes using the T1DGC data. *Diab. Obes. Metab.* **2009**, *11*, 53–59. [CrossRef]

23. Purcell, S.; Beale, B.; Todd-Brown, K.; Thomas, L.; Ferreira, M.A.R.; Bender, D.; Maller, J.; Sklar, P.; Bakker, P.I.W.; Daly, M.J.; et al. PLINK: A tool set for whole-genome association and population-based linkage analyses. *Am. J. Hum. Genet.* **2007**, *81*, 559–575. [CrossRef] [PubMed]

24. Abecasis, G.R.; Cherny, S.S.; Cookson, W.O.; Cardon, L.R. Merlin–rapid analysis of dense genetic maps using sparse gene flow trees. *Nat. Genet.* **2002**, *30*, 97–101. [CrossRef] [PubMed]

25. Erlich, H.; Valdes, A.M.; Noble, J.; Carlson, J.A.; Varney, M.; Concannon, P.; Mychaleckyj, J.C.; Todd, J.A.; Bonella, P.; Fear, A.L.; et al. HLA DR-DQ haplotypes and genotypes and type 1 diabetes risk. Analysis of the Type 1 Diabetes Genetics Consortium families. *Diabetes* **2008**, *57*, 1084–1092. [CrossRef] [PubMed]

26. Larsen, C.E.; Alford, D.R.; Trautwein, M.R.; Jalloh, Y.K.; Tarnacki, J.L.; Kunnenkeri, S.K.; Fici, D.A.; Yunis, E.J.; Awdeh, Z.L.; Alper, C.A. Dominant sequences of human major histocompatibility complex conserved extended haplotypes from *HLA-DQA2* to *DAXX*. *PLoS Genet.* **2014**, *10*, e1004637. [CrossRef] [PubMed]

27. Traherne, J.A.; Horton, R.; Roberts, A.N.; Miretti, M.M.; Hurles, M.E.; Stewart, C.A.; Ashurst, J.L.; Atrazhev, A.M.; Coggill, P.; Palmer, S.; et al. Genetic analysis of completely sequenced disease-associated MHC haplotypes identifies shuffling of segments in recent human history. *PLoS Genet.* **2006**, *2*, e9. [CrossRef]

28. Hodge, S.E.; Boehnke, M.; Spence, M.A. Loss of information due to ambiguous haplotyping of SNPs. *Nature* **1999**, *21*, 360–361. [CrossRef] [PubMed]

 *cells*

Article

HLA-E Polymorphism Determines Susceptibility to BK Virus Nephropathy after Living-Donor Kidney Transplant

Hana Rohn [1,*], Rafael Tomoya Michita [2,3], Sabine Schramm [2], Sebastian Dolff [1], Anja Gäckler [4], Johannes Korth [4], Falko M. Heinemann [2], Benjamin Wilde [4], Mirko Trilling [5], Peter A. Horn [2], Andreas Kribben [4], Oliver Witzke [1] and Vera Rebmann [2]

[1] Department of Infectious Diseases, West German Centre for Infectious Diseases (WZI), University Hospital Essen, University Duisburg-Essen, 45147 Essen, Germany
[2] Institute for Transfusion Medicine, University Hospital Essen, University Duisburg-Essen, 45147 Essen, Germany
[3] Post-Graduation Program in Genetics and Molecular Biology, Genetics Department, Universidade Federal do Rio Grande do Sul (UFRGS), Porto Alegre 91501-970, Brazil
[4] Department of Nephrology, University Hospital Essen, University Duisburg-Essen, 45147 Essen, Germany
[5] Institute for Virology, University Hospital Essen, University Duisburg-Essen, 45147 Essen, Germany
* Correspondence: hana.rohn@uk-essen.de; Tel.: +49 201 723 3394; Fax: +49 201 723 3393

Received: 22 June 2019; Accepted: 6 August 2019; Published: 7 August 2019

Abstract: Human leukocyte antigen (HLA)-E is important for the regulation of anti-viral immunity. BK polyomavirus (BKPyV) reactivation after kidney transplant is a serious complication that can result in BKPyV-associated nephropathy (PyVAN) and subsequent allograft loss. To elucidate whether HLA-E polymorphisms influence BKPyV replication and nephropathy, we determined the HLA-E genotype of 278 living donor and recipient pairs. A total of 44 recipients suffered from BKPyV replication, and 11 of these developed PyVAN. Homozygosity of the recipients for the HLA-E*01:01 genotype was associated with the protection against PyVAN after transplant ($p = 0.025$, OR 0.09, CI [95%] 0.83–4.89). Considering the time course of the occurrence of nephropathy, recipients with PyVAN were more likely to carry the HLA-E*01:03 allelic variant than those without PyVAN (Kaplan–Meier analysis $p = 0.03$; OR = 4.25; CI (95%) 1.11–16.23). Our findings suggest that a predisposition based on a defined HLA-E genotype is associated with an increased susceptibility to develop PyVAN. Thus, assessing HLA-E polymorphisms may enable physicians to identify patients being at an increased risk of this viral complication.

Keywords: BK virus; polyomavirus; nephropathy; human leukocyte antigen-E; kidney transplantation

---

1. Introduction

Human leukocyte antigen (HLA)-E belongs to the non-classical major histocompatibility complex (MHC) class Ib molecules located on chromosome 6p21.3. In terms of function, HLA-E is special within the immune system because it acts as a key player exhibiting regulatory functions both in innate and adaptive immune responses [1].

HLA-E predominantly acts as an indicator for "missing-self" by continuously presenting peptides derived from signal sequences from HLA class Ia molecules. Infected or transformed cells, which have silenced HLA-I expression to evade from canonical CD8 T-cell recognition, do not produce such signal peptides. In absence of the signal peptides, HLA-E is not stabilized and does not reach the cell surface. NK cells sense the absence of HLA-E and become activated. Recently it became apparent that HLA-E can also bind and present antigenic peptides derived from pathogens such as HIV, hepatitis B and C viruses as well as cytomegalovirus [2], a fact implying that HLA-E plays an important role in anti-viral immunity mediated by T cells [3,4]. Being the cognate ligand for the C-type lectin CD94/NKG2 receptor family, HLA-E enables natural killer (NK) cells and cytotoxic T lymphocytes to monitor cell integrity because it is recognized by either the inhibitory CD94/NKG2A or the activating CD94/NKG2C receptor [5–7]. Moreover, HLA-E can be recognized by the stimulatory α/β T-cell receptor (TCR) expressed on the CD8+ T cells, resulting in cytotoxic elimination of target cells presenting foreign peptides presented by HLA-E [8].

In contrast to the highly polymorphic HLA-A, HLA-B, or HLA-C molecules, HLA-E exhibits fewer allelic variants [9]. To date, only 27 HLA-E alleles have been reported, most of which are found in neglectable frequencies in the population or do not encode functional proteins at all [10]. The two most prevalent allotypes in the Caucasian population are HLA-E*01:01 (HLA-E107R) and HLA-E*01:03 (HLA-E107G), which are found with nearly equal frequencies [11]. Although the two corresponding proteins only differ by an arginine (HLA-E*01:01) or a glycine (HLA-E*01:03) at position 107 [12], functional differences have been reported. HLA-E*01:03 has been reported to exhibit a higher affinity for available peptides, and for this reason its cell surface expression is significantly increased compared to HLA-E*01:01 [11,13]. Studies of viral diseases have shown that the various HLA-E genotypes affect a clinical outcome [14–16].

Little is known about the relevance of HLA-E concerning viral infection and reactivation after solid organ transplant. Viruses are the leading cause of infections and mortality after transplant. In particular, the reactivation of otherwise mostly subclinical, opportunistic viruses from pre-existing latency reservoirs is clinically relevant during immunosuppression. BK polyomavirus (BKPyV), a small non-enveloped double-stranded DNA virus, has emerged as one of the most challenging pathogens after kidney transplant, causing severe allograft dysfunction and graft loss. Approximately 80% to 95% of the human population are persistently infected with BKPyV; the infection mostly occurs in healthy adults [17–19]. However, the immunosuppression necessary after kidney transplant enables the virus to reactivate. Viral replication can occur in as many as 60% of recipients [20], and can progress to the most severe form of BKPyV invasive kidney disease, polyomavirus-associated nephropathy (PyVAN).

PyVAN is linked to kidney malfunction and rejection, with a significant risk of allograft loss in as many as 60% of the cases [19,21–23]. Unfortunately, no direct antiviral treatment is approved for BKPyV replication, and the current recommendation for BKPyV management and PyVAN treatment is restricted to a reduction of immunosuppression, leading to a substantial risk of acute rejection. Little is known about the pathogenesis of BKPyV, and it is not clear which factors determine the clinical course of PyVAN. From a clinical perspective, the identification of genetic biomarkers that influence the course of BKPyV infection may allow early risk stratification to prevent the progression of PyVAN.

Thus, considering the functional differences between the HLA-E allelic variants, we hypothesized that these allelic variants may affect the clinical occurrence and onset of PyVAN after kidney transplant and may serve as a prognostic parameter for determining which patients are at risk.

2. Materials and Methods

2.1. Study Population, BKPyV and PyVAN Screening

A total of 278 living-donor kidney transplant recipients and their 278 corresponding donors from the living-donor kidney program at the University Hospital Essen, Germany, were enrolled in this retrospective study. Transplant procedures were performed between 2005 and 2017. Informed consent was obtained from all patients in accordance with the Declaration of Helsinki, and the local ethics committee approved the study (approval number 12-5312-BO). All patients underwent regular follow-up examinations. An exclusion criterion was treatment with an mTOR inhibitor, since clinical data indicate that the use of mTOR-based immunosuppressive regimens may be protective against BKPyV replication [22,24–26]. The following data were collected: demographic and transplant-related characteristics of recipients and donors (Table 1), the occurrence of BKPyV, and the occurrence of biopsy-proven PyVAN.

Table 1. Patient characteristics at baseline.

	A	B	C	
	Total	HLA-E*01:03 Carrier#	HLA-E*01:03 Non-Carrier	*p*-Value B vs. C
Donor	**N = 278**	**N = 211**	**N = 67**	
Sex (men/women)	115/163	89/122	26/41	0.67 [a]
Age (y) ±SD	51.41 ± 10.03	51.99 ± 10.06	43.30 ± 9.83	0.10 [b]
Recipient	**N = 278**	**N = 198**	**N = 80**	
Sex (men/women)	162/116	116/82	46/34	0.87 [a]
Age (y) ±SD	41.06 ± 15.43	40.46 ± 15.42	42.53 ± 15.47	0.32 [b]
KTx Related Variables				
HLA A, B mismatches, mean ± SD	1.97 ± 1.14	1.83 ± 1.1	2.31 ± 1.17	0.002 **[b]
HLA-DR mismatch, mean ± SD	1.12 ± 0.62	1.09 ± 0.60	1.19 ± 0.66	0.22 [b]
Panel of Antibodies (%)				
0%	244	172	72	0.55 [a]
1–10%	14	9	5	0.55 [a]
10–50%	14	12	2	0.36 [a]
>50%	6	5	1	0.68 [a]
Cold ischemia time, median, in minutes (range)	134.99 ± 48.90	136.07 ± 48.63	131.32 ± 48.86	0.57 [b]
Warm ischemia time, median, in minutes (range)	19.97 ± 6.87	19.38 ± 5.84	21.46 ± 8.83	0.057 [b]
Immunosuppressive Therapy				
ATG-based induction therapy, yes/no	13/265	9/189	4/76	1 [a]
CNI administration, yes/no	278/0	198/0	80/0	nd
MMF co-administration, yes/no	260/18	189/9	75/5	0.55 [a]
Steroid co-administration, yes/no	278/0	198/0	80/0	nd
PyVAN and CMV				
PyVAN, yes/no	11/267	11/187	0/80	0.031 *[a]
CMV positive recipient	159/119	116/82	43/37	0.46 [a]
CMV positive donor	177/101	128/70	49/31	0.59 [a]
CMV Infection, yes/no	38/240	32/166	6/74	0.057 [a]
PyVAN and CMV infection, yes/no	3/275	3/195	0/80	0.27 [a]
PyVAN or CMV infection vs. no PyVAN and no CMV infection	46/232	40/158	6/74	0.01 [a]

y: years; HLA: human leukocyte antigen; ATG: antithymocyte globulin; CNI: calcineurin inhibitor; MMF: mycophenolate mofetil; KTx: kidney transplant; PyVAN: BK polyomavirus associated nephropathy; CMV: cytomegalovirus; nd: not determined; SD: standard deviation. [a] Fisher's exact test; [b] Mann–Whitney U test; # HLA E*01:03 carrier: HLA E*01:03/01:03 and HLA E*01:03/01:01genotype; HLA E*01:03 non-carrier: HLA E*01:01/01:01 genotype. * $p < 0.05$, ** $p < 0.01$.

Screening for BKPyV replication after transplant evolved during the course of the study in accordance with the introduction of new clinical guidelines. Initially, only patients with graft dysfunction were tested for BKPyV replication. Starting from 2010, the screening for BKPyV replication after kidney transplant was performed according to the 2009 clinical practice guideline Kidney Disease:

Improving Global Outcomes (KDIGO). In addition to scheduled BKPyV screening, patients' serum was tested for BKPyV replication once allograft dysfunction was detected. All samples were analyzed by quantitative real-time polymerase chain reaction (RT-PCR) with a lower limit of quantification of 400 copies per milliliter. BKPyV reactivation was defined as viremia when RT-PCR detected BKPyV DNA in patients' serum.

Occurrence of PyVAN was assessed during the first three years after transplantation. PyVAN was diagnosed, if found in at least one kidney biopsy specimen judged positive by an experienced renal pathologist according to standard criteria [27]. Kidney biopsies were performed only upon clinical indication of PyVAN, like evidence of graft dysfunction or high levels of BKPyV viremia. Besides BKPyV, cytomegalovirus (CMV) represents an important viral pathogen that negatively affects allograft survival [28–31]. We have previously reported that the polymorphism of HLA-E is associated with CMV infection after kidney transplantation [16]. To evaluate possible interactions between these two viral infections with respect to the HLA-E polymorphism, we determined the rate of CMV infections for the present cohort as defined by international recommendations [32]. Standard induction immunosuppression consisted of treatment with a basiliximab-based regimen (which blocks the IL-2 receptor CD25) and a calcineurin inhibitor (CNI) in combination with steroids and an antiproliferative drug (mycophenolic acid or azathioprine).

2.2. HLA-E Genotyping

Recipients and corresponding donors were typed for HLA-E with a sequence-specific primer-polymerase chain reaction (SSP-PCR) method as described previously [33,34]. Genomic DNA was isolated from buffy-coats of peripheral blood using QIAamp® DNA Blood Mini Kit (QIAGEN GmbH, D-40724 Hilden, Germany). The SSP-PCR method was performed with the following conditions: 50 ng of genomic DNA was amplified in a final reaction volume of 10 μL containing 3 μL PCR Master Mix (Olerup® SSP AB, Stockholm, Sweden), 0.6 units of Taq DNA Polymerase (QIAGEN GmBH), 15 pmol of detection primers and 15 pmol of each positive control primer. The initial denaturation of the sequence-specific products was performed for 2 min at 94 °C, followed by a two stage PCR program: 10 cycles of 10 s at 94 °C and 20 s at 65 °C, as well as 20 relaxed cycles of 10 s at 94 °C, of 1 min at 61 °C and of 30 s at 72 °C. HLA-E alleles were identified at the resolution level of the second field. Human growth hormone was used as a positive control for PCR amplification. HLA-E genotype distributions in the two groups matched expectations according to the Hardy–Weinberg equilibrium.

2.3. Prediction of HLA-E Affinity for BKPyV Derived Peptides

The non-classical HLA-E molecule exhibits preferential binding to a highly conserved set of nonameric signal peptides derived from leader sequences of other HLA-class I molecules. However, HLA-E is also able to present pathogen-derived antigens [2], hence antigens derived from CMV, human immunodeficiency virus (HIV) and other pathogens give rise to peptides, which are loaded onto HLA-E molecules [14,35]. To our knowledge, HLA-E restricted peptides encoded by BKPyV have not been described to date. To identify peptide sequences derived from BKPyV antigens that may bind to HLA-E with high affinity and thus potentially representing immunogenic T-cell epitopes, we applied the machine-learning bioinformatics algorithm NetMHC 4.0 (http://www.cbs.dtu.dk/services/NetMHC/index.php) [36,37]. NetMHC 4.0 is a method trained on both binding affinity and eluted ligand data. The algorithm uses an allele-specific approach, whereby separate predictors are trained for each MHC allele; the input to the model is the peptide of interest (in our case derived from BKPyV) [36]. The following BKPyV-derived antigens were assessed for HLA-E binding potential: large-T and small-T antigen, the major capsid proteins VP1 and VP2, and the agnoprotein. In order to compare the affinities of putative BKPyV-derived peptides predicted by the NetMHC 4.0 algorithm, functionally well-defined HLA-E binding peptides derived from the classical HLA class I molecule HLA-G and HLA-Cw*3, the heath shock protein 60, CMV proteins (pUL40, pUL18, pUL83, and pUL123), and two other pathogen-derived peptides (viz., the HIV gag protein and mycobacterium tuberculosis enoyl-[acyl-carrier-protein] reductase) were also determined [8,38–41]. We only took "strong binding peptides" as defined by the NetMHC 4.0 algorithm into consideration. Strong binding peptides are defined by an affinity threshold of 50,000 nM as well as a percentile rank assignment within <0.5 being a predicted affinity compared to a set of 400,000 random natural peptides. Solely the HLA-E*01:01 allele is present in NetMHC 4.0; thus we were able to determine the peptide affinity for this allelic variant only.

2.4. Statistical Analysis

Statistical analyses were performed using the SPSS 21.0 software (IBM corp. Released 2012, IBM SPSS Statistic for Windows, Armonk, NY, USA). Baseline characteristics of donors and recipients were compared with two-sided Fisher's exact test or the Wilcoxon rank-sum test, as appropriate. The contribution of allelic variants as risk factors for PyVAN was evaluated by Fisher's exact test. Joint genotype analysis was performed by Mantel–Haenszel test. The occurrence of PyVAN was estimated by the Kaplan–Meier method, and estimates were compared with the log-rank test. A two-sided *p*-value of 0.05 or lower was considered statistically significant.

3. Results

3.1. HLA-E*01:01 May Be Able to Present Peptide Sequences Derived from BKPyV

We were able to identify BKPyV-derived peptides that may be presented by the HLA-E*01:01 allelic variant using the machine-learning bioinformatics algorithm NetMHC 4.0 (National Institute of Allergy and Infectious Diseases, National Institutes of Health, Bethesda, MD, USA). Tables 2–5 contains the identified high and moderate affinity peptide sequences derived from BKPyV antigens (large-T and small-t antigen, VP1, VP2 and agnoprotein) which may be presented to immune effector cells by HLA-E*01:01.

Table 2. High and moderate affinity peptide sequences derived from BK polyomavirus (BKPyV) large T antigen and small t antigen that may be presented by the HLA-E*01:01 allelic variant as identified by the machine-learning bioinformatics algorithm NetMHC4.0. Position: residue number (starting from 0); peptide: amino acid sequence of the potential ligand.

Large-T and Small-t Antigen			
Position	**Peptide**	**1 − log50k(aff)**	**Affinity (nM)**
416	NVPKRRYWL	0.202	5619.93
568	RILQSGMTL	0.173	7666.40
454	PMERLTFEL	0.165	8432.78
557	SLQNSEFLL	0.160	8874.14
570	LQSGMTLLL	0.150	9850.58

Table 3. High and moderate affinity peptide sequences derived from BKPyV major capsid protein VP1 that may be presented by the HLA-E*01:01 allelic variant as identified by the machine-learning bioinformatics algorithm NetMHC 4.0. Position: residue number (starting from 0); peptide: amino acid sequence of the potential ligand.

BK Polyomavirus Major Capsid Protein VP1			
Position	**Peptide**	**1 − log50k(aff)**	**Affinity (nM)**
18	KEPVQVPKL	0.186	6716.19
237	TNTATTVLL	0.142	10,773.00
297	NPYPISFLL	0.139	11,067.54

Table 4. High affinity peptide sequences derived from BKPyV major capsid protein VP2 that may be presented by the HLA-E*01:01 allelic variant as identified by the machine-learning bioinformatics algorithm NetMHC 4.0. Position: residue number (starting from 0); peptide: amino acid sequence of the potential ligand.

BK Polyomavirus Major Capsid Protein VP2			
Position	Peptide	1 − log50k(aff)	Affinity(nM)
292	WMLPLLLGL	0.152	9685.81

Table 5. High and moderate affinity peptide sequences derived from BKPyV agnoprotein that may be presented by the HLA-E*01:01 allelic variations as identified by the machine-learning bioinformatics algorithm NetMHC 4.0. Position: residue number (starting from 0); peptide: amino acid sequence of the potential ligand.

BK Polyomavirus Agnoprotein			
Position	Peptide	1 − log50k(aff)	Affinity (nM)
46	SXXPESVMF	0.139	11,081.08
52	VMFCEPKNL	0.139	11,162.18

The affinities of putative nonameric BKPyV-derived peptides predicted by the NetMHC 4.0 algorithm were compared to the HLA-E-binding peptide repertoire reported in the literature [8,38–41] (Supplementary Table S1a–i). Table 6 summarizes the results of the binding affinities of the identified peptide motifs with the highest affinity as predicted by the NetMHC 4.0 algorithm.

Leader peptide sequences of HLA-Cw*3, the CMV homologue pUL40 [35], and HLA-G molecules had the highest predicted binding affinity to HLA-E. The predicted binding affinities of the BKPyV-derived antigens ranged between these and other well-characterized peptide sequences. It is noteworthy that the primary anchor residues among the leader peptides sequences are largely conserved at the canonical position 2 Met (methionine) and position 9 Leu (leucine) [6,9]. Three out of four identified BKPyV HLA-E binding motifs have Leu at position 9 and the BK polyomavirus major capsid protein VP2 has also Met at position 2.

Table 6. The high affinity peptide motifs reported in the literature and derived from different self or foreign antigens as well as BKPyV-derived antigens that may be presented by HLA-E*01:01 allelic variations identified by the machine-learning bioinformatics algorithm NetMHC 4.0.

Derived from	Peptide	1 − log50k(aff)	Affinity (nM)
Human leukocyte antigen-Cw*3	VMAPRTLIL	0.589	85.69
CMV protein pUL40	VMAPRTLIL	0.589	85.69
Human leukocyte antigen-G	VMAPRTLFL	0.564	112.12
Mycobacterium tuberculosis enoyl-[acyl-carrier-protein] reductase [NADH]	RLPAKAPLL	0.547	134.77
CMV phosphorylated matrix protein pp65 (UL83)	VLPHETRLL	0.330	1405.23
CMV immediate-early protein 1 (pUL123)	VMLAKRPLI	0.303	1887.52
Human heat shock protein 60 (hsp60)	QMRPVSRVL	0.276	2530.17
CMV protein pUL18	SEPQCNPLL	0.273	2614.63
BK polyomavirus large-T and small-t antigen	NVPKRRYWL	0.202	5619.93
BK polyomavirus major capsid protein VP1	KEPVQVPKL	0.186	6716.19
Human immunodeficiency virus 1 Gag protein	RMYSPVSIL	0.173	7728.11
BK polyomavirus major capsid protein VP2	WMLPLLLGL	0.152	9685.81
BK polyomavirus agnoprotein	SXXPESVMF	0.139	11081.08

Methionine at amino acid positions 2 and leucine at position 9 amino acid have been described to be of functional relevance for HLA-E binding [9,35,42]. Peptide: amino acid sequence of the potential ligand.

3.2. HLA-E*01:01 Homozygosity Exerts Protection against PyVAN

HLA-E allelic frequencies in our cohort were similar among recipients (HLA-E*01:01 294 of 556 [52.9%]; HLA-E*01:03 262 of 556 [47.1%]) and donors (HLA-E*01:01 282 of 556 [50.7%]; HLA-E*01:03 274 of 556 [49.3%]). No other HLA-E variants were detected. The observed allelic distribution of HLA-E was in accordance with expectations indicated by the Hardy–Weinberg equilibrium.

In total, 44 recipients (15.8%) tested positive for BKPyV viremia; 23 of these exhibited high levels of BKPyV viremia ($>10^4$ copies per milliliter). Eleven BKPyV-positive recipients had at least one biopsy-proven PyVAN during the first three years after kidney transplant. Three out of the eleven patients experienced CMV replication prior to the development of a BK virus-associated nephropathy. All three patients, in which the CMV replication preceded the BK virus-associated nephropathy, were HLA-E*01:03 carriers.

With respect to HLA-E genotypes, there was no association of HLA-E genotypes of the donor and the occurrence of PyVAN (Table 7).

However, the recipient HLA-E*01:01 homozygous state was associated with protection against PyVAN, (p = 0.025, odds ratio [OR] 0.09, 95% confidence interval [CI] 0.83–4.89; Table 7). By combining BK virus nephropathy with the occurrence of CMV replication, the same observation was made (Table 8).

Table 7. Genotype distribution of allele frequencies of HLA-E polymorphism in living-donor kidney transplant recipients (A) and corresponding donors (B) with respect to BK polyomavirus nephropathy (PyVAN).

(A) Recipient HLA-E genotype	PyVAN N = 11	No PyVAN N = 267	p-Value	OR	95% CI
01:03/01:03	3 (27.3%)	61 (22.8%)	0.73	1.26	0.33–4.92
01:01/01:03	8 (72.7%)	126 (47.2%)	0.10	2.98	0.77–11.50
01:01/01:01	0 (0%)	80 (30%)	0.025 *	0.09	0.005–1.59
Allele Frequencies					
01:03	14	248	0.11	2.02	0.83–4.89
01:01	8	286			
(B) Donor HLA-E genotype	PyVAN N = 11	No PyVAN N = 267	p-Value	OR	CI (95%)
01:03/01:03	1 (9.1%)	62 (23.2%)	0.27	0.33	0.04–2.63
01:01/01:03	9 (81.8%)	139 (52.1%)	0.052	4.14	0.89–19.55
01:01/01:01	1 (9.1%)	66 (24.7%)	0.24	0.3	0.04–2.42
Allele Frequencies					
01:03	11	263	0.77	1.1	0.47–2.7
01:01	10	271			

* $p < 0.05$.

Table 8. Genotype distribution of allele frequencies of HLA-E polymorphism in living-donor kidney transplant recipients (A) and corresponding donors (B) with respect to BK polyomavirus nephropathy (PyVAN) or cytomegalovirus (CMV) infection.

(A) Recipient HLA-E genotype	PyVAN or CMV Infection N = 46	No PyVAN and no CMV Infection N = 267	*p*-Value	OR	95% CI
01:03/01:03	12 (26.1%)	52 (22.4%)	0.57	1.22	0.59–2.53
01:01/01:03	28 (60.9%)	106 (45.7%)	0.07	1.85	0.96–3.53
01:01/01:01	6 (13.0%)	74 (31.9%)	0.012 *	0.32	0.13–0.79
Allele Frequencies					
01:03	52	210	0.052	1.57	1.00–2.47
01:01	40	254			
(B) Donor HLA-E genotype	PyVAN or CMV Infection N = 46	No PyVAN and no CMV Infection N = 267	*p*-Value	OR	CI (95%)
01:03/01:03	12 (9.1%)	51 (22.0%)	0.44	1.33	0.64–2.79
01:01/01:03	26 (81.8%)	122 (52.6%)	0.75	1.17	0.62–2.22
01:01/01:01	8 (9.1%)	59 (25.4%)	0.34	0.62	0.27–1.39
Allele Frequencies					
01:03	50	224	0.31	1.3	0.81–1.99
01:01	42	240			

* p < 0.05.

Considering the time course of PyVAN occurrence, the results of Kaplan–Meier analysis combined with those of the log-rank test (Figure 1) indicate that recipients carrying the HLA-E*01:03 allele were at significantly higher risk of PyVAN during the first three years after kidney transplant as compared to non-carriers ($p = 0.03$; OR = 4.25; 95% CI 1.11–16.23). Of note, all 11 recipients with PyVAN carried at least one HLA-E*01:03 allele.

With regard to HLA-E polymorphism of donor or recipient, we found no association with the occurrence or level of BKPyV viremia.

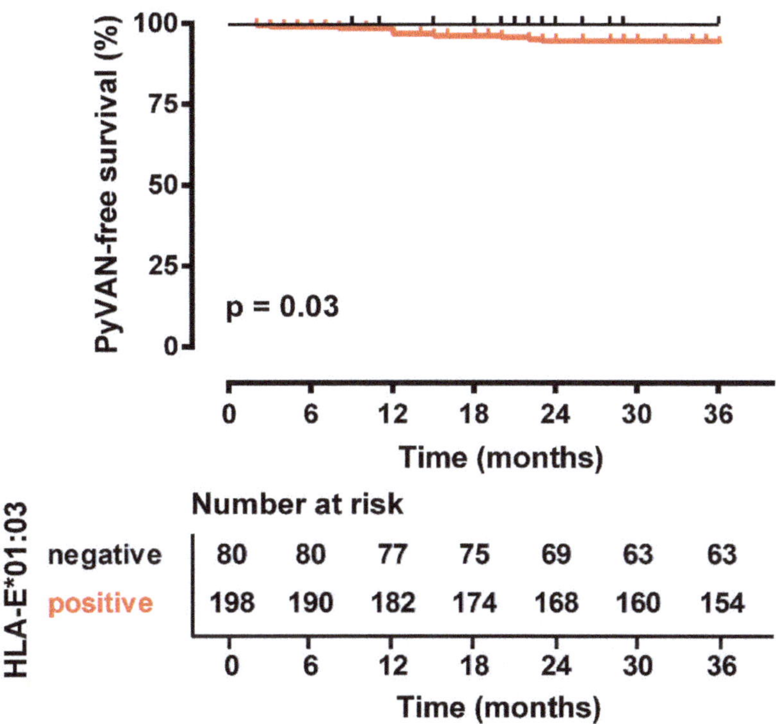

Figure 1. Association between the PyVAN-free survival and the HLA-E*01:03 carrier status of the transplant recipient during the first three years after living-donor kidney transplant. Recipients carrying the HLA-E*01:03 allelic variant were significantly more susceptible to develop PyVAN.

4. Discussion

To gain insights into BKPyV replication after kidney transplant, we evaluated the role of HLA-E genotypes in a large cohort of living kidney transplant pairs. To our knowledge, this is the first study demonstrating that the HLA-E polymorphism is associated with the pathogenesis of PyVAN after kidney transplant. Our study indicates that considering the time course of PyVAN occurrence, the prevalence of the HLA-E*01:03 allelic variant among living-donor kidney transplant recipients with PyVAN was significantly higher than that among recipients with no PyVAN. In addition, the homozygous status of the HLA-E*01:01 allele in the recipient is associated with protection against PyVAN.

This finding is in line with those of the previous studies attributing the distinctive susceptibility of patients with the HLA-E polymorphism to various viral infections. Recently, we found that HLA-E*01:03 carrier status is associated with cytomegalovirus infection after kidney transplant. Similarly, Schulte et al. reported that the homozygous HLA-E*01:01 genotype exerts a protective effect against Hepatitis C virus [16,43].

Despite the highly negative impact of BKPyV on patient and allograft survival, little is known about the immunology of BKPyV infection. Patients with a high BKPyV viral load are particularly likely to develop PyVAN and to experience allograft loss. The implementation of regular BKPyV monitoring has become a useful tool allowing the detection of patients at risk of PyVAN. The current treatment strategy against sustained BKPyV replication is a stepwise reduction or modification of the immunosuppression so that the endogenous immune control of the virus is gradually restored, thereby potentially preventing the development of PyVAN. However, this approach must be balanced against the increasing risk of allograft rejection and is not always successful. The identification of additional predictive markers indicating which patients are especially prone to PyVAN is therefore important. In combination with viral molecular monitoring, the analysis of specific immune biomarkers can make a decisive contribution in supporting the surveillance and treatment of kidney transplant recipients at risk of PyVAN. This approach will allow new preventive measures, alternative treatment strategies, and patient-tailored immunosuppression. As is the case for cytomegalovirus infection, clinical evidence indicate that the use of mTOR-based immunosuppressive regimens may be protective against BKPyV replication [24,26].

Only three patients had a CMV and BKPyV co-replication in this cohort. In such cases of combined BK virus nephropathy and CMV replication, it became evident that both events were associated with HLA-E*01:03 recipient carrier status. However, we cannot exclude that CMV additionally promotes development of BK virus nephropathy. After transplantation, CMV DNAemia has been associated with a general state of over-immunosuppression, potentially increasing risk of complications elicited by opportunistic infections [44]. Interestingly, however, Reischig et al. recently reported that the risk of BK viremia as well as PyVAN is decreased in patients who experienced CMV DNAemia [45]. At this stage, it is impossible to clarify if CMV affected the occurrence and/or severity of PyVAN in the three cases.

The host immune response is essential in controlling the viral carrier state. Increasing amount of evidences suggest that the immunomodulatory molecule HLA-E plays an important role in viral immunity. HLA-E is broadly expressed at low levels, and its expression can be up-regulated during cellular stress, such as viral infection. HLA-E is special within the immune system because it not only exhibits regulatory functions in innate and adaptive immune responses, but also promotes activating as well as inhibitory signals. These differences in HLA-E functions lead to differential effects on T- and NK-cell activity and depend on three determinants: (i) the nature of the presented peptides, (ii) the cognate receptor repertoire, and (iii) the expression level of the HLA-E alleles. HLA-E preferentially binds leader peptides sequences derived from classical HLA-class I molecules. In addition to presenting self-antigens, HLA-E can bind several nonameric peptides of viral origin [35,46,47]. It has been shown that the binding repertoire of HLA-E is broad and that the diversity of HLA-E binding motifs can considerably differ from the leader peptides sequences of classical HLA-class I molecules [35,41]. To date no study has shown that HLA-E binds BKPyV-specific peptides, however, the bioinformatics algorithm NetMHC 4.0 [36,37] has identified several peptide sequences derived from BKPyV early antigens (large-T and small-t) and late antigens (VP1 and VP2) that may bind to HLA-E with high affinity. The predicted binding affinity of the BKPyV-derived antigens in our study falls between the values of the leader peptide sequences of HLA-G and HLA-Cw*3 molecules and the HIV Gag-derived peptide. Of note the HIV Gag-peptide is a NetMHC predicted epitope being a homologous to the Simian immunodeficiency virus (SIV) Gag-derived peptide, which has recently been shown to comprise one of the two supertopes recognized by 100% of rhesus macaques vaccinated against SIV [48].

The presented peptide is of great importance because the interaction between the HLA-E-peptide and the cognate activating CD94/NKG2C or inhibitory CD94/NKG2A receptors expressed on the surface of NK cells and certain T cells has been described as peptide-sensitive [49,50]. Analysis of amino acid preferences for HLA-E binding at each anchor position demonstrated a clear preference for methionine (Met) at positions 2 and a leucine (Leu) at position 9 [9,35,42]. Three of the four BKPyV-derived peptide sequences identified in this study have a Leu at position 9, and one also has a Met at position 2, rendering HLA-E binding rather likely. Intriguingly, the activating receptor CD94/NKG2C is found less frequently and binds to the ligand HLA-E with lower affinity and stricter peptide selectivity than the inhibitory CD94/NKG2A receptor [51,52]. HLA-presented peptides are also of importance for the regulation of adaptive NK-cells expressing the activating CD94/NKG2C receptor [53–58]. Recently, Horowitz A et al. demonstrated that the binding affinity of the presented HLA-E peptide by affection of HLA-E cell surface expression impacts on the NK cell education through its influence on HLA-E cell surface disposition. As a result of enhanced HLA-E expression, NK cell education is dominated by CD94/NKG2A, whereas in individuals with peptide motifs with low binding affinity to HLA-E, NK cell education is dominated by inhibitory killer-cell immunoglobulin-like receptors (KIRs) [59].

Moreover, viral infections may imprint on the CD94/NKG2 receptor compartment, as has been shown for cytomegalovirus and Hepatitis C virus [60]. Studies using mice infected with polyomavirus (PyV) found that CD94/NKG2A expression is rapidly induced on antiviral CD8+ T-cells, thereby reducing the cytotoxic activity of PyV-specific CD8+ T-cells [61–64]. Since the HLA-E*01:03 variant consistently shows higher cell surface expression compared to HLA-E*01:01, the baseline engagement of the inhibitory CD94/NKG2A might even be higher during viral infection, thus promoting viral immune escape. With regard to HLA-E*01:01, herein described potential HLA-E*01:01-specific T cell epitopes might be recognized by CD8+ T cells leading to immune activation.

In our cohort, the recipient homozygous genotype HLA-E*01:01 was associated with protection against PyVAN, whereas HLA-E*01:03 carrier status in recipients was associated with an increased risk of PyVAN after kidney transplant. In the context of transplant, it is necessary to take recipient- and donor-derived genetic factors that may influence the outcome of viral infection into account. Although the two HLA-E alleles differ only by a single amino acid, this subtle difference has important functional implications, because HLA-E*01:03 proteins exhibit consistently higher surface expression than does HLA-E*01:01 [11,65]. This discrepancy in cell surface expression suggests differences in the effective engagement of the CD94/NKG2A or C receptors expressed on NK and T cells. With a dominant presence of the inhibitory CD94/NKG2A receptor on effector cells, the HLA-E*01:03 variant may be related to a more pronounced dampening of inhibitory receptor responses and, thus, to insufficient immune control of viruses like CMV and BKPyV. Conversely, the presence of the homozygous HLA-E*01:01 variant may result in a lower probability of interaction with its cognate inhibitory CD94/NKG2A receptor, a condition resulting in increased susceptibility to lysis mediated by effector cells.

However, this single-center analysis has its limitations due to the retrospective nature of the study and the low incidence of PyVAN in our cohort, which did not allow multivariate analysis. Additionally, our conclusions on the effects of HLA-E polymorphism on BKPyV viremia are limited due to the changes in the screening procedure for BKPyV viremia during study period. Nevertheless, our results provide a rational for future prospective clinical studies and detailed mechanistic analyses.

5. Conclusions

The data presented here suggest that the HLA-E genetic predisposition of the recipients may influence the susceptibility to PyVAN. Thus, testing for the HLA-E polymorphism may enable physicians to determine which patients are at risk of PyVAN.

Supplementary Materials: The following are available online at http://www.mdpi.com/2073-4409/8/8/847/s1, Table S1a-i: High- and moderate peptide sequences that may be presented by HLA-E*01:01 allelic variations (as identified by the machine-learning bioinformatics algorithm NetMHC 4.0) deriving from (a) CMV Protein UL-40, (b) CMV Protein UL-18, (c) CMV Phosphorylated matrix protein pp65 (UL83), (d) CMV immediate-early protein 1 (UL123), (e) Human leukocyte antigen-G, (f) Human leukocyte antigen-Cw*3, (g) Human heat shock protein (hsp60), (h) Human immundeficency virus 1 Gag protein, (i) Mycobacterium tuberculosis Enoyl -[acyl-carrier-protein]-reductase [NADH].

Author Contributions: H.R., P.A.H., A.K., O.W., V.R.: conceived and designed research. H.R., R.T.M., S.S.: performed the experiments. F.M.H.: contributed reagents. S.D., A.G., B.W., J.K.: collected and provided clinical data. H.R., R.T.M., S.D., M.T., V.R.: interpreted data and H.R., R.T.M., A.G., B.W., J.K., V.R.: performed statistical analysis. H.R. and V.R.: wrote the initial draft. H.R., R.T.M., S.S., S.D., A.G., M.T., B.W., J.K., F.M.H., P.A.H., A.K., O.W., V.R.: read and approved the final article.

Funding: We acknowledge support by the Open Access Publication Fund of the University Duisburg Essen. H.R. was supported by the "Programm zur internen Forschungsförderung Essen" (IFORES) Research fellowship program of the University Duisburg-Essen Medical School". R.M.T. was supported by the "Coordenação de Aperfeicoamento de Pessoal de Nível Superior (CAPES) Foundation, Ministry of Education of Brazil, Brasília-DF Brazil (99999.000124/2016-08)", the "Conselho Nacional de Desenvolvimento Científico e Tecnológico" (CNPq) Foundation, Ministry of Science and Technology of Brazil, Brasília-DF Brazil (142475/2015-7) and by the "Deutsche Akademische Austauschdienst" (DAAD) Scholarship of the German Federal Ministry of Education and Research. O.W. is supported by an unrestricted grant of the Rudolf-Ackermann-Stiftung (Stiftung für Klinische Infektiologie).

Acknowledgments: We thank the donors and recipients for participating in this study and Florence M. Witte, (Bluegrass Editorial Services Team, Lexington, Kentucky, USA) for editorial assistance.

Conflicts of Interest: The authors declare no conflict of interest.

References

1. Sullivan, L.C.; Clements, C.S.; Rossjohn, J.; Brooks, A.G. The major histocompatibility complex class Ib molecule HLA-E at the interface between innate and adaptive immunity. *Tissue Antigens* **2008**, *72*, 415–424. [CrossRef] [PubMed]

2. Joosten, S.A.; Sullivan, L.C.; Ottenhoff, T.H. Characteristics of HLA-E Restricted T-Cell Responses and Their Role in Infectious Diseases. *J. Immunol. Res.* **2016**, *2016*, 2695396. [CrossRef] [PubMed]

3. Romagnani, C.; Pietra, G.; Falco, M.; Mazzarino, P.; Moretta, L.; Mingari, M.C. HLA-E-restricted recognition of human cytomegalovirus by a subset of cytolytic T lymphocytes. *Hum. Immunol.* **2004**, *65*, 437–445. [CrossRef] [PubMed]

4. Petrie, E.J.; Clements, C.S.; Lin, J.; Sullivan, L.C.; Johnson, D.; Huyton, T.; Heroux, A.; Hoare, H.L.; Beddoe, T.; Reid, H.H.; et al. CD94-NKG2A recognition of human leukocyte antigen (HLA)-E bound to an HLA class I leader sequence. *J. Exp. Med.* **2008**, *205*, 725–735. [CrossRef] [PubMed]

5. Lauterbach, N.; Voorter, C.E.; Tilanus, M.G. Molecular typing of HLA-E. *Methods Mol. Biol.* **2012**, *882*, 143–158. [CrossRef] [PubMed]

6. Lee, N.; Llano, M.; Carretero, M.; Ishitani, A.; Navarro, F.; Lopez-Botet, M.; Geraghty, D.E. HLA-E is a major ligand for the natural killer inhibitory receptor CD94/NKG2A. *Proc. Natl. Acad. Sci. USA* **1998**, *95*, 5199–5204. [CrossRef] [PubMed]

7. Romagnani, C.; Pietra, G.; Falco, M.; Millo, E.; Mazzarino, P.; Biassoni, R.; Moretta, A.; Moretta, L.; Mingari, M.C. Identification of HLA-E-specific alloreactive T lymphocytes: A cell subset that undergoes preferential expansion in mixed lymphocyte culture and displays a broad cytolytic activity against allogeneic cells. *Proc. Natl. Acad. Sci. USA* **2002**, *99*, 11328–11333. [CrossRef] [PubMed]

8. Garcia, P.; Llano, M.; de Heredia, A.B.; Willberg, C.B.; Caparros, E.; Aparicio, P.; Braud, V.M.; Lopez-Botet, M. Human T cell receptor-mediated recognition of HLA-E. *Eur. J. Immunol.* **2002**, *32*, 936–944. [CrossRef]

9. Braud, V.; Jones, E.Y.; McMichael, A. The human major histocompatibility complex class Ib molecule HLA-E binds signal sequence-derived peptides with primary anchor residues at positions 2 and 9. *Eur. J. Immunol.* **1997**, *27*, 1164–1169. [CrossRef]

10. Felicio, L.P.; Porto, I.O.; Mendes-Junior, C.T.; Veiga-Castelli, L.C.; Santos, K.E.; Vianello-Brondani, R.P.; Sabbagh, A.; Moreau, P.; Donadi, E.A.; Castelli, E.C. Worldwide HLA-E nucleotide and haplotype variability reveals a conserved gene for coding and 3′ untranslated regions. *Tissue Antigens* **2014**, *83*, 82–93. [CrossRef]

11. Ulbrecht, M.; Couturier, A.; Martinozzi, S.; Pla, M.; Srivastava, R.; Peterson, P.A.; Weiss, E.H. Cell surface expression of HLA-E: Interaction with human beta2-microglobulin and allelic differences. *Eur. J. Immunol.* **1999**, *29*, 537–547. [CrossRef]

12. Strong, R.K.; Holmes, M.A.; Li, P.; Braun, L.; Lee, N.; Geraghty, D.E. HLA-E allelic variants. Correlating differential expression, peptide affinities, crystal structures, and thermal stabilities. *J. Biol. Chem.* **2003**, *278*, 5082–5090. [CrossRef] [PubMed]

13. Pyo, C.W.; Williams, L.M.; Moore, Y.; Hyodo, H.; Li, S.S.; Zhao, L.P.; Sageshima, N.; Ishitani, A.; Geraghty, D.E. HLA-E, HLA-F, and HLA-G polymorphism: Genomic sequence defines haplotype structure and variation spanning the nonclassical class I genes. *Immunogenetics* **2006**, *58*, 241–251. [CrossRef] [PubMed]

14. Lajoie, J.; Hargrove, J.; Zijenah, L.S.; Humphrey, J.H.; Ward, B.J.; Roger, M. Genetic variants in nonclassical major histocompatibility complex class I human leukocyte antigen (HLA)-E and HLA-G molecules are associated with susceptibility to heterosexual acquisition of HIV-1. *J. Infect. Dis.* **2006**, *193*, 298–301. [CrossRef] [PubMed]

15. Zidi, I.; Laaribi, A.B.; Bortolotti, D.; Belhadj, M.; Mehri, A.; Yahia, H.B.; Babay, W.; Chaouch, H.; Zidi, N.; Letaief, A.; et al. HLA-E polymorphism and soluble HLA-E plasma levels in chronic hepatitis B patients. *Hla* **2016**, *87*, 153–159. [CrossRef] [PubMed]

16. Guberina, H.; da Silva Nardi, F.; Michita, R.T.; Dolff, S.; Bienholz, A.; Heinemann, F.M.; Wilde, B.; Trilling, M.; Horn, P.A.; Kribben, A.; et al. Susceptibility of HLA-E*01:03 Allele Carriers to Develop Cytomegalovirus Replication After Living-Donor Kidney Transplantation. *J. Infect. Dis.* **2018**, *217*, 1918–1922. [CrossRef] [PubMed]

17. Hirsch, H.H.; Randhawa, P. AST Infectious Diseases Community of Practice. BK polyomavirus in solid organ transplantation-Guidelines from the American Society of Transplantation Infectious Diseases Community of Practice. *Clin. Transplant.* **2019**. [CrossRef]

18. Hirsch, H.H.; Randhawa, P. AST Infectious Diseases Community of Practice. BK polyomavirus in solid organ transplantation. *Am. J. Transplant.* **2013**, *13* (Suppl. 4), 179–188. [CrossRef] [PubMed]

19. Babel, N.; Volk, H.D.; Reinke, P. BK polyomavirus infection and nephropathy: The virus-immune system interplay. *Nat. Rev. Nephrol.* **2011**, *7*, 399–406. [CrossRef]

20. Bressollette-Bodin, C.; Coste-Burel, M.; Hourmant, M.; Sebille, V.; Andre-Garnier, E.; Imbert-Marcille, B.M. A prospective longitudinal study of BK virus infection in 104 renal transplant recipients. *Am. J. Transplant.* **2005**, *5*, 1926–1933. [CrossRef]

21. Korth, J.; Widera, M.; Dolff, S.; Guberina, H.; Bienholz, A.; Brinkhoff, A.; Anastasiou, O.E.; Kribben, A.; Dittmer, U.; Verheyen, J.; et al. Impact of low-level BK polyomavirus viremia on intermediate-term renal allograft function. *Transpl. Infect. Dis.* **2018**, *20*, e12817. [CrossRef]

22. Berger, S.P.; Sommerer, C.; Witzke, O.; Tedesco, H.; Chadban, S.; Mulgaonkar, S.; Qazi, Y.; de Fijter, J.W.; Oppenheimer, F.; Cruzado, J.M.; et al. Two-year outcomes in de novo renal transplant recipients receiving everolimus-facilitated calcineurin inhibitor reduction regimen from TRANSFORM study. *Am. J. Transplant.* **2019**. [CrossRef]

23. Budde, K.; Lehner, F.; Sommerer, C.; Reinke, P.; Arns, W.; Eisenberger, U.; Wuthrich, R.P.; Muhlfeld, A.; Heller, K.; Porstner, M.; et al. Five-year outcomes in kidney transplant patients converted from cyclosporine to everolimus: The randomized ZEUS study. *Am. J. Transplant.* **2015**, *15*, 119–128. [CrossRef] [PubMed]

24. Jouve, T.; Rostaing, L.; Malvezzi, P. Place of mTOR inhibitors in management of BKV infection after kidney transplantation. *J. Nephropathol.* **2016**, *5*, 1–7. [CrossRef] [PubMed]

25. Korth, J.; Anastasiou, O.E.; Verheyen, J.; Dickow, J.; Sertznig, H.; Frericks, N.; Bleekmann, B.; Kribben, A.; Brinkhoff, A.; Wilde, B.; et al. Impact of immune suppressive agents on the BK-Polyomavirus non coding control region. *Antivir. Res.* **2018**, *159*, 68–76. [CrossRef]

26. Acott, P.; Babel, N. BK virus replication following kidney transplant: Does the choice of immunosuppressive regimen influence outcomes? *Ann. Transplant.* **2012**, *17*, 86–99. [CrossRef]

27. Kopp, J.B. Banff Classification of Polyomavirus Nephropathy: A New Tool for Research and Clinical Practice. *J. Am. Soc. Nephrol.* **2018**, *29*, 354–355. [CrossRef]

28. Kotton, C.N.; Kumar, D.; Caliendo, A.M.; Huprikar, S.; Chou, S.; Danziger-Isakov, L.; Humar, A. The Third International Consensus Guidelines on the Management of Cytomegalovirus in Solid-organ Transplantation. *Transplantation* **2018**, *102*, 900–931. [CrossRef]

29. Sagedal, S.; Hartmann, A.; Nordal, K.P.; Osnes, K.; Leivestad, T.; Foss, A.; Degre, M.; Fauchald, P.; Rollag, H. Impact of early cytomegalovirus infection and disease on long-term recipient and kidney graft survival. *Kidney Int.* **2004**, *66*, 329–337. [CrossRef]

30. Sagedal, S.; Nordal, K.P.; Hartmann, A.; Sund, S.; Scott, H.; Degre, M.; Foss, A.; Leivestad, T.; Osnes, K.; Fauchald, P.; et al. The impact of cytomegalovirus infection and disease on rejection episodes in renal allograft recipients. *Am. J. Transplant.* **2002**, *2*, 850–856. [CrossRef]

31. Petersen, P.; Schneeberger, H.; Schleibner, S.; Illner, W.D.; Hofmann, G.O.; Land, W. Positive donor and negative recipient cytomegalovirus status is a detrimental factor for long-term renal allograft survival. *Transpl. Int.* **1994**, *7* (Suppl. 1), S336–S338. [CrossRef] [PubMed]

32. Ljungman, P.; Boeckh, M.; Hirsch, H.H.; Josephson, F.; Lundgren, J.; Nichols, G.; Pikis, A.; Razonable, R.R.; Miller, V.; Griffiths, P.D.; et al. Definitions of Cytomegalovirus Infection and Disease in Transplant Patients for Use in Clinical Trials. *Clin. Infect. Dis.* **2017**, *64*, 87–91. [CrossRef] [PubMed]

33. Grimsley, C.; Kawasaki, A.; Gassner, C.; Sageshima, N.; Nose, Y.; Hatake, K.; Geraghty, D.E.; Ishitani, A. Definitive high resolution typing of HLA-E allelic polymorphisms: Identifying potential errors in existing allele data. *Tissue Antigens* **2002**, *60*, 206–212. [CrossRef] [PubMed]

34. Guberina, H.; Rebmann, V.; Wagner, B.; da Silva Nardi, F.; Dziallas, P.; Dolff, S.; Bienholz, A.; Wohlschlaeger, J.; Bankfalvi, A.; Heinemann, F.M.; et al. Association of high HLA-E expression during acute cellular rejection and numbers of HLA class I leader peptide mismatches with reduced renal allograft survival. *Immunobiology* **2017**, *222*, 536–543. [CrossRef] [PubMed]

35. Kraemer, T.; Celik, A.A.; Huyton, T.; Kunze-Schumacher, H.; Blasczyk, R.; Bade-Doding, C. HLA-E: Presentation of a Broader Peptide Repertoire Impacts the Cellular Immune Response-Implications on HSCT Outcome. *Stem Cells Int.* **2015**, *2015*, 346714. [CrossRef] [PubMed]

36. Andreatta, M.; Nielsen, M. Gapped sequence alignment using artificial neural networks: Application to the MHC class I system. *Bioinformatics* **2016**, *32*, 511–517. [CrossRef] [PubMed]

37. Nielsen, M.; Lundegaard, C.; Worning, P.; Lauemoller, S.L.; Lamberth, K.; Buus, S.; Brunak, S.; Lund, O. Reliable prediction of T-cell epitopes using neural networks with novel sequence representations. *Protein Sci.* **2003**, *12*, 1007–1017. [CrossRef]

38. Heatley, S.L.; Pietra, G.; Lin, J.; Widjaja, J.M.; Harpur, C.M.; Lester, S.; Rossjohn, J.; Szer, J.; Schwarer, A.; Bradstock, K.; et al. Polymorphism in human cytomegalovirus UL40 impacts on recognition of human leukocyte antigen-E (HLA-E) by natural killer cells. *J. Biol. Chem.* **2013**, *288*, 8679–8690. [CrossRef]

39. Michaelsson, J.; Teixeira de Matos, C.; Achour, A.; Lanier, L.L.; Karre, K.; Soderstrom, K. A signal peptide derived from hsp60 binds HLA-E and interferes with CD94/NKG2A recognition. *J. Exp. Med.* **2002**, *196*, 1403–1414. [CrossRef]

40. Nattermann, J.; Nischalke, H.D.; Hofmeister, V.; Kupfer, B.; Ahlenstiel, G.; Feldmann, G.; Rockstroh, J.; Weiss, E.H.; Sauerbruch, T.; Spengler, U. HIV-1 infection leads to increased HLA-E expression resulting in impaired function of natural killer cells. *Antivir. Ther.* **2005**, *10*, 95–107.

41. Walters, L.C.; Harlos, K.; Brackenridge, S.; Rozbesky, D.; Barrett, J.R.; Jain, V.; Walter, T.S.; O'Callaghan, C.A.; Borrow, P.; Toebes, M.; et al. Pathogen-derived HLA-E bound epitopes reveal broad primary anchor pocket tolerability and conformationally malleable peptide binding. *Nat. Commun.* **2018**, *9*, 3137. [CrossRef] [PubMed]

42. Schulte, D.; Vogel, M.; Langhans, B.; Kramer, B.; Korner, C.; Nischalke, H.D.; Steinberg, V.; Michalk, M.; Berg, T.; Rockstroh, J.K.; et al. The HLA-E(R)/HLA-E(R) genotype affects the natural course of hepatitis C virus (HCV) infection and is associated with HLA-E-restricted recognition of an HCV-derived peptide by interferon-gamma-secreting human CD8(+) T cells. *J. Infect. Dis.* **2009**, *200*, 1397–1401. [CrossRef] [PubMed]

43. Fishman, J.A. Infection in Organ Transplantation. *Am. J. Transplant.* **2017**, *17*, 856–879. [CrossRef] [PubMed]

44. Reischig, T.; Kacer, M.; Hes, O.; Machova, J.; Nemcova, J.; Lysak, D.; Jindra, P.; Pivovarcikova, K.; Kormunda, S.; Bouda, M. Cytomegalovirus prevention strategies and the risk of BK polyomavirus viremia and nephropathy. *Am. J. Transplant.* **2019**. [CrossRef] [PubMed]

45. Celik, A.A.; Kraemer, T.; Huyton, T.; Blasczyk, R.; Bade-Doding, C. The diversity of the HLA-E-restricted peptide repertoire explains the immunological impact of the Arg107Gly mismatch. *Immunogenetics* **2016**, *68*, 29–41. [CrossRef] [PubMed]

46. Lampen, M.H.; Hassan, C.; Sluijter, M.; Geluk, A.; Dijkman, K.; Tjon, J.M.; de Ru, A.H.; van der Burg, S.H.; van Veelen, P.A.; van Hall, T. Alternative peptide repertoire of HLA-E reveals a binding motif that is strikingly similar to HLA-A2. *Mol. Immunol.* **2013**, *53*, 126–131. [CrossRef]

47. Hansen, S.G.; Wu, H.L.; Burwitz, B.J.; Hughes, C.M.; Hammond, K.B.; Ventura, A.B.; Reed, J.S.; Gilbride, R.M.; Ainslie, E.; Morrow, D.W.; et al. Broadly targeted CD8(+) T cell responses restricted by major histocompatibility complex E. *Science* **2016**, *351*, 714–720. [CrossRef]

48. Llano, M.; Lee, N.; Navarro, F.; Garcia, P.; Albar, J.P.; Geraghty, D.E.; Lopez-Botet, M. HLA-E-bound peptides influence recognition by inhibitory and triggering CD94/NKG2 receptors: Preferential response to an HLA-G-derived nonamer. *Eur. J. Immunol.* **1998**, *28*, 2854–2863. [CrossRef]

49. Stevens, J.; Joly, E.; Trowsdale, J.; Butcher, G.W. Peptide binding characteristics of the non-classical class Ib MHC molecule HLA-E assessed by a recombinant random peptide approach. *BMC Immunol.* **2001**, *2*, 5. [CrossRef]

50. Miller, J.D.; Weber, D.A.; Ibegbu, C.; Pohl, J.; Altman, J.D.; Jensen, P.E. Analysis of HLA-E peptide-binding specificity and contact residues in bound peptide required for recognition by CD94/NKG2. *J. Immunol.* **2003**, *171*, 1369–1375. [CrossRef]

51. Vales-Gomez, M.; Reyburn, H.T.; Erskine, R.A.; Lopez-Botet, M.; Strominger, J.L. Kinetics and peptide dependency of the binding of the inhibitory NK receptor CD94/NKG2-A and the activating receptor CD94/NKG2-C to HLA-E. *EMBO J.* **1999**, *18*, 4250–4260. [CrossRef]

52. Kaiser, B.K.; Barahmand-Pour, F.; Paulsene, W.; Medley, S.; Geraghty, D.E.; Strong, R.K. Interactions between NKG2x immunoreceptors and HLA-E ligands display overlapping affinities and thermodynamics. *J. Immunol.* **2005**, *174*, 2878–2884. [CrossRef]

53. Guma, M.; Angulo, A.; Vilches, C.; Gomez-Lozano, N.; Malats, N.; Lopez-Botet, M. Imprint of human cytomegalovirus infection on the NK cell receptor repertoire. *Blood* **2004**, *104*, 3664–3671. [CrossRef]

54. Guma, M.; Budt, M.; Saez, A.; Brckalo, T.; Hengel, H.; Angulo, A.; Lopez-Botet, M. Expansion of CD94/NKG2C+ NK cells in response to human cytomegalovirus-infected fibroblasts. *Blood* **2006**, *107*, 3624–3631. [CrossRef]

55. Hammer, Q.; Ruckert, T.; Borst, E.M.; Dunst, J.; Haubner, A.; Durek, P.; Heinrich, F.; Gasparoni, G.; Babic, M.; Tomic, A.; et al. Peptide-specific recognition of human cytomegalovirus strains controls adaptive natural killer cells. *Nat. Immunol.* **2018**, *19*, 453–463. [CrossRef]

56. Rolle, A.; Jager, D.; Momburg, F. HLA-E Peptide Repertoire and Dimorphism-Centerpieces in the Adaptive NK Cell Puzzle? *Front. Immunol.* **2018**, *9*, 2410. [CrossRef]

57. Rolle, A.; Meyer, M.; Calderazzo, S.; Jager, D.; Momburg, F. Distinct HLA-E Peptide Complexes Modify Antibody-Driven Effector Functions of Adaptive NK Cells. *Cell Rep.* **2018**, *24*, 1967–1976. [CrossRef]

58. Boudreau, J.E.; Hsu, K.C. Natural Killer Cell Education and the Response to Infection and Cancer Therapy: Stay Tuned. *Trends Immunol.* **2018**, *39*, 222–239. [CrossRef]

59. Horowitz, A.; Djaoud, Z.; Nemat-Gorgani, N.; Blokhuis, J.; Hilton, H.G.; Beziat, V.; Malmberg, K.J.; Norman, P.J.; Guethlein, L.A.; Parham, P. Class I HLA haplotypes form two schools that educate NK cells in different ways. *Sci. Immunol.* **2016**, *1*. [CrossRef]

60. Nattermann, J.; Feldmann, G.; Ahlenstiel, G.; Langhans, B.; Sauerbruch, T.; Spengler, U. Surface expression and cytolytic function of natural killer cell receptors is altered in chronic hepatitis C. *Gut* **2006**, *55*, 869–877. [CrossRef]

61. Moser, J.M.; Gibbs, J.; Jensen, P.E.; Lukacher, A.E. CD94-NKG2A receptors regulate antiviral CD8(+) T cell responses. *Nat. Immunol.* **2002**, *3*, 189–195. [CrossRef]

62. Moser, J.M.; Byers, A.M.; Lukacher, A.E. NK cell receptors in antiviral immunity. *Curr. Opin. Immunol.* **2002**, *14*, 509–516. [CrossRef]

63. Byers, A.M.; Kemball, C.C.; Andrews, N.P.; Lukacher, A.E. Regulation of antiviral CD8+ T cells by inhibitory natural killer cell receptors. *Microbes Infect.* **2003**, *5*, 169–177. [CrossRef]

64. Byers, A.M.; Andrews, N.P.; Lukacher, A.E. CD94/NKG2A expression is associated with proliferative potential of CD8 T cells during persistent polyoma virus infection. *J. Immunol.* **2006**, *176*, 6121–6129. [CrossRef]

65. Wagner, B.; da Silva Nardi, F.; Schramm, S.; Kraemer, T.; Celik, A.A.; Durig, J.; Horn, P.A.; Duhrsen, U.; Nuckel, H.; Rebmann, V. HLA-E allelic genotype correlates with HLA-E plasma levels and predicts early progression in chronic lymphocytic leukemia. *Cancer* **2017**, *123*, 814–823. [CrossRef]

Article

Distribution of Killer-Cell Immunoglobulin-Like Receptor Genes and Combinations of Their Human Leucocyte Antigen Ligands in 11 Ethnic Populations in China

Yufeng Yao, Lei Shi, Jiankun Yu, Shuyuan Liu, Yufen Tao and Li Shi *

Institute of Medical Biology, Chinese Academy of Medical Sciences & Peking Union Medical College, Kunming 650118, China

* Correspondence: shili@imbcams.pumc.edu.cn or shili.imb@gmail.com; Tel.: +86-871-6833-5632; Fax: +86-871-6833-4483

Received: 31 May 2019; Accepted: 9 July 2019; Published: 12 July 2019

Abstract: The aim of this study was to analyze the distribution of killer-cell immunoglobulin-like receptor (KIR) genes and their human leucocyte antigen (HLA) ligand combinations in different original ethnic populations in China, and thus, to provide relevant genomic diversity data for the future study of viral infections, autoimmune diseases, and reproductive fitness. A total of 1119 unrelated individuals from 11 ethnic populations—including Hani, Jinuo, Lisu, Nu, Bulang, Wa, Dai, Maonan, Zhuang, Tu, and Yugu—from four original groups, were included. The presence/absence of the 16 KIR loci were detected, and the KIR gene's phenotype, genotype, and haplotype A and B frequencies, as well as KIR ligand's HLA allotype and KIR–HLA pairs for each population, were calculated. Principal component analysis and phylogenetic trees were constructed to compare the characteristics of the KIR and KIR–HLA pair distributions of these 11 populations. In total, 92 KIR genotypes were identified, including six new genotypes. The KIR and its HLA ligands had a distributed diversity in 11 ethnic populations in China, and each group had its specific KIR and KIR–HLA pair profile. The difference among the KIR–HLA pairs between northern and southern groups, but not among the four original groups, may reflect strong pressure from previous or ongoing infectious diseases, which have a significant impact on KIR and its HLA combination repertoires.

Keywords: KIR; KIR–HLA pairs; ethnic populations in China

---

1. Introduction

Segregated in different chromosomes, 6p21 and 19q13.4, human leucocyte antigen (HLA) and killer-cell immunoglobulin-like receptor (KIR) genes, respectively, exhibit diverse polymorphisms and their molecular expressions interact with each other as receptor ligands to ensure the proper role of nature killer (NK) cells in modulating an immune response [1,2]. Several studies on the coinheritance of these two genetic systems have indicated that carrying the appropriate KIR–HLA combination is important for human survival [3,4]. During human migration outward from Africa and the successive colonization worldwide, the cooperative KIR haplotypes and activating KIR–HLA pairs are important for humans to adapt to quickly changing environments and to increase population reproduction [5,6].

HLA polymorphism has been well studied in worldwide populations, which makes it a genetic marker for tracing a population's origin, migration, and admixture [7,8]. Compared to HLA, KIR genes show polymorphisms, both at content and allelic levels. Among 16 identified KIR genes, *2DL1*, *2DL2*, *2DL3*, *2DL5*, *3DL1*, *3DL2*, and *3DL3* belong to inhibitory KIR genes, while *2DS1*, *2DS2*, *2DS3*, *2DS4*, *KIR2DS5*, and *3DS1* belong to activating KIR genes, and *KIR2DL4* has both inhibitory and activating capacities [9]. Four framework KIR genes—*3DL3*, *3DP1*, *2DL4*, and *3DL2*—which are located from the centromeric to the telomeric region, respectively, are observed consistently in almost all individuals [10]. On the basis of gene content, KIRs are divided into two haplotype groups: A and B. The A haplotype has a fixed gene content (*3DL3-2DL3-2DL1-3DP1-2DL4-3DL1-2DS4-3DL2*) and an activating *2DS4*. Four inhibitory KIRs (*2DL1*, *2DL3*, *3DL1*, and *3DL2*) are specific for four major HLA class I ligands (*C2*, *C1*, *Bw4*, and *A3/A11*, respectively) [11]. In contrast, haplotype B is variable both in the numbers and combinations of KIR genes and comprises several genes (*2DL2*, *2DL5*, *2DS1*, *2DS2*, *2DS3*, *2DS5*, and *3DS1*) that do not exist in the A haplotype. Among them, *2DS1* binds to *HLA-C2* but with a low affinity; *3DS1* may bind to the *HLA-Bw4* allotype, especially for *Bw4-80T*; and *2DS2* may bind to *HLA-A11* [12–14]. KIR genes, genotypes, haplotypes, and KIR–HLA pairs show a diverse distribution in different populations worldwide. Examination of KIR and ethnic populations may permit analysis of the evolutionary basis for KIR variation, allowing insight into the role of these receptors in health and disease.

The 55 officially recognized ethnic populations of China, which contribute to about 8% of the overall Chinese population, provide abundant genetic resources for KIR–HLA studies. The ethnic groups living in the south and southwest of China can be traced back to three major ancient groups: Di-Qiang, Bai-Pu, and Bai-Yue [15]. According to historical records, the ancient Di-Qiang tribe migrated from northern to southern China before the Qin dynasty in 206 BC and formed several ethnic populations who spoke the language of Tibeto-Burman, which belongs to the Sino-Tibetan linguistic family [15,16]. The ancient Baipu tribe settled down in the south and southwest of Yunnan Province in China and developed into the major Mon-Khmer speaking populations of the Austo-Asiatic linguistic family. Most of the Mon-Khmer ancient tribes migrated to the Indochina Peninsula by the end of 2000 BC, whereas others remained in the Yunnan Province [15,16]. The ancient Baiyue tribe, which was widely distributed along the southeast coast of China, migrated to Yunnan Province and the northern part of Southeast Asia 2000–3000 years ago and then later on migrated to Northern Thailand and contributed to the ancestral gene pool of the Thais [15]. They formed ethnic populations who spoke a language of Daic, part of the Sino-Tibetan linguistic family [17]. There are several ethnic populations in northwestern China, such as the Tu and Yugu. It has been suggested that the Tu population originated from an ancient Xian-Bei tribe, who constructed the Kingdom of Tuguhun in 400 AD, and most of them live in Qinhai and Gansu Provinces. The Yugu originated from Hui-Hu in 600 AD and live only in Gansu Province now. Both the Tu and Yugu integrated with Mongolian and Han populations during the following centuries; moreover, both speak a language of Mongolian belonging to the Altaic linguistic family [15,17].

In the present study, 11 ethnic populations—Hani, Jinuo, Lisu, and Nu, speaking Tibeto-Burman; Bulang and Wa, speaking Mon-Khmer; Dai, Maonan, and Zhuang, speaking Daic; and Tu and Yugu, speaking Mongolian—were selected for HLA and KIR genotyping. The KIR gene's distribution in Bulang, Nu, Yugu, and Zhuang has been reported previously [18]. The presence/absence of the 16 KIR loci were detected and the KIR gene's phenotype, genotype, and A and B haplotype frequencies, as well as the KIR ligand's HLA allotype and KIR–HLA pairs, are reported. Principal component analysis (PCA) and phylogenetic trees were constructed to compare the characteristics of the KIR and KIR–HLA pair distributions in the 11 populations.

2. Material and Methods

2.1. Subject and Samples

A total of 1119 unrelated individuals were recruited from 11 Chinese ethnic populations in China. The geographic location, sample size of each population, the language family to which they belong, and the original ancient groups they are from are listed in Figure 1 and Supplementary Table S1. These populations are descended from four ancient Chinese groups and belong to four different language subfamilies as mentioned in the introduction. The geographic origin, nationalities, and pedigree (unrelated through at least three generations) of each individual were ascertained before sampling. The present study has been approved by the Committee on the Ethics of Institute of Medical Biology, Chinese Academy of Medical Sciences, the batch number is YIKESHENGLUNZI [2012]12. All the individuals are healthy and gave written informed consent in accordance with the Declaration of Helsinki.

Figure 1. The geographic locations of the 11 ethnic populations in China. dq: Di-Qiang ancient group; bp: Baipu ancient group; by: Baiyue ancient group; m: Mongolian group; TB: Tibeto-Burman sublanguage family; MK: Mon-Khmer sublanguage family; D: Daic sublanguage family; M: Mongolian sublanguage family.

Genomic DNA was extracted from peripheral lymphocytes using a QIAamp Blood Kit (Qiagen, Hilden, Germany), in accordance with the manufacturer's protocol. DNA samples were quantified with a NanoDrop ND-1000 spectrophotometer (NanoDrop Technologies, Wilmington, WI, USA) and adjusted to a concentration of 20 ng/μL.

2.2. KIR Genotyping

The 16 KIR genes were genotyped using the Luminex MultiAnalyte Profiling System (xMAP) with a One Lambda KIR typing kit (One Lambda, Canoga Park, CA, USA), as previously reported [18]. Briefly, three separate PCR products were amplified: exon 3, exon 5, and exons 7–9. The PCR products were run on 2% agarose gel to confirm the specificity and efficiency of the reactions. Then, the PCR amplicons were denatured and hybridized with complementary 81-nucleotide oligonucleotide probes that had been immobilized on fluorescent-coated microsphere beads. At the same time, the biotinylated PCR products were labeled with phycoerythrin-conjugated streptavidin and immediately examined with the Luminex 200 system (Luminex, Austin, TX, USA). Genotype determination and data analysis were performed automatically using the LABScan 100 platform (One Lambda, Canoga Park, CA, USA) in accordance with the manufacturer's instructions.

2.3. Statistical Analysis

Hardy–Weinberg's equilibrium for each of the alleles was assessed using the Guo and Thompson method [19]. For KIR genes, the observed frequency for each KIR gene was determined via direct counting and corresponded to the ratio of the number within the population that carried the gene to the total population number. KIR locus gene frequencies (KLFs) were estimated by using the formula KLF $= 1 - \sqrt{1-f}$, where f is the observed frequency of a particular KIR sequence in a population. The genotypes were defined by referring to the Allele Frequencies website (http:

//www.allelefrequencies.net). Each genotype was named in accordance with the genotype number. The genotypes that could not be found in the database were named as unknown. Group A and B haplotypes and frequencies were predicted on the basis of a previous study [20].

HLA genotyping of 11 ethnic populations has been reported previously [21–25], and the allelic frequencies are summarized in Supplementary Table S2. The frequencies of *HLA-C1/C2* allotype, *HLA-Bw4* (*Bw4-80I* and *Bw4-80T*)/*Bw6* allotype, and *HLA-A11, -A3* were calculated using direct counting. The *HLA-C1* and *HLA-C2, HLA-A11* and *HLA-A3*, and *HLA-Bw4/Bw6* groups were used to analyze KIR–HLA combinations. The observed frequencies of KIR–HLA matched pairs were calculated using direct counting. The significance of the correlations of KIR and HLA frequencies among populations were estimated using the correlation coefficient (r) and the *t*-test was used to establish whether the correlation coefficient was significant using SPSS 16.0 [26]. The chord distance of Nei (Da distances among the populations were calculated based on 11 KIR gene (*2DL1, 2DL2, 2DL3, 3DL1, 2DS1, 2DS2, 2DS3, 2DS4, 2DS5, 3DS1*, and *2DL5*) frequencies; *HLA-A, HLA-B*, and *HLA-C* allele frequencies; or HLA–KIR combinations. A neighbor-joining (NJ) tree was constructed using Mega 7.0 software based on the DA distance [27]. Principal component analysis (PCA) was also performed based either on KIR genes, HLA alleles, or KIR–HLA combination frequencies using SPSS 16.0 software [26]. Significant differences in KIR and KIR–HLA pair frequencies between two populations were determined using a contingency test. The difference between the northern and southern groups were detected using a *t*-test with SPSS 16.0 software [26]. A value of $p < 0.05$ was considered to be statistically significant. The observed 11 KIR gene (*2DL1, 2DL2, 2DL3, 3DL1, 2DS1, 2DS2, 2DS3, 2DS4, 2DS5, 3DS1*, and *2DL5*) frequencies of 47 other populations were from previous studies and the DA distance among 58 populations were calculated (Supplementary Tables S5 and S6). The phylogenetic tree was constructed based on the DA distance using the minimum evolution method from Mega 7.0 software [28].

3. Results

3.1. KIR Gene, Genotype, and Haplotype Frequencies

The observed KIR frequencies and the estimated gene frequencies for each locus in the 11 populations are listed in Table 1. The four framework loci (*KIR3DL3, 3DP1, 2DL4*, and *3DL2*) were exhibited in all individuals in 10 populations, except one individual in Yugu, for whom *2DL4* and *3DP1* were not observed. The non-framework pseudogene *2DP1* was observed in all individuals in Hani, Nu, Dai, Zhuang, and Tu, but not in all other populations, with frequencies of 94.8%–99.1%. The frequencies of *3DL1, 2DL1*, and *2DS4* were about 90%–100% in 11 populations, with the exception in Jinuo and Bulang, which showed frequencies of 88% and 81% at *3DL1* and 88% and 79% at *2DS4*, respectively. Other activating KIRs, including *3DS1, 2DS1, 2DS3*, and *2DS5*, as well as inhibitory KIRs, including *2DL2* and *2DL5*, exhibited diverse distributions in different populations. Bulang was different compared with the other 10 populations at *3DS1* and *2DS1* ($p < 0.05$), 9 other populations except Zhuang at *2DS3*, and 9 other populations except Jinuo at *2DS4*. The following difference was among Nu and others. For all populations, the most diverse was at *2DS3* (Supplementary Table S3).

In total, 92 KIR genotypes were identified, including 6 new genotypes: 3 in Tu, 2 in Jinuo, and 1 in Wa (Table 2). Genotypes 1, 2, 4, and 8 were observed in all the populations but showed diverse frequencies. Genotype 1 was predominant in all populations except in Bulang. On the contrary, the predominant genotypes in Bulang were genotype 8 followed by genotypes 1, 2, and 75, as previously reported [18]. Originating from the same Baipu ancient group, Wa did not show a distribution similar to Bulang, with genotype 1 being the most predominant, followed by genotypes 2 and 4. The frequencies of genotype 1 were as high as 0.679 in Nu, 0.510 in Hani, 0.469 in Yugu, 0.465 in Lisu, 0.432 in Zhuang, and 0.425 in Wa.

Table 1. The observed KIR frequencies and the estimated gene frequencies for each locus in 11 ethnic populations.

	Hani (n = 145)		Jinuo (n = 94)		Lisu (n = 99)		Nu (n = 106)		Bulang (n = 106)		Wa (n = 107)		Dai (n = 112)		Maonan (n = 89)		Zhuang (n = 95)		Tu (n = 70)		Yugu (n = 96)	
	OF (%)	GF	OF (%)	GF	OF (%)	GF	OF (%)	GF	OF (%)	GF	OF (%)	GF	OF (%)	GF	OF (%)	GF	OF (%)	GF	OF (%)	GF	OF (%)	GF
3DL1	99	0.917	88	0.658	92	0.716	98	0.863	81	0.566	98	0.863	91	0.701	99	0.894	94	0.749	96	0.793	94	0.750
2DL1	100	1.000	100	1.000	99	0.899	100	1.000	98	0.863	100	1.000	97	0.836	99	0.894	100	1.000	100	1.000	95	0.772
2DL3	97	0.834	95	0.769	98	0.858	100	1.000	96	0.806	97	0.832	96	0.811	94	0.763	98	0.855	93	0.733	93	0.730
2DS4	99	0.917	88	0.658	92	0.716	98	0.863	79	0.544	98	0.863	90	0.687	99	0.894	94	0.749	97	0.831	92	0.711
2DL2	27	0.145	27	0.143	33	0.184	10	0.053	16	0.084	32	0.176	41	0.232	39	0.221	29	0.160	33	0.181	27	0.146
2DL5	30	0.165	35	0.194	34	0.190	26	0.142	70	0.451	24	0.126	53	0.312	35	0.193	46	0.267	33	0.181	45	0.257
3DS1	34	0.186	50	0.293	30	0.165	25	0.131	73	0.477	31	0.170	41	0.232	35	0.193	41	0.232	53	0.313	35	0.196
2DS1	34	0.191	44	0.249	38	0.215	23	0.120	73	0.477	34	0.187	43	0.244	30	0.165	37	0.205	39	0.216	41	0.229
2DS2	27	0.145	27	0.143	33	0.184	10	0.053	16	0.084	32	0.176	41	0.232	39	0.221	29	0.160	33	0.181	27	0.146
2DS3	34	0.191	29	0.156	8	0.041	8	0.038	48	0.280	11	0.058	28	0.150	31	0.172	37	0.205	13	0.066	21	0.110
2DS5	9	0.046	39	0.221	29	0.159	16	0.084	41	0.229	37	0.205	33	0.182	20	0.107	15	0.077	43	0.244	31	0.171
2DL4	100	1.000	100	1.000	100	1.000	100	1.000	100	1.000	100	1.000	100	1.000	100	1.000	100	1.000	100	1.000	99	0.898
3DL2	100	1.000	100	1.000	100	1.000	100	1.000	100	1.000	100	1.000	100	1.000	100	1.000	100	1.000	100	1.000	100	1.000
3DL3	100	1.000	100	1.000	100	1.000	100	1.000	100	1.000	100	1.000	100	1.000	100	1.000	100	1.000	100	1.000	100	1.000
2DP1	100	1.000	99	0.897	99	0.899	100	1.000	98	0.863	99	0.903	100	1.000	99	0.894	100	1.000	100	1.000	95	0.772
3DP1	100	1.000	100	1.000	100	1.000	100	1.000	100	1.000	100	1.000	100	1.000	100	1.000	100	1.000	100	1.000	99	0.898

OF: Observed Frequencies, GF: estimated Gene frequencies.

Table 2. KIR genotypes identified in 11 ethnic populations.

Hapl Group	Geno Group	Genotype ID	3DL1	2DL1	2DL3	2DS4	2DL2	2DL5	3DS1	2DS1	2DS2	2DS3	2DS5	2DL4	3DL2	3DL3	2DP1	3DP1	Hani Fre.	Jinuo Fre.	Lisu Fre.	Nu Fre.	Bulang Fre.	Wa Fre.	Dai Fre.	Maonan Fre.	Zhuang Fre.	Tu Fre.	Yugu Fre.
AA	AA	1																	0.510	0.287	0.465	0.679	0.208	0.425	0.286	0.393	0.432	0.271	0.469
AA	AA	180																	0.007	0.021	0.010							0.014	
AA	AA	203																						0.009					
AA	AA	332																							0.009				
Bx	AB	2																	0.034	0.117	0.101	0.113	0.170	0.113	0.080	0.056	0.074	0.143	0.094
Bx	AB	3																	0.007	0.021	0.040	0.009	0.028	0.009	0.063		0.021	0.043	0.021
Bx	AB	4																	0.069	0.074	0.091	0.057	0.028	0.085	0.107	0.135	0.095	0.143	0.052
Bx	AB	5																	0.007	0.011	0.020		0.019	0.019	0.063	0.022	0.042		0.021
Bx	AB	6																	0.021	0.011		0.009			0.018	0.067	0.011		0.042
Bx	AB	7																	0.055	0.011	0.030		0.028	0.009	0.009		0.063	0.014	
Bx	AB	8																	0.076	0.053	0.010	0.057	0.255	0.019	0.063	0.101	0.137	0.057	0.083
Bx	AB	9																	0.007		0.030	0.009		0.038	0.009	0.011			0.042
Bx	AB	11																	0.021			0.009			0.018				
Bx	AB	12																							0.009				
Bx	AB	13																	0.034		0.010				0.009	0.022	0.032	0.014	
Bx	AB	14																		0.032	0.020				0.018	0.022	0.011	0.014	
Bx	AB	15																	0.007		0.010				0.009				
Bx	AB	17																										0.014	
Bx	AB	19																							0.009				
Bx	AB	23																		0.011				0.019	0.009			0.014	
Bx	AB	25																							0.009				
Bx	AB	27																							0.009				
Bx	AB	30																						0.009	0.009	0.011			0.010
Bx	AB	31																		0.021									
Bx	AB	33																	0.007										
Bx	AB	35																		0.011				0.009	0.009				
Bx	AB	41																		0.011						0.011			
Bx	AB	44																	0.007					0.038		0.011			
Bx	AB	57																							0.009				
Bx	AB	62																		0.021						0.011			
Bx	AB	63																		0.011				0.028					
Bx	BB	68																			0.030	0.009	0.009		0.009	0.011			
Bx	BB	69																		0.011	0.020	0.009	0.038	0.009	0.027				0.031
Bx	BB	70																		0.011	0.010		0.009		0.018		0.011		
Bx	BB	71																	0.014	0.011						0.011	0.021		0.010
Bx	BB	72																					0.009						0.010
Bx	BB	73																							0.018				
Bx	BB	75																	0.007	0.053			0.104				0.032		0.010
Bx	BB	76																											0.021
Bx	BB	79																											0.010
Bx	BB	80																					0.009						0.010
Bx	BB	81																											0.010
Bx	BB	86																										0.014	
Bx	BB	89																								0.022			
Bx	BB	97																								0.011			0.010
Bx	BB	104																					0.009						

Table 2. *Cont.*

Hapl Group	Geno Group	Genotype ID	3DL1	2DL1	2DL3	2DS4	2DL2	2DL5	3DS1	2DS1	2DS2	2DS3	2DS5	2DL4	3DL2	3DL3	2DP1	3DP1	Hani Fre.	Jinuo Fre.	Lisu Fre.	Nu Fre.	Bulang Fre.	Wa Fre.	Dai Fre.	Maonan Fre.	Zhuang Fre.	Tu Fre.	Yugu Fre.
Bx	BB	106																			0.010								
Bx	BB	113																	0.007						0.009				
Bx	BB	117																		0.011			0.019		0.009		0.021		
Bx	BB	151																								0.011			
Bx	BB	154																							0.009				
Bx	BB	164																		0.010									
Bx	AB	188																							0.009				
Bx	AB	192																		0.030					0.019				
Bx	BB	194																											0.010
Bx	AB	202																		0.074	0.020		0.019	0.028	0.009	0.011		0.057	
Bx	AB	205																											0.010
Bx	AB	233																	0.014						0.009			0.014	
Bx	BB	243																							0.009				
Bx	BB	247																					0.009						
Bx	AB	260																								0.011			
Bx	AB	264																	0.007	0.011			0.019					0.014	
Bx	AB	268																							0.009				
Bx	BB	280																		0.011									
Bx	BB	289																							0.009				
Bx	BB	293																							0.009				
Bx	BB	317																		0.021								0.014	
Bx	AB	319																		0.011						0.011			
Bx	BB	320																					0.009						
Bx	BB	325																					0.009						
Bx	BB	331																							0.009				
Bx	AB	370																		0.010			0.009					0.014	
Bx	AB	372																	0.076	0.011			0.009	0.009					
Bx	BB	375																		0.011									
Bx	BB	379																										0.014	
Bx	AB	381																		0.010									
Bx	AB	382																	0.007						0.009				
Bx	BB	390																		0.010					0.009				
Bx	BB	394																										0.014	
Bx	AB	400																		0.011			0.019			0.011		0.029	0.010
Bx	BB	402																										0.014	
Bx	AB	433																				0.038							0.010
Bx	BB	466																								0.011			
Bx	AB	570																							0.009				
Bx	BB	578																							0.009				
Bx	AB	587																							0.009				
Bx	AB	n1*																										0.029	
Bx	AB	n2*																		0.011									
Bx	BB	n3*																										0.014	
Bx	BB	n4*																										0.014	
Bx	BB	n5*																		0.011									
Bx	BB	n6*																							0.009				

Fre.: Frequency. *n1–n6: unknown genotype ID.

The frequencies of the group A and B haplotypes in the 11 populations were deduced from the genotype data (Figure 2). As with most populations worldwide, the A haplotype was predominant. The frequencies of the A haplotype were around 0.657–0.830 in Nu, Hani, Wa, Lisu, Zhuang, Maonan, and Yugu, while they were around 0.576–0.593 in Tu, Jinuo, and Dai. In Bulang, the frequencies of the A and B haplotypes were almost equal (0.491 vs. 0.509). Haplotype differences were identified among Nu and 10 other populations except Jinuo; among Bulang and Hai, Lisu, Nu, Wa, Maonan, Zhuang, and Yugu; and between Hani and Jinuo, and Dai and Tu (Supplementary Table S3). The distributions of KIR genes, genotypes, and haplotypes did not show any consistency among their original ancient group or linguistic subfamily.

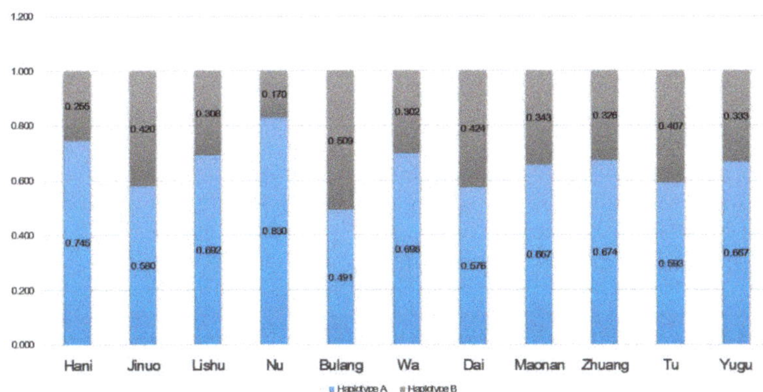

Figure 2. Haplotypes A and B frequencies in 11 ethnic populations in China.

3.2. HLA Allotype Frequencies

The frequencies of *HLA-A11/A3*, *HLA-Bw4* (*Bw4-80I* and *Bw4-80T*), and *HLA-C1* and *HLA-C2* were calculated from the *HLA-A*, *HLA-B*, and *HLA-C* allele genotyping results (Table 3). *HLA-A11/A3* were predominant in all the populations living in southern China, with frequencies higher than 0.556, and they accounted for around 80% of HLA-A alleles in Bulang, Wa, and Hani. On the contrary, *HLA-A11/A3* were around 40% in Tu and Yugu living in northern China. *HLA-Bw4* existed commonly in Tu and Yugu, with frequencies of 0.700 and 0.688, but only with frequencies of 0.283 in Bulang. *HLA-C1* was observed in all individuals in Jinuo, Lisu, Maonan, and Zhuang, with frequencies around >95% in other southern Chinese populations, but with frequencies of 0.914 and 0.833 in Tu and Yugu, respectively, from northern China. On the contrary, the frequencies of *HLA-C2* were around 50% in Tu and Yugu but was only 0.073 in Maonan. This HLA characteristic reflected the northern and southern Chinese original difference, which has been confirmed in previous studies. Therefore, we divided the present study populations into two groups: the southern group, which included Hani, Jinuo, Lisu, Nu, Bulang, Wa, Dai, Maonan, and Zhuang, and the northern group, which included Tu and Yugu. Differences between the northern and southern groups were observed (data not shown).

Table 3. HLA allotype frequencies in 11 ethnic populations.

Title	Hani (n = 145)	Jinuo (n = 94) *	Lisu (n = 99)	Nu (n = 106)	Bulang (n = 106)	Wa (n = 107)	Dai (n = 112)	Maonan (n = 89) *	Zhuang (n = 95)	Tu (n = 70)	Yugu (n = 96)
A11/A3	0.821	0.585	0.556	0.642	0.792	0.879	0.643	0.697	0.579	0.443	0.417
Bw4	0.462	0.394	0.505	0.472	0.283	0.393	0.607	0.596	0.653	0.700	0.688
Bw4-80I	0.400	0.074	0.394	0.198	0.179	0.196	0.321	0.135	0.400	0.400	0.448
Bw4-80T	0.062	0.330	0.162	0.283	0.104	0.215	0.411	0.494	0.305	0.386	0.333
HLA-C1	0.986	1.000	1.000	0.981	0.943	0.991	0.982	1.000	1.000	0.914	0.833
HLA-C2	0.145	0.258	0.162	0.208	0.406	0.262	0.179	0.073	0.221	0.443	0.542
C1/C1	0.855	0.742	0.838	0.792	0.594	0.738	0.821	0.927	0.779	0.557	0.458
C1/C2	0.131	0.258	0.162	0.189	0.349	0.252	0.161	0.073	0.221	0.357	0.375
C2/C2	0.014	0.000	0.000	0.019	0.057	0.009	0.018	0.000	0.000	0.086	0.167

* The numbers for *HLA-C1* and *HLA-C2* was 89 in Jinuo and was 82 in Maonan.

3.3. KIR–HLA Combination

The frequencies of *KIR3DL2* and *HLA-A11/A3* were calculated first. Since the interaction of *KIR2DS4* and *KIR2DS2* with *HLA-A11* has been demonstrated, their combinations have also been calculated [14,29]. The frequencies of *KIR3DL2+A11/A3* and *KIR2DS4+A11/A3* were from 0.417 to 0.879, while the frequencies of *KIR2DS2+A11* were lower, with frequencies from 0.038 to 0.299 (Table 4). In Nu, the frequency of *KIR2DS2+A11* was only 0.038, though *HLA-A*11:01* was predominant, with a frequency of 0.411(Table 4).

Table 4. Distribution of KIR and HLA-A11/A3 pairs in 11 ethnic populations.

Title	3DL2+A11/A3		2DS4+A11/A3		2DS2+A11	
	Counts	Fre.	Counts	Fre.	Counts	Fre.
Hani (n = 145)	119	0.821	119	0.821	31	0.214
Jinuo (n = 94)	55	0.585	46	0.489	12	0.128
Lisu(n = 99)	55	0.556	50	0.505	21	0.212
Nu (n = 106)	68	0.642	68	0.642	4	0.038
Bulang (n = 106)	84	0.792	65	0.613	15	0.142
Wa (n = 107)	94	0.879	92	0.860	32	0.299
Dai (n = 112)	72	0.643	66	0.589	30	0.268
Maonan (n = 89)	62	0.697	58	0.652	24	0.270
Zhuang (n = 95)	55	0.579	49	0.516	16	0.168
Tu (n = 70)	31	0.443	30	0.429	11	0.157
Yugu (n = 96)	40	0.417	38	0.396	10	0.104

Fre.: Frequency.

The individuals carrying either *KIR3DL1* or *KIR3DS1* and its *HLA-Bw4* ligands, or carrying both *KIR3DL1* and *KIR3DS1* together with its HLA-Bw4 ligands, were counted for the KIR–HLA combination (Table 5). *KIR3DL1/3DS1+Bw4* was commonly exhibited in all populations, with frequencies of 0.383–0.700, except in Bulang. The total frequency of the *KIR3DL1/3DS1+Bw4* combination was 0.283, and the frequency of either *KIR3DL1+Bw4* or *KIR3DS1+Bw4* pairs was 0.057. The frequencies of *3DL1+3DS1+Bw4* were predominant in Bulang, and *3DL1+3DS1+Bw4* and *3DL1+Bw4* were almost similar in Jinuo; however, *KIR3DL1+Bw4* was predominant in other 9 populations. In Hani and Wa, all individuals carried either *3DL1+Bw4* or *3DL1+3DS1+Bw4* together, and no one carrying only *3DS1+Bw4* was observed.

Table 5. Distributions of KIR and HLA-B pairs in 11 ethnic populations.

	3DL1+3DS1+Bw4		3DL1+Bw4		3DS1+Bw4		3DL1+3DS1+BW4 80I		3DL1+BW4 80I		3DS1+BW4 80I		3DL1+3DS1+BW4 80T		3DL1+BW4 80T		3DS1+BW4 80T	
	Counts	Fre.	Counts	Fre.	Counts	Fre.	Counts	Fre.	Counts	Fre.	Counts	Fre.	Counts	Fre.	Counts	Fre.	Counts	Fre.
Hani (n = 145)	23	0.159	44	0.303	0	0.000	21	0.145	37	0.255	0	0.000	2	0.014	9	0.062	0	0.000
Jinuo (n = 94)	15	0.160	17	0.181	4	0.043	0	0.000	6	0.064	1	0.011	15	0.160	13	0.138	2	0.021
Lisu (n = 99)	13	0.131	32	0.323	4	0.040	12	0.121	24	0.242	2	0.020	2	0.020	11	0.111	2	0.020
Nu (n = 106)	11	0.104	38	0.358	1	0.009	7	0.066	20	0.189	0	0.000	4	0.038	26	0.245	1	0.009
Bulang (n = 106)	18	0.170	6	0.057	6	0.057	10	0.094	5	0.047	3	0.028	5	0.047	3	0.028	3	0.028
Wa (n = 107)	7	0.065	35	0.327	0	0.000	2	0.019	19	0.178	0	0.000	5	0.047	18	0.168	0	0.000
Dai (n = 112)	21	0.188	38	0.339	8	0.071	14	0.125	19	0.170	2	0.018	14	0.125	24	0.214	8	0.071
Maonan (n = 89)	23	0.258	29	0.326	1	0.011	5	0.056	7	0.079	0	0.000	20	0.225	23	0.258	1	0.011
Zhuang (n = 95)	19	0.200	39	0.411	4	0.042	10	0.105	23	0.242	3	0.032	10	0.105	22	0.232	2	0.021
Tu (n = 70)	20	0.286	27	0.386	2	0.029	10	0.143	13	0.186	2	0.029	12	0.171	15	0.214	0	0.000
Yugu (n = 96)	16	0.167	46	0.479	4	0.042	9	0.094	32	0.333	2	0.021	10	0.104	20	0.208	2	0.021

Fre.: Frequency.

Further analysis of KIR with the presence of isoleucine at position 80(*Bw4-80I*) as well as with the presence of threonine at position 80(*Bw4-80T*) was performed. One individual in Dai and Jinuo with *HLA-Bw4* were unable to have their KIR ligands identified, two individuals in Lisu were not able to have their KIR ligands identified, while all the other individuals with *KIR-Bw4* ligands were identified. For the *3DL1/3DS1+Bw4* pair, the frequencies of *80I3DL1+Bw4 80I* were higher than *3DS1+Bw4 80I*. Moreover, in Hani, Nu, Wa, and Maonan, no individuals who only carried *3DS1+Bw4 80I* were observed. There were more individuals only carrying *KIR3DL1+Bw4-80I* than those only carrying *KIR3DL1+Bw4-80T* in Hani, Lisu, Bulang, and Yugu, while there were fewer in Jinuo, Dai, and Maonan, and there was no difference in Nu, Wa, Zhuang, and Tu. There was a similar finding for the *KIR3DL1+Bw4-80T* and *KIR3DS1+Bw4-80T* pairs. Further analysis of the southern and northern groups indicated that the frequencies of *KIR3DL1+Bw4* were lower in the southern group than in the northern group (0.451 ± 0.120 vs. 0.659 ± 0.018, p = 0.001).

KIR2DL2/2DL3 and its ligand *HLA-C1* specifically control NK cell response. Its combination commonly exists in all the populations, but the frequencies were higher in the southern group (0.978 ± 0.006, 95% CI: 0.966–0.988) than in the northern group (0.860 ± 0.020, 95% CI: 0.833–0.886) (p = 0.00005). *KIR2DL1* and *KIR2DS2* had a similar binding specificity for *HLA-C2*. The frequencies of *2DL1+HLA-C2* were lower in the southern group (0.211 ± 0.026, 95% CI: 0.162–0.264) than in the northern group (0.472 ± 0.021, 95% CI: 0.443–0.500) (p = 0.003). However, there was no difference in frequencies of *2DS1+HLA-C2* combinations between the southern and northern groups (Table 6).

Table 6. Distributions of KIR and HLA-C pairs in 11 ethnic populations.

	2DL2/3+HLA-C1		*2DL1+HLA-C2*		*2DS1+HLA-C2*		*2DS2+HLA-C1*	
	Counts	Fre.	Counts	Fre.	Counts	Fre.	Counts	Fre.
Hani (n = 145)	142	0.979	21	0.145	10	0.069	42	0.290
Jinuo (n = 89)	87	0.978	23	0.258	9	0.101	24	0.270
Lisu (n = 99)	98	0.990	16	0.162	5	0.051	33	0.333
Nu (n = 106)	104	0.981	22	0.208	4	0.038	11	0.104
Bulang (n = 106)	99	0.934	41	0.387	31	0.292	17	0.160
Wa (n = 107)	104	0.972	28	0.262	10	0.093	34	0.318
Dai (n = 112)	110	0.982	20	0.179	10	0.089	44	0.393
Maonan (n = 82)	81	0.988	6	0.073	2	0.024	35	0.427
Zhuang (n = 95)	95	1.000	21	0.221	11	0.116	28	0.295
Tu (n = 70)	62	0.886	31	0.443	11	0.157	22	0.314
Yugu (n = 96)	80	0.833	48	0.500	23	0.240	22	0.229

Fre.: Frequency.

The comparison between the two populations indicated that Yugu and Tu did not show any difference in any of the KIR–HLA pairs, while the other populations all showed differences of between 1 and 15 pairs (Supplementary Table S4). The *2DL2/3+HLA-C1*, *2DL1+HLA-C2*, and *2DS1+HLA-C2* pairs showed clear differences between the northern and southern groups. For other *HLA-Bw* or *HLA-A11/A3* pairs, the difference between the two populations was extensive. For example, Wa showed differences with: every other population except Lisu and Nu for carrying both *KIR3DL1+Bw4* and *KIR3DS1+Bw4* pairs; every other population except Jinuo, Nu, and Maonan for carrying both *KIR3DL1+Bw4-80I* and *KIR3DS1+Bw4-80I* pairs; and Hani, Lisu, Nu, Bulang, Dai, and Yugu for carrying *KIR3DL1+Bw4-80T* and *KIR3DS1+Bw4-80T*. For carrying *KIR3DL1+Bw4*, Bulang showed differences with every population except Jinuo, and Jinuo, Wa, Zhuang, Tu, and Yugu showed differences with at least five populations. The differences at *KIR3DL1+Bw4-80I* and *KIR3DL1+Bw4-80T* between the populations were not same as for *KIR3DL1+Bw4*. The differences in carrying *KIR3DL2+A11/A3* or *KIR2DS4+A11/A3* among the populations were similar, and Hani, Bulang, Wa, Tu, and Yugu showed differences with at least five populations.

3.4. HLA/KIR Correlation

The correlations of each KIR–HLA receptor and ligand pair were analyzed. The observed frequencies for *KIR2DL3* and *HLA-C1* ligands showed a significant correlation (r = 0.637, *p* = 0.035). Positive correlations were also observed in *KIR3DL1+HLA-Bw4*, *KIR3DL1+HLA-Bw4 80I*, *KIR3DL1+HLA-Bw4 80T*, *KIR2DL2+HLA-C1*, *KIR2DS1+HLA-C2*, and *KIR2DS2+HLA-C1* pairs, but they were not significant. Negative correlations between *KIR3DS1* and *HLA-Bw4*, and *KIR2DL1* and *HLA-C2* were observed, but they were not significant neither (Figure 3 and Supplementary Figure S1).

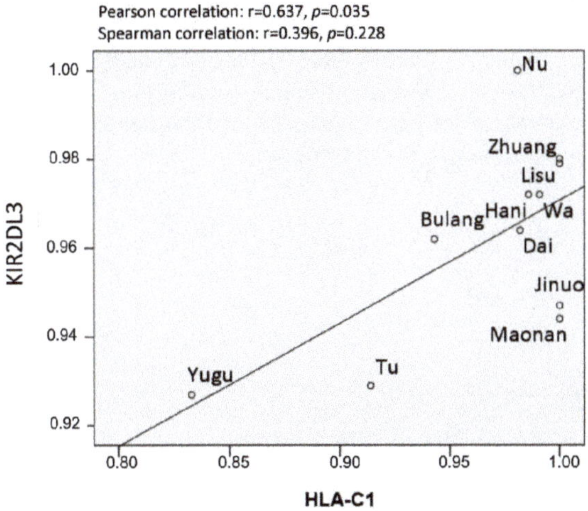

Figure 3. Correlation of KIR2DL3 and HLA-C1.

3.5. Phylogenetic Analysis

Both principal component analysis (PCA) and phylogenetic trees using KIR, HLA, and KIR–HLA combination frequencies were employed. For PCA based on 11 KIR (*3DL1, 2DL1, 2DL3, 2DS4, 2DL2, 2DL5, 3DS1, 2DS1, 2DS2, 2DS3,* and *2DS5*) plots, Bulang showed distance from the other 10 populations (Figure 4a), and the other 10 populations did not cluster with their linguistic family as based on *HLA-A*, *-B*, and *-C* (Figure 4b), which agreed with previous studies. On the PCA plots based on KIR–HLA pairs, Yugu and Nu from northern China clustered together and showed distance from the other nine populations from southern China (Figure 4c).

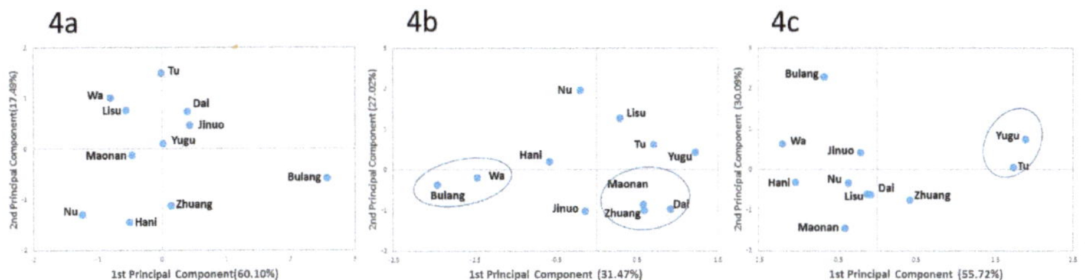

Figure 4. Principal component analysis. (**a**) PCA based on 11 KIR genes. Contributions of the first and second components were 60.10% and 17.49%, respectively. (**b**) PCA based on *HLA-A, -B,* and *-C* allele frequencies. Contributions of first and second components were 31.47% and 27.02%, respectively. (**c**) PCA based on KIR–HLA pairs. Contributions of first and second components were 55.72% and 30.09%, respectively.

On the phylogenetic tree based on KIRs, for the trees constructed either by all 11 KIR genes (Figure 5a), or by the inhibitor KIR genes or by the activating KIR genes (data not shown) had no clear clustering among the populations. For the NJ tree constructed using HLA genes, Tu and Yugu clustered together as one major branch. Bulang and Wa of Khmer clustered together, and Maonao, Zhuang, and Dai of Daic clustered together with Jinuo (Figure 5b). When 58 populations were compared on the phylogenetic tree using 11 KIRs frequencies, most populations clustered together according to their geographic location of Asian, European, African, and American (Figure 6). However, the closeness were not displayed clearly as in the NJ tree constructed by the HLA genes, in which the populations clustered according to their evolutional relationship [25]. In the Asian branch, on one hand, most Han populations in northern China clustered with Japanese and Korean, but also together with Nu, Yi, and Hani ethnic populations living in southern China. On the other hand, Yugu and Tu living in Northern China clustered with Southern Hans living in Guangdong, Hong Kong, together with Maonan, Bulang, Jinuo, etc., ethnic populations in southern China. The 11 ethnic populations in the present study still did not show a clear origin or linguistic clustering trend.

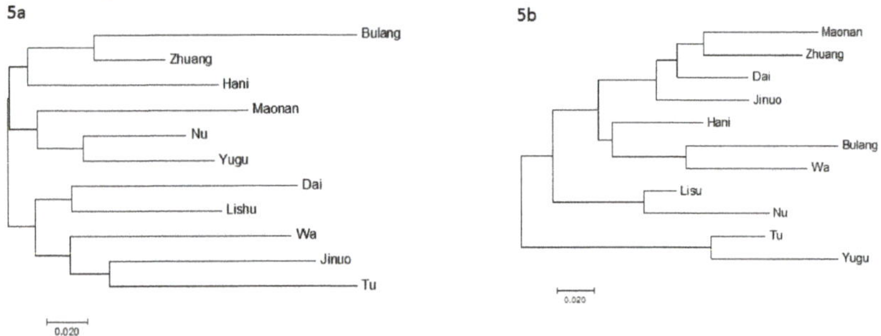

Figure 5. Neighbor-joining tree. (**a**) Neighbor-joining tree based on DA genetic distance from 11 KIR gene frequencies. The optimal tree was one with the sum of branch length = 1.166. (**b**) Neighbor-joining tree based on DA genetic distance from *HLA-A*, *-B*, and *-C* allele frequencies. The optimal tree was one with the sum of branch length = 0.875.

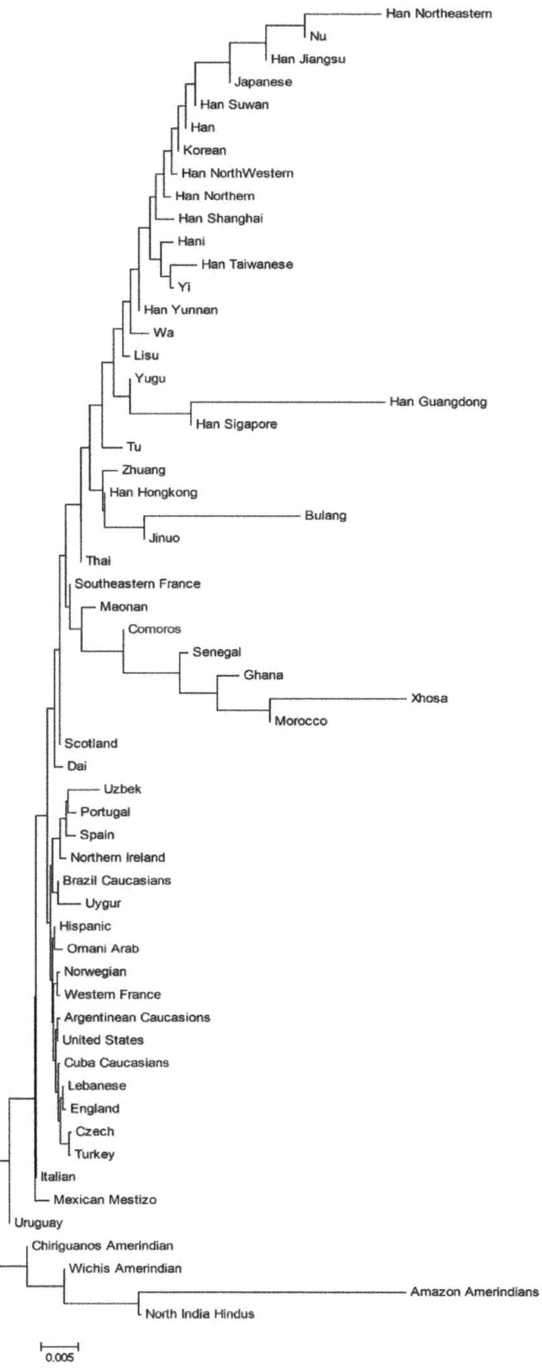

Figure 6. Neighbor-joining tree constructed using 11 KIR genes frequencies of 58 populations worldwide. The optimal tree was one with the sum of branch length = 0.159.

4. Discussion

The extensive diversity of HLA and KIR genes and their interactive roles in the immune response make these genes coevolve in genotypic combination. Disease and population studies have confirmed that they evolved together as specific KIR–HLA pairs to regulate NK cell function and play a vital role in the innate defense against pathogens and early placentation [30–32]. KIR–HLA combination studies have been performed in different populations worldwide; however, previous studies in China were limited to the Han population [4,33–38]. In the present study, we analyzed KIR and its HLA pairs in 11 ethnic populations from northern and southern China covering four different linguistic language families that represent the major origins of Chinese ethnic populations. This study not only provides useful genomic diversity data for the future study of viral infections, autoimmune diseases, and reproductive fitness among these populations, but also reveals clues about HLA and KIR interaction and coevolution under the diverse change of pathogen infections.

The coevolution of HLA and KIR has been proved by several population studies worldwide, and different KIR–HLA pair correlations were identified in different populations. In 2006, Single et al. studied the distribution of KIR and its HLA ligands in 30 populations worldwide and observed that a balancing selection acted on the negative correlation between *KIR3DS1* and *HLA-Bw4-80I* pairs [30]. In 2013, Hollenbach et al. compared KIR–HLA pairs in 105 populations worldwide and revealed a significant correlation between *KIR2DL3* and *HLA-C* ligands. However, the correlation for *KIR3DL1* and *HLA-Bw4* pairs was not significant [39]. In the Italian population, a correlation between *KIR* and *HLA-C* ligands was not observed; instead, a correlation between *KIR3DL1* and *HLA-Bw4* ligands, as well as *KIR3DL2* and *HLA-A3* and *HLA-A11* ligands, was observed [40]. In the present study, the observed frequencies for *KIR2DL3* and *HLA-C1* ligands showed a correlation (r = 0.637, *p* = 0.035) that agreed with Hollenbach et al.'s study, as well as Gendzekhadze's study in the Yucpa og South Amerindian [4]. The association of *KIR2DL3* and *HLA-C1* has also been investigated regarding the Hepatitis C virus and Malaria infection [41,42]. In Hirayasu et al.'s study, they found *KIR2DL3+HLA-C1*, but no other KIR–HLA pairs were associated with cerebral malaria, and the frequency of combination was significant lower in malaria high-endemic populations. This result suggested that natural selection has reduced the *KIR2DL3+HLA-C1* frequencies in malaria high-endemic populations to favor the development of malaria [41]. Thus, KIR–HLA coevolution may be driven by microbial pathogens, resulting in specific distributions of KIR–HLA pairs in different populations.

The genetic difference between southern and northern Chinese has been confirmed in studies of HLA [35,43], as well as immunoglobulins [44], microsatellites [45], and Y-chromosome single-nucleotide polymorphisms [46]. Furthermore, population migration from northern to southern China has frequently happened throughout Chinese history [15–17]. In the present study, *KIR2DL2/3+HLA-C1*, *KIR2DL1+HLA-C2*, and *KIR2DS1+HLA-C2* pairs showed clear differences between the northern and southern groups. The frequencies of *KIR2DL2/3+HLA-C1* pairs were significantly higher in the southern group than in the northern group (0.978 vs. 0.860, *p* = 0.00005). This difference has also been investigated in Han population. The frequencies of *KIR2DL2/3+HLA-C1* pairs were higher in two southern Chinese Han, namely Guangdong Han and Yunnan Han, than in the northern ethnic group, with frequencies of 0.981 and 0.950, respectively. In contrast, the frequencies of *KIR2DL1+HLA-C2* and *KIR2DS1+HLA-C2* pairs were lower than in the northern group, with frequencies of 0.294 and 0.351 for *KIR2DL1+HLA-C2*, and 0.100 and 0.155 for *KIR2DS1+HLA-C2* in Guangdong Han and Yunnan Han, respectively [36,38]. Furthermore, the Tu and Yugu formed a cluster and showed a distance from other ethnic populations in southern China in the PCA plot constructed using KIR–HLA pair frequencies. According to historical records, northern and southern China underwent different pathogenic pressures. Regarding malaria (a serious infectious disease prevalent in China since 2700 BC), its epidemic area was focused in southern China, whereas northern China was malaria free or had a very low incidence rate [47,48]. Historically, Yunnan Province has been the most high-risk malaria area, especially along the China–Myanmar border [49,50]. In the present study, Tu and Yugu are from northern China, while the other populations are all in southern China, the most high-risk malaria areas. Therefore, we deduced that the different distribution between the northern and southern groups in China may have been caused by severe infectious disease epidemics, such as malaria.

Except for *KIR+HLA-C* pairs, *KIR+HLA-Bw* and *KIR+HLA-A3/A11* pair differences were diverse in different populations. When considering KIR genes, genotypes, and haplotypes, more diversity was exhibited. It is interesting to note that the HLA gene distributions were in accordance with the population linguistic group and their origins. Both in the PCA plot and neighbor-joining tree constructed using HLA allele frequencies, the same linguistic origin populations clustered together. However, there were no clear cluster trends to distinguish the populations according to their origin or linguistic classification using either KIR, activated KIR, or KIR genotype frequencies among 11 ethnic populations. Moreover, in the phylogenetic tree constructed using 11 KIRs frequencies of 58 populations worldwide, there was a geographic closeness among Asians, Africans, Europeans, and Americans, while the 11 ethnic populations in the present study still did not show a clear origin or linguistic clustering trend. Compared with HLA genes, KIR genes have experienced a rapid evolution through a combination of gene duplication and nonhomologous recombination [1]. The extensive diversity of KIR genes in different populations worldwide indicates that distinct diseases have recently acted or are still acting to select on KIR repertoires [2]. This evolution was thought to be driven by the selective pressure of pathogen invasion, as well as reproduction. Moreover, haplotypes A and B are thought to have maintained a balance selection in human beings. The A haplotypes are associated with an improved response to pathogens, while B haplotypes are associated with reproductive fitness [51,52]. Previous studies have indicated that the populations are related to their geographic distribution based on KIR haplotype B but do not show a correlation based on haplotype A [53]. Moreover, it has been reported that B haplotypes are more prevalent in Australian Aborigines and Asian Indians, where the possible reason is due to these populations maybe being under strong pressure from infectious diseases [2]. In the present study, the frequencies of haplotypes A and B were almost similar to each other in Bulang, which showed a difference from other Asian populations. On the contrary, in Nu, haplotype A was as high as 0.830, which showed a significant difference from the other 10 populations. This extensive range of haplotype A in the present study, from 0.491 in Bulang to 0.830 in Nu, together with the diverse frequencies worldwide, may be the result of a founder effect, genetic drift, or natural selection [5,31]. Therefore, the distribution of KIR profiles among the present study populations could not be interpreted as a phylogenetic tree.

In conclusion, the distribution of KIR and its HLA ligands in 11 ethnic populations in China exhibited diverse characteristics, where each group had its specific KIR and KIR–HLA pair profile. The difference of KIR–HLA pairs between the northern and southern groups, but not among the four original groups, may reflect the strong pressure from previous or ongoing infectious diseases that have had a significant impact on KIR and its HLA combination repertoires.

Supplementary Materials: The Supplementary Materials are available online at http://www.mdpi.com/2073-4409/8/7/711/s1.

Author Contributions: Conceived and designed the experiments: L.S.; Sample collection: L.S. and J.Y.; Sample Genotyping: Y.Y., L.S., and Y.T.; Data Analysis: Y.Y. and S.L.; Manuscript writing: L.S.

Funding: This research was found by the National Natural Science Foundation of China (81573206) and the Special Funds for High-Level Health Talents of Yunnan Province (D-201669 and L-201615). The funders had no role in study design, data collection and analysis, decision to publish, or preparation of the manuscript.

Conflicts of Interest: The authors declare no conflict of interest. The sponsors had no role in the design, execution, interpretation, or writing of the study.

References

1. Parham, P. MHC class I molecules and KIRs in human history, health and survival. *Nat. Rev. Immunol.* **2005**, *5*, 201–214. [CrossRef] [PubMed]

2. Middleton, D.; Gonzelez, F. The extensive polymorphism of KIR genes. *Immunology* **2010**, *129*, 8–19. [CrossRef] [PubMed]

3. Du, Z.; Gjertson, D.W.; Reed, E.F.; Rajalingam, R. Receptor-ligand analyses define minimal killer cell Ig-like receptor (KIR) in humans. *Immunogenetics* **2007**, *59*, 1–15. [CrossRef] [PubMed]

4. Gendzekhadze, K.; Norman, P.J.; Abi-Rached, L.; Graef, T.; Moesta, A.K.; Layrisse, Z.; Parham, P. Co-evolution of KIR2DL3 with HLA-C in a human population retaining minimal essential diversity of KIR and HLA class I ligands. *Proc. Natl. Acad. Sci. USA* **2009**, *106*, 18692–18697. [CrossRef] [PubMed]

5. Manser, A.R.; Weinhold, S.; Uhrberg, M. Human KIR repertoires: Shaped by genetic diversity and evolution. *Immunol. Rev.* **2015**, *267*, 178–196. [CrossRef] [PubMed]

6. Guinan, K.J.; Cunningham, R.T.; Meenagh, A.; Dring, M.M.; Middleton, D.; Gardiner, C.M. Receptor systems controlling natural killer cell function are genetically stratified in Europe. *Genes Immun.* **2010**, *11*, 67–78. [CrossRef]

7. Tokunaga, K.; Imanishi, T.; Takahashi, K.; Juji, T. On the origin and dispersal of East Asian populations as viewed from HLA haplotypes. In *Prehistoric Mongoloid Dispersals*; Akazawa, T., Szathmary, E.J., Eds.; Oxford University: Oxford, UK, 1996; pp. 187–197.

8. Tokunaga, K.; Ishikawa, Y.; Ogawa, A.; Wang, H.; Mitsunaga, S.; Moriyama, S.; Lin, L.; Bannai, M.; Watanabe, Y.; Kashiwase, K.; et al. Sequence-based association analysis of HLA class I and II alleles in Japanese supports conservation of common haplotypes. *Immunogenetics* **1997**, *46*, 199–205. [CrossRef]

9. Marsh, S.G.; Parham, P.; Dupont, B.; Geraghty, D.E.; Trowsdale, J.; Middleton, D.; Vilches, C.; Carrington, M.; Witt, C.; Guethlein, L.A.; et al. Killer-cell immunoglobulin-like receptor (KIR) nomenclature report, 2002. *Tissue Antigens* **2003**, *62*, 79–86. [CrossRef]

10. Wilson, M.J.; Torkar, M.; Haude, A.; Milne, S.; Jones, T.; Sheer, D.; Beck, S.; Trowsdale, J. Plasticity in the organization and sequences of human KIR/ILT gene families. *Proc. Natl. Acad. Sci. USA* **2000**, *97*, 4778–4783. [CrossRef]

11. Rajalingam, R. Diversity of Killer Cell Immunoglobulin-Like Receptors and Disease. *Clin. Lab. Med.* **2018**, *38*, 637–653. [CrossRef]

12. Stewart, C.A.; Laugier-Anfossi, F.; Vely, F.; Saulquin, X.; Riedmuller, J.; Tisserant, A.; Gauthier, L.; Romagne, F.; Ferracci, G.; Arosa, F.A.; et al. Recognition of peptide-MHC class I complexes by activating killer immunoglobulin-like receptors. *Proc. Natl. Acad. Sci. USA* **2005**, *102*, 13224–13229. [CrossRef] [PubMed]

13. Augusto, D.G.; Lobo-Alves, S.C.; Melo, M.F.; Pereira, N.F.; Petzl-Erler, M.L. Activating KIR and HLA Bw4 ligands are associated to decreased susceptibility to pemphigus foliaceus, an autoimmune blistering skin disease. *PLoS ONE* **2012**, *7*, e39991. [CrossRef] [PubMed]

14. Liu, J.; Xiao, Z.; Ko, H.L.; Shen, M.; Ren, E.C. Activating killer cell immunoglobulin-like receptor 2DS2 binds to HLA-A*11. *Proc. Natl. Acad. Sci. USA* **2014**, *111*, 2662–2667. [CrossRef] [PubMed]

15. Guo, D.; Dong, J. *Summarization of Chinese Nationalities (Zhonghua mingzu zhishi tonglan)*; Yunnan Education Press: Kunming, China, 2000. (In Chinese)

16. You, Z. *History of Yunnan Nationalities (Yunnan Mingzu Shi)*; Yunnan University Press: Kunming, China, 1994. (In Chinese)

17. Chu, J.; Yu, J.; Huang, X.; Sun, H. China Nationalities. In *Genetic Diversity in Chinese Populations*; Jin, L., Chu, J., Eds.; Shanghai Science and Technology Press: Shanghai, China, 2006.

18. Yao, Y.; Shi, L.; Tao, Y.; Lin, K.; Liu, S.; Yu, L.; Yang, Z.; Yi, W.; Huang, X.; Sun, H.; et al. Diversity of killer cell immunoglobulin-like receptor genes in four ethnic groups in China. *Immunogenetics* **2011**, *63*, 475–483. [CrossRef] [PubMed]

19. Guo, S.W.; Thompson, E.A. Performing the exact test of Hardy-Weinberg proportion for multiple alleles. *Biometrics* **1992**, *48*, 361–372. [CrossRef] [PubMed]

20. Rajalingam, R.; Du, Z.; Meenagh, A.; Luo, L.; Kavitha, V.J.; Pavithra-Arulvani, R.; Vidhyalakshmi, A.; Sharma, S.K.; Balazs, I.; Reed, E.F.; et al. Distinct diversity of KIR genes in three southern Indian populations: Comparison with world populations revealed a link between KIR gene content and pre-historic human migrations. *Immunogenetics* **2008**, *60*, 207–217. [CrossRef] [PubMed]

21. Shi, L.; Huang, X.Q.; Shi, L.; Tao, Y.F.; Yao, Y.F.; Yu, L.; Lin, K.Q.; Yi, W.; Sun, H.; Tokunaga, K.; et al. HLA polymorphism of the Zhuang population reflects the common HLA characteristics among Zhuang-Dong language-speaking populations. *J. Zhejiang Univ. Sci. B* **2011**, *12*, 428–435. [CrossRef] [PubMed]

22. Shi, L.; Ogata, S.; Yu, J.K.; Ohashi, J.; Yu, L.; Shi, L.; Sun, H.; Lin, K.; Huang, X.Q.; Matsushita, M.; et al. Distribution of HLA alleles and haplotypes in Jinuo and Wa populations in Southwest China. *Hum. Immunol.* **2008**, *69*, 58–65. [CrossRef]

23. Shi, L.; Shi, L.; Yao, Y.F.; Matsushita, M.; Yu, L.; Huang, X.Q.; Yi, W.; Oka, T.; Tokunaga, K.; Chu, J.Y. Genetic link among Hani, Bulang and other Southeast Asian populations: Evidence from HLA -A, -B, -C, -DRB1 genes and haplotypes distribution. *Int. J. Immunogenet.* **2010**, *37*, 467–475. [CrossRef]
24. Shi, L.; Yao, Y.F.; Shi, L.; Matsushita, M.; Yu, L.; Lin, Q.K.; Tao, Y.F.; Oka, T.; Chu, J.Y.; Tokunaga, K. HLA alleles and haplotypes distribution in Dai population in Yunnan province, Southwest China. *Tissue Antigens* **2010**, *75*, 159–165. [CrossRef]
25. Yao, Y.; Shi, L.; Tao, Y.; Kulski, J.K.; Lin, K.; Huang, X.; Xiang, H.; Chu, J.; Shi, L. Distinct HLA allele and haplotype distributions in four ethnic groups of China. *Tissue Antigens* **2012**, *80*, 452–461. [CrossRef] [PubMed]
26. Lalouel, J. Distance analysis and multidimensional scaling. In *Current Developments in Anthropological Genetics, Volume 1: Theory and Methods*; Mielke, J., Crawford, M., Eds.; Plenum Press: New York, NY, USA; London, UK, 1980; pp. 209–250.
27. Tamura, K.; Dudley, J.; Nei, M.; Kumar, S. MEGA4: Molecular Evolutionary Genetics Analysis (MEGA) software version 4.0. *Mol. Biol. Evol.* **2007**, *24*, 1596–1599. [CrossRef] [PubMed]
28. Rzhetsky, A.; Nei, M. A simple method for estimating and testing minimum evolution trees. *Mol. Biol. Evol.* **1992**, *9*, 945–967.
29. Graef, T.; Moesta, A.K.; Norman, P.J.; Abi-Rached, L.; Vago, L.; Older Aguilar, A.M.; Gleimer, M.; Hammond, J.A.; Guethlein, L.A.; Bushnell, D.A.; et al. KIR2DS4 is a product of gene conversion with KIR3DL2 that introduced specificity for HLA-A*11 while diminishing avidity for HLA-C. *J. Exp. Med.* **2009**, *206*, 2557–2572. [CrossRef] [PubMed]
30. Single, R.M.; Martin, M.P.; Gao, X.; Meyer, D.; Yeager, M.; Kidd, J.R.; Kidd, K.K.; Carrington, M. Global diversity and evidence for coevolution of KIR and HLA. *Nat. Genet.* **2007**, *39*, 1114–1119. [CrossRef] [PubMed]
31. Augusto, D.G.; Petzl-Erler, M.L. KIR and HLA under pressure: Evidences of coevolution across worldwide populations. *Hum. Genet.* **2015**, *134*, 929–940. [CrossRef] [PubMed]
32. Hao, L.; Nei, M. Rapid expansion of killer cell immunoglobulin-like receptor genes in primates and their coevolution with MHC Class I genes. *Gene* **2005**, *347*, 149–159. [CrossRef] [PubMed]
33. Norman, P.J.; Hollenbach, J.A.; Nemat-Gorgani, N.; Guethlein, L.A.; Hilton, H.G.; Pando, M.J.; Koram, K.A.; Riley, E.M.; Abi-Rached, L.; Parham, P. Co-evolution of human leukocyte antigen (HLA) class I ligands with killer-cell immunoglobulin-like receptors (KIR) in a genetically diverse population of sub-Saharan Africans. *PLoS Genet.* **2013**, *9*, e1003938. [CrossRef]
34. Rajalingam, R.; Krausa, P.; Shilling, H.G.; Stein, J.B.; Balamurugan, A.; McGinnis, M.D.; Cheng, N.W.; Mehra, N.K.; Parham, P. Distinctive KIR and HLA diversity in a panel of north Indian Hindus. *Immunogenetics* **2002**, *53*, 1009–1019. [CrossRef]
35. Yao, Y.; Shi, L.; Shi, L.; Matsushita, M.; Yu, L.; Lin, K.; Tao, Y.; Huang, X.; Yi, W.; Oka, T.; et al. Distribution of HLA-A, -B, -Cw, and -DRB1 alleles and haplotypes in an isolated Han population in Southwest China. *Tissue Antigens* **2009**, *73*, 561–568. [CrossRef]
36. Zhen, J.; Wang, D.; He, L.; Zou, H.; Xu, Y.; Gao, S.; Yang, B.; Deng, Z. Genetic profile of KIR and HLA in southern Chinese Han population. *Hum. Immunol.* **2014**, *75*, 59–64. [CrossRef] [PubMed]
37. Guinan, K.J.; Cunningham, R.T.; Meenagh, A.; Gonzalez, A.; Dring, M.M.; McGuinness, B.W.; Middleton, D.; Gardiner, C.M. Signatures of natural selection and coevolution between killer cell immunoglobulin-like receptors (KIR) and HLA class I genes. *Genes Immun.* **2010**, *11*, 467–478. [CrossRef] [PubMed]
38. Shen, Y.; Cao, D.; Li, Y.; Kulski, J.K.; Shi, L.; Jiang, H.; Ma, Q.; Yu, J.; Zhou, J.; Yao, Y.; et al. Distribution of HLA-A, -B, and -C alleles and HLA/KIR combinations in Han population in China. *J. Immunol. Res.* **2014**, *2014*, 565296. [CrossRef] [PubMed]
39. Hollenbach, J.A.; Augusto, D.G.; Alaez, C.; Bubnova, L.; Fae, I.; Fischer, G.; Gonzalez-Galarza, F.F.; Gorodezky, C.; Karabon, L.; Kusnierczyk, P.; et al. 16(th) IHIW: Population global distribution of killer immunoglobulin-like receptor (KIR) and ligands. *Int. J. Immunogenet.* **2013**, *40*, 39–45. [CrossRef] [PubMed]
40. Fasano, M.E.; Rendine, S.; Pasi, A.; Bontadini, A.; Cosentini, E.; Carcassi, C.; Capittini, C.; Cornacchini, G.; Espadas de Arias, A.; Garbarino, L.; et al. The distribution of KIR-HLA functional blocks is different from north to south of Italy. *Tissue Antigens* **2014**, *83*, 168–173. [CrossRef] [PubMed]

41. Hirayasu, K.; Ohashi, J.; Kashiwase, K.; Hananantachai, H.; Naka, I.; Ogawa, A.; Takanashi, M.; Satake, M.; Nakajima, K.; Parham, P.; et al. Significant association of KIR2DL3-HLA-C1 combination with cerebral malaria and implications for co-evolution of KIR and HLA. *PLoS Pathog.* **2012**, *8*, e1002565. [CrossRef] [PubMed]

42. Khakoo, S.I.; Thio, C.L.; Martin, M.P.; Brooks, C.R.; Gao, X.; Astemborski, J.; Cheng, J.; Goedert, J.J.; Vlahov, D.; Hilgartner, M.; et al. HLA and NK cell inhibitory receptor genes in resolving hepatitis C virus infection. *Science* **2004**, *305*, 872–874. [CrossRef] [PubMed]

43. Shi, L.; Xu, S.B.; Ohashi, J.; Sun, H.; Yu, J.K.; Huang, X.Q.; Tao, Y.F.; Yu, L.; Horai, S.; Chu, J.Y.; et al. HLA-A, HLA-B, and HLA-DRB1 alleles and haplotypes in Naxi and Han populations in southwestern China (Yunnan province). *Tissue Antigens* **2006**, *67*, 38–44. [CrossRef] [PubMed]

44. Zhao, T.M.; Lee, T.D. Gm and Km allotypes in 74 Chinese populations: A hypothesis of the origin of the Chinese nation. *Hum. Genet.* **1989**, *83*, 101–110. [CrossRef] [PubMed]

45. Chu, J.Y.; Huang, W.; Kuang, S.Q.; Wang, J.M.; Xu, J.J.; Chu, Z.T.; Yang, Z.Q.; Lin, K.Q.; Li, P.; Wu, M.; et al. Genetic relationship of populations in China. *Proc. Natl. Acad. Sci. USA* **1998**, *95*, 11763–11768. [CrossRef] [PubMed]

46. Su, B.; Xiao, J.; Underhill, P.; Deka, R.; Zhang, W.; Akey, J.; Huang, W.; Shen, D.; Lu, D.; Luo, J.; et al. Y-Chromosome evidence for a northward migration of modern humans into Eastern Asia during the last Ice Age. *Am. J. Hum. Genet.* **1999**, *65*, 1718–1724. [CrossRef] [PubMed]

47. Diouf, G.; Kpanyen, P.N.; Tokpa, A.F.; Nie, S. Changing landscape of malaria in China: Progress and feasibility of malaria elimination. *Asia Pac. J. Public Health* **2014**, *26*, 93–100. [CrossRef] [PubMed]

48. Cox, F.E. History of the discovery of the malaria parasites and their vectors. *Parasites Vectors* **2010**, *3*, 5. [CrossRef] [PubMed]

49. Bi, Y.; Yu, W.; Hu, W.; Lin, H.; Guo, Y.; Zhou, X.N.; Tong, S. Impact of climate variability on Plasmodium vivax and Plasmodium falciparum malaria in Yunnan Province, China. *Parasites Vectors* **2013**, *6*, 357. [CrossRef] [PubMed]

50. Lu, L.; Yuan, X.; Cheng, C. The History, Current St at us, and Challenges of Infectious Diseases in Yunnan. *J. Kunming Med. Univ.* **2009**, *8*, 17–20. (In Chinese)

51. Hiby, S.E.; Apps, R.; Sharkey, A.M.; Farrell, L.E.; Gardner, L.; Mulder, A.; Claas, F.H.; Walker, J.J.; Redman, C.W.; Morgan, L.; et al. Maternal activating KIRs protect against human reproductive failure mediated by fetal HLA-C2. *J. Clin. Investig.* **2010**, *120*, 4102–4110. [CrossRef] [PubMed]

52. Jamil, K.M.; Khakoo, S.I. KIR/HLA interactions and pathogen immunity. *J. Biomed. Biotechnol.* **2011**, *2011*, 298348. [CrossRef] [PubMed]

53. Middleton, D.; Meenagh, A.; Moscoso, J.; Arnaiz-Villena, A. Killer immunoglobulin receptor gene and allele frequencies in Caucasoid, Oriental and Black populations from different continents. *Tissue Antigens* **2008**, *71*, 105–113. [CrossRef]

Article

Role of MHC-I Expression on Spinal Motoneuron Survival and Glial Reactions Following Ventral Root Crush in Mice

Luciana Politti Cartarozzi [1], **Matheus Perez** [2], **Frank Kirchhoff** [3] and **Alexandre Leite Rodrigues de Oliveira** [1,*]

[1] Laboratory of Nerve Regeneration, University of Campinas–UNICAMP, Cidade Universitaria "Zeferino Vaz, Rua Monteiro Lobato, 255, 13083-970 Campinas, SP, Brazil; lucartarozzi@gmail.com
[2] School of Physical Education and Sport of Ribeirao Preto, University of Sao Paulo, Av. Bandeirantes, 3900, 14040-907 Ribeirão Preto, SP, Brazil; matheusperez@usp.br
[3] Molecular Physiology, Center for Integrative Physiology and Molecular Medicine (CIPMM), University of Saarland, Building 48, 66421 Homburg, Germany; Frank.Kirchhoff@uks.eu
* Correspondence: alroliv@unicamp.br; Tel.: +55-19-3521-6295

Received: 26 February 2019; Accepted: 7 May 2019; Published: 21 May 2019

Abstract: Lesions to the CNS/PNS interface are especially severe, leading to elevated neuronal degeneration. In the present work, we establish the ventral root crush model for mice, and demonstrate the potential of such an approach, by analyzing injury evoked motoneuron loss, changes of synaptic coverage and concomitant glial responses in β2-microglobulin knockout mice (β2m KO). Young adult (8–12 weeks old) C57BL/6J (WT) and β2m KO mice were submitted to a L4–L6 ventral roots crush. Neuronal survival revealed a time-dependent motoneuron-like cell loss, both in WT and β2m KO mice. Along with neuronal loss, astrogliosis increased in WT mice, which was not observed in β2m KO mice. Microglial responses were more pronounced during the acute phase after lesion and decreased over time, in WT and KO mice. At 7 days after lesion β2m KO mice showed stronger Iba-1$^+$ cell reaction. The synaptic inputs were reduced over time, but in β2m KO, the synaptic loss was more prominent between 7 and 28 days after lesion. Taken together, the results herein demonstrate that ventral root crushing in mice provides robust data regarding neuronal loss and glial reaction. The retrograde reactions after injury were altered in the absence of functional MHC-I surface expression.

Keywords: astrogliosis; PNS/CNS interface; microglial reaction; synaptic covering; β2m knockout mice

1. Introduction

Compression of spinal roots is a common clinical incident, and the root axotomy is directly related to the modification in neuronal function [1]. After trauma, there are axonal changes that indicate dysfunction and degeneration happening both proximal and distal to the lesion [2].

Distal to the lesion, the process, called Wallerian degeneration, is characterized by the deterioration of axon fragments [3]. Cell bodies of lesioned neurons undergo several morpho-functional changes that, together, are called chromatolysis. They comprise cell body hypertrophy, nucleus displacement, Nissl substance dissociation and expressive disturbance in the expression of structural molecules related to synaptic transmission [4].

The axotomy-induced retrograde synaptic responses are known as "synaptic stripping", i.e., the extensive detachment of presynaptic terminals from perikarya and dendrites of axotomized motoneurons [5,6]. The plasticity of the nervous system ensures that structural and functional circuitry remodeling occurs after injury, in particular, with the detachment of the excitatory boutons that were in apposition with the lesioned neuron, which leads to a metabolic change and a shift from the synaptic transmission state to a recovery state [7–9]. In the acute phase after lesion, there is a significant reduction of synaptic inputs, even temporally ceasing the synaptic transmission [10,11].

The axotomy rapidly activates astrocytes and microglia in the vicinity of the affected synaptic terminals. After lesion, microglia and astrocytes quickly react through structural modifications of their cytoplasmic projections that get interposed between the axotomized-motoneuron membrane (post-synaptic membrane) and the retracting synaptic terminals [7,12–14]. Within a few days, axotomized motoneurons are surrounded by activated microglial cells which remain closely in apposition to the perikarya of injured neurons. The number of activated microglia in lesioned motor nuclei increases dramatically due to the proliferation that lasts for about two to four days post-axotomy [6].

Neurons express molecules such as the *major histocompatibility complex of class I* (MHC-I), a transmembrane complex [15] consisting of an α-chain associated with a light and obligatory extracellular chain, named β2-microglobulin (β2m) [16]. Classical MHC-I genes are well known for their role in the adaptive immune response and have their primary function related to binding peptides derived from intracellular proteolysis. The peptides that are processed intracellularly are loaded to the MHC-I molecules by the *Transporter associated with Antigen Processing 1* (TAP-1) protein for delivery to the cell surface; such peptides are recognized and identified by T-cytotoxic lymphocytes. Mice that are genetically deficient for β2m or TAP-1, lack stable cell surface expression of most MHC-I molecules [17].

MHC-I mRNA is expressed by neurons and glial cells in the olfactory system, cerebral cortex, striatum, hippocampus, and spinal cord, both during development and adulthood. MHC-I proteins are expressed on the surface of axons and dendrites and are also located pre- and post-synaptically [16,18]. In the healthy brain, MHC-I expression is regulated by neuronal activity [15,19] and its upregulation is related to critical periods where synaptic refinement occurs more substantially [15,20–22]. Therefore, MHC-I expression is directly related to synaptic re-organization of the Central Nervous System (CNS).

In the spinal cord, MHC-I and β2m are upregulated after peripheral lesion [23]. Furthermore, Oliveira, Thams, Lidman, Piehl, Hokfelt, Karre, Linda and Cullheim [9] observed more synaptic stripping after axotomy in β2m knockout (β2m KO) and in TAP-1 knockout (TAP1 KO) mice. Enhanced synapse detachment in β2m KO animals could be attributed to a preferential removal of inhibitory terminals, while excitatory terminals were removed to a similar extent in KO and wildtype (WT) mice; i.e., MHC-I molecules are important for a selective maintenance of inhibitory synaptic terminals after lesion.

We suggest that MHC-I signaling is used by neurons and glia to interact both in normal and pathological processes. MHC-I is expressed by astrocytes, microglia and at pre- and postsynaptic terminals to selectively maintain inhibitory terminals and glial processes that are putatively involved in the detachment of pre-synaptic buttons from the lesioned motoneuron cell body. Indeed, microglia and astrocytes reactivity after peripheral lesion are influenced by MHC-I modulation, also affecting the synaptic plasticity process and regenerative capacity [24,25]. However, the exact role of MHC-I in neuron-glia signaling and spatiotemporal events that outline this process is still elusive.

2. Materials and Methods

2.1. Mice and Experimental Groups

For this study, we used female C57BL/6J wild-type (WT) and B6.129P2B2mtm1Unc/J (β2m KO) mice. Mice homozygous for the β2m targeted mutation have little if any MHC-I protein expression on the cell surface.

Six to eight-week-old female mice were bred in-house, at the Laboratory of Nervous Regeneration, Institute of Biology, University of Campinas. Mice were kept in appropriate micro-isolators, under a 12 h light-dark cycle, with controlled temperature and humidity, with water and food ad libitum.

Protocols concerning the animal use and handling were approved by the Institutional Committee for Ethics in Animal Experimentation (Committee for Ethics in Animal Use—Institute of Biology—CEUA/IB/UNICAMP, Protocol number 3336-1) and were performed in accordance with the guidelines of the Brazilian College for Animal Experimentation. All experiments concerning animal experimentation.

2.2. Surgical Procedure for Ventral Root Crush (Figure 1)

Mice were anesthetized with xylazine (König, Argentina, 5 mg/kg) and ketamine (Fort Dodge, USA, 100 mg/kg. Bepanthen (Bayer, Germany) was applied in the eyes to prevent dryness. Mice were placed in a heated bed and the hair was removed from the back. In the absence of the toe-pinch withdraw, an incision parallel to the vertebral column was made at the thoracic level. Paravertebral muscles were removed to expose the lower thoracic and upper lumbar vertebrae. Laminectomy was performed to expose the spinal cord. After that, an incision was made in the dura-mater allowing to reach ventral roots correspondent to L4, L5 and L6 spinal segments that were crushed (3 times of 10 s) with number 4 forceps. After muscle and skin were sutured and mice were kept under controlled heating until completely recovered from anesthesia. A dose of Tramadol (Neoquimica, Brazil, 5 mg/Kg) was administered by gavage after surgery and for the following 3 days.

To ensure that the lesion was successfully performed, in the following days after surgery, mice were behaviorally analyzed to check the correspondent paw paralysis (Figure 2C,D). Moreover, the morphology of fixed spinal cords was analyzed, to check the gross anatomical preservation of the nervous tissue after lesion, as well as to detect trace elements of degeneration on the axotomized roots (Figure 2A,B). Only mice fitting both criteria were used for further analysis. The success ratio of the surgery was estimated at 80%.

Each strain had three time-points of analysis after surgery, days after injury (dpi), constituting the following experimental groups: WT–7 dpi (n = 4), WT–14 dpi (n = 5), WT–28 dpi (n = 4), β2m KO–7 dpi (n = 4), β2m KO–14 dpi (n = 5), β2m KO–28 dpi (n = 4).

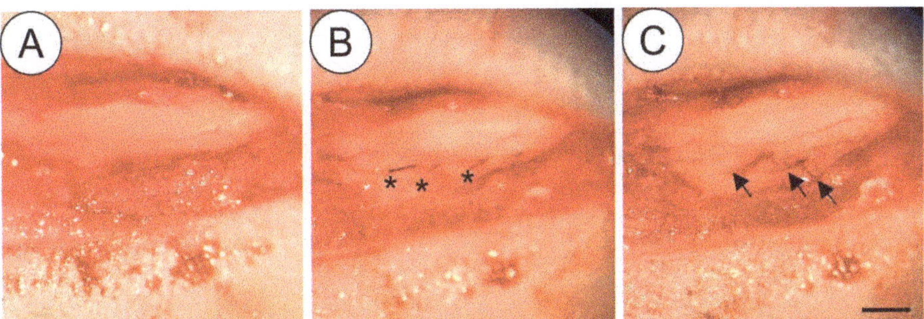

Figure 1. Details of the surgical procedure for ventral root crush in mice showing the exposed spinal cord with intact dura-mater after laminectomy (**A**), the L4, L5 and L5 ventral roots (asterisk) isolated after dura-mater longitudinal incision (**B**) Crushed roots (**C**-arrows), where it is possible to depict the persistence of the connective tissue at the injury site. Scale bar = 1 mm.

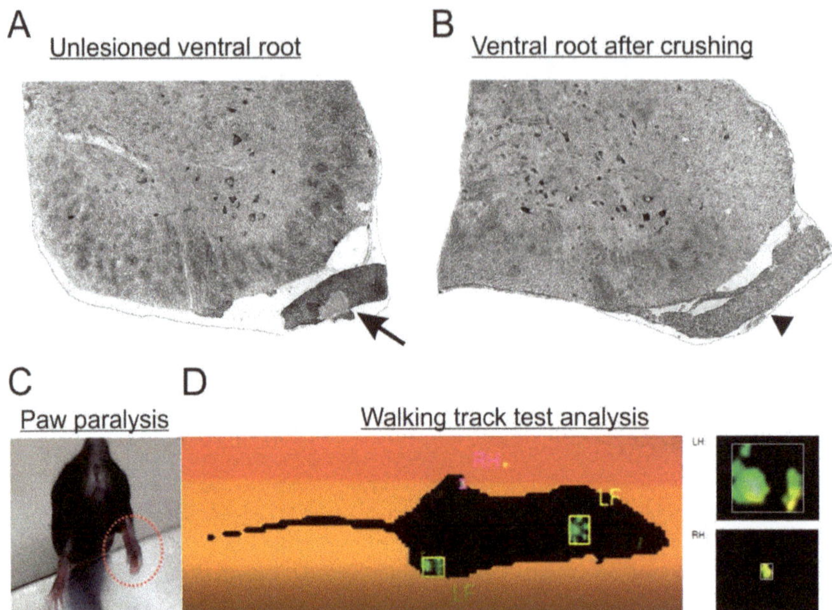

Figure 2. (**A**) Micrograph of a lumbar spinal cord transverse section showing the unlesioned ventral root morphology with intact axons (arrow). (**B**) Lesioned ventral root (arrowhead) with signs of ongoing Wallerian degeneration. (**C**) Paw paralysis behavior following L4, L5 and L6 ventral root lesion. Observe the adduction pattern of the toes combined with muscle atrophy (dotted red circle). (**D**) Gait analysis by the CatWalk system showing hindlimb loss of function ipsilateral to the lesion (RH).

2.3. Perfusion

Mice were anesthetized with an overdose of xylazine and ketamine and submitted to thoracotomy followed by transcardiac perfusion with PBS (0,1M Sodium Phosphate Buffer, PB, with 0.9% NaCl; pH 7.38). Afterwards, mice designated for the analysis of motoneuron survival and immunohistochemistry were perfused with a fixative solution (4% Formaldehyde in 0.1 M PB). After fixation, lumbar intumescences were dissected out and immersed in the same fixative solution overnight, at 4 °C, washed 3 times with 0.1 M PB, immersed in sucrose solutions (10%, 20% and 30%, 12 h each), finally soaked in Tissue-Tek, frozen in n-Hexane at controlled temperature (−32 °C to −35 °C) and stored at −20 °C.

2.4. Immunohistochemistry

Frozen 12 μm–thick serial sections were obtained using a cryostat (Microm, HM525), transferred to a gelatin-covered microscopic slide and stored at −20 °C until use.

For immunohistochemistry, microscopic slides were left at room temperature and the sections were outlined by a hydrophobic pen. Slides were transferred to a humid and light-protected chamber. Sections were immersed in 0.01 M PB (3 times, 5 min each), dried and incubated with 150 μL of blocking solution (3% Fetal Bovine Serum in 0.1 M PB) for 45 min. Subsequently, the different primary antibodies (Table 1) were diluted in incubation solution (1.5% Fetal Bovine Serum and 0.2% Tween in 0.1 M PB) and sections were incubated for 4 h or overnight.

After first incubation, slice sections were washed with 0.01 M PB and incubated with the respective secondary antibodies (cy2 anti-rabbit, cy3 anti-mouse or cy3 anti-rat; Jackson Immunoresearch, 1:500) for 45 min. As a nuclear marker, DAPI (4′,6-diamidino-2-phenylindole, 1:1000 in 0.1 M) was used. Sections were again washed with 0.01 M PB and mounted with coverslips using as medium glycerin/PB (3:1).

Table 1. Detailed description of the primary antibodies used.

Antibody	Host	Company	Code	Dilution
Anti-Iba-1	Rabbit	Wako	019-19741	1:750
Anti-GFAP	Rabbit	Abcam	ab7260	1:1500
Anti-Synaptophysin	Rabbit	Novus Biologicals	NBP2-25170	1:1000
Anti-MHC-I	Rat	BMA Biomedicals	T2105	1:100
Anti-NeuN	Mouse	Millipore	MAB377	1:500

Immunostainings were observed with an epifluorescence microscope (Leica DMB5500, Leica, Wetzlar, Germany) and documented with a digital camera (Leica DFC 345 FX), using specific filters according to the secondary antibodies or DAPI.

For quantification, three representative images from each animal of the respective experimental group were selected. The integrated density of pixels, representing protein immunolabeling, was measured in the lateral motor nucleus at anterior horn from the ipsi- and contralateral sides of the spinal cord, according to Oliveira, Thams, Lidman, Piehl, Hokfelt, Karre, Linda and Cullheim [9], using Image J software (1.51j8 version, National Institutes of Health, NIH, USA). As illustrated in Supplementary Figure S1, for GFAP and IBA-1, the whole area of the picture was quantified; for anti-synaptophysin immunolabeling, eight small areas around each motoneuron were measured (Supplementary Figure S2).

Integrated density of pixels was acquired for each animal and then the mean ± standard error of each experimental group was calculated.

2.5. Putative Motoneuron Cell Survival

For motoneuron-like cells counting, tissue specimens were processed just as described for immunohistochemistry, until sectioning and storage of the microscopic slides. Transversal sections from the lumbar intumescence were stained in 0.5% Toluidine Blue (Synth, Diadema, Brazil) for 1 min, rinsed in water, dehydrated, diaphanized and mounted with Entellan (Merck, Darmstadt, Germany) and coverslip.

Following, putative motoneurons from the lateral motor nucleus of the anterior horn of the spinal segments L4, L5 and L6 were counted in the ipsi- and contralateral sides every four slides from the specimen. In order to correct double counting, Abercrombie's formula was used [26]:

$$N = n \times t/(t + d)$$

"N" is the corrected number of neurons, "n" is the number of counted cells, "t" is the section thickness and "d" is the average diameter of the neurons. Once the neuron size difference affects the corrected number of neurons, the "d" value was calculated for each experimental group (ipsi- and contralateral).

2.6. Statistical Analysis

The data are presented as the mean ± SEM and analyzed using one-way ANOVA followed by Bonferroni post-hoc test, for multiple comparisons or two-tailed t-test; at $p < 0.05$ (*), $p < 0.01$ (**), and $p < 0.001$ (***).

3. Results

3.1. Time-Course of Motoneuron Survival, Glial Reaction, and Synaptic Covering Following VRC in C57BL/6J Mice

Neuronal survival rate was evaluated 7, 14 and 28 days after ventral root crush (VRC) by the counting of large α-motoneurons-like cells present in the ipsilateral lamina IX and compared to the contralateral side (ipsi/contralateral ratio). No statistical differences between the contralateral sides from C57BL/6J (WT) mice at the different time points were observed (mean ± standard error for different experimental groups: 7 dpi, 6.65 ± 0.17; 14 dpi, 6.71 ± 0.44; 28 dpi 7.8 ± 0.24).

Seven days after VRC, C57BL/6J mice showed already 25% of neuron loss which increased to 38 and 47% 14 and 28 days later, respectively (Figure 3). Loss of motoneuron-like cells in the lateral motor nucleus from the lumbar intumescence was statistically higher 14 and 28 days after VRC when compared to 7 dpi ($p < 0.0001$, F $_{2,11}$ = 25.83, Bonferroni post-test).

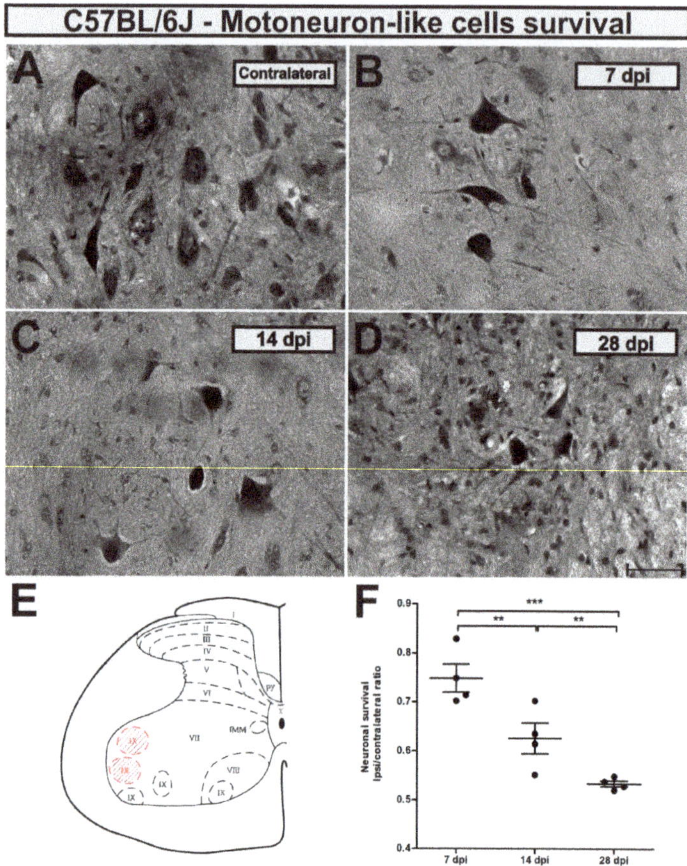

Figure 3. Time-course of motoneuron-like cell survival in the wild type (WT) mice (**A**) contralateral - unlesioned, (**B**) 7 days after injury (dpi), (**C**) 14 dpi and (**D**) 28 dpi. (**E**) Schematic spinal cord transverse section showing the laminae of Rexed distribution. Lamina IX is highlighted in red and represents the area where cell counting, and immunohistochemistry analysis were performed. (**F**) Neuronal survival at the time points after ventral root crush (ipsi/contralateral ratio); $p < 0.01$ (**), and $p < 0.001$ (***). Scale bar = 50 μm.

Description of calculated motoneuron survival means ± standard errors from C57BL/6J experimental groups: 7 dpi = 0.64 ± 0.02; 14 dpi = 0.55 ± 0.02; 28 dpi = 0.49 ± 0.02.

To analyze astroglial changes in the lesioned neuronal microenvironment we used immunohistochemical labeling of glial fibrillary acid protein (GFAP)—an astrocytic intermediary filament marker. In WT mice, astrogliosis increased in a time-dependent manner, reaching five-fold upregulation by 28 days after injury (Figure 4A,D,I,J). Comparing the time points, increased astrogliosis was statistically higher 28 dpi when compared to 7 and 14 dpi ($p = 0.0005$, F2,10 = 18.17, Bonferroni post-test).

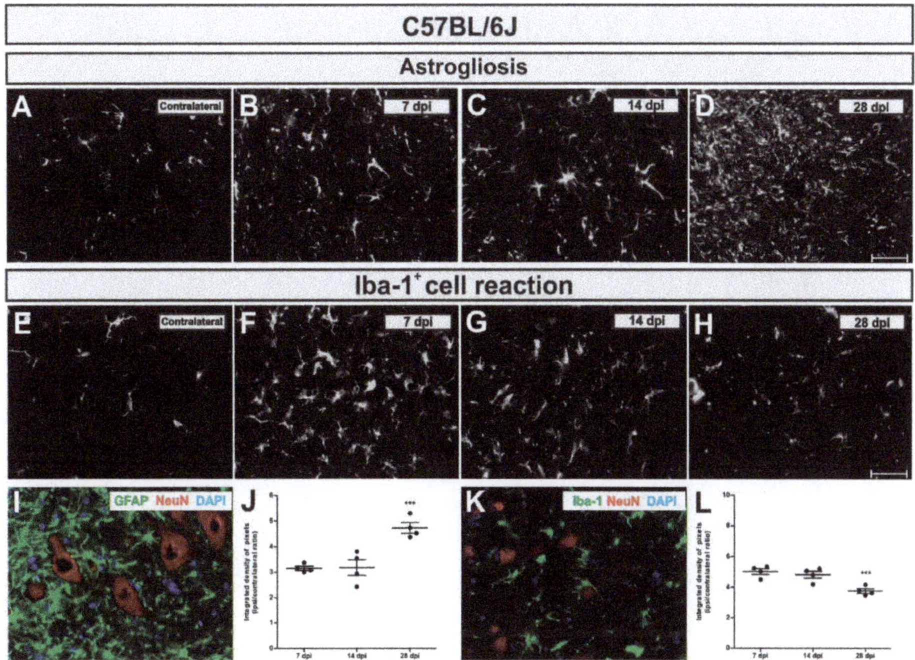

Figure 4. Time-course of astrogliosis (**A** to **D** and **I**) and Iba-1$^+$ cell reaction (**E** to **H** and **L**) in the lumbar spinal cord contralateral side and 7, 14 and 28 dpi post ventral root injury in WT mice. Note that astrogliosis increases over time. Contrarily, the Iba-1$^+$ cell reaction is more intense in the acute phase after injury. **I** and **K** illustrate the glial reaction in the ipsilateral side nearby axotomized motoneuron-like cells, positive to NeuN (in red). (**J** and **L**) show the comparative quantification among time points post injury (ipsi/contralateral ratio); $p < 0.001$ (***). Scale bar = 50 μm.

The calculated IL/CL integrated density of pixels ratio mean ± standard error from GFAP immunoreactivity in C57BL/6J - WT experimental groups provided the following results: 7 dpi = 3.14 ± 0.1; 14 dpi = 3.25 ± 0.4; 28 dpi = 4.72 ± 0.2.

To analyze microglia and macrophages changes surrounding lesioned motoneurons, we used immunolabelling of Iba-1 (Ionized calcium binding adaptor molecule–a microglia/macrophage calcium-binding marker). It is important to point out that VRC is a lesion that results in the blood brain barrier disruption and Iba-1 also labels macrophages. Thus, Iba-1 positive labelling nearby motoneurons does not exclude macrophages derived from circulating monocytes.

In C57BL/6J mice, on the contrary to what was detected for astrogliosis, the Iba-1$^+$ cells reaction was more intense in the acute phase after lesion (7 days), becoming reduced by 26% at day 28 (Figure 4E,H,K,L). The reduction in the microglial reaction in the lateral motor nucleus of WT mice at 28 dpi was significantly different compared to 7 and 14 dpi ($p = 0.0031$, F2,10 = 10.84, Bonferroni post-test).

Calculating IL/CL integrated density of pixels ratio mean ± standard error from Iba-1 immunoreactivity in C57BL/6J experimental groups led to the following data: 7 dpi = 5.02 ± 0.2; 14 dpi = 4.68 ± 0.2; 28 dpi = 3.76 ± 0.2.

To analyze synaptic changes after VRC, we used an anti-synaptophysin antibody and assessed pre-synaptic terminals in apposition to the axotomized neurons as described in Methods section.

Synaptic inputs were reduced by 27% adjacent to the putative axotomized motoneurons at day 7 after lesion, which was further reduced to 47% ($p < 0.0001$, F2,10 = 34.21, Bonferroni post-test), both 14- and 28-days post-lesion (Figure 5).

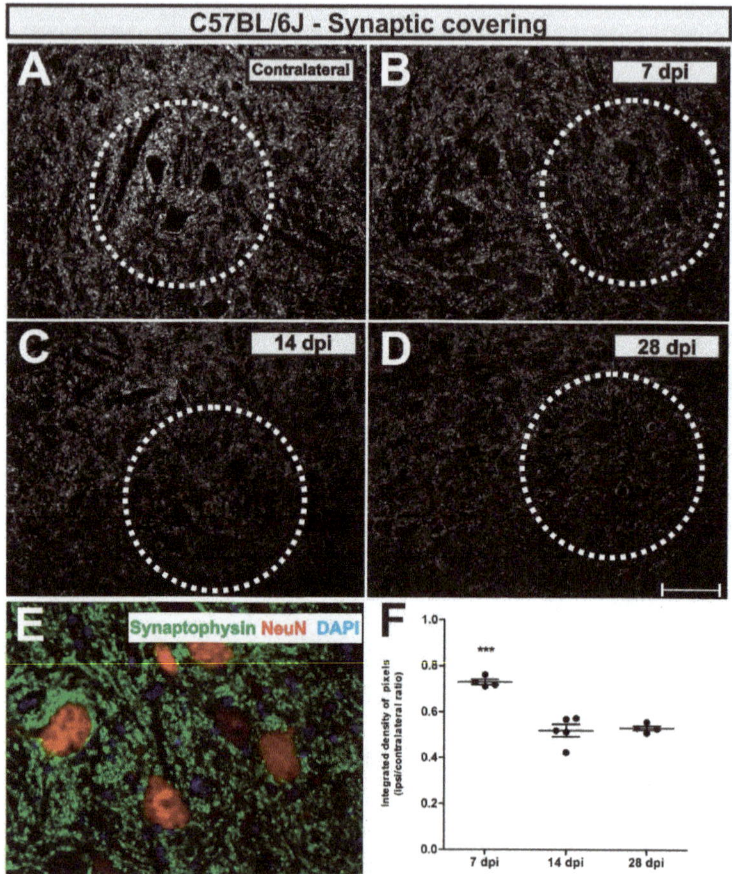

Figure 5. Synaptic covering in WT mice. (**A**) representative contralateral image, (**B**) 7 dpi, (**C**) 14 dpi and (**D**) 28 dpi. (**E**) illustrates the synaptic inputs (in green) to ipsilateral motoneuron-like cells, labelled with NeuN (in red). (**F**) Comparative quantification among time points (ipsi/contralateral ratio); $p < 0.001$ (***). Scale bar = 50 μm.

The calculated IL/CL integrated density of pixels ratio mean ± standard error from synaptophysin immunoreactivity in C57BL/6J - WT experimental groups resulted in the following values: 7 dpi = 0.73 ± 0.05; 14 dpi = 0.52 ± 0.03; 28 dpi = 0.53 ± 0.05.

3.2. Enhanced Microglial Reaction and Synaptic Stripping in β2mKO Mice after VRC

After establishing the VRC lesion model in wildtype mice, we used β2m KO mice to test glial responses and neuronal survival in a condition with impaired antigen presentation and altered neuron-glia communication. First, we monitored motoneuron-like cells survival rate at 7, 14 and 28 days after VRC. Seven days after VRC, the KO mice showed 36% of neuron loss which increased to 45 and 51% fourteen and twenty-eight days after lesion, respectively (Figure 6). Statistically, the motoneuron-like cells loss was significantly higher 28 days after VRC when compared to 7 dpi ($p = 0.0003$, F 2,10 = 20.66, Bonferroni post-test).

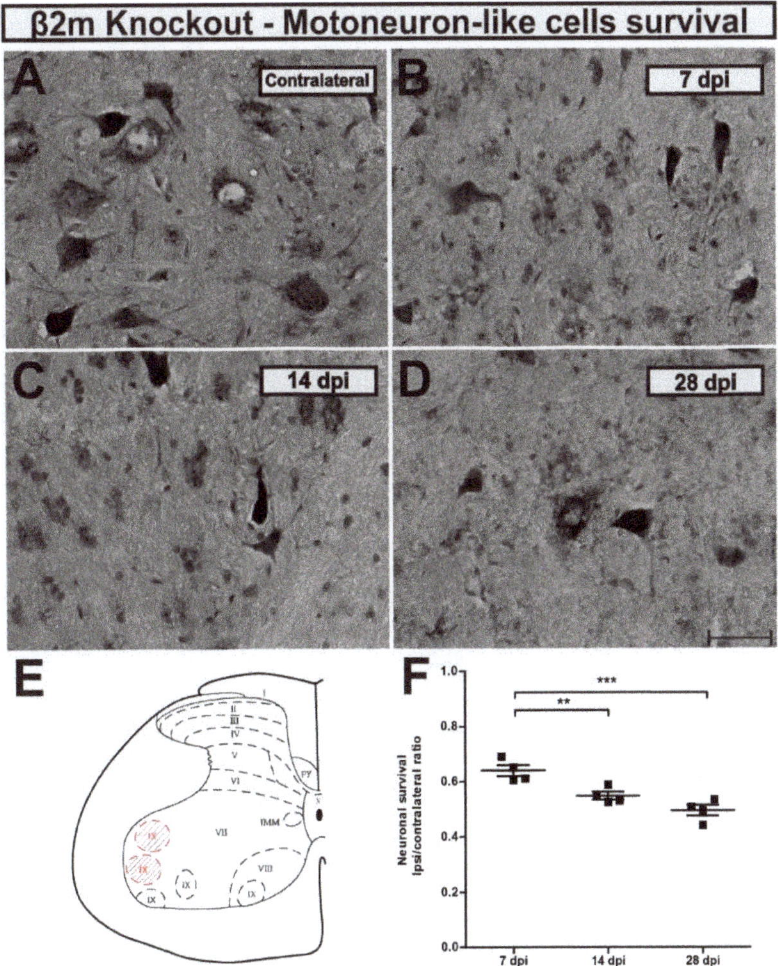

Figure 6. Time-course of motoneuron-like cell survival in the β2m knockout mice (**A**) contralateral - unlesioned, (**B**) 7 dpi, (**C**) 14 dpi and (**D**) 28 dpi. (**E**) Schematic spinal cord transverse section showing the laminae of Rexed distribution. Lamina IX is highlighted in red and represents the area where cell counting, and immunohistochemistry analysis were performed. (**F**) Neuronal survival at the time points after ventral root crush (ipsi/contralateral ratio); $p < 0.01$ (**), and $p < 0.001$ (***). Scale bar = 50 μm.

When looking at the reactive astrogliosis in β2mKO mice, we observed that GFAP was upregulated in the IL side of about 2-fold related to the CL (Figure 7A,D,I,J), but no changes among the time points were detected ($p = 0.511$, F 2,11 = 0.7138, Bonferroni post-test).

By calculating the IL/CL integrated density of pixels ratio mean ± standard error from GFAP immunoreactivity in β2m KO groups the following data was obtained: 7 dpi = 2.17 ± 0.2; 14 dpi = 2.5 ± 0.2; 28 dpi = 2.16 ± 0.2.

The Iba-1$^+$ cells reaction in β2mKO mice showed a strong transient upregulation. The Iba-1 immunolabel reached a seven-fold increase at 7 days after VRC, which was then reduced to four-fold upregulation day 28 (Figure 7E,H,K,L). When statistically compared to other time points, the increased reaction in the lateral motor nucleus from the lumbar intumescence was statistically higher 7 days after VRC when compared to 7 and 14 dpi ($p < 0.0001$, F 2,10 = 106.1, Bonferroni post-test).

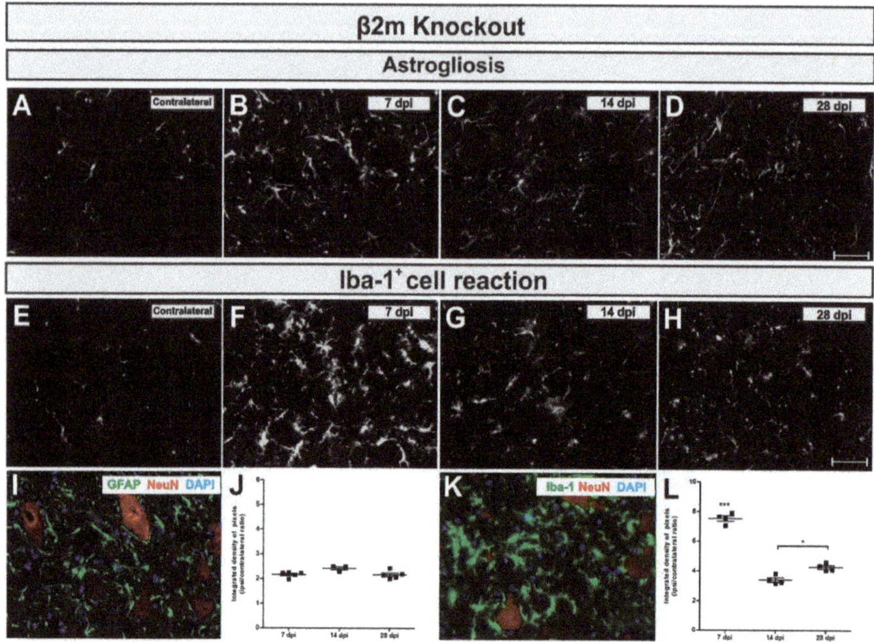

Figure 7. Time-course of astrogliosis (**A** to **D** and **I**) and Iba-1⁺ cell reaction (**E** to **H** and **L**) in the lumbar spinal cord contralateral side and 7, 14 and 28 dpi post ventral root injury in β2m knockout mice is kept constant over time and microglial reaction is more intense in the acute phase after injury. **I** and **K** illustrate the glial reaction in the ipsilateral side nearby axotomized motoneuron-like cells, positive to NeuN (in red). (**J** and **L**) show the comparative quantification among time points post injury (ipsi/contralateral ratio); $p < 0.05$ (*) and $p < 0.001$ (***). Scale bar = 50 μm.

Description of the calculated IL/CL integrated density of pixels ratio mean ± standard error from Iba-1 immunoreactivity in 7dpi in β2m KO groups: 7 dpi = 7.53 ± 0.2; 14 dpi = 3.40 ± 0.2; 28 dpi = 4.16 ± 0.2.

In β2m KO mice, the analysis of the anti-synaptophysin immunolabeling around lesioned motoneuron-like cells showed that the synaptic coverage was reduced by 48% at day 7 after lesion, reaching 56% reduction at day 28 (Figure 8). When statistically compared, the synaptic coverage was 29% lower 28 days after VRC when compared to 7 dpi ($p = 0.0013$, F 2,10 = 13.87, Bonferroni post-test).

IL/CL integrated density of pixels ratio means ± standard errors from synaptophysin immunoreactivity β2m KO groups led to the following data: 7 dpi = 0.62 ± 0.05; 14 dpi = 0.47 ± 0.03; 28 dpi = 0.44 ± 0.05.

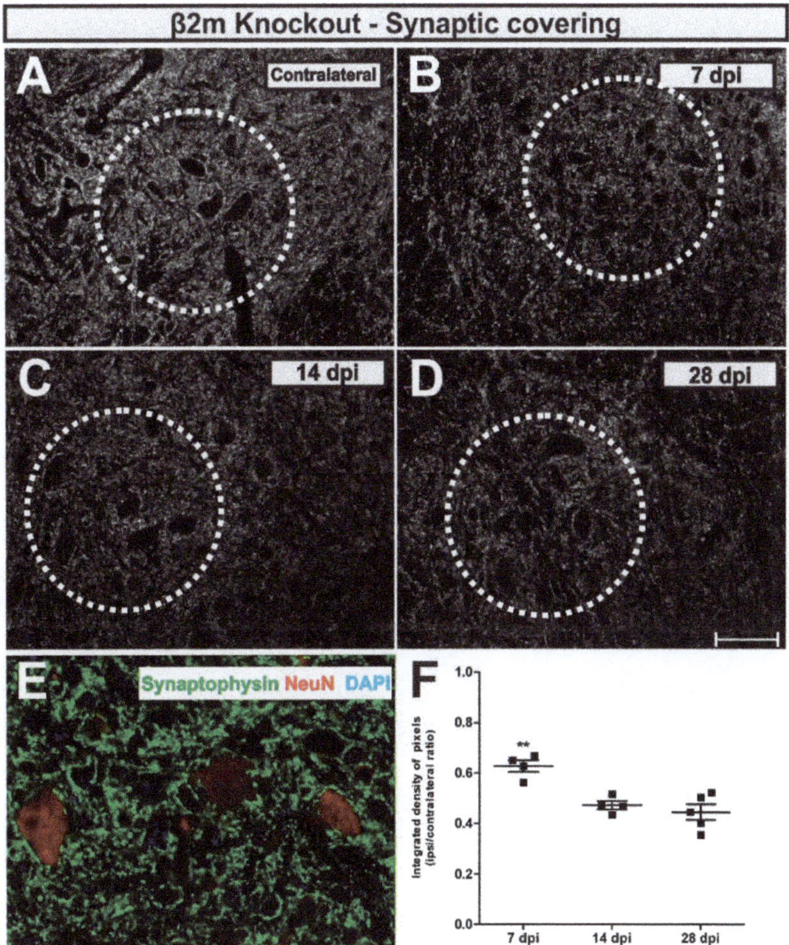

Figure 8. Synaptic covering in β2m knockout mice. (**A**) representative contralateral image, (**B**) 7 dpi, (**C**) 14 dpi and (**D**) 28 dpi. (**E**) illustrates the synaptic inputs (in green) to ipsilateral motoneuron-like cells, labelled with NeuN (in red). (**F**) Comparative quantification among time points (ipsi/contralateral ratio); $p < 0.01$ (**). Scale bar = 50 μm.

3.3. Comparison between C57BL/6J and B2m KO Responses to Injury

A qualitative MHC-I immunolabeling was performed 7 days post lesion in both C57BL/6J and β2m KO mice (Figure 9) to show that there is a low basal labelling in WT contralateral side and an upregulation in MHC-I labeling after lesion. In β2m KO mice the labelling is almost absent and the upregulation after lesion cannot be depicted.

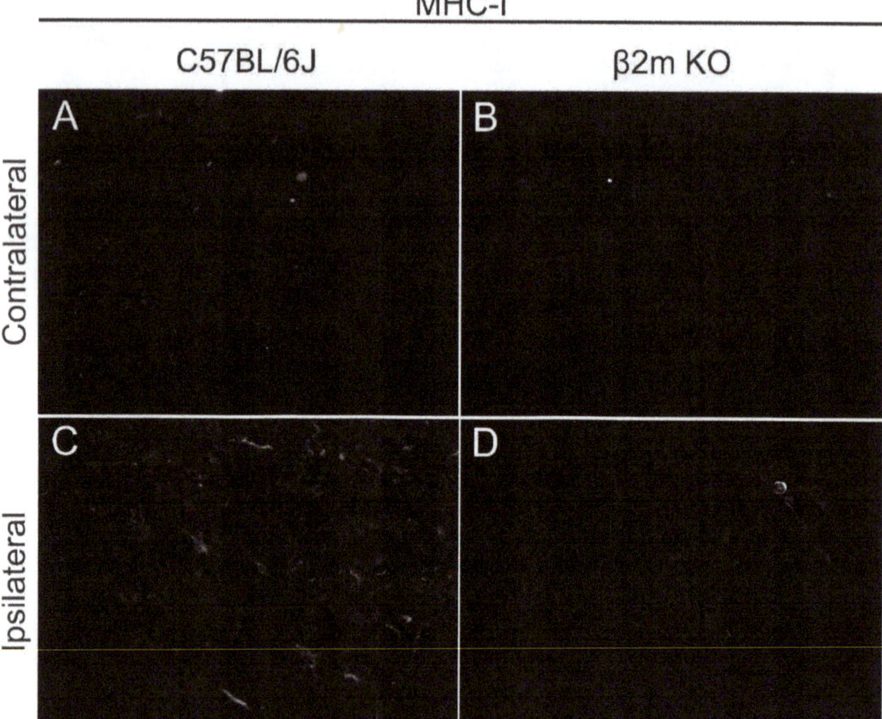

Figure 9. Major histocompatibility complex of class I (MHC-I) expression in C57BL/6J (**A,C**) and β2mKO mice (**B,D**) 7 days after VRC. Note that contralateral labeling is almost absent in C57BL/6J showing upregulation after injury. β2mKO mice showed weak staining both in the contra and ipsilateral sides. Scale bar = 25 μm.

Concerning putative motoneuron survival, when comparing the results from C57BL/6J and β2mKO mice (Figure 10), for each time point, the motoneuron loss at day 7 post-injury was sharply higher in β2m KO compared to the WT ($p = 0.0094$, $R^2 = 0.6425$, t-test), pointing out that there is an increased neuronal susceptibility in this phase in the absence of MHC-I.

Unlesioned spinal cords of C57BL/6J and β2mKO mice were labeled with GFAP and Iba-1 antibodies and no differences between the strains could be depicted (Supplementary Figure S3).

The reactive astrogliosis (Figure 10) was significantly higher ($p < 0.0001$, $R^2 = 0.95$, t-test) in β2mWT mice 7 and 28 dpi, when compared to β2m KO. On the other hand, reaction of Iba-1$^+$ cells at 7 dpi was higher in β2m KO compared to C57BL/6J ($p < 0.0001$, $R^2 = 0.9395$, t-test) and, at 14 dpi this reaction was 27% reduced in the β2m KO when compared to the WT ($p = 0.0015$, $R^2 = 0.7824$, t-test) pointing out that following the exacerbate Iba-1$^+$ cells reaction in the β2mKO at day 7 post-injury, there was an intense reduction by the 14th day in the absence of MHC I, when compared to WT.

The comparative analysis from WT and β2m KO synaptic covering after VRC showed that the synaptic loss was 16% higher in β2m KO mice 7 and 28 dpi when compared to WT ($p < 0.007$, $R^2 = 0.783$, t-test).

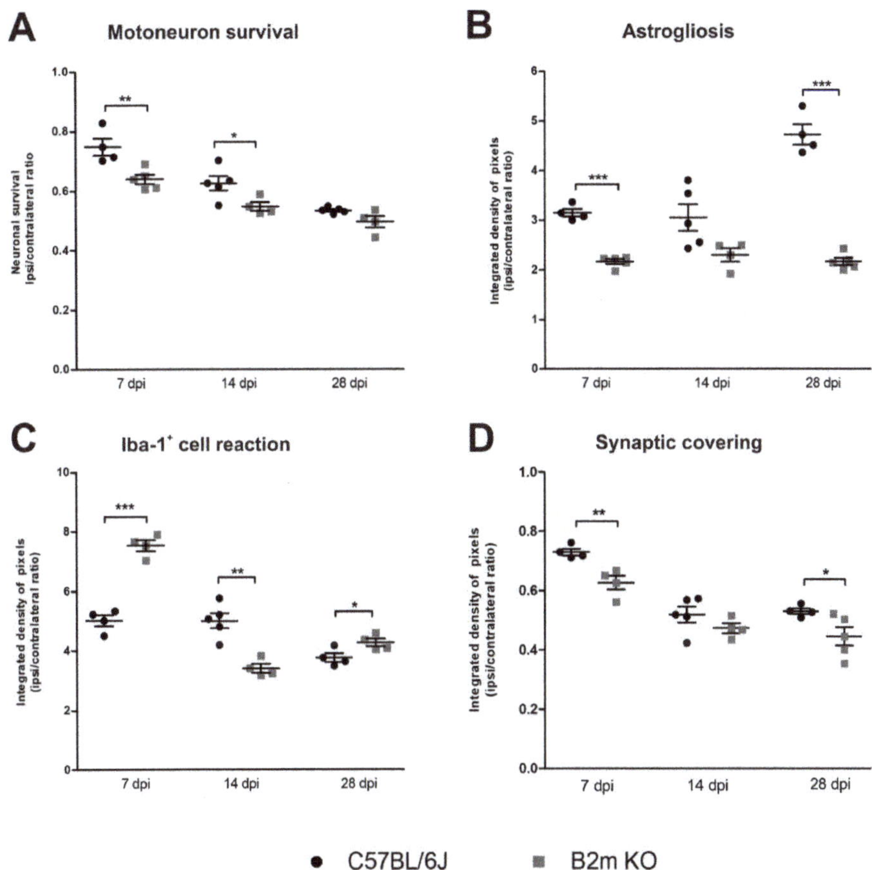

Figure 10. Comparative time course analysis of motoneuron-like cell survival (**A**), astrogliosis (**B**), Iba-1$^+$ cell reaction (**C**) and synaptic covering (**D**) in WT and β2m KO mice. Note that the progressive loss of motoneuron-like cells is equivalent in both strains, despite astrogliosis in β2m KO that did not show increasing over time. On the other hand, the Iba-1$^+$ cells reaction was more intense in β2m KO mice in the acute phase after VRC, which was coincident to a more pronounced synaptic detachment in the same strain, a difference that was detected also 28 dpi. Graphs show the comparative quantification among time points (ipsi/contralateral ratios); Significance levels: * $p < 0.5$, ** $p < 0.01$, *** $p < 0.001$.

4. Discussion

We show the time course of neuronal degeneration, synapse retraction, and glial reaction after ventral root crushing in C57BL/6J - WT mice and compared to β2mKO mice.

In the first week after lesion, 25% of putative motoneuron loss was already detected, increasing to 38% in the second week and reaching 47% in the fourth week after VRC, showing similarities with data already published regarding VRC in rats, which presented motoneuron loss up to 51%, 28 days after lesion [27].

As expected, astrogliosis increased in the ipsilateral side after lesion to the CNS/PNS interface [28,29], reaching in the mouse 3-fold upregulation between 7 to 14 days, increasing in a time-dependent manner and reaching 5-fold upregulation by 28 days after injury. On the other hand, the Iba-1$^+$ cells reaction was also induced by VRC, being more intense in the acute phase after lesion (7 days) and becoming reduced by 26% at day 28.

Concerning the synaptic coverage, synaptic inputs were reduced by 27% nearby axotomized motoneurons at day seven after lesion, and such coverage was reduced to 47% until 28 days post-injury. These data corroborate with previous results from our group [27–29] Thus, the results described so far demonstrate that ventral root crushing in mice provides robust data regarding neuronal loss and glial reaction. This allows studies with transgenic strains and/or therapeutic approaches that may, in turn, unveil strategies to improve motor recovery after proximal lesions.

In this context, we performed the same time course analysis in β2m KO mice. It is already known that MHC-I influences the interaction between pre- and post-synaptic neurons as well as between neurons and glial cells. Also, this molecule is upregulated both in the CNS and PNS as a response to axotomy [23,24,30]. The lack of MHC-I showed decreased motoneuron axonal regeneration after peripheral lesion [9]. On the other hand, the enhanced neuronal MHC-I expression has a beneficial effect following spinal cord injury, leading to the significantly better recovery of locomotor abilities [31].

Following VRC, the absence of MHC-I plays an important role, especially acutely post-injury (7 dpi). In this case, in β2mKO mice, the motoneuron-like cell loss was 33% higher at this time point, astrogliosis was decreased and the Iba-1$^+$ cells reaction was more intense when compared to WT mice.

Bombeiro, et al. [32] showed in vitro a reduction in GFAP-labelling and hypertrophic astrocytes following β2m knock down by using siRNA (small interfering RNA). In a co-culture system, the reduced expression of GFAP in knocked down astrocytes was related to maintenance of synaptophysin immunostaining in neurons. Similarly, seven days after VRC, we observed 31% less GFAP immunoreactivity in the lateral motor nucleus at lamina IX from β2m KO mice when compared to WT. This reduced reactive astrogliosis was even more accentuated 28 days post-injury, 54% lower in β2m KO when compared to C57BL/6J-WT.

It is important to point out that astrocytes play an essential role in neurotransmitter reuptake from the synaptic space and this feature is especially important under deleterious stimuli, such as a lesion. The loss of synaptic inputs from the lesioned motoneuron cell body occurs preferentially to excitatory inputs, subjecting the cells mostly to an inhibitory influence during repair [33]. In the lack of MHC-I, the selective maintenance of inhibitory synaptic terminals after axotomy is altered and a larger number of inhibitory terminals are removed [9].

Chen, et al. [34] pointed out that, in the adult brain, microglia-mediated synaptic stripping of pre-synaptic terminals is neuroprotective and the activated microglia in close apposition to neurons displaced GABAergic synapses. Chen, Jalabi, Hu, Park, Gale, Kidd, Bernatowicz, Gossman, Chen, Dutta and Trapp [34] reported that, in the adult brain, microglia-mediated synaptic stripping of pre-synaptic terminals is neuroprotective and the activated microglia in close apposition to neurons displaced GABAergic synapses. In this scenario, we propose that the higher number of inhibitory terminals retracting in β2m KO mice after lesion [9] could be due to physical displacement by microglial processes. Indeed, seven days after VRC, the Iba-1$^+$ cells reactivity detected by anti-Iba-1 immunostaining showed that β2m KO had 33% more reactive cells when compared to WT. Even though Iba-1 positive cells in this scenario include microglia and macrophages derived from circulating monocytes, it is plausible to assume that the microglial reaction is also enhanced in this situation. Thus, we hypothesize that the higher synaptic loss in β2m knockout mice, is due to the enhanced microglial activation, which is actively involved in the detaching of synaptic inputs.

Liddelow, et al. [35] showed that astrocytes with a harmful profile (defined by the authors as A1 astrocytes) are induced by factors secreted by activated microglia (namely: IL-1α, TNF and C1q). A1 astrocytes lose their normal astrocytic function, such as synaptic maintenance, and can also be induced by axotomy and neurodegenerative diseases. In a model of optic nerve crush, a fast generation of A1 astrocytes in parallel to the death of retinal ganglion cells was detected. Testing the deleterious effects of A1-like astrocytes on several sorts of neuronal types, it was detected that 20% of spinal motoneurons remained viable after submitted to A1-like astrocytes conditioned medium, while other cells, like retinal ganglion cells, did not survive at all. Moreover, γ-motoneurons were not susceptible to A1 astrocyte toxicity [35].

Reactive astrogliosis or microglial reaction to injury or neurodegenerative process did not occur in isolation but as part of a multicellular response [36]. So, in response to injury, a glial response occurs, constituting both micro- and astrogliosis and leading to a local innate inflammatory reaction in the vicinity of the axotomized nerve cells. In a model of ventral root avulsion, for example, after approximately 1 week there is a phase of microglia activation paralleled by up-regulation of astrogliosis and lymphocyte infiltration, coinciding with the beginning of the death of axotomized cells, that are about 50% gone by the third week [37]. As, after a traumatic lesion, the microenvironment changes dictate the destiny of axotomized motoneurons and the maintenance of synaptic connectivity through glial reaction, the lack of a functional MHC-I can modulate the inflammation accentuating glial reaction towards a harmful way causing ultimately a more pronounced synaptic detachment in the lesion as described here.

Supplementary Materials: The following are available online at http://www.mdpi.com/2073-4409/8/5/483/s1.

Author Contributions: L.P.C. conceived and performed experiments, analyzed data, and wrote the manuscript; M.P. performed experiments and analyzed data; A.L.R.d.O. and F.K. conceived experiments and wrote the manuscript.

Funding: This work was supported by grants from São Paulo Research Foundation (FAPESP, 2013/16134-6; 2014/06892-3), National Council for Scientific and Technological Development (CNPq; 303085/2017-7), Coordination for the Improvement of Higher Education Personnel (CAPES) and Deutsche Forschungsgemeinschaft SPP 1757 and FOR 2289.

Acknowledgments: The authors are grateful to Ricardo Gazzineli and Ana Carolina Teixeira from FIOCRUZ/MG for donating β2m KO mice.

Conflicts of Interest: The authors declare no conflict of interest.

References

1. Rothman, S.M.; Winkelstein, B.A. Chemical and mechanical nerve root insults induce differential behavioral sensitivity and glial activation that are enhanced in combination. *Brain Res.* **2007**, *1181*, 30–43. [CrossRef]
2. Kobayashi, S.; Yoshizawa, H.; Yamada, S. Pathology of lumbar nerve root compression. Part 2: morphological and immunohistochemical changes of dorsal root ganglion. *J. Orthop. Res.* **2004**, *22*, 180–188. [CrossRef]
3. Gaudet, A.D.; Popovich, P.G.; Ramer, M.S. Wallerian degeneration: Gaining perspective on inflammatory events after peripheral nerve injury. *J. Neuroinflammation* **2011**, *8*, 110. [CrossRef]
4. Zochodne, D.W. The challenges and beauty of peripheral nerve regrowth. *J. Peripher. Nerv. Syst.* **2012**, *17*, 1–18. [CrossRef]
5. Blinzinger, K.; Kreutzberg, G. Displacement of synaptic terminals from regenerating motoneurons by microglial cells. *Z. Zellforsch. Mikrosk. Anat.* **1968**, *85*, 145–157. [CrossRef]
6. Cullheim, S.; Thams, S. The microglial networks of the brain and their role in neuronal network plasticity after lesion. *Brain Res. Rev.* **2007**, *55*, 89–96. [CrossRef] [PubMed]
7. Aldskogius, H.; Liu, L.; Svensson, M. Glial responses to synaptic damage and plasticity. *J. Neurosci. Res.* **1999**, *58*, 33–41. [CrossRef]
8. Linda, H.; Shupliakov, O.; Ornung, G.; Ottersen, O.P.; Storm-Mathisen, J.; Risling, M.; Cullheim, S. Ultrastructural evidence for a preferential elimination of glutamate-immunoreactive synaptic terminals from spinal motoneurons after intramedullary axotomy. *J. Comp. Neurol* **2000**, *425*, 10–23. [CrossRef]
9. Oliveira, A.L.; Thams, S.; Lidman, O.; Piehl, F.; Hokfelt, T.; Karre, K.; Linda, H.; Cullheim, S. A role for MHC class I molecules in synaptic plasticity and regeneration of neurons after axotomy. *Proc. Natl. Acad. Sci. USA* **2004**, *101*, 17843–17848. [CrossRef]
10. Delgado-Garcia, J.M.; Del Pozo, F.; Spencer, R.F.; Baker, R. Behavior of neurons in the abducens nucleus of the alert cat–III. Axotomized motoneurons. *Neuroscience* **1988**, *24*, 143–160. [CrossRef]
11. Takata, M.; Nagahama, T. Synaptic efficacy of inhibitory synapses in hypoglossal motoneurons after transection of the hypoglossal nerves. *Neuroscience* **1983**, *10*, 23–29. [CrossRef]
12. Novikov, L.; Novikova, L.; Kellerth, J.O. Brain-derived neurotrophic factor promotes axonal regeneration and long-term survival of adult rat spinal motoneurons in vivo. *Neuroscience* **1997**, *79*, 765–774. [CrossRef]

13. Oliveira, A.L.; Langone, F. GM-1 ganglioside treatment reduces motoneuron death after ventral root avulsion in adult rats. *Neurosci. Lett.* **2000**, *293*, 131–134. [CrossRef]

14. Svensson, M.; Eriksson, P.; Persson, J.K.; Molander, C.; Arvidsson, J.; Aldskogius, H. The response of central glia to peripheral nerve injury. *Brain Res. Bull.* **1993**, *30*, 499–506. [CrossRef]

15. Boulanger, L.M.; Huh, G.S.; Shatz, C.J. Neuronal plasticity and cellular immunity: shared molecular mechanisms. *Curr. Opin. Neurobiol.* **2001**, *11*, 568–578. [CrossRef]

16. Elmer, B.M.; McAllister, A.K. Major histocompatibility complex class I proteins in brain development and plasticity. *Trends Neurosci.* **2012**, *35*, 660–670. [CrossRef]

17. Boulanger, L.M.; Shatz, C.J. Immune signalling in neural development, synaptic plasticity and disease. *Nat. Rev. Neurosci.* **2004**, *5*, 521–531. [CrossRef]

18. Ohtsuka, M.; Inoko, H.; Kulski, J.K.; Yoshimura, S. Major histocompatibility complex (Mhc) class Ib gene duplications, organization and expression patterns in mouse strain C57BL/6. *BMC Genom.* **2008**, *9*, 178. [CrossRef]

19. Corriveau, R.A.; Huh, G.S.; Shatz, C.J. Regulation of class I MHC gene expression in the developing and mature CNS by neural activity. *Neuron* **1998**, *21*, 505–520. [CrossRef]

20. Fourgeaud, L.; Boulanger, L.M. Role of immune molecules in the establishment and plasticity of glutamatergic synapses. *Eur J. Neurosci.* **2010**, *32*, 207–217. [CrossRef]

21. Huh, G.S.; Boulanger, L.M.; Du, H.; Riquelme, P.A.; Brotz, T.M.; Shatz, C.J. Functional requirement for class I MHC in CNS development and plasticity. *Science* **2000**, *290*, 2155–2159. [CrossRef]

22. Needleman, L.A.; Liu, X.B.; El-Sabeawy, F.; Jones, E.G.; McAllister, A.K. MHC class I molecules are present both pre- and postsynaptically in the visual cortex during postnatal development and in adulthood. *Proc. Natl. Acad. Sci. USA* **2010**, *107*, 16999–17004. [CrossRef]

23. Linda, H.; Hammarberg, H.; Cullheim, S.; Levinovitz, A.; Khademi, M.; Olsson, T. Expression of MHC class I and beta2-microglobulin in rat spinal motoneurons: Regulatory influences by IFN-gamma and axotomy. *Exp. Neurol.* **1998**, *150*, 282–295. [CrossRef]

24. Zanon, R.G.; Cartarozzi, L.P.; Victorio, S.C.; Moraes, J.C.; Morari, J.; Velloso, L.A.; Oliveira, A.L. Interferon (IFN) beta treatment induces major histocompatibility complex (MHC) class I expression in the spinal cord and enhances axonal growth and motor function recovery following sciatic nerve crush in mice. *Neuropathol. Appl. Neurobiol.* **2010**, *36*, 515–534. [CrossRef]

25. Zanon, R.G.; Oliveira, A.L. MHC I upregulation influences astroglial reaction and synaptic plasticity in the spinal cord after sciatic nerve transection. *Exp. Neurol.* **2006**, *200*, 521–531. [CrossRef]

26. Abercrombie, M.; Johnson, M.L. Quantitative histology of Wallerian degeneration: I. Nuclear population in rabbit sciatic nerve. *J. Anat.* **1946**, *80*, 37–50.

27. Spejo, A.B.; Carvalho, J.L.; Goes, A.M.; Oliveira, A.L. Neuroprotective effects of mesenchymal stem cells on spinal motoneurons following ventral root axotomy: Synapse stability and axonal regeneration. *Neuroscience* **2013**, *250*, 715–732. [CrossRef]

28. Rodrigues Hell, R.C.; Silva Costa, M.M.; Goes, A.M.; Oliveira, A.L. Local injection of BDNF producing mesenchymal stem cells increases neuronal survival and synaptic stability following ventral root avulsion. *Neurobiol. Dis.* **2009**, *33*, 290–300. [CrossRef]

29. Scorisa, J.M.; Zanon, R.G.; Freria, C.M.; de Oliveira, A.L. Glatiramer acetate positively influences spinal motoneuron survival and synaptic plasticity after ventral root avulsion. *Neurosci. Lett.* **2009**, *451*, 34–39. [CrossRef]

30. Bombeiro, A.L.; Thome, R.; Oliveira Nunes, S.L.; Monteiro Moreira, B.; Verinaud, L.; Oliveira, A.L. MHC-I and PirB Upregulation in the Central and Peripheral Nervous System following Sciatic Nerve Injury. *PLoS ONE* **2016**, *11*, e0161463. [CrossRef]

31. Joseph, M.S.; Bilousova, T.; Zdunowski, S.; Wu, Z.P.; Middleton, B.; Boudzinskaia, M.; Wong, B.; Ali, N.; Zhong, H.; Yong, J.; et al. Transgenic mice with enhanced neuronal major histocompatibility complex class I expression recover locomotor function better after spinal cord injury. *J. Neurosci. Res.* **2011**, *89*, 365–372. [CrossRef]

32. Bombeiro, A.L.; Hell, R.C.; Simoes, G.F.; Castro, M.V.; Oliveira, A.L. Importance of major histocompatibility complex of class I (MHC-I) expression for astroglial reactivity and stability of neural circuits in vitro. *Neurosci. Lett.* **2017**, *647*, 97–103. [CrossRef]

33. Carlstedt, T. Nerve root replantation. *Neurosurg. Clin. N Am.* **2009**, *20*, 39–50. [CrossRef]

34. Chen, Z.; Jalabi, W.; Hu, W.; Park, H.J.; Gale, J.T.; Kidd, G.J.; Bernatowicz, R.; Gossman, Z.C.; Chen, J.T.; Dutta, R.; et al. Microglial displacement of inhibitory synapses provides neuroprotection in the adult brain. *Nat. Commun.* **2014**, *5*, 4486. [CrossRef]
35. Liddelow, S.A.; Guttenplan, K.A.; Clarke, L.E.; Bennett, F.C.; Bohlen, C.J.; Schirmer, L.; Bennett, M.L.; Munch, A.E.; Chung, W.S.; Peterson, T.C.; et al. Neurotoxic reactive astrocytes are induced by activated microglia. *Nature* **2017**, *541*, 481–487. [CrossRef]
36. Sofroniew, M.V. Astrogliosis. *Cold Spring Harb. Perspect Biol.* **2014**, *7*, a020420. [CrossRef]
37. Olsson, T.; Piehl, F.; Swanberg, M.; Lidman, O. Genetic dissection of neurodegeneration and CNS inflammation. *J. Neurol. Sci.* **2005**, *233*, 99–108. [CrossRef]

Article

Genetic Association between Swine Leukocyte antigen Class II Haplotypes and Reproduction Traits in Microminipigs

Asako Ando [1], Noriaki Imaeda [2], Tatsuya Matsubara [2], Masaki Takasu [2], Asuka Miyamoto [1], Shino Oshima [1], Naohito Nishii [2], Yoshie Kametani [1], Takashi Shiina [1], Jerzy K. Kulski [1,3] and Hitoshi Kitagawa [4,*]

[1] Department of Molecular Life Science, Division of Basic Medical Science and Molecular Medicine, Tokai University School of Medicine, Isehara 259-1193, Japan

[2] Department of Veterinary Medicine, Faculty of Applied Biological Sciences, Gifu University, Gifu 501-1193, Japan

[3] Faculty of Health and Medical Sciences, UWA Medical School, The University of Western Australia, Crawley WA 6009, Australia

[4] Laboratory of Veterinary Internal Medicine, Faculty of Veterinary Medicine, Okayama University of Science, 1-3 Ikoino-oka, Imabari, Ehime 794-8555, Japan

* Correspondence: h-kitagawa@vet.ous.ac.jp; Tel.: +81-898-52-9256

Received: 28 May 2019; Accepted: 22 July 2019; Published: 26 July 2019

Abstract: The effects of swine leukocyte antigen (SLA) molecules on numerous production and reproduction performance traits have been mainly reported as associations with specific SLA haplotypes that were assigned using serological typing methods. In this study, we intended to clarify the association between SLA class II genes and reproductive traits in a highly inbred population of 187 Microminipigs (MMP), that have eight different types of SLA class II haplotypes. In doing so, we compared the reproductive performances, such as fertility index, gestation period, litter size, and number of stillbirth among SLA class II low resolution haplotypes (Lrs) that were assigned by a polymerase chain reaction-sequence specific primers (PCR-SSP) typing method. Only low resolution haplotypes were used in this study because the eight SLA class II high-resolution haplotypes had been assigned to the 14 parents or the progenitors of the highly inbred MMP herd in a previous publication. The fertility index of dams with Lr-0.13 was significantly lower than that of dams with Lr-0.16, Lr-0.17, Lr-0.18, or Lr-0.37. Dams with Lr-0.23 had significantly smaller litter size at birth than those with Lr-0.17, Lr-0.18, or Lr-0.37. Furthermore, litter size at weaning of dams with Lr-0.23 was also significantly smaller than those dams with Lr-0.16, Lr-0.17, Lr-0.18, or Lr-0.37. The small litter size of dams with Lr-0.23 correlated with the smaller body sizes of these MMPs. These results suggest that SLA class II haplotypes are useful differential genetic markers for further haplotypic and epistatic studies of reproductive traits, selective breeding programs, and improvements in the production and reproduction performances of MMPs.

Keywords: swine leukocyte antigen; reproductive performance; production trait; haplotype; micro-mini-pigs

---

1. Introduction

A Microminipigs (MMP) is a miniature pig for laboratory use with an extremely small body size developed by Fuji Micra Inc. in Japan. The body sizes, such as body weight, height, chest width, and chest circumference at 4–6 months of age were much smaller than those of young adult beagles at 10 months old [1–3]. In the population of MMPs, eleven swine leukocyte antigen (SLA) class I and II high resolution haplotypes, including three recombinant haplotypes, were identified in the 14 parents of the offspring cohorts, and the dams and sires of the inbred MMP herd, since its serendipitous establishment in Japan in 2008 [4].

The major histocompatibility complex (MHC) region has been fully sequenced for the Large White pig breed with the haplotype Hp-1.1 at was found to be consist with class I, II, and III gene regions that characterize the core architecture and organization of most mammalian MHCs, including those of primates. A total of 151 loci were annotated within the 2.4-Mb sequences, including three classical (*SLA-1, SLA-2,* and *SLA-3*), and three non-classical class I genes (*SLA-6, SLA-7,* and *SLA-8*) in the class I region, and four classical (*DRA, DRB1, DQA,* and *DQB1*), and four non-classical class II genes (*DMA, DMB, DOA,* and *DOB*) in the class II region as expressed SLA genes [5]. Based on the genomic and cDNA sequences in the SLA loci, many kinds of molecular-based SLA typing systems, including polymerase chain reaction (PCR) - sequence-specific primers (SSPs) [6,7], -fluorescently labeled sequence-specific oligonucleotide probes (SSOPs) [8], and -sequence-based typing (SBT) by traditional Sanger methods and/or next generation sequencing (NGS) [4,9–12] have been reported to assign SLA class I and II alleles. A systematic nomenclature for the SLA genes, alleles, and haplotypes are established by the SLA Nomenclature Committee, formed in 2002. There are currently 237 class I, 211 class II alleles officially designated, and registered to the Immunopolymorphism Database–Major Histocompatibility Complex (IPD-MHC) SLA sequence database (www.ebi.ac.uk/ipd/mhc/group/SLA) Release 3.3.0.0 (2019-06-13) build 785; and 29 class I and 21 class II haplotypes have been defined by means of high-resolution DNA sequencing [13]. In addition, recently 10 class I and 12 class II novel haplotypes have been defined and designated by the SLA Nomenclature Committee (Ho et al. unpublished data).

The MHC class I and class II genes are highly polymorphic and have important roles in regulating the immune system against infectious diseases [14,15] and influencing various biological traits, such as immune recognition, autoimmunity, mating preferences, and pregnancy outcomes [16]. The polymorphisms of SLA genes have enabled the analysis of associations between genotypes, alleles, haplotypes, and various infections and phenotypes, including reproductive performance and production traits [17–23]. The genotypes or SNPs at one gene locus are often used to associate DNA sequence diversity with phenotypes or disease traits. However, the single locus analysis of variants as genotypes or alleles misses the epistatic or linked effects of variants at multiple loci on phenotypes and diseases. In this regard, haplotyping of heterozygous SNPs in genomic DNA was developed as a multi-locus method, in order to study the effect of different combination of genes on physiological and disease traits, and for elucidating population structure and histories [17,24–26]. Thus, genomic information, reported as haplotypes rather than isolated genotypes or SNPs, has become increasingly important in genomic medicine and for elucidating the links between SNPs, gene regulation, and protein function [24,27].

SLA-homozygotes with alleles at each of the class I and II genes (as homozygous haplotypes) can be obtained more commonly with the establishment of new inbred colonies, like the MMP population [4], than with those in outbred colonies with high heterozygosity, such as the Chinese Wuzhishan minipigs [28]. The polymorphic SLA haplotypes and SLA-homozygotes in the highly inbred MMP population are naturally phased, and they are often simpler to analyze than the highly outbred or heterozygote breeds with respect to the comparison of the effects among SLA haplotypes on various genetic traits. For example, when comparing birth weights and weights on postnatal day 50 among eight SLA class II haplotypes in the MMP population, both of the weights in piglets, the SLA class II low resolution haplotype (Lr)-0.23 were significantly lower than those in piglets with Lr-0.17 or Lr-0.37 [3].

In humans, certain HLA antigens between couples in closely related populations may lead to infertility and miscarriages [29]. Many studies showed that the sharing of certain maternal-fetal or paternal HLA and BoLA antigens may influence fetal development and survival [29,30]. Similarly, in pigs, the influence of SLA-encoded genes on reproductive performances has been reported as the association between specific SLA haplotypes and genital tract development in males, ovulation rates, litter size, and piglet weight at birth and weaning [3,17–20]. The SLA complex also appears to influence baby pig mortality [22]. However, in most of the pig studies, SLA haplotypes were assigned using serological typing methods. Recently, we analyzed the genetic association of the relationships between the performance of swine reproduction and SLA haplotypes, assigned by SLA-DNA typing techniques in other breeds of pigs [20,22]. In selective breeding Duroc pigs with two SLA class II haplotypes, Lr-0.13 and Lr-0.30, assigned by a PCR-SSP method, Lr-0.30 was associated with higher weaning and rearing rates [20]. In addition, in a Landrace pig line selected for resistance to mycoplasmal pneumonia, PCR-SBT and PCR-SSOP methods were used for assignments of 11 SLA-class II haplotypes, and the associations between haplotypes and immune-related traits and reproductive traits, such as phagocytic activities of lymphocytes, activities of the alternative pathway of compliment, body weights and a rate of daily gain were observed [22]. In the present study, to evaluate the contribution of SLA class II genes on reproductive traits in MMPs, we compared the reproductive performances on fertility index, gestation periods, litter sizes, and number of stillbirth among SLA class II haplotypes, including SLA-homozygous individuals.

2. Materials and Methods

2.1. Animals

In this study, pigs were bred as a MMP herd at Fuji Micra Inc. (Fujinomiya, Japan) from June, 2008 to February, 2017. In the herd, the records of 2,288-cumulative matings of MMPs consisting of 129 sows and 58 boars assigned to eight different SLA class II haplotypes were utilized for measurement of reproductive performances, such as fertility indices, gestation periods, litter sizes at birth and weaning, and numbers of stillbirths per delivery. This study was approved by the Animal Care and Use Committee of Gifu University (#17042, May 26, 2017). The care and use of the laboratory animals were conducted in compliance with the guidelines of Good Laboratory Practice of Gifu University and Fuji Micra Inc.

2.2. SLA Class II Typing

SLA class II-DRB1 and DQB1 alleles were assigned by low-resolution SLA genotyping in 187 MMPs, using a PCR-SSP method, as described previously [4]. Eight types of low-resolution SLA class II haplotypes, Lr-0.7, Lr-0.11, Lr-0.13, Lr-0.16, Lr-0.17, Lr-0.18, Lr-0.23, and Lr-0.37 were determined by an analysis of the inheritance and segregation of eight and four alleles of the DRB1 and DQB1 genes, respectively, in descendants of the MMP population (Table 1). The expected high-resolution allele specificities in Table 1 are based on the 14 high resolution genotyped parents [4] of the offspring cohorts and the dams and sires of the inbred MMP herd in this study (Table 2). Consequently, we only needed to use the cheaper and more convenient low-resolution typing method to group the 187 individual MMP into their high resolution haplotypic groups for statistical comparisons and analysis.

Table 1. SLA-class II genotypes and number of SLA class II haplotypes in Microminipigs.

SLA Class II	Allele Specificity by Low Resolution Typing		* Expected Allele Specificity by High Resolution Typing		Number of Haplotypes (Frequency (%))			
haplotype	DRB1	DQB1	DRB1	DQB1	Dams		Sires	
Lr-0.7	06:XX	06:XX	06:01	06:01	2	(0.8)	1	(0.9)
Lr-0.11	09:XX	04:XX	09:01	04:01:02/04:02	19	(7.4)	10	(8.6)
Lr-0.13	04:XX	03:XX	04:03	03:03	13	(5.0)	2	(1.7)
Lr-0.16	11:XX	06:XX	11:03	06:01	16	(6.2)	7	(6.0)
Lr-0.17	08:XX	05:XX	08:01	05:01/05:02	44	(17.1)	18	(15.5)
Lr-0.18	14:XX	04:XX	14:01	04:01:02/04:02	24	(9.3)	8	(6.9)
Lr-0.23	10:XX	06:XX	10:01	06:01	79	(30.6)	42	(36.2)
Lr-0.37	07:XX	05:XX	07:01	05:01/05:02	61	(23.6)	28	(24.1)

Total number of dams and sires were 129 and 58, respectively. * Expected allele specificity by high resolution typing indicates deduced alleles by low resolution typing at two digit level in Microminipigs. DQB1*04:02 (Hp-0.7) and DQB1*04:01:02 (Hp-0.23), and DQB1*05:01 (Hp-0.17) and DQB1*05:02 (Hp-0.37) are assigned as DQB1*04XX (Lr-0.7 or Lr-0.23) and DQB1*05XX (Lr-0.17 or Lr-0.37) using a PCR-SSP method, respectively.

Table 2. Summary of reproduction traits in 187 Microminipigs.

Trait	Mean *	SE *	Number of Haplotypes **
Gestation period (day)	115.1	0.06	2760
Litter size at birth (No. of piglets)	5.48	0.04	2804
Litter size at weaning (No. of piglets)	3.88	0.04	2804
No. of stillbirth/delivery (No. of piglets)	0.97	0.03	2768

* Mean and SE indicate mean value and standard error, respectively. ** Number of haplotypes indicated as total number of haplotypes consisting of various mating combinations among 129 dams and 58 sires in each trait.

2.3. Measurement of Reproductive Performances

The fertility index was calculated as the ratio of the number of deliveries to that of matings. The indices were calculated from the data on a total number of 2288 cumulative matings in 129 dams and 58 sires associated with eight kinds of SLA class II haplotypes. Gestation periods were measured for 1380 deliveries (2760 haplotypes) that represented an overall fertility index (1380/2288) of 60.3%. To eliminate the data on premature deliveries, the gestation periods were analyzed for continuous deliveries over 100 days after copulations. Litter sizes were measured at birth including the total number of living and stillbirth newborn piglets in 1,402 deliveries and at weaning in 1384 deliveries. Abnormal piglet productions calculated as the number of stillbirths per delivery were analyzed for 1402 deliveries. To separate the influence of maternal and paternal SLA class II haplotypes on reproductive performances, the analyses of gestation periods, litter sizes at birth and weaning, and the number of stillbirth per delivery of matings were analyzed separately in dams or sires with homozygous or heterozygous haplotypes. These four reproductive performances were also analyzed in dams with homozygous haplotypes, Lr-0.11, Lr-0.17, Lr-0.23, or Lr-0.37 after 75, 78, 74, or 78 deliveries, respectively; and in sires with homozygous haplotypes, Lr-0.11, Lr-0.16, Lr-0.18, or Lr-0.23 after 141, 148, 142, or 148 deliveries, respectively (Table 2).

2.4. Statistical Analyses

Statistical comparisons were carried out by multiple group comparison with ANOVA and/or Kruskal-Walis and Sheffe's F tests (BellCurve in Excel, Social Survey Research Information Co., Ltd. Tokyo, Japan). Pairwise comparisons were adjusted for multiple tests with a Bonferroni correction. Fertility indices were evaluated by the Chi-square for independence test, using an m × n contingency table. Data are indicated as means ± standard error, and p-values of less than 0.05 were considered significant.

3. Results

3.1. Reproductive Performance in MMPs

The most frequently observed haplotype in 129 sows and 58 boars was Lr-0.23 (30.6%, and 36.2%, respectively), followed by Lr-0.37 (23.6%, and 24.1%, respectively) and Lr-0.17 (17.1%, and 15.5%, respectively). The two least frequent haplotypes were Lr-0.7 (0.8% in sows, 0.9% in boars) and Lr-0.13 (5.0% in sows, 1.7% in sires) (Table 1). Data of reproductive performance in MMPs with the lowest frequency haplotype, Lr-0.7, were excluded from all of the statistical analyses. A total of 1410 pregnancies were obtained as the result of 2,288 matings of 187 MMPs, representing a fertility index of 61.6%. The fertility index in MMPs was considerably lower than 88.4 ± 4.6 (standard deviation (SD)) in mixed breed domestic pigs in Japan [31]. Of the other reproductive performances in MMPs, the mean values of gestation period, while litter sizes at birth and weaning were 115.1 days, and 5.5 and 3.9 piglets/delivery, respectively. The gestation period in MMPs was comparable with those in other pig breeds [18,32,33]. Litter sizes at birth and weaning in MMPs were smaller than those in domestic pigs; 10.6 in commercial mixed breed pig herds in Japanese farm [34], 11.05 ± 0.77 (SD) in mixed breed domestic pigs in Japan [35], and 8.0—8.4 in Iberian pigs [36]. On the other hand, the litter size at birth in Göttingen minipigs was 5 to 6 piglets, which was slightly larger than that in MMPs [37].

However, the litter sizes in MMPs were slightly larger than that in other miniature pigs, NIBS minipigs; 4.4 ± 1.5 (SD) and 3.4 ± 1.3 (SD) piglets at birth and weaning, respectively [38], even though body sizes of the MMPs were considerably smaller than the NIBS minipigs produced in Japan [1,2] (Table 2).

3.2. Association between SLA Class II Haplotypes and Fertility Index

The mean value of the fertility index of each SLA haplotype with dams and sires were distributed across relatively wide ranges, from 52.9–73.3%, and 53.1–72.7%, respectively. The mean value of fertility index of dams was significantly lower for those with Lr-0.13 than those with Lr-0.17, Lr-0.18, or Lr-0.37 ($p < 0.05$, Figure 1A). Furthermore, the mean value of fertility index of boars with Lr-0.13 as mating partners of sows also was significantly lower than those with Lr-0.18 ($p < 0.05$, Figure 1B). Moreover, the mean values of fertility indices in mating with sires carrying Lr-0.16 or Lr-0.37 were significantly lower than those of sires carrying Lr-0.18 or Lr-0.23. Relatively high mean values of fertility indices, 73.3% and 72.7%, were observed in dams and sires as mating partners of sows with Lr-0.7, respectively. Furthermore, homozygous dams with Lr-0.11 had significantly lower fertility index than Lr-0.17 or Lr-0.37 ($p < 0.05$; Figure 2A). In contrast, homozygous sires with Lr-0.16 had significantly lower fertility index than those with Lr-0.23 ($p < 0.01$; Figure 2B).

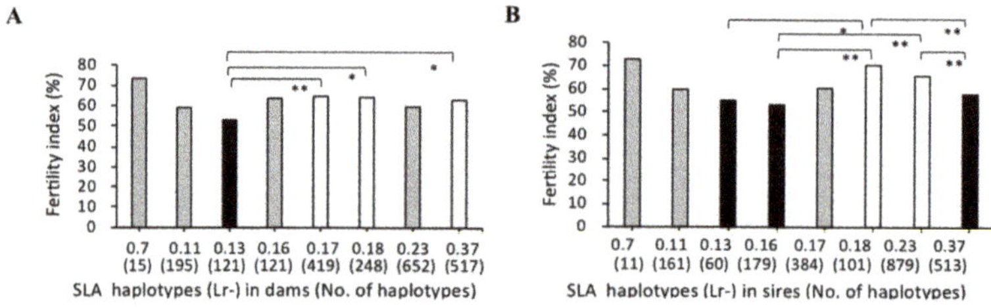

Figure 1. Comparison of the fertility indices of microminipigs (MMP) with different swine leukocyte antigen (SLA) class II haplotypes. *X*-axis shows the haplotypes of homozygous or heterozygous dams (**A**) and sires (**B**) and number of matings for each haplotype in brackets (no. of haplotypes). The *Y*-axis shows the fertility index as indicated by the ratio (%) of the number of deliveries to the number of matings, expressed as the mean value (bar). The number of haplotypes was counted as two in homozygous individuals. Black and white bars represent lower, and higher fertility indices, respectively, of the mean values, and the significant differences among haplotypes are indicated by the asterisks. Probabilities of significant differences among haplotypes are indicated by single ($p < 0.05$) and double ($p < 0.01$) asterisks. Gray bars represent mean values of fertility indices without significant differences among the haplotypes.

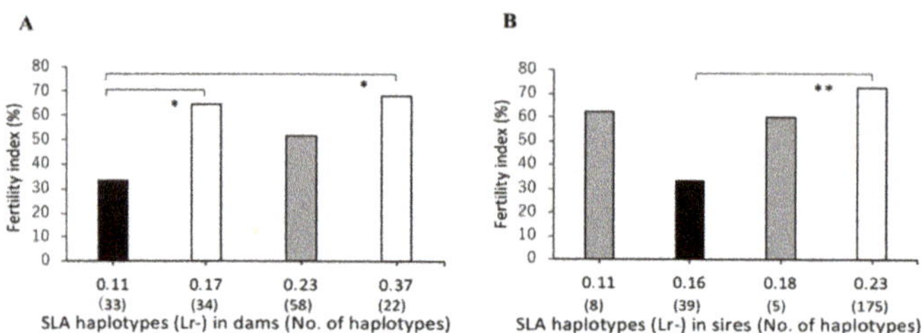

Figure 2. Comparison of fertility indices among SLA class II haplotypes in homozygous MMPs. *X*-axis shows the haplotypes of homozygous dams (**A**) and sires (**B**) and the number of matings for each haplotype in brackets (no. of haplotypes). The *Y*-axis shows the fertility index as indicated by the ratio (%) of the number of deliveries to the number of matings, expressed as the mean value (bar). Black and white bars represent lower, and higher fertility indices, respectively, of the mean values and the significant differences among haplotypes are indicated by the asterisks. Probabilities of significant differences among haplotypes are indicated by single ($p < 0.05$) and double ($p < 0.01$) asterisks. Gray bars represent the mean values of the fertility indices without significant differences among the haplotypes.

3.3. Association of SLA Class II Haplotypes and Gestation Periods

The mean values of gestation periods were within the small ranges of 115.0–116.6 or 113.8–115.6 among the eight SLA class II haplotypes for dams or sires, used as mating partners of sows, respectively, in total number of 1380 deliveries. There were no significant differences between the mean values of gestation periods among the haplotypes (Figure 3A,B). In addition, no obvious differences were observed between the mean values for gestation periods of four homozygous haplotypes, Lr-0.11, Lr-0.17, Lr-0.23, and Lr-0.37 in dams, and Lr-0.11, Lr-0.16, Lr-0.18 and Lr-0.23 in sires (Figure 4A,B). These results suggest that there are no maternal or paternal SLA class II genotype effects on gestation periods.

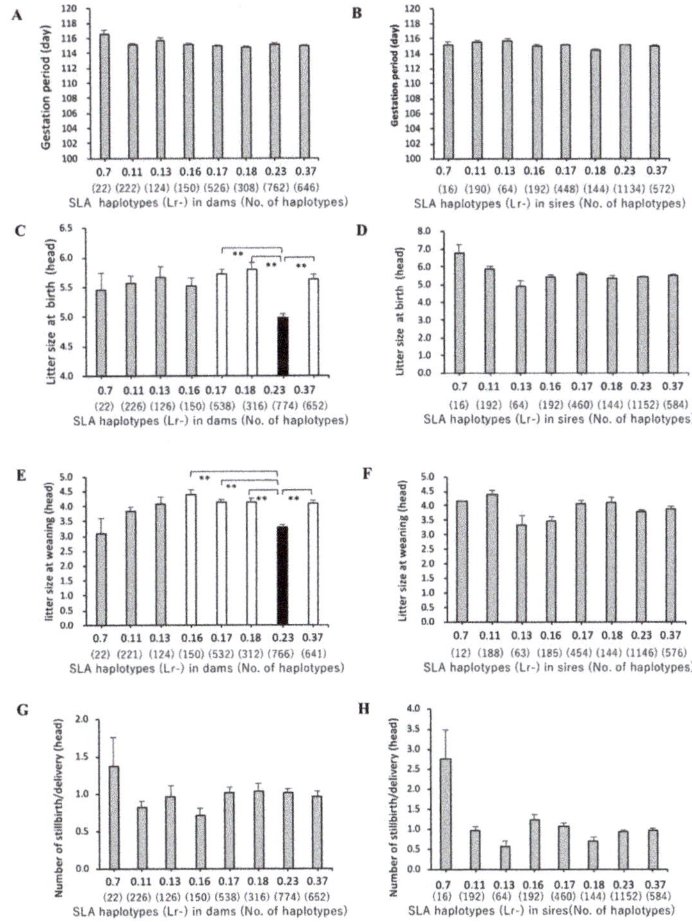

Figure 3. Comparison of reproductive performances of MMPs with different SLA class II haplotypes. *X*-axis for A to H shows the homozygous or heterozygous SLA haplotypes and the number of matings for each haplotype in brackets (no. of haplotypes). Each bar along the *X*-axis indicates the mean values and whiskers (standard errors) of gestation periods in dams (**A**) and sire (**B**), litter sizes at birth in dams (**C**) and sires (**D**), litter sizes at weaning in dams (**E**) and sires (**F**), and the number of stillbirths/delivery in dams (**G**) and sires (**H**) shown along the *Y*-axis. Black and white bars represent lower and higher haplotypes, respectively, with mean values of each trait and significant differences among haplotypes. The gray bars represent mean values of haplotypes without any significant differences among the haplotypes. The probabilities of significant differences among haplotypes are indicated by single ($p < 0.05$) and double ($p < 0.01$) asterisks.

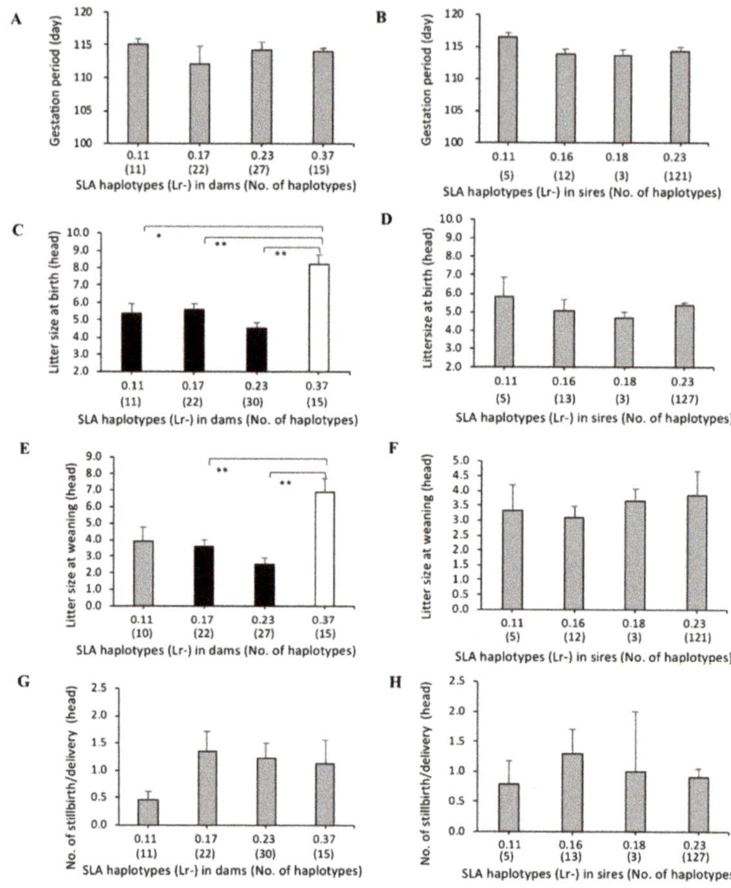

Figure 4. Comparison of reproductive performances among SLA class II homozygous haplotypes in MMPs. Haplotypes and number of deliveries for each haplotype are shown along the *X*-axis, and the bars for each homozygous haplotype show the mean values and whiskers (standard errors) of gestation periods in dams (**A**) and sire (**B**), litter sizes at birth in dams (**C**) and sires (**D**), litter sizes at weaning in dams (**E**) and sires (**F**), and the number of stillbirths/delivery in dams (**G**) and sires (**H**) along the *Y*-axis. Black and white bars represent lower and higher haplotypes, respectively, and mean values of each trait with significant differences among haplotypes. Gray bars represent the mean values of haplotypes without any significant differences among the haplotypes. The probabilities of significant differences among haplotypes are indicated by single ($p < 0.05$) and double ($p < 0.01$) asterisks.

3.4. Association between SLA Class II Haplotypes and Litter Sizes at Birth and Weaning

The dams with Lr-0.23 had the smallest litter sizes at birth compared to those with other SLA class II haplotypes. The mean value of litter sizes at birth for dams with Lr-0.23 was significantly smaller than those with Lr-0.17, Lr-0.18 or Lr-0.37 ($p < 0.01$; Figure 3C) and Lr-0.11 ($p < 0.05$; Figure 3C). In contrast, no significant differences on litter sizes at birth were detected among sires with seven SLA class II haplotypes except Lr-0.7 suggesting that there are no effects of paternal SLA types on the trait (Figure 3D). In the dams with Lr-0.23, the mean value of litter sizes at weaning rates was also significantly smaller than those with Lr-0.16, Lr-0.17, Lr-0.18, or Lr-0.37 ($p < 0.01$; Figure 3E). Moreover, the mean litter sizes at birth and weaning of piglets of dams with the homozygous Lr-0.23 haplotypes were smaller than those of dams with homozygous Lr-0.11, Lr-0.17, or Lr-0.37. Furthermore, dams with homozygous Lr-23 had significantly smaller litter sizes at birth and weaning than homozygous Lr-0.37 ($p < 0.01$; Figure 3C,E). In contrast, there were no significant differences between the mean values of litter sizes at birth and weaning among the seven haplotypes in sires (Figure 3D,F). Furthermore, there were no significant effects of SLA class II haplotypes on litter sizes in sires with homozygous Lr-0.11, Lr-0.16, Lr-0.18, or Lr-0.23 (Figure 4D,F).

3.5. Association of SLA Class II Haplotypes and Number of Stillbirths per Delivery

The largest mean values of the number of stillbirths per deliveries were observed for both the dams and sires with Lr-0.7 (Figure 3G,H). Homozygous pigs with Lr-0.7 as mating pairs could not be found in the 129 dams and fifty-eight sires. Due to the low frequencies (0.8% in dams, 0.9% in sires) of MMPs with Lr-0.7, no statistical comparisons of the mean numbers of stillbirths per delivery between Lr-0.7 and the other seven haplotypes were carried out. However, statistical analysis of these seven haplotypes showed that there were no significant differences among them for the mean values of the number of stillbirths per deliveries. Also, no significant differences among the mean numbers of stillbirths per delivery were observed in dams with homozygous Lr-0.11, Lr-0.17, Lr-0.23 or Lr-0.37, and sires with homozygous Lr-0.11, Lr-0.16, Lr-0.18 or Lr-0.23 (Figure 4G,H).

4. Discussion

To improve reproduction performances using genetic marker-assisted selection, reproductive traits, such as gestation periods and litter sizes have been analyzed in the pig populations of various breeds, including the MMPs with the extra small body sizes [31–38]. In comparing reproductive performances between the MMPs and other breeds of pigs [18,31–38], the mean gestation period, and litter sizes at birth and weaning in MMPs (Table 2) were similar to those of other breeds of domestic pigs [18,31,32], including Göttingen and NIBS minipigs [37,38]. Thus, the inbred MMP population, used in the present study, showed relatively normal reproductive traits that were useful and easier for evaluating the differential effects of SLA homozygous and heterozygous haplotypes on reproductive traits, than by using the more confounding SLA heterozygous haplotypes of most other breeds.

In this study, we haplotyped SLA class II alleles using a low resolution PCR-SSP method in well-defined MMP segregating families, consisting of 187 MMPs, and examined the association between the haplotypes and various reproduction traits, such as fertility indices, gestation periods, litter sizes at birth and weaning, and the numbers of stillbirths per delivery. The off-spring were segregated into the SLA class II low resolution haplotype families, based on the genotype results of the 129 sows and 58 boars that were assigned to eight different SLA class II haplotypes (Table 1). Moreover, we estimated the expected high-resolution allele specificities of the eight different SLA class II haplotypes (Table 1) because they could be easily inferred from the 14 high resolution genotyped parents [4] of all the offspring cohorts, and the dams and sires of the inbred MMP herd in this study (Table 2). Consequently, because of the stability of ancestral haplotypes due to low mutation rates in mammals [25], we only needed to use the much cheaper and more convenient low-resolution typing methods to group the 187 individual MMP into their low resolution (or inferred high resolution) haplotypic groups for statistical comparisons and analysis.

The significant effects of two SLA class II haplotypes, Lr-0.13 and Lr-0.18, in both dams and sires showed lower, and higher fertility indices, respectively. In our present data, we could not determine whether much lower or higher fertility indices would be observed in mating pairs with homozygous haplotypes Lr-0.13 or Lr-0.18, respectively. Dams with Lr-0.17, Lr-0.18, or Lr-0.37 tended to have a relatively high fertility index, litter sizes at birth and weaning, and low number of stillbirths per delivery, suggesting relatively good performances of reproduction traits. In general, these haplotypes were at relatively high frequency in the MMP population (Table 1). Moreover, both sires and dams with the Lr-0.18 haplotype showed similar trends for the four reproduction traits. In contrast, sires and dams exhibited an opposite trend for associations between Lr-37 and fertility index; sires with Lr-0.37 had relatively low fertility index suggesting that the contribution of Lr-0.37 on fertility may be different between sows and boars in a MMP herd. On the other hand, the large numbers of stillbirths per delivery observed in sires with Lr-0.7 are consistent with the production of low frequencies of pigs with this haplotype in the MMP population (Table 1).

The Lr-0.23 haplotype has the highest frequency of 36.2% in dams of the MMP population (Table 1), and they also had the smallest litter size at birth compared to the dams with the other seven haplotypes (Figure 3C). Furthermore, litter sizes at weaning of the dams with Lr-0.23 were significantly smaller than those of dams with Lr-0.16, Lr-0.17, Lr-0.18, or Lr-0.37. In our previous analyses of association between SLA haplotypes and body weights in MMPs, piglets with the Lr-0.23 had lower body weights at birth (0.415 kg) and on postnatal day 50 (3.14 kg) than those with the other SLA haplotypes, although they had no significant differences in daily gains (DGs) in comparison to those with the other haplotypes [3]. These data on low body weights and DGs of MMPs with Lr-0.23 characterize their small body size and slow growth rates; although their small litter size at weaning might simply reflect their small size at birth. Nevertheless, in the MMP population, Lr-0.23 is one of the genetic markers both for small body size and small litter size. Taken together, the associations between specific SLA haplotypes and reproduction traits showed that SLA class II haplotypes differ among the various traits of reproduction performance in MMPs, and they might be useful genetic markers for the improvement of production and reproduction performances in selective breeding programs.

A large litter size is commonly used as a measure of successful breeding and animal production [39,40]. Although animal production management strives for large litter sizes, there can be serious limitations and biological problems with large litter sizes, such as increased numbers of stillbirths. In some cases, moderately low litter sizes can be of greater benefit to the overall animal production than large litter sizes and benefits, such as greater lactation efficiency for the mother and lower nutritional requirements for the litter and the mother [40]. There are many biological and genetic factors involved in litter size and/or embryonic and foetal growth rates [39,41,42]. Here, we focused on the effect of MHC class II haplotypes on reproductive traits and found that the embryonic survival and litter size increases with the probability that the embryo receives the haplotype from the dam, rather than the boar. In this regard, the MHC-DRB1 and -DQB1 genes expressed by placental macrophages [43,44] might affect embryonic growth rates and litter size as a consequence of gene activations under various regulatory scenarios, such as methylation or gene activation due to stress or infections. An alternative explanation for the correlation between MHC class II haplotypes and litter size is the haplotypic linkage between MHC class II alleles and various other gene alleles located within and outside the SLA region. The linkage between the MHC class I and class II alleles was not investigated in this study, but MHC class I is known to mediate immunological tolerance, leucocyte recruitment, and the development of the trophoblast during pregnancy, and rejection during parturition in the bovine [30]. The MHC also has effects on placental leucocyte recruitment during gestation and parturition in the mouse [44]. There are numerous other genes within the MHC class III or extended class II region that have a possible role in the reproduction and lactation processes. For example, butyrophilin is involved genetically and biologically with lactation [45], RING1 with polycomb function during oogenesis [46], KIFC1 with oocyte meiosis [47], and POU1F5 (OCT4) transcription factor in stem cell modulation and embryonic development [48].

The polymorphic MHC class II haplotypes also might be linked with one or other of the SNPs of 376 functional genes outside the MHC region that were associated significantly with reproductive traits in Large White pigs [49]. Multiple genes were associated with some swine production traits and many of them were mapped on several chromosomes including *Sus scrofa* chromosome (SSC) 7 (Animal Quantitative Trait Loci (QTL) Database, https://www.animalgenome.org/QTLdb). QTL detection analyses for traits on growth and fatness indicated that the SLA region on SSC7 was excluded as a candidate region [50]. However, in a herd of Meishan/Large White pigs, Wei et al. mapped QTL influencing growth traits near the SLA region [51]. Taken together, these reports suggest that the SLA genes or haplotypes associate indirectly with those traits. Nevertheless, to better define the SLA or other loci within the SLA region that are responsible for productive and reproductive traits, further association analyses will need to be carried out using next generation sequencing (NGS) techniques in MMPs.

For a few decades, several authors reported that SLA haplotypes are associated with various traits or measures of growth and reproduction performances, suggesting the possibility of SLA molecules as important determinants on many economical traits [17–19,52]. Because it is difficult to define the most likely corresponding locus for productive or reproductive performance within the SLA region simply by the data analyses of serologically assigned SLA haplotypes. Many different molecular-based SLA typing techniques were developed and used for the analyses of associations of production and reproduction traits and SLA alleles or haplotypes [3,22,23]. These association data often indicated that SLA alleles or haplotypes are useful genetic markers to achieve improvements in pig breeding programs. These studies also highlighted the advantage of using well-defined, genetically conserved haplotypes over just using trait-associated SNPs for linking the MHC with reproduction, and possibly other medical and phenotypic traits of interest. Since our association studies were analyzed using only a few SLA haplotypic class II loci in a limited pig population and breed, similar haplotype studies will need to be expanded to other MHC loci and other breeds, populations and species for comparison and confirmation that the MHC haplotypic markers might be associated productively with animal breeding performance.

5. Conclusions

In our study of the unique inbred MMP population, association analyses between specific SLA haplotypes and reproduction traits exhibited significant effects of SLA class II homozygous haplotypes on reproduction performances, that differed among fertility index, litter sizes, and number of stillbirths. The present study forms a basis for a broader and more detailed investigation of the differential effects of the MHC class I, II and III genes as well as the many other genes outside the MHC complex that are known to influence reproduction in pigs and various other mammalian species.

Author Contributions: Conceptualization, A.A. and H.K.; Data curation, T.M. and M.T.; Formal analysis, N.I., T.M. and N.N.; Investigation, A.A., Y.K. and H.K.; Methodology, A.M. and S.O.; Project administration, A.A., J.K.K., T.S. and H.K.; Writing—original draft, A.A. and H.K.; Writing—review & editing, J.K.K., T.S., A.A. and H.K.

Acknowledgments: We thank Takashi Nishimura and Toshiaki Nishimura (Fuji Micra Inc., Fujinomiya, Japan) for providing us with recorded information on reproduction performances for MMPs.

Conflicts of Interest: We certify that there is no conflict of interest with any financial organization regarding the material discussed in the manuscript.

References

1. Kaneko, N.; Itoh, K.; Sugiyama, A.; Izumi, Y. Microminipig, a non-rodent experimental animal optimized for life science research: Preface. *J. Pharmacl. Sci.* **2011**, *115*, 112–114. [CrossRef]

2. Takasu, M.; Tsuji, E.; Imaeda, N.; Kaneko, N.; Matsubara, T.; Maeda, M.; Ito, Y.; Shibata, S.; Ando, A.; Nishii, N.; et al. Body and, major organ sizes of young mature Microminipigs determined by computed tomography. *Lab. Anim.* **2015**, *49*, 65–70. [CrossRef] [PubMed]

3. Matsubara, T.; Takasu, M.; Imaeda, N.; Nishii, N.; Takashima, S.; Nishimura, T.; Nishimura, T.; Shiina, T.; Ando, A.; Kitagawa, H. Genetic association of swine leukocyte antigen class II haplotypes and body weight in Microminipigs. *Asian Aust. J. Animal Sci.* **2018**, *31*, 163–166. [CrossRef] [PubMed]

4. Ando, A.; Imaeda, N.; Ohshima, S.; Miyamoto, A.; Kaneko, N.; Takasu, M.; Shiina, T.; Kulski, J.K.; Inoko, H.; Kitagawa, H. Characterization of swine leucocyte antigen alleles and haplotypes on a novel miniature pig line, Microminipig. *Anim. Gen.* **2015**, *45*, 791–798. [CrossRef] [PubMed]

5. Renard, C.; Hart, E.; Sehra, H.; Beasley, H.; Coggill, P.; Howe, K.; Harrow, J.; Gilbert, J.; Sims, S.; Rogers, J.; et al. The genomic sequence and analysis of the swine major histocompatibility complex. *Genomics* **2006**, *88*, 96–110. [CrossRef] [PubMed]

6. Ho, C.S.; Lunney, J.K.; Franzo-Romain, M.H.; Martens, G.W.; Lee, Y.J.; Lee, J.H.; Wysocki, M.; Rowland, R.R.; Smith, D.M. Molecular characterization of swine leucocyte antigen class I genes in outbred pig populations. *Anim. Genet.* **2009**, *40*, 468–478. [CrossRef] [PubMed]

7. Ho, C.S.; Lunney, J.K.; Franzo-Romain, M.H.; Martens, G.W.; Rowland, R.R.; Smith, D.M. Molecular characterization of swine leucocyte antigen class II genes in outbred pig populations. *Anim. Genet.* **2010**, *41*, 428–432. [CrossRef] [PubMed]

8. Ando, A.; Shigenari, A.; Ota, M.; Sada, M.; Kawata, H.; Azuma, F.; Kojima-Shibata, C.; Nakajoh, M.; Suzuki, K.; Uenishi, H.; et al. SLA-DRB1 and -DQB1 genotyping by the PCR-SSOP-Luminex method. *Tissue Antigens* **2011**, *78*, 49–55. [CrossRef] [PubMed]

9. Kita, Y.F.; Ando, A.; Tanaka, K.; Suzuki, S.; Ozaki, Y.; Uenishi, H.; Inoko, H.; Kulski, J.K.; Shiina, T. Application of high-resolution, massively parallel pyrosequencing for estimation of haplotypes and gene expression levels of swine leukocyte antigen (SLA) class I genes. *Immunogenetics* **2012**, *64*, 187–199. [CrossRef] [PubMed]

10. Le, M.; Choi, H.; Choi, M.K.; Cho, H.; Kim, J.H.; Seo, H.G.; Cha, S.Y.; Seo, K.; Dadi, H.; Park, C. Development of a simultaneous high resolution typing method for three SLA class II genes, SLA-DQA, SLA-DQB1, and SLA-DRB1 and the analysis of SLA class II haplotypes. *Gene* **2015**, *564*, 228–232. [CrossRef]

11. Sørensen, M.R.; Ilsøe, M.; Strube, M.L.; Bishop, R.; Erbs, G.; Hartmann, S.B.; Jungersen, G. Sequence-Based Genotyping of Expressed Swine Leukocyte Antigen Class I Alleles by Next-Generation Sequencing Reveal Novel Swine Leukocyte Antigen Class I Haplotypes and Alleles in Belgian, Danish, and Kenyan Fattening Pigs and Göttingen Minipigs. *Front. Immunol.* **2017**, *8*, 701. [CrossRef] [PubMed]

12. Lee, C.; Moroldo, M.; Perdomo-Sabogal, A.; Mach, N.; Marthey, S.; Lecardonnel, J.; Wahlberg, P.; Chong, A.Y.; Estellé, J.; Ho, S.Y.W.; et al. Inferring the evolution of the major histocompatibility complex of wild pigs and peccaries using hybridisation DNA capture-based sequencing. *Immunogenetics* **2018**, *70*, 401–417. [CrossRef] [PubMed]

13. Ho, C.S.; Lunney, J.K.; Ando, A.; Rogel-Gaillard, C.; Lee, J.H.; Schook, L.B.; Smith, D.M. Nomenclature for factors of the SLA system, update 2008. *Tissue Antigens* **2009**, *73*, 307–315. [CrossRef] [PubMed]

14. Blackwell, J.M.; Jamieson, S.E.; Burgner, D. HLA and infectious diseases. *Clin. Microbiol. Rev.* **2009**, *22*, 370–385. [CrossRef] [PubMed]

15. Gutiérrez, S.E.; Esteban, E.N.; Lützelschwab, C.M.; Juliarena, M.A. Chapter 6, Major Histocompatibility Complex-associated resistance to infectious diseases: The case of Bovine leukemia virus infection. In *Trends and Advances in Veterinary Genetics*; Abubakar, M., Ed.; IntechOpen Limited: London, UK, 2017; pp. 101–126. [CrossRef]

16. Sommer, S. The importance of immune gene variability (MHC) in evolutionary ecology and conservation. *Front. Zool.* **2005**, *2*, 16. [CrossRef] [PubMed]

17. Renard, C.; Vaiman, M. Possible relationships between SLA and porcine reproduction. *Reprod. Nutr. Dev.* **1989**, *29*, 569–576. [CrossRef]

18. Gautschi, C.; Gaillard, C. Influence of major histocompatibility complex on reproduction and production traits in swine. *Anim. Genet.* **1989**, *21*, 161–170. [CrossRef]

19. Conley, A.J.; Jung, Y.C.; Schwartz, N.K.; Warner, C.M.; Rothschild, M.F.; Ford, S.P. Influence of SLA haplotype on ovulation rate and litter size in miniature pigs. *J. Reprod. Fertil.* **1988**, *82*, 595–601. [CrossRef]

20. Imaeda, N.; Ando, A.; Takasu, M.; Matsubara, T.; Nishii, N.; Takashima, S.; Shigenari, A.; Shiina, T.; Kitagawa, H. Influences of swine leukocyte antigen haplotypes on serum antigen titers against swine erysipelas vaccine and traits of reproductive ability and meat production in a SLA-defined Duroc pigs. *J. Vet. Med. Sci.* **2018**, *80*, 1662–1668. [CrossRef]

21. Vaiman, M.; Chardon, P.; Rothschild, M.F. Porcine major histocompatibility complex. *Rev. Sci. Tech. Off. Int. Epiz.* **1998**, *17*, 95–107. [CrossRef]

22. Ando, A.; Shigenari, A.; Kojima-Shibata, C.; Nakajoh, M.; Suzuki, K.; Kitagawa, H.; Shiina, T.; Inoko, H.; Uenishi, H. Association of swine leukocyte antigen class II haplotypes and immune-related traits in a swine line selected for resistance to mycoplasmal pneumonia. *Comp. Immunol. Microbiol. Infect. Dis.* **2016**, *48*, 33–40. [CrossRef] [PubMed]

23. Zhang, S.; Yang, J.; Wang, L.; Li, Z.; Pang, P.; Li, F. SLA-11 mutations are associated with litter size traits in Large White and Chinese DIV pigs. *Anim. Biotech.* **2018**. [CrossRef] [PubMed]

24. Glusman, G.; Cox, H.C.; Roach, J.C. Whole-genome haplotyping approaches and genomic medicine. *Genome Med.* **2014**, *6*, 73. [CrossRef] [PubMed]

25. Sánchez-Molano, E.; Tsiokos, D.; Chatziplis, D.; Jorjani, H.; Degano, L.; Diaz, C.; Rossoni, A.; Schwarzenbacher, H.; Seefried, F.; Varona, L.; et al. A practical approach to detect ancestral haplotypes in livestock populations. *BMC Genet.* **2016**, *7*, 91. [CrossRef] [PubMed]

26. Huang, M.; Tu, J.; Lu, Z. Recent Advances in Experimental Whole Genome Haplotyping Methods. *Int. J. Mol. Sci.* **2017**, *18*, 1944. [CrossRef] [PubMed]

27. Murphy, N.M.; Burton, M.; Powell, D.R.; Rossello, F.J.; Cooper, D.; Chopra, A.; Hsieh, M.J.; Sayer, D.C.; Gordon, L.; Pertile, M.D.; et al. Haplotyping the human leukocyte antigen system from single chromosomes. *Sci. Rep.* **2016**, *6*, 30381. [CrossRef] [PubMed]

28. Min, F.; Pan, J.; Wang, X.; Chen, R.; Wang, F.; Luo, S.; Ye, J. Biological characteristics of captive Chinese Wuzhishan minipigs (*Sus scrofa*). *Int. Sch. Res. Notices* **2014**, 761257. [CrossRef]

29. Agenor, A.; Bhattacharya, S. Infertility and miscarriage: Common pathways in manifestation and management. *Womens Health* **2015**, *11*, 527–541. [CrossRef]

30. Rapacz-Leonard, A.; Dąbrowska, M.; Janowski, T. Major histocompatibility complex I mediates immunological tolerance of the trophoblast during pregnancy and may mediate rejection during parturition. *Mediat. Inflamm.* **2014**, *2014*, 579279. [CrossRef]

31. Suzuki, K.; Somei, H. Present condition and opinions of artificial insemination in pig farming. *Bull. Chiba Prefect. Livestock Exptl. Station* **1982**, *6*, 39–43. (In Japanese)

32. Rothkötter, H.J.; Sowa, E.; Pabst, R. The pig as a model of developmental immunology. *Hum. Exp. Toxicol.* **2002**, *21*, 533–536. [CrossRef] [PubMed]

33. Tsumagari, S. *Textbook of Theriogenology*, 2nd ed.; Mori, J., Kanagawa, H., Hamana, K., Eds.; Buneido Co., Ltd.: Tokyo, Japan, 2002; pp. 159–160. ISBN 4-8300-3184-0. (In Japanese)

34. Koketsu, Y. Reproductive productivity measurements in Japanese swine breeding herds. *J. Vet. Med. Sci.* **2002**, *64*, 195–198. [CrossRef] [PubMed]

35. Yamane, I.; Ishizaki, S.; Yamazaki, H. Parameters Contributing to Improved Reproductive Performance at Farrow-to-Finish Swine Farms in Japan. *J. Jpn. Vet. Med. Assoc.* **2014**, *67*, 177–182. (In Japanese) [CrossRef]

36. Casellas, J.; Ibáñez-Escriche, N.; Varona, L.; Rosas, J.P.; Noguera, J.I. Inbreeding depression load for litter size in *Entrepelado* and *Retinto* Iberian pig varieties. *J. Anim. Sci.* **2019**, skz084. [CrossRef]

37. Schuleri, K.H.; Boyle, A.J.; Centola, M.; Amado, L.C.; Evers, R.; Zimmet, J.M.; Evers, K.S.; Ostbye, K.M.; Scorpio, D.G.; Hare, J.M.; et al. The adult Göttingen Minipig as a model for chronic heart failure after myocardial infarction: Focus on cardiovascular imaging and regenerative therapies. *Comp. Med.* **2008**, *58*, 568–579. [PubMed]

38. Saitoh, T. Utilization of a swine other than food. *All About Swine* **2009**, *35*, 14–20. (In Japanese)

39. Lawlor, P.G.; Lynch, P.B. A review of factors influencing litter size in Irish sows. *Ir. Vet. J.* **2007**, *60*, 359–366. [CrossRef]

40. Rutherford, K.M.D.; Baxter, E.M.; D'Eath, R.B.; Turner, S.P.; Arnott, G.; Roehe, R.; Ask, B.; Sandoe, P.; Moustsen, V.A.; Thorup, F.; et al. The welfare implications of large litter size in the domestic pig I: Biological factors. *Anim. Welfare* **2013**, *22*, 199–218. [CrossRef]

41. Chen, P.; Baas, T.J.; Mabry, J.W.; Koehler, K.J.; Dekkers, J.C. Genetic parameters and trends for litter traits in U.S.; Yorkshire, Duroc, Hampshire, and Landrace pigs. *J. Anim. Sci.* **2003**, *81*, 46–53. [CrossRef]

42. Kwon, S.G.; Hwang, J.H.; Park, D.H.; Kim, T.W.; Kang, D.G.; Kang, K.H.; Kim, I.S.; Park, H.C.; Na, C.S.; Ha, J.; et al. Identification of differentially expressed genes associated with litter size in Berkshire pig placenta. *PLoS ONE* **2016**, *11*, e0153311. [CrossRef]

43. Athanassakis-Vassiliadis, I.; Thanos, D.; Papamatheakis, J. Induction of class II major histocompatibility complex antigens in murine placenta by 5-azacytidine and interferon-γ involves different cell populations. *Eur. J. Immunol.* **1989**, *19*, 2341–2348. [CrossRef]

44. Kieckbusch, J.; Balmas, E.; Hawkes, D.A.; Colucci, F. Disrupted PI3K p110δ signaling dysregulates maternal immune cells and increases fetal mortality in mice. *Cell Rep.* **2015**, *13*, 2817–2828. [CrossRef] [PubMed]

45. Stefferl, A.; Schubart, A.; Storch, M.; Amini, A.; Mather, I.; Lassmann, H.; Linington, C. Butyrophilin, a milk protein, modulates the encephalitogenic T cell response to myelin oligodendrocyte glycoprotein in experimental autoimmune encephalomyelitis. *J. Immunol.* **2000**, *165*, 2859–2865. [CrossRef] [PubMed]

46. Posfai, E.; Kunzmann, R.; Brochard, V.; Salvaing, J.; Cabuy, E.; Roloff, T.C.; Liu, Z.; Tardat, M.; van Lohuizen, M.; Vidal, M.; et al. Polycomb function during oogenesis is required for mouse embryonic development. *Genes Dev.* **2012**, *26*, 920–932. [CrossRef] [PubMed]

47. Camlin, N.J.; McLaughlin, E.A.; Holt, J.E. Motoring through: The role of kinesin superfamily proteins in female meiosis. *Hum. Reprod. Update* **2017**, *23*, 409–420. [CrossRef] [PubMed]
48. Wu, G.; Schöler, H.R. Role of Oct4 in the early embryo development. *Cell Regen.* **2014**, *3*, 7. [CrossRef] [PubMed]
49. Wang, Y.; Ding, X.; Tan, Z.; Xing, K.; Yang, T.; Wang, Y.; Sun, D.; Wang, C. Genome-wide association study for reproductive traits in a Large White pig population. *Anim. Genet.* **2018**, *49*, 127–131. [CrossRef] [PubMed]
50. Demeure, O.; Sanchez, M.P.; Riquet, J.; Iannuccelli, N.; Demars, J.; Fève, K.; Kernaleguen, L.; Gogué, J.; Billon, Y.; Caritez, J.C.; et al. Exclusion of the swine leukocyte antigens as candidate region and reduction of the position interval for the *Sus scrofa* chromosome 7 QTL affecting growth and fatness. *J. Anim. Sci.* **2005**, *83*, 1979–1987. [CrossRef]
51. Wei, W.H.; Skinner, T.M.; Anderson, J.A.; Southwood, O.I.; Plastow, G.; Archibald, A.L.; Haley, C.S. Mapping QTL in the porcine MHC region affecting fatness and growth traits in a Meishan/Large White composite population. *Anim. Genet.* **2011**, *42*, 83–85. [CrossRef]
52. Lunney, J.K.; Ho, C.S.; Wysocki, M.; Smith, D.M. Molecular genetics of the swine major histocompatibility complex, the SLA complex. *Dev. Comp. Immunol.* **2009**, *33*, 362–374. [CrossRef]

Article

Genetic Diversity and Differentiation at Structurally Varying MHC Haplotypes and Microsatellites in Bottlenecked Populations of Endangered Crested Ibis

Hong Lan [1,2], Tong Zhou [1], Qiu-Hong Wan [1,*] and Sheng-Guo Fang [1,*]

[1] MOE Key Laboratory of Biosystems Homeostasis & Protection, State Conservation Centre for Gene Resources of Endangered Wildlife, College of Life Sciences, Zhejiang University, Hangzhou 310058, China; amber_lan@163.com (H.L.); zhoutong2016@zju.edu.cn (T.Z.)

[2] Department of Agriculture, Zhejiang Open University, Hangzhou 310012, China

* Correspondence: qiuhongwan@zju.edu.cn (Q.-H.W.); sgfanglab@zju.edu.cn (S.-G.F.)

Received: 19 December 2018; Accepted: 23 April 2019; Published: 25 April 2019

Abstract: Investigating adaptive potential and understanding the relative roles of selection and genetic drift in populations of endangered species are essential in conservation. Major histocompatibility complex (MHC) genes characterized by spectacular polymorphism and fitness association have become valuable adaptive markers. Herein we investigate the variation of all MHC class I and II genes across seven populations of an endangered bird, the crested ibis, of which all current individuals are offspring of only two pairs. We inferred seven multilocus haplotypes from linked alleles in the Core Region and revealed structural variation of the class II region that probably evolved through unequal crossing over. Based on the low polymorphism, structural variation, strong linkage, and extensive shared alleles, we applied the MHC haplotypes in population analysis. The genetic variation and population structure at MHC haplotypes are generally concordant with those expected from microsatellites, underlining the predominant role of genetic drift in shaping MHC variation in the bottlenecked populations. Nonetheless, some populations showed elevated differentiation at MHC, probably due to limited gene flow. The seven populations were significantly differentiated into three groups and some groups exhibited genetic monomorphism, which can be attributed to founder effects. We therefore propose various strategies for future conservation and management.

Keywords: MHC; genetic drift; haplotype; crested ibis; founder effect; bottleneck; conservation genetics; selection

---

1. Introduction

Bottlenecked species are prone to genetic drift and depletion of adaptive variation [1], which may raise the probability of extinction owing to inbreeding depression and reduced adaptability [2]. Elucidating the mechanism influencing adaptive genetic variation and differentiation in endangered species is thus informative for evolutionary processes and critical for conservation biology [3,4].

The major histocompatibility complex (MHC), an essential component of the vertebrate immune system, is an ideal fitness-relevant marker [5]. This highly polymorphic genetic region contains multigene family members involved in presenting pathogen-derived peptides to T-cells and triggering an adaptive immune reaction [6]. MHC class I molecules, consisting of an α chain and an associated β2-microglobulin, present intracellular antigens to CD8+ T cells, while MHC class II molecules, consisting of α and β chains encoded by separate MHC genes, present extracellular pathogens to CD4+ T cells [6,7].

The remarkable level of polymorphism in MHC is thought to result from sexual selection, maternal–fetal interactions, pathogen-mediated selection (PMS), and demographic processes such as genetic drift [7–10]. There are three nonexclusive mechanisms for PMS: (1) heterozygotes may be favored because they can recognize a greater range of antigens than homozygotes (heterozygote advantage); (2) rare alleles may have a selective advantage as few pathogens are adapted to them (frequency-dependent selection); (3) spatial/temporal heterogeneity in the pathogen load and variation may maintain MHC diversity at the global scale (fluctuating selection) [11]. Evaluating the relative roles of natural selection and genetic drift in driving the maintenance or loss of MHC diversity is challenging, and discordant results have been reported. Some studies found similar population structures between MHC and neutral loci, and it was proposed that genetic drift may overwhelm selection in shaping MHC variation, especially in small and bottlenecked populations [1,12,13]. Elevated MHC differentiation across populations has also been reported, supporting the dominant role of fluctuating selection [14–16]. Moreover, limited gene flow among populations can promote local adaptation to consistent pathogen communities, and this spatial variation in pathogen selection regimes may also create more structure in MHC genes [8,17,18].

Previous studies of MHC variation were mainly focused at the allele level. However, particularly in birds and reptiles, alleles from different MHC loci are sometimes highly similar and even identical to each other because of frequent gene duplication, conversion, and recombination [15,19–21], and the gene copy numbers may also vary among individuals [21–25]. These issues all imply that genotypes and variation estimated on the basis of alleles alone would be unreliable and potentially biased [21]. Even when the MHC loci are reliably separated, analysis based on valid genotypes alone, without considering the linkage among loci (which is often strong), may also bias the results [21]. For example, different results were obtained on the basis of MHC linkage groups compared to individual alleles in populations of the Egyptian vulture [26]. Similarly, certain infections in badgers were associated with MHC class II–class I haplotypes rather than allele presence [27]. Levels of individual fitness can be differentiated more profoundly using multilocus haplotypes. Thus, population analysis using multilocus MHC haplotype with consideration of the sharing of alleles, the copy number variation (CNV), and the linkage among loci may provide more comprehensive knowledge and credible evaluation of the adaptive variation and species fitness. Moreover, it is essential to examine both MHC class I and II genes, in recognition of their different functions, although only a few studies do this. To the best of our knowledge, no avian population study has adopted the MHC class I–class II multilocus haplotype as an adaptive marker.

The crested ibis (*Nipponia nippon*, Pelecaniformes; hereafter referred to as *Nini*), which is listed as "Endangered" according to IUCN criteria [28], was once believed to be extinct in the wild until a tiny population of seven birds (two breeding pairs and three nestlings) was rediscovered in 1981 at Yangxian, China [29]. Since then, the Chinese government have made great efforts to protect this wild population (in-situ conservation) and have also established several captive populations at different sites (ex-situ conservation) (Figure 1, Table 1). Surprisingly, the total number of crested ibis has now increased from seven to over 2000. However, previous studies using mitochondrial DNA and microsatellite markers on small *Nini* populations revealed extremely low levels of polymorphism [30,31]. By analyzing the genomes of 57 historic and 8 modern samples, a recent study revealed ancestral loss of genetic variation and high deleterious mutation owing to genetic drift and inbreeding depression in current populations [32]. There have been a growing number of reports regarding various infections [33–36] and high mortality [29,33] in this bird, indicating frequent threats from pathogens and undetermined fitness. However, there has not yet been any large-scale genetic assessment on *Nini* populations. It is imperative that such an assessment is completed, preferably by using MHC markers characterized by high polymorphism and fitness relatedness. This could provide insight into the adaptive diversity and differentiation in the recovering populations, thereby better evaluating species fitness and guiding future conservation. Fortunately, the gene architecture of the MHC region has been well illustrated in the crested ibis: ~500 kb MHC region contains five class I genes including one major locus (*UAA*), three minor loci (*UBA*, *UCA1*, and *UCA2*), and one non-classical locus (*UDA*), but the *UAA* locus is located beyond the Class I Region and gathered with the class II genes in the compact Core

Region; all these MHC loci were expressed, and different expression levels were found among class I genes [20]. The class IIα and IIβ genes are alternately organized into one to four elementary "αβ" units [20,37,38]. Phylogenetic analysis on the four "αβ" units identified two clusters [20] corresponding to the two reported ancestral avian MHC-IIβ lineages (DAB1 and DAB2) generated by duplication [19]. These results provided a solid foundation for our study. Here, we investigated the variation of all *Nini*-MHC class I and II genes across seven main populations. We further characterized the MHC haplotypes consisting of all class II genes and the major class I gene in the Core Region, and compared the genetic variation and population differentiation at MHC haplotypes with those expected from microsatellites. We aimed to (1) infer the evolutionary history of the structurally varying MHC haplotypes; (2) determine the relative contribution of genetic drift and selective pressures to MHC loci in the bottlenecked populations; (3) interpret the current population structure and provide corresponding conservation and management strategies.

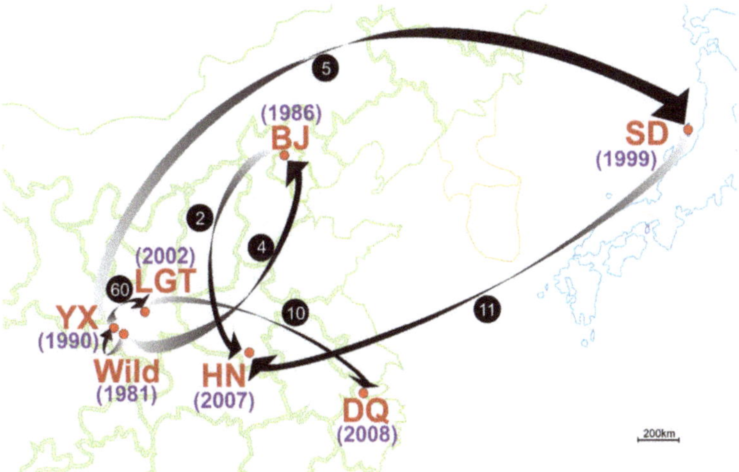

Figure 1. Map of the distribution and establishment of seven crested ibis populations (represented by red dots) in China (green lines) and Japan (blue lines). The SD population is from Sado Island, Japan, while the other populations are all from China. The year that each population was established is given beside the population name. Abbreviations for the six captive populations are: YX, Yangxian; LGT, Louguantai; DQ, Deqing; BJ, Beijing; SD, Sado; HN, Henan. The arrows represent the movement of founders with numbers indicated in the solid black circles.

Table 1. Summary of the establishment of seven populations of crested ibis.

Population	Year Established	Founder Size	Sample Size	Sampling Year	Population Size Around (Sampling Year)	Original Population of Founders
Wild	1981	6 [b]	8	1990–1993	17 (1993)	-
YX	1990	unknown	54/65 *	1997–2006	145 (2004)	Wild
LGT	2002	60	48/50 *	2006	185 (2004)	YX
DQ	2008	10	45	2008–2012	45 (2012)	LGT
BJ [a]	1986	4	35	2011	35 (2011)	Wild
SD	1999	5	18	2011–2012	120 (2011)	YX
HN	2007	13	86	2012	86 (2012)	11 from SD, 2 from BJ

[a] Four individuals were introduced from SD into BJ in 2002, and two of these participated in breeding of eight "hybrid" generations. [b] The wild population was developed from six of the seven rediscovered birds, as one nestling fell off the nest and was later captively bred. * Numbers before and after the slash represent samples used in MHC and microsatellite genotyping, respectively. Population abbreviations are as in Figure 1.

2. Materials and Methods

2.1. Sampling and DNA Extraction

We collected 307 samples including 141 feather, 130 blood, 25 muscle, and 11 eggshell samples from one wild and six captive populations (Table 1, Table S1). Blood samples were obtained from a wing vein by pricking the skin carefully and collecting a trace of blood on a cotton ball; muscle samples were obtained from dead birds; feather and eggshell samples were collected during molting and breeding times, respectively. All sampling from live birds was approved by the ethics committees of the relevant Breeding Centers, conducted in accordance with the guidelines, and supervised by the technical staff. Genomic DNA was extracted using the standard phenol/chloroform method [39].

2.2. MHC Genotyping

We analyzed the polymorphic MHC peptide-binding regions (PBR), which are coded by exons 2 and 3 of class I genes and exon 2 of class II genes, using single-strand conformation polymorphism and heteroduplex (SSCP-HD) analysis [40] according to the protocol described by Zhu, et al. [41]. For class I genotyping, given the highly similar introns flanking PBR exons among the five loci [20], we mainly used universal primers and focused on the heteroduplex (HD) banding pattern. An SSCP-HD profile containing the alleles of five class I loci was constructed from the positive bacterial artificial chromosome (BAC) clones stored in our laboratory [20,42], and was verified by profile reconstitution [40]. The profile was supplemented with novel alleles obtained by extracting and cloning new HD bands from genotyping gels (Figure S1a,b). For *UBA* and *UCA2*, which involved a single nucleotide substitution on exon 3, we designed locus-specific primers inside exon 3 and performed precise genotyping focusing on single-strand band patterns (Figure S1c,d). Meanwhile, the exon 3 sequences of *UCA1*01* and *UCA2*01* were identical. Thus, to avoid the interference from *UCA1*01*, we also utilized a SNP specific for *UCA2*01* on intron 1 to differentiate between *UCA2*01/*02* and *UCA2*02/*02* through sequencing polymerase chain reaction (PCR) products. For class II genotyping, the eight genes were classified according to their orthologies [20]: For *DAA* and *DAB*, respective locus-specific primers were designed (IIA1e2, IIB1e2) (Figure S1e,f). For *DBAs* (*DBA1*, *DBA2*, and *DBA3*) and *DBBs* (*DBB1*, *DBB2*, and *DBB3*), given the extensive sharing of alleles and identical introns flanking exon 2 among genes of each orthology, orthology-specific primers were designed (IIA2e2, IIB2e2) (Figure S1g,h). Primers for genotyping are illustrated in Table S2. To exclude the possibility of amplification failure of potential alleles, representative genotypes (3 samples for each) were collected for each genotyping and amplified by ≥8 additional valid primers designed for different regions, and the PCR products were sequenced and verified. For *DAB*, an SSCP profile of reference bands was constructed by cloning different genotypes (Figure S1f) [40]. The PCR-SSCP analysis was performed ≥3 times for each primer pair per DNA sample. For cloning and sequencing, we randomly selected ≥5 positive clones per sequence, which were then subjected to Sanger sequencing in both directions by Majorbio (Shanghai, China) using an ABI 3730 automated sequencer. For samples with no PCR product from some primers (putative gene deletion), we extracted DNA twice and designed ≥10 additional sets of primers on both coding (exons 1, 2, 3, and 4) and noncoding (introns 1 and 2) regions to exclude the possibility of poor template quality or PCR failure. All obtained sequences of *Nini*-MHC PBR exons were deposited in GenBank (accession numbers: MK829161–MK829185).

2.3. Haplotype Analysis

In this study, we mainly analyzed the *Nini*-MHC haplotypes combining all class II loci and the major class I locus in the Core Region. Linkage disequilibrium (LD) between polymorphic loci (gametic phase unknown) was tested by Arlequin 3.5 [43]. Individual haplotypes were inferred from genotypes both manually and by multiple programs. We first inferred all potential haplotypes from the individuals with 0–1 heterozygous loci (i.e., homozygous at *DAB* and/or *UAA*). For individuals with ≥2 heterozygous loci (i.e., heterozygous at both *DAB* and *UAA*), most haplotypes were deduced using pedigree information provided by reserves (assuming Mendelian inheritance), and the remainder lacking parentage data were phased according to the most frequent haplotypes in their respective populations. Notably, although the crested ibis is strictly monogamous, captive cages are sometimes occupied by >1 breeding pairs (2–3 in most cases), creating the potential for recording false pedigrees. Thus, we verified the pedigrees using microsatellite data, and corrected the paternities of three suspect individuals among candidate parents. Moreover, individual haplotypes were estimated by Haplotype Inference (gametic phase unknown) implemented in Arlequin 3.5 using both Excoffier-Laval-Balding (ELB) and Expectation-Maximization (EM) algorithms (identical results were obtained), and by PHASE 2.1 using a Bayesian statistical method credited with higher accuracy than the EM algorithm [44]. LD coefficients between linked alleles were further computed based on the estimated haplotypes in Arlequin (gametic phase known). Both intra-population and intra-species (combining population data together) analyses were performed.

To determine the genomic structure of the class II region ("*COL11A2–BRD2*" fragment) for each haplotype, we randomly selected twelve homozygotes from different populations as follows: four HT02/HT02, three HT03/HT03, one HT07/HT07, two HT04/HT04, and two HT05/HT05 (there were no homozygotes of HT01 and HT06). The BAC clone containing HT01 [20] and an HT01/HT02 heterozygote were chosen as positive controls. We first amplified the full-length sequences of IIβ genes using two sets of primers (3UT1 and 3UT2) (Table S2), and PCR products were cloned and sequenced. Then we amplified the six overlapped segments using the corresponding primers P1–P4, P6, and P7 reported previously [20], and the PCR products of twelve homozygotes were cloned and subjected to primer-walking Sanger sequencing (for both clones and PCR products) by Majorbio (Shanghai, China). Each segment was verified by ≥3 independent rounds of long and accurate PCR (La-PCR).

2.4. Microsatellite Genotyping

After validity assessment of amplification, we first chose 12 reported microsatellite loci for genotyping [30,45]. Each forward primer was labeled with one of the three fluorescent dyes (FAM, HEX, TAMRA), and the triplex PCR products were resolved on the ABI 3730xl DNA analyzer (Applied Biosystems, Foster City, CA, USA). GeneMapper 3.7 (Applied Biosystems) was used to score the genotypes. PCR and genotyping were performed ≥3 times per sample. MICROCHECKER 2.2.1 [46] was used to check for genotyping errors and/or null alleles. All pairs of loci were tested for LD using the exact test (Bonferroni correction) in GENEPOP 4.0 [47]. LD tests were also performed between microsatellite loci and MHC haplotype. One locus was excluded because of its significant LD with another locus in certain populations. Genotypes of 11 loci were used for population analysis (Nn01, Nn03, Nn04, Nn12, Nn16, Nn17, Nn18, Nn21, Nn25, Nn26 [30], and NnNF5 [45]).

2.5. Sequence Analysis

Sequences of the *Nini*-IIβ exon 2 obtained above were aligned using the ClustalW algorithm with manual modifications in MEGA 6 [48]. We used two methods to detect historical positive selection on exon 2 of *DAB* and *DBB*. As the exons 2 of *DBB1* and *DBB2* are derived from *DBB3* [20], we integrated the three *DBB* loci as one locus with four alleles (*DBB1*01*, *DBB2*01*, *DBB2*02*, *DBB3*01*). The selection parameter ω, which estimates the ratio of the nonsynonymous substitution rate (dN) to the synonymous substitution rate (dS), was calculated, and the Z-test with Jukes-Cantor correction was performed in MEGA 6 [48]. Given the limited variation, we input each sequence a number of times equal to its frequency in the primary data. We also used omegaMap 0.5 [49] to detect positive selection under the influence of recombination by simultaneously estimating ω and the recombination rate ρ along the sequence. Analyses of positive selection were performed using a set of objective priors: ω and ρ following an inverse distribution (0.02–500); mutation rate (μ; starting from 0.1), transition/transversion rate (κ; starting from 3.0), and deletion rate (φ; starting from 0.1) following an improper inverse distribution. A variable model was used with a block length of 2 codons. Two independent Markov Chain Monte Carlo (MCMC) chains were performed for 1,000,000 iterations with a thinning interval of 100, and the first 10% of iterations were abandoned as a burn-in. The two chains with high convergence were merged to infer the posterior distribution of ω and ρ. Sites with posterior probability ≥95% were deemed to be under positive selection. R code was used to interpret the test output and generate plots.

Recombination on the full-length sequences of IIβ genes was detected by the RDP beta 4.16 package [50] using seven methods: RDP, GENECONV, Bootscan, MaxChi, Chimaera, SiScan, and 3Seq. A preliminary scan for recombination was performed with $p \leq 0.05$ for multiple comparisons. Recombination events identified by at least two methods were then verified and refined by several methods according to the manual, including examination of the plots, matrices, and phylogenetic trees.

2.6. Population Analysis

Haplotype/allele frequencies, and observed (H_O) and expected (H_E) heterozygosities were obtained from Arlequin 3.5. Allelic richness (AR) was calculated using FSTAT 2.9.3 [51], and the correlation between MHC and microsatellites was evaluated using Pearson product-moment correlation in SPSS 20.0. We used Wilcoxon's test in the program BOTTLENECK [52] to assess evidence of population bottlenecks at microsatellites after Bonferroni correction, with 2000 replicates under the Stepwise Mutation Model.

STRUCTURE 2.3.4 [53] was used to explore population structure using a Bayesian clustering algorithm. Preliminary analysis using standard structure models without information on the sampling location failed to offer a distinct signal of structure, although the pairwise F_{ST} showed significant differences among populations. Thus, the admixture model with sampling location as a prior (LOCPRIOR) was used to detect the population structure. The parameter α was inferred from the data and λ was set to 1. An MCMC chain of 1,000,000 generations was run with an initial burn-in of 100,000 generations. Seven independent replicates of K = 1–7 were performed, and the final number of clusters was determined by simultaneously evaluating posterior probability and the delta K using the online Structure Harvester program [54]. The estimates of r, which evaluates the amount of information carried by the sampling locations, were 0.19 and 0.26 for MHC and microsatellites, respectively, indicating that the sampling locations are informative (near 1 or <1). The plot of the Q matrix was generated by distruct 1.1 [55]. In addition, neighbor-joining (NJ) trees for both markers were constructed to evaluate the genetic distances among populations in Populations 1.2.32 [56] using chord distance. Node supports were evaluated with 2000 bootstrap replicates over individuals.

As expectations for R_{ST} may suffer from higher sampling variances under a random mutation process [57], we first performed a size permutation test implemented in SPAGeDi 1.4 [58] to check whether the microsatellite allele sizes carried relevant information on genetic differentiation. As the result of this test was nonsignificant, we then calculated pairwise F_{ST} values for both MHC and microsatellites in Arlequin 3.5. Analysis of molecular variance (AMOVA) was conducted on 17 potential groupings with 50,000 permutations in Arlequin 3.5.

G'_{ST} was adopted to compare the population structures inferred by MHC and microsatellites because it can control differences in variation among genetic markers [59]. G'_{ST} was calculated in R code and 95% confidence intervals (CIs) for pairwise microsatellite G'_{ST} values were generated from 1000 bootstrap replications. Then we compared MHC-G'_{ST} to the 95% CIs of microsatellite-G'_{ST}, and the outliers represent population structures that are significantly different from that expected under neutrality. The correlation between MHC-G'_{ST} and microsatellite-G'_{ST} was tested using Pearson product-moment correlation in SPSS 20.0.

3. Results

3.1. Characterization of Multilocus MHC Haplotypes

The MHC class I genotyping showed that *UAA*, *UBA*, and *UCA2* were dimorphic whereas *UCA1* and *UDA* were monomorphic among the crested ibis studied. *UBA*02* and *UCA2*02* were newly discovered in this study, but differed from *UBA**01 and *UCA2**01, respectively, by only a single silent nucleotide substitution on exon 3. Thus, only *UAA* was dimorphic at the protein level. The allele frequencies of *UAA**01/*02, *UBA**01/*02, and *UCA2**01/*02 were 0.604/0.396, 0.914/0.086, and 0.500/0.500, respectively. Notably, *UBA**02 was only present with *UCA2**01, and these two loci showed significant linkage disequilibrium ($p = 0.000$). However, this linkage was not significantly supported between *UAA* and *UBA–UCA2* ($p = 0.083$), in accordance with the reported physical distance [20]. Six class I haplotypes were inferred from 213 individuals with 0–1 heterozygous locus (Figure S2).

In the class II exon 2 genotyping, we obtained 1 allele for *DAA* (1 sequence per individual), 1 sequence for *DBAs* (0–1 sequences per individual), 4 alleles for *DAB* (1–2 sequences per individual), and 2 sequences for *DBBs* (one is from *DBB1*, and the other is from *DBB2*/*DBB3*) (0–2 sequences per individual). Three *DBA* genes share the same exon 2; *DBB2* and *DBB3* share the same exon 2. Accordingly, all the class II loci except *DAB* were monomorphic. We also found that *DBAs* and *DBBs* were absent in individuals with *DAB**04, whereas *DBB1**01 was only present with *DAB**01.

The multilocus MHC haplotype of the Core Region consisted of all class II loci and the major class I locus [20]. Among these loci, only *DAB* and *UAA* were polymorphic, and they were in significant linkage disequilibrium (Table S3). Accordingly, we first read all seven haplotypes (HTs01–07) (Figure 2) from 177 homozygotes, and we found strong linkage between certain alleles: *DAB**01 was 100% present with *UAA**01; *DAB**02 was 98.8% present with *UAA**01; *DAB**03 was 88.1% present with *UAA**02; *DAB**04 was almost equally present with *UAA**01 and *02. This is consistent with the results of LD test between alleles (Table S3). We further phased 89 heterozygotes according to their verified parentages. For the 28 heterozygotes without pedigree information (from Wild, YX and LGT), HT02 and HT03 were favored in the presence of *DAB**02 and *DAB**03, respectively, considering their higher frequencies against HT06 and HT07 in all populations (e.g., *DAB**02/*04–*UAA**01/*02: HT02/HT05, *DAB**02/*03–*UAA**01/*02: HT02/HT03). The manually inferred haplotypes were consistent with the estimates from PHASE and Arlequin (intra-species level) for each individual. When estimating at intra-population level in Arlequin, three individuals (2 from SD and 1 from BJ) showed inconsistent results with the above-inferred haplotypes. However, given their unusually low phase-frequencies (0.517–0.580, overall level: >0.900), these results should be considered invalid and were likely biased by the relative frequencies of HT04 and HT05. We detected nine recombination events between class II genes and *UAA* in parentages: two HT06 were recombined from HT02, three HT07 from HT03, two HT04 from HT05, and two HT05 from HT04.

In structural analysis of the class II region using homozygotes, the sequencing results of full-length IIβ sequences showed that *DBB2* and *DBB3* were differentiated by their downstream sequences, and two full-length *DBB2* alleles varying in exon 3 were found, in HT01 and HTs02/06 (Table S4); the electrophoresis of six overlapped long-range segments showed various band patterns among haplotypes (Figure S3), and assembly of the six segments showed that IIα and IIβ genes were alternately organized on each haplotype, with four types of haplotype structure (Table S4). Compared to the four "αβ" units (*DA, DB1, DB2,* and *DB3*) in HT01, there are only three (*DA, DB2,* and *DB3*) in HTs02/06, two (*DA* and *DB3*) in HTs03/07, and one (*DA*) in HTs04/05 (Figure 2). Intriguingly, each type can be distinguished by a unique allele of *DAB*. For HTs02/06, HTs03/07, and HTs04/05, haplotypes within each pair differ by alleles of *UAA*. HT01 was the haplotype of the target BAC clone reported previously [20], and the class II regions of HTs02/06, HTs03/07, and HTs04/05 were identical to the three reported *Nini* class II haplotypes HP3, HP2, and HP1, respectively [38].

3.2. Positive Selection and Recombination on Nini-MHC-IIβ

The alignment of the *Nini*-IIβ exon 2 amino acid sequences clearly indicated two lineages (*DAB* and *DBBs*) by the first 13 residues (grey regions, Figure 3). Selection analysis on the exon 2 using MEGA revealed a trend towards positive selection on *DAB* (ABS: dN = 0.116, dS = 0.029, $p = 0.054$; non-ABS: dN = 0.020, dS = 0.012, $p = 0.171$) and significant positive selection on *DBB* (ABS: dN = 0.021, dS = 0.000, $p = 0.004$; non-ABS: dN = 0.003, dS = 0.001, $p = 0.031$). All the substitutions on ABS of *DBB* were nonsynonymous. When considering the recombination, omegaMap revealed a mean value of $\omega = 14.04$ per codon and identified fourteen codons under positive selection on the exon 2 of *DAB* (Figure 3 and Figure S4a), of which eight (57%) were in the predicted ABS and the rest were in the immediate vicinity. A higher mean value of ω was obtained ($\omega = 19.40$ per codon) for *DBB* and thirteen positively selected sites were identified (Figure 3 and Figure S4b), of which eight (61.5%) were in the predicted ABS and the rest were adjacent.

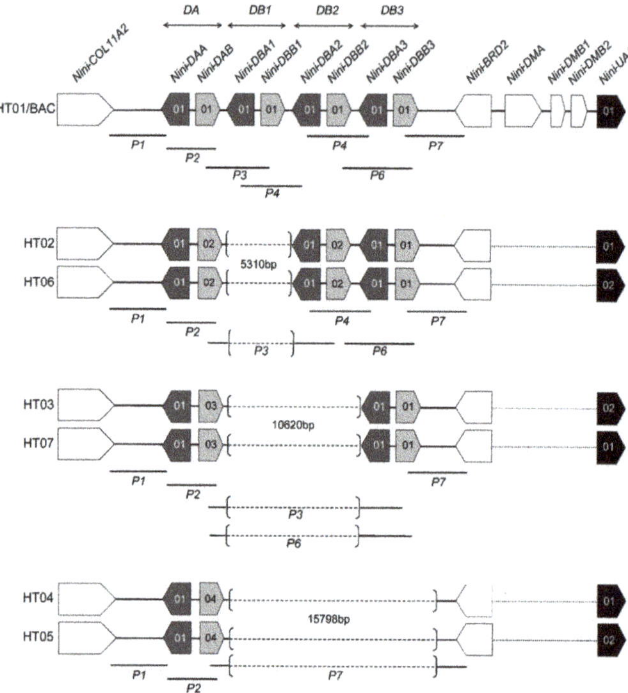

Figure 2. Genomic structures of seven crested ibis MHC haplotypes. A total of seven haplotypes were inferred from genotype data, and their genomic structures of class II region were further determined by La-PCR using representative homozygotes. Dark and light grey boxes indicate the IIα and IIβ genes, respectively, with numbers indicating different alleles on the full-length level. Black boxes indicate the major class I gene (*UAA*) and white boxes indicate other functional genes [20]. Haplotypes are aligned, with bracketed dash lines representing deleted regions (the lengths were denoted aside) in relative haplotypes. Dotted lines between *BRD2* and *UAA* represent the unsequenced region. Locations of the La-PCR fragments amplified by the six reported sets of primers (P1–P4, P6, and P7) [20] are marked below the haplotypes.

RDP analysis on the eight full-length IIβ sequences detected five significant recombination events with the most likely parental sequences of *DAB**03 and *DBB3**01 (Table S5). This role of recombination in shaping the IIβ sequences was also indicated by their mosaic structures (Figure 4a). Along the whole sequences, most of the nucleotide variation existed in introns 1 and 5, 3′UTR, and exons 2 and 3. All these regions were genetically dimorphic, except exon 2 which was phylogenetically dimorphic (i.e., two lineages). Thus, we designated these dimorphic regions on the parental sequences as "type A" (blue) and "type B" (red) (Figure 4). Accordingly, *DAB**01, *02, and *03 contained all "type A" regions, whereas *DBB3**01 contained all "type B" regions. The other four IIβ sequences showed mixed patterns.

Introns 1, 5, and 3'UTR were far longer in "type A" than in "type B" (intron 1: 660 vs. 285 bp, intron 5: 171 vs. 84 bp, 3'UTR: 244 vs. 199 bp).

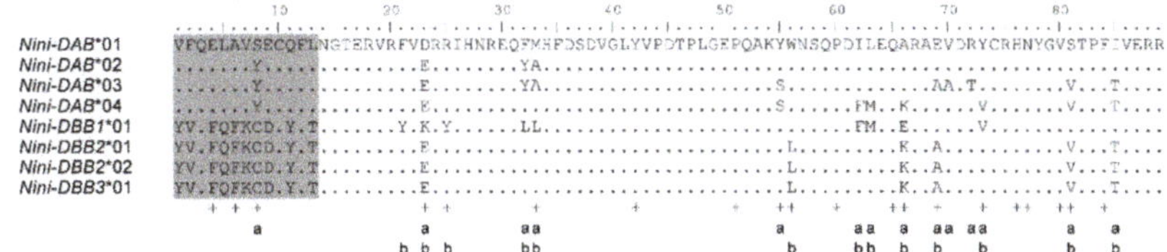

Figure 3. Amino acid alignment of the exon 2 of full-length *Nini*-IIβ sequences. Dots indicate identity to *DAB**01. Crosses represent putative antigen binding sites (ABS) inferred from Jardetzky, et al. [60] and Stern, et al. [61]. "a" and "b" denote sites under positive selection for *DAB* and *DBB*s, respectively, as estimated by omegaMap.

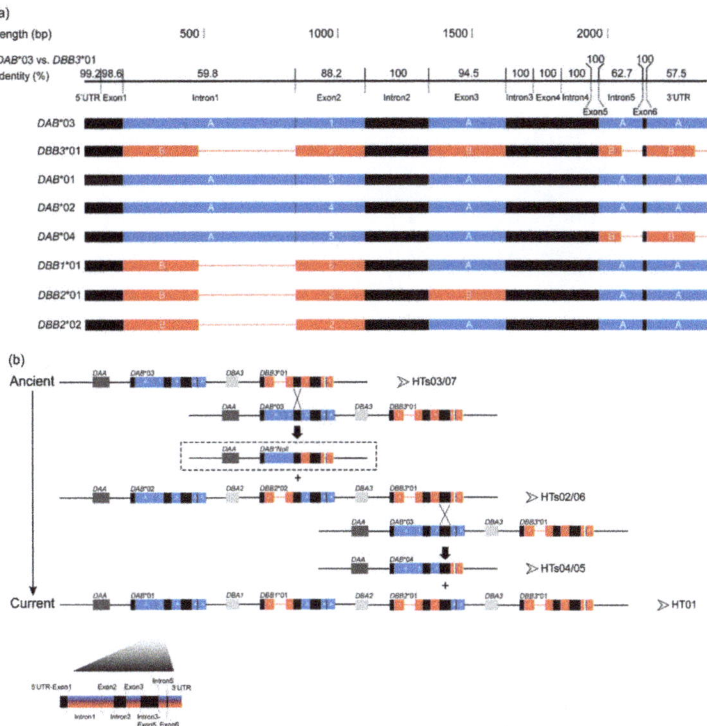

Figure 4. Comparison of the full-length IIβ sequences (**a**) and evolutionary model of the class II regions in seven MHC haplotypes (**b**) in crested ibis. Blue (A) and red (B) boxes respectively indicate "type A" and "type B" regions represented by *DAB**03 and *DBB3**01. Numbers on exon 2 represent different nucleotide sequences. Black boxes indicate highly similar homologous regions which provide good opportunities for unequal crossing over between two ancestral IIβ genes. Dotted lines indicate alignment gaps. (**a**) Sequence identity (%) for each gene region was calculated in Lasergene software (DNASTAR Inc., Madison, WI) for *DAB**03 vs. *DBB3**01. (**b**) Dark and light grey boxes represent *DAA* and *DBA*s, respectively. The scale for intragenic regions is shown in the lower left corner. Breakpoints of recombination (as indicated by RDP) are shown by black crosses (intron 2 and intron 3–exon 5). The haplotype framed by the dashed-line was not detected in this study.

3.3. Genetic Variation within Populations

Each of the seven populations carried 5 or 6 haplotypes (Table 2). In general, HT02 (33.2%) and HT03 (30.8%) were most common. HT01 (3.9%), HT06 (0.3%), and HT07 (4.3%) were relatively rare and restricted to certain populations. HT01 was only present in group 1, and the only two HT06 detected were products of recombination. Among the four types of haplotypes, HTs02/06 were most prevalent in Wild, LGT, and DQ, while HTs03/07 were most common in BJ, SD, and HN. The majority of individuals in BJ contained HTs03/07. HTs04/05 occurred at higher frequencies in YX and HN than in the other populations.

Table 2. Summary of genetic variation and haplotype frequencies across seven populations of crested ibis.

	Population		Wild	YX	LGT	DQ	BJ	SD	HN	Total
Sample size (MHC/microsatellite)			8/8	54/65	48/50	45/45	35/35	18/18	86/86	294/307
	Homozygote [1]		5	34	34	23	27	9	45	177
	Heterozygote [2]		3	20	14	22	8	9	41	117
		HT01	6.3	7.4	5.2	10.0	0.0	0.0	0.0	3.9
		HT02	43.8	28.7	44.8	51.1	14.3	25.0	28.5	33.2
		HT03	25.0	26.9	21.9	18.9	60.0	36.1	32.0	30.8
MHC	Haplotype frequency (%)	HT04	12.5	19.4	15.6	6.7	2.9	22.2	32.0	18.5
		HT05	12.5	17.6	11.5	12.2	8.6	5.6	1.2	9.0
		HT06	0.0	0.0	1.0	1.1	0.0	0.0	0.0	0.3
		HT07	0.0	0.0	0.0	0.0	14.3	11.1	6.4	4.3
		HTs02/06	43.8	28.7	45.8	52.2	14.3	25.0	28.5	33.5
		HTs03/07	25.0	26.9	21.9	18.9	74.3	47.2	38.4	35.1
		HTs04/05	25.0	37.0	27.1	18.9	11.5	27.8	33.2	27.5
	H_O		0.63	0.78	0.75	0.71	0.54*	0.78	0.81	0.71
	H_E		0.76	0.78	0.72	0.68	0.60	0.76	0.71	0.75
	AR		5.00	4.67	4.59	4.60	4.09	4.61	3.84	4.78
	H_O		0.43	0.50	0.46	0.44	0.49	0.55	0.39	0.45
Microsatellite	H_E		0.49	0.45	0.44	0.43	0.46	0.47	0.37	0.44
	AR		2.27	2.22	2.22	2.14	2.14	2.17	2.13	2.23

[1] Homozygous at *DAB* and/or *UAA*; [2] Heterozygous at both *DAB* and *UAA*; AR, allelic richness; H_E, expected heterozygosity (haplotype level for MHC); H_O, observed heterozygosity (haplotype level for MHC). * significant deficiency of heterozygotes at $P < 0.05$.

Of the eleven microsatellite loci, eight were dimorphic and three were trimorphic (Table S6). No population exhibited a significant deficit or excess of heterozygotes for both MHC and microsatellites, except BJ which showed significant heterozygote deficiency for MHC ($p = 0.015$). For both markers, allelic richness (AR) was highest and lowest in Wild and HN, respectively, and decreased progressively in more recently established populations (order of establishment: Wild→ YX→ LGT→ DQ, Wild→ BJ→ HN, Wild→ YX→ SD→ HN). There was a significant correlation (r = 0.805, $p = 0.029$) between the AR values of MHC and microsatellites. The BOTTLENECK analysis supported a significant historical bottleneck in all populations ($p < 0.005$).

3.4. Patterns of Population Differentiation

Clustering analysis using STRUCTURE revealed that both MHC and microsatellite captured the best population genetic structure when K = 3 (Figure S5). According to the proportions of memberships from the three genetic clusters, both analyses allocated individuals from Wild, YX, LGT, and DQ into group 1, individuals from BJ into group 2, and individuals from SD and HN into group 3 (Figure 5). Memberships in group 2 and group 3 were mainly from a single cluster, whereas group 1 contained more even memberships from all three clusters. The NJ trees constructed from MHC and microsatellites shared similar pattern of three major clusters and similar bootstrap supports (Figure S6). The first cluster was "(SD, HN) (BJ)" while the second was "(DQ, LGT) (YX)". The Wild formed the third cluster as a basal branch, indicating its ancestral position.

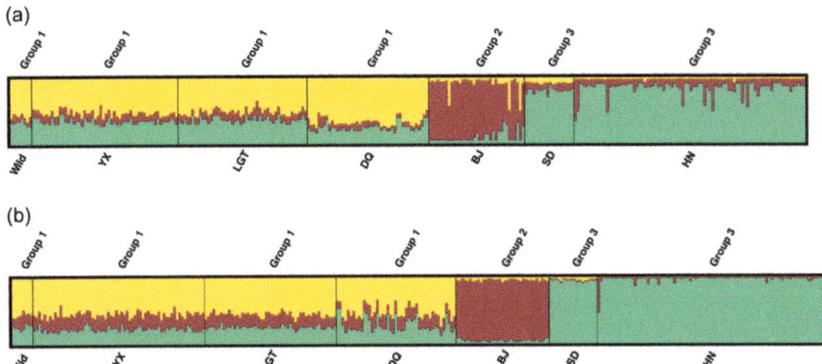

Figure 5. Bayesian clustering analysis (K = 3) using (**a**) MHC haplotypes and (**b**) microsatellite markers. Each individual is represented by a vertical bar, vertically partitioned into segments with lengths proportional to the individual's estimated membership fraction of each of the three inferred clusters indicated by different colors. The seven populations are separated by black lines and can be divided into three groups based on the proportion of membership of the three clusters.

F_{ST} values between pairwise populations ranged from −0.036 to 0.199 for MHC and −0.014 to 0.099 for microsatellites (Table S7). AMOVA analysis of 17 potential groupings indicated the highest F_{CT} values and most significant *p* values when the grouping was "(Wild, LGT, DQ) (YX) (BJ) (SD, HN)" for MHC and "(Wild, YX, LGT, DQ) (BJ) (SD, HN)" for microsatellites (Table S8). Notably, "(Wild, YX, LGT, DQ) (BJ) (SD, HN)" was the second most likely pattern for MHC, and only slightly less likely than the best grouping (F_{CT}: 7.29 vs. 7.81, *P* value: 0.010 vs. 0.005) (Table S8). For both markers, most of the variation was found within populations; nonetheless, the level of differentiation among groups was higher than that among populations within groups (Table 3). Similarly, *t*-tests showed that pairwise F_{ST} values among groups were significantly higher than those within groups (MHC: $t = 2.751$, d.f. = 19, $p = 0.040$; microsatellite: $t = 5.064$, d.f. = 19, $p = 0.050$) (Table S7).

Table 3. Results from AMOVA of MHC haplotypes and microsatellite loci.

Locus	Grouping	Source of Variation	d.f.	SS	Variance Components	Percentage of Variation	Fixation Index	*P* Value
MHC	(Wild, LGT, DQ) (YX) (BJ) (SD, HN)	Among groups	3	12.800	0.02996	7.81	0.07810	0.00484
		Among populations within groups	3	0.633	−0.00250	−0.65	−0.00706	0.81299
		Within populations	581	206.899	0.35611	92.84	0.07159	0.00000
Microsatellite	(Wild, YX, LGT, DQ) (BJ) (SD, HN)	Among groups	2	49.130	0.10917	5.10	0.05102	0.00897
		Among populations within groups	4	17.424	0.03245	1.52	0.01598	0.00002
		Within populations	607	1212.932	1.99824	93.38	0.06618	0.00000

Significance values are derived from 1023 permutations.

The global G'_{ST} for MHC haplotypes (0.381) was significantly higher than that for microsatellites (0.126, 99% CI = 0.061–0.212). Meanwhile, 10 of the 21 pairwise MHC-G'_{ST} values fell above the 95% CIs of the microsatellites-G'_{ST} (Figure 6), and 6 pairs were associated with BJ. Three pairwise MHC-G'_{ST} values fell below the 95% CIs of the microsatellites-G'_{ST}, and these were all pairs from the group 1 suggested by STRUCTURE (W–Y, W–L, and W–D). The mantel test indicated significant correlation between pairwise MHC-G'_{ST} and microsatellites-G'_{ST} estimates ($r = 0.660$, $P = 0.001$).

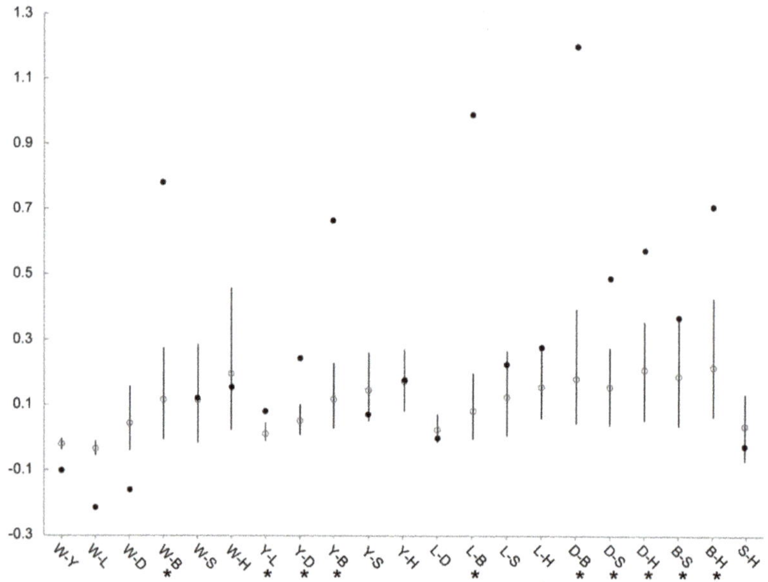

Figure 6. Comparison between pairwise G'_{ST} for MHC haplotypes and microsatellite loci among seven populations of the crested ibis. Pairwise G'_{ST} with 95% CIs for microsatellite loci is indicated by white circles with corresponding error bars. Black circles indicate pairwise MHC-G'_{ST} values. Pairwise populations with significantly higher MHC-G'_{ST} than microsatellite-G'_{ST} values are marked by asterisks. Populations are W, Wild; Y, YX; L, LGT; D, DQ; B, BJ; S, SD, and H, HN.

4. Discussion

The *Nini*-MHC haplotypes are characterized by different copy numbers of IIαβ units. Evidence for CNVs in MHC class II regions between individuals has been increasingly reported in birds [21–25]. For example, one to two IIαβ units were found across ten MHC class II haplotypes of the Oriental stork [24]; in quail, five MHC-IIβ haplotypes were flexibly organized with 1–7 transcribed IIβ loci [25]. CNV, which emerges faster than other types of mutation, plays a substantial role in generating MHC variation both within and among species [62]. In birds, passerines have strikingly more MHC gene copies than non-passerines. This avian CNV was thought to be correlated with life-history traits (lifespan and migratory behavior) and differences in exposure to pathogens [63]. The evolution of MHC-CNV can be explained by the birth-and-death model that posits that new genes in the multigene family are generated from successive duplication and eventually either maintained or lost [64]. Generally, there are three molecular dynamic mechanisms that may produce CNV: recombination between homologous sequences (unequal crossing over), replication slippage, and retrotransposition [65]. Given the CNV and large-tract intragenic recombination in the MHC class II regions among *Nini* haplotypes, we inferred an evolutionary model using the unequal crossing over mechanism. The previous study on HT01 proposed that *DAA/DAB* and *DBA3/DBB3* pairs belonging to the two ancestral avian IIαβ lineages might represent the two primitive *Nini*-IIαβ dyads [20]. This speculation is strongly supported by the RDP result, regarding *DAB*03* and *DBB3*01* on HTs03/07 as the two parental sequences. The other five haplotypes all contain 1–2 recombinant IIβ sequences. Accordingly, we strongly suggest that HTs03/07 represent the ancestral *Nini* haplotypes. Based on the RDP results, we hypothesized that HTs02/06, HTs04/05, and HT01 were successively generated from HTs03/07 through two rounds of unequal crossing over events between *DAB* and *DBB3* (Figure 4b). In addition, the recombination in the "*BRD2–DMB2*" region further generated three new haplotypes with either allele of *UAA*. In the studied populations, HT06 and HT07 occupy strikingly lower frequencies than HT02 and HT03, respectively.

The only two cases of HT06 (found in LGT and DQ) were recombined from parental HT02. In contrast, most cases of HT07 (found in BJ, SD, and HN) can be traced back to one founder (HT03/HT07) of BJ, born in the wild in 1985. Thus, we suggest an earlier generation and/or higher probability of presence in the initial seven birds for HT07 than HT06. Nevertheless, in view of the higher genetic diversity found in historical samples [32], other unknown MHC haplotypes might have existed prior to the species bottleneck and therefore the haplotype evolution might be more complex than we expected.

Despite structural diversity of *Nini*-MHC haplotypes, most loci were monomorphic. The class II genes showed higher polymorphism than class I genes, and this is thought to be associated with the stronger selection and gene conversion at class II genes (in non-passerines) [9]. Given the low polymorphism of *Nini-UAA*, the crested ibis is predicted to be at high risk for infections by intracellular pathogens, such as bird flu and Newcastle virus [35], and this deserves great attention in conservation and management. The recent genome-wide study on the crested ibis revealed that the contemporary population has only retained a small amount of the ancestral genetic variation [32]. Concordantly, the historical presence of positive selection on *Nini*-IIβ genes (as suggested by MEGA and omegaMap) indicated that the ancestral MHC variation should be high [12,66], and therefore the current low polymorphism is probably caused by genetic drift through species bottleneck. Additionally, the genetic drift still has a predominant role over selection in shaping MHC variation within and across the recovering populations. This is strongly supported by the significant correlations in AR values and pairwise G'_{ST} values between MHC and microsatellites, as well as the consistent population structures at MHC and microsatellites in most differentiation analyses. The only exception is that AMOVA supported the differentiation of one more group with MHC than with microsatellites; however, this result was not informative as there was also a pronounced support for an identical grouping between the two markers (Table S8). A previous simulation on bottlenecked populations found that the relative roles of selection and drift in driving MHC diversity depend on the timescales. Initially, selection was not effective in maintaining high polymorphism, but after ~40 post-bottleneck generations, selection overwhelmed the drift and restored variation to pre-bottleneck levels [67]. Nevertheless, high MHC diversity generated by strong selection pressures was reported in bottlenecked fox populations within only ~10–20 generations from 10 individuals [68]. Thus, we call for long-term monitoring of the adaptive diversity in *Nini* populations. With such dramatic loss of MHC variation, average fitness should decline substantially below the pre-bottleneck level, so why did this species successfully thrive within just 38 years? One possibility is that the captive environments under considerable care may serve as a good shelter for this bird, and thus prevent exposure to potential threats such as competitors, predators, and especially the human activities that were considered to be the dominant cause of this species retrogression [32]. Alternatively, the stress from pathogens may be moderate at the present; in this case, the crested ibis is still at risk from catastrophic deterioration caused by novel parasites. Nevertheless, there is little doubt that this bird still maintains a robust reproductive rate.

The global G'_{ST} values suggested elevated differentiation at MHC relative to microsatellites. However, this trend appears to be a by-product of strong MHC differentiation between only a few populations. This claim can be strongly supported by the comparisons of pairwise G'_{ST} between two markers (Figure 6). Most outliers with significantly higher MHC-G'_{ST} values were associated with BJ. BJ was highly differentiated at MHC from all the other six populations, and obtained extremely high MHC-G'_{ST} with the four populations from group 1 (as suggested by STRUCTURE). This might be associated with the unusually high frequency of HT03 (60.0%) and low frequency of HT02 (14.3%) in BJ compared to those in other six populations (Table 2). Given the similar habitat and management across captive populations, the partially higher differentiation at MHC might arise from restricted gene flow translated into different adaptations to local pathogen communities (e.g., reference [18]), rather than fluctuating selection under spatial/temporal heterogeneity in pathogen communities. This is consistent with the demographic history of BJ: as the first captive population, it was established in 1986, and has been almost completely isolated for a long time with restricted in-migrants. This could also explain the significantly lower MHC-G'_{ST} than microsatellite-G'_{ST} values for three pairs from group 1 (Figure 6).

The founder effect resulted from improper management is a major cause of genetic drift in conservation programs [69–72]. Its negative impact is associated with the number and genetic makeup of founders, number of generations (time between founding and sampling), gene flow, reproductive variance, etc [69,71,73]. Ex-situ daughter populations of the same origin may evolve divergently under respective founder effects [71,72]. Thus, it is necessary to investigate and interpret the population structure of the crested ibis after 38 years of conservation practices, thereby evaluating and guiding the captive management [74]. By referring to population histories, we found the current population structure (as suggested by STRUCTURE) could be attributed to inconsistent founder effects across populations. (1) YX (sampled at 7–16 years since establishment) was the second captive population formed with several wild birds, and continuously received injured or sick individuals and abandoned eggs (i.e., gene flow) from the nearby wild population [29]. After 12 years, YX exported 60 individuals as founders to establish LGT (sampled 4 years later). DQ, as the youngest captive population derived from LGT, was sampled within only two generations. Accordingly, these three populations suffered from a relatively weak founder effect and captured richest genetic variation from Wild. Nevertheless, the sample size of Wild is small and sampling time is early (because only 17 wild individuals existed at sampling and unbiased sampling is difficult after its range expansion). Currently, the wild individuals account for nearly half of the total birds. Thus, it is possible that the wild population has already become genetically differentiated from captive populations, especially given that wild individuals are prone to perform mate choice for MHC dissimilarity [7]. (2) In BJ, despite four birds introduced from SD in 2002, most individuals are descendants of a single breeding pair with high reproductive rate but low MHC dissimilarity (HT03/HT03 and HT03/HT07), and BJ had developed for 25 years prior to sampling. Therefore, its founder effect is expected to be extreme, and this was reflected by its isolation with monomorphic genetic resource. Although YX and BJ both originated from the wild around the same time, their different management histories have led to a great distance between BJ (but not YX) and Wild in the NJ trees. (3) Similar to BJ, most individuals in SD (sampled at 12–13 years since establishment) are descendants of two founders from YX. After 8 years of development, 11 descendants were chosen as the main founders of HN. Therefore, the small founder size of group 2 led to its genetic isolation from YX. Notably, the decreasing AR values with establishment of new populations indicate a trend towards continuous genetic drift through the founder effect and inbreeding depression. Accordingly, we call for serious consideration of the anthropogenic founder effect in future conservation management.

Various markers with different characteristics have been developed for conservation of crested ibis over the past decade. It is important that appropriate markers are chosen for different management issues. The microsatellite markers [30,45] plus the more powerful genome-wide STR markers [75] provide a DNA identification profiling platform, and can be mostly used in demographic management, such as pedigree construction, individual identity and paternity testing. The defensin markers can be applied to improvement of innate immune response among embryos and nestlings [42]. Alternatively, the MHC haplotypes provide a crucial index of individual fitness and population viability. Construction of MHC haplotype profiles, by genotyping *DAB* and *UAA*, was recommended for all reserves. This has already been applied in DQ, as an important reference for conservation strategies, such as artificial pairing, individual exchange, and reintroduction. Moreover, new populations should be established from enough founders with maximum MHC dissimilarity. In order to mitigate the founder effect, we call for appropriate individual exchange between captive populations [70,73]. For example, group 1, which contained the only birds with HT01 and an abundance of HTs02/06, should have more genetic connectivity with groups 2/3 which suffer from limited genetic resources and are rich in HTs03/07. Recently, several reserves have successfully reintroduced captive birds into the wild, providing a giant step forward. However, the low level of adaptive diversity implies a survival risk for these birds with exposure to more diverse pathogen faunas but less artificial care. Similar consideration was also proposed based on the genome-wide diversity loss [32]. Thus, it is crucial to persistently monitor their genetic diversity and population structure using MHC haplotypes and neutral markers.

Associations between certain MHC haplotypes and infections have been increasingly identified in birds [76–78]. In the crested ibis, the structural and genetic variation across haplotypes strongly implies the immunological divergence. HT01, which holds the maximum gene copies including a unique one (*DBB1*), may be the most immunologically competitive haplotype, whereas HTs04/05 may be most susceptible. Nevertheless, the immunity can be confounded by various factors, especially when gene copies function at different levels. A small-scale analysis on the correlation between *Nini*-MHC haplotypes and offspring mortality by our laboratory found a potentially higher mortality in HT03/HT04 heterozygotes [79]. However, the offspring might not have died from infections, so further investigation of *Nini*-MHC haplotype–pathogen relationships should be undertaken for improved conservation. Besides, the MHC haplotypes are ideal markers for studying mate-choice mechanism.

Here we demonstrated an application of multilocus MHC haplotype in population genetic analysis, which particularly suits to populations with low polymorphism, such as bottlenecked or isolated populations [27,40]. Loci with limited alleles are not informative and can sometimes produce inaccurate analysis results, especially in the presence of CNV and/or shared alleles. In contrast, the MHC haplotype integrating overall variation may provide a more precise assessment of fitness. Individual MHC genes are often in tight linkage that is undesirable in some analyses [21], hence the use of haplotypes recoded from the linked region was suggested as a valid solution (e.g., STRUCTURE [53]). However, we acknowledge that haplotype inference may be hindered for populations with masses of alleles, and given the extensive repeats and high GC-content in the MHC region, the characterization of CNV and shared alleles among haplotypes using La-PCR may be challenging without any prior MHC genomic information. In view of this, the recently-proposed family-assisted inference of the MHC genetic architecture can be an alternative to achieve accurate haplotyping [21]. Notably, whole-genome sequencing has played a promising role in conservation of endangered species including the crested ibis [32,75]. Various estimates can be obtained based on the genome-wide SNPs, including overall genetic diversity, deleterious mutations, phylogenetic relationships, and even population structure [32]. The noncoding SNPs were also considered as suitable neutral markers in MHC variation analysis owing to their similar mutation mechanism [4]. Nevertheless, the assembly of the complicated MHC region remains a challenge for next-generation genome sequencing methods [20,24,37], particularly in non-galliformes, therefore limiting in-depth research on this region.

Supplementary Materials: The following are available online at http://www.mdpi.com/2073-4409/8/4/377/s1. Figure S1. SSCP genotyping patterns. The upper and lower regions show the single-strand and heteroduplex regions, respectively. Arrows indicate the specific bands among genotypes in two class I SSCP-HD profiles. For each photograph, the left part shows representative genotypes, and the right part shows partial genotyping results in population surveys. (a) Exon 2 of class I genes. Numbers 1 to 3 represent *UAA**01/*01, *UAA**02/*02, and *UAA**01/*02. (b) Exon 3 of class I genes. Numbers 1 to 3 represent *UAA**01/*02, *UAA**02/*02, and *UAA**01/*01. (c) Exon 3 of *UBA*. Numbers 1 to 3 represent *UBA**02/*02, *UBA**01/*01, and *UBA**01/*02. (d) Exon 3 of *UCA2*. Number 1 represents *UCA2**01/*01, and number 2 represents *UCA2**01/*02 or *UCA2**02/*02 (see Methods for further differentiation). (e) Exon 2 of *DAA*. (f) Exon 2 of *DAB*. Numbers 1 to 9 represent *DAB**02/*02, *DAB**03/03, *DAB**04/*04, *DAB**01/*02, *DAB**01/*03, *DAB**01/*04, *DAB**02/*03, *DAB**02/*04, and *DAB**03/*04. *DAB**01/*01 was not found. (g) Exon 2 of *DBAs*. (h) Exon 2 of *DBBs*. Number 1 represents the sequence from *DBB2/DBB3*, and number 2 represents the mixed sequences from *DBB1* and *DBB2/DBB3*. The reference band of *DAB**04 was added to each lane to strengthen the bands of *DBB1* with low proportion in mixed PCR products. Figure S2. Six MHC class I haplotypes of crested ibis. Black boxes indicate the five class I genes, with numbers representing alleles of each gene. White boxes indicate other functional genes, and the dotted oval represents the gap between the Core Region and the Class I Region [20]. Figure S3. La-PCR results for six segments of the class II region (P1–P4, P6, and P7) in representative samples. The ladder with reference band sizes (bp) is shown on the left, and the bright reference bands are labelled in red. Figure S4. Spatial variation in the logarithm of the selection parameter (ω) across the exon 2 of (a) *Nini-DAB* and (b) *Nini-DBB* as estimated with omegaMap. The site-wise mean (solid line) and 95% confidence intervals (grey shaded area) are shown. Values of $\log(\omega) > 0$ imply positive selection. The plots of posterior probability and selection parameters overlapped, so only one plot is shown for each analysis. Figure S5. Estimating the true number of clusters with ΔK for (a) MHC haplotypes and (b) microsatellite loci. The uppermost level of structure is the true number of clusters. Figure S6. Neighbor-joining trees of seven crested ibis populations for (a) MHC haplotypes and (b) microsatellites. Bootstrap values were computed over 2000 replicates and are shown as percentages. Table S1. Details of the crested ibis samples analyzed in this study. Table S2. Primers for PBR exon genotyping of *Nini*-MHC genes and amplification of full-length IIβ sequences in haplotype analysis. Table S3. Results of linkage disequilibrium (LD) test from Arlequin. Table S4. Summary of the *Nini*-IIβ sequences amplified by 3UT1 and 3UT2 and long-range segments amplified by P1–P4, P6, and P7 for each haplotype. Table S5. Recombination events between the full-length sequences of *Nini*-IIβ genes, identified using RDP. Table S6. Allele frequencies (%) of microsatellite loci across seven populations of crested ibis. Table S7. Pairwise F_{ST} values between populations for MHC (below diagonal) and microsatellites (above diagonal). Table S8. Results of the hierarchical AMOVA for the crested ibis.

Author Contributions: Conceptualization, S.-G.F. and Q.-H.W.; Methodology, Q.-H.W. and H.L.; Experiment and Analysis, H.L., T.Z., Q.-H.W. and S.-G.F.; Resources, S.-G.F.; Writing–Original Draft Preparation, H.L., Q.-H.W. and S.-G.F.; Writing–Review & Editing, H.L., T.Z., Q.-H.W. and S.-G.F.; Supervision, S.-G.F. and Q.-H.W.; Project Administration and Funding Acquisition, S.-G.F.

Funding: This research was supported by the National Key Program (2016YFC0503200) from the Ministry of Science and Technology of China, a special grant from the State Forestry Administration, and the Fundamental Research Funds for the Central Universities of China.

Conflicts of Interest: The authors declare no competing interests.

References

1. Radwan, J.; Biedrzycka, A.; Babik, W. Does reduced MHC diversity decrease viability of vertebrate populations? *Biol. Conserv.* **2010**, *143*, 537–544. [CrossRef]

2. Mills, L.S. *Conservation of Wildlife Populations: Demography, Genetics, and Management*, 1st ed.; Blackwell Publishing: Malden, MA, USA, 2007.

3. Crandall, K.A.; Bininda-Emonds, O.R.P.; Mace, G.M.; Wayne, R.K. Considering evolutionary processes in conservation biology. *Trends Ecol. Evol.* **2000**, *15*, 290–295. [CrossRef]

4. Alcaide, M. On the relative roles of selection and genetic drift in shaping MHC variation. *Mol. Ecol.* **2010**, *19*, 3842–3844. [CrossRef] [PubMed]

5. Potts, W.K.; Wakeland, E.K. Evolution of diversity at the major histocompatibility complex. *Trends Ecol. Evol.* **1990**, *5*, 181–187. [CrossRef]

6. Klein, J. *Natural History of the Major Histocompatibility Complex*; John Wiley and Sons: New York, NY, USA, 1986; pp. 685–687.

7. Piertney, S.B.; Oliver, M.K. The evolutionary ecology of the major histocompatibility complex. *Heredity* **2006**, *96*, 7–21. [CrossRef]

8. Hedrick, P.W. Pathogen resistance and genetic variation at MHC loci. *Evolution* **2002**, *56*, 1902–1908. [CrossRef] [PubMed]

9. Minias, P.; Pikus, E.; Whittingham, L.A.; Dunn, P.O. A global analysis of selection at the avian MHC. *Evolution* **2018**, *72*, 1278–1293. [CrossRef] [PubMed]

10. Winternitz, J.C.; Minchey, S.G.; Garamszegi, L.Z.; Huang, S.; Stephens, P.R.; Altizer, S. Sexual selection explains more functional variation in the mammalian major histocompatibility complex than parasitism. *Proc. R. Soc. B: Biol. Sci.* **2013**, *280*, 20131605. [CrossRef] [PubMed]

11. Spurgin, L.G.; Richardson, D.S. How pathogens drive genetic diversity: MHC, mechanisms and misunderstandings. *Proc. R. Soc. B: Biol. Sci.* **2010**, *277*, 979–988. [CrossRef] [PubMed]

12. Radwan, J.; Kawalko, A.; Wojcik, J.M.; Babik, W. MHC-DRB3 variation in a free-living population of the European bison, *Bison bonasus. Mol. Ecol.* **2007**, *16*, 531–540. [CrossRef] [PubMed]

13. Babik, W.; Pabijan, M.; Radwan, J. Contrasting patterns of variation in MHC loci in the Alpine newt. *Mol. Ecol.* **2008**, *17*, 2339–2355. [CrossRef] [PubMed]

14. Loiseau, C.; Richard, M.; Garnier, S.; Chastel, O.; Julliard, R.; Zoorob, R.; Sorci, G. Diversifying selection on MHC class I in the house sparrow (*Passer domesticus*). *Mol. Ecol.* **2009**, *18*, 1331–1340. [CrossRef] [PubMed]

15. Ekblom, R.; Sæther, S.A.; Jacobsson, P.; Fiske, P.; Sahlman, T.; Grahn, M.; Kålås, J.A.; Höglund, J. Spatial pattern of MHC class II variation in the great snipe (*Gallinago media*). *Mol. Ecol.* **2007**, *16*, 1439–1451. [CrossRef] [PubMed]

16. Wang, S.; Liu, C.; Wilson, A.B.; Zhao, N.; Li, X.; Zhu, W.; Gao, X.; Liu, X.; Li, Y. Pathogen richness and abundance predict patterns of adaptive MHC variation in insular amphibians. *Mol. Ecol.* **2017**, *26*, 4671–4685. [CrossRef] [PubMed]

17. Hill, A.V.S. HLA associations with malaria in Africa: Some implications for MHC evolution. In *Molecular Evolution of the Major Histocompatibility Complex*; Klein, J., Klein, D., Eds.; Springer: Berlin/Heidelberg, Germany, 1991; pp. 403–420.

18. Alcaide, M.; Edwards, S.V.; Negro, J.J.; Serrano, D.; Tella, J.L. Extensive polymorphism and geographical variation at a positively selected MHC class II B gene of the lesser kestrel (*Falco naumanni*). *Mol. Ecol.* **2008**, *17*, 2652–2665. [CrossRef] [PubMed]

19. Burri, R.; Salamin, N.; Studer, R.A.; Roulin, A.; Fumagalli, L. Adaptive divergence of ancient gene duplicates in the avian MHC class II beta. *Mol. Biol. Evol.* **2010**, *27*, 2360–2374. [CrossRef] [PubMed]

20. Chen, L.C.; Lan, H.; Sun, L.; Deng, Y.L.; Tang, K.Y.; Wan, Q.H. Genomic organization of the crested ibis MHC provides new insight into ancestral avian MHC structure. *Sci. Rep.* **2015**, *5*, 7963. [CrossRef]

21. Gaigher, A.; Burri, R.; Gharib, W.; Taberlet, P.; Roulin, A.; Fumagalli, L. Family-assisted inference of the genetic architecture of major histocompatibility complex variation. *Mol. Ecol. Resour.* **2016**, *16*, 1353–1364. [CrossRef] [PubMed]

22. Alcaide, M.; Munoz, J.; Martínez-de la Puente, J.; Soriguer, R.; Figuerola, J. Extraordinary MHC class II B diversity in a non-passerine, wild bird: The Eurasian Coot Fulica atra (Aves: Rallidae). *Ecol. Evol.* **2014**, *4*, 688–698. [CrossRef]

23. Eimes, J.A.; Bollmer, J.L.; Whittingham, L.A.; Johnson, J.A.; Van Oosterhout, C.; Dunn, P.O. Rapid loss of MHC class II variation in a bottlenecked population is explained by drift and loss of copy number variation. *J. Evol. Biol.* **2011**, *24*, 1847–1856. [CrossRef]

24. Tsuji, H.; Taniguchi, Y.; Ishizuka, S.; Matsuda, H.; Yamada, T.; Naito, K.; Iwaisaki, H. Structure and polymorphisms of the major histocompatibility complex in the Oriental stork, *Ciconia boyciana. Sci. Rep.* **2017**, *7*, 42864. [CrossRef]

25. Hosomichi, K.; Shiina, T.; Suzuki, S.; Tanaka, M.; Shimizu, S.; Iwamoto, S.; Hara, H.; Yoshida, Y.; Kulski, J.K.; Inoko, H.; et al. The major histocompatibility complex (Mhc) class IIB region has greater genomic structural flexibility and diversity in the quail than the chicken. *BMC Genom.* **2006**, *7*, 322. [CrossRef] [PubMed]

26. Agudo, R.; Alcaide, M.; Rico, C.; Lemus, J.A.; Blanco, G.; Hiraldo, F.; Donázar, J.A. Major histocompatibility complex variation in insular populations of the Egyptian vulture: Inferences about the roles of genetic drift and selection. *Mol. Ecol.* **2011**, *20*, 2329–2340. [CrossRef] [PubMed]

27. Sin, Y.W.; Annavi, G.; Dugdale, H.L.; Newman, C.; Burke, T.; MacDonald, D.W. Pathogen burden, co-infection and major histocompatibility complex variability in the European badger (*Meles meles*). *Mol. Ecol.* **2014**, *23*, 5072–5088. [CrossRef] [PubMed]

28. The IUCN Red List of Threatened Species (Version 2018-1). Available online: www.iucnredlist.org (accessed on 12 June 2018).

29. Ding, C.Q. *Research on the Crested Ibis*; Shanghai Scientific and Technological Education Publishing: Shanghai, China, 2004; p. 388.

30. He, L.P.; Wan, Q.H.; Fang, S.G.; Xi, Y.M. Development of novel microsatellite loci and assessment of genetic diversity in the endangered Crested Ibis, *Nipponia Nippon. Conserv. Genet.* **2006**, *7*, 157–160. [CrossRef]

31. He, X.L.; Ding, C.Q.; Han, J.L. Lack of structural variation but extensive length polymorphisms and heteroplasmic length variations in the mitochondrial DNA control region of highly inbred crested ibis, *Nipponia nippon. PLoS ONE* **2013**, *8*. [CrossRef] [PubMed]

32. Feng, S.; Fang, Q.; Barnett, R.; Li, C.; Han, S.; Kuhlwilm, M.; Zhou, L.; Pan, H.; Deng, Y.; Chen, G.; et al. The genomic footprints of the fall and recovery of the crested ibis. *Curr. Biol.* **2019**, *29*, 340–349.e347. [CrossRef]

33. Fan, G.L.; Yang, Z.Q.; Gao, G.G.; Bin, J.J.; Qiao, H.L.; Huang, Z.X.; Hou, Y.F. The histopathological observations of young crested ibis infected with *Escherichia coli. Chinese. J. Zool.* **2004**, *39*, 44–46.

34. Xi, Y.M.; Wood, C.; Lu, B.Z.; Zhang, Y.M. Prevalence of a septicemia disease in the crested ibis (*Nipponia nippon*) in China. *Avian Dis.* **2007**, *51*, 614–617. [CrossRef]

35. Chen, S.L.; Hao, H.F.; Liu, Q.T.; Wang, R.; Zhang, P.; Wang, X.L.; Du, E.Q.; Yang, Z.Q. Phylogenetic and pathogenic analyses of two virulent Newcastle disease viruses isolated from Crested Ibis (*Nipponia nippon*) in China. *Virus Genes.* **2013**, *46*, 447–453. [CrossRef]

36. Zhang, X.; Qiao, J.Y.; Wu, X.M.; Ma, Q.Y.; Hu, H.; Wang, J.; Che, L.F. Ascaris spp. and Capillaria caudinflata infections in captive-bred crested ibis (*Nipponia nippon*) in China. *Zoo Biol.* **2015**, *34*, 80–84. [CrossRef] [PubMed]

37. Chang, L.; He, S.; Mao, D.; Liu, Y.; Xiong, Z.; Fu, D.; Li, B.; Wei, S.; Xu, X.; Li, S.; et al. Signatures of crested Ibis MHC revealed by recombination screening and short-reads assembly strategy. *PLoS ONE* **2016**, *11*, e0168744. [CrossRef] [PubMed]

38. Taniguchi, Y.; Matsumoto, K.; Matsuda, H.; Yamada, T.; Sugiyama, T.; Homma, K.; Kaneko, Y.; Yamagishi, S.; Iwaisaki, H. Structure and polymorphism of the major histocompatibility complex class II region in the Japanese crested ibis, *Nipponia nippon*. *PLoS ONE* **2014**, *9*, e108506. [CrossRef] [PubMed]

39. Sambrook, J.; Russell, D.W. *Molecular Cloning: A Laboratory Manual*, 3rd ed.; Cold Spring Harbor Laboratory Press: New York, NY, USA, 2001.

40. Wan, Q.H.; Zhang, P.; Ni, X.W.; Wu, H.L.; Chen, Y.Y.; Kuang, Y.Y.; Ge, Y.F.; Fang, S.G. A novel HURRAH protocol reveals high numbers of monomorphic MHC class II loci and two asymmetric multi-locus haplotypes in the Père David's deer. *PLoS ONE* **2011**, *6*, e14518. [CrossRef]

41. Zhu, Y.; Sun, D.D.; Ge, Y.F.; Yu, B.; Chen, Y.Y.; Wan, Q.H. Isolation and characterization of class I MHC genes in the giant panda (*Ailuropoda melanoleuca*). *Chin. Sci. Bull.* **2013**, *58*, 2140–2147. [CrossRef]

42. Lan, H.; Chen, H.; Chen, L.C.; Wang, B.B.; Sun, L.; Ma, M.Y.; Fang, S.G.; Wan, Q.H. The first report of a Pelecaniformes defensin cluster: Characterization of β-defensin genes in the crested ibis based on BAC libraries. *Sci. Rep.* **2014**, *4*, 6923. [CrossRef]

43. Excoffier, L.; Lischer, H.E.L. Arlequin suite ver 3.5: A new series of programs to perform population genetics analyses under Linux and Windows. *Mol. Ecol. Resour.* **2010**, *10*, 564–567. [CrossRef]

44. Stephens, M.; Donnelly, P. A comparison of bayesian methods for haplotype reconstruction from population genotype data. *Am. J. Hum. Genet.* **2003**, *73*, 1162–1169. [CrossRef]

45. Ji, Y.J.; Liu, Y.D.; Ding, C.Q.; Zhang, D.X. Eight polymorphic microsatellite loci for the critically endangered crested ibis, *Nipponia nippon* (Ciconiiformes: Threskiornithidae). *Mol. Ecol. Notes.* **2004**, *4*, 615–617. [CrossRef]

46. Van Oosterhout, C.; Hutchinson, W.F.; Wills, D.P.M.; Shipley, P. MICRO-CHECKER: Software for identifying and correcting genotyping errors in microsatellite data. *Mol. Ecol. Notes.* **2004**, *4*, 535–538. [CrossRef]

47. Raymond, M.; Rousset, F. GENEPOP (Version 1.2): Population genetics software for exact tests and ecumenicism. *J. Hered.* **1995**, *86*, 248–249. [CrossRef]

48. Tamura, K.; Stecher, G.; Peterson, D.; Filipski, A.; Kumar, S. MEGA6: Molecular evolutionary genetics analysis version 6.0. *Mol. Biol. Evol.* **2013**, *30*, 2725–2729. [CrossRef] [PubMed]

49. Wilson, D.J.; McVean, G. Estimating diversifying selection and functional constraint in the presence of recombination. *Genetics* **2006**, *172*, 1411–1425. [CrossRef]

50. Martin, D.P.; Lemey, P.; Lott, M.; Moulton, V.; Posada, D.; Lefeuvre, P. RDP3: A flexible and fast computer program for analyzing recombination. *Bioinformatics* **2010**, *26*, 2462–2463. [CrossRef]

51. Goudet, J. FSTAT, a Program to Estimate and Test Gene Diversities and Fixation Indices (ver 2.9.3.2). Available online: http://www2.unil.ch/popgen/softwares/fstat.htm (accessed on 1 February 2018).

52. Piry, S.; Luikart, G.; Cornuet, J.M. Bottleneck: A computer program for detecting recent reductions in the effective population size using allele frequency data. *J. Hered.* **1999**, *90*, 502–503. [CrossRef]

53. Pritchard, J.K.; Stephens, M.; Donnelly, P. Inference of population structure using multilocus genotype data. *Genetics* **2000**, *155*, 945–959. [PubMed]

54. Earl, D.A.; von Holdt, B.M. STRUCTURE HARVESTER: A website and program for visualizing STRUCTURE output and implementing the Evanno method. *Conserv. Genet. Resour.* **2012**, *4*, 359–361. [CrossRef]

55. Rosenberg, N.A. Distruct: A program for the graphical display of population structure. *Mol. Ecol. Notes.* **2004**, *4*, 137–138. [CrossRef]

56. Langella, O. POPULATIONS (version 1.2.31). Available online: http://bioinformatics.org/populations/ (accessed on 3 May 2018).

57. Balloux, F.; Goudet, J. Statistical properties of population differentiation estimators under stepwise mutation in a finite island model. *Mol. Ecol.* **2002**, *11*, 771–783. [CrossRef]

58. Hardy, O.J.; Vekemans, X. SPAGeDi: A versatile computer program to analyse spatial genetic structure at the individual or population levels. *Mol. Ecol. Notes.* **2002**, *2*, 618–620. [CrossRef]

59. Hedrick, P.W. A standardized genetic differentiation measure. *Evolution* **2005**, *59*, 1633–1638. [CrossRef] [PubMed]

60. Jardetzky, T.S.; Brown, J.H.; Gorga, J.C.; Stern, L.J.; Urban, R.; Chi, Y.I.; Stauffacher, C.; Strominger, J.L.; Wiley, D.C. Three-dimensional structure of the human class II histocompatibility antigen HLA-DR1. *Nature* **1994**, *368*, 711–718. [CrossRef]

61. Stern, L.J.; Brown, J.H.; Jardetzky, T.S.; Gorga, J.C.; Urban, R.G.; Strominger, J.L.; Wiley, D.C. Crystal structure of the human class II MHC protein HLA-DR1 complexed with an influenza virus peptide. *Nature* **1994**, *368*, 215–221. [CrossRef] [PubMed]

62. Hastings, P.J.; Lupski, J.R.; Rosenberg, S.M.; Ira, G. Mechanisms of change in gene copy number. *Nat. Rev. Genet.* **2009**, *10*, 551–564. [CrossRef] [PubMed]

63. Minias, P.; Pikus, E.; Whittingham, L.A.; Dunn, P.O. Evolution of copy number at the MHC varies across the avian tree of life. *Genome Biol Evol.* **2018**, *11*, 17–28. [CrossRef]

64. Nei, M.; Gu, X.; Sitnikova, T. Evolution by the birth-and-death process in multigene families of the vertebrate immune system. *Proc. Natl. Acad. Sci. USA.* **1997**, *94*, 7799–7806. [CrossRef]

65. Schrider, D.R.; Hahn, M.W. Gene copy-number polymorphism in nature. *Proc. R. Soc. B: Biol. Sci.* **2010**, *277*, 3213–3221. [CrossRef]

66. Ploshnitsa, A.I.; Goltsman, M.E.; Macdonald, D.W.; Kennedy, L.J.; Sommer, S. Impact of historical founder effects and a recent bottleneck on MHC variability in Commander Arctic foxes (*Vulpes lagopus*). *Ecol. Evol.* **2012**, *2*, 165–180. [CrossRef] [PubMed]

67. Ejsmond, M.J.; Radwan, J. MHC diversity in bottlenecked populations: A simulation model. *Conserv. Genet.* **2011**, *12*, 129–137. [CrossRef]

68. Aguilar, A.; Roemer, G.; Debenham, S.; Binns, M.; Garcelon, D.; Wayne, R.K. High MHC diversity maintained by balancing selection in an otherwise genetically monomorphic mammal. *Proc. Natl. Acad. Sci. USA* **2004**, *101*, 3490–3494. [CrossRef]

69. Miller, K.A.; Nelson, N.J.; Smith, H.G.; Moore, J.A. How do reproductive skew and founder group size affect genetic diversity in reintroduced populations? *Mol. Ecol.* **2009**, *18*, 3792–3802. [CrossRef] [PubMed]

70. Shen, F.; Zhang, Z.; He, W.; Yue, B.; Zhang, A.; Zhang, L.; Hou, R.; Wang, C.; Watanabe, T. Microsatellite variability reveals the necessity for genetic input from wild giant pandas (*Ailuropoda melanoleuca*) into the captive population. *Mol. Ecol.* **2009**, *18*, 1061–1070. [CrossRef] [PubMed]

71. Winton, C.L.; Plante, Y.; Hind, P.; McMahon, R.; Hegarty, M.J.; McEwan, N.R.; Davies-Morel, M.C.; Morgan, C.M.; Powell, W.; Nash, D.M. Comparative genetic diversity in a sample of pony breeds from the UK and North America: A case study in the conservation of global genetic resources. *Ecol. Evol.* **2015**, *5*, 3507–3522. [CrossRef]

72. Zhang, Q.; Ji, Y.; Zeng, Z. Influence of the founder effect on genetic diversity of translocated populations: An example from Hainan Eld's Deer. *Chinese. J. Zool.* **2007**, *42*, 54.

73. Vonholdt, B.M.; Stahler, D.R.; Bangs, E.E.; Smith, D.W.; Jimenez, M.D.; Mack, C.M.; Niemeyer, C.C.; Pollinger, J.P.; Wayne, R.K. A novel assessment of population structure and gene flow in grey wolf populations of the Northern Rocky Mountains of the United States. *Mol. Ecol.* **2010**, *19*, 4412–4427. [CrossRef] [PubMed]

74. Hinkson, K.M.; Henry, N.L.; Hensley, N.M.; Richter, S.C. Initial founders of captive populations are genetically representative of natural populations in critically endangered dusky gopher frogs, *Lithobates sevosus*. *Zoo Biol.* **2016**, *35*, 378–384. [CrossRef] [PubMed]

75. Li, S.; Li, B.; Cheng, C.; Xiong, Z.; Liu, Q.; Lai, J.; Carey, H.V.; Zhang, Q.; Zheng, H.; Wei, S. Genomic signatures of near-extinction and rebirth of the crested ibis and other endangered bird species. *Genome. Biol.* **2014**, *15*, 557. [CrossRef]

76. Taylor, R.L., Jr. Major histocompatibility (B) complex control of responses against Rous sarcomas. *Poult. Sci.* **2004**, *83*, 638–649. [CrossRef] [PubMed]

77. Banat, G.R.; Tkalcic, S.; Dzielawa, J.A.; Jackwood, M.W.; Saggese, M.D.; Yates, L.; Kopulos, R.; Briles, W.E.; Collisson, E.W. Association of the chicken MHC B haplotypes with resistance to avian coronavirus. *Dev. Comp. Immunol.* **2013**, *39*, 430–437. [CrossRef] [PubMed]

78. Cotter, P.F.; Taylor, R.L., Jr.; Abplanalp, H. B-complex associated immunity to Salmonella enteritidis challenge in congenic chickens. *Poult. Sci.* **1998**, *77*, 1846–1851. [CrossRef]

79. Deng, Y.L. Paternity Testing and Correlation Analysis Between MHC and Offspring Mortality of Deqing Captive Crested Ibises. Master's Thesis, Zhejiang University, Zhejiang, China, 2015.

Article

Reproductive Strategy Inferred from Major Histocompatibility Complex-Based Inter-Individual, Sperm-Egg, and Mother-Fetus Recognitions in Giant Pandas (*Ailuropoda melanoleuca*)

Ying Zhu [1,†], Qiu-Hong Wan [1,†], He-Min Zhang [2] and Sheng-Guo Fang [1,*]

[1]   MOE Key Laboratory of Biosystems Homeostasis & Protection, State Conservation Centre for Gene
     Resources of Endangered Wildlife, College of Life Sciences, Zhejiang University, Hangzhou 310058, China;
     so_zy2003@126.com (Y.Z.); qiuhongwan@zju.edu.cn (Q.-H.W.)
[2]   China Conservation and Research Center for the Giant Panda, No. 98 Tongjiang Road,
     Dujiangyan 611800, China; panda_zhanghm@163.com
[*]   Correspondence: sgfanglab@zju.edu.cn; Tel.: +86-571-88206472
[†]   These authors contributed equally to this work.

Received: 13 December 2018; Accepted: 13 March 2019; Published: 19 March 2019

Abstract: Few major histocompatibility complex (MHC)-based mate choice studies include all MHC genes at the inter-individual, sperm-egg, and mother-fetus recognition levels. We tested three hypotheses of female mate choice in a 17-year study of the giant panda (*Ailuropoda melanoleuca*) while using ten functional MHC loci (four MHC class I loci: *Aime*-C, *Aime*-F, *Aime*-I, and *Aime*-L; six MHC class II loci: *Aime*-DRA, *Aime*-DRB3, *Aime*-DQA1, *Aime*-DQA2, *Aime*-DQB1, and *Aime*-DQB2); five super haplotypes (SuHa, SuHaI, SuHaII, DQ, and DR); and, seven microsatellites. We found female choice for heterozygosity at *Aime*-C, *Aime*-I, and DQ and for disassortative mate choice at *Aime*-C, DQ, and DR at the inter-individual recognition level. High mating success occurred in MHC-dissimilar mating pairs. No significant results were found based on any microsatellite parameters, suggesting that MHCs were the mate choice target and there were no signs of inbreeding avoidance. Our results indicate *Aime*-DQA1- and *Aime*-DQA2-associated disassortative selection at the sperm-egg recognition level and a possible *Aime*-C- and *Aime*-I-associated assortative maternal immune tolerance mechanism. The MHC genes were of differential importance at the different recognition levels, so all of the functional MHC genes should be included when studying MHC-dependent reproductive mechanisms.

Keywords: MHC-I- and MHC-II-dependent inter-individual recognition; MHC-II-associated sperm-egg recognition; MHC-I-based mother-fetus recognition; giant panda; long-fragment super haplotype

1. Introduction

Understanding the genetic basis and the driving forces of mate choice in animals has always been a major goal of evolutionary ecologists [1,2]. It has been proposed that females prefer males who can maximize their reproductive success and increase offspring quality/fitness [3,4]. Immunocompetence is undoubtedly an essential index of an individual's fitness [5,6] and major histocompatibility complex (MHC) genes are suitable candidates in investigating the genetic basis underlying mate choice decisions, as MHC molecules can recognize and present antigens to T cells and trigger immune reactions [7–10]. A growing body of evidence shows the association between pathogen resistance and MHC haplotypes or alleles [11–14].

In recent years, an increasing number of studies have focused on MHC-associated mate choice, including three non-exclusive hypotheses that could explain the MHC-based mate choice. Firstly, heterozygous advantage: according to this hypothesis, the choosy sex could obtain additive benefits from mating with MHC-heterozygous mates whose disease resistance might be inherited by their offspring [15,16]. For example, in tuco-tucos (*Ctenomys* spp.), females prefer MHC-heterozygous males [17]. In the scarlet rosefinch (*Carpodacus erythrinus*), social males with low MHC heterozygosity are cheated on by their females more frequently than highly MHC-heterozygous males [18]. Secondly, the genetic compatibility hypothesis: the choosy sex is assumed to select MHC-dissimilar partners, resulting in the production of offspring with diverse genotypes that can recognize a broad array of pathogens and hence increase their fitness [15,16,19]. In the grey mouse lemur (*Microcebus murinus*), the fathers have lower allele sharing and a greater amino acid distance to the mother than the randomly assigned males [20]. A study on blue petrels (*Halobaena caerulea*) revealed that females mated more with functionally (not evolutionary) MHC-dissimilar males than with random males [21]. Similar patterns have been observed in great frigatebirds (*Fregata minor*) [22], the Chinese rose bitterling (*Rhodeus ocellatus*) [23], the pot-bellied seahorse (*Hippocampus abdominalis*) [24], the fat-tailed dwarf lemur (*Cheirogaleus medius*) [25], and mandrills (*Mandrillus* spp.) [26]. Furthermore, to acquire the best immunogenetic composition for their offspring, females would choose mates with the most appropriate MHC diversity (maximum or intermediate), which is also referred to as the optimal hypothesis, and it is an extension of the genetic compatibility hypothesis [10]. Studies in sticklebacks (family Gasterosteidae) have found that females with many alleles prefer males with few alleles, and vice versa, in order to obtain an optimal level of MHC diversity in their offspring for resistance against parasites and pathogens [27,28]. Thirdly, the inbreeding avoidance hypothesis: the choosy sex is expected to seek dissimilar partners, not only with regard to MHC, but also genome-wide, in order to gain fitness benefits [29,30].

Mate choice results from non-random reproductive investment, which could happen at the precopulatory stage or/and at the postcopulatory stage. In the precopulatory stage (or individual recognition level), the choosy sex uses visual, acoustic, or odor cues to choose mates [15]. In the postcopulatory stage, the females increase offspring quality/fitness by their eggs differentiating between sperm during fertilization (sperm-egg recognition), and by following a differential allocation strategy during embryo implantation (mother-fetus recognition) [16]. Human leukocyte antigen (HLA) class I and class II molecules are expressed on sperm cell surfaces [31–33], making the complementary cryptic female choice for MHC genotypes possible at the gamete level. Sperm selection that targets different levels of MHC diversity has been reported in several species. In some rodent species, females refuse to accept MHC-similar sperm [34], and in humans, females that mate with a male who has the same HLA haplotype tend to have a greater chance of spontaneous abortion [35–37]. In red junglefowl (*Gallus gallus*), the eggs favor sperm that is from MHC-dissimilar males [38]. In contrast, a fertilization advantage for MHC-similar mates has been observed in Atlantic salmon (*Salmo salar*) [39], Chinook salmon (*Oncorhynchus tshawytscha*) [40], and guppies (*Poecilia reticulata*) [41]. A recent study of the three-spined stickleback (*Gasterosteus aculeatus*) revealed that sperm selection resulted in offspring with an intermediate level of MHC diversity [42].

The giant panda (*Ailuropoda melanoleuca*) is an endangered species and its captive breeding has always been focused in China. Giant pandas usually have polyandrous/polygynous multiple mating systems [43]. Females are very choosy, and they have significantly higher copulation and birth rates when paired with preferred males [44]. In breeding programs, natural mating as well as artificial insemination are adopted to increase reproductive success [45]. Although the number of female giant pandas that are available to breed has increased, the fertilization rate is still low (~50% for nearly 20 years). Furthermore, female giant pandas have spontaneous abortions when they are inseminated by sperm from a male that they do not like [46]. Therefore, there may be an MHC-based mechanism that determines mate choice and fertilization in giant pandas.

It is best to investigate all MHC genes concerning mate choice, as MHC class I and class II molecules mainly present intracellular and extracellular pathogen-derived antigens, respectively [9], and the selection of MHC genes might influence mate choice results [47]. Few studies have focused on a large region of the MHC due to a lack of structural knowledge and an effective genotyping method [48], including a giant panda mate choice study that only used three MHC class II genes [49]. In previous studies, we characterized six functional MHC class II genes (*Aime*-DRA, *Aime*-DRB3, *Aime*-DQA1, *Aime*-DQA2, *Aime*-DQB1, and *Aime*-DQB2) [50,51] and four classical MHC class I genes (*Aime*-C, *Aime*-F, *Aime*-I, and *Aime*-L) [52], and we developed their genotyping protocols in the giant panda. Therefore, our previous studies provide a good foundation to investigate the relationship between a large number of MHC genes and female choice at the inter-individual, sperm-egg, and mother-fetus recognition levels in giant pandas.

In the present study, we took advantage of multiple years of observations and the possession of genetic data of a captive population in the Wolong Chinese Research and Conservation Center for the Giant Panda. We aimed to: (1) test three mate choice hypotheses at the individual recognition level (MHC-heterozygous choice, MHC-compatibility choice, and inbreeding avoidance) while using 10 MHC genes and seven microsatellites. If the heterozygote advantage is the main driving force of female mate choice, more diverse males should be preferred than less diverse ones, regardless of the females' MHC genotypes. If female choice favors the production of offspring with high fitness that is based on MHC compatibility, we expect that females would choose MHC-dissimilar partners that provide the most appropriate level of MHC dissimilarity (maximum or intermediate) in the offspring. If mate choice aims to avoid inbreeding, then we expect genome-wide dissimilarity between partners (MHC and microsatellites); (2) test whether cryptic female choice for MHC compatibility occurs at the gamete level; and, (3) characterize successful embryo implantations by comparing the MHC genotypes of mothers and their offspring.

2. Materials and Methods

2.1. Study Species and Behavioral Observations of Inter-Individual Recognition

Giant pandas were housed in the Wolong Chinese Research and Conservation Center for the Giant Panda. Female giant pandas come into estrus from February to June [43], when their appetite decreases, males' urination frequency increases, and they rub their genitalia against the ground or walls [43,46]. Two rutting females and males are usually put into two adjacent cages. Neighboring pandas have access to each other through a cage fence with full sight, hearing, and smell, but limited touch. Female pandas face multiple males that are consecutively presented. Veterinarians use pre-mating behaviors to determine when the males are sent to females for mating. If a female is interested in a male, then they usually respond by sniffing and pushing the fence, before the male is sent to the female's cage to copulate. If a female is uninterested, then she does not respond or behaves aggressively. Female mate choice at the inter-individual recognition level was determined by naturally analyzing mating pairs that were recorded in a studbook [53]. There were 33 females and 21 males in the breeding program from 1991 and 2008, except for 1994, but we failed to obtain samples from five females and four males. Therefore, for the inter-individual recognition study, we included 182 natural mating events that involved 28 females and 17 males.

2.2. Sperm-Egg and Mother-Fetus Recognition

Female giant pandas have an annual estrus cycle with spontaneous ovulation [54,55]. They naturally mate with multiple males and require artificial insemination with males that they mate with and with males that they do not mate with but have good-quality sperm. Sperm from all males have a chance to access the egg, but only one of them is successful through a mechanism at the gamete stage, and sperm-egg recognition may occur at the gamete stage. Over 17 years, a total of 80 offspring (zygotes observed) were produced. Other zygotes were combinations of egg-to-sperm haplotypes, except for offspring haplotypes. The information that is required for natural mating and artificial insemination was acquired from the studbook and SPARKS 1.5 [53].

In addition, we compared the difference between the zygotes observed and other zygotes (combinations of egg-to-sperm haplotypes, except for offspring haplotypes) to mothers in successful embryo implantation events, and called it the "mother-fetus recognition level".

2.3. DNA Extraction, MHC and Microsatellite Genotyping, and the Definition of a Super Haplotype

We collected 110 blood and the fecal samples. Blood samples were obtained during a routine medical examination and preserved in liquid nitrogen. Fecal samples were less than two days old and stored in 95% ethanol. Genomic DNA extraction from the blood and fecal samples was conducted, as described by Wan et al. [56].

We used seven microsatellite loci (*Aim-3*, *Aim-5*, *Aim-10*, *Aim-11*, *Aim-13*, *Aim-14*, and *Aim-16*) that performed well in a previous paternity test [57] to conduct paternity analysis for 113 individuals. Polymerase chain reaction (PCR) amplification and genotyping mirrored that described by Zhang et al. [58].

We used 10 functional *Aime*-MHC loci, including four class I loci (*Aime*-C, *Aime*-F, *Aime*-I, and *Aime*-L) and six class II loci (*Aime*-DRA, *Aime*-DRB3, *Aime*-DQA1, *Aime*-DQA2, *Aime*-DQB1, and *Aime*-DQB2). We genotyped all individuals at the polymorphic MHC class I exon 2–3 regions and MHC class II exon 2. The primer sets, PCR amplification, and genotyping were as described in two previous studies [52,59].

A physical MHC map revealed that as well as four MHC class I genes, there were also six MHC class II genes linked together [52,60]. In addition to using the above-mentioned independent MHC loci, we analyzed the allele linkage relationships of four MHC class I genes, six MHC class II genes, all MHC genes, genes in the DQ region, and genes in the DR region. For example, we identified homozygotes in four MHC class I genes of the offspring and then inferred the linkage relationship between the mother and father according to Mendel's law. We named the linkage of four MHC class I genes as SuperHaplotypeI (SuHaI). Using the same procedure, we named the linkage relationships between six MHC class II genes and all MHC genes SuHaII and SuHa, respectively. Genes in the DQ and DR regions were simply called DQ and DR, respectively.

2.4. Paternity Test

We performed a paternity analysis with seven microsatellite loci while using an exclusive method that was based on Mendel's law, and MHC genotype data confirmed the results. Information on mothers was obtained from the studbook [53].

2.5. Data Analysis

2.5.1. Female Mate Choice at the Inter-Individual Recognition Level

Three mate choice hypotheses, heterozygote advantage, genetic compatibility, and inbreeding avoidance were tested at the inter-individual recognition level.

We tested whether females prefer heterozygous males by utilizing three parameters: (1) The number of heterozygotes (H_{obs}) at MHC loci in males. We used 0 and 1 to represent homozygote and heterozygote, respectively; (2) Multilocus heterozygosity (MLH) in males, i.e., the proportion of heterozygous microsatellite loci that accounted for the total number of microsatellite loci; (3) d^2 value (microsatellites), i.e., the genetic distance between the two alleles. We calculated the d^2 for each individual according to the formula $d^2 = 1/n \, \Sigma^n \, (a_i - a_j)^2$, where a_i and a_j refer to the repeat length of two individual alleles [61].

We tested whether female giant pandas prefer MHC-dissimilar or -similar males (choice for genetic compatibility) with three parameters: (1) The number of MHC alleles that is shared between females and males as "Nas", which is twice the number of alleles that is shared by females and males divided by the total number of females and males [62]; (2) The pairwise functional amino acid distance between MHC alleles of females and males, calculated as "Faadis" = $D_{ab} + D_{aB} + D_{AB} + D_{ab}$, where A, a, B, and b are four alleles in two mates [63]. Each amino acid was characterized by five physicochemical variables: z1 (hydrophobicity), z2 (steric bulk), z3 (polarity), and z4 and z5 (electronic effects) [64]. The functional amino acid distance was the Euclidean metric of the two vectors, which consists of two alleles from a female and male [65]. We not only considered the distance at all sites, but also at the antigen binding sites (ABSs), as it is ABSs that determine pathogen binding and recognition and they are considered a functional region [14,66,67]. We defined ABSs according to human sequences [68].

We tested the inbreeding avoidance hypothesis using Queller and Goodnight's relatedness (microsatellites) [69], as well as Nas and Faadis. Queller and Goodnight's relatedness was calculated in SPAGeDi v 1.4 [70] to estimate the genetic similarity between female and male giant pandas. If mate choice aims to avoid inbreeding, we would expect to see choice that is based on compatibility at the genome level (MHC and microsatellites), or female mate choice was targeted to MHCs.

We adopted three approaches to test each hypothesis. Firstly, we performed a randomization test, which is a nonparametric approach that is based on Monte Carlo sampling, to test whether female giant pandas randomly choose their partners for mating. We simulated the natural mating scenario (same accessible males for each female in the respective year), with the exception that females randomly chose males. We generated a null distribution by allowing each female to randomly choose 10,000 times between all males in the respective year. We then compared the mean of the values that were obtained from 182 naturally mated pairs or males involved in natural mating with a set of randomly matched pairs. We calculated the exact p values as twice the proportion of the simulations, which gave higher values than those observed. If females preferred heterozygous males, we expected that the mean values (H_{obs}, MLH, and d^2) of males that were naturally mated would be significantly higher than those of the randomly assigned males. If females preferred MHC-dissimilar males, we expected that the mean Nas of naturally mated pairs would be significantly lower than that of the randomly assigned pairs, or that the Faadis of naturally mated pairs would be significantly higher than that of the randomly assigned pairs. If females aim to avoid inbreeding, then we expected that the mean values (Faadis and relatedness) would be higher than those of randomly assigned pairs. The randomization tests were conducted in ResamplingStats v 4.0 (ResamplingStats Inc., Arlington, TX, USA).

Secondly, we used a paired Student's t-test (or Wilcoxon test if the model assumptions were not met) to test whether the natural mating group exhibited higher heterozygosity and/or greater MHC divergence than non-mating group by comparing the mean values of males that are involved in natural mating and those of males that were not involved in mating. H_{obs}, MLH, and d^2 were used to test the heterozygote advantage hypothesis, and Nas, Faadis, and relatedness were used to test the compatibility and inbreeding avoidance hypotheses.

We employed a model-based approach to explore the association between genetic variables and the probability of natural mating success. We performed a generalized linear mixed model (GLMM) with a binomial distribution and logic link function for 956 paring events, and the dependent variable was coded as 1 for a natural mating event and 0 for those that did not. The MHC genetic variables (H_{obs}, Nas, and Faadis) were included as the fixed effects and female ID (to avoid female pseudoreplication), male ID, and year (to account for differences among breeding years) were included as random effects.

2.5.2. Sperm-Egg Recognition Level

We used a randomization test, a paired Student's t-test, and a GLMM to test the MHC compatibility hypothesis at the sperm-egg recognition level.

Randomization was based on the data obtained (same males for fertilization or natural mating in the respective year), with the exception that the eggs randomly chose sperm. We generated a null distribution by allowing each egg to randomly choose 10,000 times between all of the sperm of the respective year. Subsequently, we compared the mean MHC divergence from the zygotes that were observed (Faadis = D_{AB}, where A and B are two haplotypes in two gametes, N = 65) with a distribution of egg-to-sperm divergence from randomly assigned zygotes. The exact p values were calculated, as described above.

Paired Student t-tests (or Wilcoxon tests if the model assumptions were not met) were used to test whether the observed combination of eggs and sperm indicated greater MHC divergence by comparing the mean values of the zygotes observed (offspring) with other zygotes.

We used a GLMM with a binomial distribution and logic link to test whether MHC divergence between eggs and sperm influenced breeding success (successful births). The dependent variable was coded as 1 for a breeding event and 0 for a non-breeding event. Egg-to-sperm MHC divergence (Faadis) was included as a fixed effect and female ID, male ID, and year were included as the random effects.

The paired Student t-tests, Wilcoxon tests, and GLMMs were conducted in SPSS v. 20.0 (SPSS Inc., Chicago, IL, USA).

2.5.3. Mother-Fetus Recognition Level

We used paired Student t-tests or Wilcoxon tests, as appropriate, to test whether the zygotes observed (offspring) had lower MHC divergence (Faadis) from the mother than other zygotes (combinations of egg-to-sperm haplotypes, except for offspring haplotypes) by comparing the mean Faadis of zygotes observed to the mother with other zygotes to the mother.

A GLMM with a binomial distribution and logic link was used to explore the association between genetic variables and the probability of breeding success. The dependent variable was coded as 1 for a breeding event and 0 for a non-breeding event. The MHC divergence of zygotes from the mother was included as a fixed effect, and the random effects mirrored those described above.

We adjusted the p values with a false discovery rate of 20%, following the Benjamini–Hochberg procedure for multiple testing [71,72].

2.6. Declarations Ethics Statement

All blood and fecal samples were collected from captive pandas that were housed in the Wolong Chinese Research and Conservation Center for the Giant Panda. Blood samples were collected during routine examinations with permission from the China Giant Panda Protection and Management Office. We obtained specific permission from the China Research and Conservation Center for the Giant Panda to take fecal samples from captive individuals during the non-breeding season. Permission to use the samples was given by the State Conservation Center for Gene Resources of Endangered Wildlife of China, where they were deposited.

3. Results

3.1. Microsatellite and MHC Diversity

The male MLH ranged from 0.286 to 1 (median = 0.714), d^2 ranged from 0.714 to 41.571 (median = 25.857), and Queller and Goodnight's relatedness ranged from -0.484 to 0.829 (median = -0.070).

A total of 47 MHC sequences were isolated, which are the same as the published sequences, and included 22 MHC class I alleles and 25 MHC class II alleles: seven *Aime*-C (*Aime*-C*01–08, JX987000–JX987005, and JX987007), one *Aime*-F (*Aime*-F*01 and JX9870008), seven *Aime*-I (*Aime*-I*01–07 and JX987009–JX987015), seven *Aime*-L (*Aime*-L*01–06 and JX987016–JX987021), seven DQA1 (DQA1*01–07), three DQA2 (DQA2*01–03), 6 DQB1 (DQB1*01–06), seven DRB3 (DRB3*01–05, 07–08), one DRA, and one DQB2. We obtained only one allele at *Aime*-F, *Aime*-DRA, and *Aime*-DQB2, so we excluded these three loci from the analysis.

Furthermore, we obtained 36 SuHa, 20 SuHaI, 18 SuHaII, 18 DQ, and seven DR. The linkage relationships of SuHaII, SuHaI, SuHa, and DQ are presented in Figures S1–S4, respectively. The number of super haplotypes in the DR region was the same as the allele number at DRB3, as there were only two loci in the DR region, with DRA being a homozygote.

The mean number of variable amino acid sites in MHC class II genes was higher than that in MHC class I genes with respect to the whole exon and all of the ABS sites (12.8% vs. 11.5%, respectively, and 41.0% vs. 25.7%, respectively). This was found by comparing the variable amino acid sites in SuHaI and SuHaII (Table S1). However, the mean difference between pairwise alleles in MHC class I genes was greater than that in MHC class II genes (Table S1), as was the case for SuHaI and SuHaII. Moreover, the number of variable amino acid sites and the difference between the pairwise alleles in the DR region were both greater than those in the DQ region.

3.2. Inter-Individual Recognition

The number of males that were accessible to females ranged from three to 11 during the 17-year period, and females naturally mated with 1–4 males.

3.2.1. Heterozygosity Advantage

Males Involved in Natural Mating Versus Randomly Assigned Males

Concerning super haplotypes, males that were involved in natural mating had significantly more heterozygotes at SuHa, SuHaII, and DQ than the randomly assigned males (P_{SuHa} = 0.009, P_{SuHaII} = 0.011, and P_{DQ} = 0.001; Figure 1a). Regarding individual loci, males that are involved in natural mating had more heterozygotes at *Aime*-C, *Aime*-I, and *Aime*-DQB1 than randomly assigned males (P_C = 0.020, P_I = 0.000, and P_{DQB1} = 0.000; Figure 1b). However, there were no significant differences in heterozygosity at any other super haplotype or locus between the males that naturally mated and randomly assigned males (Figure 1). Furthermore, we found no significant difference in MLH or d^2 between males that naturally mated and those that were randomly assigned (Table 1).

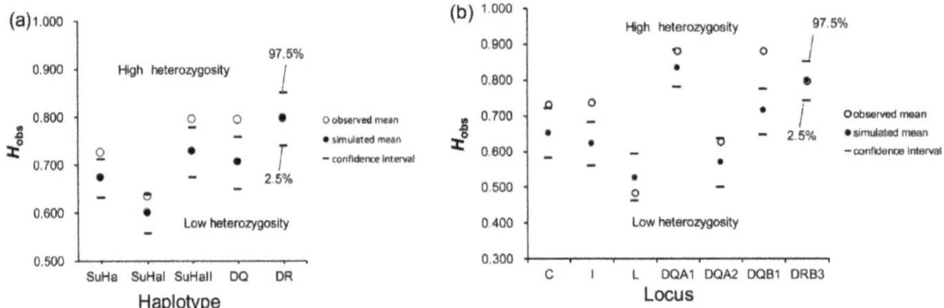

Figure 1. Mean individual major histocompatibility complex heterozygosity (H_{obs}) of males involved in natural mating and randomly assigned males. (**a**) five super haplotypes; and, (**b**) seven individual loci. Two-tailed 95% confidence intervals are indicated by black lines.

Table 1. Microsatellite results from randomization and paired Student *t*-tests.

		Compatibility Test	Heterozygote Advantage Test	
		Relatedness	MLH	d^2
Randomization	Observed	−0.087	0.751	25.961
Test [a]	Simulated mean	−0.065	0.744	27.491
	95% CI	[−0.090, −0.040]	[0.713, 0.775]	[24.319, 30.648]
	P	0.079	0.651	0.345
Paired test [b]	Potential fathers	−0.093	0.782	25.157
	Other males	−0.061	0.741	27.480
	Statistics	$t = 1.592$	$Z = 0.556$	$t = 0.913$
	P	0.115	0.578	0.364

[a] Comparison between the natural mating group and the randomly assigned group. [b] Comparison between the natural mating group and the non-mating group.

Males Involved in Natural Mating Versus Natural Non-Mating Males

Males that were involved in natural mating had a higher proportion of heterozygotes at SuHa, SuHaI, SuHaII, and SuHaDQ than those natural non-mating males (males not involved in mating, $P_{SuHa} = P_{SuHaI} = P_{SuHaII} = P_{DQ} = 0.000$; Figure 2a). This pattern was found at all loci, except *Aime*-L and *Aime*-DRB ($P_C = 0.000$, $P_I = 0.000$, $P_{DQA1} = 0.001$, $P_{DQA2} = 0.007$, and $P_{DQB1} = 0.000$; Figure 2b). We found no significant differences in MLH or d^2 between males that were involved in natural mating and other males (Table 1).

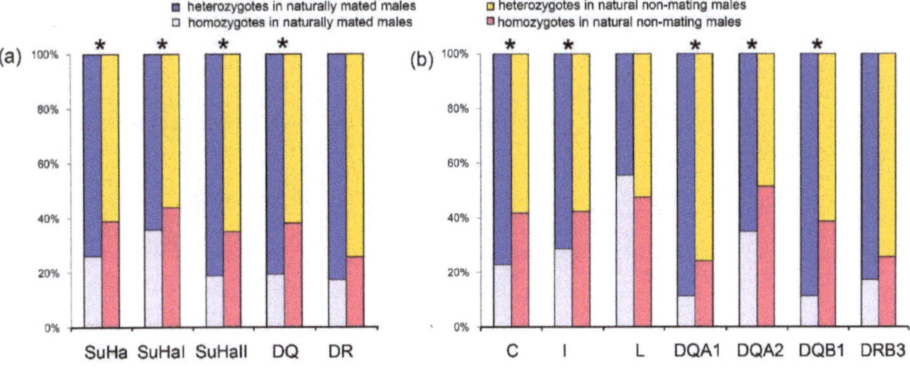

Figure 2. Proportions of heterozygotes and homozygotes in males involved in natural mating and natural non-mating males. (**a**) five super haplotypes; and, (**b**) seven individual loci. Asterisk shows results with *p* values that are smaller than 0.05.

Relationship between Male MHC Heterozygosity and Natural Mating Success

Being encouraged by the results above, we combined seven polymorphic MHC loci (DQA1, DQA2, DQB1, DRB3, *Aime*-C, *Aime*-I, and *Aime*-L) to model the relationship between male MHC heterozygosity and natural mating success, and found a nonsignificant relationship between overall male MHC heterozygosity and natural mating success (F = 2.963, *df* = 954, P = 0.86). Female ID, male ID, and year did not have any effect on natural mating success.

3.2.2. Genetic Compatibility and Inbreeding Avoidance

Naturally Mated Pairs versus Randomly Assigned Pairs

There was a significant difference in Nas at SuHa, SuHaII, DQ, and DR between naturally mated pairs and randomly assigned pairs ($P_{SuHa} = 0.006$, $P_{SuHaII} = 0.000$, $P_{DQ} = 0.000$, and P_{DR}

= 0.000; Figure 3a). Males that were involved in natural mating had significantly lower Nas at *Aime*-DQA1, DQA2, DQB1, and DRB3 than the randomly assigned pairs (P_{DQA1} = 0.038, P_{DQA2} = 0.001, P_{DQB1} = 0.000, and P_{DRB3} = 0.000; Figure 3b). However, we did not find any significant differences in Nas at SuHaI or other loci (Figure 3).

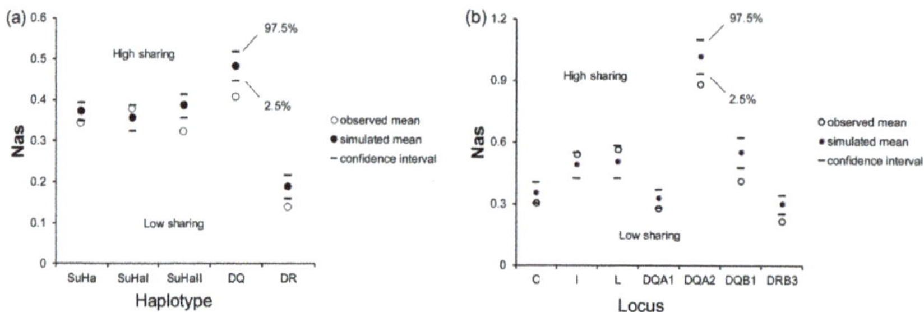

Figure 3. Mean allele sharing values (Nas) in females and males involved in natural mating and in females and randomly assigned males. (**a**) five super haplotypes; and, (**b**) seven individual loci. Two-tailed 95% confidence intervals are indicated by black lines.

Naturally mated pairs had significantly higher Faadis at SuHa, SuHaII, DQ, DR, *Aime*-C, DQA1, DQA2, DQB1, and DRB3 than randomly assigned pairs with respect to all sites and the ABS sites (Table 2). However, we did not find a significant difference in Faadis at SuHaI, *Aime*-I, or *Aime*-L between the pairs with respect to all sites and ABS sites (Table 2). In addition, there was no significant difference between the naturally mated pairs and randomly assigned pairs in genetic relatedness at microsatellites (Table 1).

Table 2. Functional amino acid distances for female genetic compatibility between the observed pairs and randomly assigned pairs.

Locus	Region	Simulated Mean [95% CI]	Observe Mean	*P*
SuHa	ABS	123.873 [121.759, 125.899]	127.832	**0.001**
	ALL	136.412 [134.088, 138.677]	140.728	**0.000**
SuHaI	ABS	93.169 [90.939, 95.359]	94.326	0.297
	ALL	101.310 [98.883, 103.635]	102.285	0.416
SuHaII	ABS	73.374 [71.216, 75.579]	78.033	**0.000**
	ALL	82.600 [80.143, 85.028]	87.462	**0.000**
DQ	ABS	53.815 [51.715, 55.876]	58.000	**0.000**
	ALL	57.319 [55.059, 59.577]	61.433	**0.001**
DR	ABS	45.051 [43.434, 46.662]	47.498	**0.002**
	ALL	53.531 [51.588, 55.516]	56.510	**0.003**
C	ABS	60.340 [58.005, 62.674]	62.840	**0.036**
	ALL	63.839 [61.359, 66.265]	66.431	**0.038**
I	ABS	46.787 [44.406, 49.218]	45.218	0.196
	ALL	50.388 [47.956, 52.819]	48.681	0.172
L	ABS	33.315 [31.085, 35.588]	32.624	0.548
	ALL	39.862 [37.267, 42.449]	38.584	0.340
DQA1	ABS	32.307 [30.740, 33.938]	35.587	**0.000**
	ALL	35.121 [30.740,33.938]	38.400	**0.000**
DQA2	ABS	11.596 [10.674, 12.542]	13.076	**0.001**
	ALL	11.603 [10.674, 12.542]	13.076	**0.001**
DQB1	ABS	32.821 [30.624, 35.005]	36.002	**0.005**
	ALL	34.555 [32.231, 36.817]	37.924	**0.003**
DRB3	ABS	45.051 [43.434, 46.662]	47.498	**0.002**
	ALL	53.531 [51.588, 55.516]	56.510	**0.003**

Note: Bolded figures indicate a significant difference after false discovery rate correction. ABS, antigen binding site.

Naturally Mated Pairs versus Non-Mating Pairs

The comparison between naturally mated pairs and non-mating pairs that are based on allele sharing, functional amino acid distances, and genetic relatedness was the same as between naturally mated pairs and randomly assigned pairs (Figures 4 and 5 and Tables 1 and 3). We found no evidence to support the inbreeding avoidance hypothesis when comparing the results of the MHC genes (significant) and microsatelites (nonsignificant).

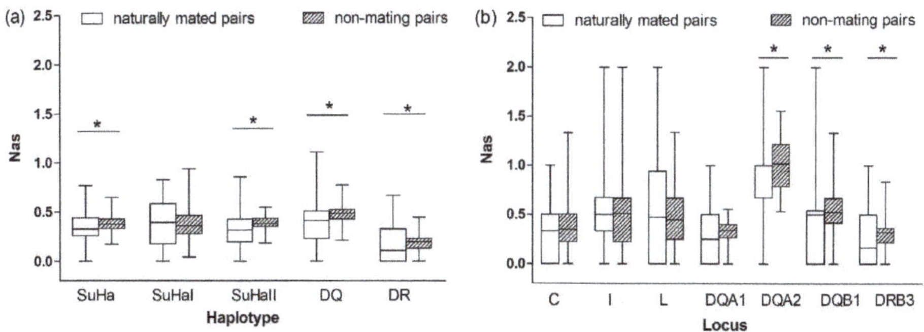

Figure 4. Proportion of allele sharing (Nas) in naturally mated pairs and non-mating pairs (**a**) five super haplotypes, and (**b**) seven individual loci. Asterisk shows results with p values that are smaller than 0.05.

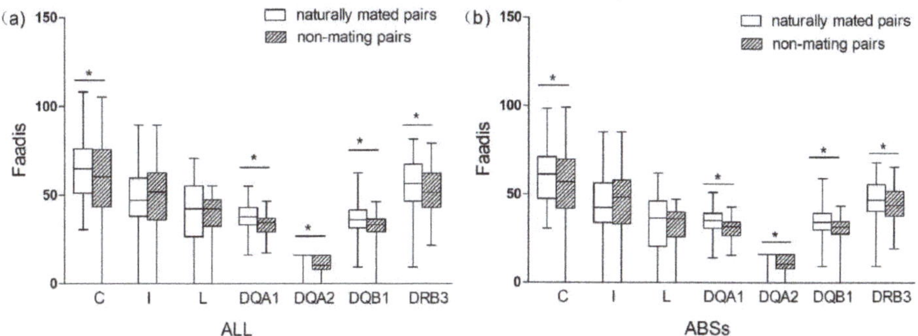

Figure 5. Functional amino acid distances (Faadis) in naturally mated pairs and non-mating pairs. (**a**) functional amino acid distance of all sites; and, (**b**) functional amino acid distance of antigen binding sites (ABSs). Asterisk shows results with p values that are smaller than 0.05.

Table 3. Mean functional amino acid distances at five super haplotypes in naturally mated pairs and non-mating pairs.

Locus		Natural Mated Pairs	Non-Mating Pairs	Statistics	P
SuHa	ABS	126.209	119.677	t = 3.310	**0.001**
	ALL	138.990	132.018	t = 3.298	**0.001**
SuHaI	ABS	92.315	90.284	t = 1.054	0.294
	ALL	100.215	98.262	t = 0.932	0.353
SuHaII	ABS	77.548	70.363	t = 3.960	**0.000**
	ALL	86.828	79.532	t = 3.630	**0.000**
DQ	ABS	57.961	50.322	Z = −4.604	**0.000**
	ALL	61.375	53.769	Z = −4.435	**0.000**
DR	ABS	46.879	44.284	t = 2.298	**0.024**
	ALL	55.732	52.649	t = 2.235	**0.027**

Note: Bolded figures indicate a significant difference after false discovery rate correction. ABS, antigen binding site.

Relationship between MHC Divergence between Females and Males and Natural Mating Success

The model-based analysis confirmed the randimization results. We found a negative relationship between allele sharing females and males and natural mating success at SuHa, SuhaII, DQ, and DR. Separately examining the MHC loci revealed that the overall negative trend might have been caused by lower allele sharing at DQA1, DQA2, DQB1, and DRB3 between the mated pairs (Figure 6). A positive relationship was found between the functional amino acid distance of females to males and natural mating success at SuHa, SuhaII, and DQ, and at three individual loci (DQA1, DQA2, and DQB1; Table 4). The model-based analysis indicated that higher natural mating success resulted from more MHC-dissimilar mating pairs. These results support the MHC genetic compatibility hypothesis.

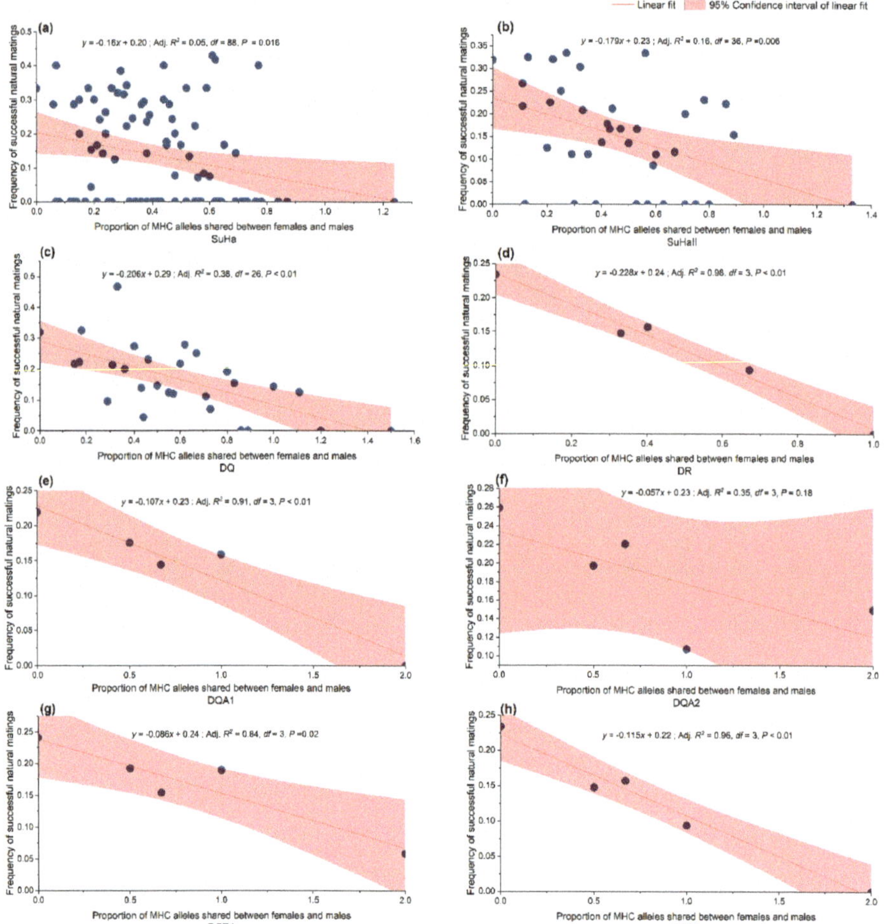

Figure 6. Relationship between shared major histocompatibility complex (MHC) alleles and frequency of successful natural mating. (**a**) SuHa, including four MHC I loci (*Aime*-C, *Aime*-F, *Aime*-I, and *Aime*-L) and six MHC II loci; (**b**) SuHaII, including six MHC II loci (DQA1, DQA2, DQB1, DQB2, DRB3, and DRA); (**c**) DQ, including DQB1, DQB2, DQA1, and DQA2; and, (**d**) DR, including DRB3 and DRA; (**e**–**h**) Four individual loci.

Table 4. Association between natural mating success and major histocompatibility complex functional amino acid distances of females and males.

	ABS		ALL	
	F	**P**	**F**	**P**
SuHa	9.004	**0.003**	8.518	**0.004**
SuHaI	1.241	0.266	0.919	0.338
SuHaII	10.933	**0.001**	8.634	**0.003**
DQ	11.649	**0.001**	10.11	**0.002**
DR	3.33	0.068	2.861	0.091
C	3.849	0.050	3.498	0.062
I	0.406	0.524	0.475	0.491
L	0.122	0.727	0.469	0.494
DQA1	10.915	**0.001**	9,465	**0.002**
DQA2	4.688	**0.031**	4.688	**0.031**
DQB1	6.953	**0.009**	6.88	**0.009**
DRB3	3.33	0.068	2.861	0.091

Note: There were 954 degrees of freedom for each genetic variable tested. Bolded figures indicate a significant difference after false discovery rate correction. ABS, antigen binding site.

3.3. Sperm-Egg Recognition

We successfully assigned 16 fathers to 80 offspring, including 30 twins. In 65 cases, adult females accepted sperm from 1–4 adult males, including males that are involved in natural mating and artificial insemination.

3.3.1. Observed Zygotes (Offspring) Versus Randomly Assigned Zygotes

We found no significant difference between the observed zygotes and the randomly assigned zygotes in Faadis (egg-to-sperm MHC divergence) at all types of super haplotype and all MHC loci, except for DQA1 and DQA2 (Table S2). The Faadis of the observed zygotes at DQA1 and DQA2 was outside the 97.5% hypothesis distribution with respect to all sites and the ABS sites (Figure 7, Table S2).

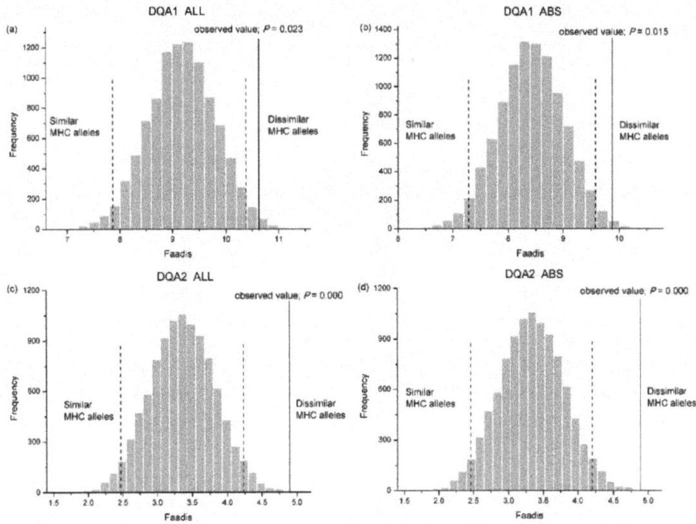

Figure 7. Frequency distributions of functional amino acid distances (Faadis) between eggs and randomly assigned sperm after 10,000 simulations. (**a**) whole exon2 of *Aime*-DQA1; (**b**) antigen binding site (ABS) within exon2 of *Aime*-DQA1; (**c**) whole exon2 of *Aime*-DQA2; and, (**d**) ABS within exon2 of *Aime*-DQA2. Two-tailed 95% confidence intervals are indicated by black dashed lines. MHC, major histocompatibility complex.

3.3.2. Observed Zygotes (Offspring) Versus Other Zygotes

The results between the observed zygotes and other zygotes (combinations of egg-to-sperm haplotypes, except for offspring haplotypes)) were similar to those between the observed zygotes and randomly assigned zygotes (Table 5, Figure 8). Observed zygotes had higher Faadis values at DQA1 and DQA2 than other zygotes at all sites and ABS sites ($P_{\text{DQA1-ABS}} = 0.008$, $P_{\text{DQA1-ALL}} = 0.012$, $P_{\text{DQA2-ABS}} = 0.000$, and $P_{\text{DQA2-ALL}} = 0.000$; Figure 8).

Table 5. Mean functional amino acid distances at five super haplotypes in observed zygotes and other zygotes.

Locus		Observed Zygotes	Other Zygotes	Statistics	P
SuHa	ABS	31.119	31.153	Z = −1.184	0.237
	ALL	34.056	34.383	t = −0.228	0.820
SuHaI	ABS	22.290	23.649	Z = −0.096	0.924
	ALL	24.082	25.684	Z = −0.027	0.978
SuHaII	ABS	19.959	18.785	Z = −1.696	0.090
	ALL	22.136	21.261	Z = −1.108	0.268
DQ	ABS	15.164	13.717	Z = −2.624	**0.009**
	ALL	15.997	14.677	Z = −2.526	**0.012**
DR	ABS	11.857	11.633	Z = −0.304	0.761
	ALL	13.924	13.943	Z = −0.023	0.982

Note: Bolded figures indicate a significant difference after false discovery rate correction. ABS, antigen binding site.

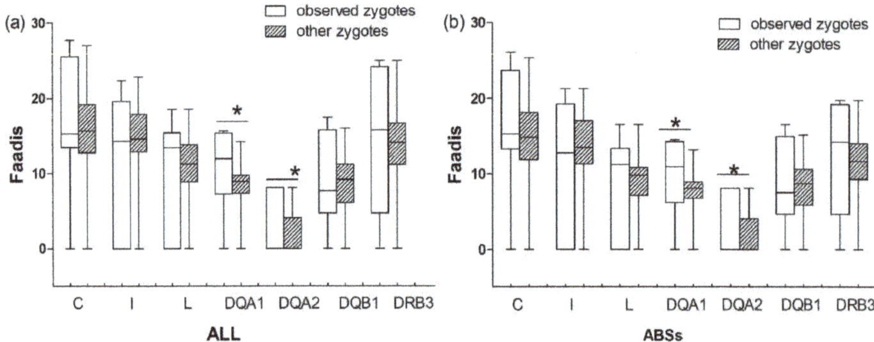

Figure 8. Functional amino acid distance (Faadis) for genetic compatibility at seven major histocompatibility complex loci between observed zygotes and other zygotes. (**a**) functional amino acid distance of all sites; and, (**b**) functional amino acid distance of antigen binding sites (ABSs). Asterisk shows results with *p* values that are smaller than 0.05.

3.3.3. Relationship between Breeding Success and MHC Divergence of Zygotes

The model-based analysis revealed that a large MHC functional amino acid distance between eggs and sperm at DQ (mainly at DQA1 and DQA2) and *Aime*-C resulted in higher breeding success, indicating the preference for maximum MHC divergence between eggs and sperm (Table 6).

Table 6. Association between breeding success and major histocompatibility complex functional amino acid distances between egg and sperm haplotypes.

		ABS		ALL	
	df	**F**	***P***	**F**	***P***
SuHa	472	0.671	0.413	0.989	0.320
SuHaI	467	1.827	0.177	2.072	0.151
SuHaII	453	0.531	0.466	0.120	0.730
DQ	453	0.001	0.982	12.32	**0.000**
DR	344	0.007	0.935	0.101	0.751
C	283	0.205	0.651	0.135	0.714
I	256	5.440	0.020	5.749	0.017
L	252	0.902	0.343	1.136	0.288
DQA1	**399**	7.058	**0.008**	6.325	**0.012**
DQA2	**146**	16.090	**0.000**	16.09	**0.000**
DQB1	242	0.107	0.743	0.072	0.788
DRB3	344	0.007	0.935	0.101	0.751

Note: Bolded figures indicate a significant difference after false discovery rate correction. ABS, antigen binding site.

3.4. Mother-Fetus Recognition

There was no significant difference in Faadis between the observed zygotes and their mothers and other zygotes and their mothers at all super haplotypes and all loci, except for *Aime*-C and *Aime*-I (Figure 9, Table S3). The observed zygotes had significantly lower Faadis to mothers at *Aime*-C and *Aime*-I than other zygotes ($p < 0.05$, Figure 9).

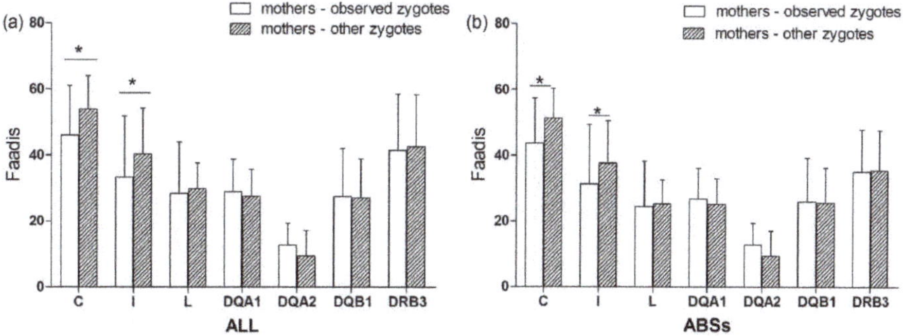

Figure 9. Functional amino acid distances (Faadis) at seven major histocompatibility complex loci between mothers to observed zygotes (offspring) and between mothers to other zygotes. (**a**) functional amino acid distances at all sites; and, (**b**) functional amino acid distances at antigen binding sites (ABSs). Asterisk shows results with *p* values that are smaller than 0.05.

We did not find a significant relationship between breeding success and the functional amino acid distance of mothers to a combination of egg and sperm haplotypes at any MHC loci (Table S4).

4. Discussion

4.1. Female Choice at the Inter-Individual Recognition Level

We found that female giant pandas preferred males with MHC-heterozygous and MHC-dissimilar genotypes, which supports the heterozygosity advantage and disassortative choice of the compatibility hypotheses, respectively. It has been proposed that females usually use male ornaments as a cue for choosing MHC-heterozygous males, and olfaction as a cue for choosing MHC-dissimilar males [15]. Body color varies among giant pandas, and it may reflect condition [73–75]. Therefore, we propose that female giant pandas assess males' physical condition in order to choose MHC-heterozygous mates by evaluating the depth of body color. During the mating season, giant pandas secrete musk from their anal glands and in their urine, which is highly odorous [43], thus it is possible that females use odor to choose among male MHC genotypes. Animals can simultaneously use two cues, and the two strategies are not mutually exclusive, particularly in species with both good vision and olfaction [19,76,77]. A recent study on giant pandas reported multimodal signal behavior between females and males, including olfaction, vision, and hearing [55], supporting the use of two strategies. It has been reported that female pandas use odor cues prior to face-to-face meetings, so we propose that females choose MHC-dissimilar mating partners based on odor cues alone, and they use body color depth to visually choose MHC-heterozygous males. A "multiple-cue strategy" was also found in a study of female mice that changed strategy when the males' urinary scent-marking rate changed [78], and in lizards that use coloration as a mate choice cue at long distances, but use odor at short distances [79]. Using multiple cues to choose a mate may be more common than expected, as females gain more information and reduce their selection costs [80].

The preference for MHC-dissimilar mates might be due to inbreeding avoidance in giant panda populations. In general, if mate choice aims to avoid inbreeding, then we expected to see significantly lower relatedness in the observed pairs than in randomly assigned pairs [20,63,81]. However, our results revealed no significant difference between observed pairs and randomly assigned pairs (Table 1), suggesting that female giant pandas do not attempt to avoid inbreeding. Avoiding inbreeding may not be as important to females as maximizing the number of MHC polymorphisms in their offspring in order to resist pathogens. In addition, we did not find any evidence for the heterozygote advantage hypothesis that is based on the MLH and d^2 results, suggesting that overall genetic diversity has no effect on mate choice. These findings indicate that female mate choice targets functional MHC genes rather than other regions, or it is a byproduct of inbreeding avoidance.

Some studies have reported that giant pandas are susceptible to parasites and viruses [82–85], suggesting that the MHC genes are important in female mate choice at the inter-individual recognition level. Furthermore, our results show that females favored partners that were the most MHC-dissimilar to themselves, resulting in offspring with high heterozygosity. Whether offspring have high immunocompetence should be addressed in future studies regarding the relationship between MHC heterozygosity and immunocompetence in the giant panda.

4.2. Cryptic Female Choice at the Sperm-Egg Recognition Level

Female giant pandas usually mate with multiple males, possibly to increase fertilization success and offspring genetic quality, as has been found in other species [3,86,87]. Our results revealed that sperm and eggs do not randomly combine, and that sperm from zygotes observed was more dissimilar to eggs at DQA1 and DQA2 than sperm from other zygotes (Figures 7 and 8 and Table 5). Two mechanisms may explain this: cryptic female choice or sperm competition [15,16,20], but we could not identify which was the most important. Nevertheless, combined sperm and eggs had the maximum functional amino acid distance, which supports the female MHC-disassortative choice of compatibility hypothesis that is described above.

4.3. Mother-Fetus Recognition Level

The observed zygotes were more similar to mothers at *Aime*-C and *Aime*-I than the randomly assigned zygotes and other zygotes (Figure 9). These findings suggest that the observed zygotes had higher compatibility at *Aime*-C and *Aime*-I, which could decrease or block graft rejections from the mother's immune system. MHC class I molecules play an important role in the immune reaction between the mother and fetus, e.g., HLA-G molecules are present at the mother-child interface of trophoblastic cells and protect the fetus from the lytic activity of maternal uterine natural killer cells [88,89]. The identification of an HLA-G ortholog in the giant panda would elucidate the effects of mother-fetus immunity in this species; however, MHC class I genes have similar loci, making HLA-G orthologs difficult to identify in the giant panda.

4.4. Hierarchical and Cooperative Effects

From the super haplotype level to individual loci, female mate choice was hierarchical in nature. For example, we found that SuHa, which represents all of the functional MHC genes in giant pandas, predicted MHC-heterozygous female choice. However, when separately analyzing SuHaI and SuHaII, the effect could only be observed in SuHaII. When we then excluded DR from SuHaII and analyzed DQ alone, the heterozygosity advantage could still be detected. In contrast, the heterozygosity advantage could not be detected when only analyzing DR. Finally, when we focused on the individual locus level, only one out of three DQ genes (DQB1) still predicted female choice. A similar pattern was found for the genetic compatibility hypothesis.

Alternatively, multiple MHC genes that act cooperatively can explain the above example. The effect of DQ seemed to be larger than that of DQB1 alongside DQA1 and DQA2. The integration of the loci or super haplotypes had a greater effect than the sum of the loci or super haplotypes.

The SuHa results were inconsistent with those of SuHaII with respect to inter-individual recognition, suggesting that SuHaII is more important than SuHaI. This may have been caused by greater variation in SuHaII than in SuHaI.

4.5. Relative Importance of MHC-I, MHC-II, DQ, and DR

MHC class I, MHC class II, DQ, and DR were differentially important at the inter-individual, gamete, and mother-fetus recognition levels. More MHC II genes predicted compatible female mate choice than the MHC I genes, e.g., naturally mated males were more dissimilar to their partners at SuHaII but not at SuHaI. Furthermore, all four of the MHC class II genes (DQA1, DQA2, DQB1, and DRB3) predicted compatible female mate choice, while only one out of three MHC class I genes had a significant result. At the gamete level, the sperm were more dissimilar to eggs at DQ than at DR. At the mother-fetus recognition level, the observed zygotes (offspring) were more similar to their mothers at MHC class I genes, but not at MHC class II genes.

Genes that predict female mate choice or gamete selection may be more important than other MHC loci that are involved in pathogen resistance, as revealed by many studies in which distinct molecules that are coded by different alleles recognize specific pathogens, and their ability to resist pathogens differs [11–14], such as the DQ region. The DQ region in the giant panda differs to that in other mammals, because it contains more genes and alleles than the DR region [59]. DRB3 exhibited the most variation among the seven individual polymorphic loci. MHC polymorphisms may be driven by sexual selection [10], which is in line with our results at DQ and DRB3.

Our findings suggest that MHC loci do not play equal roles in female mate choice at the inter-individual recognition level or at other levels, and that targeted MHC loci may be key for female mate choice at the individual recognition level. Therefore, more MHC loci should be isolated and large MHC regions surveyed.

5. Conclusions

Our results for *Aime*-C, *Aime*-I, and DQ support the heterozygosity hypothesis, while the results for *Aime*-C, DQ, and DR support the genetic compatibility hypothesis. Comparisons of combined sperm and other sperm-to-eggs revealed that sperm competition or *Aime*-DQA1- and DQA2-associated gamete selection occurred. The comparison of zygotes observed (offspring) and other zygotes revealed the possible *Aime*-C- and *Aime*-I-associated maternal immune tolerance mechanisms. We suggest that captive breeding programs should consider the MHC constitution. Our study provides a good foundation for studying the relationships between the MHC constitution, individual fitness (lifetime reproductive success), and mate choice cues in the giant panda.

Supplementary Materials: The following are available online at http://www.mdpi.com/2073-4409/8/3/257/s1. Figure S1, Allele linkage relationships of six major histocompatibility complex (MHC) class II genes for SuHaII. Asterisk indicates recombinational SuHaII. Figure S2, Allele linkage relationships of four major histocompatibility complex (MHC) class I genes for SuHaI. Figure S3, Allele linkage relationships between SuHaI and SuHaII for SuHa. Figure S4, Allele linkage relationships of DQ genes. Asterisk indicates recombinational DQ super haplotype. Table S1, Genetic variation in major histocompatibility complex (MHC) class I and class II molecules and super haplotypes. Table S2, Functional amino acids in observed zygotes and randomly assigned zygotes. Table S3, Mean functional amino acid distances at five super haplotypes between mothers and observed zygotes and between mothers and other zygotes. Table S4, Association between breeding success and major histocompatibility complex (MHC) functional amino acid distances of the mother and a combination of egg and sperm haplotypes.

Author Contributions: Conceptualization, S.-G.F. and Q.-H.W.; Methodology, Q.-H.W. and Y.Z.; Experiment and Analysis, Y.Z, Q.-H.W. and S.-G.F.; Resources, H.-M.Z.; Writing–Original Draft Preparation, Y.Z, Q.-H.W. and S.-G.F.; Writing–Review & Editing, Y.Z., Q.-H.W. and S.-G.F.; Supervision, S.-G.F. and Q.-H.W.; Project Administration and Funding Acquisition, S.-G.F.

Funding: This work was supported by a National Key Program (2016YFC0503200) from the Ministry of Science and Technology of China, a special grant for the giant panda from the State Forestry Administration of the People's Republic of China, and the Fundamental Research Funds for the Central Universities of the People's Republic of China.

Acknowledgments: We thank China Research and Conservation Center for the giant panda (Wolong) for precious samples of giant panda.

Conflicts of Interest: The authors declare no conflict of interest.

References

1. Andersson, M.B. *Sexual Selection*; Princeton University Press: Princeton, NJ, USA, 1994.
2. Andersson, M.; Simmons, L.W. Sexual selection and mate choice. *Trends Ecol. Evol.* **2006**, *21*, 296–302. [CrossRef] [PubMed]
3. Jennions, M.D.; Petrie, M. Why do females mate multiply? A review of the genetic benefits. *Biol. Rev.* **2000**, *75*, 21–64. [CrossRef]
4. Searcy, W.A. The evolutionary effects of mate selection. *Annu. Rev. Ecol. Syst.* **1982**, *13*, 57–85. [CrossRef]
5. Folstad, I.; Karter, A.J. Parasites, bright males, and the immunocompetence handicap. *Am. Nat.* **1992**, 603–622. [CrossRef]
6. Hamilton, W.D.; Zuk, M. Heritable true fitness and bright birds: A role for parasites? *Science* **1982**, *218*, 384–387. [CrossRef]
7. Klein, J. *The Natural History of the Major Histocompatibility Complex*; Wiley & Sons: New York, NY, USA, 1986.
8. Sommer, S. The importance of immune gene variability (MHC) in evolutionary ecology and conservation. *Front. Zool.* **2005**, *2*, 16. [CrossRef] [PubMed]
9. Piertney, S.B.; Oliver, M.K. The evolutionary ecology of the major histocompatibility complex. *Heredity* **2006**, *96*, 7–21. [CrossRef] [PubMed]
10. Milinski, M. The major histocompatibility complex, sexual selection, and mate choice. *Annu. Rev. Ecol. Evol. Syst.* **2006**, *37*, 159–186. [CrossRef]
11. de Campos-Lima, P.O.; Gavioli, R.; Zhang, Q.J.; Wallace, L.E.; Dolcetti, R.; Rowe, M.; Rickinson, A.B.; Masucci, M.G. HLA-A11 epitope loss isolates of Epstein-Barr virus from a highly A11+ population. *Science* **1993**, *260*, 98–100. [CrossRef] [PubMed]
12. Grimholt, U.; Larsen, S.; Nordmo, R.; Midtlyng, P.; Kjoeglum, S.; Storset, A.; Saebø, S.; Stet, R.J.M. MHC polymorphism and disease resistance in Atlantic salmon (*Salmo salar*); facing pathogens with single expressed major histocompatibility class I and class II loci. *Immunogenetics* **2003**, *55*, 210–219. [CrossRef]
13. Harf, R.; Sommer, S. Association between major histocompatibility complex class II DRB alleles and parasite load in the hairy-footed gerbil, *Gerbillurus paeba*, in the southern Kalahari. *Mol. Ecol.* **2004**, *14*, 85–91. [CrossRef] [PubMed]
14. Langefors, Å.; Lohm, J.; Grahn, M.; Andersen, Ø.; von Schantz, T. Association between major histocompatibility complex class IIB alleles and resistance to Aeromonas salmonicida in Atlantic salmon. *Proc. R. Soc. Lond. B Biol. Sci.* **2001**, *268*, 479–485. [CrossRef] [PubMed]
15. Mays, H.L.; Hill, G.E. Choosing mates: Good genes versus genes that are a good fit. *Trends Ecol. Evol.* **2004**, *19*, 554–559. [CrossRef] [PubMed]

16. Neff, B.D.; Pitcher, T.E. Genetic quality and sexual selection: An integrated framework for good genes and compatible genes. *Mol. Ecol.* **2005**, *14*, 19–38. [CrossRef] [PubMed]

17. Cutrera, A.P.; Fanjul, M.S.; Zenuto, R.R. Females prefer good genes: MHC-associated mate choice in wild and captive tuco-tucos. *Anim. Behav.* **2012**, *83*, 847–856. [CrossRef]

18. Promerová, M.; Vinkler, M.; Bryja, J.; Poláková, R.; Schnitzer, J.; Munclinger, P.; Albrecht, T. Occurrence of extra-pair paternity is connected to social male's MHC-variability in the scarlet rosefinch *Carpodacus erythrinus*. *J. Avian Biol.* **2011**, *42*, 5–10. [CrossRef]

19. Tregenza, T.; Wedell, N. Genetic compatibility, mate choice and patterns of parentage: Invited review. *Mol. Ecol.* **2000**, *9*, 1013–1027. [CrossRef] [PubMed]

20. Schwensow, N.; Eberle, M.; Sommer, S. Compatibility counts: MHC-associated mate choice in a wild promiscuous primate. *Proc. R. Soc. B Biol. Sci.* **2008**, *275*, 555–564. [CrossRef]

21. Strandh, M.; Westerdahl, H.; Pontarp, M.; Canback, B.; Dubois, M.P.; Miquel, C.; Taberlet, P.; Bonadonna, F. Major histocompatibility complex class II compatibility, but not class I, predicts mate choice in a bird with highly developed olfaction. *Proc. R. Soc. B Biol. Sci.* **2012**, *279*, 4457–4463. [CrossRef]

22. Juola, F.A.; Dearborn, D.C. Sequence-based evidence for major histocompatibility complex-disassortative mating in a colonial seabird. *Proc. R. Soc. B Biol. Sci.* **2012**, *279*, 153–162. [CrossRef]

23. Reichard, M.; Spence, R.; Bryjova, A.; Bryja, J.; Smith, C. Female Rose Bitterling Prefer MHC-Dissimilar Males: Experimental Evidence. *PLoS ONE* **2012**, *7*. [CrossRef] [PubMed]

24. Bahr, A.; Sommer, S.; Mattle, B.; Wilson, A.B. Mutual mate choice in the potbellied seahorse (*Hippocampus abdominalis*). *Behav. Ecol.* **2012**, *23*, 869–878. [CrossRef]

25. Schwensow, N.; Fietz, J.; Dausmann, K.; Sommer, S. MHC-associated mating strategies and the importance of overall genetic diversity in an obligate pair-living primate. *Evol. Ecol.* **2008**, *22*, 617–636. [CrossRef]

26. Setchell, J.; Charpentier, M.; Abbott, K.; Wickings, E.; Knapp, L. Opposites attract: MHC-associated mate choice in a polygynous primate. *J. Evol. Biol.* **2010**, *23*, 136–148. [CrossRef] [PubMed]

27. Reusch, T.B.H.; Haberli, M.; Aeschlimann, P.B.; Milinski, M. Female sticklebacks count alleles in a strategy of sexual selection explaining MHC polymorphism. *Nature* **2001**, *414*, 300–302. [CrossRef] [PubMed]

28. Aeschlimann, P.B.; Haberli, M.A.; Reusch, T.B.H.; Boehm, T.; Milinski, M. Female sticklebacks *Gasterosteus aculeatus* use self-reference to optimize MHC allele number during mate selection. *Behav. Ecol. Sociobiol.* **2003**, *54*, 119–126. [CrossRef]

29. Arkush, K.D.; Giese, A.R.; Mendonca, H.L.; McBride, A.M.; Marty, G.D.; Hedrick, P.W. Resistance to three pathogens in the endangered winter-run chinook salmon (*Oncorhynchus tshawytscha*): Effects of inbreeding and major histocompatibility complex genotypes. *Can. J. Fish. Aquat. Sci.* **2002**, *59*, 966–975. [CrossRef]

30. Meagher, S.; Penn, D.J.; Potts, W.K. Male-Male Competition Magnifies Inbreeding Depression in Wild House Mice. *Proc. Natl. Acad. Sci. USA* **2000**, *97*, 3324–3329. [CrossRef]

31. MartinVilla, J.M.; Luque, I.; MartinezQuiles, N.; Corell, A.; Regueiro, J.R.; Timon, M.; ArnaizVillena, A. Diploid expression of human leukocyte antigen class I and class II molecules on spermatozoa and their cyclic inverse correlation with inhibin concentration. *Biol. Reprod.* **1996**, *55*, 620–629. [CrossRef]

32. Martin-Villa, J.M.; Longas, J.; Arnaiz-Villena, A. Cyclic expression of HLA class I and II molecules on the surface of purified human spermatozoa and their control by serum inhibin B levels. *Biol. Reprod.* **1999**, *61*, 1381–1386. [CrossRef]

33. Paradisi, R.; Neri, S.; Pession, A.; Magrini, E.; Bellavia, E.; Ceccardi, S.; Flamigni, C. Human leukocyte antigen II expression in sperm cells: Comparison between fertile and infertile men. *Arch. Androl.* **2000**, *45*, 203–213. [PubMed]

34. Rulicke, T.; Chapuisat, M.; Homberger, F.R.; Macas, E.; Wedekind, C. MHC-genotype of progeny influenced by parental infection. *Proc. R. Soc. B Biol. Sci.* **1998**, *265*, 711–716. [CrossRef] [PubMed]

35. Knapp, L.A.; Ha, J.C.; Sackett, G.P. Parental MHC antigen sharing and pregnancy wastage in captive pigtailed macaques. *J. Reprod. Immunol.* **1996**, *32*, 73–88. [CrossRef]

36. Ober, C.; Weitkamp, L.R.; Cox, N.; Dytch, H.; Kostyu, D.; Elias, S. HLA and mate choice in humans. *Am. J. Hum. Genet.* **1997**, *61*, 497–504. [CrossRef]

37. Schacter, B.; Weitkamp, L.R.; Johnson, W.E. Parental HLA compatibility, fetal wastage and neural-tube defects-evidence for a T/T-linke locus in humans. *Am. J. Hum. Genet.* **1984**, *36*, 1082–1091.

38. Lovlie, H.; Gillingham, M.A.F.; Worley, K.; Pizzari, T.; Richardson, D.S. Cryptic female choice favours sperm from major histocompatibility complex-dissimilar males. *Proc. R. Soc. B Biol. Sci.* **2013**, *280*, 20131296. [CrossRef]

39. Yeates, S.; Einum, S.; Fleming, I.; Megens, H.-J.; Hindar, K.; Holt, W.; Van Look, K.; Gage, M. Atlantic salmon eggs favour sperm in competition that have similar major histocompatibility alleles. *Proc. R. Soc. Lond. B Biol. Sci.* **2009**, *276*, 559–566. [CrossRef]

40. Gessner, C.; Nakagawa, S.; Zavodna, M.; Gemmell, N.J. Sexual selection for genetic compatibility: The role of the major histocompatibility complex on cryptic female choice in Chinook salmon (*Oncorhynchus tshawytscha*). *Heredity* **2017**, *118*. [CrossRef]

41. Gasparini, C.; Congiu, L.; Pilastro, A. Major histocompatibility complex similarity and sexual selection: Different does not always mean attractive. *Mol. Ecol.* **2015**, *24*, 4286–4295. [CrossRef]

42. Lenz, T.L.; Hafer, N.; Samonte, I.E.; Yeates, S.E.; Milinski, M. Cryptic haplotype-specific gamete selection yields offspring with optimal MHC immune genes. *Evolution* **2018**, *72*, 2478–2490. [CrossRef]

43. Hu, J.C. *Research on the Giant Panda*; Shanghai Publishing House of Science and Technology: Shanghai, China, 2001.

44. Martin-Wintle, M.S.; Shepherdson, D.; Zhang, G.; Zhang, H.; Li, D.; Zhou, X.; Li, R.; Swaisgood, R.R. Free mate choice enhances conservation breeding in the endangered giant panda. *Nat. Commun.* **2015**, *6*. [CrossRef] [PubMed]

45. Zhang, H.M.; Wang, P.Y. *The Study on Reproduction in Giant Panda*; China FORESTRY publishing House: Bei Jing, China, 2003.

46. Feng, W.H.; Zhang, A.J. *The Research on Breeding and Disease of Giant Panda*; Sichuan Publishing House of Science and Technology: Cheng Du, China, 1991.

47. Huchard, E.; Baniel, A.; Schliehe-Diecks, S.; Kappeler, P.M. MHC-disassortative mate choice and inbreeding avoidance in a solitary primate. *Mol. Ecol.* **2013**, *22*, 4071–4086. [CrossRef] [PubMed]

48. Babik, W. Methods for MHC genotyping in non-model vertebrates. *Mol. Ecol. Resour.* **2010**, *10*, 237–251. [CrossRef] [PubMed]

49. Yu, L.; Nie, Y.; Yan, L.; Hu, Y.; Wei, F. No evidence for MHC-based mate choice in wild giant pandas. *Ecol. Evol.* **2018**, *8*, 8642–8651. [CrossRef] [PubMed]

50. Wan, Q.H.; Zeng, C.J.; Ni, X.W.; Pan, H.J.; Fang, S.G. Giant panda genomic data provide insight into the birth-and-death process of mammalian major histocompatibility complex class II genes. *PLoS ONE* **2009**, *4*, e4147. [CrossRef] [PubMed]

51. Wan, Q.H.; Zhang, P.; Ni, X.W.; Wu, H.L.; Chen, Y.Y.; Kuang, Y.Y.; Ge, Y.F.; Fang, S.G. A novel HURRAH protocol reveals high numbers of monomorphic MHC class II loci and two asymmetric multi-locus haplotypes in the Père David's deer. *PLoS ONE* **2011**, *6*, e14518. [CrossRef]

52. Zhu, Y.; Sun, D.D.; Ge, Y.F.; Yu, B.; Chen, Y.Y.; Wan, Q.H. Isolation and characterization of class I MHC genes in the giant panda (*Ailuropoda melanoleuca*). *Chin. Sci. Bull.* **2012**, *57*, 1–8. [CrossRef]

53. Xie, Z.; Gipps, J. *The 2009 International Studbood for Giant Panda (Ailuropoda melanoleuca)*; Chinese Association of Zoological Garden: Beijing, China, 2009.

54. Durrant, B.S.; Olson, M.A.; Amodeo, D.; Anderson, A.; Russ, K.D.; Campos-Morales, R.; Gual-Sill, F.; Garza, J.R. Vaginal cytology and vulvar swelling as indicators of impending estrus and ovulation in the giant panda (*Ailuropoda melanoleuca*). *Zoo Biol.* **2003**, *22*, 313–321. [CrossRef]

55. Owen, M.A.; Swaisgood, R.R.; McGeehan, L.; Zhou, X.P.; Lindburg, D.G. Dynamics of Male-Female Multimodal Signaling Behavior across the Estrous Cycle in Giant Pandas (*Ailuropoda melanoleuca*). *Ethology* **2013**, *119*, 869–880. [CrossRef]

56. Wan, Q.H.; Zhu, L.; Wu, H.; Fang, S.G. Major histocompatibility complex class II variation in the giant panda (*Ailuropoda melanoleuca*). *Mol. Ecol.* **2006**, *15*, 2441–2450. [CrossRef]

57. Li, D.; Cui, H.; Wang, C.; Ling, S.; Huang, Z.; Zhang, H. A fast and effective method to perform paternity testing for Wolong giant pandas. *Chin. Sci. Bull.* **2011**, *56*, 2559–2564. [CrossRef]

58. Zhang, H.M.; Guo, Y.; Li, D.S.; Wang, P.Y.; Fang, S.G. Sixteen novel microsatellite loci developed for the giant panda (*Ailuropoda melanoleuca*). *Conserv. Genet.* **2009**, *10*, 589–592. [CrossRef]

59. Chen, Y.Y.; Zhu, Y.; Wan, Q.H.; Lou, J.K.; Li, W.J.; Ge, Y.F.; Fang, S.G. Patterns of adaptive and neutral diversity identify the Xiaoxiangling Mountains as a refuge for the giant panda. *PLoS ONE* **2013**. [CrossRef]

60. Zeng, C.J.; Pan, H.J.; Gong, S.B.; Yu, J.Q.; Wan, Q.H.; Fang, S.G. Giant panda BAC library construction and assembly of a 650-kb contig spanning major histocompatibility complex class II region. *BMC Genom.* **2007**, *8*, 315. [CrossRef]

61. Coltman, D.; Slate, J. Microsatellite measures of inbreeding: A meta-analysis. *Evolution* **2003**, *57*, 971–983. [CrossRef]

62. Wetton, J.H.; Carter, R.E.; Parkin, D.T.; Walters, D. Demographic study of a wild house sparrow population by DNA fingerprinting. *Nature* **1987**, *327*, 147–149. [CrossRef]

63. Landry, C.; Garant, D.; Duchesne, P.; Bernatchez, L. 'Good genes as heterozygosity': The major histocompatibility complex and mate choice in Atlantic salmon (*Salmo salar*). *Proc. R. Soc. Lond. B Biol. Sci.* **2001**, *268*, 1279–1285. [CrossRef]

64. Sandberg, M.; Eriksson, L.; Jonsson, J.; Sjöström, M.; Wold, S. New chemical descriptors relevant for the design of biologically active peptides. A multivariate characterization of 87 amino acids. *J. Med. Chem.* **1998**, *41*, 2481–2491. [CrossRef]

65. Huchard, E.; Weill, M.; Cowlishaw, G.; Raymond, M.; Knapp, L.A. Polymorphism, haplotype composition, and selection in the Mhc-DRB of wild baboons. *Immunogenetics* **2008**, *60*, 585–598. [CrossRef]

66. Hill, G.E. Plumage coloration is a sexually selected indicator of male quality. *Nature* **1991**, *350*, 337–339. [CrossRef]

67. Kurtz, J.; Wegner, K.M.; Kalbe, M.; Reusch, T.B.; Schaschl, H.; Hasselquist, D.; Milinski, M. MHC genes and oxidative stress in sticklebacks: An immuno-ecological approach. *Proc. Rroc. Soc. B Biol. Sci.* **2006**, *273*, 1407–1414. [CrossRef] [PubMed]

68. Bjorkman, P.J.; Saper, M.A.; Samraoui, B.; Bennett, W.S.; Strominger, J.L.; Wiley, D.C. The foreigh antigen bing site and T cell recognition regions of class I histocompatibility antigens. *Nat. Immunol.* **1987**, *329*, 512–518.

69. Queller, D.C.; Goodnight, K.F. Estimating relatedness using genetic markers. *Evolution* **1989**, 258–275. [CrossRef] [PubMed]

70. Hardy, O.J.; Vekemans, X. SPAGeDi: A versatile computer program to analyse spatial genetic structure at the individual or population levels. *Mol. Ecol. Notes* **2002**, *2*, 618–620. [CrossRef]

71. Benjamini, Y.; Hochberg, Y. Controlling the False Discovery Rate: A Practical and Powerful Approach to Multiple Testing. *J. R. Stat. Soc. Ser. B Stat. Methodol.* **1995**, *57*, 289–300. [CrossRef]

72. Yekutieli, D.; Benjamini, Y. Resampling-based false discovery rate controlling multiple test procedures for correlated test statistics. *J. Stat. Plan. Infer.* **1999**, *82*, 171–196. [CrossRef]

73. McGraw, K.J.; Hill, G.E. Differential effects of endoparasitism on the expression of carotenoid-and melanin-based ornamental coloration. *Proc. R. Soc. Lond. B Biol. Sci.* **2000**, *267*, 1525–1531. [CrossRef] [PubMed]

74. Fitze, P.S.; Richner, H. Differential effects of a parasite on ornamental structures based on melanins and carotenoids. *Behav. Ecol.* **2002**, *13*, 401–407. [CrossRef]

75. Siefferman, L.; Hill, G.E. Structural and melanin coloration indicate parental effort and reproductive success in male eastern bluebirds. *Behav. Ecol.* **2003**, *14*, 855–861. [CrossRef]

76. Evans, J.; Magurran, A. Multiple benefits of multiple mating in guppies. *Proc. Natl. Acad. Sci. USA* **2000**, *97*, 10074–10076. [CrossRef]

77. Colegrave, N.; Kotiaho, J.S.; Tomkins, J.L. Mate choice or polyandry: Reconciling genetic compatibility and good genes sexual selection. *Evol. Ecol. Res.* **2002**, *4*, 911–917.

78. Roberts, S.C.; Gosling, L.M. Genetic similarity and quality interact in mate choice decisions by female mice. *Nat. Genet.* **2003**, *35*, 103–106. [CrossRef] [PubMed]

79. Lopez, P.; Martin, J. Pheromonal recognition of females takes precedence over the chromatic cue in male Iberian wall lizards Podarcis hispanica. *Ethology* **2001**, *107*, 901–912. [CrossRef]

80. Candolin, U. The use of multiple cues in mate choice. *Biol. Rev.* **2003**, *78*, 575–595. [CrossRef]

81. Huchard, E.; Knapp, L.A.; Wang, J.; Raymond, M.; Cowlishaw, G. MHC, mate choice and heterozygote advantage in a wild social primate. *Mol. Ecol.* **2010**, *19*, 2545–2561. [CrossRef] [PubMed]

82. Feng, W.H.; Wang, R.L.; Zhong, S.M.; Ye, Z.Y.; Cui, X.Z.; Zeng, J.H. *Analysis on the Dead Cause of the Anatomical Carcass of Giant Panda (Ailuropoda Melanoleuca)*; Sichuan Scientific & Technical Publishers: Chengdu, China, 1991; pp. 244–248.

83. Ye, Z.Y. *The Control of the Diseases of Giant Panda in Field: Report of 50 Cases*; Sichuan Scientific & Technical Publishers: Chengdu, China, 1991.

84. Mainka, S.A.; Qiu, X.M.; He, T.M.; Appel, M.J. Serologic survey of giant pandas (*Ailuropoda melanoleuca*), and domestic dogs and cats in the Wolong Reserve, China. *J. Wildl. Dis.* **1994**, *30*, 86–89. [CrossRef]

85. Qin, Q.; Li, D.S.; Zhang, H.M.; Hou, R.; Zhang, Z.H.; Zhang, C.L.; Zhang, J.G.; Wei, F.W. Serosurvey of selected viruses in captive giant pandas (*Ailuropoda melanoleuca*) in China. *Vet. Microbiol.* **2010**, *142*, 199–204. [CrossRef]

86. Fedorka, K.M.; Mousseau, T.A. Material and genetic benefits of female multiple mating and polyandry. *Anim. Behav.* **2002**, *64*, 361–367. [CrossRef]

87. Yasui, Y. A" good-sperm" model can explain the evolution of costly multiple mating by females. *Am. Nat.* **1997**, *149*, 573–584. [CrossRef]

88. Schmidt, C.M.; Orr, H.T. Maternal/fetal interactions: The role of the MHC class I molecule HLA-G. *Crit. Rev. Immunol.* **1992**, *13*, 207–224.

89. Rouas-Freiss, N.; GoncAlves, R.M.-B.; Menier, C.; Dausset, J.; Carosella, E.D. Direct evidence to support the role of HLA-G in protecting the fetus from maternal uterine natural killer cytolysis. *Proc. Natl. Acad. Sci. USA* **1997**, *94*, 11520–11525. [CrossRef] [PubMed]

Communication

Discovery of a Novel MHC Class I Lineage in Teleost Fish which Shows Unprecedented Levels of Ectodomain Deterioration while Possessing an Impressive Cytoplasmic Tail Motif

Unni Grimholt [1,*], Kentaro Tsukamoto [2], Keiichiro Hashimoto [2] and Johannes M. Dijkstra [2,*]

[1]   Fish Health Research Group, Norwegian Veterinary Institute, Ullevaalsveien 68, 0454 Oslo, Norway
[2]   Institute for Comprehensive Medical Science, Fujita Health University, Toyoake, Aichi 470-1192, Japan
*   Correspondence: Unni.Grimholt@vetinst.no (U.G.); Dijkstra@fujita-hu.ac.jp (J.M.D.);
    Tel.: +047-908-24-112 (U.G.); +81-562-93-9381 (J.M.D.)

Received: 14 August 2019; Accepted: 4 September 2019; Published: 9 September 2019

Abstract: A unique new nonclassical MHC class I lineage was found in Teleostei (teleosts, modern bony fish, e.g., zebrafish) and Holostei (a group of primitive bony fish, e.g., spotted gar), which was designated "H" (from "hexa") for being the sixth lineage discovered in teleosts. A high level of divergence of the teleost sequences explains why the lineage was not recognized previously. The spotted gar H molecule possesses the three MHC class I consensus extracellular domains $\alpha 1$, $\alpha 2$, and $\alpha 3$. However, throughout teleost H molecules, the $\alpha 3$ domain was lost and the $\alpha 1$ domains showed features of deterioration. In fishes of the two closely related teleost orders Characiformes (e.g., Mexican tetra) and Siluriformes (e.g., channel catfish), the H ectodomain deterioration proceeded furthest, with H molecules of some fishes apparently having lost the entire $\alpha 1$ or $\alpha 2$ domain plus additional stretches within the remaining other ($\alpha 1$ or $\alpha 2$) domain. Despite these dramatic ectodomain changes, teleost H sequences possess rather large, unique, well-conserved tyrosine-containing cytoplasmic tail motifs, which suggests an important role in intracellular signaling. To our knowledge, this is the first description of a group of MHC class I molecules in which, judging from the sequence conservation pattern, the cytoplasmic tail is expected to have a more important conserved function than the ectodomain.

Keywords: major histocompatibility complex; MHC; evolution; nonclassical; fish

---

1. Introduction

1.1. The Structure and Function of Classical MHC-I

Classical MHC class I (MHC-I) molecules consist of extracellular $\alpha 1$, $\alpha 2$, and $\alpha 3$ domains, plus a connecting peptide (CP)/transmembrane (TM)/Cytoplasmic (CY) region. Such a "heavy chain" molecule forms a complex together with a single domain molecule β_2-microglobulin (β_2-m) and a peptide ligand, and this pMHC-I complex is presented at the surface of cells for screening by CD8[+] T cells [1,2]. This system helps CD8[+] T cells to detect virus-infected and cancerous cells, which can then be eliminated.

The $\alpha 3$ and β_2-m domains are typical immunoglobulin superfamily (IgSF) structures, while the $\alpha 1$ and $\alpha 2$ domains form a unique structure of two anti-parallel helical structures on top of a β-sheet, which, in the case of classical MHC-I, forms a groove in which peptides of ~9 amino acids length can be bound [1]. Classical MHC molecules are renown for extensive polymorphism, which resides mostly in the peptide-binding domains and affects the sets of peptides presented by different allelic MHC molecules [3,4]. This polymorphism is believed to increase the resistance of a population against pathogens. Throughout jawed vertebrates (Gnathostomata), polymorphic sequences of classical MHC-I can be found [5], and for teleost fish, the ability of such molecules to bind β_2-m and peptide ligand was proven and their ability to stimulate cytotoxic CD8$^+$ T cells suggested by a variety of data ([6–8], reviewed in [9]).

1.2. Classical and Nonclassical MHC-I Can Show Differences in Domain Organization

At various times during evolution, classical MHC-I genes duplicated and new copies diverged into "nonclassical" MHC-I genes [10,11]. Depending on the respective molecule, nonclassical MHC-I molecules have retained similarity more or less with classical MHC-I and perform a wide variety of functions within and outside the immune system [12]. The evolutionary younger nonclassical MHC-I molecules especially tend to share features with the classical molecules, such as peptide binding, whereas many of the more ancient nonclassical lineages exhibit more diverged functions and properties [10,12]. The number of molecular domains can also differ from the classical MHC-I situation. For example, ZAG (zinc-α2-glycoprotein, alias AZGP1; [13]) is a soluble mammalian nonclassical MHC-I molecule which consists of α1, α2, and α3 domains, which does not bind β_2-m and can bind fatty acids [14,15]. Another example is represented by EPCR (endothelial protein C receptor, alias PROCR; [16]) which is a nonclassical MHC-I molecule found in birds, reptiles, and mammals (reviewed in [10]), and which is a transmembrane molecule that promotes protein C activation (reviewed in [17]) and possesses only α1 and α2 ectodomains that form a hydrophobic groove that can bind hydrophobic molecules [18]. EPCR molecules belong to the same nonclassical MHC-I lineage as CD1 molecules [10,16], which are also found in mammals [19], birds, and reptiles (reviewed in [10]), although CD1 are transmembrane molecules that retained the α1, α2, and α3 ectodomain organization and β_2-m binding ability of classical MHC-I. Like EPCR, CD1 molecules have a groove for binding hydrophobic molecules [20], but in the case of CD1, they present these to T cells (reviewed in [21]). Other nonclassical MHC-I transmembrane molecules in mammals without an α3 domain and without a β_2-m partner are members of the RAET/ULBP (retinoic acid early transcripts/UL16 binding proteins) family [22]. RAET/ULBP molecules tend to have closed grooves, are not involved in presentation of small ligands, and can interact with NKG2D receptors on natural killer (NK) cells (reviewed in [12]); while some members of this family have transmembrane domains, others are associated with the membrane by means of a GPI anchor (reviewed in [23]). Although the evidence is somewhat thin [10], it has been proposed that the RAET/ULBP family forms a phylogenetic group together with the MIC/MILL family of NKG2D binding nonclassical MHC-I molecules which typically do possess an α3 domain [24]. Therefore, besides the EPCR/CD1 situation, the RAET/ULBP/MIC/MILL group may provide another example of related molecules with and without an α3 domain.

1.3. MHC-I in Teleost Fish and the Target of the Present Study

In teleost fish, there are currently five described MHC class I lineages denoted U, Z, S, L, and P, and all essentially contain transmembrane heavy chain molecules with the canonical three extracellular domains organization ([25–29]; reviewed in [30]), although at least at the genetic level, some variation in the number of domains can be found (e.g., [30–32]). Defined by polymorphism, expression pattern, and expectations for peptide binding ability, all identified teleost classical MHC-I genes belong to the U lineage, while the U lineage also contains nonclassical MHC-I genes (reviewed in [30]). Among teleost MHC-I, only classical U has been well studied at the genetic, structural, and functional levels (reviewed in [9]), whereas for Z, S, L [25,27–29], and P [30], only genetic information is available. Among the nonclassical lineages, only Z lineage members are expected to bind (probably N-terminally modified) peptides in a way reminiscent of the peptide binding mode of classical MHC-I [28,30].

In the present study, we describe a sixth MHC-I lineage in teleosts. Previously, this lineage had not been noted in teleost fish because of the high level of divergence. We designated the lineage H, after the Greek hexa for six. Teleost H sequences are highly unusual in showing an unprecedented level of deterioration of their ectodomains, while having a large, unique, well-conserved motif in their cytoplasmic tails.

2. Materials and Methods

2.1. Datamining

A mixture of annotated and un-annotated MHC-I sequences were identified using various blastn and tblastn searches of Ensembl and NCBI databases using evolutionary diverged, as well as species-specific, sequences. Genes, genomic regions, and regional genes were identified using either the Ensembl blast browser: https://www.ensembl.org/index.html [33] (Nile tilapia Orenil1.0; stickleback BROAD S1; tetraodon TETRAODON 8.0; Spotted gar LepOcu 1) or the NCBI genome browser https://www.ncbi.nlm.nih.gov/genome for the remaining species with available genomes (see Supplementary Table S1 for further details). Some open reading frames were predicted using FGENESH [34], aligning genomic and expressed sequences using Splign (https://www.ncbi.nlm.nih.gov/sutils/splign/splign.cgi) and, in some other cases, exons and exon-intron junctions were defined using ORF finder in the Sequence Manipulation suite [35], followed by manual inspection. Deduced amino acid sequences are presented in Supplementary Text S1, while data on genomic sequences and matching expressed sequences are compiled in Supplementary Table S1.

2.2. Usage of the Word MHC-I

Our usage of the word "MHC-I" is based on phylogeny [10] and not on location (in the human genome) or function. Therefore, we do not use the distinction between "MHC-I" and "MHC-I-like" as used by some other researchers because we find that to be troublesome when discussing deep MHC evolution across species borders. Among MHC-I molecules, we distinguish between classical and nonclassical based on known or expected presence of classical functions.

2.3. Experimental Analysis of Mexican Tetra HAA Transcript Sequences

Total RNA was isolated using TRIzol (Gibco) from the gill and mixed internal organs of a Mexican tetra (*Astyanax mexicanus*) purchased from a pet shop. Animal handling and experiments were in agreement with regulations at Fujita Health University. Total RNA was reverse–transcribed into cDNA using ReverTra Ace (TOYOBO, Osaka, Japan), and the coding sequence of *Asme-HAA* gene was amplified by Ex-Taq HS (Takara Bio, Shiga, Japan) using primers *Asme-HAA 5′UTR.F1* (5′-AAATCATACCTGGGGTCAGCTGTTA-3′) and *Asme-HAA 3′UTR.R1* (5′-GCGAAGCACAACCACATGGTCATGA-3′) designed at 5′ and 3′ untranslated regions, respectively, of the Transcriptome Shotgun Assembly sequence report GFIF01006274. The PCR conditions were: denaturation at 98 °C for 30 s, 40 cycles of denaturation at 98 °C for 10 s, annealing at 54 °C for 30 s, and elongation at 72 °C for 90 s, and final elongation at 72 °C for 3 min. The PCR products were cloned into pGEM-T Easy vector (Promega, Madison, WI, USA) and sequencing reactions were performed with BigDye Terminator v3.1 Sequencing Standard kit (Applied Biosystems, Foster City, CA, USA), and the nucleotide sequences were determined using 3130xl Genetic Analyzer (Applied Biosystems). For both the gill and the mixed internal organ samples, multiple clones were determined to exclude PCR and sequencing artefacts, and for both samples, two different sequences were amplified that only show a single silent nucleotide exchange in the Asme-HAA coding sequence, *Asme-HAA*01* and *Asme-HAA*02* (Supplementary Text S2), which were deposited at GenBank as accessions LC494124 and LC494125.

2.4. Phylogenetic Analysis

Alignments of deduced MHC-I amino acid sequences were made by hand based on considerations on sequence similarity, relatedness of the sequences, and structure as described previously [10]. The transmembrane domain of HLA-A2 was predicted by TMPRED software, https://embnet.vital-it.

ch/software/TMPRED_form.html [36]. A phylogenetic tree using the best conserved domain, namely the α2 domain, was inferred using the Neighbor-Joining method [37] with bootstrap testing according to Felsenstein [38]. During the bootstrapping process, 132 failed, so bootstrap values are based on only 868 replicates. Thus, some pairwise distances could not be estimated, such as between sequence Icfu-HAA and Taru-TR6. The evolutionary distances were computed using the p-distance method [39]. The lineage clustering was supported by another phylogenetic tree constructed using Maximum Likelihood method based on the optimal JTT matrix-based model [40] (Supplementary Figure S1). Evolutionary analyses were conducted in MEGA7 [41].

2.5. Synonymous Versus Non-Synonymous Nucleotide Substitution Rates

The teleost HAA sequence fragments encoding the unique cytoplasmic tail motif were analyzed by the Synonymous Non-synonymous Analysis Program, SNAP v2.1.1, https://www.hiv.lanl.gov/content/sequence/SNAP/SNAP.html [42]. A similar analysis was performed for the available parts of the full-length coding sequences of characiform *HAA*.

2.6. Expression Analyses

Transcriptional values (Reads Per Kilobase per Million mapped reads or RPKM) were calculated using CLC Genomic Workbench 6.0.5 (https://www.qiagenbioinformatics.com/products/clc-genomics-workbench). Reads were mapped with high stringency, i.e., greater than 95% identity over more than 90% of the total length of the query read. Only open reading frame query sequence was used for each gene. Transcriptomes used in this study were: Atlantic salmon (*Salmo salar*) gills (SRR1422858), head kidney (SRR1422860), gut (SRR1422859), ovary (SRR1422871), testis (SRR1422872), spleen (SRR1422870), heart (SRR1422862), brain (SRR1422856), nose (SRR1422867), liver (SRR1422865), skin (SRR1422869), eye (SRR1422857) [43]; Northern pike (*Esox Lucius*) gills (SRR1533653), kidney (SRR1533657), intestine (SRR1533659), ovary (SRR1533651), testis (SRR1533661) [44]; Zebrafish (*Danio rerio*) gills (SRR1524239), kidney (SRR1524243), intestine (SRR1524245), ovary (SRR1524248), testis (SRR1524249); Spotted gar (*Lepososteus oculatus*) gills (SRR1524251), kidney (SRR1524255), intestine (SRR1524257), ovary (SRR1524259), and testis (SRR1524260).

3. Results

3.1. Identification of H Lineage Sequences in Holostei and Teleostei

H lineage sequences were found in Teleostei and Holostei by blast similarity searches using a spotted gar (*Lepisosteus oculatus*) sequence that we previously reported as *LO1* (*Lepisosteus oculatus* sequence 1) [30] and which we now recognize as a member of the previously unknown H lineage and therefore renamed *Leoc-HAA* (nomenclature as suggested in [45]). The sequences presented in the current study were found in public databases (Supplementary Text S2) and, in the case of *Asme-HAA* of Mexican tetra (*Astyanax mexicanus*), also confirmed by experiments (Supplementary Text S2). Phylogeny of the fish clades to which the investigated species belong is shown in Figure 1. Finding of H lineage members in both Holostei and Teleostei implies that the lineage is more than 300 million years old ([46]; Figure 1). We were unable to find H lineage members in other clades of species.

3.2. Genomic Positions of Detected H Lineage Genes Reveal Orthology

Genomic positions of representative H lineage genes are shown in Figure 2 and reveal orthology of the *HAA* genes in teleost fish and spotted gar. In the present study, in cases where the genomic location is not known, the first H lineage genes detected for a species are also named *HAA* (Supplementary Text S1, Table S1). Salmonid fishes experienced a whole genome duplication early in their evolution (e.g., [43,47]), which explains the gene duplication and the presence of *HAA* and a similar *HBA* gene in similar genetic surroundings (Figure 2 and Supplementary Table S1). In Atlantic salmon and rainbow trout (Figure 2 and Supplementary Text S1), but potentially not in coho salmon (*Oncorhynchus kisutch*; Supplementary Text S1), the *HBA* became a probable pseudogene. In common carp (*Cyprinus carpio*) two H lineage loci, *HAA* and *HBA*, are situated closely together (Table S1), indicating their origin by tandem duplication.

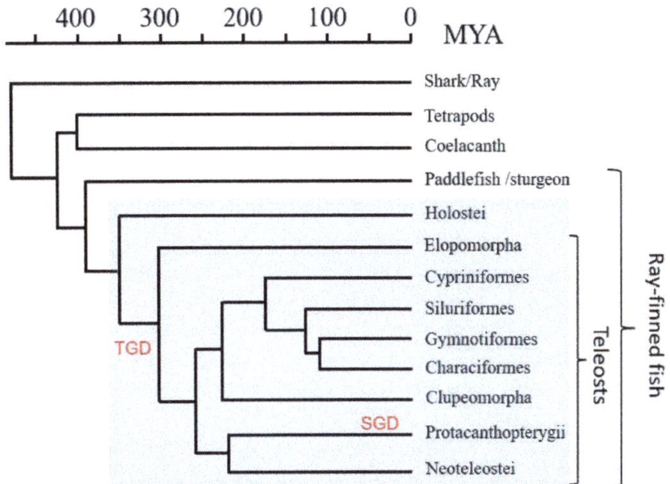

Figure 1. Phylogeny of the fish clades investigated in this study according to Near et al. [46]. A timescale is depicted in millions of years ago (MYA) above the figure. As for the investigated fish species: Holostei include bowfin and spotted gar; Elopomorpha include European eel and American eel; Cypriniformes include common carp, zebrafish, and horned golden-line barbel; Siluriformes include channel catfish, blue catfish, southern catfish, and amur catfish; Gymnotiformes include glass knifefish; Characiformes include Mexican tetra, tambaqui, and pacu; Clupeomorpha include Atlantic herring, Hilsa ilisa, sardine, and allis shad; Protacanthopterygii include Atlantic salmon, rainbow trout, coho salmon, and Northern pike; and Neoteleostei include Nile tilapia, stickleback, tetraodon, yellow croaker, red-lip croaker, guppy, medaka, and turquoise killifish. Clades possessing HAA gene sequences are boxed in blue. The teleost specific third whole genome duplication (TGD) and the salmonid specific fourth whole genome duplication (SGD) are shown.

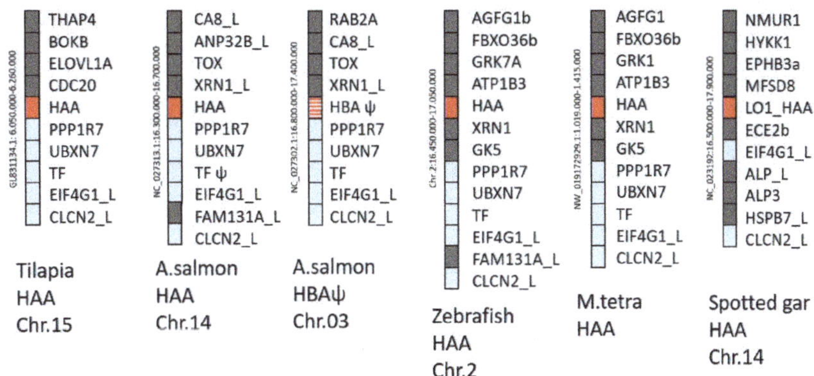

Figure 2. Genomic locations of H lineage loci in representative teleost fishes and spotted gar. H lineage genes and their flanking genes in Tilapia (*Oreochromis niloticus*), Atlantic salmon (*Salmo salar*), Zebrafish

(*Danio rerio*), Mexican tetra (*Astyanax mexicanus*), and Spotted gar (*Lepisosteus oculatus*) are shown as boxes. In Atlantic salmon, the region is duplicated and found on two different chromosomes. Red boxes represent intact H lineage genes, striped red represents a probable H lineage pseudogene, and light blue shading is used for non-MHC genes present in three or more species, while grey shading represents non-MHC genes showing poorer conservation in this region. The database sequence position of the depicted genomic region is shown in small font on the left side of each region, and if known the chromosome number is shown at the bottom of the figure. The ψ symbol indicates a probable pseudogene.

The H lineage loci appear not to be linked with the classical *Mhc* region (e.g., compare the H loci positions in Figure 2 with the positions of the classical MHC-I loci in reference [30]), which is quite common among nonclassical MHC-I (e.g., [10,30]).

3.3. Intron-Exon Organization of H Lineage Genes and Losses of Ectodomain Exons

Comparison of available genomic and cDNA information (Supplementary Table S1) allowed analysis of intron-exon organization. Intron-exon organizations of representative H lineage genes are shown in Figure 3, and sequences encoded by the α1, α2, and α3 exons are separately aligned in Figure 4. Spotted gar *Leoc-HAA* encodes all domains of a consensus MHC-I molecule, including α1, α2, α3, and CP/TM/CY domains, and the intron-exon organization is as commonly found among MHC-I genes (Figure 3). In contrast, cDNA analysis indicates that teleost fish H lineage genes do not possess α3 domain exon sequences (Figure 4 and Supplementary Text S1) and analysis of the genomic region sequences confirms this absence (Figure 3). Furthermore, in neither cDNA (Figure 4; Supplementary Text S2) nor genomic DNA (Figure 3) could an α1 domain exon sequence be found for Mexican tetra *Asme-HAA*, and this α1 absence was confirmed at the cDNA level for another characiform fish, pacu (*Piaractus mesopotamicus*) (Figure 4; Supplementary Text S1). Characiformes are related with Siluriformes and Gymnotiformes (Figure 1; [46]), and in Siluriformes, the deterioration of a consensus type MHC-I ectodomain is also pronounced. For example, in the two investigated fish of the genus *Silurus*, Southern catfish (*Silurus meridionalis*) and Amur catfish (*Silurus asotus*), the α2 exon appears to be entirely lacking according to investigated cDNA sequences (Figure 4; Supplementary Text S1); however, genomic information for these *Silurus* H genes is absent and, theoretically, there might be additional transcripts from the same genes that do include an α2 exon. Yet, for the related siluriform species channel catfish (*Ictalurus punctatus*), both cDNA and genomic sequence information is available (Supplementary Table S1), providing solid evidence that large parts of both α1 and α2 exon consensus were lost (Figures 3 and 4), and this probably relates to the same lack of importance of the ectodomains as reflected in the complete loss of α1 exon sequences in characiform H and of α2 exon sequences in *Silurus* H. Lack of conservation pressure for maintaining H lineage ectodomains may also explain the short α1 and α2 sequences of *Eivi-HAA* of glass knifefish (*Eigenmannia virescens*) belonging to Gymnotiformes (Figure 4), although the fragment losses are less extreme compared to those in Characiformes and Siluriformes. Teleost fishes other than Characiformes/Siluriformes/Gymnotiformes (C/S/G) possess H lineage sequences with an α2 length that is quite similar to MHC-I consensus, but their α1 sequence length is considerably shorter than the consensus, although not as short as in H sequences in C/S/G fish (Figures 3 and 4).

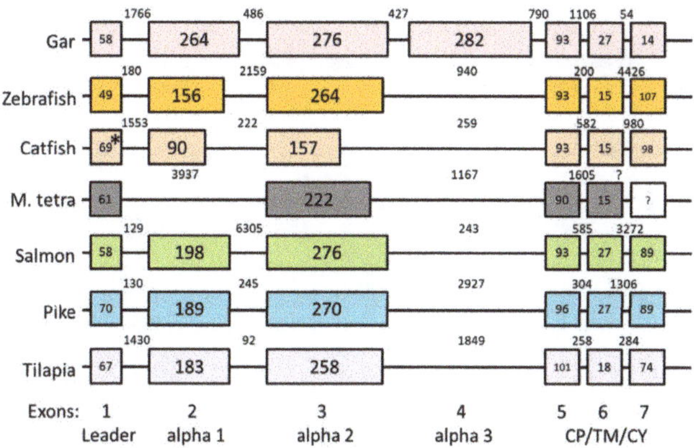

Figure 3. Exon-intron organization of selected *HAA* genes. Exons are shown as differently colored boxes for H lineage genes in Spotted gar (*Lepisosteus oculatus*), Zebrafish (*Danio rerio*), Channel catfish (*Ictalurus punctatus*), Mexican tetra (*Astyanax mexicanus*), Atlantic salmon (*Salmo salar*), Northern pike (*Esox Lucius*), and Tilapia (*Oreochromis niloticus*). The last exon of Mexican tetra found in different cDNA sequences was not present in the currently assembled GenBank genome sequence (Supplementary Text S1 and Text S2) and thus is presented as an uncolored box with question marks as to exon and intron sizes. Because of small differences between genomic and cDNA sequence reports, it was impossible to correctly align the first exon-intron boundary of the catfish *HAA* gene sequence, which is highlighted by an asterisk. Numbers within blocks indicate nucleotide lengths of the coding parts of exons, and numbers between blocks indicate intron lengths. Numbers at the bottom of the figure refer to MHC-I exon consensus numbers, with a description of their encoded domains.

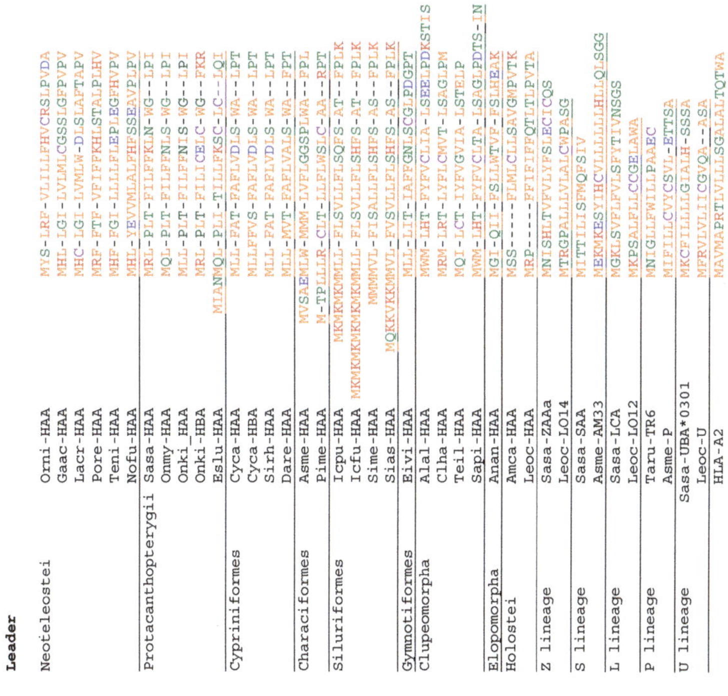

Figure 4. *Cont.*

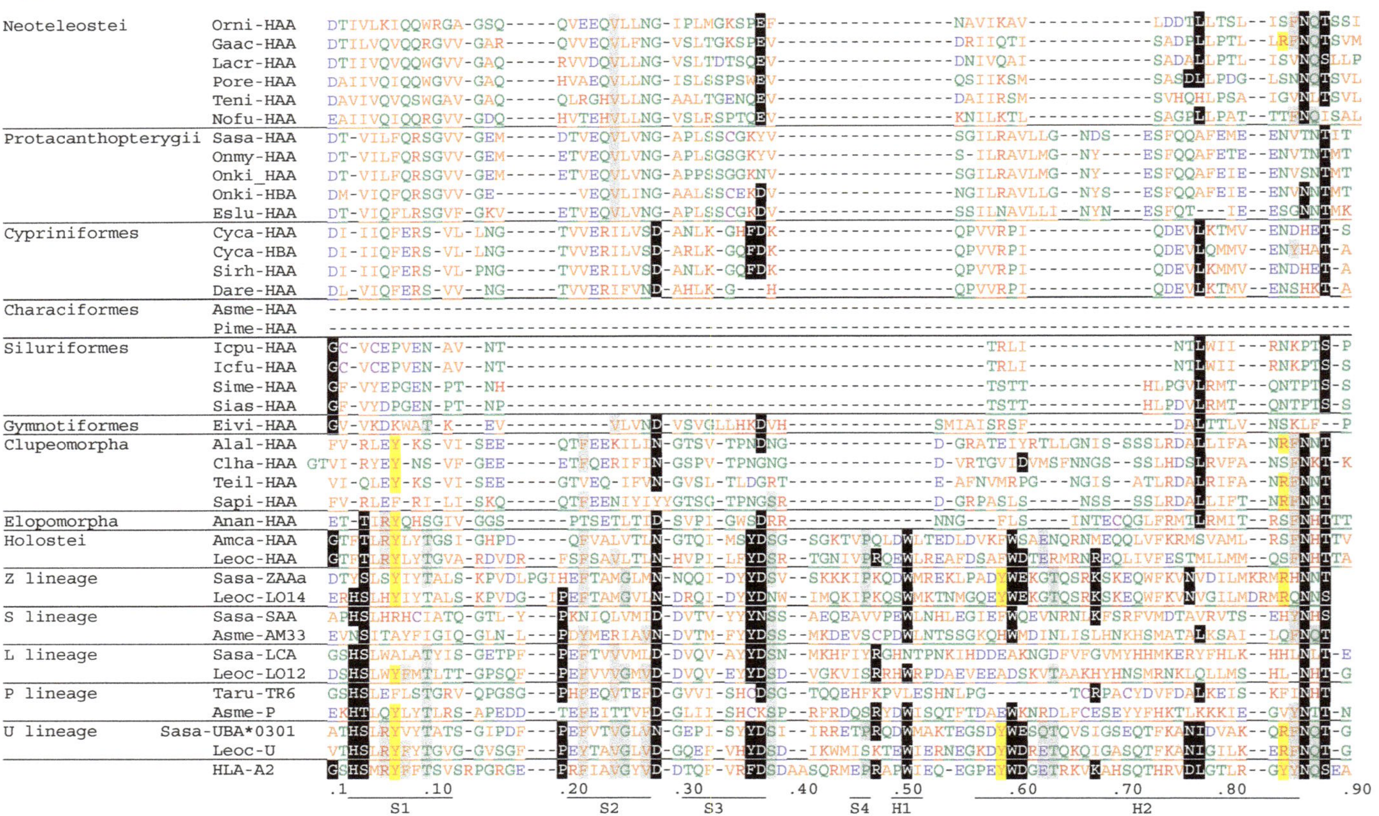

Figure 4. *Cont.*

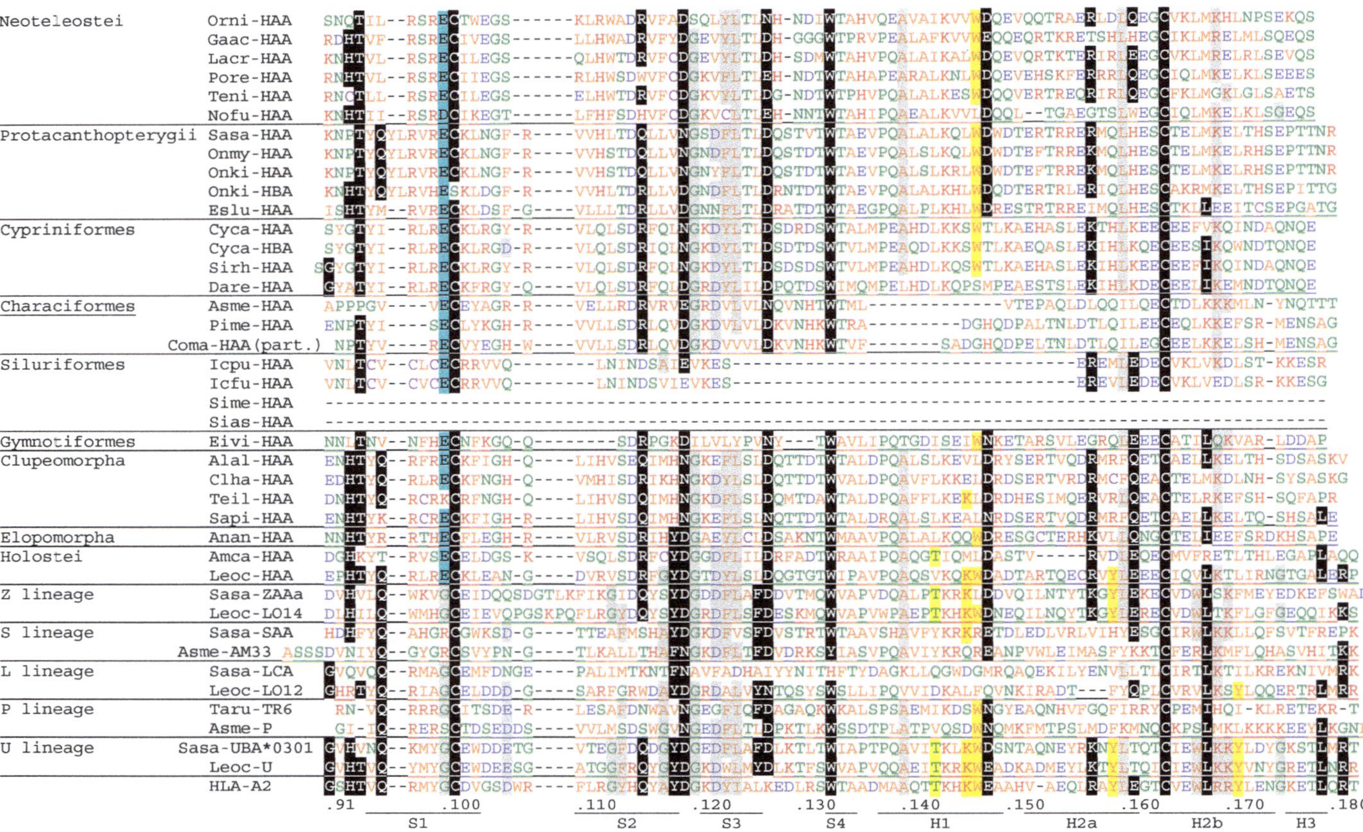

Figure 4. *Cont.*

Figure 4. *Cont.*

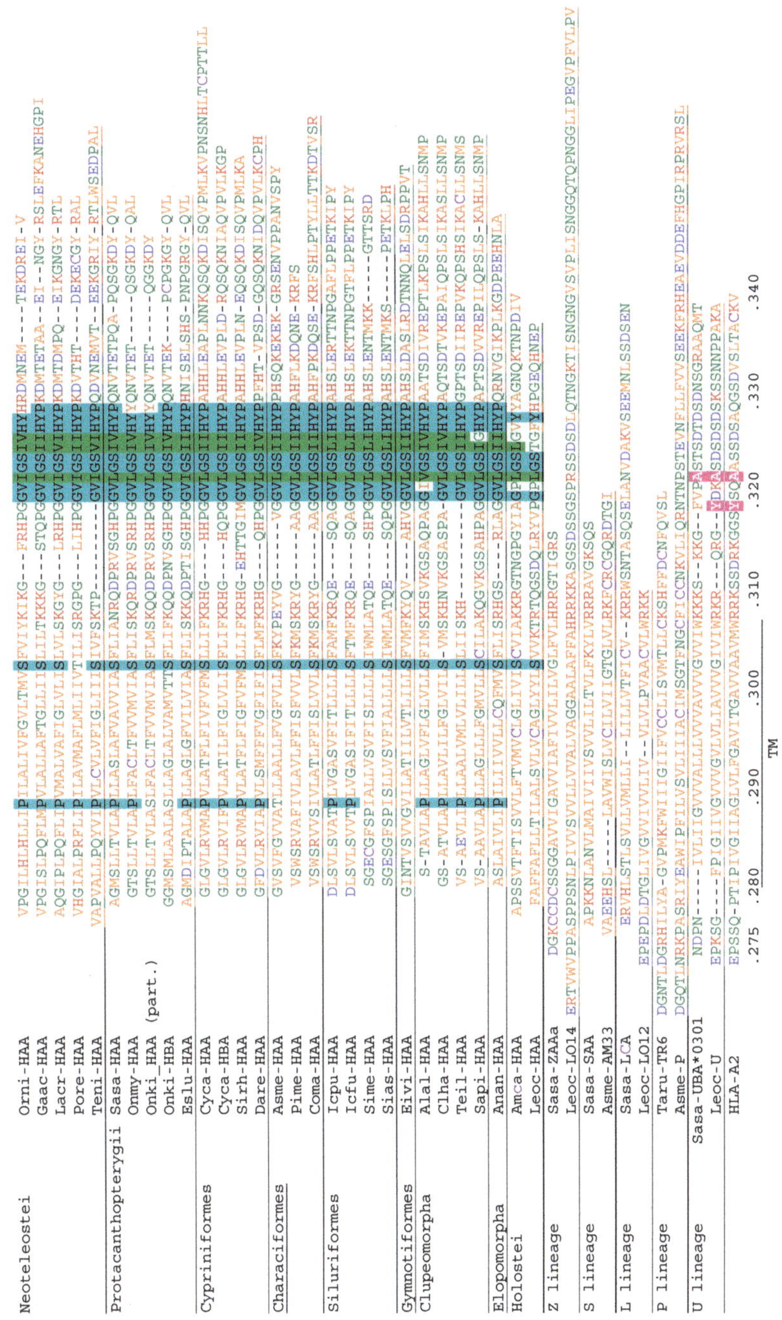

Figure 4. Alignment of deduced H lineage amino acid sequences with other representative MHC-I sequences. Color shading of residues highlights conserved features and is explained in the main text. The sequences are divided into leader, α1, α2, α3, and CP/TM/CY regions, the borders of the first four domains being defined by (expected) exon borders. Numbers under the alignment are based on residue

positions in mature HLA-A2 protein, and β-strand (S) and helix (H) structural indications are based on the pHLA-A2 structure in the PDB database accession 3PWN. Non-shaded cysteines are in purple font. Font colors of non-shaded residues are based on reference [48], with basic residues in red, acidic residues in blue, and green residues being more hydrophilic than the orange ones. The (part.) indications refer to incomplete sequence information. H sequence references are as follows: Orni (*Oreochromis niloticus*, Nile tilapia) -HAA (TSA: GBAZ01123113); Gaac (*Gasterosteus aculeatus*, Stickleback) -HAA (DW655318); Lacr (*Larimichthys crocea*, Yellow croaker) –HAA (XP_010741942.1); Pore (*Poecilia reticulata*, Guppy) -HAA (XP_008432358.1 and TSA: GFHH01045885); Teni (*Tetraodon nigroviridis*, Tetraodon) -HAA (CAG07665.1); Nofu (*Nothobranchius furzeri*, Turquoise killifish) -HAA (JZ213307); Sasa (*Salmo salar*, Atlantic salmon) -HAA (XP_013995094.1); Onmy (*Oncorhynchus mykiss*, Rainbow trout) -HAA (XP_021468778.1); Onki (*Oncorhynchus kisutch*, Coho salmon) -HAA (TSA: GDQG01022519.1) and -HBA (TSA: GDQG01022514.1); Eslu (*Esox Lucius*, Northern pike) -HAA (XP_010881869); Cyca (Cyprinus carpio, Common carp) -HAA (KTG41314) and -HBA (KTG33590); Sirh (*Sinocyclocheilus rhinocerous*, Horned golden-line barbel) -HAA (Genomic sequence NW_015649561.1:87.947-91.348); Dare (*Danio rerio*, Zebrafish) -HAA (XM_003197841); Asme (*Astyanax mexicanus*, Mexican tetra) -HAA (LC494124); Pime (*Piaractus mesopotamicus*, Pacu) -HAA (Bioproject PRJEB6656:); Coma (*Colossoma macropomum*, tambaqui) -HAA(TSA: GGHL01056846 and Bioproject PRJNA292457); Icpu (*Ictalurus punctatus*, Channel catfish) -HAA (TSA: JT437950); Icfu (*Ictalurus furcatus*, Blue catfish) -HAA (Bioproject PRJNA195453); Sime (*Silurus meridionalis*, Southern catfish) -HAA (Bioproject PRJNA427243); Sias (*Silurus asotus*, Amur catfish) -HAA (TSA: GHGF01004423); Eivi (*Eigenmannia virescens*, Glass knifefish) -HAA (TSA: GGGZ01064726); Alal (*Alosa*, Allis shad) -HAA (TSA: GETY01043622); Clha (*Clupea harengus*, Atlantic herring) -HAA (Genomic sequence NW_012220971.1: 1.071.225-1.073.475); Teil (*Tenualosa ilisha*, Hilsa ilisa) -HAA (QYSC01123722.1: 356.174-357.933); Sapi (*Sardina pilchardus*, Sardine) -HAA (TSA: GGSC01229082); Amca (*Amia calva*, Bowfin) -HAA (TSA: GEUG01019669); Leoc (*Lepisosteus oculatus*, Spotted gar) -HAA (XP_015216910). The references of the other sequences are Sasa (*Salmo salar*, Atlantic salmon) -SAA (ACY30362.1), -LCA (XP_013983104.1), -ZAAa (ACX35596.1), -UBA (*0301, XP_014032819); Leoc (*Lepisosteus oculatus*, Spotted gar) -Z (LO14, TSA: GFIM01040660), -L (LO12, JH591577:52,184-56,541), -U (TSA: GFIM01032149); Asme (*Astyanax mexicanus*, Mexican tetra) -S (AM33, ENSAMXG00000017444), -P (TSA: GFIF01000014); Taru (*Takifugu rubripes*, Fugu) -P (Scaffold_497:44,782-49,340); and Human HLA-A2 (AAA76608.2). See Supplementary Table S1 and Text S1 for more details.

3.4. Deduced Amino Acid Sequences of Teleost H Lineage Molecules Reveal Deterioration of Ectodomains and the Possession of an Unusual Cytoplasmic Tail Motif

In Figure 4, the deduced amino acid sequences of H lineage sequences are compared with representative sequences of other MHC-I lineages found in teleost fish, and with the human classical MHC-I molecule HLA-A2. Black shading highlights residues or sets of similar residues that probably were also present before the evolutionary separation of the MHC classes I and II [10,49], gray shading highlights residues or sets of similar residues that are common in and rather specific for classical MHC-I ([10] and our ongoing investigations), and yellow shading highlights conserved residues which in classical MHC-I are involved in binding the peptide termini [3,5]. The depicted Atlantic salmon Sasa-UBA*0301 sequence is an allele of the polymorphic classical *UBA* locus [30,50]. From its length and shading pattern in Figure 4, it is readily concluded that spotted gar Leoc-HAA shows an overall similarity to classical MHC-I and may build a similar structure, whereas it lost four of the classical MHC-I residues used for binding of peptide termini and possibly does not possess a groove for peptide binding. Holostei are classified into the orders Lepisosteiformes and Amiiformes, represented by spotted gar and bowfin (*Amia calva*), respectively. Although an α3 domain is present in spotted gar Leoc-HAA, in bowfin Amca-HAA and teleost H lineage sequences, the α3 domain is absent (Figure 2), which suggests that the α3 domain was independently lost in the bowfin and teleost fish ancestors. However, the absence of an α3 domain in bowfin Amca-HAA would need confirmation at the genomic level because the single available cDNA sequence might represent only one of multiple splicoform variants.

Because of the large degree of diversification between the sequences, some regions of the Figure 4 alignment are only tentative. Except for the apparent and differential losses of ectodomain parts, it is difficult to find common features that collectively distinguish the H lineage ectodomains from other MHC-I lineages. The only readily distinguishable specific feature may be the acidic residue at position 100, shaded blue in Figure 4. Nonetheless, phylogenetic tree analysis shows that the H lineage α2 domain sequences do share overall specific similarity as the H sequences form a single cluster (Figure 5; we refrained from such analysis for the α1 domain because of uncertainty about the correct alignment).

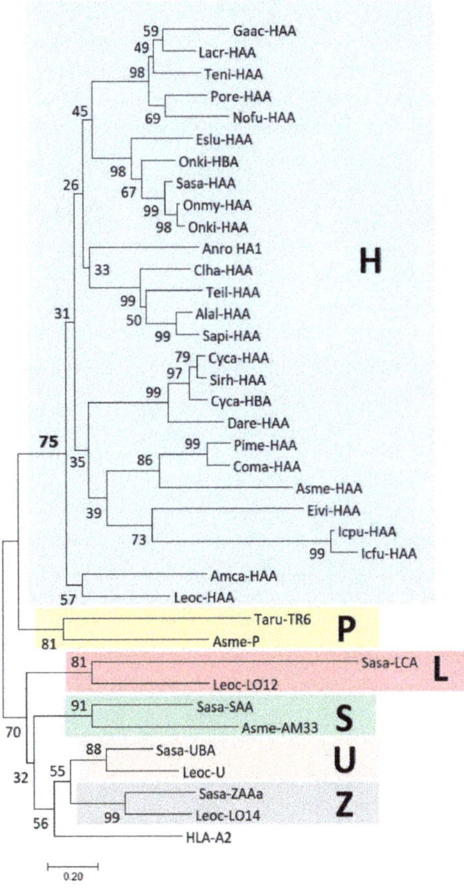

Figure 5. Phylogenetic tree based on α2 domain amino acid sequences of representative H lineage and other MHC-I molecules as aligned in Figure 4. Clusters with the six different MHC class I lineages found in teleosts are shown with colored boxes. The evolutionary history was inferred using the Neighbor-Joining method [37]. The optimal tree with the sum of branch length = 15.21024459 is shown. The percentage of replicate trees in which the associated taxa clustered together in the bootstrap test (1000 replicates) are shown next to the branches [38]. The tree is drawn to scale, with branch lengths in the same units as those of the evolutionary distances used to infer the phylogenetic tree. The evolutionary distances were computed using the Poisson correction method [39] and are in the units of the number of amino acid substitutions per site. The analysis involved 38 amino acid sequences. All ambiguous positions were removed for each sequence pair. There were a total of 104 positions in the final dataset. Evolutionary analyses were conducted in MEGA7 [41]. Sequence references can be found in legend to Figure 4 and in Supplementary Text S1.

Among classical MHC-I cytoplasmic tails, the YXXΛ motif (X denotes any possible residue) is rather well conserved (shaded magenta in Figure 4), although in teleost fish the tyrosine is commonly replaced by phenylalanine (Figure 4; [51]). The YXXΛ motif in classical MHC-I plays a role in endocytic trafficking of surface MHC-I molecules and the loading with exogenous antigens [52,53]. These residues are not present in the H lineage sequences.

However, in sharp contrast to the diversification of their ectodomains, the teleost H lineage sequences show remarkable conservation of a large motif in the cytoplasmic tail GV(I/L)GS(I/L/V)(I/V)HYP. In Figure 4, the single residues within this motif in teleost H are shaded blue and the variable residues are shaded green. We are not aware of a similar cytoplasmic tail motif in other proteins, but the conserved tyrosine suggests involvement in intracellular pathways by means of phosphorylation. The conservation of hydrophobic residues at some positions within the cytoplasmic tail motif suggests interaction with another as yet unidentified protein. Equally, the conservation of a hydrophilic serine at position 304 in the teleost H lineage transmembrane domain sequences, and maybe also the partial conservation of a proline at position 290, suggest interaction of the TM domain with some other molecule. Examples of molecules believed to use a conserved serine/threonine or proline within the transmembrane domain for intermolecular protein binding are CD74 [54,55] and Igα [56], respectively. The identity of the molecules putatively interacting with the HAA TM/CY regions can only be guessed. In H molecules of Holostei, only parts of the teleost H lineage GV(I/L)GS(I/L/V)(I/V)HYP cytoplasmic tail motif are found, but, importantly, the tyrosine within the motif is conserved, as is the unusual serine within the TM domain. Thus, unique features of the H lineage transmembrane and cytoplasmic tail domains have been conserved over 300 million years, underlining their probable importance.

3.5. Nucleotide Sequences Encoding the Teleost H Lineage Cytoplasmic Tail Motif Indicate Purifying Selection

A possible explanation for the deterioration of the H lineage ectodomains in teleosts could be that the proteins lost their function and that the teleost H genes and gene transcripts are only nonfunctional remnants of an evolutionary past, or, alternatively, that teleost H genes only have a function at the transcript level. Because a lack of protein function is reflected in the rates of synonymous versus non-synonymous (ds versus dn) substitutions that accumulate in a gene, the nucleotide sequences encoding the cytoplasmic tail GV(I/L)GS(I/L/V)(I/V)HYP motif were compared between all teleost H lineage molecules shown in Figure 4 and between the H lineage molecules of only Characiformes/Siluriformes/Gymnotiformes (C/S/G) (Supplementary Text S3A–C). The results in Supplementary Text S3 show that the ds/dn ratio among H lineage molecules of all compared teleost aligned is 14.60 and that, among the H lineage molecules of the C/S/G teleost subgroup, it is 20.03. These numbers indicate purifying selection at the amino acid level and suggest that H lineage proteins are functional in teleosts, including in C/S/G fish.

The characiform *HAA* sequences *Asme-HAA*, *Pime-HAA*, and *Coma-HAA* are sufficiently similar and dissimilar for allowing reliable alignment and meaningful ds/dn analysis for the available part of their full-length coding sequences, and the calculated ds/dn ratios were 2.00 for *Asme/Pime*, 2.35 for *Asme/Coma*, and 1.71 for *Pime/Coma* (Supplementary Text S3D). This is additional evidence for purifying selection at the amino acid level in characiform H lineage sequences, even after losses of both the α1 and α3 domains.

3.6. Expression Pattern

The tissue distribution of H lineage transcripts was determined by analysis of transcriptome data. Initially, we investigated a panel with a wide variety of Atlantic salmon tissue transcriptomes available in GenBank [43]. This analysis showed highest expression in Atlantic salmon ovary with low to medium expression levels in other tissues, disregarding liver and skin where the expression levels were insignificant (Table 1). The expression levels are comparable to those of other non-classical Atlantic salmon MHC-I genes, such as *UDA*, *LDA*, and *ZBAa*, and are generally much lower than found for the classical *UBA* gene (Table 1, data from [30]). We then analyzed the expression of *HAA* genes in transcriptome data publicly available for several ray-finned fish species and found an overall low to medium expression level (Table 1). The high expression level found in Atlantic salmon ovary was neither seen in ovary samples from the other species shown in Table 1 nor in rainbow trout ovary (GenBank ERR324375; rainbow trout data not shown). However, in general, it is difficult to compare transcription values between animals and samples as the biological age and status of the animals, providing the transcriptomes are not well defined. From Table 1, it can be concluded that H lineage genes are transcribed in most tissues in species ranging from spotted gar to Atlantic salmon, but the detected expression pattern is not helpful for predicting a specific function.

Table 1. *HAA* transcription in various tissues of several teleost fish species. Transcriptional values are given as RPKM, i.e., Reads Per Kilobase per Million mapped reads. See Materials and Methods section for dataset references. * Tissue denoted kidney is defined as head kidney in Atlantic salmon but as kidney in the other species. ** Intestine/gut is defined as gut in Atlantic salmon but as intestine in the other species. Atlantic salmon RPKM values in grey colored cells for the classical *UBA* locus and the nonclassical *UDA*, *LDA*, and *ZBAa* genes originate from our previous study [30].

Tissue\Gene	Sasa-HAA	Eslu-HAA	Dare-HAA	Leoc-HAA	Sasa-UBA	Sasa-UDA	Sasa-LDA	Sasa-ZBAa
Gills	9.48	1.57	3.50	0.95	253.57	5.65	3.36	43.46
Kidney *	11.49	1.79	4.11	3.55	66.97	4.05	4.17	9.42
Intestine/gut **	16.47	1.90	4.41	11.09	361.72	4.27	1.86	18.57
Ovary	56.84	2.66	8.65	1.35	1.25	4.29	6.08	0.11
Testis	10.49	1.54	3.18	2.07	60.84	6.31	1.14	2.70
Spleen	9.62	n/a	n/a	n/a	260.43	6.63	2.72	19.47
Heart	2.85	n/a	n/a	n/a	16.79	1.35	1.24	8.11
Brain	5.61	n/a	n/a	n/a	17.65	1.75	0.30	3.34
Nose	2.40	n/a	n/a	n/a	60.70	5.44	2.02	8.61
Liver	0.94	n/a	n/a	n/a	12.92	0.78	1.23	4.30
Skin	0.17	n/a	n/a	n/a	4.84	6.31	0.09	0.08
Eye	1.77	n/a	n/a	n/a	11.75	0.79	0.15	2.76
Query Length (bp)	741	741	693	1056	1068	1068	1080	1128

4. Discussion

In the present study, a sixth lineage of teleost MHC-I sequences is presented and designated H. H lineage sequences are also found in Holostei, and spotted gar HAA at least looks structurally similar to classical sequences. Spotted gar HAA has all domains found in classical MHC-I, and those domains are of similar length and have many of the MHC-I and MHC-I/II characteristic residues (gray and black shading in Figure 4), although several of the classical residues for binding of peptide termini (yellow shading in Figure 4) were lost. In contrast to the spotted gar HAA sequence, the teleost H lineage sequences show an unprecedented deterioration of the canonical MHC-I ectodomain structure, with a loss of the α3 domain and a seemingly random loss of stretches and residues within the α1 and α2 domains, or even of the entire α1 or α2 domains. Although firm conclusions cannot be drawn, the impression from the sequence comparisons is that, in teleost H lineage molecules, the ectodomain lost most of its function. In contrast, unique residues in the transmembrane domain and cytoplasmic tail which are found in Holostei H molecules are also conserved in teleosts and suggest binding of a yet unknown partner molecule and intracellular signaling through tyrosine phosphorylation. Especially among teleost H molecules, the conservation of a large cytoplasmic motif including the mentioned tyrosine is impressive. We are not aware of any previous descriptions of a group of MHC molecules with such deteriorated ectodomains or with such an impressive cytoplasmic tail motif. Regarding the often posed question of how the unique MHC peptide-binding domain structure emerged in evolution the teleost H sequences are quite interesting, as they seem to provide evidence that also partial MHC structures can be stable. Therefore, future studies should not only investigate the function but also the structures of the H lineage molecules.

Supplementary Materials: The following are available online at http://www.mdpi.com/2073-4409/8/9/1056/s1, Text S1: Gene specifics and sequences; Text S2: Mexican tetra amplification; Text S3: Substitution rates; Table S1. Genomes, genomic location and expressed match; and Figure S1. Maximum Likelihood phylogenetic tree.

Author Contributions: Conceptualization, U.G. and J.M.D.; methodology, J.M.D.; validation, U.G., K.H. and J.M.D.; formal analysis, U.G. and J.M.D.; investigation, K.T and J.M.D.; writing—original draft preparation, U.G. and J.M.D.; writing—review and editing, U.G. and J.M.D.; visualization, U.G., K.T. and J.M.D.; supervision, U.G. and K.H.; project administration, U.G. and K.H.; funding acquisition, U.G. and K.H.

Funding: This study was funded by a research grant from the Norwegian Research Council (Grant # 274635) and by Japan Society for the Promotion of Science Grants-in-Aid for Scientific Research Grant Number JP26440201.

Conflicts of Interest: The authors declare no conflict of interest.

References

1. Bjorkman, P.J.; Saper, M.A.; Samraoui, B.; Bennett, W.S.; Strominger, J.L.; Wiley, D.C. Structure of the human class I histocompatibility antigen, HLA-A2. *Nature* **1987**, *329*, 506–512. [CrossRef] [PubMed]

2. Neefjes, J.; Jongsma, M.L.; Paul, P.; Bakke, O. Towards a systems understanding of MHC class I and MHC class II antigen presentation. *Nat. Rev. Immunol.* **2011**, *11*, 823–836. [CrossRef] [PubMed]

3. Madden, D.R. The three-dimensional structure of peptide-MHC complexes. *Annu.Rev.Immunol.* **1995**, *13*, 587–622. [CrossRef] [PubMed]

4. Rammensee, H.G. Chemistry of peptides associated with MHC class I and class II molecules. *Curr. Opin. Immunol.* **1995**, *7*, 85–96. [CrossRef]

5. Hashimoto, K.; Okamura, K.; Yamaguchi, H.; Ototake, M.; Nakanishi, T.; Kurosawa, Y. Conservation and diversification of MHC class I and its related molecules in vertebrates. *Immunol. Rev.* **1999**, *167*, 81–100. [CrossRef] [PubMed]

6. Chen, W.; Jia, Z.; Zhang, T.; Zhang, N.; Lin, C.; Gao, F.; Wang, L.; Li, X.; Jiang, Y.; Li, X.; et al. MHC class I presentation and regulation by IFN in bony fish determined by molecular analysis of the class I locus in grass carp. *J. Immunol.* **2010**, *185*, 2209–2221. [CrossRef]

7. Chen, Z.; Zhang, N.; Qi, J.; Chen, R.; Dijkstra, J.M.; Li, X.; Wang, Z.; Wang, J.; Wu, Y.; Xia, C. The Structure of the MHC Class I Molecule of Bony Fishes Provides Insights into the Conserved Nature of the Antigen-Presenting System. *J. Immunol.* **2017**, *199*, 3668–3678. [CrossRef]

8. Dijkstra, J.M.; Fischer, U.; Sawamoto, Y.; Ototake, M.; Nakanishi, T. Exogenous antigens and the stimulation of MHC class I restricted cell-mediated cytotoxicity: Possible strategies for fish vaccines. *Fish Shellfish Immunol.* **2001**, *11*, 437–458. [CrossRef]

9. Yamaguchi, T.; Dijkstra, J.M. Major Histocompatibility Complex (MHC) Genes and Disease Resistance in Fish. *Cells* **2019**, *8*, 378. [CrossRef]

10. Dijkstra, J.M.; Yamaguchi, T.; Grimholt, U. Conservation of sequence motifs suggests that the nonclassical MHC class I lineages CD1/PROCR and UT were established before the emergence of tetrapod species. *Immunogenetics* **2018**, *70*, 459–476. [CrossRef]

11. Hughes, A.L.; Nei, M. Evolution of the major histocompatibility complex: Independent origin of nonclassical class I genes in different groups of mammals. *Mol. Bio. Evol.* **1989**, *6*, 559–579.

12. Adams, E.J.; Luoma, A.M. The adaptable major histocompatibility complex (MHC) fold: Structure and function of nonclassical and MHC class I-like molecules. *Annu. Rev. Immunol.* **2013**, *31*, 529–561. [PubMed]

13. Araki, T.; Gejyo, F.; Takagaki, K.; Haupt, H.; Schwick, H.G.; Burgi, W.; Marti, T.; Schaller, J.; Rickli, E.; Brossmer, R.; et al. Complete amino acid sequence of human plasma Zn-alpha 2-glycoprotein and its homology to histocompatibility antigens. *Proc. Natl. Acad. Sci. USA* **1988**, *85*, 679–683. [CrossRef] [PubMed]

14. Delker, S.L.; West, A.P., Jr.; McDermott, L.; Kennedy, M.W.; Bjorkman, P.J. Crystallographic studies of ligand binding by Zn-alpha2-glycoprotein. *J. Struct. Biol.* **2004**, *148*, 205–213. [CrossRef] [PubMed]

15. Sanchez, L.M.; Chirino, A.J.; Bjorkman, P. Crystal structure of human ZAG, a fat-depleting factor related to MHC molecules. *Science* **1999**, *283*, 1914–1919. [PubMed]

16. Fukudome, K.; Esmon, C.T. Identification, cloning, and regulation of a novel endothelial cell protein C/activated protein C receptor. *J. Boil. Chem.* **1994**, *269*, 26486–26491.

17. Mohan Rao, L.V.; Esmon, C.T.; Pendurthi, U.R. Endothelial cell protein C receptor: A multiliganded and multifunctional receptor. *Blood* **2014**, *124*, 1553–1562. [CrossRef] [PubMed]

18. Oganesyan, V.; Oganesyan, N.; Terzyan, S.; Qu, D.; Dauter, Z.; Esmon, N.L.; Esmon, C.T. The crystal structure of the endothelial protein C receptor and a bound phospholipid. *J. Boil. Chem.* **2002**, *277*, 24851–24854.

19. Calabi, F.; Milstein, C. A novel family of human major histocompatibility complex-related genes not mapping to chromosome 6. *Nature* **1986**, *323*, 540–543. [CrossRef]

20. Zeng, Z.; Castano, A.R.; Segelke, B.W.; Stura, E.A.; Peterson, P.A.; Wilson, I.A. Crystal structure of mouse CD1: An MHC-like fold with a large hydrophobic binding groove. *Science* **1997**, *277*, 339–345.

21. Van Rhijn, I.; Godfrey, D.I.; Rossjohn, J.; Moody, D.B. Lipid and small-molecule display by CD1 and MR1. *Nat. Rev. Immunol.* **2015**, *15*, 643–654. [CrossRef] [PubMed]

22. Zou, Z.; Nomura, M.; Takihara, Y.; Yasunaga, T.; Shimada, K. Isolation and characterization of retinoic acid-inducible cDNA clones in F9 cells: A novel cDNA family encodes cell surface proteins sharing partial homology with MHC class I molecules. *J. Biochem.* **1996**, *119*, 319–328. [CrossRef] [PubMed]

23. Fernandez-Messina, L.; Reyburn, H.T.; Vales-Gomez, M. Human NKG2D-ligands: Cell biology strategies to ensure immune recognition. *Front. Immunol.* **2012**, *3*, 299. [CrossRef] [PubMed]

24. Kasahara, M.; Yoshida, S. Immunogenetics of the NKG2D ligand gene family. *Immunogenetics* **2012**, *64*, 855–867. [CrossRef] [PubMed]

25. Dijkstra, J.M.; Katagiri, T.; Hosomichi, K.; Yanagiya, K.; Inoko, H.; Ototake, M.; Aoki, T.; Hashimoto, K.; Shiina, T. A third broad lineage of major histocompatibility complex (MHC) class I in teleost fish; MHC class II linkage and processed genes. *Immunogenetics* **2007**, *59*, 305–321. [CrossRef]

26. Grimholt, U.; Hordvik, I.; Fosse, V.M.; Olsaker, I.; Endresen, C.; Lie, O. Molecular cloning of major histocompatibility complex class I cDNAs from Atlantic salmon (*Salmo salar*). *Immunogenetics* **1993**, *37*, 469–473. [CrossRef] [PubMed]

27. Hashimoto, K.; Nakanishi, T.; Kurosawa, Y. Isolation of carp genes encoding major histocompatibility complex antigens. *Proc. Natl. Acad. Sci. USA* **1990**, *87*, 6863–6867. [CrossRef]

28. Kruiswijk, C.P.; Hermsen, T.T.; Westphal, A.H.; Savelkoul, H.F.; Stet, R.J. A novel functional class I lineage in zebrafish (*Danio rerio*), carp (*Cyprinus carpio*), and large barbus (*Barbus intermedius*) showing an unusual conservation of the peptide binding domains. *J. Immunol.* **2002**, *169*, 1936–1947. [CrossRef]

29. Shum, B.P.; Rajalingam, R.; Magor, K.E.; Azumi, K.; Carr, W.H.; Dixon, B.; Stet, R.J.; Adkison, M.A.; Hedrick, R.P.; Parham, P. A divergent non-classical class I gene conserved in salmonids. *Immunogenetics* **1999**, *49*, 479–490. [CrossRef]

30. Grimholt, U.; Tsukamoto, K.; Azuma, T.; Leong, J.; Koop, B.F.; Dijkstra, J.M. A comprehensive analysis of teleost MHC class I sequences. *BMC Evol. Boil.* **2015**, *15*. [CrossRef]

31. Dijkstra, J.M.; Kiryu, I.; Yoshiura, Y.; Kumanovics, A.; Kohara, M.; Hayashi, N.; Ototake, M. Polymorphism of two very similar MHC class Ib loci in rainbow trout (*Oncorhynchus mykiss*). *Immunogenetics* **2006**, *58*, 152–167. [CrossRef] [PubMed]

32. Dirscherl, H.; Yoder, J.A. Characterization of the Z lineage Major histocompatability complex class I genes in zebrafish. *Immunogenetics* **2013**, *66*, 185–198. [CrossRef] [PubMed]

33. Aken, B.L.; Achuthan, P.; Akanni, W.; Amode, M.R.; Bernsdorff, F.; Bhai, J.; Billis, K.; Carvalho-Silva, D.; Cummins, C.; Clapham, P.; et al. Ensembl 2017. *Nucleic Acids Res.* **2017**, *45*, D635–D642. [CrossRef]

34. Salamov, A.A.; Solovyev, V.V. Ab initio gene finding in Drosophila genomic DNA. *Genome Res.* **2000**, *10*, 516–522. [CrossRef] [PubMed]

35. Stothard, P. The sequence manipulation suite: JavaScript programs for analyzing and formatting protein and DNA sequences. *BioTechniques* **2000**, *28*, 1102–1104. [CrossRef] [PubMed]

36. Hofmann, K. TMBASE—A database of membrane spanning protein segments. *Biol.Chem. Hoppe-Seyel* **1993**, *374*, 166.

37. Saitou, N.; Nei, M. The neighbor-joining method: A new method for reconstructing phylogenetic trees. *Mol. Boil. Evol.* **1987**, *4*, 406–425.

38. Felsenstein, J. Confidence limits on phylogenies: An approach using the bootstrap. *Evolution* **1985**, *39*, 783–791. [CrossRef] [PubMed]

39. Zuckerkandl, E.; Pauling, L. Evolutionary divergence and convergence in proteins. In *Evolving Genes and Proteins*; Bryson, V., Vogel, H.J., Eds.; Academic Press: New York, NY, USA, 1965; Volume 33, pp. 1870–1874.

40. Jones, D.T.; Taylor, W.R.; Thornton, J.M. The rapid generation of mutation data matrices from protein sequences. *Comput. Appl. Biosci.* **1992**, *8*, 275–282. [CrossRef]

41. Kumar, S.; Stecher, G.; Tamura, K. MEGA7: Molecular Evolutionary Genetics Analysis Version 7.0 for Bigger Datasets. *Mol. Boil. Evol.* **2016**, *33*, 1870–1874. [CrossRef]

42. Korber, B. HIV Signature and Sequence Variation Analysis. In *Computational Analysis of HIV Molecular Sequences*; Rodrigo, A.G., Learn, G.H., Eds.; Kluwer Academic Publishers: Dordrecht, Netherlands, 2000; pp. 55–72.

43. Lien, S.; Koop, B.F.; Sandve, S.R.; Miller, J.R.; Kent, M.P.; Nome, T.; Hvidsten, T.R.; Leong, J.S.; Minkley, D.R.; Zimin, A.; et al. The Atlantic salmon genome provides insights into rediploidization. *Nature* **2016**, *533*, 200–205. [CrossRef] [PubMed]

44. Rondeau, E.B.; Minkley, D.R.; Leong, J.S.; Messmer, A.M.; Jantzen, J.R.; von Schalburg, K.R.; Lemon, C.; Bird, N.H.; Koop, B.F. The genome and linkage map of the northern pike (*Esox lucius*): Conserved synteny revealed between the salmonid sister group and the Neoteleostei. *PLoS ONE* **2014**, *9*, e102089. [CrossRef] [PubMed]

45. Klein, J.; Bontrop, R.E.; Dawkins, R.L.; Erlich, H.A.; Gyllensten, U.B.; Heise, E.R.; Jones, P.P.; Parham, P.; Wakeland, E.K.; Watkins, D.I. Nomenclature for the major histocompatibility complexes of different species: A proposal. *Immunogenetics* **1990**, *31*, 217–219. [CrossRef] [PubMed]

46. Near, T.J.; Eytan, R.I.; Dornburg, A.; Kuhn, K.L.; Moore, J.A.; Davis, M.P.; Wainwright, P.C.; Friedman, M.; Smith, W.L. Resolution of ray-finned fish phylogeny and timing of diversification. *Proc. Natl. Acad. Sci. USA* **2012**, *109*, 13698–13703. [CrossRef] [PubMed]

47. Kodama, M.; Brieuc, M.S.; Devlin, R.H.; Hard, J.J.; Naish, K.A. Comparative mapping between Coho Salmon (Oncorhynchus kisutch) and three other salmonids suggests a role for chromosomal rearrangements in the retention of duplicated regions following a whole genome duplication event. *G3 (Bethesda)* **2014**, *4*, 1717–1730. [CrossRef] [PubMed]

48. Hopp, T.P.; Woods, K.R. Prediction of protein antigenic determinants from amino acid sequences. *Proc. Natl. Acad. Sci. USA* **1981**, *78*, 3824–3828. [CrossRef] [PubMed]

49. Kaufman, J. Vertebrates and the evolution of Major Histocompatibility Complex (MHC) class I and class II molecules. *Verh.Dtsch.Zool.Ges.* **1988**, *81*, 131–144.

50. Grimholt, U.; Drablos, F.; Jorgensen, S.M.; Hoyheim, B.; Stet, R.J. The major histocompatibility class I locus in Atlantic salmon (*Salmo salar* L.): Polymorphism, linkage analysis and protein modelling. *Immunogenetics* **2002**, *54*, 570–581. [CrossRef]

51. Aoyagi, K.; Dijkstra, J.M.; Xia, C.; Denda, I.; Ototake, M.; Hashimoto, K.; Nakanishi, T. Classical MHC class I genes composed of highly divergent sequence lineages share a single locus in rainbow trout (*Oncorhynchus mykiss*). *J. Immunol.* **2002**, *168*, 260–273. [CrossRef]

52. Basha, G.; Lizee, G.; Reinicke, A.T.; Seipp, R.P.; Omilusik, K.D.; Jefferies, W.A. MHC class I endosomal and lysosomal trafficking coincides with exogenous antigen loading in dendritic cells. *PLoS ONE* **2008**, *3*, e3247. [CrossRef]

53. Lizee, G.; Basha, G.; Jefferies, W.A. Tails of wonder: Endocytic-sorting motifs key for exogenous antigen presentation. *Trends Immunol.* **2005**, *26*, 141–149. [CrossRef] [PubMed]

54. Ashman, J.B.; Miller, J. A role for the transmembrane domain in the trimerization of the MHC class II-associated invariant chain. *J. Immunol.* **1999**, *163*, 2704–2712. [PubMed]

55. Dijkstra, J.M.; Yamaguchi, T. Ancient features of the MHC class II presentation pathway, and a model for the possible origin of MHC molecules. *Immunogenetics* **2019**, *71*, 233–249. [CrossRef] [PubMed]

56. Gottwick, C.; He, X.; Hofmann, A.; Vesper, N.; Reth, M.; Yang, J. A symmetric geometry oftransmembrane domains inside the B cell antigen receptor complex. *Proc. Natl. Acad. Sci. USA* **2019**, *116*, 13468–13473. [CrossRef] [PubMed]

MDPI
St. Alban-Anlage 66
4052 Basel
Switzerland
Tel. +41 61 683 77 34
Fax +41 61 302 89 18
www.mdpi.com

Cells Editorial Office
E-mail: cells@mdpi.com
www.mdpi.com/journal/cells